Fundamentals of Heat Transfer

Frank P. Incropera

David P. DeWitt

School of Mechanical Engineering
Purdue University

Fundamentals of Heat Transfer

John Wiley & Sons

New York   Chichester   Brisbane   Toronto

Chapter opening art   David A. Hruby

Library of Congress Cataloging in Publication Data:

Incropera, Frank P.
    Fundamentals of heat transfer.

    Includes bibliographies and index.
    1. Heat-Transmission. I. DeWitt, David P.,
1934–   joint author. II. Title.
QC320.I45   536'.2   80-17209
ISBN 0-471-42711-X

Printed in the United States of America

10 9 8 7 6 5 4 3 2

Preface

Heat transfer processes have always been an integral part of our environment, and with each passing year they are becoming more relevant to our technology. If one wishes to work on the many traditional and newly emerging problems that involve such processes, two requirements must be met: the physical features of the processes must be understood, and the facility to quantify the processes must be developed. There are two primary objectives in this text:

1. To instill within the student an appreciation for the physical effects that underlie heat transfer phenomena.
2. To develop methodologies that allow the student to solve a wide variety of practical problems.

It is implicit in the first objective that, when confronted with a problem, the student should be able to identify whether there exists a heat transfer component. If such a component does exist, the student should be able to ascertain its particular nature. For example, is conduction important? If so, may one-dimensional, steady-state conditions be assumed, or is the problem characterized by multidimensional and/or transient effects? Is radiation relevant, and if so, is it significant compared to any convection effects that may exist? A major objective is to condition the student to undertake such a mental analysis for each and every problem.

It is implicit in the second objective that the student be able to discern appropriate simplifying assumptions, identify the relevant dependent and independent variables, and choose or develop appropriate expressions for evaluating the dependent variables. In short, the student must be able to effect the kind of *engineering analysis* which, although not always exact, often provides useful information concerning the performance and/or design of a particular system or process.

This book's development has been strongly motivated by the foregoing objectives and by the fact that it is intended for use in a first course on the subject. Priorities have been established accordingly. For example, although considerable attention is given to the physical processes that underlie boundary layer behavior, the development of detailed boundary layer solutions is deemphasized. Emphasis is instead placed on the empirical results needed to solve practical problems involving convection heat and mass transfer.

We assume that the student has had a prior course in thermodynamics and has had or is concurrently taking a separate course in fluid mechanics. Hence the text does not provide an integrated treatment of convection heat transfer and momentum transfer (fluid mechanics). Instead, it reinforces an existing background in the fluid mechanics area. Moreover, through its treatment of conservation principles, particularly the first law of thermodynamics, the text also reinforces the student's background in thermodynamics. General — yet simple — formulations of the energy conservation requirement are provided early in the text and are used as a basis for systematically treating all subsequent applications, whether to a differential or to a finite control volume. The importance of the first law to heat transfer analysis is firmly established through numerous examples and problems.

The extent to which numerical methods are treated should be noted. Since such methods often provide the only means of obtaining accurate solutions to conduction problems, considerable attention has been given to the manner in which finite-difference equations may be formulated and to methods that may be used for their solution. We have not, however, included any computer programs or subroutines that could be used to effect these solutions, since such software is readily available in most computer libraries and/or easily written by the student.

Although this text is concerned primarily with the subject of heat transfer, it does include more than just a casual description of mass transfer effects and problems. Mass transfer is, of course, a complex subject that could not be considered in detail in an introductory text. However, its strong analogy to heat transfer suggests that it is appropriate to introduce the subject in a text of this nature. Accordingly, mass transfer effects are considered *to whatever extent they can be inferred by analogy to heat transfer.*

The manner in which mass transfer is considered varies according to the mass transfer mode. Because engineering applications of mass transfer by *diffusion* are limited, all topics related to this mode have been treated separately and confined to a single chapter at the end of the text. Hence any instructor wishing to ignore such topics may readily do so. In contrast, because engineering applications of *convection* mass transfer are extensive and because the process can make a significant contribution to total energy exchange at a vapor–liquid interface, its treatment has been *integrated* with that of convection heat transfer. This integration has been facilitated by the strong analogy between convection heat and mass transfer effects.

The specific contents of the text are as follows. Chapter 1 develops the physical basis of heat transfer by conduction, convection and radiation and emphasizes the important role played by the energy conservation law. Through carefully selected examples and problems, the student is encouraged to develop the facility to identify the relevant heat transfer processes for a variety of practical situations. Chapter 2 provides a detailed introduction to the conduction process. The nature and origins of Fourier's law are discussed, and general forms of the conduction heat equation are developed. The physical significance of each of the terms in the equation is emphasized, along with the nature of the simplifications which may often be made. Chapters 3, 4, and 5 are structured along traditional lines, with methods of solution developed for one-dimensional, two-dimensional, and transient conduction problems.

Chapter 6 introduces energy and species transfer by convection and presents the major physical concepts related to this transfer. The concepts of the velocity, thermal, and concentration boundary layers are developed, and the related conservation equations are derived. The physical significance of the various terms in the equations and their relation to the wall shear stress and to heat and mass transfer across the boundary layers are emphasized. The various similarity parameters relevant to convection transfer are obtained by non-dimensionalizing the conservation equations, and the important role that they play in the generalization of heat and mass transfer results is developed. The dimensionless forms of the conservation equations are also used to infer important boundary layer analogies. Finally, the effect of turbulence on convective transfer is considered.

In many respects Chapter 6 is a cornerstone to the treatment of convection transfer. It provides the student with firm conceptual underpinnings before providing the tools needed to perform engineering calculations. Although subsequent chapters may be considered without the benefit of Chapter 6, this course of action is not recommended. The development of engineering methodologies to the deliberate exclusion of underlying physical principles is incompatible with sound engineering education.

Subsequent chapters deal with specific convection effects. Chapters 7 and 8 treat the unique features associated with convection heat and mass transfer in external and internal flows, respectively. Computational tools are stressed. Chapters 9 and 10 deal with the special cases of free convection and heat transfer with phase change, respectively, while Chapter 11 deals with heat transfer equipment.

The fundamentals of radiative transfer are introduced in Chapter 12. The physical basis of surface emission, absorption, and reflection is discussed, and expressions for the related properties are developed. Spectral and directional effects are considered, and the concepts of the blackbody and the diffuse-gray surface are developed. In writing this chapter, we have been strongly influenced by our belief that the subject of radiation cannot be developed properly without recognizing and dealing with its directional nature. For this reason the concept of radiation intensity is introduced early in the chapter and is used to develop all directional aspects of surface emission, irradiation, and radiative properties. While such an approach has the disadvantage of requiring a significant effort from the student in the early going, it has the distinct advantage of allowing a complete and realistic description of surface radiative processes. In Chapter 13 we emphasize predicting radiative exchange between surfaces in a non-participating medium, although attention is also given to volumetric effects in absorbing–emitting media.

Chapter 14 deals exclusively with multimode heat transfer effects. A variety of examples related to industrial and environmental problems are considered to illustrate the manner in which previously developed tools may be systematically integrated for effective problem solving. Emphasis is placed on engineering problems for which the student must establish objectives, make appropriate assumptions, and develop solutions that best respond to the objectives. The material of this chapter can contribute greatly to developing student confidence in using the tools of heat transfer analysis, and the instructor is strongly urged to include it in the course requirements.

Although the manner in which this text is used for a particular heat transfer course will depend largely on the time available and the preferences of the instructor, we offer the following suggestions. If the course is restricted to a 10-week session (30 class periods), it is impossible to do justice to all topics in the text. A reasonable course outline would consider the following portions of the text: Chapter 1 (2 periods), Chapter 2 (1 period), Chapter 3 (4 periods), Sections 5.1 to 5.6 (3 periods), Chapter 6 (2 periods), Chapters 7 and 8 (5 periods), Chapter 11 (2 periods), Chapter 12 (5 periods), and Sections 13.1 to 13.3 (3 periods). Three periods could then be reserved for exams and contingencies. The foregoing coverage would exclude two-dimensional conduction (Chapter 4), finite-difference approaches to transient conduction (Section 5.7), free convection (Chapter 9), convection with phase change (Chapter 10), volumetric radiation phenomena (Section 13.4), multimode heat transfer effects (Chapter 14), and mass transfer by diffusion (Chapter 15).

Coverage of many of the foregoing topics could, of course, be included if 15 weeks (45 class periods) are available for the course. For example, Chapter 4 and Section 5.7 could be considered, with considerable attention given to numerical methods of solving conduction problems. Chapters 9 and 10 could also be considered, along with Sections 13.4 and Chapter 14. In the event that two 10-week courses are used to cover the subject (60 class periods), more time would be available for all of the foregoing topics, as well as to mass transfer by diffusion (Chapter 15) and analytical methods of solving conduction problems (Appendix C) and boundary layer problems (Appendices D and E).

The text examples and problems may be divided into two categories. In one category, problems are somewhat routine and appear with the introduction of new material. They familiarize the student with basic computational procedures and with the units and orders of magnitude of the relevant variables. The second, larger category of problems contributes more significantly to the objectives of the text. Many of these problems are couched in terms of actual engineering systems and motivate the student by demonstrating the connection between transport phenomena and contemporary problems in science and technology. Specific systems relate to industrial production and processing, energy production and conversion, and environmental protection and control. Solutions may not be obvious and will generally require that the tools of the course be implemented in a thoughtful and systematic fashion. In fact, many of the problems were designed to provide the student with opportunities to make decisions and to develop judgmental skills.

In connection with solving problems, we have stressed a particular methodology that requires that considerable thought be given to the problem before its solution is effected. The physical system must be described by a schematic, on which all relevant processes are identified. Assumptions must be clearly described, and conservation laws and rate equations must be applied in a systematic way.

Because transition from the English to the SI (Systéme Internationale) system of units is proceeding rapidly, the SI system is used exclusively in this text. Conversion factors are provided for use of the English system.

This text is an outgrowth of many years of teaching the subject of heat transfer. In its preparation we have strived to remain conscious of student learning needs and difficulties and to move carefully and systematically toward achieving the stated objectives. In many respects the book reflects perspectives toward the subject that have been developed at Purdue over a long period of time. We are indebted to our many colleagues who have contributed to this development. We are particularly grateful to Professor P. E. Liley, who did much to insure the inclusion of reliable thermophysical property values, and to Professor R. W. Fox, who exhaustively reviewed two drafts of the manuscript, provided many valuable suggestions, and was a constant source of encouragement.

Finally, we thank the many students who offered constructive criticisms of the text during its trial use in the classroom.

West Lafayette, Indiana

Frank P. Incropera
David P. DeWitt

Contents

Symbols

A	area, m^2
A_c	cross-sectional area, m^2
A_s	surface area, m^2
a	acceleration, m/s^2
Bi	Biot number
C	molar concentration, $kmol/m^3$; heat capacity rate, W/K
C_D	drag coefficient
C_f	friction coefficient
C_t	thermal capacitance, J/kg
c	specific heat, $J/kg \cdot K$; speed of light, m/s
c_p	specific heat at constant pressure, $J/kg \cdot K$
c_v	specific heat at constant volume, $J/kg \cdot K$
D	diameter, m
D_{AB}	binary mass diffusion coefficient, m^2/s
D_h	hydraulic diameter, m
E	thermal (sensible) internal energy, J; electric potential, V; emissive power, W/m^2
Ec	Eckert number
$\dot{E}_g$	rate of energy generation, W
$\dot{E}_{in}$	rate of energy transfer into a control volume, W
$\dot{E}_{out}$	rate of energy transfer out of control volume, W
$\dot{E}_{st}$	rate of increase of energy stored within a control volume, W
e	thermal internal energy per unit mass, J/kg; surface roughness, m
F	force, N; heat exchanger correction factor; fraction of blackbody radiation in a wavelength band; view factor
Fo	Fourier number
f	friction factor; similarity variable
G	irradiation, W/m^2
Gr	Grashof number

Gz	Graetz number
g	gravitational acceleration, m/s^2
g_c	gravitational constant, $1 \ kg \cdot m/N \cdot s^2$ or $32.17 \ ft \cdot lb_m/lb_f \cdot s^2$
h	convection heat transfer coefficient, $W/m^2 \cdot K$; Planck's constant
h_{fg}	latent heat of vaporization, J/kg
h_m	convection mass transfer coefficient, m/s
h_{rad}	radiation heat transfer coefficient, $W/m^2 \cdot K$
I	electric current, A; radiation intensity, $W/m^2 \cdot sr$
i	electric current density, A/m^2; enthalpy per unit mass, J/kg
J	radiosity, W/m^2
J_i^*	diffusive molar flux of species i relative to the mixture molar average velocity, $kmol/s \cdot m^2$
j_i	diffusive mass flux of species i relative to the mixture mass average velocity, $kg/s \cdot m^2$
j_H	Colburn j factor for heat transfer
j_m	Colburn j factor for mass transfer
k	thermal conductivity, $W/m \cdot K$; Boltzmann's constant
k_0	zero-order, homogeneous reaction rate constant, $kmol/s \cdot m^3$
k_1	first-order, homogeneous reaction rate constant, $1/s$
k_1''	first-order, homogeneous reaction rate constant, m/s
L	characteristic length, m
Le	Lewis number
M	mass, kg; number of heat transfer lanes in a flux plot; reciprocal of the Fourier number for finite-difference solutions
$\dot{M}_i$	rate of transfer of mass for species i, kg/s
$\dot{M}_{i,g}$	rate of increase of mass of species i due to chemical reactions, kg/s
$\dot{M}_{in}$	rate at which mass enters a control volume, kg/s
$\dot{M}_{out}$	rate at which mass leaves a control volume, kg/s
$\dot{M}_{st}$	rate of increase of mass stored within a control volume, kg/s
$\mathscr{M}_i$	molecular weight of species i, $kg/kmol$
m	mass, kg
$\dot{m}$	mass flowrate, kg/s
m_i	mass fraction of species i, ρ_i/ρ
N	number of temperature increments in a flux plot; number of rows in a tube bank; number of surfaces in an enclosure
Nu	Nusselt number
NTU	number of transfer units
N_i	molar transfer rate of species i relative to fixed coordinates, $kmol/s$
N_i''	molar flux of species i relative to fixed coordinates, $kmol/s \cdot m^2$
$\dot{N}_i$	molar rate of increase of species i per unit volume due to chemical reactions, $kmol/s \cdot m^3$
$\dot{N}_i''$	surface reaction rate of species i, $kmol/s \cdot m^2$
n_i''	mass flux of species i relative to fixed coordinates, $kg/s \cdot m^2$
$\dot{n}_i$	mass rate of increase of species i per unit volume due to chemical reactions, $kg/s \cdot m^3$

P_L, P_T	dimensionless longitudinal and transverse pitch of a tube bank
P	perimeter, m; general fluid property designation
Pe	Peclet number ($RePr$)
Pr	Prandtl number
p	pressure, N/m^2
Q	heat transfer, J
q	heat transfer rate, W
$\dot{q}$	rate of energy generation per unit volume, W/m^3
q'	heat transfer rate per unit length, W/m
q''	heat flux, W/m^2
R	cylinder radius, m
$\mathscr{R}$	universal gas constant
Ra	Rayleigh number
Re	Reynolds number
R_c	thermal contact resistance, K/W
R_e	electric resistance, Ω
R_f	fouling factor, m$^2 \cdot$K/W
R_m	mass transfer resistance, s/m^3
$R_{m,n}$	residual for the m,n nodal point
R_t	thermal resistance, K/W
r_o	cylinder or sphere radius, m
r, ϕ, z	cylindrical coordinates
r, θ, ϕ	spherical coordinates
S	solubility, kmol/m$^3 \cdot$atm; shape factor for two-dimensional diffusion, m
S_c	solar constant
Sc	Schmidt number
Sh	Sherwood number
St	Stanton number
S_D, S_L, S_T	diagonal, longitudinal and transverse pitch of a tube bank, m
T	temperature, K
t	time, s
U	overall heat transfer coefficient, W/m$^2 \cdot$K
u, v, w	mass average fluid velocity components, m/s
u^*, v^*, w^*	molar average velocity components, m/s
V	volume, m^3; fluid velocity, m/s
v	specific volume, m^3/kg
$\dot{W}$	rate at which work is performed, W
X, Y, Z	components of the body force per unit volume, N/m^3
x, y, z	rectangular coordinates, m
x_c	critical location for transition to turbulence, m
$x_{\text{fd},c}$	concentration entry length, m
$x_{\text{fd},h}$	hydrodynamic entry length, m
$x_{\text{fd},t}$	thermal entry length, m
x_i	mole fraction of species i, C_i/C

Greek Letters

α	thermal diffusivity, m²/s; absorptivity
β	volumetric thermal expansion coefficient, 1/K
Γ	mass flowrate per unit width in film condensation, kg/s·m
δ	hydrodynamic boundary layer thickness, m
δ_c	concentration boundary layer thickness, m
δ_t	thermal boundary layer thickness, m
ε	emissivity; porosity of a packed bed; heat exchanger effectiveness
ε_f	fin effectiveness
ε_H	turbulent diffusivity for heat transfer, m²/s
ε_M	turbulent diffusivity for momentum transfer, m²/s
ε_m	turbulent diffusivity for mass transfer, m²/s
η	similarity variable
η_f	fin efficiency
θ	zenith angle, rad; temperature difference, K
κ	absorption coefficient, 1/m
λ	wavelength, μm
μ	viscosity, kg/s·m
v	kinematic viscosity, m²/s; frequency of radiation, 1/s
ρ	mass density, kg/m³; reflectivity
σ	Stefan-Boltzmann constant; electrical conductivity, 1/Ω·m; normal viscous stress, N/m²; surface tension, N/m
Φ	viscous dissipation function, 1/s²
ϕ	azimuthal angle, rad
ψ	stream function, m²/s
τ	shear stress, N/m²; transmissivity
ω	solid angle, sr

Subscripts

A,B	species in a binary mixture
abs	absorbed
am	arithmetic mean
b	base of an extended surface; blackbody
c	cross-sectional; concentration; cold fluid
cr	critical insulation thickness
cond	conduction
conv	convection
CF	counterflow
D	diameter; drag
dif	diffusion
e	excess; emission
evap	evaporation
f	fluid properties; fin conditions; saturated liquid conditions
fd	fully developed conditions

g	saturated vapor conditions
H	heat transfer conditions
h	hydrodynamic; hot fluid
i	general species designation; inner surface of an annulus; initial condition; tube inlet condition; incident radiation
L	based on characteristic length
l	saturated liquid conditions
lm	log mean condition
M	momentum transfer condition
m	mass transfer condition; mean value over a tube cross section
max	maximum fluid velocity
o	center or midplane condition; tube outlet condition; outer
R	reradiating surface
r,ref	reflected radiation
rad	radiation
S	solar conditions
s	surface conditions; solid properties
sat	saturated conditions
sky	sky conditions
sur	surroundings
t	thermal
tr	transmitted
v	saturated vapor conditions
x	local conditions on a surface
λ	spectral
∞	free stream conditions

Superscripts

′	fluctuating quantity
*	molar average; dimensionless quantity

Overbar

‾	surface average conditions; time mean

Fundamentals
of Heat Transfer

1   Introduction

From the study of thermodynamics, you have learned that energy can be transferred by interactions of a system with its surroundings. These interactions are called work and heat. However, thermodynamics deals with the end states of the process during which an interaction occurs and provides no information concerning the nature of the interaction or the time rate at which it occurs. The objective of this text is to extend thermodynamic analysis through study of the *mechanisms* of heat transfer and through development of relations to calculate heat transfer *rates*. In this chapter we lay the foundation for much of the material treated in the text. We do so by raising several questions. *What is heat transfer? How is it transferred? Why is it important to study it?* In answering these questions, we will begin to appreciate the physical mechanisms that underlie heat transfer processes and the relevance of these processes to our industrial and environmental problems.

1.1   WHAT AND HOW?

A simple, yet general, definition provides sufficient response to the question: what is heat transfer?

> *Heat transfer (or, heat) is energy in transit as the result of a temperature difference.*

Whenever there exists a temperature difference in a medium or between media, heat transfer must occur.

As shown in Figure 1.1, we refer to different types of heat transfer processes as *modes*. When a temperature gradient exists in a stationary medium, which may be a solid or a fluid, we use the term *conduction* to refer to the heat transfer that will occur across the medium. In contrast, the term *convection* refers to heat

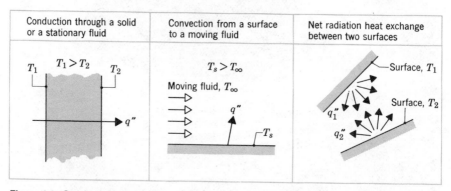

Conduction through a solid or a stationary fluid	Convection from a surface to a moving fluid	Net radiation heat exchange between two surfaces

Figure 1.1   Conduction, convection, and radiation heat transfer modes.

transfer that will occur between a surface and a moving fluid when they are at different temperatures. The third mode of heat transfer is termed *thermal radiation*. All surfaces of finite temperature emit energy in the form of electromagnetic waves. Hence, in the absence of an intervening medium, there is net heat transfer by radiation between two surfaces at different temperatures.

1.2 PHYSICAL ORIGINS AND RATE EQUATIONS

As engineers it is important that we understand the physical mechanisms that underlie the heat transfer modes and that we be able to use the rate equations that quantify the amount of energy being transferred per unit time.

1.2.1 Conduction

At mention of the word conduction, we should immediately conjure up concepts of *atomic* and *molecular activity*, for it is processes at these levels that sustain this mode of heat transfer. Conduction may be viewed as the transfer of energy from the more energetic to the less energetic particles of a substance due to interactions between the particles.

The physical mechanism of conduction is most easily explained by considering a gas and using ideas familiar from your thermodynamics background. Consider a gas in which there exists a temperature gradient and assume that there is *no* bulk motion. The gas may occupy the space between two surfaces that are maintained at different temperatures, as shown in Figure 1.2. We associate the temperature at any point with the energy stored by the gas molecules in the vicinity of the point. This energy is related to the random translational motion, as well as to the internal rotational and vibrational

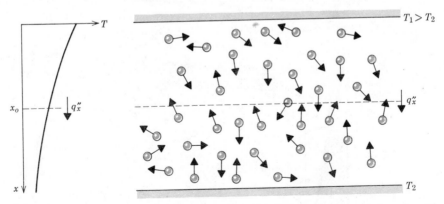

Figure 1.2 Association of conduction heat transfer with diffusion of energy due to molecular activity.

motions, of the molecules. Moreover, higher temperatures are associated with higher molecular energies, and when neighboring molecules collide, as they are constantly doing, a transfer of energy from the more energetic to the less energetic molecules must occur. In the presence of a temperature gradient, energy transfer by conduction must then occur in the direction of decreasing temperature. This transfer is evident from Figure 1.2. The hypothetical plane at x_o is constantly being crossed by molecules from above and below due to their *random* motion. However, molecules from above are associated with a larger temperature than those from below, in which case there must be a *net* transfer of energy in the positive x direction. Because of its association with this random motion, we may speak of heat transfer by conduction as a *diffusion* of energy.

The situation is much the same in liquids, although the molecules are more closely spaced and the molecular interactions stronger and more frequent. Similarly, in a solid, conduction may be attributed to molecular activity in the form of lattice vibrations. The modern view is to ascribe the energy transfer to *lattice waves* induced by atomic motion. In a nonconductor, the energy transfer is exclusively via these lattice waves; in a conductor it is also due to the translational motion of the free electrons. We treat the important properties associated with conduction phenomena in Chapter 2 and in Appendix A.

Examples of conduction heat transfer are legion. The exposed end of a metal spoon suddenly immersed in a cup of hot coffee will eventually be warmed due to the conduction of energy through the spoon. On a winter day there is significant energy loss from a heated room to the outside air. This loss is principally due to conduction heat transfer through the composite wall (plaster, insulation, and wood or brick) that separates the room air from the outside air.

It is possible to quantify heat transfer processes in terms of appropriate *rate equations*. These equations may be used to compute the amount of energy being transferred per unit time. For heat conduction, the rate equation is known as *Fourier's law*. For the one-dimensional plane wall shown in Figure 1.3, having a temperature distribution $T(x)$, the rate equation is expressed as

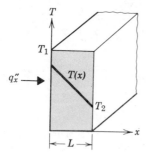

Figure 1.3 One-dimensional heat transfer by conduction (diffusion of energy).

$$q_x'' = -k\frac{dT}{dx} \tag{1.1}$$

The *heat flux* q_x'' (W/m²) is the heat transfer rate in the *x* direction *per* unit area perpendicular to the direction of transfer, and it is proportional to the temperature gradient, dT/dx, in this direction. The proportionality constant k is a *transport* property known as the *thermal conductivity* (W/m·K) and is a characteristic of the wall material. The minus sign is a consequence of the fact that heat is transferred in the direction of decreasing temperature. Under the conditions shown in Figure 1.3, wherein the temperature distribution is linear, the temperature gradient may be expressed as

$$\frac{dT}{dx} = \frac{T_2 - T_1}{L}$$

and the heat flux is then

$$q_x'' = -k\frac{(T_2 - T_1)}{L}$$

or

$$q_x'' = k\frac{(T_1 - T_2)}{L} = k\frac{\Delta T}{L} \tag{1.2}$$

Equation 1.2 represents the simplest form of the conduction rate equation. Despite this simplicity, however, it finds frequent application in relating the overall temperature difference, ΔT, across a medium to the rate at which heat is transferred by conduction through the medium.

EXAMPLE 1.1

The wall of an industrial furnace is constructed from 0.15 m thick fire clay brick having a thermal conductivity of 1.7 W/m·K. Measurements made during steady-state operation reveal temperatures of 1400 K and 1150 K at the inner and outer surfaces, respectively. What is the rate of heat loss through a wall which is 0.5 m by 3 m on a side?

SOLUTION

KNOWN:

Steady-state conditions with prescribed wall thickness, area, thermal conductivity, and surface temperatures.

FIND:

Wall heat loss.

SCHEMATIC:

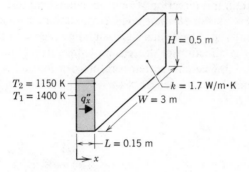

$H = 0.5$ m

$T_2 = 1150$ K
$T_1 = 1400$ K

q_x''

$k = 1.7$ W/m·K

$W = 3$ m

$L = 0.15$ m

x

ASSUMPTIONS:

1. Steady-state conditions.
2. One-dimensional conduction through the wall.
3. Constant properties.

ANALYSIS:

Since heat transfer through the wall is only by conduction, the heat flux may be determined from Fourier's law. Using Equation 1.2

$$q_x'' = k\frac{\Delta T}{L} = 1.7 \text{ W/m·K} \times \frac{250 \text{ K}}{0.15 \text{ m}} = 2833 \text{ W/m}^2$$

The heat flux represents the rate of heat transfer through a section of unit area. The total heat loss is then

$$q_x = (HW)q_x'' = (0.5 \text{ m} \times 3.0 \text{ m}) 2833 \text{ W/m}^2$$

$$q_x = 4250 \text{ W}$$ ◁

COMMENTS:

1. Note direction of heat flow.
2. Note distinction between heat flux and heat rate.

1.2.2 Convection

The convection heat transfer *mode* is comprised of *two mechanisms*. In addition to energy transfer due to *random molecular motion* (diffusion), there is also energy being transferred by the *bulk*, or *macroscopic, motion* of the fluid. This fluid motion is associated with the fact that, at any instant, large numbers of molecules are moving collectively or as aggregates. Such motion, in the presence of a temperature gradient, will give rise to heat transfer. Because the molecules in the aggregate retain their random motion, the total heat transfer is then due to a superposition of energy transport by the random motion of the molecules and by the bulk motion of the fluid. It is customary to use the term *convection* when referring to this cumulative transport.

We are especially interested in the convection heat transfer which occurs between a fluid in motion and a bounding surface, when the two are at different temperatures. Consider fluid flow over the heated surface of Figure 1.4. A consequence of the fluid-surface interaction is the development of a region in the fluid over which its velocity varies from zero at the surface to a finite value, u_∞, associated with the flow. This region of the fluid is known as the *hydrodynamic*, or *velocity, boundary layer*. Moreover, if the surface and flow temperatures differ, there will be a region of the fluid over which the temperature varies from T_s at $y = 0$ to T_∞ in the outer flow. This region, called the *thermal boundary layer*, may be smaller, larger or the same size as that over which the velocity varies. In any case, if $T_s > T_\infty$, convection heat transfer will occur between the surface and the outer flow.

The convection heat transfer mode is sustained by both random molecular motion and by the bulk motion of the fluid within the boundary layer. The contribution due to random molecular motion (diffusion) generally dominates near the surface where the fluid velocity is low. In fact, at the interface between the surface and the fluid, ($y = 0$), heat is transferred only by this mechanism. The contribution due to bulk fluid motion originates from the fact that the boundary layers *grow* as the flow progresses in the x direction. Appreciation of boundary layer phenomena is essential to understanding convection heat transfer. It is for

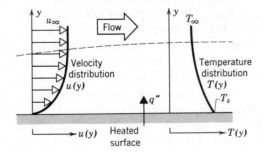

Figure 1.4 Boundary layer development in convection heat transfer.

this reason that the discipline of fluid mechanics will play a vital role in our later analysis of the convection mechanisms.

Convection heat transfer may be categorized according to the nature of the flow. We speak of *forced convection* when the flow is caused by some external means, such as by a fan, a pump, or by atmospheric winds. In contrast, for *free* (or *natural*) *convection* the flow is induced by buoyancy forces in the fluid. These forces arise from density variations that always accompany temperature variations in the fluid. An example is the free convection heat transfer that occurs from a hot pavement to the surrounding atmosphere on a still day. Air that is in contact with the hot pavement has a lower density than that of the cooler air above the pavement. Hence, a circulation pattern exists in which the warm air moves up from the pavement and the cooler air moves downward. However, in the presence of atmospheric winds, heat transfer from the pavement to the air is likely to be dominated by forced convection, even though the free convection mode still exists. Note that, because of free convection, heat transfer exclusively by conduction rarely occurs in a fluid.

We have described the convection heat transfer mode as energy transfer occurring within a fluid due to the combined effects of conduction and gross fluid motion. In general, the energy that is being transferred is the *sensible*, or internal thermal, energy of the fluid. However, there are convection processes for which there is, in addition, *latent* heat exchange. This latent heat exchange is generally associated with a phase change between the liquid and vapor states of the fluid. Two special cases of interest in this text are *boiling* and *condensation*.

Regardless of the particular nature of the convection heat transfer mode, the appropriate rate equation is of the form

$$q'' = h(T_s - T_\infty) \tag{1.3}$$

where q'', the convective *heat flux* (W/m^2), is proportional to the difference between the surface and fluid temperatures, T_s and T_∞, respectively. This expression is known as *Newton's law of cooling*, and the proportionality constant, h ($W/m^2 \cdot K$), is referred to as the *heat transfer coefficient*, the *film conductance*, or the *film coefficient*. It attempts to encompass, in a single quantity, all of the effects that influence the convection mode. The heat transfer coefficient depends on conditions in the boundary layer, which depend on the surface geometry, the nature of the fluid motion, and a number of the fluid thermodynamic and transport properties. Moreover, any study of convection ultimately reduces to a study of the means by which h may be determined. Although consideration of these means is deferred to Chapter 6, it is important to note that convection heat transfer will frequently appear as a boundary condition in the solution of conduction problems (Chapters 2 to 5). In the solution of such problems we presume h to be known, using typical values given in Table 1.1. Note the range of values associated with the various convection processes.

Table 1.1 Typical values of the convection heat transfer coefficient

	h (W/m$^2 \cdot$K)
Free convection	5–25
Forced convection	
Gases	25–250
Liquids	50–20,000
Convection with phase change	
Boiling or condensation	2,500–100,000

1.2.3 Radiation

Thermal radiation is energy emitted by matter that is at a finite temperature. Although we focus primarily on radiation from solid surfaces, emission may also occur from liquids and gases. Regardless of the form of matter, the emission may be attributed to changes in the electron configurations of the constituent atoms or molecules. Moreover, the energy of the radiation field is transported by electromagnetic waves (or alternatively, photons) and originates at the expense of the internal energy of the matter that is emitting. Note carefully that the transfer of energy by conduction or convection requires the presence of a material medium while radiation transfer does not. In fact, radiation transfer occurs most efficiently in a vacuum.

The *maximum* flux (W/m^2) at which radiation may be emitted from a surface is given by the *Stefan-Boltzmann law*

$$q'' = \sigma T_s^4 \tag{1.4}$$

where T_s is the *absolute* temperature (K) of the surface and σ is the *Stefan-Boltzmann constant* ($\sigma = 5.67 \times 10^{-8}$ W/m$^2 \cdot$K^4). Such a surface is called an ideal radiator or *blackbody*. The heat flux emitted by a real surface is less than that of the ideal radiator and is given by

$$q'' = \varepsilon \sigma T_s^4 \tag{1.5}$$

where ε is a radiative property of the surface called the *emissivity*. This property indicates how efficiently the surface emits compared to an ideal radiator.

Equation 1.5 determines the rate at which energy is *emitted* by a surface. Determination of the *net* rate at which *radiation* is *exchanged between surfaces* is generally a good deal more complicated. However, a special case that occurs very frequently in practice involves the net exchange between a small surface and a much larger surface that completely surrounds the smaller one (Figure 1.5). The surface and the surroundings are separated by a gas that has no effect on the radiation transfer. The net rate of radiation heat exchange, $q(W)$,

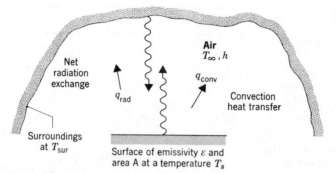

Figure 1.5   Radiation exchange between a surface and its surroundings.

between the surface and its *surroundings*, may then be expressed as

$$q = \varepsilon A \sigma (T_s^4 - T_{sur}^4) \tag{1.6}$$

In this expression, A is the surface area and ε is its emissivity, while T_{sur} is the temperature of the surroundings. For this special case, the emissivity and the area of the surroundings do not influence the net heat exchange rate.

There are many applications for which it is convenient to express the net radiation heat exchange in the form

$$q_{rad} = h_r A (T_s - T_{sur}) \tag{1.7}$$

where h_r is a *radiation heat transfer coefficient* defined as

$$h_r \equiv \varepsilon \sigma (T_s + T_{sur})(T_s^2 + T_{sur}^2) \tag{1.8}$$

Here we have modeled the radiation mode in a manner similar to convection. In this sense we have *linearized* the radiation rate equation, making the heat rate proportional to a temperature difference rather than to the difference between two temperatures to the fourth power. Note, however, that h_r depends strongly on temperature, while the temperature dependence of the convection heat transfer coefficient, h, is generally weak.

The surface within the surroundings may also simultaneously transfer heat by convection to the adjoining gas (Figure 1.5). The total rate of heat transfer *from* the surface is then the sum of the heat rates due to the two modes. That is,

$$q = q_{conv} + q_{rad}$$

or,

$$q = hA(T_s - T_\infty) + \varepsilon A \sigma (T_s^4 - T_{sur}^4) \tag{1.9}$$

Note that the convection heat transfer rate, q_{conv}, is simply the product of the flux given by Equation 1.3 and the surface area.

EXAMPLE 1.2

An uninsulated steam pipe passes through a room in which the air and walls are at a temperature of 25°C. The outside diameter of the pipe is 70 mm, and its surface temperature and emissivity are 200°C and 0.8, respectively. If the coefficient associated with free convection heat transfer from the surface to the air is 15 W/m$^2\cdot$K, what is the rate of heat loss from the surface per unit length of pipe?

SOLUTION

KNOWN:

Uninsulated pipe of prescribed diameter, emissivity and surface tempera-ture in a room with fixed wall and air temperature.

FIND:

Pipe heat loss per unit length, q' (W/m).

SCHEMATIC:

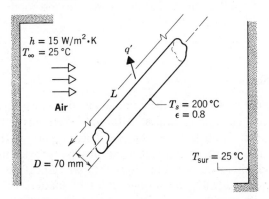

$h = 15$ W/m$^2\cdot$K
$T_\infty = 25$ °C

Air

q'

L

$D = 70$ mm

$T_s = 200$ °C
$\epsilon = 0.8$

$T_{sur} = 25$ °C

ASSUMPTIONS:

1. Steady-state conditions.
2. Radiation exchange between the pipe and the room is between a small surface enclosed within a much larger surface.

ANALYSIS:

Heat loss from the pipe is by convection to the room air and by radiation exchange with the walls. Hence, from Equation 1.9.

$$q = h(\pi DL)(T_s - T_\infty) + \varepsilon(\pi DL)\sigma(T_s^4 - T_{sur}^4)$$

The heat loss per unit length of pipe is then

$$q' = (q/L) = 15 \text{ W/m}^2 \cdot \text{K} \; (\pi \times 0.07 \text{ m})(200 - 25)°\text{C}$$
$$+ 0.8(\pi \times 0.07 \text{ m})5.67 \times 10^{-8} \text{ W/m}^2 \cdot \text{K}^4 \; (473^4 - 298^4)\text{K}^4$$

$$q' = 577 \text{ W/m} + 421 \text{ W/m}$$

$$q' = 998 \text{ W/m} \qquad\qquad\qquad \triangleleft$$

COMMENTS:

1. Note that temperature may be expressed in °C or K units when evaluating the temperature difference for a convection (or conduction) heat transfer rate. However, temperature must be expressed in kelvins (K) when evaluating a radiation transfer rate.

2. In this situation the radiation and convection heat transfer rates are comparable because T_s is large compared to T_{sur} and the coefficient associated with free convection is small. For more moderate values of T_s and the larger values of h associated with forced convection, the effect of radiation may often be neglected. The radiation heat transfer coefficient may be computed from Equation 1.8, and for the conditions of this problem its value is $h_r = 11 \text{ W/m}^2 \cdot \text{K}$.

1.2.4   Relationship to Thermodynamics

At this point it is important to take note of the fundamental differences that exist between heat transfer and thermodynamics. Although thermodynamics is concerned with the heat interaction and the vital role which it plays in the first and second laws, it considers neither the basic mechanisms that provide for heat exchange nor the methods that exist for computing the *rate* of heat exchange. Thermodynamics is concerned with *equilibrium* states of matter, where an equilibrium state necessarily precludes the existence of a temperature gradient. Although thermodynamics may be used to determine the amount of energy required in the form of heat for a system to pass from one equilibrium state to another, it does not acknowledge that *heat transfer is inherently a nonequilibrium process*. In order for heat transfer to occur, there must be a temperature gradient and hence thermodynamic nonequilibrium. The discipline of heat transfer therefore seeks to do what thermodynamics is inherently unable to do. It seeks to quantify the *rate* at which heat transfer occurs in terms of the degree of thermal nonequilibrium. This is done through the rate equations for the three modes, expressed by Equations 1.1, 1.3, and 1.6 and summarized in Table 1.5.

1.3 THE CONSERVATION OF ENERGY REQUIREMENT

The subjects of thermodynamics and heat transfer are highly complementary. For example, heat transfer is an extension of thermodynamics in that it considers the *rate* at which energy is transported. Moreover, in many heat transfer analyses the first law of thermodynamics (the *law of conservation of energy*) plays an important role. It is therefore useful to consider general statements for this law.

1.3.1 Conservation of Energy for a Control Volume

In our application of the conservation laws, we first need to identify the control volume, a fixed region of space bounded by a control surface through which energy and matter may pass. With respect to a control volume, a form of the energy conservation requirement that is most useful for heat transfer analyses may be stated as follows.

> *The rate at which thermal and mechanical energy enters a control volume minus the rate at which this energy leaves the control volume must equal the rate at which this energy is stored in the control volume.*

If the inflow of energy exceeds the outflow, there will be an increase in the amount of energy stored (or accumulated) in the control volume; if the converse is true, there will be a decrease in energy storage. If the inflow of energy equals the outflow, then a *steady state* condition must prevail in which there will be no change in the amount of energy stored in the control volume.

Consider applying energy conservation to the system shown in Figure 1.6. The first step is to carefully identify the system boundaries associated with the control volume surface. We have done this in the figure by drawing a dashed line around the actual system. The next step is to identify the energy terms. In general these may include energy entering and leaving the system *through* the control surface, $\dot{E}_{in}$ and $\dot{E}_{out}$, as well as energy generation, $\dot{E}_g$, and storage, $\dot{E}_{st}$, *within* the control volume. A general form of the energy conservation requirement may then be expressed on a *rate* basis as

$$\dot{E}_{in} + \dot{E}_g - \dot{E}_{out} = \dot{E}_{st} \tag{1.10}$$

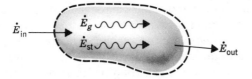

Figure 1.6 Conservation of energy for a control volume.

In words this relation says that energy inflow and generation act to increase the amount of energy stored within the control volume, whereas outflow acts to decrease the stored energy.

The inflow and outflow rate terms, $\dot{E}_{in}$ and $\dot{E}_{out}$, are *surface phenomena*. That is, they are associated exclusively with processes occurring at the control surface, and the rate at which they occur is proportional to surface area. The most common situation will involve energy inflow and outflow due to heat transfer by the conduction, convection, and/or radiation modes. In a situation involving fluid flow across the control surface, $\dot{E}_{in}$ and $\dot{E}_{out}$ would also include the energy transported into and out of the system with the fluid. This energy may be composed of potential, kinetic, and thermal forms. However, for most of the heat transfer problems encountered in this text, the potential and kinetic energy forms are generally negligible. The inflow and outflow terms may also include work interactions.

The *thermal energy generation term*, $\dot{E}_g$, is associated with the conversion from some other energy form (chemical, electrical, electromagnetic or nuclear) to thermal energy. It is a *volumetric phenomenon*. That is, it occurs within the system, and its rate is proportional to the system volume. For example, an exothermic chemical reaction may be occurring in the system, converting chemical to thermal energy. The net effect is an increase in the thermal energy of the system. Conversely, if the reaction were endothermic, thermal energy would be required to effect the conversion, and there would be a decrease in the thermal energy of the system. We could then speak of a thermal energy sink, in which case the generation term $\dot{E}_g$ would be negative. Another source of thermal energy would be the conversion from electrical energy that occurs due to resistance heating when an electric current is passed through a conductor.

It is important not to confuse the physical process of *energy storage* with that of *energy generation*. Although energy generation may certainly contribute to energy storage, the two processes are fundamentally different. Energy storage is also a *volumetric phenomenon*, but it is simply associated with an increase ($\dot{E}_{st} > 0$) or decrease ($\dot{E}_{st} < 0$) in the energy of the matter occupying the control volume. Under steady-state conditions there is, of course, no storage term ($\dot{E}_{st} = 0$).

EXAMPLE 1.3

A long conducting rod of diameter D and electrical resistance per unit length R'_e is initially in thermal equilibrium with the ambient air and its surroundings. This equilibrium is disturbed when an electrical current I is passed through the rod. Develop an equation that could be used to compute the variation of the rod temperature with time during passage of the current.

SOLUTION

KNOWN:

Temperature of a rod of prescribed diameter and electrical resistance changes with time due to passage of an electrical current.

FIND:

Equation that governs temperature change with respect to time for the rod.

SCHEMATIC:

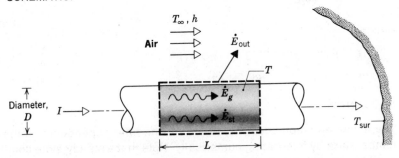

ASSUMPTIONS:

1. End effects are negligible.
2. At any time t the temperature of the rod is uniform.
3. Constant properties.
4. Radiation exchange between the outer surface of the rod and the surroundings is between a small surface and a large enclosure.

ANALYSIS:

Applying Equation 1.10 to a control volume of length L about the rod, it follows that

$$\dot{E}_g - \dot{E}_{out} = \dot{E}_{st}$$

where energy generation is due to the electric resistance heating,

$$\dot{E}_g = I^2 R'_e L$$

energy outflow is due to convection and net radiation from the surfaces, Equations 1.3 and 1.6, respectively,

$$\dot{E}_{\text{out}} = h(\pi DL)(T - T_\infty) + \varepsilon\sigma(\pi DL)(T^4 - T_{\text{sur}}^4)$$

and the change in energy storage is due to the temperature change,

$$\dot{E}_{\text{st}} = \frac{d}{dt}(\rho VcT)$$

The term $\dot{E}_{\text{st}}$ is associated with the change in the internal energy of the rod, where ρ and c are the mass density and specific heat, respectively, of the rod material and V is the volume of the rod, $V = (\pi D^2/4)L$. Substituting the rate equations into the energy balance, it follows that

$$I^2 R'_e L - h(\pi DL)(T - T_\infty) - \varepsilon\sigma(\pi DL)(T^4 - T_{\text{sur}}^4)$$
$$= \rho c(\pi D^2/4)L\frac{dT}{dt}$$

Hence

$$\frac{dT}{dt} = \frac{I^2 R'_e - \pi Dh(T - T_\infty) - \pi D\varepsilon\sigma(T^4 - T_{\text{sur}}^4)}{\rho c(\pi D^2/4)} \qquad \triangleleft$$

COMMENTS:

The above equation could be solved for the time dependence of the rod temperature by integrating numerically. Note that a steady-state condition would eventually be reached for which $dT/dt = 0$. The rod temperature is then determined by an algebraic equation of the form

$$\pi Dh(T - T_\infty) + \pi D\varepsilon\sigma(T^4 - T_{\text{sur}}^4) = I^2 R'_e$$

1.3.2 The Surface Energy Balance

We will frequently have occasion to apply the conservation of energy requirement at the surface of a medium. In this special case the control surface includes no mass or volume and appears as shown in Figure 1.7. Accordingly, the generation and storage terms of the conservation expression, Equation 1.10, are no longer relevant and it is only necessary to deal with surface phenomena. For this case the conservation requirement then becomes

$$\dot{E}_{\text{in}} - \dot{E}_{\text{out}} = 0 \tag{1.11}$$

Even though thermal energy generation may be occurring in the medium, the process would not affect the energy balance at the control surface. Note that this conservation requirement holds for both *steady state* and *transient* conditions.

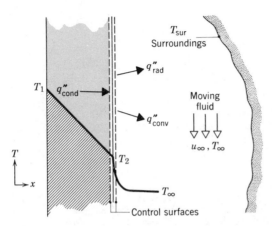

Figure 1.7 The energy balance for conservation of energy at the surface of a medium.

In Figure 1.7 three heat transfer terms are shown for the control surface. On a unit area basis they are conduction from the medium *to* the control surface, q''_{cond}, convection *from* the surface to a fluid, q''_{conv}, and net radiation exchange from the surface to the surroundings, q''_{rad}. The energy balance then takes the form

$$q''_{cond} - q''_{conv} - q''_{rad} = 0 \tag{1.12}$$

and we can express each of the terms according to the appropriate rate equations, Equations 1.2, 1.3, and 1.6.

EXAMPLE 1.4

The hot combustion gases of a furnace are separated from the ambient air and its surroundings, which are at 25°C, by a brick wall 0.15 m thick. The brick has a thermal conductivity of 1.2 W/m·K and a surface emissivity of 0.8. Under steady-state conditions an outer surface temperature of 100°C is measured. Free convection heat transfer in the air adjoining this surface is characterized by a convection coefficient of $h = 20$ W/m^2·K. What is the brick inner surface temperature?

SOLUTION

KNOWN:

Outer surface temperature of a furnace wall of prescribed thickness, thermal conductivity and emissivity.

FIND:

Wall inner surface temperature.

SCHEMATIC:

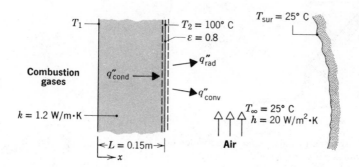

ASSUMPTIONS:

1. Steady-state conditions.
2. One-dimensional heat transfer by conduction across the wall.
3. Radiation exchange between the outer surface of the wall and the surroundings is between a small surface and a large enclosure.

ANALYSIS:

The inside surface temperature may be obtained by performing an energy balance at the outer surface. From Equation 1.11

$$\dot{E}_{in} - \dot{E}_{out} = 0$$

it follows that, on a unit area basis,

$$q''_{cond} - q''_{conv} - q''_{rad} = 0$$

or, rearranging and substituting from Equations 1.2, 1.3, and 1.6,

$$k\frac{T_1 - T_2}{L} = h(T_2 - T_\infty) + \varepsilon\sigma(T_2^4 - T_{sur}^4)$$

Therefore, substituting the appropriate numerical values

$$1.2 \text{ W/m}\cdot\text{K} \frac{(T_1 - 373)\text{K}}{0.15 \text{ m}} = 20 \text{ W/m}^2\cdot\text{K} \, (373 - 298)\text{K}$$

$$+ 0.8(5.67 \times 10^{-8} \text{ W/m}^2\cdot\text{K}^4)(373^4 - 298^4)\text{K}^4$$

$$= 1500 \text{ W/m}^2 + 520 \text{ W/m}^2 = 2020 \text{ W/m}^2$$

Solving for T_1,

$$T_1 = 373 \text{ K} + \frac{0.15 \text{ m}}{1.2 \text{ W/m} \cdot \text{K}} (2020 \text{ W/m}^2)$$

$$T_1 = 625 \text{ K} = 352°\text{C} \qquad \triangleleft$$

COMMENTS:

1. Note that the contribution of radiation to heat transfer from the outer surface is significant. The relative contribution would diminish, however, with increasing h and/or decreasing T_2.

2. When using energy balances involving radiation exchange and other modes, it is good practice to express all temperatures in kelvin units. This practice is *necessary* when the unknown temperature appears in the radiation term and in one or more of the other terms.

1.3.3 Application of the Conservation Laws: Methodology

In addition to being familiar with the transport rate equations described in Section 1.2, the heat transfer analyst must be able to work with the energy conservation requirements of Equations 1.10 to 1.12. The application of these balances is simplified if a few basic rules are followed.

1. The system and its boundaries must be identified. The appropriate control volume must be established, with the control surface represented by a dashed line on a schematic of the system.

2. The *physics* of the problem should be carefully considered and the relevant energy processes identified. Each of these processes should be shown on the control volume by an appropriately labeled arrow.

3. The conservation equation(s) should then be written, and appropriate rate expressions should be substituted for the terms in the equation(s).

In connection with applying the energy conservation equation, (1.10), note that this application may be in terms of a *finite* control volume or a *differential* (infinitesimal) control volume. In the first case an equation would be obtained that governs the gross behavior of the system. In contrast, for the second case a differential equation would be obtained that could be solved for conditions at each point in the system. Both types of control volumes are used extensively in this text.

1.4   ANALYSIS OF HEAT TRANSFER PROBLEMS: METHODOLOGY

A major objective of this text is to prepare you to solve engineering problems that involve heat transfer processes. To this end numerous problems are provided at the end of each chapter. In working these problems you will gain a deeper appreciation for the fundamentals of the subject, and you will gain confidence in your ability to apply these fundamentals to the solution of engineering problems.

It is our conviction that there exists a preferred approach to problem solving. In contrast to a somewhat random, "plug-and-chug" approach, we advocate the use of a systematic procedure characterized by a prescribed format. We consistently employ the procedure in all of the examples of this text, and we require that it be used by our students in their solution of problems. It consists of the following steps.

1. KNOWN: After carefully reading the problem, state briefly and concisely what is known about the problem. Do not repeat the problem statement.

2. FIND: State briefly and concisely what must be found.

3. SCHEMATIC: Draw a schematic of the physical system to be considered, carefully identifying its boundaries. If application of the conservation laws is anticipated, represent the required control surface by dashed lines on the schematic. *Think* about the physical processes that characterize the problem and identify relevant heat transfer processes by appropriately labeled arrows on the schematic.

4. ASSUMPTIONS: List all simplifying assumptions that you feel to be pertinent to the problem.

5. PROPERTIES: Compile material property values needed for subsequent calculations and identify the table from which they are obtained.

6. ANALYSIS: Begin your analysis of the problem. Apply appropriate conservation laws and introduce rate equations as needed. Develop the analysis as completely as possible before substituting numerical values. Perform the calculations needed to obtain the desired results.

7. COMMENTS: Where appropriate, discuss the results. Such a discussion may include a summary of key conclusions, an inference of trends, and a critique of the original assumptions.

The importance of following steps 1 through 4 should not be underestimated. These steps provide the opportunity to systematically think about a problem before actually effecting its solution.

1.5 RELEVANCE OF HEAT TRANSFER

In the vernacular of the time, heat transfer is indeed a relevant subject, not to mention an inherently fascinating part of the engineering sciences. We will devote much time to acquiring an understanding of heat transfer effects and to developing the skills needed to predict heat transfer rates. What is the value of this knowledge and to what kinds of problems may it be applied?

Heat transfer phenomena play an important role in many industrial and environmental problems. As an example, consider the vital area of energy production and conversion. There is not a single application in this area that does not involve heat transfer effects in some way. In the generation of electrical power, whether it be through nuclear fission or fusion, the combustion of fossil fuels, magnetohydrodynamic processes, or the use of geothermal energy sources, there are numerous heat transfer problems that must be solved. These problems involve conduction, convection, and radiation processes and relate to the design of systems such as boilers, condensers, and turbines. One is often confronted with the need to maximize heat transfer rates and to maintain the integrity of materials in high-temperature environments. One is also confronted with the thermal pollution problem associated with the discharge of large amounts of waste heat from a power plant to the environment. Numerous heat transfer considerations relate to the design of cooling towers to alleviate the environmental problems associated with this discharge.

On a smaller scale there are many heat transfer problems related to the development of solar energy conversion systems for space heating and air conditioning, as well as for electric power production. Heat transfer processes also affect the performance of propulsion systems, such as the internal combustion and rocket engines. Heat transfer problems arise in the design of conventional space and water heating systems, in the design of incinerators and cryogenic storage equipment, in the cooling of electronic equipment, in the design of refrigeration and air conditioning systems, and in many other industrial problems. Heat transfer processes are also relevant to air and water pollution and strongly influence local and global climate.

1.6 UNITS AND DIMENSIONS

The physical quantities of heat transfer are specified in terms of *dimensions*, which are measured in terms of units. Four *basic* dimensions are required for the development of heat transfer, and they include the length (L), mass (M), time (t), and temperature (T). All other physical quantities of interest may be related to these four basic dimensions.

In the United States it has been customary to measure dimensions in terms of an *English system of units*, for which the *base units* are

DIMENSION		UNIT
Length (L)	→	foot (ft)
Mass (M)	→	pound mass (lb_m)
Time (t)	→	second (s)
Temperature (T)	→	degree Fahrenheit (°F)

The units required to specify other physical quantities may then be inferred from the above group. For example, the dimension of force is related to mass through Newton's second law of motion

$$F = \frac{1}{g_c} Ma \tag{1.13}$$

where the acceleration, a, has units of feet per square second and g_c is a proportionality constant. If this constant is arbitrarily set equal to unity and made *dimensionless*, the dimensions of force are $(F) = (M) \cdot (L)/(t)^2$ and the unit of force is

$$1 \text{ poundal} = 1 \text{ } lb_m \cdot ft/s^2$$

Alternatively, one could work with a system of basic dimensions that includes both mass and force. However, in this case the proportionality constant must have the dimensions $(M) \cdot (L)/(F) \cdot (t)^2$. Moreover, if one defines the pound force (lb_f) as a unit of force that will accelerate one pound mass by 32.17 ft/s², the proportionality constant must be of the form

$$g_c = 32.17 \text{ } lb_m \cdot ft/lb_f \cdot s^2$$

The units of work may be inferred from its definition as the product of a force times a distance, in which case the units are ft·lb_f. The units of work and energy are, of course, equivalent, although it is customary to use the British thermal unit (Btu) as the unit of thermal energy. One British thermal unit will raise the temperature of 1 lb_m of water at 68°F by 1°F. It is equivalent to 778.16 ft·lb_f, which is termed the *mechanical equivalent of heat*.

In recent years there has been a strong trend toward worldwide usage of a standard set of units. In 1960 the SI (Systéme International d'Unites) system of units was defined by the Eleventh General Conference on Weights and Measures and recommended as a worldwide standard. In response to this trend, the American Society of Mechanical Engineers has required the use of SI units in all of its publications since July 1, 1974. For this reason and because it is operationally more convenient than the English system, the SI system will be used for the calculations of this text. However, because it is recognized that, for some time to come, engineers will also have to work with results expressed in the

English system, you should be able to convert from one system to the other. For your convenience conversion factors are provided on the inside back cover of the text.

The SI *base* units required for this text are summarized in Table 1.2. With regard to these units note that 1 mol is the amount of substance that has as many atoms or molecules as there are atoms in 12 g of carbon 12; this is the gram-mole (mol). Although the mol has been recommended as the unit quantity of matter for the SI system, it is more consistent to work with the kmol (the kg-mole). One kmol is simply the amount of substance that has as many atoms or molecules as there are atoms in 12 kg of carbon 12. So long as the use is consistent within a given problem, no difficulties arise in using either the mol or kmol. The molecular weight of a substance is the mass associated with a mol or kmol. For oxygen, as an example, the molecular weight, $\mathcal{M}$, is 16 g/mol or 16 kg/kmol.

Also note that, although the SI unit of temperature is the kelvin, use of the Celsius temperature scale remains widespread. Zero on the Celsius scale (0°C) is equivalent to 273.15 K on the thermodynamic scale,[1] in which case

$$T(\text{K}) = T(°\text{C}) + 273.15$$

However, temperature *differences* are equivalent for the two scales and may be denoted as °C or K. Note also that, although the SI unit of time is the second, other units of time (minute, hour, and day) are so common that their use with the SI system is generally accepted.

The SI units comprise a rigorously coherent form of the metric system. By this it is meant that all remaining units may be derived from the base units using

Table 1.2 SI base and supplementary* units

QUANTITY AND SYMBOL	UNIT AND SYMBOL
Length (L)	Meter (m)
Mass (M)	Kilogram (kg)
Concentration (C)	Mole (mol)
Time (t)	Second (s)
Electric current (I)	Ampere (A)
Thermodynamic temperature (T)	Kelvin (K)
Plane angle* (θ)	Radian (rad)
Solid angle* (ω)	Steradian (sr)

[1] The degree symbol is retained for designation of the Celsius temperature (°C) to avoid confusion with use of C for the unit of electrical charge (coulomb). It is not necessary to use the degree symbol with the kelvin.

Table 1.3 SI derived units for selected quantities

QUANTITY	NAME AND SYMBOL	FORMULA	EXPRESSION IN SI BASE UNITS
Force	Newton (N)	$m \cdot kg/s^2$	$m \cdot kg/s^2$
Pressure and Stress	Pascal (Pa)	N/m^2	$kg/m \cdot s^2$
Energy	Joule (J)	$N \cdot m$	$m^2 \cdot kg/s^2$
Power	Watt (W)	J/s	$m^2 \cdot kg/s^3$

formulas which do not involve any numerical factors. *Derived* units for selected quantities are listed in Table 1.3. Note that force is measured in newtons, where a 1-N force will accelerate a 1-kg mass at 1 m/s^2. Hence 1 N = 1 kg·m/s^2. The unit of pressure is N/m^2 and is often referred to as the pascal. In the SI system there is one unit of energy (thermal, mechanical, or electrical). The SI unit of energy is the joule (J), where 1 J = 1 N·m. The unit for energy rate, or power, is then J/s, where one joule per second is equivalent to one watt (1 J/s = 1 W). Since it is frequently necessary to work with extremely large or small numbers, a set of standard prefixes has been introduced to simplify matters (Table 1.4). For example, a megawatt is 1 MW = 10^6 W, and a micrometer is 1 μm = 10^{-6} m.

1.7 SUMMARY

Although much of the material of this chapter will subsequently be discussed in greater detail, you should now have a reasonable overview of heat transfer. You should be aware of the several modes of transfer and their physical origins. Moreover, given a physical situation, you should be able to perceive the relevant transport phenomena. The importance of developing this facility must not be

Table 1.4 Multiplying prefixes

PREFIX	ABBREVIATION	MULTIPLIER
Pico	p	10^{-12}
Nano	n	10^{-9}
Micro	μ	10^{-6}
Milli	m	10^{-3}
Centi	c	10^{-2}
Hecto	h	10^2
Kilo	k	10^3
Mega	M	10^6
Giga	G	10^9
Tera	T	10^{12}

underestimated. You will be devoting much time to acquiring the tools needed to calculate heat transfer phenomena. However, before you can begin to use these tools to solve practical problems, you must have the intuition to determine what is happening physically. In short, you must be able to look at a problem and identify the pertinent transport phenomena. You will find examples and problems at the end of this chapter that should help you to begin developing this intuition.

You should also appreciate the significance of the rate equations and feel comfortable in using them to compute transport rates. These equations are summarized in Table 1.5 and *should be committed to memory*. You must also recognize the importance of the conservation laws and the need to carefully identify control volumes. With the rate equations, the conservation laws may be used to solve numerous problems involving heat transfer.

Table 1.5 Summary of heat transfer rate processes

MODE	MECHANISM	RATE EQUATION	EQUATION NUMBER	TRANSPORT PROPERTY OR COEFFICIENT
Conduction	Diffusion of energy due to random molecular motion	$q_x'' \ (\text{W/m}^2) = -k\dfrac{dT}{dx}$	(1.1)	$k \ (\text{W/m·K})$
Convection	Diffusion of energy due to random molecular motion plus energy transfer due to bulk motion	$q'' \ (\text{W/m}^2) = h(T_s - T_\infty)$	(1.3)	$h \ (\text{W/m}^2\text{·K})$
Radiation	Energy transfer by electromagnetic waves	$q \ (\text{W}) = \varepsilon A\sigma(T_s^4 - T_{sur}^4)$ or $q \ (\text{W}) = h_r A(T_s - T_{sur})$	(1.6) (1.7)	$h_r \ (\text{W/m}^2\text{·K}), \varepsilon$

EXAMPLE 1.5

A closed container filled with hot coffee is in a room whose air and walls are at a fixed temperature. Identify all heat transfer processes that contribute to cooling of the coffee. Comment on features that would contribute to a superior container design.

SOLUTION

KNOWN:

Hot coffee is separated from its cooler surroundings by a plastic flask, an air space, and a plastic cover.

FIND:

Relevant heat transfer processes.

SCHEMATIC:

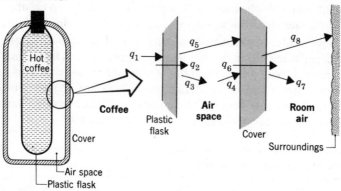

Pathways for energy transfer from the coffee are as follows.

q_1: free convection from the coffee to the flask

q_2: conduction through the flask

q_3: free convection from the flask to the air

q_4: free convection from the air to the cover

q_5: radiation exchange between the outer surface of the flask and the inner surface of the cover

q_6: conduction through the cover

q_7: free convection from the cover to the room air

q_8: radiation exchange between the outer surface of the cover and the surroundings

COMMENTS:

Design improvements are associated with (1) use of aluminized (low emissivity) surfaces for the flask and cover and (2) evacuating the air space or use of a filler material to retard free convection.

PROBLEMS

1.1 A heat rate of 3 kW is conducted through a section of an insulating material of cross-sectional area $10 \, m^2$ and of thickness 2.5 cm. If the inner (hot) surface temperature is at 415°C and the thermal conductivity of the material is 0.2 $W/m \cdot K$, what is the outer surface temperature?

1.2   A concrete wall, which has a surface area of 30 m² and is 0.30 m thick, separates warm room air from cold ambient air. The inner surface of the wall is known to be at a temperature of 25°C, while the outer surface is at −15°C. The thermal conductivity of the concrete is 1 W/m·K.

a) Referring to Equation 1.1, describe the conditions that must be satisfied in order for the temperature distribution in the wall to be linear, as shown in Figure 1.3.

b) What is the heat loss through the wall?

1.3   The heat flux through a wood slab 50 mm thick, whose inner and outer surface temperatures are 40°C and 20°C, respectively, has been determined to be 40 W/m². What is the thermal conductivity of the wood?

1.4   The inner and outer surface temperatures of a glass window 5 mm thick are 15°C and 5°C, respectively. What is the heat loss through a window that is 1 m by 3 m on a side? The thermal conductivity of glass is 1.4 W/m·K.

1.5   The convection heat transfer coefficient between a surface at 40°C and ambient air at 20°C is 20 W/m²·K. Calculate the heat flux leaving the plate by convection.

1.6   Air at 300°C flows over a flat plate of dimensions 0.50 m by 0.25 m. If the convection heat transfer coefficient is 250 W/m²·K, determine the heat transfer rate from the air to one side of the plate when the plate is maintained at 40°C.

1.7   A water-cooled, spherically shaped object of diameter 10 mm and $\varepsilon = 0.9$ is maintained at 80°C when placed in a large vacuum oven whose walls are maintained at 400°C. What is the heat transfer rate from the oven walls to the object?

1.8   A surface of area equal to 0.5 m², emissivity equal to 0.8, and temperature equal to 150°C is placed in a large, evacuated chamber whose walls are maintained at a temperature of 25°C. What is the rate at which radiation is emitted by the surface? What is the net rate at which radiation is exchanged between the surface and the chamber walls?

1.9   The variation of temperature with position in a wall is shown below for a specific time, t_1, during a transient (time-varying) process.

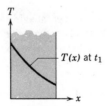

$T(x)$ at t_1

Is the wall being heated or cooled?

1.10 A solid aluminum sphere of emissivity ε is initially at an elevated temperature and is cooled by placing it in a chamber. The walls of the chamber are maintained at a lower temperature, and a cold gas is circulated through the chamber. Obtain an equation that could be used to predict the variation of the aluminum temperature with time during the cooling process. Do not attempt to solve.

1.11 An aluminum plate 4 mm thick is mounted in a horizontal position, and its bottom surface is well insulated. A special, thin coating is applied to the top surface such that it absorbs 80 per cent of any incident solar radiation, while having an emissivity of $\varepsilon = 0.25$. The density and specific heat of aluminum are known to be $\rho = 2700$ kg/m^3 and $c = 900$ J/kg·K, respectively.

   a) Consider conditions for which the plate is at a temperature of 25°C and its top surface is suddenly exposed to ambient air at $T_\infty = 20°C$ and to solar radiation that provides an incident flux of 900 W/m^2. The convection heat transfer coefficient between the surface and the air is $h = 20$ W/m^2·K. What is the initial rate of change of the plate temperature?

   b) What will be the equilibrium temperature of the plate when steady-state conditions are reached?

1.12 The coating on a plate is to be cured by exposure to an infrared lamp providing a radiant flux of 2000 W/m^2 incident upon the coating. The coating absorbs 80 per cent of the incident radiant flux and has an emissivity, $\varepsilon = 0.50$. The surroundings are large compared to the plate and at a temperature of 30°C. The convection heat transfer coefficient between the plate surface and the ambient air at 20°C is 15 W/m^2·K. Determine the temperature of the plate.

1.13 The backside of a metallic plate is perfectly insulated while the front side absorbs a solar radiant flux of 800 W/m^2. The convection coefficient between the plate and the ambient air is 12 W/m^2·K.

   a) Neglecting radiation exchange with the surroundings, calculate the temperature of the plate under steady-state conditions if the ambient air temperature is 20°C.

   b) For the same ambient air temperature, calculate the temperature of the plate if its surface emissivity is 0.8 and the temperature of the surroundings is also 20°C.

1.14 The heat flux by diffusion through a plane wall to a surface is known to be 400 W/m^2. Determine the surface temperature for each of the following conditions.

   a) Convection between the surface and an air stream at 20°C with a heat transfer coefficient, $h = 10$ W/m^2·K.

   b) The same convection process occurs along with radiative heat transfer between the surface and cold surroundings at $-150°C$, with a radiative heat transfer coefficient of $h_r = 5$ W/m^2·K.

1.15 A surface whose temperature is maintained at 400°C is separated from an airflow by a layer of insulation 25 mm thick for which the thermal conductivity is 0.1 W/m·K. If the air temperature is 35°C and the convection coefficient between the air and the outer surface of the insulation is 500 W/m^2·K, what is the temperature of this outer surface?

1.16 An experiment to determine the convection coefficient associated with airflow over the surface of a thick steel casting involves insertion of thermocouples in the casting at distances of 10 mm and 20 mm from the surface along a hypothetical line normal to the surface. The steel has a thermal conductivity of 15 W/m·K. If the thermocouples measure temperatures of 50°C and 40°C in the steel when the air temperature is 100°C, what is the convection coefficient?

1.17 Plate glass at 600°C is cooled by passing air over its surfaces such that the convection heat transfer coefficient is $h = 5$ W/m²·K. To prevent cracking, it is known that the temperature gradient must not exceed 15°C/mm at any point in the glass during the cooling process. If the thermal conductivity of the glass is 1.4 W/m·K and its surface emissivity is 0.8, what is the lowest temperature of the air that can be initially used for the cooling?

1.18 Consider a flat plate solar collector operating under steady-state conditions. Solar radiation is incident, per unit surface area of the collector, at a rate q''_S (W/m²). The cover glass is completely transparent to this radiation, and the fraction of the radiation which is absorbed by the blackened absorber plate is designated as α (the absorptivity). The fraction of the radiation not absorbed by the absorber plate (1 − α) may be assumed to be reflected through the cover plate and back to the atmosphere and space.

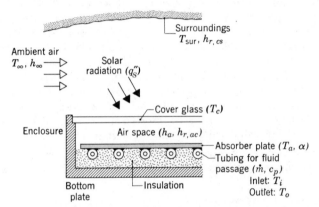

Useful energy is obtained from the collector by passing a working fluid through copper tubing that is brazed to the underside of the absorber plate. The tubing forms a serpentine arrangement for which fluid, at a constant flow rate $\dot{m}$ and specific heat c_p, is heated from an inlet temperature T_i to an outlet temperature T_o. Although the bottom of the collector may be assumed to be perfectly insulated (no heat loss), there will be heat loss from the absorber plate due to convection across the air space and radiation exchange with the cover plate. Assuming that the absorber and cover plates have uniform temperatures T_a and T_c, respectively, the parallel convection and radiation heat fluxes may be expressed as $h_a(T_a - T_c)$ and $h_{r,ac}(T_a - T_c)$, respectively. The quantity h_a is the convection heat transfer coefficient associated with the air space, while $h_{r,ac}$ is the radiation heat transfer coefficient associated with the absorber plate–cover plate combination. The cover glass also transfers heat by convection to the ambient air, $h_\infty(T_c - T_\infty)$, and exchanges energy in the form of radiation with its surroundings, $h_{r,cs}(T_c - T_{sur})$. The effective temperature, T_{sur}, of the sky and surrounding surfaces seen by the cover glass is generally less than the ambient air temperature.

a) Write an equation for the rate at which useful energy, $q_u(W)$, is collected by the working fluid, expressing your result in terms of $\dot{m}$, c_p, T_i, and T_o.

b) Perform an energy balance on the absorber plate. Use this balance to obtain an expression for q_u in terms of q_S'', α, T_a, T_c, h_a, $h_{r,ac}$, and A (the surface area of the absorber and cover plates).

c) Perform an energy balance on the cover plate.

d) Perform an overall energy balance on the entire collector, working with a control volume about the collector. Compare your result with that obtained in parts b) and c).

e) The collector efficiency η is defined as the ratio of the useful heat collected to the rate at which solar energy is incident on the collector. Obtain an expression for η.

f) Comment on what effect the value of $\dot{m}$ will have on T_a, T_o, and η. What would happen to T_a if the cover plate is removed?

1.19 In analyzing the performance of a thermal system, it is essential that the engineer be able to identify the relevant heat transfer processes. Only then can the system behavior be properly quantified. For the following systems identify the pertinent processes, designating them by an appropriately labeled arrow on a sketch of the system. Answer any additional questions which appear in the problem statement. Note: In considering problems involving heat transfer in the natural environment (outdoors), recognize that solar radiation is comprised of long and short wavelength components. If this radiation is incident on a *semitransparent medium*, such as water or glass, two things will happen to the nonreflected portion of the radiation. The long wavelength component will be absorbed at the surface of the medium, whereas the short wavelength component will be transmitted by the surface.

a) Identify the heat transfer processes that determine the temperature of an asphalt pavement on a summer day. Write an energy balance for the surface of the pavement.

b) Consider an exposed portion of your body (e.g., your forearm with a short-sleeved shirt) while you are sitting in a room. Identify all heat transfer processes that occur at the surface of your skin.

   In the interest of conserving fuel and funds, an engineer's spouse insists on keeping the thermostat of their home at 15°C (59°F) throughout the winter months. The engineer is able to tolerate this condition if the outside ambient air temperature exceeds −10°C (14°F) but complains of being cold if the ambient temperature falls much below this value. Is the engineer imagining things?

c) Consider an incandescent light source that consists of a tungsten filament enclosed in an air-filled glass bulb. Assuming steady-state operation with the filament at a temperature of approximately 2900 K, list all of the pertinent heat transfer processes for (i) the filament and (ii) the glass bulb.

d) In recent years there has been considerable interest in developing building materials that have improved insulating qualities. The development of such materials would do much to enhance energy conservation by reducing space heating requirements.

   It has been suggested that superior structural and insulating qualities could be obtained by using the composite shown on the opposite page. The material consists of a honeycomb, with cells of square cross section, sandwiched between solid slabs. The cells are filled with air, and the slabs, as well as the honeycomb matrix, are fabricated from plastics of low thermal conductivity. Identify all heat transfer processes pertinent to the performance of the composite. Suggest ways in which this performance could be enhanced.

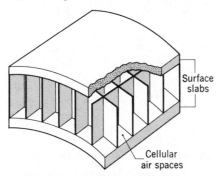

e) The number of panes in a window can strongly influence the heat loss from a heated room to the outside ambient air. Compare the single- and double-paned units shown below by identifying relevant heat transfer processes for each case.

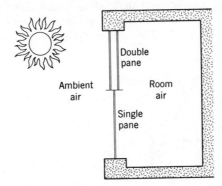

f) In a typical flat plate solar collector, energy is collected by a working fluid that is circulated through tubes that are in good contact with the back face of an absorber plate. The back face is insulated from the surroundings, and the absorber plate receives solar radiation on its front face, which is typically covered by one or more transparent plates. Identify the relevant heat transfer processes, first for the absorber plate with no cover plate and then for the absorber plate with a single cover plate.

g) Evacuated tube solar collectors are capable of improved performance relative to flat plate collectors. The design consists of an inner tube enclosed within an outer tube that is transparent to solar radiation. The annular space between the tubes is evacuated. The outer, opaque surface of the inner tube absorbs solar radiation, and a working fluid is passed through the tube to collect the solar energy. The collector design generally consists of a row of such tubes arranged in front of a reflecting panel.

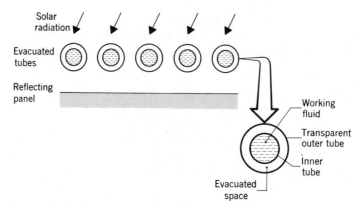

Identify all heat transfer processes relevant to the performance of this device.

h) A solar energy collector design that has been used for agricultural applications is shown as follows. Air is blown through a long duct whose cross section is in the form of an equilateral triangle. One side of the triangle is comprised of a double-paned, semitransparent cover, while the other two sides are constructed from aluminum sheets painted flat black on the inside and covered on the outside with a layer of styrofoam insulation. During sunny periods air entering the system is heated for delivery to either a greenhouse, grain drying unit, or a storage system.

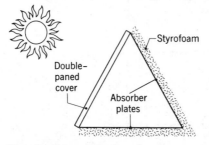

Identify all heat transfer processes associated with the cover plates, the absorber plate(s), and the air.

i) A thermocouple junction is used to measure the temperature of a hot gas stream flowing through a channel by inserting the junction into the mainstream of the gas. The surface of the channel is cooled such that its temperature is well below that of the gas. Identify the heat transfer processes associated with the junction surface. Will the junction sense a temperature that is less than, equal to, or greater than the gas temperature? A radiation shield is a small, open-ended tube that encloses the thermocouple junction, yet allows for passage of the gas through the tube. How does use of such a shield improve the accuracy of the temperature measurement?

j) A double-glazed, glass fire screen is inserted between a woodburning fireplace and the interior of a room. The screen consists of two vertical glass plates that are separated by a space through which room air may flow (the space is open at the top and bottom). Identify the heat transfer processes associated with the fire screen.

2 Introduction to Conduction

Recall that conduction (heat transfer by diffusion) refers to the transport of energy in a medium due to a temperature gradient, and the physical mechanism is that of random atomic or molecular activity. In this chapter we consider in greater detail the conduction rate equation and the relationship of energy conservation to the conduction process. In Chapters 3 to 5 we consider applications involving various geometrical and time-dependent conditions.

2.1 THE CONDUCTION RATE EQUATION

Although the conduction (energy diffusion) rate equation, Fourier's law, was introduced in Section 1.2, it is now appropriate to consider its origin. Fourier's law is a phenomenological law; that is, the law is developed from observed phenomena rather than being derived from first principles. Hence, we view the rate equation as a generalization based upon much experimental evidence. For example, consider the conduction experiment of Figure 2.1. A cylindrical rod of known material is insulated on its lateral surface, while its end faces are maintained at different temperatures with $T_1 > T_2$. The temperature difference causes conduction heat transfer in the positive x direction. We are able to measure the heat transfer rate, q_x, and we seek to determine how q_x depends on the variables ΔT, the temperature difference; Δx, the rod length; and A, the cross-sectional area.

We might imagine first holding ΔT and Δx constant and varying A. If we did so, we would find that q_x is directly proportional to A. Similarly, holding ΔT and A constant, we would observe that q_x varies inversely with Δx. Finally, holding A and Δx constant, we would find that q_x is directly proportional to ΔT. The collective effect is then

$$q_x \propto A \frac{\Delta T}{\Delta x}$$

In changing the material (e.g., from a metal to a plastic), we would find that the above proportionality remains valid. However, we would also find that, for equal values of A, Δx, and ΔT, the value of q_x would be smaller for the plastic than for the metal. This suggests that the proportionality may be converted to

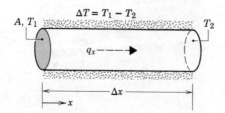

Figure 2.1 Heat conduction experiment.

an equality by introducing a coefficient that is a measure of the material behavior. Hence, we write

$$q_x = kA \frac{\Delta T}{\Delta x}$$

where k, the *thermal conductivity* (W/m·K), is an important *property* of the material. Evaluating this expression in the limit as $\Delta x \to 0$, we obtain for the heat *rate*

$$q_x = -kA \frac{dT}{dx} \tag{2.1}$$

or for the heat *flux*

$$q_x'' = \frac{q_x}{A} = -k \frac{dT}{dx} \tag{2.2}$$

Recall that the minus sign is necessitated by the fact that heat is always transferred in the direction of decreasing temperature.

Fourier's law, as written in Equation 2.2, implies that the heat flux is a directional quantity. In particular the direction of q_x'' is *normal* to the cross-sectional area, A. Or, more generally, the direction of heat flow will always be normal to a surface of constant temperature, called an *isothermal* surface. Figure 2.2 illustrates the direction of heat flow, q_x'', in a one-dimensional

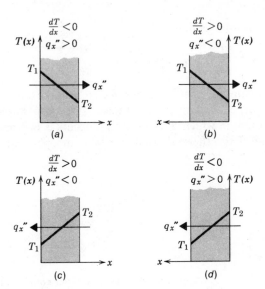

Figure 2.2 The relationship between coordinate system, heat flow direction, and temperature gradient in one dimension.

coordinate system when the gradient, dT/dx, is positive or negative. You should study this figure carefully to recognize the relationship between the temperature gradient, the direction of heat flow and the space coordinate. For example, in (d) we have a negative temperature gradient causing a heat flow in the positive x direction. Note also that the isothermal surfaces are planes normal to the x direction.

Recognizing that the heat flux is a vector quantity, we can write a more general statement of the conduction rate equation (*Fourier's law*) as follows:

$$q'' = -k\,\nabla T = -k\left(i\frac{\partial T}{\partial x} + j\frac{\partial T}{\partial y} + k\frac{\partial T}{\partial z}\right) \tag{2.3}$$

where ∇ is the three-dimensional del operator and T is the scalar temperature field. It is implicit in Equation 2.3 that the heat flux vector is in a direction perpendicular to the isothermal surfaces. An alternative form of Fourier's law is therefore

$$q''_n = -k\frac{\partial T}{\partial n} \tag{2.4}$$

where q''_n is the heat flux in a direction n which is normal to an *isotherm*, as shown for the two-dimensional case in Figure 2.3. The heat transfer is sustained by a temperature gradient along n. Note also that the heat flux vector can be resolved into components such that, in cartesian coordinates, the general expression for q'' is

$$q'' = i\,q''_x + j\,q''_y + k\,q''_z \tag{2.5}$$

where, from Equation 2.3, it follows that

$$q''_x = -k\frac{\partial T}{\partial x} \qquad q''_y = -k\frac{\partial T}{\partial y} \qquad q''_z = -k\frac{\partial T}{\partial z} \tag{2.6}$$

Each of these expressions relates the heat flux *across a surface* to the temperature gradient in a direction perpendicular to the surface. It is also implicit in Equation 2.3 that the medium in which the conduction occurs is *isotropic*. For such a medium the value of the thermal conductivity is independent of the coordinate direction.

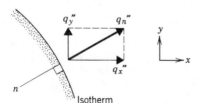

Figure 2.3   The heat flux vector normal to an isotherm in a two-dimensional coordinate system.

Since Fourier's law provides the basis of conduction heat transfer, its key features are summarized as follows. It is *not* an expression that may be derived from first principles; it is instead a generalization based on experimental evidence. It is also an expression that *defines* an important material property, the thermal conductivity. In addition, Fourier's law is a vector expression indicating that the heat flux is normal to an isotherm and in the direction of decreasing temperature. Finally, note that Fourier's law applies for all matter regardless of its state: solid, liquid, or gas.

EXAMPLE 2.1

Referring to Figure 2.2 consider a plane wall of 100 mm thickness and thermal conductivity $k = 100$ W/m·K. Steady-state conditions are known to exist with $T_1 = 400$ K and $T_2 = 600$ K. Determine the heat flux q''_x and the temperature gradient dT/dx for the coordinate systems shown in Figure 2.2c and d.

SOLUTION

KNOWN:

Temperature difference across a plane wall.

FIND:

The heat flux, q''_x, and temperature gradient, dT/dx, for the two coordinate systems.

SCHEMATIC:

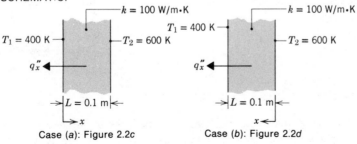

Case (a): Figure 2.2c   Case (b): Figure 2.2d

ASSUMPTIONS:

1. One-dimensional conduction.
2. Constant properties.
3. No internal heat generation.

ANALYSIS:

The rate equation for this situation is given by Equation 2.2

$$q_x'' = -k\frac{dT}{dx}$$

For the prescribed conditions (one-dimensional, steady-state conduction with no heat generation), conservation of energy requires that the heat flux, q_x'', be a constant, independent of x. Moreover, since the thermal conductivity is assumed to be constant, it follows that the temperature gradient, dT/dx, is also independent of x. Hence the temperature distribution is linear with

Case (a): $\dfrac{dT}{dx} = \dfrac{T_2 - T_1}{L} = (600 - 400)\,\text{K}/0.1\,\text{m} = 2000\,\text{K/m}$ ◁

Case (b): $\dfrac{dT}{dx} = \dfrac{T_1 - T_2}{L} = (400 - 600)\,\text{K}/0.1\,\text{m} = -2000\,\text{K/m}$ ◁

Using these gradients to calculate the heat flux,

Case (a): $q_x'' = -100\,\text{W/m}\cdot\text{K} \times 2000\,\text{K/m} = -200\,\text{kW/m}^2$ ◁

Case (b): $q_x'' = -100\,\text{W/m}\cdot\text{K} \times (-2000\,\text{K/m}) = +200\,\text{kW/m}^2$ ◁

COMMENTS:

1. Carefully note the temperature gradient and the direction of q_x'' for each case.
2. Recognize that Fourier's law involves directional quantities.

2.2 THE THERMAL PROPERTIES OF MATTER

Using the heat diffusion law necessitates knowledge of the thermal conductivity. This property, which is referred to as a *transport property*, provides an indication of the rate at which energy is transferred by the diffusion process. It depends on the physical structure of matter, atomic and molecular, which is related to the state of the matter. In this section we consider various forms of matter, identifying important aspects of their behavior and presenting typical property values.

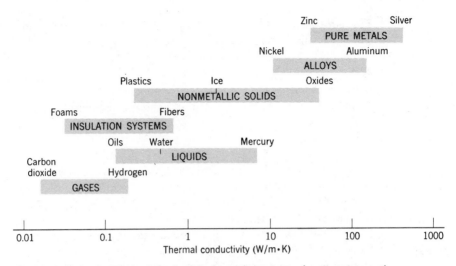

Figure 2.4 Range of thermal conductivity for various states of matter at normal temperatures and pressure.

2.2.1 Thermal Conductivity

From Fourier's law, Equation 2.6, the thermal conductivity is defined as

$$k \equiv -q_x''/\partial T/\partial x$$

It follows that, for a prescribed temperature gradient, the conduction heat flux increases with increasing thermal conductivity. Recalling the physical mechanism associated with conduction (Section 1.2.1), it also follows that, in general, the thermal conductivity of a solid is larger than that of a liquid, which is larger than that of a gas. As illustrated in Figure 2.4, the thermal conductivity of a solid may be more than four orders of magnitude larger than that of a gas. This trend is due largely to the difference in the intermolecular spacing for the two states.

The Solid State

In the modern view of materials, a solid may be comprised of free electrons and of atoms bound in a periodic arrangement called the lattice. Accordingly, transport of thermal energy is due to two effects: the migration of free electrons and lattice vibrational waves. These effects are additive, such that the thermal conductivity, k, is the sum of the electronic component, k_e, and the lattice component, k_l

$$k = k_e + k_l$$

For *pure metals*, k_e is much larger than k_l and is determined by the Wiedemann-Franz-Lorenz law [1]

$$k_e = \frac{L_0 T}{\rho_e}$$

The Lorenz number, L_0, is a constant, and the electrical resistivity, ρ_e, is composed of two parts

$$\rho_e = \rho_0 + \rho' T$$

The ρ_0 term is the residual resistivity at absolute zero and is associated with imperfections in the lattice. The second term is the intrinsic resistivity, and it is the dominant term at temperatures in excess of 100 K. It follows that, to a first approximation, k_e is independent of temperature.

For *nonmetallic* solids, k is determined primarily by k_l, which depends on the frequency of interactions between the atoms of the lattice. Due to the fact that this frequency increases with temperature, the value of k_l increases with increasing temperature for temperatures in excess of 100 K. The regularity of the lattice arrangement has an important effect on k_l, with crystalline (well-ordered) materials like quartz having a higher thermal conductivity than amorphous materials like glass. In fact, at temperatures near 100 K, the value of k_l for crystalline, nonmetallic solids can be quite large, with values for solids such as diamond and beryllium oxide exceeding those of good conductors, such as copper.

For *alloys* the magnitude of k_e is less than that of pure metals, which are of lower electrical resistivity, ρ_e, and the contribution of k_l to the thermal conductivity is no longer negligible. The net effect is that, in general, the thermal conductivity increases with increasing temperature.

The behavior of these three solid forms can be seen in Figure 2.5. Values for selected materials of technical importance are also provided in Table A.1 (metallic solids) and Tables A.2 and A.3 (nonmetallic solids). More detailed treatments of thermal conductivity, including low temperature (<100 K) effects, are available in the literature [1].

Insulation Systems

Thermal insulations are comprised of low thermal conductivity materials fabricated or combined in a manner to achieve an even lower system thermal conductivity. In *fiber*, *powder*, and *flake* type insulations, the solid material is finely dispersed throughout an airspace. The effective conductivity of the system depends on several characteristics, which include the thermal conductivity and surface radiative properties of the solid material and the nature and volumetric fraction of the air or void space. A special parameter of the system is its bulk

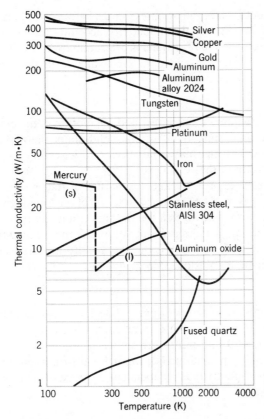

Figure 2.5 The temperature dependence of the thermal conductivity of selected solids.

density (solid mass/total volume), which depends strongly on the manner in which the solid material is interconnected.

If small voids or hollow spaces are formed by bonding or fusing portions of the solid material, a rigid matrix is created. When these spaces are sealed from each other, the system is referred to as a cellular insulation. Examples of such rigid insulations are *foamed* systems, particularly those made from plastic and glass materials. *Reflective* insulations are composed of multilayered, parallel, thin sheets or foils of high reflectivity, which are spaced so as to reflect radiant heat back to its source. The spacing between the foils is designed to restrict the motion of air, and in high performance insulations, the space is even evacuated. In all types of insulation, evacuation of the air in the void space will reduce the effective thermal conductivity of the system.

It is important to recognize that heat transfer through any of these insulation systems may include several modes: conduction through the solid materials; conduction or convection through the air in the void spaces; and, if the temperature is sufficiently high, radiation exchange between the surfaces of the solid matrix. The *effective* thermal conductivity accounts for all of these effects, and values for selected insulation systems are summarized in Table A.3. Additional background information and data are available in the literature [2, 3].

The Fluid State

Since the intermolecular spacing is much larger and the motion of the molecules is more random for the fluid state than for the solid state, thermal energy transport is less effective. The thermal conductivity of gases and liquids is therefore generally smaller than that of solids.

The effect of temperature, pressure and the nature of the species on the thermal conductivity of a gas may be explained in terms of the kinetic theory of gases [4]. From this theory it is known that the thermal conductivity is directly proportional to the number of particles per unit volume, n, the mean molecular speed, $\bar{c}$, and the mean free path, λ, which is the average distance traveled by a molecule before experiencing a collision. Hence

$$k \propto n\bar{c}\lambda$$

Because $\bar{c}$ increases with increasing temperature and decreasing molecular mass, the thermal conductivity of a gas increases with increasing temperature and decreasing molecular weight. These trends are shown in Figure 2.6. However, because n and λ are directly and indirectly proportional to the gas pressure, respectively, the thermal conductivity is independent of pressure. This assump-

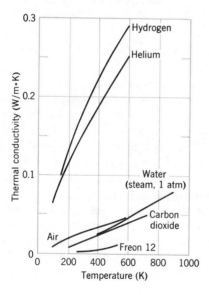

Figure 2.6 The temperature dependence of the thermal conductivity of selected gases at normal pressures.

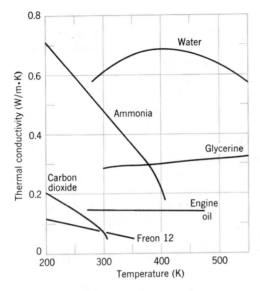

Figure 2.7 The temperature dependence of the thermal conductivity of selected nonmetallic liquids under saturated conditions.

tion is appropriate for the gas pressures of interest in this text. Accordingly, although the values of k presented in Table A.4 pertain to atmospheric pressure or the saturation pressure corresponding to the prescribed temperature, they may be used over a much wider range.

Molecular conditions associated with the liquid state are more difficult to describe and physical mechanisms for explaining the thermal conductivity are not well understood [5]. As shown in Figure 2.7, the thermal conductivity of nonmetallic liquids generally decreases with increasing temperature, the notable exceptions being glycerine and water. This property is insensitive to pressure except near the critical point. Also, it generally follows that thermal conductivity decreases with increasing molecular weight. Values of the thermal conductivity are generally tabulated as a function of temperature for the saturated state of the liquid. Tables A.5 and A.6 present such data for several common liquids.

Metallic liquids are useful as heat transfer media in high heat flux situations, as occur in nuclear power plants. The thermal conductivity of such liquids is given in Table A.7. Note that the values are much larger than for the nonmetallic liquids [6].

2.2.2 Other Relevant Properties

In our subsequent analysis of heat transfer problems, it will be necessary to use many properties of matter. These properties are generally referred to as *thermophysical* properties and include two distinct categories, *transport* and *thermodynamic* properties. The transport properties include the diffusion rate coefficients such as k, the thermal conductivity (for heat transfer) and v, the kinematic viscosity (for momentum transfer). Thermodynamic properties, on the other hand, define the equilibrium state of a system; density and specific heat are two such properties used extensively in thermodynamic analysis. Such properties are provided in the tables of Appendix A. At this point, it would be useful for you to take some time to become acquainted with the information contained in these tables.

Often the accuracy of engineering calculations depends on the accuracy with which the thermophysical properties are known. Numerous examples could be cited wherein flaws in equipment or process design or failure to meet performance specifications were attributable to misinformation associated with the selection of key property values used in the initial system analysis. Selection of reliable property data is an integral part of any careful engineering analysis. The casual use of data from the literature or handbooks, which has not been well characterized or evaluated, is to be avoided. Recommended data values for many thermophysical properties can be obtained from Reference 7. This reference, available in most institutional libraries, was prepared by the Thermophysical Properties Research Center (TPRC) at Purdue University. This center maintains a continuing program to provide current, comprehensive coverage on thermophysical properties [8].

EXAMPLE 2.2

The thermal diffusivity, α, is the controlling transport property for transient diffusion and is equal to $k/\rho c_p$. Using the appropriate values of k, ρ, and c_p from Appendix A, calculate α for the following materials at the prescribed temperatures:

　　Pure aluminum, 300 K and 700 K
　　Silicon carbide, 1000 K
　　Paraffin, 300 K

SOLUTION

KNOWN:

Definition of the thermal diffusivity, α.

FIND:

Numerical values of the thermal diffusivity for selected materials and temperatures.

PROPERTIES:

Table A.1, pure aluminum (300 K):

$$\left.\begin{array}{l} \rho \ = 2702 \ \text{kg/m}^3 \\[8pt] c_p = 903 \ \text{J/kg} \cdot \text{K} \\[8pt] k \ = 237 \ \text{W/m} \cdot \text{K} \end{array}\right\} \quad \alpha = \frac{k}{\rho c_p} = \frac{237 \ \text{W/m} \cdot \text{K}}{2702 \ \text{kg/m}^3 \times 903 \ \text{J/kg} \cdot \text{K}}$$
$$= 97.1 \times 10^{-6} \ \text{m}^2/\text{s}$$

Table A.1, pure aluminum (700 K):

$\rho \ = 2702 \ \text{kg/m}^3$ at 300 K

$c_p = 1090 \ \text{J/kg} \cdot \text{K}$ at 700 K (by linear interpolation)

$k \ = 225 \ \text{W/m} \cdot \text{K}$ at 700 K (by linear interpolation)

Hence

$$\alpha \ = \frac{k}{\rho c_p} = \frac{225 \ \text{W/m} \cdot \text{K}}{2702 \ \text{kg/m}^3 \times 1090 \ \text{J/kg} \cdot \text{K}} = 76 \times 10^{-6} \ \text{m}^2/\text{s}$$

Table A.2, silicon carbide (1000 K):

$$\left.\begin{array}{ll} \rho \ = 3160 \ \text{kg/m}^3 & \text{at 300 K} \\[8pt] c_p = 1195 \ \text{J/kg} \cdot \text{K} & \text{at 1000 K} \\[8pt] k \ = 87 \ \text{W/m} \cdot \text{K} & \text{at 1000 K} \end{array}\right\} \quad \alpha = \frac{87 \ \text{W/m} \cdot \text{K}}{3160 \ \text{kg/m}^3 \times 1195 \ \text{J/kg} \cdot \text{K}}$$
$$= 23 \times 10^{-6} \ \text{m}^2/\text{s}$$

Table A.3, paraffin (300 K):

$$\left.\begin{array}{l} \rho \ = 900 \ \text{kg/m}^3 \\[8pt] c_p = 2890 \ \text{J/kg} \cdot \text{K} \\[8pt] k \ = 0.020 \ \text{W/m} \cdot \text{K} \end{array}\right\} \quad \alpha = \frac{k}{\rho c_p} = \frac{0.020 \ \text{W/m} \cdot \text{K}}{900 \ \text{kg/m}^3 \times 2890 \ \text{J/kg} \cdot \text{K}}$$
$$= 7.7 \times 10^{-9} \ \text{m}^2/\text{s}$$

COMMENTS:

1. Note temperature dependence of the thermophysical properties of aluminum and silicon carbide. For example, for silicon carbide, $\alpha(1000 \ \text{K}) \approx 0.1 \times \alpha(300 \ \text{K})$; hence properties of this material have a strong temperature dependence.

2. The physical interpretation of α is that it provides a measure of heat transport (k) relative to energy storage (ρc_p). In general, metallic

materials have higher α, while nonmetallic solids — like paraffin — have lower values of α.

3. Linear interpolation of property values is generally acceptable for engineering purposes.

4. Use of the low-temperature (300 K) density at higher temperatures ignores thermal expansion effects and is also acceptable for engineering purposes.

2.3 THE HEAT DIFFUSION EQUATION

A major objective in a conduction analysis is to determine the *temperature field* in a medium resulting from conditions that are imposed on the boundaries of the medium. That is, we wish to know the *temperature distribution*, which represents how temperature varies with position in the medium. Once this distribution is known, the heat transfer rate at any point in the medium or on its surface may be computed from Fourier's law. Other important quantities of interest may also be determined. For the solid medium, knowledge of the temperature distribution could be used to ascertain structural integrity through determination of thermal stresses, expansions, and deflections. The temperature distribution could also be used to optimize the thickness of an insulating material or to determine the compatibility of special coatings or adhesives used with the material.

We now consider the manner in which the temperature distribution can be determined. The approach follows the methodology described in Section 1.3.3 of applying the energy conservation requirement. That is, we will define a differential control volume, identify the relevant energy transfer processes, and introduce the appropriate rate equations. The result will be a differential equation whose solution provides the temperature distribution in the medium.

Consider a homogeneous medium in which temperature gradients exist and the temperature distribution, $T(x, y, z)$, is expressed in cartesian coordinates. Following the methodology of applying conservation of energy, Section 1.3.3, we first define an infinitesimally small (differential) control volume, $dx \cdot dy \cdot dz$, as shown in Figure 2.8. The second step is to consider the energy processes that are relevant to this control volume. If there are temperature gradients, conduction heat transfer will occur across each of the control surfaces. The conduction heat rates perpendicular to each of the control surfaces at the x, y, and z coordinate locations are indicated by the terms q_x, q_y, and q_z, respectively. The conduction heat rates at the opposite surfaces can then be expressed as

$$q_{x+dx} = q_x + \frac{\partial q_x}{\partial x} \, dx \qquad (2.7a)$$

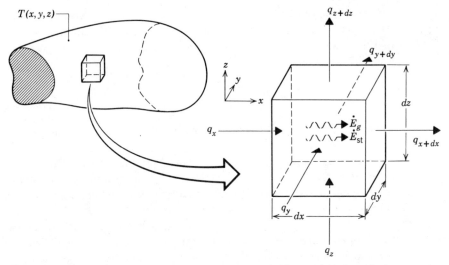

Figure 2.8 Differential control volume, *dxdydz*, for conduction analysis in cartesian coordinates.

$$q_{y+dy} = q_y + \frac{\partial q_y}{\partial y} dy \tag{2.7b}$$

$$q_{z+dz} = q_z + \frac{\partial q_z}{\partial z} dz \tag{2.7c}$$

In words, Equation 2.7a simply states that the x component of the heat transfer rate at $x + dx$ is equal to the value of this component at x plus the amount by which it changes with respect to x times dx. This condition is graphically illustrated in Figure 2.9 for the x-coordinate direction. In writing these equations, we are acknowledging that the components of the conduction rate must vary continuously with x, y, and z.

Within the medium there may also be an *energy source* term associated with the generation of thermal energy. This term is represented as

$$\dot{E}_g = \dot{q} \, dxdydz \tag{2.8}$$

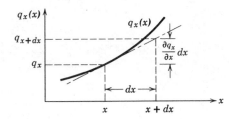

Figure 2.9 Graphical illustration of Equation 2.7a expressing the change in the heat transfer rate as a function of distance.

where $\dot{q}$ is the rate at which energy is generated per unit volume of the medium (W/m^3). In addition there may occur changes in the amount of the internal thermal energy stored by the material in the control volume. We express this *energy storage* term as

$$\dot{E}_{st} = \rho c_p \frac{\partial T}{\partial t} dxdydz \qquad (2.9)$$

where $\rho c_p \, \partial T / \partial t$ is the time rate of change of the internal energy of the medium per unit volume.

Once again it is important to note that the terms $\dot{E}_g$ and $\dot{E}_{st}$ represent different physical processes. The energy generation term, $\dot{E}_g$, is a manifestation of some energy conversion process involving thermal energy on one hand and chemical, electrical, or nuclear energy on the other. The term is positive (a *source*) if thermal energy is being generated in the material at the expense of some other energy form; it is negative (a *sink*) if thermal energy is being consumed. In contrast the energy storage term, $\dot{E}_{st}$, refers specifically to the rate of change of internal energy stored by the matter.

The third step in the methodology outlined in Section 1.3.3 is to express conservation of energy using the foregoing rate equations. On a *rate* basis, the general form of the conservation of energy requirement is

$$\dot{E}_{in} + \dot{E}_g - \dot{E}_{out} = \dot{E}_{st} \qquad (1.10)$$

Hence, recognizing that the conduction rates constitute the energy inflow, $\dot{E}_{in}$, and outflow, $\dot{E}_{out}$, and substituting Equations 2.8 and 2.9 we obtain

$$q_x + q_y + q_z + \dot{q} \, dxdydz - q_{x+dx} - q_{y+dy}$$
$$- q_{z+dz} = \rho c_p \frac{\partial T}{\partial t} dxdydz \qquad (2.10)$$

Substituting from Equations 2.7 we then obtain

$$-\frac{\partial q_x}{\partial x} dx - \frac{\partial q_y}{\partial y} dy - \frac{\partial q_z}{\partial z} dz + \dot{q} dxdydz = \rho c_p \frac{\partial T}{\partial t} dxdydz \qquad (2.11)$$

The conduction heat rates may be evaluated from Fourier's law,

$$q_x = -k \, dydz \frac{\partial T}{\partial x} \qquad (2.12a)$$

$$q_y = -k \, dxdz \frac{\partial T}{\partial y} \qquad (2.12b)$$

$$q_z = -k \, dxdy \frac{\partial T}{\partial z} \qquad (2.12c)$$

where each heat flux component of Equation 2.6 has been multiplied by the appropriate control surface (differential) area to obtain the heat transfer rate. Substituting Equations 2.12 into Equation 2.11 and dividing out the dimensions of the control volume ($dx\,dy\,dz$), we obtain

$$\frac{\partial}{\partial x}\left(k\frac{\partial T}{\partial x}\right) + \frac{\partial}{\partial y}\left(k\frac{\partial T}{\partial y}\right) + \frac{\partial}{\partial z}\left(k\frac{\partial T}{\partial z}\right) + \dot{q} = \rho c_p \frac{\partial T}{\partial t} \qquad \text{HEAT EQUATION} \tag{2.13}$$

Equation 2.13 is the general form, in cartesian coordinates, of the *heat diffusion equation.* This equation, usually known as the *heat equation*, provides the basic tool for heat conduction analysis. From its solution, we can obtain the temperature distribution, $T(x,y,z)$, as a function of time. The apparent complexity of this expression should not obscure the fact that it describes an important physical condition, that is, conservation of energy. You should have a clear understanding of the physical significance of each term appearing in the equation. For example, the term $\partial/\partial x(k\,\partial T/\partial x)$ describes the *net* conduction heat flux *into* the control volume for the x-coordinate direction. That is,

$$\frac{\partial}{\partial x}\left(k\frac{\partial T}{\partial x}\right) = q''_x - q''_{x+dx} \tag{2.14}$$

with similar expressions applying for the fluxes in the y and z directions. In words, the heat equation, Equation 2.13, therefore states that *at any point in a medium the net conduction heat rate into a unit volume plus the volumetric rate of thermal energy generation must equal the rate of change of thermal energy stored within the volume.*

It is often possible to work with simplified versions of Equation 2.13. For example, if the thermal conductivity is a constant independent of position or temperature, the heat equation may be expressed as

$$\frac{\partial^2 T}{\partial x^2} + \frac{\partial^2 T}{\partial y^2} + \frac{\partial^2 T}{\partial z^2} + \frac{\dot{q}}{k} = \frac{1}{\alpha}\frac{\partial T}{\partial t} \tag{2.15}$$

where $\alpha = k/\rho c_p$ is the *thermal diffusivity*. This important thermophysical property is the ratio of the thermal conductivity, k, of the medium to the thermal capacitance, ρc_p. A large value of α (large k and/or low ρc_p) implies that a medium is more effective in transferring energy by conduction than it is in storing energy.

Additional simplifications of the general form of the heat equation are often possible. For example, under *steady-state* conditions, there can be no change in the amount of energy storage and hence Equation 2.13 reduces to

$$\frac{\partial}{\partial x}\left(k\frac{\partial T}{\partial x}\right) + \frac{\partial}{\partial y}\left(k\frac{\partial T}{\partial y}\right) + \frac{\partial}{\partial z}\left(k\frac{\partial T}{\partial z}\right) + \dot{q} = 0 \qquad (2.16)$$

Moreover, if the heat transfer is *one-dimensional* (e.g., in the x direction) and there is *no energy generation*, Equation 2.16 reduces to

$$\frac{d}{dx}\left(k\frac{dT}{dx}\right) = 0 \qquad (2.17)$$

The important implication of this result is that *under steady-state, one-dimensional conditions with no energy generation*, the heat flux is a constant in the direction of transfer $(dq_x''/dx = 0)$.

The heat equation may also be expressed in cylindrical and spherical coordinates. The differential control volumes for these two coordinate systems are shown in Figures 2.10 and 2.11. Applying the methodology used for the cartesian coordinates, the general forms of the heat equation are,

Cylindrical Coordinates:

$$\frac{1}{r}\frac{\partial}{\partial r}\left(kr\frac{\partial T}{\partial r}\right) + \frac{1}{r^2}\frac{\partial}{\partial \phi}\left(k\frac{\partial T}{\partial \phi}\right)$$

$$+ \frac{\partial}{\partial z}\left(k\frac{\partial T}{\partial z}\right) + \dot{q} = \rho c_p \frac{\partial T}{\partial t} \qquad (2.18)$$

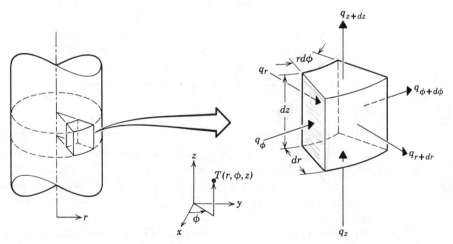

Figure 2.10 Differential control volume, $dr \cdot rd\phi \cdot dz$, for conduction analysis in cylindrical coordinates, (r,ϕ,z).

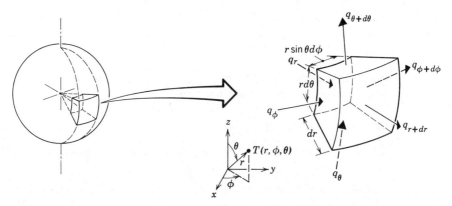

Figure 2.11   Differential control volume, $dr \cdot r\sin\theta d\phi \cdot rd\theta$, for conduction analysis in spherical coordinates, (r, ϕ, θ).

Spherical Coordinates:

$$\frac{1}{r^2}\frac{\partial}{\partial r}\left(kr^2\frac{\partial T}{\partial r}\right) + \frac{1}{r^2\sin^2\theta}\frac{\partial}{\partial \phi}\left(k\frac{\partial T}{\partial \phi}\right)$$

$$+ \frac{1}{r^2\sin\theta}\frac{\partial}{\partial \theta}\left(k\sin\theta\frac{\partial T}{\partial \theta}\right) + \dot{q} = \rho c_p\frac{\partial T}{\partial t} \qquad (2.19)$$

Since it is important that you be able to apply conservation principles to differential control volumes, you should attempt to derive one of these expressions (see Problems 2.23 and 2.24).

EXAMPLE 2.3

The temperature distribution across a wall 1 m thick at a certain instant of time is given as

$$T(x) = a + bx + cx^2$$

where T is in °C and x is in m, while $a = 900$°C, $b = -300$°C/m, and $c = -50$°C/m². A uniform heat generation, $\dot{q} = 1000$ W/m³, is present in the wall of area 10 m² having the properties $\rho = 1600$ kg/m³, $k = 40$ W/m·K, and $c_p = 4$ kJ/kg·K.

1. Determine the rate of heat transfer entering the wall $(x = 0)$ and leaving the wall $(x = 1$ m$)$.

2. Determine the rate of change of energy storage in the wall.

3. Determine the time rate of temperature change at $x = 0$, 0.25, and 0.5 m.

SOLUTION

KNOWN:

Temperature distribution, $T(x)$, at a given instant of time, t, within a one-dimensional wall with uniform heat generation.

FIND:

1. Heat rates entering, q_{in} ($x = 0$), and leaving, q_{out} ($x = 1$), the wall.
2. Rate of change of energy storage in the wall, $\dot{E}_{st}$.
3. Time rate of temperature change at $x = 0$, 0.25, and 0.5 m.

SCHEMATIC:

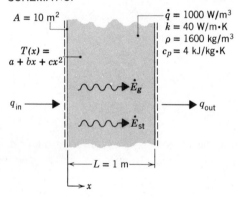

$A = 10\ m^2$

$T(x) = a + bx + cx^2$

$\dot{q} = 1000\ W/m^3$
$k = 40\ W/m\cdot K$
$\rho = 1600\ kg/m^3$
$c_p = 4\ kJ/kg\cdot K$

q_{in}

$\dot{E}_g$

$\dot{E}_{st}$

q_{out}

$L = 1\ m$

x

ASSUMPTIONS:

1. One-dimensional conduction in the x direction.
2. Homogeneous medium with constant properties.
3. Uniform internal heat generation, $\dot{q}$ (W/m³).

ANALYSIS:

1. Recall that once the temperature distribution is known for a medium, it is a simple matter to determine the conduction heat transfer rate at any point in the medium, or at its surfaces, by using Fourier's law. Hence, the desired heat rates may be determined by using the prescribed temperature distribution with Equation 2.1. Accordingly,

$$q_{in} = q_x(0) = -kA\frac{\partial T}{\partial x}\bigg|_{x=0} = -kA(b + 2cx)_{x=0}$$

$$q_{\text{in}} = -bkA = 300\ ^{\circ}\text{C/m} \times 40\ \text{W/m}\cdot\text{K} \times 10\ \text{m}^2$$

$$q_{\text{in}} = 120\ \text{kW} \qquad \triangleleft$$

Similarly,

$$q_{\text{out}} = q_x(L) = -kA \left.\frac{\partial T}{\partial x}\right|_{x=L} = -kA(b + 2cx)_{x=L}$$

$$q_{\text{out}} = -(b + 2cL)kA = -[-300\ ^{\circ}\text{C/m}$$

$$+ 2(-50\ ^{\circ}\text{C/m}^2) \times 1\ \text{m}] \times 40\ \text{W/m}\cdot\text{K} \times 10\ \text{m}^2$$

$$q_{\text{out}} = 160\ \text{kW} \qquad \triangleleft$$

2. The rate of change of energy storage in the wall, $\dot{E}_{\text{st}}$, may be determined by applying an overall energy balance to the wall. Using Equation 1.10 for a control volume about the wall,

$$\dot{E}_{\text{in}} + \dot{E}_g - \dot{E}_{\text{out}} = \dot{E}_{\text{st}}$$

where $\dot{E}_g = \dot{q}AL$, it follows that

$$\dot{E}_{\text{st}} = \dot{E}_{\text{in}} + \dot{E}_g - \dot{E}_{\text{out}} = q_{\text{in}} + \dot{q}AL - q_{\text{out}}$$

$$\dot{E}_{\text{st}} = 120\ \text{kW} + 1000\ \text{W/m}^3 \times 10\ \text{m}^2 \times 1\ \text{m} - 160\ \text{kW}$$

$$\dot{E}_{\text{st}} = -30\ \text{kW} \qquad \triangleleft$$

3. The time rate of change of the temperature at any point in the medium may be determined from the heat equation, Equation 2.15, rewritten as

$$\frac{\partial T}{\partial t} = \frac{k}{\rho c_p} \frac{\partial^2 T}{\partial x^2} + \frac{\dot{q}}{\rho c_p}$$

From the prescribed temperature distribution, it follows that

$$\frac{\partial^2 T}{\partial x^2} = \frac{\partial}{\partial x}\left(\frac{\partial T}{\partial x}\right) = \frac{\partial}{\partial x}(b + 2cx) = 2c = 2(-50\ ^{\circ}\text{C/m}^2) = -100\ ^{\circ}\text{C/m}^2$$

Note that this derivative is independent of position in the medium. Hence the time rate of temperature change is also independent of position and is given by

$$\frac{\partial T}{\partial t} = \frac{40\ \text{W/m}\cdot\text{K}}{1600\ \text{kg/m}^3 \times 4\ \text{kJ/kg}\cdot\text{K}} \times (-100\ ^{\circ}\text{C/m}^2)$$

$$+ \frac{1000\ \text{W/m}^3}{1600\ \text{kg/m}^3 \times 4\ \text{kJ/kg}\cdot\text{K}}$$

$$\frac{\partial T}{\partial t} = -6.25 \times 10^{-4} \ ^{\circ}C/s + 1.56 \times 10^{-4} \ ^{\circ}C/s$$

$$\frac{\partial T}{\partial t} = -4.69 \times 10^{-4} \ ^{\circ}C/s \qquad \qquad \triangleleft$$

COMMENTS:

1. From the above result it is evident that the temperature at every point within the wall is decreasing with time.

2. Recognize that Fourier's law can always be used to compute the conduction heat rate from knowledge of the temperature distribution, even for unsteady conditions with internal heat generation.

2.4   BOUNDARY AND INITIAL CONDITIONS

To determine the temperature distribution in a medium, it is necessary to solve the appropriate form of the heat equation. However, such a solution depends on the physical conditions existing at the *boundaries* of the medium and, if the situation is time dependent, on conditions existing in the medium at some *initial time*. With regard to the *boundary conditions*, there exist several common physical effects, which are simply expressed in mathematical form. Because the heat equation is second order in the spatial coordinates, two boundary conditions must be expressed for each coordinate needed to describe the system. Because the equation is first order in time, however, only one condition, referred to as the *initial condition*, must be specified.

The three kinds of boundary conditions that are commonly encountered in heat transfer are summarized in Table 2.1. The conditions are specified at the surface $x = 0$ for a one-dimensional system. The heat transfer is in the positive x direction with the temperature distribution, which may be time dependent, designated as $T(x,t)$. The first condition corresponds to a situation for which the surface is maintained at a fixed or constant temperature, T_s. The second condition corresponds to the existence of a fixed or constant heat flux, q_s'', at the surface. Note that this heat flux is related to the temperature gradient at the surface by Fourier's law, Equation 2.6, which may be expressed as

$$q_x''(0) = -k \left. \frac{\partial T}{\partial x} \right|_{x=0}$$

A special case of this condition corresponds to the *perfectly insulated*, or *adiabatic*, surface for which $\partial T/\partial x|_{x=0} = 0$. The third condition corresponds to the existence of convection heating (or cooling) at the surface. Note the

Table 2.1 Boundary conditions for the heat diffusion equation at the surface ($x = 0$)

1. Constant surface temperature

$$T(0,t) = T_s \qquad (2.20)$$

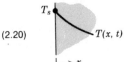

2. Constant surface heat flux

   (a) Finite heat flux

$$-k\left.\frac{\partial T}{\partial x}\right|_{x=0} = q''_s \qquad (2.21)$$

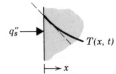

   (b) Adiabatic or insulated surface

$$\left.\frac{\partial T}{\partial x}\right|_{x=0} = 0 \qquad (2.22)$$

3. Convection surface condition

$$-k\left.\frac{\partial T}{\partial x}\right|_{x=0} = h[T_\infty - T(0,t)] \qquad (2.23)$$

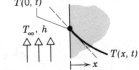

relationship of this last condition to the surface energy balance discussed in Section 1.3.2.

Physical situations in which the foregoing boundary conditions are realized will be illustrated by the examples and problems of this text. However, at this time you should attempt to conceptualize your own examples for each of these conditions.

EXAMPLE 2.4

A long copper bar of rectangular cross section, whose width w is much greater than its thickness L, is maintained in contact with an icebath at its lower surface, and the temperature throughout the bar is approximately equal to that of the ice, T_o. Suddenly, an electric current is passed through the bar and an airstream of temperature T_∞ is passed over the top surface, while the bottom surface continues to be maintained at T_o by the icebath. Obtain the differential equation and the boundary and initial conditions that could be solved to determine the temperature as a function of position and time in the bar.

SOLUTION

KNOWN:

Copper bar initially in thermal equilibrium with an icebath is suddenly heated by passage of an electric current.

FIND:

Differential equation and boundary and initial conditions needed to determine temperature as a function of position and time within the bar.

SCHEMATIC:

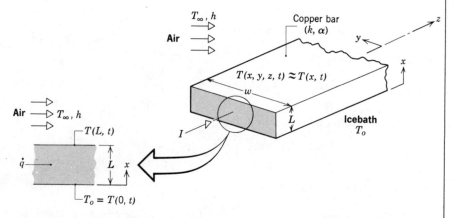

ASSUMPTIONS:

1. Since $w \gg L$, side effects are negligible and heat transfer within the bar is primarily one dimensional in the x direction.
2. Uniform volumetric heat generation, $\dot{q}$.
3. Constant properties.

ANALYSIS:

The temperature distribution is governed by the heat equation, (2.13), which, for the one-dimensional and constant property conditions of the present problem, reduces to

$$\frac{\partial^2 T}{\partial x^2} + \frac{\dot{q}}{k} = \frac{1}{\alpha}\frac{\partial T}{\partial t} \qquad \qquad \lhd \quad (1)$$

where the temperature is a function of position and time, $T(x,t)$. Recognize that this differential equation is second order in the spatial coordinate, x, and first order in time, t. Hence there must be two boundary conditions for the x direction and one condition, termed the initial condition, for time. The boundary condition appropriate for the bottom surface corresponds to case 1 of Table 2.1. In particular, since the temperature of this surface is maintained at a value, T_o, which is fixed with time, it follows that

$$T(0,t) = T_o \qquad \lhd \quad (2)$$

In contrast the convection surface condition, case 3 of Table 2.1, is appropriate for the top surface. Hence

$$-k\left.\frac{\partial T}{\partial x}\right|_{x=L} = h[T(L,t) - T_\infty] \qquad \lhd \quad (3)$$

The initial condition is inferred from recognition of the fact that, prior to the change in conditions, the bar is at a uniform temperature T_o. Hence

$$T(x,0) = T_o \qquad \lhd \quad (4)$$

If T_o, T_∞, $\dot{q}$, and h are known, Equations 1 to 4 may be solved to obtain the temperature distribution $T(x,t)$ following the change in conditions.

COMMENTS:

1. Recognize that the temperature of the top surface, $T(L,t)$, will change with time. This temperature is an unknown and may only be obtained after finding $T(x,t)$.

2. How do you expect the temperature to vary with x at different times following the change in conditions?

2.5 SUMMARY

The primary purpose of this chapter has been to obtain an improved understanding of the conduction rate equation and to become familiar with the heat diffusion equation. You should know the origin and implications of Fourier's law and you should understand the key thermal properties and how they vary for different substances. You should also know the physical meaning of each term appearing in the heat diffusion equation. To what form does this equation reduce for simplified conditions, and what kinds of boundary conditions may be used for its solution? In short you must now have a fundamental understanding of the conduction process and its mathematical description. In the following three chapters we undertake conduction analyses for numerous systems and conditions.

REFERENCES

1. Klemens, P. G., "Theory of the Thermal Conductivity of Solids" in R. P. Tye, Ed., *Thermal Conductivity*, Vol. 1, Academic Press, London, 1969.

2. Mallory, John F., *Thermal Insulation*, Reinhold Book Corp., New York, 1969.

3. American Society of Heating, Refrigerating and Air Conditioning Engineers (ASHRAE), *Handbook of Fundamentals*, Chapters 17 and 31, ASHRAE, New York, 1972.

4. Vincenti, W. G. and C. H. Kruger, Jr., *Introduction to Physical Gas Dynamics*, Wiley, New York, 1965.

5. McLaughlin, E., "Theory of the Thermal Conductivity of Fluids" in R. P. Tye, Ed., *Thermal Conductivity*, Vol. 2, Academic Press, London, 1969.

6. Foust, O. J., Ed., "Sodium Chemistry and Physical Properties," *Sodium-NaK Engineering Handbook*, Vol. 1, Gordon and Breach, Science Publishers, Inc., New York, 1972.

7. Touloukian, Y. S. and C. Y. Ho, Eds., *Thermophysical Properties of Matter, The TPRC Data Series*, 13 volumes on thermophysical properties: thermal conductivity, specific heat, thermal radiative, thermal diffusivity, and thermal linear expansion, Plenum, New York, 1970 through 1977.

8. Center for Information and Numerical Data Analysis and Synthesis (CINDAS), Purdue University, 2595 Yeager Road, West Lafayette, IN 47906.

PROBLEMS

2.1 Consider steady-state, one-dimensional heat conduction through the axisymmetric shape shown below.

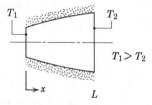

Assuming constant properties and no internal heat generation, sketch the temperature distribution on $T - x$ coordinates. Briefly explain the shape of the curve shown.

2.2 A hot water pipe with outside radius, r_1, has a temperature T_1. A thick insulation applied to reduce the heat loss has an outer radius r_2 and temperature T_2. On $T - r$ coordinates, sketch the temperature distribution in the insulation for one-dimensional, steady-state heat transfer. Give a brief explanation, justifying the shape of the curve shown.

2.3 A spherical shell with inner radius, r_1, and outer radius, r_2, has surface temperatures T_1 and T_2, respectively, where $T_1 > T_2$. Sketch the temperature distribution on $T - r$ coordinates assuming steady-state, one-dimensional conduction with constant properties. Briefly justify the shape of the curve shown.

2.4   Consider steady-state, one-dimensional heat conduction through the symmetric shape shown in the sketch.

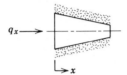

Assuming there is no internal heat generation, derive an expression for the thermal conductivity $k(x)$ for these conditions: $A(x) = (1 - x)$, $T(x) = 300(1 - 2x - x^3)$ and $q = 6000$ W where A has the units (m^2), T(K), and x(m).

2.5   A solid, truncated cone serves as a support for a system that maintains the top (truncated) face of the cone at a temperature T_1, while the base of the cone is at a temperature $T_2 < T_1$.

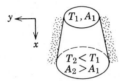

The thermal conductivity of the solid depends on temperature according to the relation $k = k_o + aT$, where a is a positive constant, and the sides of the cone are well insulated. Do the following quantities increase, decrease or remain the same with increasing x: the heat transfer rate, q_x, the heat flux, q_x'', the thermal conductivity, k, and the temperature gradient, dT/dx?

2.6   To determine the effect of the temperature dependence of the thermal conductivity on the temperature distribution in a solid, consider a material for which this dependence may be represented as

$$k = k_o + aT$$

where k_o is a positive constant and a is a coefficient that may be positive or negative. Sketch the steady-state temperature distribution associated with heat transfer in a plane wall for three cases corresponding to $a > 0$, $a = 0$, and $a < 0$.

2.7   One-dimensional, steady-state conduction without heat generation occurs in the system shown below. The thermal conductivity is 25 W/m·K and the thickness, L, is 0.5 m.

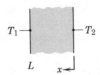

Determine the unknown quantities for each case in the following table (next page), and then sketch the temperature distribution indicating the direction of the heat flux.

CASE	T_1	T_2	$\dfrac{dT}{dx}$ (K/m)	q_x'' (W/m^2)
1	400 K	300 K		
2	100°C		−250	
3	80°C		+200	
4		−5°C		4000
5	30°C			−3000

2.8 Consider steady-state conditions for one-dimensional conduction in the system below having a thermal conductivity $k = 50$ W/m·K and a thickness $L = 0.25$ m, with no internal heat generation.

Determine the heat flux and the unknown quantity for each case and make a sketch of the temperature distribution indicating the direction of the heat flux.

CASE	T_1(°C)	T_2(°C)	$\dfrac{dT}{dx}$ (K/m)
1	50	−20	
2	−30	−10	
3	70		160
4		40	−80
5		30	200

2.9 A solid cylindrical rod of length $L = 0.1$ m and diameter $D = 25$ mm is well insulated on its side, while its end faces are maintained at temperatures of 100°C and 0°C. What is the rate of heat transfer through the rod if it is constructed from (a) pure copper, (b) aluminum alloy 2024-T6, (c) AISI 302 stainless steel, (d) silicon nitride, (e) wood (oak), (f) magnesia, 85 per cent, and (g) pyrex?

2.10 A one-dimensional system without heat generation has a thickness of 20 mm with surfaces maintained at temperatures of 275 K and 325 K. Determine the heat flux through the system if it is constructed from these materials: (a) pure aluminum, (b) plain carbon steel, (c) AISI 316, stainless steel, (d) pyroceram, (e) Teflon, and (f) concrete.

2.11 A TV advertisement by a well-known insulation manufacturer states: it isn't the thickness of the insulating material that counts, it's the R-value. The ad then goes on to show that to obtain an R-value of 19, you need 18 ft of rock, 15 in. of wood or just 6 in. of their fiberglass insulation. Is this advertisement technically reasonable? In case you are like most TV viewers, you don't know the R-value is defined as L/k where L (in.) is the thickness of the insulation and k (Btu·in./hr·ft^2·°F) is the thermal conductivity of the material.

2.12 The Prandtl number (Pr) is a dimensionless parameter used in convection analysis and is defined as $\mu c_p/k$. Using the thermophysical properties μ, c_p, and k from the appropriate tables, calculate the Prandtl number for these fluids: air at 1 atm and 300 K, engine oil at 60°C, water at 373 K, and liquid sodium at 93°C. Verify your calculations by comparison with the tabulated values of Pr.

2.13 The Grashof number is a dimensionless parameter used in free convection analysis and is defined as

$$Gr \equiv \frac{g\beta\Delta T L^3}{v^2}$$

where ΔT is the temperature difference between the surface of an object and the fluid, L is the characteristic length of the object, and β is the volumetric expansion coefficient defined as

$$\beta \equiv -\frac{1}{\rho}\left(\frac{\partial \rho}{\partial T}\right)_p$$

Consider an object having a characteristic length of 0.25 m and a situation for which the temperature difference is 25°C.

a) Show that the volumetric expansion coefficient for perfect gases is $\beta = 1/T$ where T, the temperature, is in absolute units (K, kelvins).

b) Using the required thermophysical properties evaluated at 350 K, calculate the Grashof number for free convection with these fluids: air, hydrogen, water, and ethylene glycol. Assume the pressure is 1 atm.

2.14 The Rayleigh number, Ra, is a dimensionless parameter used in free convection analysis and is defined as

$$Ra \equiv Gr \cdot Pr$$

where Pr, the Prandtl number, and Gr, the Grashof number, are identified and defined in the previous two problems. Determine the Rayleigh number for free convection between an object of characteristic length (L) 0.01 m and the fluids specified for which the temperature difference (ΔT) is 30°C. Consider these fluids and the temperatures at which their properties should be evaluated: (a) air (1 atm, 400 K), (b) helium (1 atm, 400 K), (c) glycerin (12°C) and (d) water (37°C).

2.15 To a good approximation, the transport properties dynamic viscosity, μ, and thermal conductivity, k, are independent of pressure. The kinematic viscosity, v, defined as μ/ρ, depends upon the density, ρ, and therefore may have a significant pressure dependence. Estimate the kinematic viscosity of the following gases for the conditions shown and compare with tabulated values for the 1 atm condition: (a) air (10 atm, 350 K), (b) carbon dioxide (5 atm, 400 K), and (c) hydrogen (2 atm, 300 K).

2.16 In the following two-dimensional body, the gradient at surface A is found to be $\partial T/\partial y = 30$ K/m. What are $\partial T/\partial y$ and $\partial T/\partial x$ at the surface B?

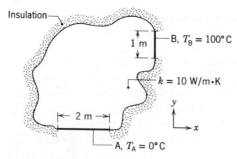

2.17 At a given instant of time the temperature distribution within an infinite homogeneous body is given by the function

$$T(x, y, z) = x^2 - 2y^2 + z^2 - xy + 2yz$$

Assuming constant properties and no internal heat generation, determine the regions where the temperature changes with time.

2.18 Uniform internal heat generation at $\dot{q}_1 = 5 \times 10^7$ W/m³ is occurring in a cylindrical nuclear reactor fuel rod of 50 mm diameter, and under steady-state conditions the temperature distribution is of the form

$$T(r) = a + br^2$$

where T is in °C and r is in m, while $a = 800$°C and $b = -4.167 \times 10^5$ °C/m². The fuel rod properties are $k = 30$ W/m·K, $\rho = 1100$ kg/m³, and $c_p = 800$ J/kg·K.

a) What is the rate of heat transfer per unit length of the rod at $r = 0$ (the centerline) and at $r = 25$ mm (the surface)?

b) If the reactor power level is suddenly increased to $\dot{q}_2 = 10^8$ W/m³, what is the initial time rate of temperature change at $r = 0$ and $r = 25$ mm?

2.19 The steady-state temperature distribution in a one-dimensional wall of thermal conductivity 50 W/m·K and thickness 50 mm is observed to be

$$T(°C) = a + bx^2$$

where $a = 200$°C, $b = -2000$°C/m², and x has units (m).

a) What is the heat generation rate, $\dot{q}$, in the wall?

b) Determine the heat fluxes at the two wall faces. In what manner are these heat fluxes related to the heat generation rate?

2.20 The cylindrical system shown on the opposite page has negligible variation of temperature in the r and z directions. Assume that $\Delta r = r_o - r_i$ is small compared to r_i and denote the length in the z direction, normal to the page, as L.

a) Beginning with a properly defined control volume and considering energy generation and storage effects, derive the differential equation that prescribes the variation in temperature with the angular coordinate ϕ. Compare your result with Equation 2.18.

b) For steady-state conditions with no internal heat generation and constant properties, determine the temperature distribution $T(\phi)$ in terms of the constants T_1, T_2, r_i, and r_o. Is this distribution linear in ϕ?

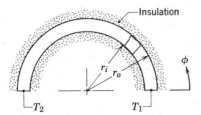

c) For the conditions of part (b) write the expression for the heat rate, q_ϕ.

2.21 Beginning with a properly defined differential control volume element, derive the heat diffusion equation for a one-dimensional, cylindrical, radial coordinate system with internal heat generation. Compare your result with Equation 2.18.

2.22 Beginning with a properly defined differential control volume element, derive the heat diffusion equation for a one-dimensional, spherical, radial coordinate system with internal heat generation. Compare your result with Equation 2.19.

2.23 Derive the heat diffusion equation, Equation 2.18, for cylindrical coordinates beginning with the differential control volume shown in Figure 2.10.

2.24 Derive the heat diffusion equation, Equation 2.19, for spherical coordinates beginning with the differential control volume shown in Figure 2.11.

2.25 In Example 2.4, we considered a copper bar that was initially at a uniform temperature and suddenly was heated by the passage of an electric current. Recall that the emphasis of the problem was to identify boundary and initial conditions that must be used in the conduction analysis.

 a) On $T - x$ coordinates, sketch the temperature distributions for the following conditions: initial condition ($t \leq 0$), steady-state condition ($t \to \infty$), and for two intermediate time conditions. Assume that the electric current is sufficiently large that the bar outer surface ($x = L$) will eventually be hotter than the air (T_∞).

 b) On $q_x'' - t$ coordinates, sketch the heat flux at the bar faces. That is, show qualitatively how $q_x''(0, t)$ and $q_x''(L, t)$ vary with time.

2.26 The one-dimensional system of mass M with constant properties and no internal heat generation shown in the figure is initially at a uniform temperature, T_i. The electrical heater is suddenly energized providing a uniform heat flux, q_o'', at the surface $x = 0$. The boundaries at $x = L$ and elsewhere are perfectly insulated.

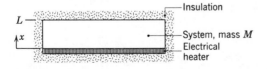

 a) Write the differential equation and identify the boundary and initial conditions that could be used to determine the temperature as a function of position and time in the system.

 b) On $T - x$ coordinates, sketch the temperature distributions for the initial condition ($t \leq 0$) and for several times after the heater is energized. Will a steady-state temperature distribution ever be reached?

c) On $q_x'' - t$ coordinates, sketch the heat flux, $q_x''(x, t)$, at the planes $x = 0$, $x = L/2$, and $x = L$ as a function of time.

d) After a period of time has elapsed, designated as t_e, the heater power is switched off. Assuming that the insulation is perfect, the system will eventually reach a uniform temperature T_e. Derive an expression that can be used to determine T_e as a function of the parameters q_o'', t_e, T_i and the system characteristics M, c_p and A_s (the heater surface area).

2.27 The plane wall with constant properties and no internal heat generation shown in the figure is initially at a uniform temperature T_i. Suddenly the surface at $x = L$ is exposed to a heating process with a fluid at T_∞ having a convection heat transfer coefficient h. The boundary at $x = 0$ is perfectly insulated.

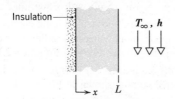

a) Write the differential equation and identify the boundary and initial conditions that could be used to determine the temperature as a function of position and time in the wall.

b) On $T - x$ coordinates, sketch the temperature distributions for the following conditions: initial condition ($t \leq 0$), steady-state condition ($t \to \infty$), and for two intermediate time conditions.

c) On $q_x'' - t$ coordinates, sketch the heat flux at the locations $x = 0$ and $x = L$. That is, show qualitatively how $q_x''(0, t)$ and $q_x''(L, t)$ vary with time.

d) Write an expression for the total heat transferred to the wall per unit volume of the wall (J/m^3).

3 One-Dimensional, Steady-State Conduction

In this chapter we treat situations for which heat is transferred by diffusion under *one-dimensional*, *steady-state* conditions. The term one-dimensional refers to the fact that only one coordinate is needed to describe the spatial variation of the dependent variables. Hence in a *one-dimensional* system, temperature gradients exist along only a single coordinate direction, and heat transfer occurs exclusively in that direction. The system is characterized by *steady-state* conditions if all of its features are time invariant. Despite their inherent simplicity, one-dimensional, steady-state models may be used to accurately represent numerous engineering systems.

We begin our consideration of one-dimensional, steady-state conduction by discussing heat transfer with no internal generation (Sections 3.1 to 3.3). The objective is to determine the temperature distribution and heat transfer rate in common geometries. The concept of thermal resistance (analogous to electrical resistance) is introduced as an aid to solving conduction heat transfer problems. The effect of internal heat generation on the temperature distribution and heat rate is then treated (Section 3.4). Finally, conduction analysis is used to describe the performance of extended surfaces or fins, wherein the role of convection at the external boundary must be considered (Section 3.5).

3.1 THE PLANE WALL

For one-dimensional conduction in a plane wall, temperature is a function of only the x coordinate and heat is transferred exclusively in this direction. In Figure 3.1*a*, a plane wall separates two fluids of different temperatures. Heat transfer occurs by convection from the hot fluid at $T_{\infty,1}$ to one surface of the wall at $T_{s,1}$, by conduction through the wall, and by convection from the other surface of the wall at $T_{s,2}$ to the cold fluid at $T_{\infty,2}$.

We begin by considering conditions *within* the wall. We will first determine the temperature distribution, from which we can then obtain the conduction heat transfer rate.

3.1.1 Temperature Distribution   *SOLVE HEAT EQUATION USING B.C.*

The temperature distribution in the wall can be determined by solving the heat diffusion equation with the proper boundary conditions. For steady-state conditions with no distributed source or sink of energy within the wall, the appropriate form of the heat equation, Equation 2.17, is

$$\frac{d}{dx}\left(k\frac{dT}{dx}\right) = 0 \qquad \textit{CONDUCTION (X-dir only)} \tag{3.1}$$

If the thermal conductivity of the wall material is assumed to be constant, the equation may be integrated twice, case 1 of Appendix B.1, to obtain the *general*

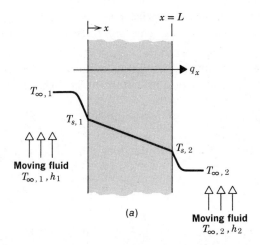

Figure 3.1 Heat transfer through a plane wall. (*a*) Temperature distribution. (*b*) Equivalent thermal circuit.

solution

$$T(x) = C_1 x + C_2 \qquad (3.2)$$

To obtain the constants of integration, C_1 and C_2, boundary conditions must be introduced. We choose to work with the constant surface conditions

$$T(0) = T_{s,1} \qquad T(L) = T_{s,2}$$

Applying the condition at $x = 0$ to the general solution, it follows that

$$T_{s,1} = C_2$$

Similarly, at $x = L$

$$T_{s,2} = C_1 L + C_2 = C_1 L + T_{s,1}$$

in which case

$$\frac{T_{s,2} - T_{s,1}}{L} = C_1$$

Substituting into the general solution the temperature distribution is then

$$T(x) = (T_{s,2} - T_{s,1})\frac{x}{L} + T_{s,1} \qquad (3.3)$$

From this result it is evident that, *for one-dimensional, steady-state conduction in a plane wall with no heat generation and constant thermal conductivity, the temperature varies linearly with* x.

Now that we have the temperature distribution, we may use Fourier's law, Equation 2.1, to determine the conduction heat transfer rate. That is,

$$q_x = -kA \frac{dT}{dx} = \frac{kA}{L}(T_{s,1} - T_{s,2}) \tag{3.4}$$

Note that A is the area of the wall *normal* to the direction of heat transfer and that the heat flux is

$$q''_x = q_x/A = \frac{k}{L}(T_{s,1} - T_{s,2}) \tag{3.5}$$

Note also that the heat flux is a constant independent of x. This result is consistent with Equation 3.1, which implies that there is no change in the heat flux for the x direction.

In the foregoing paragraphs we have used the standard approach to solving conduction problems. That is, the appropriate form of the heat equation is solved for the temperature distribution, which is then used with Fourier's law to determine the heat transfer rate. This approach may be used for *any* conduction problem. However, for certain restricted conditions an alternative approach may also be used (Section 3.2).

3.1.2 Thermal Resistance

At this point we note that a useful and very important concept is suggested by Equation 3.4. In particular, there exists an analogy between the diffusion of heat and electrical charge. Just as an electrical resistance may be associated with the conduction of electricity, a thermal resistance may be associated with the conduction of heat. From the form of the rate equation for the plane wall, Equation 3.4, it is evident that the *thermal resistance for conduction* is

$$R_{t,\,cond} \equiv \frac{(T_{s,1} - T_{s,2})}{q_x} = \frac{L}{kA} \tag{3.6}$$

where resistance is defined as the driving potential divided by the transfer rate. Similarly, for electrical conduction in the same system, Ohm's law provides an electrical resistance of the form

$$R_e = \frac{(E_{s,1} - E_{s,2})}{I} = \frac{L}{\sigma A} \tag{3.7}$$

The analogy between Equations 3.6 and 3.7 is obvious. Of course, a thermal resistance may also be associated with heat transfer by convection at a surface, and, from Newton's law of cooling,

$$q = hA(T_s - T_\infty) \tag{3.8}$$

it is of the form

$$R_{t,\,conv} \equiv \frac{(T_s - T_\infty)}{q} = \frac{1}{hA} \tag{3.9}$$

As we will find, circuit representations provide a useful tool for both conceptualizing and quantifying heat transfer problems. The *equivalent thermal circuit* for the plane wall with convection surface conditions is shown in Figure 3.1b. The heat transfer rate may be determined from separate consideration of each element in the network. That is,

$$q_x = \frac{T_{\infty,1} - T_{s,1}}{(1/h_1 A)} = \frac{T_{s,1} - T_{s,2}}{(L/kA)} = \frac{T_{s,2} - T_{\infty,2}}{(1/h_2 A)} \tag{3.10}$$

HEAT TRANSFER IS THE SAME THROUGHOUT

In terms of the *overall temperature difference*, $T_{\infty,1} - T_{\infty,2}$, and the *total resistance* to heat transfer, R_{tot}, the heat transfer rate may also be expressed as

$$q_x = \frac{T_{\infty,1} - T_{\infty,2}}{R_{tot}} \tag{3.11}$$

where R_{tot} is obtained by recognizing that the conduction and convection resistances of the circuit are in series and hence may be summed, giving

$$R_{tot} = \frac{1}{h_1 A} + \frac{L}{kA} + \frac{1}{h_2 A} \tag{3.12}$$

CONV. COND. CONV.

3.1.3 The Composite Wall

Equivalent thermal circuits may also be used for more complex systems, such as *composite walls*. Such walls may involve any number of series and parallel thermal resistances due to layers of different materials. Consider the composite wall of Figure 3.2. The one-dimensional heat transfer rate for this system may be expressed as

$$q_x = \frac{T_{\infty,1} - T_{\infty,4}}{\Sigma R_t} \tag{3.13}$$

where $(T_{\infty,1} - T_{\infty,4})$ is the *overall* temperature difference and the summation is over all thermal resistances. Hence,

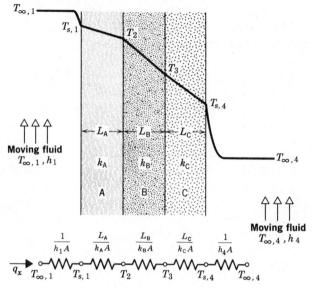

Figure 3.2   Equivalent thermal circuit for a composite wall.

$$q_x = \frac{T_{\infty,1} - T_{\infty,4}}{[(1/h_1 A) + (L_A/k_A A) + (L_B/k_B A) + (L_C/k_C A) + (1/h_4 A)]} \quad (3.14)$$

Alternatively, the heat transfer rate can be related to the temperature difference and resistance associated with each element. For example,

$$q_x = \frac{T_{\infty,1} - T_{s,1}}{(1/h_1 A)} = \frac{T_{s,1} - T_2}{(L_A/k_A A)} = \frac{T_2 - T_3}{(L_B/k_B A)} = \cdots \quad (3.15)$$

With composite systems it is often convenient to work with an *overall heat transfer coefficient, U,* which is defined by an expression analogous to Newton's law of cooling. Accordingly,

$$q_x \equiv UA \, \Delta T \quad (3.16)$$

where ΔT is the overall temperature difference. Note that the overall heat transfer coefficient is related to the total thermal resistance. From Equations 3.13 and 3.16 we see that $UA = 1/R_{tot}$, in which case, for the composite wall of Figure 3.2,

$$U = \frac{1}{R_{tot} A} = \frac{1}{[(1/h_1) + (L_A/k_A) + (L_B/k_B) + (L_C/k_C) + (1/h_4)]} \quad (3.17)$$

In general, we may write

$$R_{\text{tot}} = \Sigma R_t = \frac{\Delta T}{q} = \frac{1}{UA} \tag{3.18}$$

3.1.4 Contact Resistance

Although neglected to now, it is important to recognize that, in composite systems, the temperature drop across the interface between materials may be appreciable. This temperature change may be attributed to what is known as the *thermal contact resistance*, $R_{t,c}$. The effect is shown in Figure 3.3, and $R_{t,c}$ may be defined as

$$R_{t,c} = \frac{T_A - T_B}{q_x} \tag{3.19}$$

The existence of a finite contact resistance is due principally to surface roughness effects. Contact spots are interspersed with voids which are, in most instances, air filled. Heat transfer is therefore due to conduction across the actual contact area and to conduction (or natural convection) and radiation across the voids. The contact resistance may therefore be viewed as two parallel resistances: that due to the contact spots and that due to the voids. If the contact area is small, as it is for rough surfaces, the major contribution to the resistance is made by the voids. The resistance decreases with decreasing surface roughness and increasing joint pressure.

Although theories have been developed for the prediction of $R_{t,c}$, the most reliable results are those which have been obtained experimentally. The effect of loading on metallic interfaces can be seen in Table 3.1, which presents the

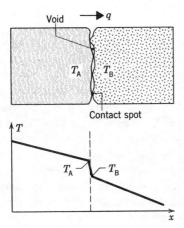

Figure 3.3 Temperature drop due to thermal contact resistance.

Table 3.1 Approximate range of thermal resistance values for metallic interfaces under vacuum conditions [1]

	THERMAL RESISTANCE, $R''_{t,c} \times 10^4$ (m²·K/W)	
Contact pressure	100 kN/m²	10,000 kN/m²
Stainless steel	6–25	0.7–4.0
Copper	1–10	0.1–0.5
Magnesium	1.5–3.5	0.2–0.4
Aluminum	1.5–5.0	0.2–0.4

approximate range of thermal resistance values under vacuum conditions. The variation of thermal resistance for an aluminum interface due to interfacial fluids of different thermal conductivity is shown in Table 3.2. A more detailed discussion of thermal contact resistance is given by Fried [1].

Table 3.2 Variation of thermal resistance for aluminum-aluminum interface (10 μm surface roughness) under 10^5 N/m² contact pressure with different interfacial fluids [1]

FLUID	THERMAL RESISTANCE, $R''_{t,c} \times 10^4$ (m²·K/W)
Air	2.75
Helium	1.05
Hydrogen	0.720
Silicone oil	0.525
Glycerine	0.265

EXAMPLE 3.1

A leading manufacturer of household appliances is proposing a self-cleaning oven design that involves use of a composite window separating the oven cavity from the room air. The composite is to consist of two high-temperature plastics (A and B) of thicknesses $L_A = 2L_B$ and thermal conductivities $k_A = 0.15$ W/m·K and $k_B = 0.08$ W/m·K. During the self-cleaning process, the oven wall and air temperatures, T_w and T_a, are 400°C, while the room air temperature is $T_\infty = 25$°C. The inside convection and radiation heat transfer coefficients, h_i and h_r, respectively, as well as the outside convection coefficient, h_o, are each approximately equal to 25 W/m²·K. What is the minimum window thickness, $L = L_A + L_B$, needed to insure a temperature that is less than or equal to 50°C at the outer surface of the window? This temperature must not be exceeded for safety reasons.

SOLUTION

KNOWN:

The properties and relative dimensions of plastic materials used for a composite oven window, and conditions associated with self-cleaning operation.

FIND:

Composite thickness, $L_A + L_B$, needed to insure safe operation.

SCHEMATIC:

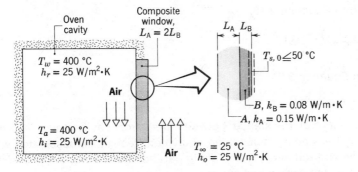

ASSUMPTIONS:

1. Steady-state conditions.
2. One-dimensional conduction through the window.
3. Negligible contact resistance.
4. Negligible radiation absorption *within* the window; hence no internal heat generation (radiation exchange between window and oven walls occurs at the window inner surface).
5. Negligible radiation exchange between window outer surface and surroundings.
6. Each plastic is homogeneous with constant properties.

ANALYSIS:

The thermal circuit can be constructed by recognizing that there is resistance to heat flow associated with convection at the outer surface, conduction in the plastics, and convection and radiation at the inner surface. Accordingly, the circuit and the resistances are of the following form.

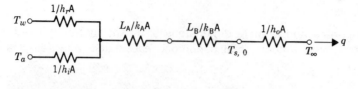

Since the outer surface temperature of the window, $T_{s,0}$, is prescribed, the required window thickness may be obtained by applying an energy balance at this surface. That is, from Equation 1.11

$$\dot{E}_{in} = \dot{E}_{out}$$

where, from Equation 3.18,

$$\dot{E}_{in} = q = \frac{T_a - T_{s,0}}{\Sigma R_t}$$

and from Equation 3.8

$$\dot{E}_{out} = q = h_o A(T_{s,0} - T_\infty)$$

The total thermal resistance between the oven cavity and the outer surface of the window includes an equivalent resistance associated with convection and radiation, which act in parallel at the inner surface of the window, and the conduction resistances of the window materials. Hence

$$\Sigma R_t = \left(\frac{1}{1/h_i A} + \frac{1}{1/h_r A}\right)^{-1} + \frac{L_A}{k_A A} + \frac{L_B}{k_B A}$$

or

$$\Sigma R_t = \frac{1}{A}\left(\frac{1}{h_i + h_r} + \frac{L_A}{k_A} + \frac{L_A}{2k_B}\right)$$

Substituting into the energy balance, it follows that

$$\frac{T_a - T_{s,0}}{(h_i + h_r)^{-1} + (L_A/k_A) + (L_A/2k_B)} = h_o(T_{s,0} - T_\infty) \qquad \dot{E}_{IN} = \dot{E}_{OUT}$$

Hence, solving for L_A,

$$L_A = \frac{(1/h_o)(T_a - T_{s,0})/(T_{s,0} - T_\infty) - (h_i + h_r)^{-1}}{(1/k_A + 1/2k_B)}$$

$$L_A = \frac{0.04 \text{ m}^2 \cdot \text{K/W}\left(\dfrac{400 - 50}{50 - 25}\right) - 0.02 \text{ m}^2 \cdot \text{K/W}}{(1/0.15 + 1/0.16) \text{ m} \cdot \text{K/W}}$$

$$L_A = 0.0418 \text{ m}$$

Hence

$$L_B = \frac{L_A}{2} = 0.0209 \text{ m}$$

and

$$L = L_A + L_B = 0.0627 \text{ m} = 62.7 \text{ mm} \qquad \triangleleft$$

COMMENTS:

Note that the self-cleaning operation is a transient process, as far as the thermal response of the window is concerned, and it is possible that steady-state conditions may not be reached in the time required for cleaning. However, the steady-state condition provides the maximum possible value of $T_{s,0}$ and hence is well suited for the design calculation.

3.2 AN ALTERNATIVE CONDUCTION ANALYSIS

At this point we note that the conduction analysis of Section 3.1 was performed in the standard manner. That is, the heat equation was first solved to obtain the temperature distribution, Equation 3.3, and Fourier's law was then applied to obtain the heat transfer rate, Equation 3.4. However, an alternative approach may be used for the conditions presently of interest. Considering conduction in the system of Figure 3.4, we recognize that, for *steady-state conditions* with *no heat generation* and *no heat loss from the sides*, the heat transfer rate q_x must be a constant independent of x. That is, for any differential element dx, $q_x = q_{x+dx}$. This condition is, of course, a consequence of the energy conservation requirement, and it must apply even if the area varies with position, $A(x)$, and the thermal conductivity varies with temperature, $k(T)$. Moreover, even though the

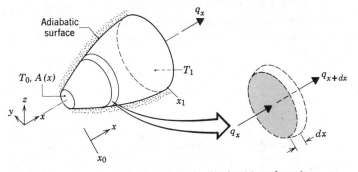

Figure 3.4 System with a constant conduction heat transfer rate.

temperature distribution may be two dimensional, varying with x and y, it is often reasonable to neglect the y variation and to assume a *one-dimensional* distribution in x.

For the above conditions it is possible to work exclusively with Fourier's law when performing a conduction analysis. In particular, since the conduction rate is a *constant*, the rate equation may be *integrated*, despite the fact that neither the rate nor the temperature distribution are known. Consider Fourier's law, Equation 2.1, which may be applied to the system of Figure 3.4. Although we may have no knowledge of the value of q_x or the form of $T(x)$, we do know that q_x is a constant. Hence we may express Fourier's law in the integral form

$$q_x \int_{x_0}^{x} \frac{dx}{A(x)} = - \int_{T_0}^{T} k(T)dT \qquad (3.20)$$

The cross-sectional area may be a known function of x, and the material thermal conductivity may vary with temperature in a known manner. If the integration is performed from a point x_0 at which the temperature T_0 is known, the resulting equation provides the functional form of $T(x)$. Moreover, if the temperature $T = T_1$ at some $x = x_1$ is also known, integration between x_0 and x_1 provides an expression from which q_x may be computed. Note that, if the area A is uniform and k is independent of temperature, Equation 3.20 reduces to

$$\frac{q_x \Delta x}{A} = - k \, \Delta T \qquad (3.21)$$

where $\Delta x = x_1 - x_0$ and $\Delta T = T_1 - T_0$.

We frequently elect to solve diffusion problems by working with integrated forms of the diffusion rate equations. However, the limiting conditions for which this may be done should be firmly fixed in our minds: *steady-state* and *one-dimensional* transfer with *no heat generation*.

EXAMPLE 3.2

The following diagram shows a conical section fabricated from pyroceram. It is of circular cross section with the diameter $D = ax$, where $a = 0.25$. The small end is located at $x_1 = 50$ mm and the large end at $x_2 = 250$ mm. The end temperatures are $T_1 = 400$ K and $T_2 = 600$ K, while the lateral surface is well insulated.

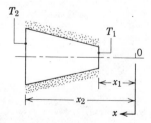

1. Derive an expression for the temperature distribution $T(x)$ in symbolic form, assuming one-dimensional conditions. Sketch the temperature distribution.
2. Calculate the heat rate, q_x, through the cone.

SOLUTION

KNOWN:

Conduction in a circular conical section having a diameter, $D = ax$, where $a = 0.25$.

FIND:

1. Temperature distribution, $T(x)$.
2. Heat transfer rate, q_x.

SCHEMATIC:

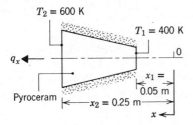

ASSUMPTIONS:

1. Steady-state conditions.
2. One-dimensional conduction in the x direction.
3. No internal heat generation.
4. Constant properties.

PROPERTIES:

Table A.2, Pyroceram (500 K): $k = 3.46$ W/m·K

ANALYSIS:

1. Since heat conduction occurs under steady-state, one-dimensional conditions with no internal heat generation, the heat transfer rate, q_x, is a constant independent of x. Accordingly, Fourier's law,

Equation 2.1, may be used to determine the temperature distribution

$$q_x = -kA\frac{dT}{dx}$$

With $A = \pi D^2/4 = \pi a^2 x^2/4$ and separating variables

$$\frac{4q_x dx}{\pi a^2 x^2} = -k\,dT$$

Integrating from x_1 to any x within the cone, and recalling that q_x and k are constants, it follows that

$$\frac{4q_x}{\pi a^2}\int_{x_1}^{x}\frac{dx}{x^2} = -k\int_{T_1}^{T}dT$$

with the result

$$\frac{4q_x}{\pi a^2}\left(-\frac{1}{x}+\frac{1}{x_1}\right) = -k(T-T_1)$$

or solving for T

$$T(x) = T_1 - \frac{4q_x}{\pi a^2 k}\left(\frac{1}{x_1}-\frac{1}{x}\right)$$

Although q_x is a constant, it is as yet an unknown. However, it may be determined by evaluating the above expression at $x = x_2$ where $T(x_2) = T_2$. Hence

$$T_2 = T_1 - \frac{4q_x}{\pi a^2 k}\left(\frac{1}{x_1}-\frac{1}{x_2}\right)$$

and solving for q_x

$$q_x = \frac{\pi a^2 k(T_1-T_2)}{4[(1/x_1)-(1/x_2)]}$$

Substituting for q_x into the expression for $T(x)$, the temperature distribution becomes

$$T(x) = T_1 + (T_1-T_2)\left[\frac{(1/x)-(1/x_1)}{(1/x_1)-(1/x_2)}\right] \qquad \triangleleft$$

From this result, temperature may be calculated as a function of x and the distribution is as shown below.

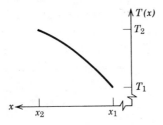

Note that, since $dT/dx = -4q_x/k\pi a^2 x^2$ from Fourier's law, it follows that the slope of the temperature distribution must decrease with increasing x. That is, $x^2(dT/dx)$ is a constant.

2. Substituting numerical values into the foregoing result for the heat transfer rate, it follows that

$$q_x = \frac{\pi(0.25)^2 \times 3.46 \text{ W/m} \cdot \text{K} \ (400 - 600)\text{K}}{4\left(\dfrac{1}{0.05 \text{ m}} - \dfrac{1}{0.25 \text{ m}}\right)}$$

$$q_x = -2.12 \text{ W} \qquad\qquad \triangleleft$$

COMMENTS:

1. Recognize that Fourier's law, Equation 2.1, is for a one-dimensional system. However, it can also be used when the heat flow is only approximately one dimensional.

2. Recognize that, when the parameter a increases, the one-dimensional assumption becomes less appropriate. That is, the assumption improves when the cross-sectional area change with distance is less pronounced.

3.3 RADIAL SYSTEMS

Cylindrical and spherical systems often experience temperature gradients in only the radial direction and may therefore be treated as one dimensional. Moreover, under steady-state conditions with no heat generation, such systems may be analyzed by using the standard method, which begins with the appropriate form of the heat equation, or the alternative method, which begins with the appropriate form of Fourier's law. In this section, the cylindrical system is analyzed by means of the standard method and the spherical system by means of the alternative method.

3.3.1 The Cylinder   — STANDARD METHOD

A common example is the hollow cylinder, whose inner and outer surfaces are exposed to fluids at different temperatures (Figure 3.5). For steady-state conditions, with no heat generation, the appropriate form of the heat equation, Equation 2.18, is

$$\frac{1}{r}\frac{d}{dr}\left(kr\frac{dT}{dr}\right)=0 \tag{3.22}$$

where, for the moment, k is treated as a variable. The physical significance of this result becomes evident, if we also consider the appropriate form of Fourier's law. The rate at which energy is conducted across any cylindrical surface in the solid may be expressed as

$$q_r = -kA\frac{dT}{dr} = -k(2\pi rL)\frac{dT}{dr} \tag{3.23}$$

where $A = 2\pi rL$ is the area normal to the direction of heat transfer. From Equations 3.22 and 3.23 it follows that the conduction *heat transfer rate*, q_r, (*not* the heat flux q_r'') is a *constant in the radial direction*.

We may determine the temperature distribution in the cylinder by solving Equation 3.22. Assuming the value of k to be constant, Equation 3.22 may be integrated twice, case 3 of Appendix B.1, to obtain the general solution

$$T(r) = C_1 \ln r + C_2 \qquad T \text{ dist. IN cylinder} \tag{3.24}$$

To obtain the constants of integration, C_1 and C_2, we introduce the following

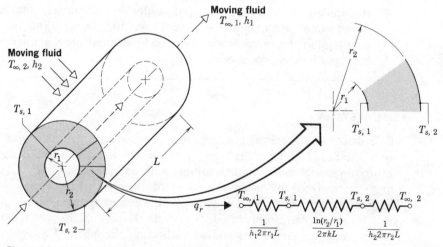

Figure 3.5 Hollow cylinder with convective surface conditions.

boundary conditions.

$$T(r_1) = T_{s,1} \qquad T(r_2) = T_{s,2}$$

Applying these conditions to the general solution, we then obtain

$$T_{s,1} = C_1 \ln r_1 + C_2$$

$$T_{s,2} = C_1 \ln r_2 + C_2$$

Solving for C_1 and C_2 and substituting back into the general solution, we then obtain

$$T(r) = \frac{T_{s,1} - T_{s,2}}{\ln(r_1/r_2)} \ln\left(\frac{r}{r_2}\right) + T_{s,2} \tag{3.25}$$

Note that the temperature distribution associated with radial conduction through a cylinder is logarithmic and not linear, as it is for the plane wall under the same conditions.

If the temperature distribution, Equation 3.25, is now used with Fourier's law, Equation 3.23, we obtain the following expression for the heat transfer rate

$$q_r = \frac{2\pi L k(T_{s,1} - T_{s,2})}{\ln(r_2/r_1)} \tag{3.26}$$

From this result it is evident that, for conduction in hollow cylinders, the thermal resistance is of the form

$$R_{t,\text{cond}} = \frac{\ln(r_2/r_1)}{2\pi L k} \tag{3.27}$$

Note that, since the value of q_r is independent of r, the foregoing result could have been obtained by using the alternative method, that is, by integrating Equation 3.23.

Consider now the composite system of Figure 3.6. Recalling how we treated the composite plane wall and neglecting the interfacial contact resistances, the heat transfer rate may be expressed as

$$q_r = \frac{T_{\infty,1} - T_{\infty,4}}{\dfrac{1}{2\pi r_1 L h_1} + \dfrac{\ln(r_2/r_1)}{2\pi k_A L} + \dfrac{\ln(r_3/r_2)}{2\pi k_B L} + \dfrac{\ln(r_4/r_3)}{2\pi k_C L} + \dfrac{1}{2\pi r_4 L h_4}} \tag{3.28}$$

The foregoing result may also be expressed in terms of an overall heat transfer coefficient. That is,

$$q_r \equiv \frac{T_{\infty,1} - T_{\infty,4}}{R_{\text{tot}}} = U_1 A_1 (T_{\infty,1} - T_{\infty,4}) \tag{3.29}$$

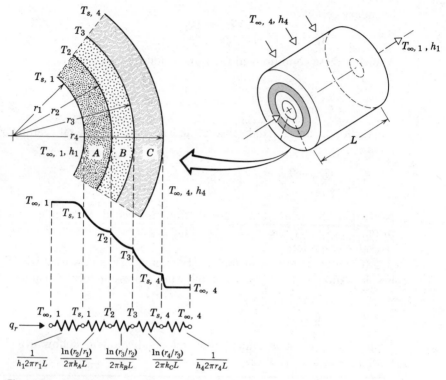

Figure 3.6   Temperature distribution for a composite cylindrical wall.

where $A_1 = 2\pi r_1 L$ and

$$U_1 = \cfrac{1}{\cfrac{1}{h_1} + \cfrac{r_1}{k_A}\ln\cfrac{r_2}{r_1} + \cfrac{r_1}{k_B}\ln\cfrac{r_3}{r_2} + \cfrac{r_1}{k_C}\ln\cfrac{r_4}{r_3} + \cfrac{r_1}{r_4}\cfrac{1}{h_4}} \qquad (3.30)$$

Equation 3.29 defines U in terms of the inside surface area, A_1, of the composite cylinder. This definition is *arbitrary*, and the overall coefficient may also be defined in terms of A_4 or any of the intermediate areas. Note that

$$U_1 A_1 = U_2 A_2 = U_3 A_3 = U_4 A_4 = (\Sigma R_t)^{-1} \qquad (3.31)$$

and the specific forms of U_2, U_3, and U_4 may be inferred from Equation 3.28.

EXAMPLE 3.3

The possible existence of an optimum insulation thickness for radial systems is suggested by the fact that there are competing effects associated with an increase in this thickness. In particular, although the conduction resistance increases with the addition of insulation, the convection resistance decreases due to increasing outer surface area. Hence there may exist an insulation thickness that minimizes heat loss by maximizing the total resistance to heat transfer. Resolve this question by considering the following system.

1. A thin-walled copper tube of radius r_i is used to transport a low temperature refrigerant and is at a temperature, T_i, which is less than that of the ambient air at T_∞ around the tube. Is there an optimum thickness associated with application of insulation to the tube?

2. Confirm the above result by computing the total thermal resistance per unit length of tube for a 10-mm diameter tube having the following insulation thicknesses: 0, 2, 5, 10, 20, and 40 mm. The insulation is composed of cellular glass, and the outer surface convection coefficient is 5 W/m²·K.

SOLUTION

KNOWN:

Radius, r_i, and temperature, T_i, of a thin-walled copper tube to be insulated from the ambient air.

FIND:

1. Whether there exists an optimum insulation thickness that minimizes the heat transfer rate.

2. Thermal resistance associated with using cellular glass insulation of varying thickness.

SCHEMATIC:

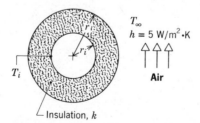

T_∞
$h = 5$ W/m²·K

Air

T_i

Insulation, k

ASSUMPTIONS:

1. Steady-state conditions.

2. One-dimensional heat transfer in the radial (cylindrical) direction.

3. Negligible tube wall thermal resistance.

4. Insulation has constant properties.

5. Negligible radiation exchange between insulation outer surface and surroundings.

PROPERTIES:

Table A.3, cellular glass (285 K, assumed): $k = 0.055 \; W/m \cdot K$.

ANALYSIS:

1. The resistance to heat transfer between the refrigerant and the air is dominated by conduction in the insulation and convection in the air. The thermal circuit is therefore

where the conduction and convection resistances per unit length follow from Equations 3.27 and 3.9, respectively. The total thermal resistance per unit length of tube is then

$$R'_{tot} = \frac{\ln(r/r_i)}{2\pi k} + \frac{1}{2\pi rh}$$

where the rate of heat transfer per unit length of tube is

$$q'_r = \frac{T_\infty - T_i}{R'_{tot}}$$

An optimum insulation thickness would be associated with the value of r which minimized q'_r or maximized R'_{tot}. Such a value could be obtained from the requirement that

$$\frac{dR'_{tot}}{dr} = 0$$

Hence

$$\frac{1}{2\pi kr} - \frac{1}{2\pi r^2 h} = 0$$

or

$$r = \frac{k}{h}$$

To determine whether the foregoing result maximizes or minimizes the total resistance, the second derivative must be evaluated. Hence

$$\frac{d^2 R'_{tot}}{dr^2} = -\frac{1}{2\pi k r^2} + \frac{1}{\pi r^3 h}$$

or, at $r = k/h$

$$\frac{d^2 R'_{tot}}{dr^2} = \frac{1}{\pi (k/h)^2} \left(\frac{1}{k} - \frac{1}{2k} \right) = \frac{1}{2\pi k^3/h^2} > 0$$

Since this result is always positive, it follows that $r = k/h$ is the insulation radius for which the total resistance is a minimum, not a maximum. Hence an *optimum* insulation thickness *does not exist*.

From the above result it makes more sense to think in terms of a *critical insulation radius*

$$r_{cr} \equiv \frac{k}{h}$$

below which q'_r increases with increasing r and above which q'_r decreases with increasing r.

2. With $h = 5$ W/m²·K and $k = 0.055$ W/m·K, the critical radius is

$$r_{cr} = \frac{0.055 \text{ W/m·K}}{5 \text{ W/m}^2 \cdot \text{K}} = 0.011 \text{ m}$$

Hence $r_{cr} > r_i$ and heat transfer will increase with the addition of insulation up to a thickness of

$$r_{cr} - r_i = (0.011 - 0.005) \text{ m} = 0.006 \text{ m}$$

The thermal resistances corresponding to the prescribed insulation thicknesses may be calculated and are summarized as follows.

INSULATION THICKNESS $(r - r_i)$ (mm)	INSULATION RADIUS r (m)	THERMAL RESISTANCES (m·K/W)		
		R'_{cond}	R'_{conv}	R'_{tot}
0	0.005	0	6.37	6.37
2	0.007	0.97	4.55	5.52
5	0.010	2.00	3.18	5.18
6	$r_{cr} = 0.011$	2.28	2.89	5.17
10	0.015	3.18	2.12	5.30
20	0.025	4.66	1.27	5.93
40	0.045	6.35	0.71	7.06

COMMENTS:

1. The effect of the critical radius is revealed by the fact that, even for 20 mm of insulation, the total resistance is still not as large as the value for no insulation.

2. Note that the problem of lowering the total resistance through the application of insulation only exists for small tubes and low convection coefficients, such that $r_{cr} > r_i$. For moderate to large values of r_i and/or h, $r_i > r_{cr}$ and the addition of any amount of insulation will increase the total resistance.

3.3.2 The Sphere

Now consider applying the alternative method to analyzing conduction in the hollow sphere of Figure 3.7. For the differential control volume of the figure, energy conservation requires that $q_r = q_{r+dr}$ for steady-state, one-dimensional conditions with no heat generation. The appropriate form of Fourier's law is

$$q_r = -kA\frac{dT}{dr} = -k(4\pi r^2)\frac{dT}{dr} \tag{3.32}$$

where $A = 4\pi r^2$ is the area normal to the direction of heat transfer.

Acknowledging that q_r is a constant, independent of r, Equation 3.32 may be expressed in the integral form

$$\frac{q_r}{4\pi}\int_{r_1}^{r_2}\frac{dr}{r^2} = -\int_{T_{s,1}}^{T_{s,2}} k(T)dT \tag{3.33}$$

Assuming constant k, we then obtain

$$q_r = \frac{4\pi k(T_{s,1} - T_{s,2})}{(1/r_1) - (1/r_2)} \tag{3.34}$$

Remembering that the thermal resistance is defined as the temperature difference divided by the heat transfer rate, we obtain

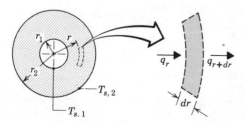

Figure 3.7 Conduction in a spherical shell.

$$R_{t, \text{cond}} = \frac{1}{4\pi k}\left(\frac{1}{r_1} - \frac{1}{r_2}\right) \tag{3.35}$$

Note that the temperature distribution and Equations 3.34 and 3.35 could have been obtained by using the standard approach, which would begin with the appropriate form of the heat equation.

Spherical composites may be treated in much the same way as composite walls and cylinders, where appropriate forms of the total resistance and overall heat transfer coefficient may be determined.

EXAMPLE 3.4

A spherical, thin-walled metallic container is used to store liquid nitrogen at 77 K. The container has a diameter of 0.5 m and is covered with an evacuated, reflective insulation system composed of silica powder. The insulation is 25 mm thick, and its outer surface is exposed to ambient air at 300 K. The convection coefficient is known to be 20 W/m² · K. The latent heat of vaporization and density of liquid nitrogen are 2×10^5 J/kg and 804 kg/m³, respectively.

1. What is the rate of heat transfer to the liquid nitrogen?

2. What is the rate of liquid boil-off?

SOLUTION

KNOWN:

Liquid nitrogen is stored in a spherical container that is insulated and exposed to ambient air.

FIND:

1. The rate of heat transfer to the nitrogen.

2. The mass rate of nitrogen boil-off.

SCHEMATIC:

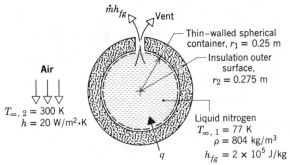

ASSUMPTIONS:

1. Steady-state conditions.
2. One-dimensional transfer in the radial direction.
3. Negligible resistance to heat transfer through the container wall and from the container to the nitrogen.
4. Constant properties.
5. Negligible radiation exchange between outer surface of insulation and surroundings.

PROPERTIES:

Table A.3, evacuated silica powder (300 K): $k = 0.0017$ W/m·K.

ANALYSIS:

1. The thermal circuit involves a conduction and convection resistance in series and is of the form

where, from Equation 3.35,

$$R_{t,\,\mathrm{cond}} = \frac{1}{4\pi k}\left(\frac{1}{r_1} - \frac{1}{r_2}\right)$$

and from Equation 3.9

$$R_{t,\,\mathrm{conv}} = \frac{1}{h4\pi r_2^2}$$

The rate of heat transfer to the liquid nitrogen is then

$$q = \frac{T_{\infty,2} - T_{\infty,1}}{(1/4\pi k)\,[(1/r_1) - (1/r_2)] + (1/h4\pi r_2^2)}$$

Hence

$$q = [(300 - 77)\ \mathrm{K}] \Bigg/ \Bigg[\frac{1}{4\pi(0.0017\ \mathrm{W/m\cdot K})} \left(\frac{1}{0.25\ \mathrm{m}} - \frac{1}{0.275\ \mathrm{m}}\right) + \frac{1}{(20\ \mathrm{W/m^2 \cdot K})4\pi(0.275\ \mathrm{m})^2} \Bigg]$$

$$q = \frac{223}{17.02 + 0.05}\ \mathrm{W}$$

$q = 13.06$ W   ◁

2. Performing an energy balance for a control volume about the nitrogen, it follows from Equation 1.11 that

$$\dot{E}_{in} - \dot{E}_{out} = 0$$

where $\dot{E}_{in} = q$ and $\dot{E}_{out} = \dot{m}h_{fg}$ is associated with the loss of latent energy due to boiling. Hence

$$q - \dot{m}h_{fg} = 0$$

and the boil-off, $\dot{m}$, is

$$\dot{m} = \frac{q}{h_{fg}}$$

$$\dot{m} = \frac{13.06 \text{ J/s}}{2 \times 10^5 \text{ J/kg}}$$

$$\dot{m} = 6.53 \times 10^{-5} \text{ kg/s}$$

The loss per day is

$$\dot{m} = 6.53 \times 10^{-5} \text{ kg/s} \times 3600 \text{ s/h} \times 24 \text{ h/d}$$

$$\dot{m} = 5.64 \text{ kg/d}$$ ◁

or on a volumetric basis

$$\dot{V} = \frac{\dot{m}}{\rho} = \frac{5.64 \text{ kg/d}}{804 \text{ kg/m}^3}$$

$$\dot{V} = 0.007 \text{ m}^3/\text{d} = 7 \text{ l/d}$$

COMMENTS:

1. $R_{t,\,conv} \ll R_{t,\,cond}$
2. With a container volume of $(4/3)(\pi r_1^3) = 0.065$ m^3 = 65 l, the daily loss amounts to (7 l/65 l) 100 percent = 10.8 percent of capacity.

3.4 DIFFUSION WITH THERMAL ENERGY GENERATION

In the preceding section we considered diffusion problems for which the temperature distribution in a medium was determined solely by conditions at the boundaries of the medium. We now want to consider the additional effect on the temperature distribution of processes that may be occurring *within* the medium. In particular, we wish to consider situations for which thermal energy is being *generated* due to *conversion* from some other energy form.

A common thermal energy generation process involves the conversion from *electrical to thermal energy* in a current-carrying medium (*ohmic* or *resistance heating*). The rate at which energy is generated by passing a current I through a medium of electrical resistance, R_e, is

$$\dot{E}_g = I^2 R_e \tag{3.36}$$

If this power generation (W) occurs uniformly throughout the medium of volume V, the volumetric generation rate (W/m^3) is then

$$\dot{q} \equiv \dot{E}_g/V = I^2 R_e/V \tag{3.37}$$

Energy generation may also occur due to the deceleration and absorption of neutrons in the fuel element of a nuclear reactor or due to exothermic chemical reactions occurring within a medium. Endothermic reactions would, of course, have the inverse effect (a thermal energy sink) of converting thermal energy to chemical bonding energy. Finally, a conversion from electromagnetic to thermal energy may occur due to the absorption of radiation within the medium. The process may occur, for example, due to the absorption of gamma rays in external nuclear reactor components (cladding, thermal shields, pressure vessels, etc.) or due to the absorption of visible radiation in a semitransparent medium. Remember not to confuse energy generation with energy storage (Section 1.3.1).

3.4.1   The Plane Wall

Consider the plane wall of Figure 3.8*a*, in which there is uniform energy generation per unit volume ($\dot{q}$ is constant) and the surfaces are maintained at $T_{s,1}$ and $T_{s,2}$. For constant thermal conductivity, k, the appropriate form of the heat equation, Equation 2.16 is

$$\frac{d^2 T}{dx^2} + \frac{\dot{q}}{k} = 0 \tag{3.38}$$

and the general solution, case 2 of Appendix B.1, is

$$T = -\frac{\dot{q}}{2k}x^2 + C_1 x + C_2 \tag{3.39}$$

where C_1 and C_2 are the constants of integration. For the prescribed boundary conditions,

$$T(-L) = T_{s,1} \qquad T(L) = T_{s,2}$$

the constants may be evaluated and are of the form

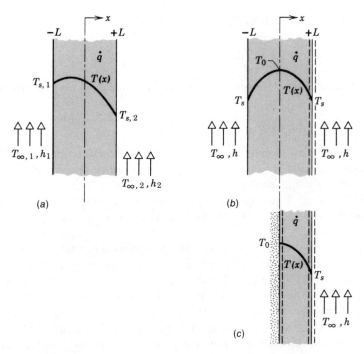

Figure 3.8 Conduction in a plane wall with uniform heat generation (a) Asymmetrical boundary conditions. (b) Symmetrical boundary conditions. (c) Adiabatic surface at midplane.

$$C_1 = \frac{T_{s,2} - T_{s,1}}{2L}$$

$$C_2 = \frac{\dot{q}}{2k}L^2 + \frac{T_{s,1} + T_{s,2}}{2}$$

in which case the temperature distribution is

$$T(x) = \frac{\dot{q}L^2}{2k}\left(1 - \frac{x^2}{L^2}\right) + \frac{T_{s,2} - T_{s,1}}{2}\frac{x}{L} + \frac{T_{s,1} + T_{s,2}}{2} \tag{3.40}$$

The heat flux at any point in the wall may, of course, be determined by using Equation 3.40 with Fourier's law. Note, however, that with generation the heat flux is no longer independent of x.

The preceding result simplifies when both surfaces are maintained at a common temperature, $T_{s,1} = T_{s,2} \equiv T_s$. The temperature distribution is then *symmetrical* about the midplane, Figure 3.8b, and is given by

$$T(x) = \frac{\dot{q}L^2}{2k}\left(1 - \frac{x^2}{L^2}\right) + T_s \tag{3.41}$$

The maximum temperature exists at the midplane

$$T(0) \equiv T_0 = \frac{\dot{q}L^2}{2k} + T_s \tag{3.42}$$

in which case the temperature distribution, Equation 3.41, may be expressed as

$$\frac{T(x) - T_0}{T_s - T_0} = \left(\frac{x}{L}\right)^2 \tag{3.43}$$

It is important to note that at the plane of symmetry in Figure 3.8b, the temperature gradient is zero, $(dT/dx)_{x=0} = 0$. Accordingly, this plane may be represented by the *adiabatic* surface shown in Figure 3.8c. One implication of this result is that Equation 3.41 applies to plane walls which are perfectly insulated on one side ($x = 0$) and maintained at a fixed temperature, T_s, on the other side ($x = L$).

Note that to use the foregoing results the surface temperature(s), T_s, must be known. However, a common situation is one for which it is the temperature of an adjoining fluid, T_∞, and not T_s, which is known. It then becomes necessary to relate T_s to T_∞. This relation may be developed by applying a surface energy balance. Consider the surface at $x = L$ for the symmetrical plane wall, Figure 3.8b, or the insulated plane wall, Figure 3.8c. Neglecting radiation and substituting the appropriate rate equations, the energy balance given by Equation 1.11 reduces to

$$-k\frac{dT}{dx}\bigg|_{x=L} = h(T_s - T_\infty) \qquad \text{ENERGY BALANCE} \tag{3.44}$$

Substituting from Equation 3.41 to obtain the temperature gradient at $x = L$, it follows that

$$T_s = T_\infty + \frac{\dot{q}L}{h} \tag{3.45}$$

Hence T_s may be computed from knowledge of T_∞, $\dot{q}$, L, and h.

Equation 3.45 may also be obtained by applying an *overall* energy balance to the plane wall of Figure 3.8b or 3.8c. For example, relative to a control volume about Figure 3.8c, the rate at which energy is generated within the wall must be balanced by the rate at which energy leaves via convection at the boundary. Equation 1.10 reduces to

$$\dot{E}_g = \dot{E}_{out} \tag{3.46}$$

or, for a unit surface area,

$$\dot{q}L = h(T_s - T_\infty) \tag{3.47}$$

Solving for T_s, Equation 3.45 is obtained.

EXAMPLE 3.5

A plane wall is a composite of two materials, A and B. The wall of material A has heat generation, $\dot{q} = 1.5 \times 10^6$ W/m^3, $k_A = 75$ W/m·K and thickness $L_A = 50$ mm. The wall material B has no generation, with $k_B = 150$ W/m·K and thickness $L_B = 20$ mm. The noncontact surface of material A is well insulated, while the noncontact surface of material B is cooled by a water stream with $T_\infty = 30°$C and $h = 1000$ W/m^2·K.

1. Sketch the temperature distribution which exists in the composite under steady-state conditions.

2. Determine the temperature T_0 of the insulated surface and the temperature T_2 of the cooled surface.

SOLUTION

KNOWN:

Plane wall of material A with internal heat generation is insulated on one side and bounded by a second wall of material B, which is without heat generation and is subjected to convection cooling.

FIND:

1. Sketch of steady-state temperature distribution in the composite.

2. Inner and outer surface temperatures of the composite.

SCHEMATIC:

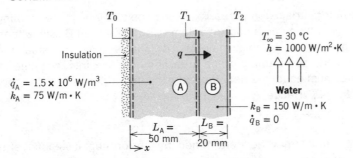

ASSUMPTIONS:

1. Steady-state conditions.

2. One-dimensional conduction in x direction.

3. Negligible contact resistance between walls.

4. Inner surface of A is adiabatic.

5. Constant properties for materials A and B.

ANALYSIS:

1. From the prescribed physical conditions, the temperature distribution in the composite is known to have the following features:

   a) Parabolic in material A.

   b) Zero slope at insulated boundary.

   c) Linear in material B.

   d) Slope change $= k_B/k_A = 2$ at interface.

   The temperature distribution in the water is characterized by

   e) Large gradients near the surface.

   These features are shown below.

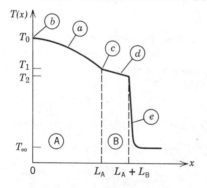

2. The outer surface temperature, T_2, may be obtained by performing an energy balance on a control volume about material B. Since there is no generation in this material, it follows that, for steady-state conditions and a unit surface area, the heat flux into the material at $x = L_A$ must equal the heat flux from the material due to convection at $x = L_A + L_B$. Hence

$$q'' = h(T_2 - T_\infty) \qquad (1)$$

The heat flux q'' may be determined by performing a second energy balance on a control volume about material A. In particular since the surface at $x = 0$ is adiabatic, there is no inflow and the rate at which energy is generated must equal the outflow. Accordingly, for a unit surface area,

$$\dot{q}L_A = q'' \qquad (2)$$

Combining Equations 1 and 2, the outer surface temperature is

$$T_2 = T_\infty + \frac{\dot{q}L_A}{h}$$

$$T_2 = 30°C + \frac{1.5 \times 10^6 \text{ W/m}^3 \times 0.05 \text{ m}}{1000 \text{ W/m}^2 \cdot \text{K}}$$

$$T_2 = 105°C \qquad \triangleleft$$

From Equation 3.42 the temperature at the insulated surface is

$$T_0 = \frac{\dot{q}L_A^2}{2k_A} + T_1 \tag{3}$$

where T_1 may be obtained from the following thermal circuit.

That is,

$$T_1 = T_\infty + (R''_{\text{cond, B}} + R''_{\text{conv}})q''$$

where the resistances for a unit surface area are

$$R''_{\text{cond, B}} = \frac{L_B}{k_B} \qquad R''_{\text{conv}} = \frac{1}{h}$$

Hence

$$T_1 = 30°C + \left(\frac{0.02 \text{ m}}{150 \text{ W/m} \cdot \text{K}} + \frac{1}{1000 \text{ W/m}^2 \cdot \text{K}} \right) 1.5 \times 10^6 \text{ W/m}^3 \times 0.05 \text{ m}$$

$$T_1 = 30°C + 85°C = 115°C$$

Substituting into Equation 3,

$$T_0 = \frac{1.5 \times 10^6 \text{ W/m}^3 \, (0.05 \text{ m})^2}{2 \times 75 \text{ W/m} \cdot \text{K}} + 115°C$$

$$T_0 = 25°C + 115°C$$

$$T_0 = 140°C \qquad \triangleleft$$

COMMENTS:

1. Recognize that material A, having heat generation, cannot be represented by a thermal circuit element.

2. Note that $[(T_2 - T_\infty)/(T_1 - T_2)] = (R_{\text{conv}}/R_{\text{cond, B}}) = 7.5$.

3.4.2   Radial Systems

Heat generation may occur in a variety of radial geometries. Consider the long, solid cylinder of Figure 3.9, which could represent a current-carrying wire or a fuel element in a nuclear reactor. For steady-state conditions the rate at which heat is generated within the cylinder must equal the rate at which heat is convected from the surface of the cylinder to a moving fluid. This condition allows.the surface temperature to be maintained at a fixed value of T_s.

To determine the temperature distribution in the cylinder, we begin with the appropriate form of the heat equation. For constant thermal conductivity, k, Equation 2.18 reduces to

$$\frac{1}{r}\frac{d}{dr}\left(r\frac{dT}{dr}\right)+\frac{\dot{q}}{k}=0 \tag{3.48}$$

Separating variables and assuming uniform generation, this expression may be integrated to obtain

$$r\frac{dT}{dr}=-\frac{\dot{q}}{2k}r^2+C_1 \tag{3.49}$$

Repeating the procedure, the general solution for the temperature distribution, case 4 of Appendix B.1, becomes

$$T(r)=-\frac{\dot{q}}{4k}r^2+C_1\ln r+C_2 \tag{3.50}$$

To obtain the constants of integration, C_1 and C_2, we apply the boundary conditions

$$\left.\frac{dT}{dr}\right|_{r=0}=0 \qquad T(r_o)=T_s$$

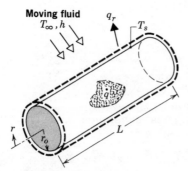

Figure 3.9   Conduction in a solid cylinder with uniform heat generation.

The first condition results from the symmetry of the situation. That is, for the solid cylinder the centerline is a line of symmetry for the temperature distribution and the temperature gradient must be zero. Recall that similar conditions existed at the midplane of a wall having symmetrical boundary conditions (Figure 3.8b). From the above symmetry condition at $r = 0$ and Equation 3.49, it is evident that $C_1 = 0$. Using the surface boundary condition at $r = r_o$ with Equation 3.50, we then obtain

$$C_2 = T_s + \frac{\dot{q}}{4k} r_o^2 \tag{3.51}$$

The temperature distribution is therefore

$$T(r) = \frac{\dot{q}r_o^2}{4k}\left(1 - \frac{r^2}{r_o^2}\right) + T_s \tag{3.52}$$

Evaluating Equation 3.52 at the centerline and dividing the result into Equation 3.52, we obtain the temperature distribution in nondimensional form

$$\frac{T(r) - T_s}{T_o - T_s} = 1 - \left(\frac{r}{r_o}\right)^2 \tag{3.53}$$

where T_o is the centerline temperature. The heat rate at any radius in the cylinder may, of course, be evaluated by using Equation 3.52 with Fourier's law.

To relate the surface temperature, T_s, to the temperature, T_∞, of the moving fluid, either a surface energy balance or an overall energy balance may be used. Choosing the second approach and using the appropriate rate equations with Equation 3.46, we obtain

$$\dot{q}(\pi r_o^2 L) = h(2\pi r_o L)(T_s - T_\infty)$$

or

$$T_s = T_\infty + \frac{\dot{q}r_o}{2h} \tag{3.54}$$

The foregoing approach may also be used to obtain the temperature distribution in solid spheres and in cylindrical and spherical shells for a variety of boundary conditions.

EXAMPLE 3.6

Consider a long solid tube, insulated at the outer radius, r_o, and cooled at the inner radius, r_i, with uniform heat generation, $\dot{q}$(W/m³), within the solid.

1. Obtain the general solution for the temperature distribution in the tube.

2. In a practical application a limit would be placed on the maximum

temperature which is permissible at the insulated surface $(r = r_o)$. Specifying this limit as T_o, identify appropriate boundary conditions that could be used to determine the arbitrary constants appearing in the general solution. Determine these constants and the corresponding form of the temperature distribution.

3. Determine the heat removal rate per unit length of tube.

4. If the coolant is available at a temperature T_∞, obtain an expression for the convection coefficient that would have to be maintained at the inner surface to allow for operation at prescribed values of T_o and $\dot{q}$.

SOLUTION

KNOWN:

Solid tube with uniform heat generation is insulated at the outer surface and cooled at the inner radius.

FIND:

1. General solution for the temperature distribution, $T(r)$.

2. Appropriate boundary conditions and the corresponding form of the temperature distribution.

3. Heat removal rate.

4. Convection coefficient at the inner surface.

SCHEMATIC:

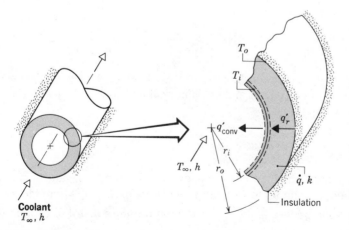

ASSUMPTIONS:

1. Steady-state conditions.
2. One-dimensional radial conduction.
3. Constant properties.
4. Uniform volumetric heat generation.
5. Outer surface is adiabatic.

ANALYSIS:

1. To determine $T(r)$, the appropriate form of the heat equation, Equation 2.18, must be solved. For the prescribed conditions, this equation reduces to

$$\frac{d}{dr}\left(r\frac{dT}{dr}\right) = -\frac{\dot{q}}{k}r$$

Separating variables and integrating, we obtain

$$r\frac{dT}{dr} = -\frac{\dot{q}}{2k}r^2 + C_1 \tag{1}$$

Separating variables,

$$dT = -\frac{\dot{q}}{2k}r\,dr + \frac{C_1}{r}\,dr$$

and integrating once again, it follows that

$$T(r) = -\frac{\dot{q}}{4k}r^2 + C_1\ln r + C_2 \tag{2}$$

where C_1 and C_2 are the arbitrary constants of integration.

2. Two boundary conditions are needed to evaluate C_1 and C_2, and in this problem it is appropriate to specify both conditions at r_o. Invoking the prescribed temperature limit,

$$T(r_o) = T_o \tag{3}$$

Moreover, with the outer surface being adiabatic, $q_r(r_o) = 0$, in which case, from Fourier's law, Equation 3.23,

$$\left.\frac{dT}{dr}\right|_{r_o} = 0 \tag{4}$$

Using Equations 2 and 3, it follows that

$$T_o = -\frac{\dot{q}}{4k}\,r_o^2 + C_1\ln r_o + C_2 \tag{5}$$

Similarly, from Equations 1 and 4

$$0 = -\frac{\dot{q}}{2k}\,r_o^2 + C_1 \tag{6}$$

Hence, from Equation 6,

$$C_1 = \frac{\dot{q}}{2k}\,r_o^2 \tag{7}$$

and from Equation 5

$$C_2 = T_o + \frac{\dot{q}}{4k}\,r_o^2 - \frac{\dot{q}}{2k}\,r_o^2\ln r_o \tag{8}$$

Substituting Equations 7 and 8 into the general solution, Equation 2, it then follows that

$$T(r) = T_o + \frac{\dot{q}}{4k}\,(r_o^2 - r^2) - \frac{\dot{q}}{2k}\,r_o^2\ln\frac{r_o}{r} \tag{9}$$

3. The heat removal rate may be determined by obtaining the conduction rate at r_i or by evaluating the total generation rate for the tube. From Fourier's law

$$q_r' = -k2\pi r\frac{dT}{dr}$$

Hence substituting from Equation 9 and evaluating the result at r_i,

$$q_r'(r_i) = -k2\pi r_i\left(-\frac{\dot{q}}{2k}\,r_i + \frac{\dot{q}}{2k}\frac{r_o^2}{r_i}\right)$$

$$q_r'(r_i) = -\pi\dot{q}(r_o^2 - r_i^2) \tag{10}$$

Alternatively, because the tube is insulated at r_o, the rate at which heat is generated in the tube must equal the rate of removal at r_i. That is, for a control volume about the tube, the energy conservation requirement, Equation 1.10, reduces to $\dot{E}_g - \dot{E}_{out} = 0$, where $\dot{E}_g = \dot{q}\pi(r_o^2 - r_i^2)L$ and $\dot{E}_{out} = q'(r_i)L$. Hence

$$q_r'(r_i) = \pi\dot{q}(r_o^2 - r_i^2) \tag{11}$$

4. Applying the energy conservation requirement, Equation 1.11, to the inner surface, it follows that

$$q_r'(r_i) = q_{conv}'$$

or

$$\pi \dot{q}(r_o^2 - r_i^2) = h2\pi r_i(T_i - T_\infty)$$

Hence

$$h = \frac{\dot{q}(r_o^2 - r_i^2)}{2r_i(T_i - T_\infty)} \tag{12}$$

where T_i may be obtained by evaluating Equation 9 at $r = r_i$.

COMMENTS:

Note that, through application of Fourier's law in Part (3), the sign on $q_r'(r_i)$ was found to be negative, Equation 10, implying that heat flow is in the negative r direction. However, in applying the energy balance, we acknowledged that heat flow was *out of* the wall and hence obtained a positive sign for q_r' in Equation 11.

3.4.3 Application of Resistance Concepts

We conclude our discussion of heat generation effects with a word of caution. In particular, when such effects are present, the heat transfer rate is not a constant, independent of the spatial coordinate. Accordingly, it would be *incorrect* to use the thermal resistance concepts and the related heat rate equations developed in Sections 3.1 and 3.3.

3.5 HEAT TRANSFER FROM EXTENDED SURFACES

The term *extended surface* is commonly used in reference to a solid which experiences energy transfer by conduction within its boundaries, as well as energy transfer by convection (and/or radiation) between its boundaries and the surroundings. Such a system is shown schematically in Figure 3.10. A strut is

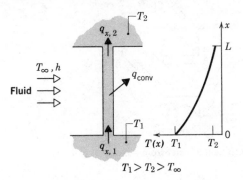

Figure 3.10 Combined conduction and convection in a structural element.

used to provide mechanical support to two walls that are at different temperatures. A temperature gradient in the x direction sustains heat transfer by
conduction internally, at the same time that there is energy transfer by
convection from the surface.

Although there are many different situations that involve combined
conduction-convection effects, the most frequent application is one in which an
extended surface is used specifically to *enhance* the heat transfer rate between a
solid and an adjoining fluid. Such an extended surface is termed a *fin*.

Consider the plane wall of Figure 3.11a. If T_s is fixed, there are two ways in
which the heat transfer rate may be increased. The convection coefficient, h,
could be increased by increasing the fluid velocity, and/or the fluid temperature,
T_∞, could be reduced. However, many situations would be encountered in which
increasing h to the maximum possible value is either insufficient to obtain the
desired heat transfer rate or the associated costs are prohibitive. Such costs are
related to the blower or pump power requirements needed to increase h through
increased fluid motion. Moreover, the second option of reducing T_∞ is often
impractical. Examining Figure 3.11b, however, we see that there exists a third
option. That is, the heat transfer rate may be increased by increasing the surface
area across which the convection occurs. This may be done by employing *fins*
that *extend* from the wall into the surrounding fluid. Note that the thermal
conductivity of the fin material has a strong effect on the temperature
distribution along the fin and therefore influences the degree to which the heat
transfer rate is enhanced.

You are already familiar with several fin applications. Consider the
arrangement for cooling engine heads on motorcycles and lawnmowers or for
cooling electric power transformers. Consider also the tubes with attached fins
used to promote heat exchange between air and the working fluid of an air

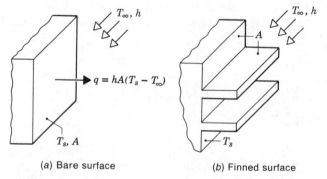

(a) Bare surface   (b) Finned surface

Figure 3.11 Use of fins to enhance heat transfer from a plane
wall.

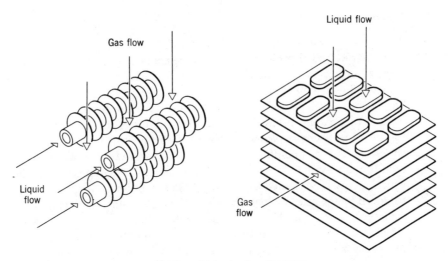

Figure 3.12 Schematic of typical finned-tube heat exchangers.

conditioner. Two such finned-tube arrangements which are commonly used are shown in Figure 3.12.

Different fin configurations are illustrated in Figure 3.13. A *straight fin* is any extended surface that is attached to a *plane wall*. It may be of uniform cross-sectional area, or its cross-sectional area may vary with the distance x from the wall. An *annular fin* is one that is circumferentially attached to a cylinder, and its cross section varies with radius from the centerline of the cylinder. The foregoing fin types have rectangular cross sections, whose area may be expressed as a product of the fin thickness t and the width w for straight fins or the circumference $2\pi r$ for annular fins. In contrast a *pin fin*, or *spine*, is an extended surface of circular cross section. Pin fins may also be of uniform or nonuniform cross section.

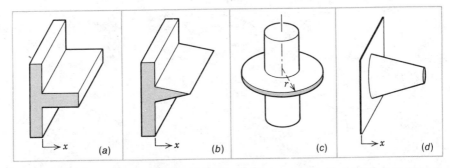

Figure 3.13 Fin configurations. (*a*) Straight fin of uniform cross section. (*b*) Straight fin of nonuniform cross section. (*c*) Annular fin. (*d*) Pin fin.

3.5.1 A General Conduction Analysis

As engineers we are primarily interested in knowing the extent to which particular extended surfaces or fin arrangements could improve heat *dissipation* from a surface to the surrounding fluid. To determine the heat transfer rate associated with a fin, we must first obtain the temperature distribution along the fin. As we have done for previous systems, we begin by performing an energy balance on an appropriate differential element. Consider the extended surface of Figure 3.14. The analysis is simplified if certain assumptions are made. We choose to assume one-dimensional conditions in the longitudinal (x) direction, even though conduction within the fin is actually two dimensional. The rate at which energy is convected to the fluid from any point on the fin surface must be balanced by the rate at which energy reaches that point due to conduction in the transverse direction. However, in practice the fin is thin, in which case temperature changes in the longitudinal direction are much larger than those in the transverse direction. Hence we may assume one-dimensional conduction in the x direction. We will consider steady-state conditions and also assume that the thermal conductivity is constant, that radiation from the surface is negligible, that heat generation effects are absent, and that the convection heat transfer coefficient h is uniform over the surface.

Applying the conservation of energy requirement, Equation 1.10, to the differential element of Figure 3.14, we obtain

$$q_x = q_{x+dx} + dq_{conv} \tag{3.55}$$

From Fourier's law we know that

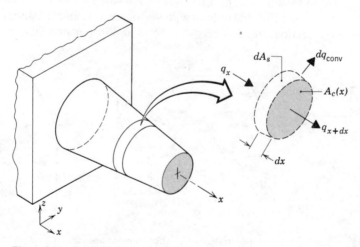

Figure 3.14   Energy balance for an extended surface.

$$q_x = -k A_c \frac{dT}{dx} \tag{3.56}$$

where A_c is the cross-sectional area, which may vary with x.
Since the conduction heat rate at $x + dx$ may be expressed as

$$q_{x+dx} = q_x + \frac{dq_x}{dx} dx \tag{3.57}$$

it follows that

$$q_{x+dx} = -k A_c \frac{dT}{dx} - k \frac{d}{dx}\left(A_c \frac{dT}{dx} \right) dx \tag{3.58}$$

The convection heat transfer rate may be expressed as

$$dq_{conv} = h\, dA_s (T - T_\infty) \tag{3.59}$$

where dA_s is the *surface* area of the differential element. Substituting the foregoing rate equations into the energy balance, Equation 3.55, we obtain

$$\frac{d}{dx}\left(A_c \frac{dT}{dx} \right) - \frac{h\, dA_s}{k\, dx}(T - T_\infty) = 0$$

or

$$\frac{d^2 T}{dx^2} + \left(\frac{1}{A_c}\frac{dA_c}{dx} \right)\frac{dT}{dx} - \left(\frac{1}{A_c}\frac{h}{k}\frac{dA_s}{dx} \right)(T - T_\infty) = 0 \tag{3.60}$$

This result provides a general form of the energy equation for one-dimensional conditions in an extended surface or fin. Its solution would provide the temperature distribution, which could then be used with Equation 3.56 to calculate the conduction rate at any x.

3.5.2 Fins of Uniform Cross-Sectional Area

To solve Equation 3.60 it is necessary to be more specific about the geometry. We begin with the simplest case of straight and pin fins of uniform cross section (Figure 3.15). Each fin is attached to a base surface of temperature $T(0) = T_b$ and extends into a fluid of temperature T_∞.

For the prescribed fins, A_c is a constant and $A_s = Px$, where A_s is the surface area measured from the base to x and P is the fin perimeter. Accordingly, $dA_c/dx = 0$ and $dA_s/dx = P$, and Equation 3.60 reduces to

$$\frac{d^2 T}{dx^2} - \frac{hP}{kA_c}(T - T_\infty) = 0 \tag{3.61}$$

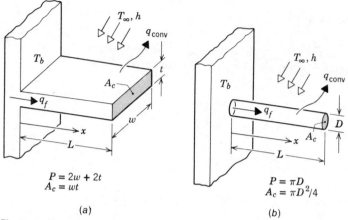

Figure 3.15   Fins of uniform cross-sectional area. (a) Straight fin. (b) Pin fin.

To simplify the form of this equation, we transform the dependent variable by defining an *excess temperature*, θ, as

$$\theta(x) \equiv T(x) - T_{\infty} \tag{3.62}$$

where, since T_{∞} is a constant, $d\theta/dx = dT/dx$. Substituting Equation 3.62 into (3.61), we then obtain

$$\frac{d^2\theta}{dx^2} - m^2\theta = 0 \tag{3.63}$$

where

$$m^2 \equiv \frac{hP}{kA_c} \tag{3.64}$$

Equation 3.63 is a linear, homogeneous, second-order differential equation with constant coefficients. Its general solution is of the form, case 7 of Appendix B.1,

$$\theta(x) = C_1 e^{mx} + C_2 e^{-mx} \tag{3.65}$$

By substitution it may be readily verified that Equation 3.65 is indeed a solution to Equation 3.63.

To evaluate the constants, C_1 and C_2, of Equation 3.65, it is necessary to specify appropriate boundary conditions. One such condition may be specified in terms of the temperature at the *base* of the fin ($x = 0$).

$$\boxed{\theta(0) = T_b - T_\infty \equiv \theta_b} \quad \text{BC. I} \tag{3.66}$$

The second condition, specified at the fin tip ($x = L$), may correspond to any one of four different physical conditions.

The first situation, case A, considers convection heat transfer from the fin tip. Applying an energy balance to a control surface about this tip, Figure 3.16, we obtain

$$hA_c[T(L) - T_\infty] = -kA_c \, dT/dx|_{x=L}$$

or

$$h\theta(L) = -kd\theta/dx|_{x=L} \tag{3.67}$$

That is, the rate at which energy is transferred to the fluid by convection from the tip must equal the rate at which energy reaches the tip by conduction through the fin. Substituting Equation 3.65 into (3.66) and (3.67), we obtain, respectively,

$$\theta_b = C_1 + C_2 \tag{3.68}$$

and

$$h(C_1 e^{mL} + C_2 e^{-mL}) = km(C_2 e^{-mL} - C_1 e^{mL})$$

Solving for C_1 and C_2, it may be shown, after some manipulation, that

$$\frac{\theta}{\theta_b} = \frac{\cosh m(L - x) + (h/mk)\sinh m(L - x)}{\cosh mL + (h/mk)\sinh mL} \tag{3.69}$$

The form of this temperature distribution is shown schematically in Figure 3.16. Note that the magnitude of the temperature gradient decreases with increasing

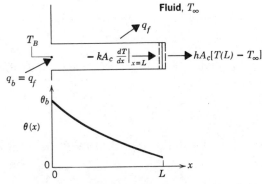

Figure 3.16 Conduction and convection in a fin of uniform cross-sectional area.

x. This trend is a consequence of the reduction in the conduction heat transfer, $q_x(x)$, with increasing x due to continuous convection losses from the fin surface.

We are also interested in the total heat transferred by the fin. From Figure 3.16 it is evident that the fin heat transfer rate, q_f, may be evaluated in two alternative ways, both of which involve use of the temperature distribution. The simpler procedure, and the one which we will use, involves applying Fourier's law at the fin base. That is,

$$q_f = q_b = -kA_c\, dT/dx|_{x=0} = -kA_c\, d\theta/dx|_{x=0} \tag{3.70}$$

Hence, knowing the temperature distribution, $\theta(x)$, q_f may be evaluated, giving

$$q_f = \sqrt{hPkA_c}\; \theta_b\, \frac{\sinh mL + (h/mk)\cosh mL}{\cosh mL + (h/mk)\sinh mL} \tag{3.71}$$

Note that conservation of energy dictates that the rate at which heat is transferred by convection from the fin must equal the rate at which it is conducted through the base of the fin. Accordingly, the alternative formulation for q_f is

$$q_f = \int_{A_f} h[T(x) - T_\infty]dA_s$$

$$q_f = \int_{A_f} h\theta(x)dA_s \tag{3.72}$$

where A_f is the *total*, including the tip, *fin surface area*.

The second tip condition, case B, corresponds to the assumption that the convective heat loss from the fin tip is negligible, in which case the tip may be treated as adiabatic and

$$d\theta/dx|_{x=L} = 0 \tag{3.73}$$

Substituting from Equation 3.65 and dividing by *m*, we then obtain

$$C_1 e^{mL} - C_2 e^{-mL} = 0$$

Using this expression with Equation 3.68 to solve for C_1 and C_2 and substituting the results into (3.65), we obtain

$$\frac{\theta}{\theta_b} = \frac{\cosh m(L - x)}{\cosh mL} \tag{3.74}$$

Using this temperature distribution with Equation 3.70, the fin heat transfer rate is then

$$q_f = \sqrt{hPkA_c}\; \theta_b \tanh mL \tag{3.75}$$

In the same manner, we can obtain the fin temperature distribution and heat transfer rate for the situation, case C, where the temperature is prescribed at the fin tip. That is, the second boundary condition is $\theta(L) = \theta_L$, and the resulting expressions are of the form

$$\frac{\theta}{\theta_b} = \frac{(\theta_L/\theta_b)\sinh mx + \sinh m(L-x)}{\sinh mL} \tag{3.76}$$

$$q_f = \sqrt{hPkA_c}\,\theta_b\frac{(\cosh mL - \theta_L/\theta_b)}{\sinh mL} \tag{3.77}$$

The *very long fin*, case D, is an interesting extension of these results. In particular, as $L \to \infty$, $\theta_L \to 0$ and it is easily verified that

$$\frac{\theta}{\theta_b} = e^{-mx} \tag{3.78}$$

$$q_f = \sqrt{hPkA_c}\,\theta_b \tag{3.79}$$

The foregoing results are summarized in Table 3.3. Note that a table of hyperbolic functions is provided in Appendix B.2.

Analysis of fin thermal behavior is, of course, a good deal more complex if the fin is of nonuniform cross section. For such cases the second term of

---

Table 3.3 Temperature distribution and heat loss for fins of uniform cross section

CASE	TIP CONDITION ($x = L$)	TEMPERATURE DISTRIBUTION, θ/θ_b	FIN HEAT TRANSFER RATE, q_f
A	Convection heat transfer: $h\theta(L) = -kd\theta/dx\vert_{x=L}$	$\dfrac{\cosh m(L-x) + (h/mk)\sinh m(L-x)}{\cosh mL + (h/mk)\sinh mL}$ (3.69)	$M\dfrac{\sinh mL + (h/mk)\cosh mL}{\cosh mL + (h/mk)\sinh mL}$ (3.71)
B	Adiabatic: $d\theta/dx\vert_{x=L} = 0$	$\dfrac{\cosh m(L-x)}{\cosh mL}$ (3.74)	$M\tanh mL$ (3.75)
C	Prescribed temperature: $\theta(L) = \theta_L$	$\dfrac{(\theta_L/\theta_b)\sinh mx + \sinh m(L-x)}{\sinh mL}$ (3.76)	$M\dfrac{(\cosh mL - \theta_L/\theta_b)}{\sinh mL}$ (3.77)
D	Infinite fin ($L \to \infty$): $\theta(L) = 0$	e^{-mx} (3.78)	M (3.79)

$\theta \equiv T - T_\infty$ $\qquad m^2 \equiv hP/kA_c$

$\theta_b = \theta(0) = T_b - T_\infty$ $\qquad M \equiv \sqrt{hPkA_c}\,\theta_b$

Equation 3.60 must be retained, and the solutions are no longer in the form of simple exponential or hyperbolic functions. The development of such solutions is beyond the scope of this text (although working results are presented in the following section), and the interested student is referred to Schneider [3].

EXAMPLE 3.7

A very long rod 25 mm in diameter has one end maintained at 100°C. The surface of the rod is exposed to ambient air at 25°C with a convection heat transfer coefficient of 10 W/m²·K.
1. If the rod is fabricated from pure copper, what is the heat loss from the rod? If the rod is fabricated from stainless steel, type AISI 316, what is its heat loss?
2. Estimate how long the rods of these two different materials must be in order to be considered infinite.

SOLUTION

KNOWN:

A long circular fin exposed to ambient air.

FIND:

1. Heat loss, q, when rod is fabricated from copper or stainless steel.
2. How long rods must be to assume they are infinitely long.

SCHEMATIC:

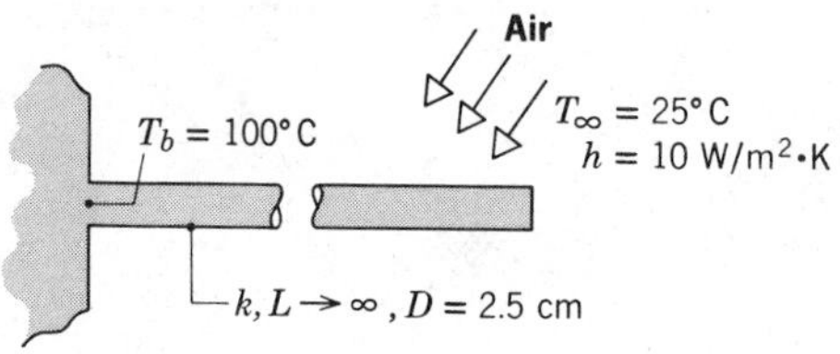

$T_b = 100°C$

Air

$T_\infty = 25°C$
$h = 10$ W/m²·K

$k, L \to \infty, D = 2.5$ cm

ASSUMPTIONS:

1. Steady-state conditions.
2. One-dimensional conduction along the rod.
3. Constant properties.
4. No internal heat generation.

5. Negligible radiation exchange with surroundings.

6. Uniform heat transfer coefficient.

7. Infinitely long rod.

PROPERTIES:

Table A.1, copper $[T = (T_b + T_\infty)/2 = 62.5°C \approx 335$ K]: $k = 398$ W/m·K.
Table A.1, stainless steel, AISI 316 (335 K): $k = 14$ W/m·K.

ANALYSIS:

1. From Equation 3.79 the heat loss is

$$q = \sqrt{hPkA_c}\, \theta_b$$

Hence for copper,

$$q = [10 \text{ W/m}^2 \cdot \text{K} \times \pi \times 0.025 \text{ m} \times 398 \text{ W/m} \cdot \text{K}$$

$$\times \frac{\pi}{4}(0.025 \text{ m})^2]^{1/2} (100 - 25)°C$$

$$q = 29.4 \text{ W} \qquad \triangleleft$$

and for stainless steel,

$$q = [10 \text{W/m}^2 \cdot \text{K} \times \pi \times 0.025 \text{ m} \times 14 \text{ W/m} \cdot \text{K}$$

$$\times \frac{\pi}{4}(0.025 \text{ m})^2]^{1/2} (100 - 25)°C$$

$$q = 5.5 \text{ W} \qquad \triangleleft$$

2. Since there is no heat loss from the tip of an infinitely long rod, an estimate of the validity of this approximation may be made by comparing (3.75) and (3.79). To a satisfactory approximation, the expressions provide equivalent results if tanh $mL \geq 0.99$ or $mL \geq 2.65$. Hence a rod may be assumed to be infinitely long if

$$L \geq L_\infty \equiv 2.65/m = 2.65 \left(\frac{kA_c}{hP}\right)^{1/2}$$

For copper,

$$L_\infty = 2.65 \left[\frac{398 \text{ W/m} \cdot \text{K} \times (\pi/4)(0.025 \text{ m})^2}{10 \text{ W/m}^2 \cdot \text{K} \times \pi(0.025 \text{ m})}\right]^{1/2} = 1.32 \text{ m} \qquad \triangleleft$$

For stainless steel,

$$L_\infty = 2.65 \left[\frac{14 \text{ W/m·K} \times (\pi/4)(0.025 \text{ m})^2}{10 \text{ W/m}^2\text{·K} \times \pi(0.025 \text{ m})} \right]^{1/2} = 0.25 \text{ m} \qquad \triangleleft$$

COMMENTS:

The foregoing results suggest that the fin heat transfer rate may be accurately predicted from the infinite fin approximation if $mL \gtrsim 2.65$. However, if the infinite fin approximation is to accurately predict the temperature distribution, $T(x)$, a larger value of mL would be required.

3.5.3 Fin Performance

Recall that fins are used to increase the heat transfer from a surface by increasing the effective surface area. However, the fin itself represents a conduction resistance to heat transfer from the original surface. For this reason, there is no assurance that the heat transfer rate will be increased through the use of fins. An assessment of this matter may be made by evaluating the *fin effectiveness*, ε_f. It is defined as the *ratio of the fin heat transfer rate to the heat transfer rate that would exist without the fin.* Therefore

$$\varepsilon_f = \frac{q_f}{h A_{c,b} \theta_b} \qquad (3.80)$$

where $A_{c,b}$ is the fin cross-sectional area at the base. In any rational design the value of ε_f should be as large as possible, and in general, the use of fins may rarely be justified unless $\varepsilon_f \gtrsim 2$.

Subject to any one of the four tip conditions that have been considered, the effectiveness for a fin of uniform cross section may be obtained by dividing the appropriate expression for q_f in Table 3.3 by $h A_{c,b} \theta_b$. For the infinite fin approximation, (case D), the result is

$$\varepsilon_f = (kP/hA_c)^{1/2} \qquad (3.81)$$

and several important trends may be inferred. Obviously, fin effectiveness is enhanced by the choice of a material of high thermal conductivity. Aluminum alloys and copper come to mind. However, although copper is superior from the standpoint of thermal conductivity, aluminum alloys are the more common choice due to additional benefits related to lower cost and weight. Fin effectiveness is also enhanced by increasing the ratio of the perimeter to the cross-sectional area. For this reason the use of *thin*, but closely spaced, fins is preferred in most engineering applications. Equation 3.81 also suggests that the use of fins can better be justified under conditions for which the convection coefficient h is small. Hence from Table 1.1 it is evident that the need for fins may be stronger when the fluid is a gas rather than a liquid and particularly when the surface heat transfer is by *free* convection. Hence, if fins are to be used on a surface separating a gas and a liquid, they are generally placed on the gas side, the side having the lower convection coefficient.

Another measure of fin thermal performance is provided by the *fin efficiency*, η_f. The maximum driving potential for convection is the temperature difference between the base ($x = 0$) and the fluid, $\theta_b = T_b - T_\infty$. Hence, the maximum rate at which a fin could dissipate energy is the rate which would exist *if* the entire fin surface were at the base temperature. However, since any fin is characterized by a finite conduction resistance, a temperature gradient must exist along the fin and the above condition is an idealization. A logical definition of fin efficiency is therefore

$$\eta_f \equiv \frac{q_f}{q_{max}} = \frac{q_f}{hA_f\theta_b} \qquad (3.82)$$

where A_f is the total surface area of the fin. From its definition, it is evident that knowledge of η_f may be used as a basis for computing the actual fin heat transfer rate.

Equation 3.82 has proven to be particularly useful for treating fins of nonuniform cross-sectional area. Since none of the results presented in Table 3.3 pertain to such fins, it is necessary to return to Equation 3.60 in order to determine the solution for the particular form of $A_c(x)$ and $A_s(x)$. For numerous fin configurations, such solutions are conveniently presented in graphical form [3]. In Figure 3.17 the fin efficiency is plotted for three common *straight fin* profiles: the previously considered *rectangular fin* of uniform cross-sectional area, Figure 3.13a; as well as the *triangular fin* ($y \sim x$), Figure 3.13b, and the *parabolic fin* ($y \sim x^2$), which are of nonuniform cross-sectional area. In Figure 3.18 results are presented for the *circular* (or *annular*) *fin* of *rectangular* profile, Figure 3.13c. Note that the foregoing results are presented in terms of a corrected length L_c. To use these results with Equation 3.82, it is necessary to calculate the maximum heat transfer rate in the following manner. For the rectangular, triangular and parabolic fins,

$$q_{max} \equiv hPL_c\theta_b \qquad (3.83a)$$

while, for the annular fin,

$$q_{max} \equiv 2\pi h(r_{2,c}^2 - r_1^2)\theta_b \qquad (3.83b)$$

Extended surfaces find use in numerous engineering applications. For comprehensive discussions of thermal effects, the treatments by Schneider [3] and by Kern and Kraus [4] should be consulted.

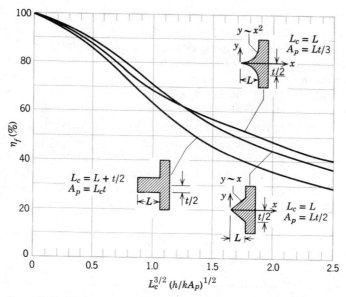

Figure 3.17 Efficiency ot straight fins (rectangular, triangular, and parabolic profiles).

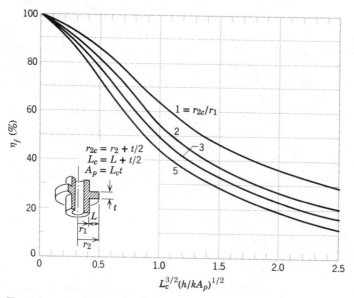

Figure 3.18 Efficiency of annular fins of rectangular profile.

EXAMPLE 3.8

An engine head on a motorcycle is constructed of 2024-T6 aluminum alloy. The head is in the form of a cylinder that has a height of $H = 0.15$ m and an outside diameter of $D = 50$ mm. Under typical operating conditions the outer surface of the head is at a temperature of 500 K and is exposed to ambient air at 300 K, with a convection coefficient of 50 W/m²·K. Annular fins of rectangular profile are typically added to the head in order to increase heat transfer to the surroundings. Assume that five such fins, which are of thickness $t = 6$ mm, length $L = 20$ mm and equally spaced, are added. What is the increase in heat transfer from the head due to addition of the fins?

SOLUTION

KNOWN:

Operating conditions of a finned motorcycle head.

FIND:

Increase in heat transfer associated with using fins.

SCHEMATIC:

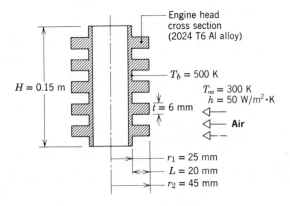

ASSUMPTIONS:

1. Steady-state conditions.
2. One-dimensional radial conduction in the fins.
3. Constant properties.
4. No internal heat generation.

5. Negligible radiation exchange with the surroundings.
6. Uniform convection coefficient over the outer surface (with or without fins).

PROPERTIES:

Table A.1, 2024-T6 aluminum ($T = 400$ K): $k = 186$ W/m·K.

ANALYSIS:

With the fins in place, the heat transfer rate is

$$q = q_f + q_o$$

From Equation 3.82, the fin heat transfer rate is

$$q_f = N\eta_f q_{max}$$

where N is the number of fins and from Equation 3.83b

$$q_{max} = 2\pi h(r_{2,c}^2 - r_1^2)(T_b - T_\infty)$$

The heat transfer from the exposed cylinder surface is

$$q_o = hA_o(T_b - T_\infty)$$

where

$$A_o = (H - Nt)(2\pi r_1)$$

Hence

$$q = N\eta_f h2\pi(r_{2,c}^2 - r_1^2)(T_b - T_\infty)$$
$$+ h(H - Nt)(2\pi r_1)(T_b - T_\infty)$$

The fin efficiency may be obtained from Figure 3.18 with

$$r_{2c} = r_2 + \frac{t}{2} = 0.048 \text{ m}, \qquad L_c = L + \frac{t}{2} = 0.023 \; m$$

$$\frac{r_{2c}}{r_1} = 1.92, \qquad A_p = L_c t = 1.38 \times 10^{-4} \, m^2,$$

$$L_c^{3/2}\left(\frac{h}{kA_p}\right)^{1/2} = 0.15$$

Hence from Figure 3.18,

$$\eta_f \approx 0.95$$

It follows that

$$q = 5\{0.95 \times 50 \text{ W/m}^2 \cdot \text{K} \times 2\pi[(0.048^2 - 0.025^2)\text{m}^2] \times (500 - 300)\text{K}\}$$
$$+ 50 \text{ W/m}^2 \cdot \text{K} \ (0.15 - 5 \times 0.006)(2\pi \times 0.025)\text{m}^2 \times (500 - 300)\text{K}$$

Hence

$$q = 5\{100.22\} \text{ W} + 188.50 \text{ W}$$

or

$$q = 690 \text{ W}$$ ◁

Without the fins, the heat transfer rate is

$$q_{wo} = hA_{wo}(T_b - T_\infty)$$

where

$$A_{wo} = H \times 2\pi r_1$$

Hence

$$q_{wo} = 50 \text{ W/m}^2 \cdot \text{K} \ (0.15 \times 2\pi \times 0.025)\text{m}^2 \ (200 \text{ K})$$

or

$$q_{wo} = 236 \text{ W}$$ ◁

COMMENTS:

Although the fins significantly increase heat dissipation from the head, considerable improvement could still be obtained by increasing the number of fins (by reducing both t and the separation between fins).

3.6 SUMMARY

Despite the inherent mathematical simplicity, one-dimensional, steady-state heat transfer occurs in numerous engineering applications. Although one-dimensional, steady-state conditions may not apply exactly, the assumptions may often be made to obtain results of reasonable accuracy. You should therefore be thoroughly familiar with the means by which such problems are treated. In particular you should be comfortable with the use of equivalent heat transfer circuits and with the expressions for the conduction resistances that pertain to each of the three common geometries. You should also be familiar with how the heat equation and Fourier's law may be used to obtain temperature distributions and the corresponding fluxes. The implications of an internally distributed source of energy should also be clearly understood. Finally, you should appreciate the important role that extended surfaces can play in the design of thermal systems and should have the facility to effect design and performance calculations for such surfaces.

REFERENCES

1. Fried, E., "Thermal Conduction Contribution to Heat Transfer at Contacts" in R. P. Tye, Ed., *Thermal Conductivity*, Vol. 2, Academic Press, London, 1969.
2. Schneider, P. J., "Conduction, Interface Resistance" in W. M. Rohsenow and J. P. Hartnett, Eds., *Handbook of Heat Transfer*, McGraw-Hill, New York, 1973.
3. Schneider, P. J., *Conduction Heat Transfer*, Addison-Wesley, Cambridge, Mass., 1955.
4. Kern, D. Q. and A. D. Kraus, *Extended Surface Heat Transfer*, McGraw-Hill, New York, 1972.

PROBLEMS

3.1 The rear window of an automobile is defogged by passing warm air at 40°C over its inner surface, and the associated convection coefficient is 30 W/m²·K. Under conditions for which the outside ambient air temperature is −10°C and the associated convection coefficient is 65 W/m²·K, what are the inner and outer surface temperatures of the window? The window glass is 4 mm thick.

3.2 The rear window of an automobile is defogged by attaching a thin, transparent, film-type heating element to its inner surface. By electrically heating this element, a uniform heat flux may be established at the inner surface. What is the electrical power that must be provided per unit window area in order to maintain an inner surface temperature of 15°C when the interior air temperature and convection coefficient are 25°C and 10 W/m²·K, respectively, and the exterior (ambient) air temperature and convection coefficient are −10°C and 65 W/m²·K, respectively? The window glass is 4 mm thick.

3.3 In a manufacturing process, a transparent film is being bonded to a substrate as shown in the following sketch. To cure the bond at a temperature, T_0, a radiant source is used to provide a heat flux q_0'' (W/m²), all of which is absorbed at the bonded surface. The back of the substrate is maintained at T_1 while the free surface of the film is exposed to air at T_∞ and a convection heat transfer coefficient, h.

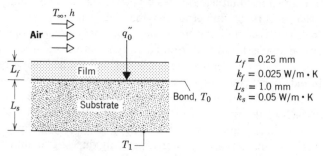

$L_f = 0.25$ mm
$k_f = 0.025$ W/m·K
$L_s = 1.0$ mm
$k_s = 0.05$ W/m·K

a) Show the thermal circuit representing the steady-state heat transfer situation. Be sure to label *all* elements, nodes, heat rates, etc. Leave in symbolic form.

b) Assume the following conditions: $T_\infty = 20°C$, $h = 50$ W/m²·K, and $T_1 = 30°C$. Calculate the heat flux, q_0'', that is required in order to maintain the bonded surface at $T_0 = 60°C$.

3.4 Both copper and stainless steel (AISI 304) are being considered as a wall material for a liquid-cooled rocket nozzle. The cooled exterior of the wall is maintained at 150°C, while combustion gases within the nozzle are at 2750°C. The gas side heat transfer coefficient is known to be $h_i = 2 \times 10^4$ W/m²·K, and the radius of the nozzle is much larger than the wall thickness. Thermal limitations dictate that the temperature of the copper must not exceed 540°C, while that of the steel must not exceed 980°C. What is the maximum wall thickness that could be employed for each of the two materials? If the nozzle is constructed with the maximum wall thickness, which material would be preferred?

3.5 One-dimensional heat conduction occurs under steady-state conditions through a plane wall of thickness L and thermal conductivity k. The inner surface of the wall is maintained at a temperature $T_{s,1}$, while the outer surface is exposed to air at T_∞. Heat transfer between the outer surface and the air is characterized by a convection coefficient h.

a) Assuming radiation effects to be negligible, sketch and label the equivalent thermal circuit for the system. Obtain an expression for the ratio of temperature differences, $(T_{s,1} - T_{s,2})/(T_{s,2} - T_\infty)$, where $T_{s,2}$ is the outer surface temperature. Defining the dimensionless parameter, hL/k, as the Biot number, sketch the temperature distribution in the wall for $Bi = 0.1$, 1.0, and 10.0. If $k = 1.2$ W/m·K, $h = 12$ W/m²·K, $L = 0.055$ m, $T_{s,1} = 350°C$, and $T_\infty = 50°C$, what are the values of q'', Bi, $(T_{s,1} - T_{s,2})/(T_{s,2} - T_\infty)$, and $T_{s,2}$?

b) Now account for the effects of radiation exchange between the outer surface and surroundings at $T_{sur} = T_\infty$, with h_r designated as the radiation transfer coefficient. Sketch and label the equivalent thermal circuit for the system. If $h_r = 12$ W/m²·K, what are the values of q'', $(T_{s,1} - T_{s,2})/(T_{s,2} - T_\infty)$, and $T_{s,2}$?

3.6 The *wind chill*, which is experienced on a cold, windy day, is related to increased heat transfer from exposed human skin to the surrounding atmosphere. Consider a layer of fatty tissue that is 3 mm thick and whose interior surface is maintained at a temperature of 36°C. On a calm day the convection heat transfer coefficient at the outer surface is 25 W/m²·K, but with 30 mph winds it reaches 65 W/m²·K. In both cases the ambient air temperature is −15°C.

a) What is the ratio of the heat loss per unit area from the skin for the calm day to that for the windy day?

b) What will be the skin outer surface temperature for the calm day? For the windy day?

c) What temperature would the air have to assume on the calm day to produce the same heat loss occurring with the air temperature at −15°C on the windy day?

3.7 A thermopane window consists of two pieces of glass 7 mm thick that enclose an air space 7 mm thick. The window separates room air at 20°C from outside ambient air at $-10°C$. The convection coefficient associated with the inner (room side) surface is $10 \ W/m^2 \cdot K$, while that associated with the outer (ambient air) side is $80 \ W/m^2 \cdot K$. What is the heat loss through a window which is 0.8 m long by 0.5 m wide? The air enclosed between the panes may be assumed to be stagnant.

3.8 The wall of a building is a composite consisting of a 100-mm layer of common brick, a 100-mm layer of glass fiber (paper faced, $28 \ kg/m^3$), a 10-mm layer of gypsum plaster (vermiculite), and a 6-mm layer of pine panel. If the inside convection coefficient is $10 \ W/m^2 \cdot K$ and the outside convection coefficient is $70 \ W/m^2 \cdot K$, what are the total resistance and the overall coefficient for heat transfer?

3.9 The composite wall of an oven consists of three materials, two of which are of known thermal conductivity, $k_A = 20 \ W/m \cdot K$ and $k_C = 50 \ W/m \cdot K$, and known thickness, $L_A = 0.30$ m and $L_C = 0.15$ m. The third material, B, which is sandwiched between materials A and C, is of known thickness, $L_B = 0.15$ m, but unknown thermal conductivity, k_B.

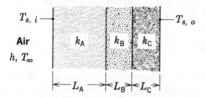

Under steady-state operating conditions, measurements reveal an outer surface temperature of $T_{s,o} = 20°C$, an inner surface temperature of $T_{s,i} = 600°C$, and an oven air temperature of $T_\infty = 800°C$. The inside convection coefficient is known to be $h = 25 \ W/m^2 \cdot K$. What is the value of k_B?

3.10 A house has composite walls of wood, fiberglass insulation, and dry wall as indicated in the following sketch. On a cold winter day the convection heat transfer coefficients are $h_o = 60 \ W/m^2 \cdot K$ and $h_i = 30 \ W/m^2 \cdot K$. The total wall surface area is $350 \ m^2$.

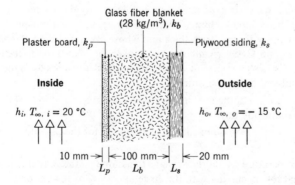

a) Determine a symbolic expression for the total thermal resistance of the walls including inside and outside convection effects for the prescribed conditions.

b) Determine the total heat loss through the walls.

c) If the wind were blowing violently, raising h_o to $300 \text{ W/m}^2 \cdot \text{K}$, determine the percentage increase in the heat loss.

d) What is the controlling resistance that determines the amount of heat flow through the walls?

3.11 Consider the composite wall of the foregoing problem under conditions for which the inside air is still characterized by $T_{\infty,i} = 20°\text{C}$ and $h_i = 30 \text{ W/m}^2 \cdot \text{K}$. However, now consider more realistic conditions for which the outside air is characterized by a diurnal (time) varying temperature of the form

$$T_{\infty,o}(\text{K}) = 273 + 5 \sin\left(\frac{2\pi}{24}t\right) \qquad 0 \le t \le 12h$$

$$T_{\infty,o}(\text{K}) = 273 + 11 \sin\left(\frac{2\pi}{24}t\right) \qquad 12 \le t \le 24h$$

with $h_o = 60 \text{ W/m}^2 \cdot \text{K}$. Estimate the daily heat loss through the wall if its total surface area is 200 m^2.

3.12 A composite wall of height H and of unit length normal to the page is insulated at its ends and is comprised of four different materials, arranged as shown in the diagram.

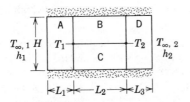

a) Sketch the thermal circuit of the system.

b) Consider a wall for which $H = 3 \text{ m}$, $H_B = H_C = 1.5 \text{ m}$, $L_1 = L_3 = 0.05 \text{ m}$, $L_2 = 0.10 \text{ m}$, $k_A = k_D = 50 \text{ W/m} \cdot \text{K}$, $k_B = 10 \text{ W/m} \cdot \text{K}$, and $k_C = 1 \text{ W/m} \cdot \text{K}$. Under conditions for which $T_{\infty,1} = 200°\text{C}$, $h_1 = 50 \text{ W/m}^2 \cdot \text{K}$, $T_{\infty,2} = 25°\text{C}$, and $h_2 = 10 \text{ W/m}^2 \cdot \text{K}$, what is the rate of heat transfer through the wall? What are the interface temperatures, T_1 and T_2?

3.13 Consider a composite wall that includes an 8-mm-thick hardwood siding, 40 mm by 130 mm hardwood studs on 0.65 m centers with glass fiber insulation (paper faced, 28 kg/m³), and a 12-mm layer of gypsum (vermiculite) wall board.

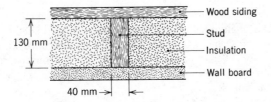

What is the thermal resistance associated with a wall that is 2.5 m high by 6.5 m wide (having 10 studs, each 2.5 m high)?

3.14 For the flat plate solar collector of Problem 1.18, sketch the thermal circuit that connects the absorber plate to the collector surroundings. Write an expression for each of the resistances appearing in the circuit.

3.15 A composite wall separates combustion gases at 2600°C from a liquid coolant at 100°C, with gas and liquid side convection coefficients of 50 and 1000 W/m²·K, respectively. The wall is comprised of a layer of beryllium oxide on the gas side 10 mm thick and a slab of stainless steel (AISI 304) on the liquid side 20 mm thick. The contact resistance between the oxide and the steel is 0.05 m²·K/W. What is the heat loss per unit surface area of the composite? Sketch the temperature distribution from the gas to the liquid.

3.16 Two stainless steel plates 10 mm thick are subjected to a contact pressure of 1 bar under vacuum conditions for which there is an overall temperature drop of 100°C across the plates. What is the heat flux through the plates? What is the temperature drop across the contact plane?

3.17 Consider a plane composite wall that is composed of two materials of thermal conductivities $k_A = 0.1$ W/m·K and $k_B = 0.04$ W/m·K and thicknesses $L_A = 10$ mm and $L_B = 20$ mm. The contact resistance at the interface between the two materials is known to be 0.30 m²·K/W. Material A adjoins a fluid at 200°C for which $h = 10$ W/m²·K, and material B adjoins a fluid at 40°C for which $h = 20$ W/m²·K.

a) What is the rate of heat transfer through a wall that is 2 m high by 2.5 m wide?

b) Sketch the temperature distribution.

3.18 The thermal conductivity of metal samples may be determined by means of the cut-bar method. The method involves working with a standard sample, whose thermal conductivity, k_S, is known and arranging it in series with other samples whose thermal conductivities are desired. Consider the following stack comprised of four materials whose thermal conductivities are to be determined and a standard sample whose thermal conductivity is known.

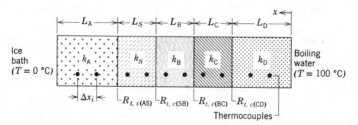

Each sample is cylindrical, and the stack is formed by arranging the cylinders in series (axially). The longitudinal surface of the stack is well insulated, while the ends are maintained at 100°C and 0°C. Different conditions associated with the sample interfaces provide for different contact resistances. Thermocouples are inserted in each of the samples to measure the temperature difference ΔT_i across a fixed distance Δx_i in the sample.

a) Obtain an expression for an unknown sample thermal conductivity, k_i, in terms of the known thermal conductivity, k_S, and the temperature differences ΔT_i and ΔT_S.

b) Consider a situation for which the samples are of 25 mm diameter and consist of high purity aluminum (A), Armco iron (S), AISI 304 stainless steel (B), 70 percent Cu–30 percent Zn brass (C), and high purity copper (D). The sample lengths are known to be $L_A = 65$ mm, $L_S = L_B = L_C = 40$ mm, and $L_D = 65$ mm. An air gap exists at the aluminum to iron interface for which $R_{t,c(AS)} = 0.9$ K/W, and a vacuum grease interface exists between the iron and the steel with $R_{t,c(SB)} = 0.15$ K/W. Soldered joints exist between the steel and the brass and between the brass and the copper for which $R_{t,c(BC)} = 0.12$ K/W and $R_{t,c(CD)} = 0.04$ K/W, respectively. Obtaining all property values from Table A.1, calculate the rate of heat transfer through the stack. Calculate all interface temperatures and plot the temperature distribution across the entire stack.

3.19 The following diagram shows a conical section fabricated from pure aluminum. It is of circular cross section with the diameter, D, given by

$$D = ax^{1/2}$$

where $a = 0.5$ m$^{1/2}$. The small end is located at $x_1 = 25$ mm and the large end at $x_2 = 125$ mm. The end temperatures are $T_1 = 600$ K and $T_2 = 400$ K, while the lateral surface is well insulated.

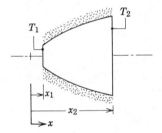

a) Derive an expression for the temperature distribution $T(x)$ in symbolic form, assuming one-dimensional conditions. Sketch the temperature distribution.

b) Calculate the heat rate, q_x.

3.20 A truncated solid cone is of circular cross section, and its diameter is related to the axial coordinate by an expression of the form $D = ax^{3/2}$, where $a = 1.0$ m$^{-1/2}$.

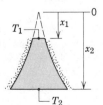

The sides are well insulated, while the top surface of the cone at x_1 is maintained at T_1 and the bottom surface at x_2 is maintained at T_2.

a) Obtain an expression for the temperature distribution, $T(x)$.

b) What is the rate of heat transfer across the cone if it is constructed of pure aluminum with $x_1 = 0.075$ m, $T_1 = 100°C$, $x_2 = 0.225$ m, and $T_2 = 20°C$.

3.21 From Figure 2.5 it is evident that, over a wide temperature range, the temperature dependence of the thermal conductivity of many solids may be approximated by a linear expression of the form

$$k = k_o + aT$$

where k_o is a positive constant and a is a coefficient that may be positive or negative. Obtain an expression for the heat flux across a plane wall whose inner and outer surfaces are maintained at T_0 and T_1, respectively. Sketch the forms of the temperature distribution corresponding to $a > 0$, $a = 0$, and $a < 0$.

3.22 Consider a tube wall of inner and outer radii, r_i and r_o, whose temperatures are maintained at T_i and T_o, respectively. The thermal conductivity of the cylinder is temperature dependent and may be represented by an expression of the form

$$k = k_o(1 + aT)$$

where k_o and a are constants. Obtain an expression for the heat transfer per unit length of the tube. What is the thermal resistance of the tube wall?

3.23 The thermal conductivity of a material varies with temperature according to the relation

$$k = k_o(1 + aT^2)$$

where k_o and a are constants. Obtain an expression for the thermal resistance per unit surface area of a plane wall constructed from this material.

3.24 A steam pipe of 0.12 m outside diameter is insulated with a 20-mm-thick layer of calcium silicate. If the inner and outer surfaces of the insulation are at temperatures of $T_{s,1} = 800$ K and $T_{s,2} = 490$ K, respectively, what is the heat loss per unit length of the pipe?

3.25 A long cylindrical rod of radius 10 cm consists of a nuclear reacting material ($k = 0.5$ W/m·K) generating 24,000 W/m³ uniformly throughout its volume. This rod is encapsulated within another cylinder whose outer radius is 20 cm and that has a thermal conductivity of 4 W/m·K. The outer surface is surrounded by a fluid at 100°C, and the convection coefficient between the surface and the fluid is 20 W/m²·K. Find the temperatures at the interface between the two cylinders and at the outer surface.

3.26 A special coating, which is applied to the inner surface of a plastic tube, is cured by placing a cylindrical radiation heat source within the tube. (Diagram at top of p. 125.) The space between the tube and the source is evacuated, and the source delivers a uniform heat flux, q_i'', which is absorbed at the inner surface of the tube. The outer surface of the tube is maintained at a uniform temperature, $T_{s,2}$.

Develop an expression for the temperature distribution, $T(r)$, in the tube wall in

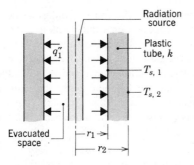

terms of q_1'', $T_{s,2}$, r_1, r_2, and k. If the inner and outer tube radii are $r_1 = 25$ mm and r_2 = 38 mm, respectively, what is the power required per unit length of the radiation source to maintain the inner surface at $T_{s,1} = 150°C$, while the outer surface is fixed at $T_{s,2} = 25°C$? The tube wall thermal conductivity is $k = 10$ W/m·K.

3.27 An electrical current of 700 A flows through a stainless steel cable having a diameter of 5 mm and an electrical resistance of 6×10^{-4} Ω/m (per meter of cable length). The cable is in an environment having a temperature of 30°C, and the total coefficient associated with convection and radiation between the cable and the environment is approximately 25 W/m²·K.

a) If the cable is bare, what is its surface temperature?

b) If a very thin coating of electrical insulation is applied to the cable, with a contact resistance of 0.02 m²·K/W, what are the insulation and cable surface temperatures?

c) There is some concern about the ability of the insulation to withstand elevated temperatures. What thickness of this insulation ($k = 0.5$ W/m·K) will yield the lowest value of the maximum insulation temperature? What is the value of the maximum temperature when this thickness is used?

3.28 The radiation heat gauge shown in the following diagram is made from constantan metal foil, which is coated black and is in the form of a circular disc of radius R and thickness t. The gauge is located in an evacuated enclosure.

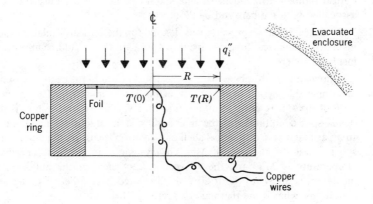

The incident radiation flux absorbed by the foil, q_i'', diffuses toward the outer circumference and into the larger copper ring, which acts as a heat sink at the constant temperature $T(R)$. Two copper lead wires are attached to the center of the foil and to the ring to complete a thermocouple circuit that allows for measurement of the temperature difference between the foil center and the foil edge, $\Delta T = T(0) - T(R)$. Obtain the differential equation that determines the temperature distribution in the foil, $T(r)$, under steady-state conditions. Solve this equation to obtain an expression relating ΔT to q_i''. You may neglect radiation exchange between the foil and its surroundings.

3.29 A 0.20-m diameter steel pipe is used to transport saturated steam at a pressure of 20 bars in a room for which the air temperature is 25°C and the convection heat transfer coefficient at the outer surface of the pipe is 20 W/m²·K.

a) What is the heat loss per unit length from the bare pipe (no insulation)? What is the heat loss per unit length if a 50-mm thick layer of insulation (magnesia, 85 percent) is added? The steel and magnesia may each be assumed to have an emissivity of 0.8.

b) The costs associated with generating the steam and installing the insulation are known to be \$4/10⁹ J and \$100/m of pipe length, respectively. If the steam line is to operate 7500 h/yr, how many years are needed to pay back the initial investment in insulation?

3.30 Steam at a temperature of 250°C flows through a steel pipe (AISI 1010) of 60 mm inside diameter and 75 mm outside diameter. The convection coefficient between the steam and the inner surface of the pipe is 500 W/m²·K, while that between the outer surface of the pipe and the surroundings is 25 W/m²·K. The pipe emissivity is 0.8, and the temperature of the air and the surroundings is 20°C. What is the heat loss per unit length of pipe?

3.31 A bakelite coating is to be used with a 10-mm diameter conducting rod, whose surface is maintained at 200°C by passage of an electrical current. The rod is in a fluid at 25°C, and the convection coefficient is 140 W/m²·K. What is the critical radius associated with the coating? What is the heat transfer rate per unit length for the bare rod and for the rod with a coating of bakelite that corresponds to the critical radius? How much bakelite should be added to reduce the heat transfer associated with the bare rod by 25 percent?

3.32 In Example 3.3, an expression was derived for the critical insulation radius for an insulated, cylindrical tube. Derive the expression that would be appropriate for an insulated sphere.

3.33 A hollow aluminum sphere, with an electrical heater in the center, is used in tests to determine the thermal conductivity of insulating materials. The inner and outer radii of the sphere are 0.15 m and 0.18 m, respectively, and testing is done under steady-state conditions with the inner surface of the aluminum maintained at 250°C. In a particular test, a spherical shell of insulation is cast on the outer surface of the sphere to a thickness of 0.12 m. The system is in a room for which the air temperature is 20°C and the convection coefficient at the outer surface of the insulation is 30 W/m²·K. If 80 W is dissipated by the heater under steady-state conditions, what is the thermal conductivity of the insulation?

3.34 The energy transferred from the anterior chamber of the eye through the cornea varies considerably depending on whether or not a contact lens is worn. Treat the eye as a spherical system and assume the system to be at steady state. The convection coefficient, h_o, is unchanged with and without the contact lens in place. The cornea and the lens cover one-third of the spherical surface area.

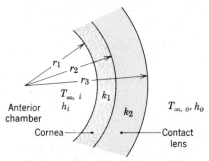

Values of the parameters representing this situation are as follows:

$r_1 = 10.2$ mm $r_2 = 12.7$ mm $r_3 = 16.5$ mm
$T_{\infty,i} = 37°C$ $T_{\infty,o} = 21°C$
$k_1 = 0.35$ W/m·K $k_2 = 0.80$ W/m·K
$h_i = 12$ W/m²·K $h_o = 6$ W/m²·K

a) Construct the thermal circuits, labeling all potentials and flows for the systems excluding the contact lens and including the contact lens. Write resistance elements in terms of appropriate parameters.

b) Determine the heat loss from the anterior chamber with and without the contact lens in place.

c) Discuss the implication of your results.

3.35 The outer surface of a hollow sphere of radius r_2 is subjected to a uniform heat flux, q_2''. The inner surface at r_1 is held at a constant temperature $T_{s,1}$.

a) Develop an expression for the temperature distribution, $T(r)$, in the sphere wall in terms of q_2'', $T_{s,1}$, r_1, r_2, and the thermal conductivity of the wall material, k.

b) If the inner and outer tube radii are $r_1 = 50$ mm and $r_2 = 100$ mm, respectively, what heat flux q_2'' is required to maintain the outer surface at $T_{s,2} = 50°C$ while the inner surface is at $T_{s,1} = 20°C$. The thermal conductivity of the wall material is $k = 10$ W/m·K.

3.36 The steady-state temperature distribution in a composite plane wall of three different materials, each of constant thermal conductivity, is shown below.

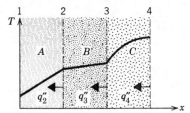

a) Comment on the relative magnitudes of q_2'' and q_3'' and of q_3'' and q_4''.

b) Comment on the relative magnitudes of k_A and k_B and of k_B and k_C.

3.37 A plane wall of thickness 0.1 m and thermal conductivity 25 W/m·K having uniform volumetric heat generation of 0.3 MW/m³ is insulated on one side while the other side is exposed to a fluid at 92°C. The convection heat transfer coefficient between the wall and the fluid is 500 W/m²·K. Determine the maximum temperature in the wall.

3.38 When passing an electrical current I, a copper bus bar of rectangular cross section (6 mm × 150 mm) experiences uniform heat generation at a rate $\dot{q}(\text{W/m}^3)$ given by

$$\dot{q} = aI^2$$

where $a = 0.015$ W/m³·A². If the bar is in ambient air with $h = 5$ W/m²·K and its maximum temperature must not exceed that of the air by more than 30°C, what is the allowable current capacity for the bus bar?

3.39 A semiconductor material of thermal conductivity $k = 2$ W/m·K and electrical resistivity $\rho = 2 \times 10^{-5}$ Ω·m is used to fabricate a cylindrical rod of 10 mm diameter and 40 mm length. The longitudinal surface of the rod is well insulated, while the ends are maintained at temperatures of 100°C and 0°C. If the rod carries a current of 10 A, what is its midpoint temperature? What is the heat rate at each of the ends?

3.40 A rear window defroster in an automobile consists of uniformly distributed high resistance wires molded into the glass. When power is applied to the wires, uniform heat generation may be assumed to occur within the window. During operation, heat that is generated is transferred by convection from both the interior and exterior surfaces of the window. However, the convection coefficient on the warmer, interior side, h_i, is less than that on the exterior side, h_o, due to the effects of vehicle speed and atmospheric winds. On the same coordinate system, sketch the steady-state temperature distributions that would exist in the glass *before* the defroster has been turned on and *after* the defroster has been on for some time.

3.41 A nuclear fuel element of thickness 2L is covered with a steel cladding of thickness b. Heat generated within the nuclear fuel at a rate $\dot{q}$ is removed by a fluid at T_∞, which adjoins one surface and is characterized by a convection coefficient h. The other surface is well insulated, and the fuel and steel have thermal conductivities of k_f and k_s, respectively.

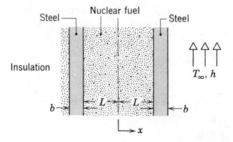

a) Obtain an equation for the temperature distribution, $T(x)$, in the nuclear fuel. Express your results in terms of $\dot{q}$, k_f, L, b, k_s, h, and T_∞.

b) Sketch the temperature distribution, $T(x)$, for the entire system.

3.42 Consider a plane wall of thickness L, which is to act as shielding for a nuclear reactor. The inner surface $(x = 0)$ receives gamma radiation that is partially absorbed within the shielding and has the effect of an internally distributed heat source. In particular heat is generated per unit volume within the shielding according to the relation

$$\dot{q}(x) = q_o'' \alpha e^{-\alpha x}$$

where q_o'' is the incident radiation flux and α is a property (the absorption coefficient) of the shielding material.

a) If the inner $(x = 0)$ and outer $(x = L)$ surfaces of the shielding are maintained at temperatures of T_1 and T_2, respectively, what is the form of the temperature distribution in the shielding?

b) Obtain an expression that could be used to determine the x location in the shield at which the temperature is a maximum.

3.43 A quartz window of thickness L serves as a viewing port in a furnace used for annealing steel. The inner surface $(x = 0)$ of the window is irradiated with a uniform heat flux, q_o'', due to emission from hot gases in the furnace. A fraction, β, of this radiation may be assumed to be absorbed at the inner surface, while the remaining radiation is partially absorbed as it passes through the quartz. The volumetric heat generation due to this absorption may be described by an expression of the form

$$\dot{q}(x) = (1 - \beta)q_o'' \alpha e^{-\alpha x}$$

where α is the absorption coefficient of the quartz. Convection heat transfer occurs from the outer surface $(x = L)$ of the window to ambient air at T_∞ and is characterized by the convection coefficient h. Convection and radiation emission from the inner surface may be neglected, along with radiation emission from the outer surface. Determine the temperature distribution in the quartz, expressing your result in terms of the foregoing parameters.

3.44 A copper cable of 30 mm diameter has an electrical resistance of 5×10^{-3} Ω/m and is used to carry an electrical current of 250 A. The cable is exposed to ambient air at 20°C, and the associated convection coefficient is 25 $\text{W/m}^2 \cdot \text{K}$. What are the surface and centerline temperatures of the copper?

3.45 The cross section of a long cylindrical fuel element in a nuclear reactor is shown below.

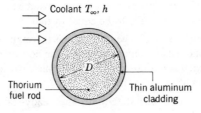

Coolant T_∞, h

D

Thorium fuel rod

Thin aluminum cladding

Energy generation occurs uniformly in the thorium fuel rod, which is of diameter $D = 25$ mm and is wrapped in a thin aluminum cladding. It is proposed that, under steady-state conditions, the system operate with a generation rate of $\dot{q} = 7 \times 10^8$ W/m^3 and cooling system characteristics of $T_\infty = 95°$C and $h = 7000$ W/m$^2 \cdot$K. Is this proposal satisfactory?

3.46 A homeowner, whose water pipes have frozen during a period of cold weather, decides to melt the ice by passing an electric current I through the pipe wall. The inner and outer radii of the wall are designated as r_1 and r_2, respectively, and its electrical resistance per unit length is designated as R'_e (Ω/m). The pipe is well insulated on the outside, and during melting the ice (and water) in the pipe remain at a constant temperature, T_m, associated with the melting process.

a) Assuming that steady-state conditions are reached shortly after application of the current, determine the form of the steady-state temperature distribution, $T(r)$, in the pipe wall during the melting process.

b) Develop an expression for the time, t_m, required to completely melt the ice. Calculate this time for $I = 100$ A, $R'_e = 0.30$ Ω/m, and $r_1 = 50$ mm.

3.47 Copper tubing is joined to the absorber of a flat plate solar collector as shown.

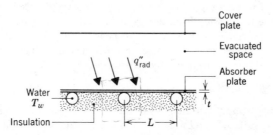

The aluminum alloy (2024-T6) absorber plate is of thickness $t = 6$ mm and well insulated on its bottom. The top surface of the plate is separated from a transparent cover plate by an evacuated air space. The tubes are spaced a distance of $L = 0.20$ m from each other, and water is circulated through the tubes in order to remove the collected energy. The water may be assumed to be at a uniform temperature of $T_w = 60°$C. Under steady-state operating conditions for which the *net* radiation heat flux to the surface is $q''_{rad} = 800$ W/m^2, what is the maximum temperature on the plate and the heat transfer rate per unit length of tube? Note that q''_{rad} represents the net effect of solar radiation absorption by the absorber plate and radiation exchange between the absorber and cover plates. You may assume the temperature of the absorber plate directly above a tube to be equal to that of the water.

3.48 A thin flat plate of length L, thickness t, and width $W \gg L$ is thermally joined to two large heat sinks that are maintained at a temperature T_o. The bottom of the plate is well insulated, while the net heat flux to the top surface of the plate is known to have a uniform value of q''_o. (Diagram at top of p. 131.)

a) Derive the differential equation which determines the steady-state temperature distribution, $T(x)$, in the plate.

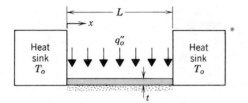

b) Solve the foregoing equation for the temperature distribution, and obtain an expression for the rate of heat transfer from the plate to the heat sinks.

3.49 Consider the flat plate of the foregoing problem, but with the bottom surface no longer insulated. Convection heat transfer is now allowed to occur between this surface and a fluid at T_∞, with a convection coefficient h.

    a) Derive the differential equation that determines the steady-state temperature distribution, $T(x)$, in the plate.

    b) Solve the foregoing equation for the temperature distribution, and obtain an expression for the rate of heat transfer from the plate to the heat sinks.

3.50 A thin metallic wire of thermal conductivity k, diameter D, and length $2L$ is annealed by passing an electrical current through the wire to induce a uniform volumetric heat generation $\dot{q}$. The ambient air around the wire is at a temperature T_∞, while the ends of the wire at $x = \pm L$ are also maintained at T_∞. Heat transfer from the wire to the air is characterized by the convection coefficient h. Obtain an expression for the steady-state temperature distribution, $T(x)$, along the wire.

3.51 A long, circular aluminum rod is attached at one end to a heated wall and transfers heat by convection to a cold fluid.

    a) If the diameter of the rod is tripled, by how much would the rate of heat removal change?

    b) If a copper rod of the same diameter is used in place of the aluminum, by how much would the rate of heat removal change?

3.52 A brass rod 100 mm long and 5 mm in diameter extends horizontally from a casting at 200°C. The rod is in an air environment with $T_\infty = 20°C$ and $h = 30$ W/m²·K. What is the temperature of the rod at distances of 25 mm, 50 mm, and 100 mm from the casting?

3.53 Two long copper rods of diameter $D = 1$ cm are soldered together end-to-end, with the solder having a melting point of 650°C. The rods are in air at 25°C with a convection coefficient of 10 W/m²·K. What is the minimum power input needed to effect the soldering?

3.54 Circular copper rods of diameter $D = 1$ mm and length $L = 25$ mm are used to enhance heat transfer from a surface which is maintained at $T_{s,1} = 100°C$. One end of the rod is attached to this surface (at $x = 0$), while the other end ($x = 25$ mm) is joined to a second surface, which is maintained at $T_{s,2} = 0°C$. Air flowing between the surfaces (and over the rods) is also at a temperature of $T_\infty = 0°C$, and an average convection coefficient of $\bar{h} = 100$ W/m²·K is maintained.

    a) What is the rate of heat transfer by convection from a single copper rod to the air?

b) What is the total rate of heat transfer from a 1 m by 1 m section of the surface at 100°C, if a bundle of the rods is installed on 4 mm centers?

3.55 Consider two long, slender rods of the same diameter but different materials. One end of each rod is attached to a base surface maintained at 100°C while the surface of the rods are exposed to ambient air at 20°C. By traversing the length of each rod with a thermocouple, it was observed that the temperatures of the rods were equal at the positions $x_A = 0.15$ m and $x_B = 0.075$ m where x is measured from the base surface. If the thermal conductivity of rod A is known to be $k_A = 70$ W/m·K, determine the value of k_B for rod B.

3.56 An experimental arrangement for measuring the thermal conductivity of solid materials involves the use of two long rods that are equivalent in every respect, except that one is fabricated from a standard material of known thermal conductivity, k_A, while the other is fabricated from the material whose thermal conductivity, k_B, is desired. Both rods are attached at one end to a heat source of fixed temperature T_b, are exposed to a fluid of temperature T_∞, and are instrumented with thermocouples to measure the temperature at a fixed distance x_1 from the heat source. If the standard material is aluminum, with $k_A = 200$ W/m·K, and measurements reveal values of $T_A = 75°C$ and $T_B = 60°C$ at x_1 for $T_b = 100°C$ and $T_\infty = 25°C$, what is the thermal conductivity, k_B, of the test material?

3.57 A disc-shaped transistor, which is mounted in an insulating medium, dissipates 0.25 W during steady-state operation. To reduce the temperature of the transistor, it is proposed that a hollow copper tube be attached to the transistor as shown below.

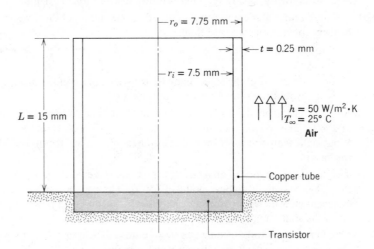

The outer surface of the tube is exposed to ambient air at $T_\infty = 25°C$ and with a convection coefficient of $h = 50$ W/m²·K. As a first approximation, heat transfer from the interior surface of the tube and from the exposed surface of the transistor may be neglected. What is the temperature of the transistor with the fin? What is the temperature of the transistor without the fin if h and T_∞ remain the same?

3.58 Determine the percentage increase in heat transfer associated with attaching aluminum fins of rectangular profile to a plane wall. The fins are 50 mm long, 0.5 mm thick, and are equally spaced at a distance of 4 mm (250 fins per meter). The convection coefficient associated with the bare wall is 40 $W/m^2 \cdot K$, while that resulting from attachment of the fins is 30 $W/m^2 \cdot K$.

3.59 Consider the use of straight, stainless steel (304) fins of rectangular and triangular profiles on a plane wall whose temperature is 100°C. The adjoining fluid is at 20°C, and the associated convection coefficient is 75 $W/m^2 \cdot K$. Each fin is 6 mm thick and has a length of 20 mm. Compare the efficiency, the effectiveness, and the heat loss per unit width associated with the two kinds of fins.

3.60 Aluminum fins of triangular profile are attached to a plane wall whose surface temperature is 250°C. The fin base thickness is 2 mm, and its length is 6 mm. The system is in ambient air at a temperature of 20°C, and the surface convection coefficient is 40 $W/m^2 \cdot K$.

a) What are the fin efficiency and effectiveness?

b) What is the heat dissipated per unit width by a single fin?

3.61 An annular aluminum fin of rectangular profile is attached to a circular tube having an outside diameter of 25 mm and a surface temperature of 250°C. The fin is 1 mm thick and 10 mm long, and the temperature and convection coefficient associated with the adjoining fluid are 25°C and 25 $W/m^2 \cdot K$, respectively.

a) What is the heat loss per fin?

b) If 200 such fins are spaced at 5 mm increments along the tube length, what is the heat loss per meter of tube length?

3.62 Annular aluminum fins of rectangular profile are attached to a circular tube having an outside diameter of 50 mm and an outer surface temperature of 200°C. The fins are 4 mm thick and have a length of 15 mm. The system is in ambient air at a temperature of 20°C, and the surface convection coefficient is 40 $W/m^2 \cdot K$.

a) What are the fin efficiency and effectiveness?

b) If there are 125 such fins per meter of tube length, what is the rate of heat transfer per unit length of tube?

3.63 Annular aluminum fins that are 2 mm thick and 15 mm long are installed on an aluminum tube of 30 mm diameter. The contact thermal resistance between a fin and the tube is known to be 2×10^{-4} $m^2 \cdot K/W$. If the tube wall is at 100°C and the adjoining fluid is at 25°C, with a convection coefficient of 75 $W/m^2 \cdot K$, what is the rate of heat transfer from a single fin? What would be the rate of heat transfer if the contact resistance could be eliminated?

4 Two-Dimensional, Steady-State Conduction

To this point we have restricted our attention to conduction problems in which the potential gradient is significant for only one coordinate direction. However, in many cases such problems are grossly oversimplified if a one-dimensional treatment is used, and it is necessary to account for multidimensional effects. In this chapter, we consider several techniques for treating two-dimensional systems under steady-state conditions.

4.1 ALTERNATIVE APPROACHES

Consider a system in which two-dimensional conduction effects are important (Figure 4.1). With two surfaces insulated and the other surfaces maintained at different temperatures, $T_1 > T_2$, heat transfer by conduction will occur from surface 1 to 2. According to Fourier's law, Equation 2.3 or 2.4, the local heat flux in the solid is a vector that is everywhere perpendicular to lines of constant temperature (*isotherms*). The directions of the heat flux vector are represented by the *heat flow lines* of Figure 4.1, and the vector itself is the resultant of heat flux components in the x and y directions. These components are determined by Equation 2.6.

Recall that in any conduction analysis, there exist two major objectives. The first objective is to determine the temperature distribution in the medium, which for the present problem necessitates determining $T(x, y)$. This objective is generally achieved by solving the appropriate form of the heat equation. For two-dimensional, steady-state conditions with no generation and constant thermal conductivity, this form is, from Equation 2.16,

$$\frac{\partial^2 T}{\partial x^2} + \frac{\partial^2 T}{\partial y^2} = 0 \qquad (4.1)$$

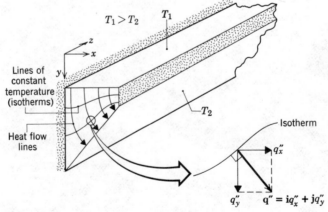

Figure 4.1 Two-dimensional conduction.

If Equation 4.1 can be solved for $T(x, y)$, it is then a simple matter to satisfy the second major objective, which is to determine the heat flux components q_x'' and q_y'' by applying the rate equations (2.6). Methods for solving Equation 4.1 include the use of *analytical, graphical,* and *numerical (finite-difference)* approaches, as well as an electrical analog.

The analytical method involves effecting an exact mathematical solution to Equation 4.1. The problem is more difficult than those considered in Chapter 3, since it now involves a partial, rather than an ordinary, differential equation. However, a solution may often be effected by the method of *separation of variables,* and the major advantage is the fact that knowledge of the temperature may be obtained for *any* point of interest in the medium. In contrast the other methods can only provide approximate results for *discrete* points in the medium. Unfortunately, however, exact solutions may only be obtained for simple geometries and special boundary conditions, and even for such cases the results appear in terms of complicated mathematical series and functions. An example of the use of the separation of variables method to obtain an exact solution for a particular two-dimensional problem is provided in Appendix C. The material is presented only to convey a feeling for the nature of the method. The mathematics are generally beyond the scope of this text and are left to more advanced treatments of the subject [1–5].

Approximate methods of solving Equation 4.1, which are without the mathematical difficulties of the analytical method, are treated in the following sections. The graphical, or flux-plotting, method (Section 4.2) may be used to obtain a rough estimate of the temperature field and hence the heat transfer rate. The accuracy of this method may be improved through the use of an experimental electrical analog (Section 4.3). However, the preferred and most widely used approach is one that employs finite-difference procedures (Sections 4.4 and 4.5). Such procedures may be used with a computer to provide accurate results for any multidimensional problem.

4.2 THE GRAPHICAL METHOD

The graphical method may be employed for two-dimensional problems involving adiabatic and/or isothermal boundaries. Because the approach demands considerable patience and an artistic flair (not to mention the use of durable paper and a good eraser), it has declined in popularity over the past three decades, having been superseded largely by computer solutions based on finite-difference procedures. Despite its limited accuracy, however, the method still finds application as a means for obtaining a first estimate of the temperature distribution. It is also useful for developing one's intuition for the nature of the temperature field and heat flow in a system.

4.2.1 Methodology of Constructing a Flux Plot

The rationale for the graphical method comes from the fact that lines of constant temperature must be perpendicular to lines that indicate the direction of heat flow (Figure 4.1). The objective of the graphical method is one of systematically constructing such a network of isotherms and heat flow lines. This network, commonly termed a *flux plot*, is used to infer the temperature distribution and the rate of heat flow through the system.

Consider a square, two-dimensional channel whose inner and outer surfaces are maintained at T_1 and T_2, respectively. A cross section of the channel is shown in Figure 4.2a. A procedure for constructing the flux plot, a portion of which is shown in Figure 4.2b, is enumerated as follows.

1. The first step in any flux plot should be one of *identifying all relevant lines of symmetry*. Such lines are determined by thermal, as well as geometrical, conditions. For the square channel of Figure 4.2a, such lines include the designated vertical, horizontal, and diagonal lines. For this system it is therefore possible to consider only one-eighth of the configuration, as shown in Figure 4.2b.

2. *Lines of symmetry are adiabatic* in the sense that there can be no heat transfer in a direction perpendicular to the lines. They are therefore heat flow lines and should be treated as such. Since there is no heat flow in a

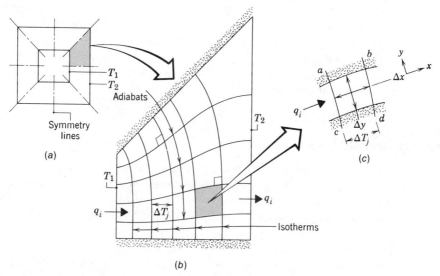

Figure 4.2 Two-dimensional conduction in a square channel of length, *l*. (a) Symmetry planes. (b) Flux plot. (c) Typical curvilinear square.

direction perpendicular to a heat flow line, such a line can be termed an *adiabat*.

3. After all known lines of constant temperature associated with the system boundaries are identified, an attempt should be made to sketch *uniformly spaced* lines of constant temperature within the system. Note that *isotherms should always be perpendicular to adiabats*.

4. The heat flow lines should then be drawn with an eye toward creating a network of *curvilinear squares*. This is done by having the *heat flow lines and isotherms intersect at right angles* and by requiring that *all sides of each square be of approximately the same length*. It is often impossible to satisfy this second requirement exactly, and it is more realistic to strive for equivalence between the sums of the opposite sides of each square, as shown in Figure 4.2c. Assigning the x coordinate to the direction of heat flow and the y coordinate to the direction normal to this flow, the requirement may be expressed as

$$\Delta x \equiv \frac{ab + cd}{2} \approx \Delta y \equiv \frac{ac + bd}{2} \tag{4.2}$$

It is difficult to create a satisfactory network of curvilinear squares in the first attempt, and numerous iterations must often be made. This trial-and-error process involves adjusting the isotherms and adiabats until satisfactory curvilinear squares are obtained for most of the network[1]. Once the flux plot is obtained, it may be used to infer the temperature distribution in the medium. From a simple analysis, the heat transfer rate may then be obtained.

4.2.2 Determination of the Heat Transfer Rate

The rate at which energy is conducted through a *lane*, which is the region between adjoining adiabats, is designated as q_i. If the flux plot is properly constructed, the value of q_i will be the same for all lanes and the total heat transfer rate may be expressed as

$$q = \sum_{i=1}^{M} q_i = M q_i \tag{4.3}$$

where M is the *number of lanes* associated with the plot. From the curvilinear square of Figure 4.2c and the application of Fourier's law, q_i may be expressed as

---

[1] In certain regions, such as corners, it may be impossible to approach the curvilinear square requirements. However, such difficulties will generally have little effect on the overall accuracy of the results obtained from the flux plot.

Table 4.1 Conduction shape factors for selected two-dimensional systems $[q = Sk(T_1 - T_2)]$

SYSTEM	SCHEMATIC	RESTRICTIONS	SHAPE FACTOR
Isothermal sphere buried in a semiinfinite medium		$z > D/2$	$\dfrac{2\pi D}{1 - D/4z}$
Horizontal isothermal cylinder of length L buried in a semiinfinite medium		$L \gg D$ $L \gg D$ $z > 3D/2$	$\dfrac{2\pi L}{\cosh^{-1}(2z/D)}$ $\dfrac{2\pi L}{\ln(4z/D)}$
Vertical cylinder in a semiinfinite medium		$L \gg D$	$\dfrac{2\pi L}{\ln(4L/D)}$
Conduction between two cylinders of length L in infinite medium		$L \gg D_1, D_2$ $L \gg w$	$\dfrac{2\pi L}{\cosh^{-1}\left(\dfrac{4w^2 - D_1^2 - D_2^2}{2D_1 D_2}\right)}$
Horizontal circular cylinder of length L midway between parallel planes of equal length and infinite width		$z > D/2$	$\dfrac{2\pi L}{\ln(8z/D)}$

System	Schematic	Restrictions	Shape Factor
Circular cylinder of length L in a square solid of equal length	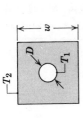	$w > D$	$\dfrac{2\pi L}{\ln(1.08\,w/D)}$
Plane wall	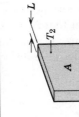	One-dimensional conduction	$\dfrac{A}{L}$
Conduction through the edge of adjoining walls	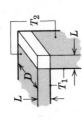	$D > L/5$	$0.54D$
Conduction through corner of three walls with a temperature difference of ΔT_{1-2} across the walls		$L \ll$ length and width of wall	$0.15L$
Disc of diameter D and T_1 on a semiinfinite medium of thermal conductivity k and T_2	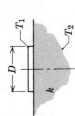	None	$2D$

$$q_i \approx kA_i \frac{\Delta T_j}{\Delta x} \approx k(\Delta y \cdot l) \frac{\Delta T_j}{\Delta x} \tag{4.4}$$

where ΔT_j is the temperature difference between successive isotherms, A_i is the conduction heat transfer area for the lane, and l is the length of the channel normal to the page. However, if the flux plot is properly constructed, the temperature increment is the same for all adjoining isotherms, and the overall temperature difference between boundaries, ΔT_{1-2}, may be expressed as

$$\Delta T_{1-2} = \sum_{j=1}^{N} \Delta T_j = N \Delta T_j \tag{4.5}$$

where N is the total number of temperature increments. Combining Equations 4.3, 44, and 4.5 and recognizing that $\Delta x \approx \Delta y$ for curvilinear squares, we obtain

$$q \approx \frac{Ml}{N} k \Delta T_{1-2} \tag{4.6}$$

The manner in which a flux plot may be used to obtain the heat transfer rate for a two-dimensional system is evident from Equation 4.6. The ratio of the number of heat flow lanes to the number of temperature increments (the value of M/N) may be obtained from the plot. Recall that specification of N is based on step 3 of the foregoing procedure, and the value, which is an integer, may be made large or small depending on the desired accuracy. The value of M is then a consequence of following step 4. Note that M is not necessarily an integer, since a fractional lane may be needed to arrive at a satisfactory network of curvilinear squares. For the network of Figure 4.2b it is evident that $N = 6$ and $M = 5$. Of course, as the network, or *mesh*, of curvilinear squares is made finer, N and M increase and the value of M/N should become more accurate.

4.2.3   The Conduction Shape Factor

Equation 4.6 may be used to define the *shape factor*, S, of a two-dimensional system. That is, the heat transfer rate may be expressed as

$$q = S \, k \, \Delta T_{1-2} \tag{4.7}$$

where

$$S \equiv \frac{Ml}{N} \tag{4.8}$$

Shape factors have been obtained for numerous two-dimensional systems, and results are summarized in Table 4.1 for some common configurations. In each case, two-dimensional conduction is presumed to occur between boundaries that are maintained at uniform temperatures, with $\Delta T_{1-2} \equiv T_1 - T_2$. Results are available for many configurations not included in the table [1, 6–9].

EXAMPLE 4.1

A hole of diameter $D = 0.25$ m is drilled through the center of a solid block of square cross section with $w = 1$ m on a side. The hole is drilled along the length, $L = 2$ m, of the block, which has a thermal conductivity of $k = 150$ W/m·K. A warm fluid passing through the hole maintains an inner surface temperature of $T_1 = 75°C$, while the outer surface of the block is kept at $T_2 = 25°C$.

1. Using the flux plot method, determine the shape factor for the system.
2. What is the rate of heat transfer through the block?

SOLUTION:

KNOWN:

Dimensions and thermal conductivity of a block with a circular hole drilled along its length.

FIND:

1. Shape factor.
2. Heat transfer rate for prescribed surface temperatures.

SCHEMATIC:

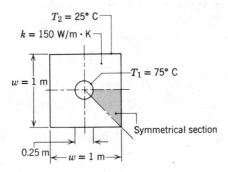

ASSUMPTIONS:

1. Steady-state conditions.
2. Two-dimensional conduction.
3. Constant properties.

ANALYSIS:

1. The flux plot may be simplified by identifying lines of symmetry and reducing the system to the one-eighth section shown in the schematic. Using a fairly coarse grid involving $N = 6$ temperature increments, the flux plot was generated and the resulting network of curvilinear squares is shown below.

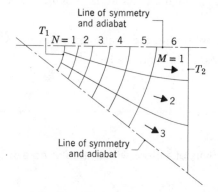

With the number of heat flow lanes for the section corresponding to $M = 3$, it follows from Equation 4.8 that the shape factor for the entire block is

$$S = 8\frac{MI}{N} = 8\frac{3 \times 2 \text{ m}}{6} = 8 \text{ m} \qquad \triangleleft$$

where the factor of 8 results from the number of symmetrical sections. The accuracy of this result may be determined by referring to Table 4.1, where, for the prescribed system, it follows that

$$S = \frac{2\pi L}{\ln(1.08\,w/D)} = \frac{2\pi \times 2 \text{ m}}{\ln(1.08 \times 1 \text{ m}/0.25 \text{ m})} = 8.59 \text{ m}$$

Hence the result of the flux plot underpredicts the shape factor by approximately 7 percent.

2. Using $S = 8.59$ with Equation 4.7, the heat rate is

$$q = Sk(T_1 - T_2)$$

$$q = 8.59 \text{ m} \times 150 \text{ W/m} \cdot \text{K} (75\text{--}25)^\circ\text{C}$$

$$q = 64.4 \text{ kW} \qquad \qquad \triangleleft$$

COMMENTS:

The accuracy of the flux plot may be improved by using a finer grid (increasing the value of N).

4.3 THE ELECTRICAL ANALOG

As discussed in Chapter 3, steady-state electrical conduction in a homogeneous medium of uniform electrical conductivity is analogous to steady-state heat conduction in a homogeneous medium of uniform thermal conductivity. Just as this analogy was applied to one-dimensional systems in Chapter 3, it may also be applied to multidimensional transport.

For steady, two-dimensional electrical conduction, the Laplace equation governs the distribution of the electric potential, E, and is given by

$$\frac{\partial^2 E}{\partial x^2} + \frac{\partial^2 E}{\partial y^2} = 0 \tag{4.9}$$

Since this expression is of precisely the same form as the two-dimensional, steady-state heat equation, the solutions to Equations 4.1 and 4.9 for analogous boundary conditions must also be of the same form. That is, the two-dimensional temperature distribution, $T(x, y)$, is of the same form as the electric potential field, $E(x, y)$, and lines of constant potential are identical to lines of constant temperature.

The foregoing analogy provides the basis for an experimental procedure that may be used to increase the accuracy associated with the flux-plotting method of the preceding section. The procedure first involves constructing a geometric model of the two-dimensional system using one of several readily available conductive papers. A schematic of the measurement system is shown in Figure 4.3, where the model is a one-eighth section of the channel of Figure 4.2. Edges intended to simulate isothermal boundaries are made highly conductive through application of either copper strips or a special silver paint. These edges then serve as electrodes, across which a potential difference, ΔE_{1-2}, may be applied. Lines of symmetry, or adiabatic boundaries, are left as untreated edges of the conductive paper. A d-c voltmeter may then be used to trace lines of constant potential across the paper. Alternatively, as shown in Figure 4.3, a null detector may be used for the same purpose. After selecting a particular potential difference, $\Delta E = E - E_2$, a probe, or stylus, may be moved across the paper to trace out the corresponding line of constant potential.

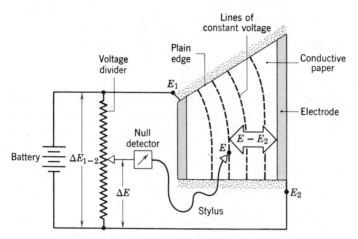

Figure 4.3   Electrical analog for two-dimensional heat conduction.

By analogy, lines of constant temperature may be inferred from the lines of constant electrical potential obtained using the above procedure. Quantitatively, this analog takes the form

$$\frac{\Delta E}{\Delta E_{1-2}} = \frac{E - E_2}{E_1 - E_2} = \frac{T - T_2}{T_1 - T_2} \tag{4.10}$$

In this way, lines of constant temperature may be determined more accurately (and frequently more quickly) than by means of the iterative flux plotting method of the preceding section. Heat flow lines may then be drawn perpendicular to the isotherms, and with efforts made to construct curvilinear squares, the resulting mesh may be used to infer the shape factor.

The electrical analog often provides a simple, yet reliable, means of solving steady, two-dimensional conduction problems, and its value should not be underestimated (even in this age of the high-speed digital computer). A more extensive description of the method is provided by Schneider [2].

4.4   FINITE-DIFFERENCE EQUATIONS

As discussed in Section 4.1, analytical methods may be used, in certain cases, to effect exact mathematical solutions to steady, two-dimensional conduction problems. These solutions have been generated for an assortment of simple geometries and boundary conditions and are well documented in the literature [1–5]. However, more often than not, a two-dimensional problem will involve a geometry and/or boundary conditions that preclude such a solution. In these cases, the best alternative is often one which involves the use of *numerical*, or *finite-difference*, techniques. These techniques are particularly well suited for use with high-speed digital computers.

4.4.1 The Nodal Network

In contrast to an analytical solution, which allows for temperature determination at *any* point of interest in a medium, a numerical solution enables determination of the temperature at only *discrete* points. The first step in any numerical analysis must therefore be one of selecting these points. Referring to Figure 4.4, this is done by subdividing the medium of interest into a number of small regions and assigning to each a reference point that is at its center. The reference point is frequently termed a *nodal point* (or simply a *node*), and the aggregate of points is termed a *nodal network* or *mesh*. The nodal points are designated by a numbering scheme which, for a two-dimensional system, may take the form shown in Figure 4.4a. The x and y locations are designated by the *m* and *n* indices, respectively.

$$\frac{\partial T}{\partial x}\bigg|_{m-1/2,n} = \frac{T_{m,n} - T_{m-1,n}}{\Delta x}$$

$$\frac{\partial T}{\partial x}\bigg|_{m+1/2,n} = \frac{T_{m+1,n} - T_{m,n}}{\Delta x}$$

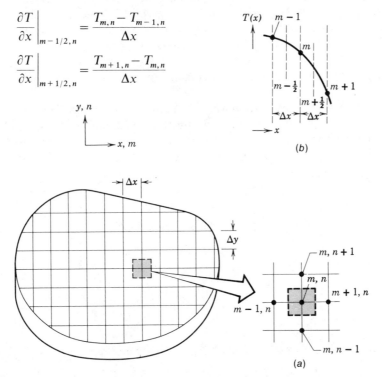

Figure 4.4 Two-dimensional conduction. (*a*) Nodal network. (*b*) Finite-difference approximation.

It is important to note that each node represents a certain region, and its temperature is a measure of the *average* temperature of the region. For example, the temperature of the node *m,n* of Figure 4.4*a* may be viewed as the average temperature of the surrounding shaded area. In addition the selection of nodal points is rarely arbitrary, depending often on matters such as geometric convenience and the desired accuracy. The numerical accuracy of calculations which are performed depends strongly on the number of designated nodal points. If this number is small (*a coarse mesh*), as the case will be if the calculations must be performed by hand, the accuracy will be limited. However, if a digital computer is used, a large number of points may be chosen (a *fine mesh*), and higher accuracy will be obtained.

4.4.2 Finite-Difference Form of the Heat Diffusion Equation

Determination of the temperature distribution numerically dictates that an appropriate conservation equation be written for *each* of the points in the nodal network. The resulting set of equations may then be simultaneously solved for the temperature at each node. For *any interior* node of a two-dimensional system with no generation and uniform thermal conductivity, the *exact* form of the energy conservation requirement is given by the heat equation, Equation 4.1. However, if the system is characterized in terms of a nodal network, it is necessary to work with an *approximate*, or *finite-difference*, form of this equation.

A finite-difference equation which is suitable for the interior nodes of a two-dimensional system may be inferred directly from Equation 4.1. Consider the second derivative, $\partial^2 T / \partial x^2$. From Figure 4.4*b*, the value of this derivative at the *m,n* nodal point may be approximated as

$$\left. \frac{\partial^2 T}{\partial x^2} \right|_{m,n} \approx \frac{\partial T/\partial x|_{m+1/2,n} - \partial T/\partial x|_{m-1/2,n}}{\Delta x} \qquad (4.11)$$

The temperature gradients may in turn be expressed as a function of the nodal temperatures. That is,

$$\left. \frac{\partial T}{\partial x} \right|_{m+1/2,n} \approx \frac{T_{m+1,n} - T_{m,n}}{\Delta x} \qquad (4.12)$$

$$\left. \frac{\partial T}{\partial x} \right|_{m-1/2,n} \approx \frac{T_{m,n} - T_{m-1,n}}{\Delta x} \qquad (4.13)$$

Substituting Equations 4.12 and 4.13 into 4.11, we obtain

$$\left. \frac{\partial^2 T}{\partial x^2} \right|_{m,n} \approx \frac{T_{m+1,n} + T_{m-1,n} - 2T_{m,n}}{(\Delta x)^2} \qquad (4.14)$$

Proceeding in a similar fashion, it is readily shown that

$$\left.\frac{\partial^2 T}{\partial y^2}\right|_{m,n} \approx \frac{\partial T/\partial y|_{m,n+1/2} - \partial T/\partial y|_{m,n-1/2}}{\Delta y}$$

$$\approx \frac{T_{m,n+1} + T_{m,n-1} - 2T_{m,n}}{(\Delta y)^2} \tag{4.15}$$

Using a network for which $\Delta x = \Delta y$ and substituting Equations 4.14 and 4.15 into 4.1, we then obtain

$$T_{m,n+1} + T_{m,n-1} + T_{m+1,n} + T_{m-1,n} - 4T_{m,n} = 0 \tag{4.16}$$

Hence for the m,n node the heat equation, which is an *exact differential equation,* is reduced to an *approximate algebraic equation.* This approximate, finite-difference form of the heat equation may be applied to any interior node which is equidistant from its four neighboring nodes. It simply requires that the sum of the temperatures associated with the neighboring nodes be equal to four times the temperature of the node of interest.

4.4.3 The Energy Balance Method

The fact that Equation 4.16 is a statement of the energy conservation requirement may be verified by performing an energy balance on an interior node. Consider the m,n node of Figure 4.5. For steady, two-dimensional conditions with no generation, energy exchange is influenced only by conduction between m,n and its four neighboring nodes. For a control volume about the m,n nodal region, the energy conservation requirement may be expressed as

$$\sum_{i=1}^{4} q_{(i)\to(m,n)} = 0 \tag{4.17}$$

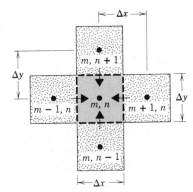

Figure 4.5 Conduction to an interior node from its adjoining nodes.

where the index i refers to the neighboring nodes and $q_{(i) \to (m, n)}$ is the conduction rate between nodes. This expression states that the total rate at which energy is conducted *into* the node, as shown in Figure 4.5, must be zero. The result is equivalent to requiring that the energy inflow and outflow rates be balanced.

To evaluate the conduction rate terms, we will *assume* that conduction transfer occurs exclusively through *lanes* that are oriented in either the x or y direction. Simplified forms of Fourier's law may therefore be used. For example, the rate at which energy is transferred by conduction from node $m - 1, n$ to m, n may be expressed as

$$q_{(m - 1, n) \to (m, n)} = k(\Delta y \cdot 1) \frac{(T_{m - 1, n} - T_{m, n})}{\Delta x} \tag{4.18}$$

The quantity $(\Delta y \cdot 1)$ is the heat transfer area, where unit depth is assumed, and the term $(T_{m - 1, n} - T_{m, n})/\Delta x$ is the approximate temperature gradient at the boundary between the two nodes. The remaining conduction rates may then be expressed as

$$q_{(m + 1, n) \to (m, n)} = k(\Delta y \cdot 1) \frac{(T_{m + 1, n} - T_{m, n})}{\Delta x} \tag{4.19}$$

$$q_{(m, n + 1) \to (m, n)} = k(\Delta x \cdot 1) \frac{(T_{m, n + 1} - T_{m, n})}{\Delta y} \tag{4.20}$$

$$q_{(m, n - 1) \to (m, n)} = k(\Delta x \cdot 1) \frac{(T_{m, n - 1} - T_{m, n})}{\Delta y} \tag{4.21}$$

Summing Equations 4.18 through 4.21, in compliance with the conservation requirement, Equation 4.17, and remembering that $\Delta x = \Delta y$, it is easily shown that Equation 4.16 will result.

Note that the foregoing energy balance may readily be extended to include the effects of an internally distributed energy source. The rate at which energy is generated within the m, n nodal region is $\dot{q}(\Delta x \cdot \Delta y \cdot 1)$, and when combined with the net rate at which energy is conducted to this node, the finite-difference equation becomes

$$T_{m, n + 1} + T_{m, n - 1} + T_{m + 1, n} + T_{m - 1, n} + \frac{\dot{q}(\Delta x \cdot \Delta y)}{k} - 4T_{m, n} = 0 \tag{4.22}$$

With little difficulty the finite-difference equation for an interior node may also be extended to account for three-dimensional conduction effects.

It is important to note that a finite-difference equation is needed for each nodal point at which the temperature is unknown. However, it is not always possible to classify all such points as interior, and hence to use Equation 4.16 or 4.22. For example, the temperature may be unknown at an insulated surface or at a surface which is exposed to convective conditions. For points on such surfaces, the finite-difference equation must be obtained by applying the energy balance method.

To illustrate this method, consider the node corresponding to the internal corner of Figure 4.6. This node may be assumed to represent the three-quarter shaded section and to exchange energy by convection with an adjoining fluid at T_∞. Conduction to the nodal region (m,n) will occur along four different lanes from neighboring nodes in the solid. The conduction heat rates, q_{cond}, may be expressed as

$$q_{(m-1,n)\to(m,n)} = k(\Delta y \cdot 1)\frac{(T_{m-1,n} - T_{m,n})}{\Delta x} \tag{4.23}$$

$$q_{(m,n+1)\to(m,n)} = k(\Delta x \cdot 1)\frac{(T_{m,n+1} - T_{m,n})}{\Delta y} \tag{4.24}$$

$$q_{(m+1,n)\to(m,n)} = k\left(\frac{\Delta y}{2} \cdot 1\right)\frac{(T_{m+1,n} - T_{m,n})}{\Delta x} \tag{4.25}$$

$$q_{(m,n-1)\to(m,n)} = k\left(\frac{\Delta x}{2} \cdot 1\right)\frac{(T_{m,n-1} - T_{m,n})}{\Delta y} \tag{4.26}$$

Note, that the areas for conduction from nodal regions $m-1,n$ and $m,n+1$ are proportional to Δy and Δx, respectively, whereas conduction from $m+1,n$ and $m,n-1$ occurs along lanes of half-width, $\Delta y/2$ and $\Delta x/2$, respectively.

Conditions in the nodal region m,n are also influenced by convective exchange with the fluid, and this exchange may be viewed as occurring along

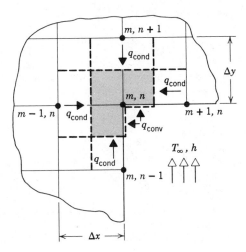

Figure 4.6 Formulation of the finite-difference equation for an internal corner of a solid with surface convection.

Table 4.2 Summary of nodal finite-difference equations

CONFIGURATION	FINITE-DIFFERENCE EQUATION FOR $\Delta x = \Delta y$

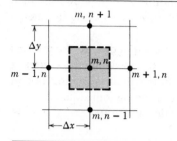

$$T_{m,n+1} + T_{m,n-1} + T_{m+1,n} + T_{m-1,n}$$
$$-4T_{m,n} = 0 \qquad (4.16)$$

1. Interior node.

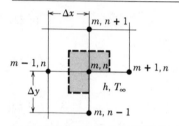

$$2(T_{m-1,n} + T_{m,n+1}) + (T_{m+1,n} + T_{m,n-1})$$
$$+2\frac{h\Delta x}{k}T_\infty - 2\left(3 + \frac{h\Delta x}{k}\right)T_{m,n} = 0 \qquad (4.28)$$

2. Node at an internal corner with convection.

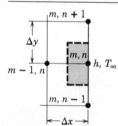

$$(2T_{m-1,n} + T_{m,n+1} + T_{m,n-1}) + \frac{2h\Delta x}{k}T_\infty$$
$$-2\left(\frac{h\Delta x}{k} + 2\right)T_{m,n} = 0 \qquad (4.29)^{\text{a}}$$

3. Node at a plane surface with convection.

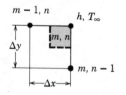

$$(T_{m,n-1} + T_{m-1,n}) + 2\frac{h\Delta x}{k}T_\infty$$
$$-2\left(\frac{h\Delta x}{k} + 1\right)T_{m,n} = 0 \qquad (4.30)$$

4. Node at an external corner with convection.

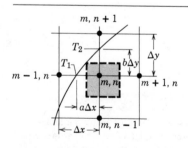

$$\frac{2}{a+1}T_{m+1,n} + \frac{2}{b+1}T_{m,n-1}$$
$$+\frac{2}{a(a+1)}T_1 + \frac{2}{b(b+1)}T_2$$
$$-\left(\frac{2}{a} + \frac{2}{b}\right)T_{m,n} = 0 \qquad (4.31)$$

5. Node near a curved surface maintained at a nonuniform temperature.

---

[a]To obtain the finite-difference equation for an adiabatic surface (or surface of symmetry), simply set h equal to zero.

half-lanes in the x and y directions. The total convection rate, q_{conv}, may be expressed as

$$q_{(\infty) \to (m,n)} = h\left(\frac{\Delta x}{2} \cdot 1\right)(T_\infty - T_{m,n}) + h\left(\frac{\Delta y}{2} \cdot 1\right)(T_\infty - T_{m,n})$$

(4.27)

Implicit in this expression is the assumption that the exposed surfaces of the corner are at a uniform temperature corresponding to the nodal temperature, $T_{m,n}$. This assumption is consistent with the concept that the entire nodal region is characterized by a single temperature, which represents an average of the actual temperature distribution in the region. In the absence of transient, three-dimensional, and generation effects, conservation of energy requires that the sum of Equations 4.23 to 4.27 must be zero. Summing these equations and rearranging, we therefore obtain

$$T_{m-1,n} + T_{m,n+1} + \frac{1}{2}(T_{m+1,n} + T_{m,n-1}) + \frac{h\Delta x}{k}T_\infty - \left(3 + \frac{h\Delta x}{k}\right)T_{m,n} = 0$$

(4.28)

where again the mesh is such that $\Delta x = \Delta y$.

Nodal energy-balance equations pertinent to several common geometries are presented in Table 4.2. Note that the retention of equal spatial increments ($\Delta x = \Delta y$) in the presence of irregular boundaries (e.g., the curved surface of case 5 in Table 4.2) constitutes a complicating feature. Finite-difference formulations for other cases are discussed by Schneider [2] and Dusinberre [10].

EXAMPLE 4.2

Using the energy balance method, derive the finite-difference equation for the m,n nodal point located on a plane insulated surface of a medium with uniform heat generation.

SOLUTION

KNOWN:

Network of points adjoining a nodal point on an insulated surface.

FIND:

Finite-difference equation for the nodal point.

SCHEMATIC:

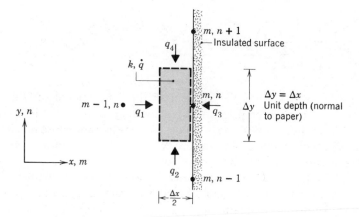

ASSUMPTIONS:

1. Steady-state conditions.
2. Two-dimensional conduction.
3. Constant properties.
4. Uniform internal heat generation.

ANALYSIS:

Applying the energy conservation requirement, Equation 4.17, to the control surface about the region ($\Delta x/2 \cdot \Delta y \cdot 1$) associated with the m,n node, it follows that, with volumetric heat generation at a rate $\dot{q}$,

$$q_1 + q_2 + q_3 + q_4 + \dot{q}\left(\frac{\Delta x}{2} \cdot \Delta y \cdot 1\right) = 0$$

where

$$q_1 = k(\Delta y \cdot 1)\frac{(T_{m-1,n} - T_{m,n})}{\Delta x}$$

$$q_2 = k\left(\frac{\Delta x}{2} \cdot 1\right)\frac{(T_{m,n-1} - T_{m,n})}{\Delta y}$$

$$q_3 = 0$$

$$q_4 = k\left(\frac{\Delta x}{2} \cdot 1\right)\frac{(T_{m,n+1} - T_{m,n})}{\Delta y}$$

Substituting into the energy balance and dividing by $k/2$, it follows that

$$2T_{m-1,n} + T_{m,n-1} + T_{m,n+1} - 4T_{m,n} + \frac{\dot{q}(\Delta x \cdot \Delta y)}{k} = 0 \qquad \lhd$$

COMMENTS:

The same result could be obtained by using the symmetry condition, $T_{m+1,n} = T_{m-1,n}$, with the finite-difference equation, Equation 4.22, for an interior nodal point. If $\dot{q} = 0$, the desired result could also be obtained by setting $h = 0$ in Equation 4.29.

4.5 FINITE-DIFFERENCE SOLUTIONS

Once the nodal network has been established and an appropriate finite-difference equation has been written for each of the nodes, the temperature distribution may be determined. Let us consider three methods which may be used for this purpose.

4.5.1 The Relaxation Method

A traditional technique of solving a system of finite-difference equations is known as the *relaxation method*. It is suitable for use with a coarse mesh and permits the solution to be effected by means of hand calculations.

The first step in the relaxation method is one of *guessing* a value for each of the unknown temperatures in the nodal network. However, if the assumed temperatures are then substituted into the finite-difference equation for a particular node, it is unlikely that this equation will be satisfied. For example, if assumed values of $T_{m,n+1}$, $T_{m,n-1}$, $T_{m+1,n}$, $T_{m-1,n}$ and $T_{m,n}$ were substituted into Equation 4.16 for an interior node, it is unlikely that the left-hand side would reduce to zero. Instead it would assume some value, $R_{m,n}$, termed the *residual* of the node, and the equation may be expressed as

$$T_{m,n+1} + T_{m,n-1} + T_{m+1,n} + T_{m-1,n} - 4T_{m,n} = R_{m,n} \qquad (4.32)$$

Hence the assumed temperatures will provide residuals for each node at which the temperature is unknown. The relaxation method is, then, a procedure that calls for systematically altering the assumed temperatures in a way which forces the residuals to some small value close to zero. Application of the method is facilitated by the following procedure.

1. After identifying lines of symmetry and establishing the nodal network for the system, the appropriate finite-difference equation is written for each node at which the temperature is unknown. The residual for the node should appear on the right-hand side of each equation.

2. Values are then assumed for each of the unknown nodal temperatures. To minimize subsequent computations, efforts are made to guess values that are close to the actual temperatures. A flux plot may often be advantageously used to establish a reasonable first estimate.

3. Using the *assumed* temperatures, the residual is then calculated at each node for which the actual temperature is desired.

4. The largest residual is identified, and the relaxation process is initiated. This is done by adjusting the nodal temperature corresponding to the largest residual by an amount which will *relax* this residual to zero. Since changing the temperature of one node will alter the residuals of neighboring nodes, the values of these residuals must be recomputed.

5. The node now having the largest residual is identified, and step 4 is repeated. The process continues until all residuals have been relaxed to values that are small enough to insure achievement of the desired accuracy.

Note that several techniques are available for increasing the speed of the relaxation process. They involve the use of overrelaxation, group relaxation, and block relaxation and are described by Razelos [11]. Note also that, for any numerical solution scheme, improved accuracy may be obtained by refining the nodal network (decreasing mesh size). However, any move in this direction must come at the expense of increased computation time.

EXAMPLE 4.3

A large industrial furnace is supported on a long column of fire clay brick, which is 1 m × 1 m on a side. During steady-state operation, installation is such that three surfaces of the column are maintained at 500 K while the remaining surface is exposed to an airstream for which $T_\infty = 300$ K and $h = 10$ W/m$^2 \cdot$K. Using the relaxation method with a grid of $\Delta x = \Delta y = 0.25$ m, determine the two-dimensional temperature distribution in the column and the heat rate to the airstream per unit length of column.

SOLUTION

KNOWN:

Dimensions and surface conditions of a support column.

FIND:

Temperature distribution and heat rate per unit length.

SCHEMATIC:

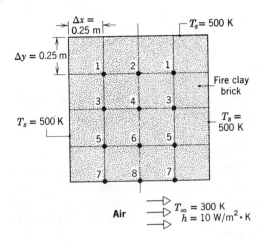

ASSUMPTIONS:

1. Steady-state conditions.
2. Two-dimensional conduction.
3. Constant properties.
4. No internal heat generation.

PROPERTIES:

Table A.3, fire clay brick ($T \approx 478$ K): $k = 1$ W/m·K.

ANALYSIS:

The prescribed grid consists of 12 nodal points at which the temperature is unknown. However, the number of unknowns is reduced to 8 through symmetry, in which case the temperature of nodal points to the left of the symmetry line must equal the temperature of those to the right.

Nodes 1, 3, and 5 are interior points for which the residual equations may be inferred from Equation 4.16. Hence

Node 1: $T_2 + T_3 + 1000 - 4T_1 = R_1$

Node 3: $T_1 + T_4 + T_5 + 500 - 4T_3 = R_3$

Node 5: $T_3 + T_6 + T_7 + 500 - 4T_5 = R_5$

Equations for points 2, 4, and 6 may be obtained in a like manner or, since they lie on a symmetry adiabat, by using Equation 4.29 with $h = 0$. Hence

Node 2: $2T_1 + T_4 + 500 - 4T_2 = R_2$

Node 4: $T_2 + 2T_3 + T_6 - 4T_4 = R_4$

Node 6: $T_4 + 2T_5 + T_8 - 4T_6 = R_6$

From Equation 4.29 and the fact that $(h\Delta x/k) = 2.5$, it also follows that

Node 7: $2T_5 + T_8 + 2000 - 9T_7 = R_7$

Node 8: $2T_6 + 2T_7 + 1500 - 9T_8 = R_8$

Having the required finite-difference equations, the relaxation calculations may be initiated. The calculations are expedited by using the relaxation table on the following page, which allocates two columns for each node, one for the temperature and the other for the residual. The calculations proceed as follows.

1. For each node, assumed temperatures and the corresponding residuals obtained from the foregoing equations are tabulated in the first row.
2. Since the residual for node 7 is the largest, the next step involves *relaxing* this residual. Reducing the temperature from 370 to 350 K, it follows that $R_7 = 0$. This change also alters the residuals of the adjoining nodes, R_5 and R_8. All changes are recorded in the second row.
3. The largest residual is then R_8, and the process is repeated.
4. The calculations are continued until all residuals are reduced to values that are low enough to insure the desired accuracy. The final results appearing in the relaxation table are accurate to within approximately ± 5 K and are summarized as follows.

$T_1 \approx 487\,\text{K}, \qquad T_2 \approx 482\,\text{K}, \qquad T_3 \approx 470\,\text{K}, \qquad T_4 \approx 457\,\text{K}$

$T_5 \approx 435\,\text{K}, \qquad T_6 \approx 414\,\text{K}, \qquad T_7 \approx 356\,\text{K}, \qquad T_8 \approx 337\,\text{K}$ $\triangleleft$

The heat rate from the column to the airstream may be computed from the expression

$$(q/L) \approx 2h[(\Delta x/2)(T_s - T_\infty) + \Delta x(T_7 - T_\infty) + (\Delta x/2)(T_8 - T_\infty)]$$

where the factor of two outside the brackets originates from the symmetry condition. Hence

$$(q/L) \approx 2 \times 10\,\text{W/m}^2 \cdot \text{K}\,[(0.125\ \text{m}\,(200\ \text{K})$$

$+0.25$ m $(56$ K$) + 0.125$ m $(37$ K$)]$

or

$(q/L) \approx 873$ W/m   ◁

COMMENTS:

Recognize that even if all the residuals are relaxed to zero, the resulting temperatures do not constitute an exact solution to the problem. Recall that finite-difference solutions are inherently approximate, and the exact solution may only be approached by going to a finer grid size.

Relaxation Table: Example 4.3

T_1	R_1	T_2	R_2	T_3	R_3	T_4	R_4	T_5	R_5	T_6	R_6	T_7	R_7	T_8	R_8
480	−10	470	10	440	50	430	20	400	100	390	20	370	−180	350	−130
									80				350	0		−170
											1		−19	331	1
					70			420	0		41		21		
	8			458	−2		56		18						
			24		12	444	0				55				
							14		32	404	−1				29
					20			428	0		7		37		
									4			354	1		37
											11		5	335	1
	14	476	0		26										
	20			464	2		26		10						
			6		8	450	2				17				
485	0		16		13										
							6		14	408	1				9
	4	480	0				10								
					17			432	−2		9		13		
	8			468	1		18		2						
			5		6	455	−2				14				
							2		6	412	−2				17
											0		15	337	−1
									8			356	−3		3
487	0		9		8										
	2	482	1				4								
	4			470	0		8		10						
					3			435	−2		6		3		
			3		5	457	0				8				
							2		0	414	0				7

4.5.2 The Gauss-Seidel Iteration

Although the relaxation method is suitable for use with simpler problems that involve a limited number of nodes and hence are amenable to hand calculations, other methods are preferred for more complex problems. Remember that, in using the finite-difference approximation, the problem is reduced to one of solving a set of algebraic equations, where the number of equations corresponds to the number of nodes at which the temperature is an unknown. One method for obtaining the simultaneous solution to a set of equations is the Gauss-Seidel iteration method.

In writing the set of finite-difference equations, it is convenient to identify the nodes by a single integer subscript, rather than by the double subscript (m,n) previously used. It follows that, for N nodes of unknown temperature, there will be N finite-difference equations. The procedure for obtaining a solution for T_1, T_2, T_i, ... T_N is then as follows:

1. The finite-difference equation for each node is written in the explicit form for the temperature of that node. That is, for node i, T_i is expressed explicitly in terms of T_j, the neighboring nodal temperatures, and other parameters of the grid network or the boundary conditions. Hence, there will be N equations of the form

$$T_i = f(T_j, \ldots)$$

2. An initial set of values for the temperatures, T_i, of the N nodes is assumed. Subsequent computational requirements may be reduced by selecting values based upon rational estimates or rough calculations.

3. New values of T_i are then calculated from the finite-difference equations using the initial set. This step is the first iteration.

4. The iteration procedure is continued by calculating new values of $T_{i,n+1}$ from the $T_{j,n}$ values of the previous iteration step. Note the use of the subscript to designate the iteration step.

5. The iteration procedure is repeated until the difference in successive values for the nodal temperatures is less than or equal to a specified value δ.

$$T_{i,n+1} - T_{i,n-1} \leq \delta$$

The constant δ represents an uncertainty in the temperature that is deemed to be acceptable in your solution.

EXAMPLE 4.4

Consider the column of the industrial furnace described in Example 4.3. Using the same grid size, find the two-dimensional temperature distribution using the Gauss-Seidel iteration method.

SOLUTION

FIND:

Temperature distribution using the Gauss-Seidel iteration method.

ANALYSIS:

Referring to the solution of Example 4.3, we know that since there are N = 8 nodes of unknown temperature, there must be eight finite-difference equations. Setting the residuals, R_i, equal to zero and expressing each nodal equation explicitly for the temperature of that node, the equations reduce to

$$T_1 = 0.25T_2 + 0.25T_3 + 250$$

$$T_2 = 0.50T_1 + 0.25T_4 + 125$$

$$T_3 = 0.25T_1 + 0.25T_4 + 0.25T_5 + 125$$

$$T_4 = 0.25T_2 + 0.50T_3 + 0.25T_6$$

$$T_5 = 0.25T_3 + 0.25T_6 + 0.25T_7 + 125$$

$$T_6 = 0.25T_4 + 0.50T_5 + 0.25T_8$$

$$T_7 = 0.2222T_5 + 0.1111T_8 + 222.22$$

$$T_8 = 0.2222T_6 + 0.2222T_7 + 166.67$$

Having the finite-difference equations in the required form, the iteration procedure may be implemented by using a table which has one column for the iteration (step) number and a column for each of the nodes labeled as T_i. The calculations proceed as follows:

1. For each node, the initial temperature estimate is entered on the row $n = 0$. Values are selected rationally in order to reduce the number of required iterations.

2. Using the N finite-difference equations and values of T_i from the first row, the new values of T_i are calculated for the first iteration ($n = 1$). These new values are entered on the next row.

3. This procedure is repeated to calculate $T_{i,n+1}$ from the previous values of $T_{i,n}$ until the temperature difference between iterations meets the prescribed maximum δ.

4. Note that, for iteration 9, the change in temperature between iterations is no larger than 1.2°C (node 4). For iteration 14, this change is reduced to 0.2°C (node 6).

n	T_1	T_2	T_3	T_4	T_5	T_6	T_7	T_8
0	480	470	440	430	400	390	370	350
1	477.5	472.5	452.5	435.0	425.0	395.0	350.0	335.5
2	481.3	472.5	459.4	443.1	426.1	405.1	353.9	332.2
3	483.0	476.4	462.6	449.1	429.6	406.9	353.8	335.3
4	484.8	478.8	465.4	452.1	430.8	410.9	354.9	335.7
5	486.1	480.4	466.9	455.9	432.8	412.4	355.2	336.8
6	486.4	482.0	468.7	456.7	433.6	414.6	355.8	337.2
7	487.7	482.4	469.2	458.5	434.8	415.3	356.0	337.8
8	487.9	483.5	470.3	459.0	435.1	416.5	356.4	338.0
9	488.5	483.7	470.5	460.2	435.8	416.8	356.5	338.4
10	488.6	484.3	471.1	460.4	436.0	417.6	356.7	338.5
11	488.9	484.4	471.3	461.0	436.4	417.7	356.7	338.7
12	488.9	484.7	471.6	461.2	436.4	418.1	356.8	338.7
13	489.1	484.8	471.6	461.5	436.6	418.2	356.8	338.8
14	489.1	484.9	471.8	461.6	436.7	418.4	356.9	338.9

COMMENTS:

1. Recognize that, even if the temperature change between iterations is reduced to zero, the resulting temperatures do not constitute an exact solution to the problem. The resulting temperatures, however, would be an exact solution to the finite-difference equations which, at best, approximate the physical problem. A finer grid size would provide a more accurate temperature distribution.

2. This iteration procedure with the finite-difference equations as written above can readily be implemented on a digital computer. If such an approach is taken, increasing the number of nodes requires little additional effort. Alternatively, you may find it convenient to use a programmable hand calculator to perform the successive, repetitive calculations of $T_{i,n}$.

4.5.3 The Matrix Inversion Method

Another method for obtaining the simultaneous solution to a set of equations is by *matrix inversion*. Rather than being an iterative method, it provides an exact solution of the unknown temperatures. Using the single integer subscript notation, the procedure for setting up a matrix inversion begins by expressing the finite-difference equations in the form

$$a_{11}T_1 + a_{12}T_2 + a_{13}T_3 + \ldots + a_{1N}T_N = C_1$$
$$a_{21}T_1 + a_{22}T_2 + a_{23}T_3 + \ldots + a_{2N}T_N = C_2$$
$$\vdots \qquad \vdots \qquad \vdots \qquad \vdots \qquad \vdots \qquad \vdots$$
$$a_{N1}T_1 + a_{N2}T_2 + a_{N3}T_3 + \ldots + a_{NN}T_N = C_N$$

where the quantities a_{11}, a_{12}, ..., C_1, ... are known coefficients involving quantities such as Δx, k, h, and T_∞. Using matrix notation, these equations may be expressed as

$$[A][T] = [C] \tag{4.33}$$

where

$$A \equiv \begin{bmatrix} a_{11} & a_{12} & \cdots & a_{1N} \\ a_{21} & a_{22} & \cdots & a_{2N} \\ \vdots & \vdots & & \vdots \\ a_{N1} & a_{N2} & \cdots & a_{NN} \end{bmatrix}, \quad T \equiv \begin{bmatrix} T_1 \\ T_2 \\ \vdots \\ T_N \end{bmatrix}, \quad C \equiv \begin{bmatrix} C_1 \\ C_2 \\ \vdots \\ C_N \end{bmatrix}$$

Moreover, for such problems, the solution may be represented as

$$[T] = [A]^{-1}[C] \tag{4.34}$$

where $[A]^{-1}$ is the inverse of $[A]$ and may be expressed as

$$[A]^{-1} \equiv \begin{bmatrix} b_{11} & b_{12} & \cdots & b_{1N} \\ b_{21} & b_{22} & \cdots & b_{2N} \\ \vdots & \vdots & & \vdots \\ b_{N1} & b_{N2} & \cdots & b_{NN} \end{bmatrix}$$

In expanded form Equation 4.34 may then be expressed as

$$\begin{bmatrix} T_1 = b_{11}C_1 + b_{12}C_2 + \ldots + b_{1N}C_N \\ T_2 = b_{21}C_1 + b_{22}C_2 + \ldots + b_{2N}C_N \\ \vdots \qquad \vdots \qquad \vdots \\ T_N = b_{N1}C_1 + b_{N2}C_2 + \ldots + b_{NN}C_N \end{bmatrix}$$

and the problem reduces to one of determining $[A]^{-1}$. That is, if $[A]$ is inverted, its elements b_{11}, b_{12}, ... may be determined and the unknown temperatures may be computed from the above expressions.

Matrix inversion may readily be effected on a digital computer by using one of numerous routines available for this purpose and hence provides a convenient means of solving complicated two-dimensional conduction problems. For simpler problems it is even possible to obtain solutions by using programmable, hand calculators. However, in problems involving a large number of unknown nodal temperatures, for example, in excess of 100, iterative numerical procedures may be more efficient from the standpoint of requiring less computer storage space and time. A more detailed description of relaxation, iterative procedures, and matrix inversion may be found in the literature [2], [4], and [10–12].

EXAMPLE 4.5

Consider the column of the industrial furnace described in Example 4.3. Using the same grid size, find the two-dimensional temperature distribution using the matrix inversion method.

SOLUTION

FIND:

Temperature distribution using the matrix inversion method.

ANALYSIS:

Referring to the solution of Example 4.3, note that there are $N = 8$ nodes of unknown temperatures and hence there will be eight finite-difference equations. These equations are the residual equations (with $R_N = 0$), which, when rearranged and written in sequence ($N = 1, 2 \ldots 8$), have the form

$$-4T_1 + T_2 + T_3 + 0 + 0 + 0 + 0 + 0 = -1000$$
$$2T_1 - 4T_2 + 0 + T_4 + 0 + 0 + 0 + 0 = -500$$
$$T_1 + 0 - 4T_3 + T_4 + T_5 + 0 + 0 + 0 = -500$$
$$0 + T_2 + 2T_3 - 4T_4 + 0 + T_6 + 0 + 0 = 0$$
$$0 + 0 + T_3 + 0 - 4T_5 + T_6 + T_7 + 0 = -500$$
$$0 + 0 + 0 + T_4 + 2T_5 - 4T_6 + 0 + T_8 = 0$$
$$0 + 0 + 0 + 0 + 2T_5 + 0 - 9T_7 + T_8 = -2000$$
$$0 + 0 + 0 + 0 + 0 + 2T_6 + 2T_7 - 9T_8 = -1500$$

In matrix notation, following Equation 4.33, these equations are of the form $[A][T] = [C]$, where

$$[A] = \begin{bmatrix} -4 & 1 & 1 & 0 & 0 & 0 & 0 & 0 \\ 2 & -4 & 0 & 1 & 0 & 0 & 0 & 0 \\ 1 & 0 & -4 & 1 & 1 & 0 & 0 & 0 \\ 0 & 1 & 2 & -4 & 0 & 1 & 0 & 0 \\ 0 & 0 & 1 & 0 & -4 & 1 & 1 & 0 \\ 0 & 0 & 0 & 1 & 2 & -4 & 0 & 1 \\ 0 & 0 & 0 & 0 & 2 & 0 & -9 & 1 \\ 0 & 0 & 0 & 0 & 0 & 2 & 2 & -9 \end{bmatrix} \quad [C] = \begin{bmatrix} -1000 \\ -500 \\ -500 \\ 0 \\ -500 \\ 0 \\ -2000 \\ -1500 \end{bmatrix}$$

Using a standard matrix inversion routine, it is a simple matter to find the inverse of $[A]$, $[A]^{-1}$, giving

$$[T] = [A]^{-1}[C]$$

where

$$[T] = \begin{bmatrix} T_1 \\ T_2 \\ T_3 \\ T_4 \\ T_5 \\ T_6 \\ T_7 \\ T_8 \end{bmatrix} = \begin{bmatrix} 489.30 \\ 485.15 \\ 472.07 \\ 462.01 \\ 437.95 \\ 418.74 \\ 356.99 \\ 339.05 \end{bmatrix} {}^\circ C \qquad \triangleleft$$

COMMENTS:

Compare these results with the values from the relaxation method and from the Gauss-Seidel iteration method, Examples 4.3 and 4.4, respectively. While the matrix inversion method is the most precise method, since the residuals at each node are zeros, it is still an approximate method.

4.6 SUMMARY

You should now have an appreciation for the nature of a two-dimensional conduction problem and the methods which are available for solving such problems. When confronted with a two-dimensional problem, you should first determine whether an exact solution is known. This may be done by examining one or more of the many excellent references in which exact solutions to the heat equation are compiled [1, 3, 4, 5]. You may also want to determine whether the shape factor is known for the system of interest [6–9]. However, conditions will often be such that the use of a shape factor or an exact solution is not possible, and it is necessary to use a finite-difference solution. You should appreciate the inherent nature of the *discretization process*, and you should know how to formulate the finite-difference equations for the discrete points of a nodal network. Although you may find it convenient to solve these equations using the relaxation method for a coarse mesh, you should always be prepared to use standard computer subroutines involving matrix inversion or iterative techniques.

REFERENCES

1. Carslaw, H. S. and J. C. Jaeger, *Conduction of Heat in Solids*, 2nd Ed., Oxford University Press, London, 1959.
2. Schneider, P. J., *Conduction Heat Transfer*, Addison-Wesley, Cambridge, Mass., 1955.
3. Arpaci, V. S., *Conduction Heat Transfer*, Addison-Wesley, Reading, Mass., 1966.
4. Özisik, M. N., *Boundary Value Problems of Heat Conduction*, International Textbook Co., Scranton, Pa., 1968.
5. Myers, G. E., *Analytical Methods in Conduction Heat Transfer*, McGraw-Hill, New York, 1971.
6. Sunderland, J. E. and K. R. Johnson, "Shape Factors for Heat Conduction through Bodies with Isothermal or Convective Boundary Conditions," *Trans. ASHRAE, 10*, 237–241, 1964.
7. Kutateladze, S. S., *Fundamentals of Heat Transfer*, Academic Press, New York, 1963.
8. General Electric Co. (Corporate Research and Development), *Heat Transfer Data Book*, Section 502, General Electric Company, Schenectady, New York, 1973.
9. Hahne, E. and U. Grigull, *International Journal of Heat and Mass Transfer, 18*, 751–767, 1975.
10. Dusinberre, G. M., *Heat Transfer Calculations by Finite-Difference*, International Textbook Co., Scranton, Pa., 1961.
11. Razelos, P., "Methods of Obtaining Approximate Solutions" in W. M. Rohsenow and J. P. Hartnett, Eds., *Handbook of Heat Transfer*, McGraw-Hill, New York, 1973.
12. Adams, J. A. and D. F. Rogers, *Computer Aided Heat Transfer Analysis*, McGraw-Hill, New York, 1973.

PROBLEMS

4.1   In the method of separation of variables (Appendix C) for two-dimensional, steady-state conduction, the separation constant λ^2 in Equations C.5 and C.6 must be a positive constant. Show that a negative or zero value of λ^2 will result in solutions that cannot satisfy the prescribed boundary conditions.

4.2   A two-dimensional rectangular plate is subjected to the uniform temperature boundary conditions shown in the following diagram. Using the results of the exact

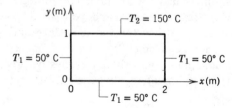

solution for the heat diffusion equation presented in Appendix C, calculate the temperature at the midpoint (1, 0.5) by considering the first five nonzero terms of the infinite series that must be evaluated. Assess the error resulting from using only the first three terms of the infinite series.

4.3 Two thin strips of materials A and B with thermal conductivities k_A and k_B, respectively, where $k_A = 2k_B$, are soldered together as shown in the following sketch. The strips are insulated on their faces and exposed edges, while their ends are maintained at temperatures T_1 and T_2, where $T_1 > T_2$. For steady-state conditions, sketch the isotherms that pass through the points indicated in the sketch. What are the special features of the isotherms in the vicinity of the interface?

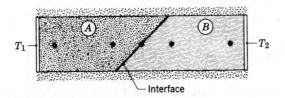

4.4 A long furnace, constructed from refractory brick with a thermal conductivity of 1.2 W/m·K, has the cross section shown below with inner and outer surface temperatures of 600°C and 60°C, respectively. Determine the shape factor and the heat transfer rate per unit length using the flux plot method. Be sure to take advantage of symmetry wherever possible.

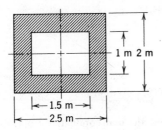

4.5 A hot pipe is imbedded eccentrically as shown below in a material of thermal conductivity 0.5 W/m·K. Using the flux plot method, determine the shape factor and the heat transfer per unit length when the pipe and outer surface temperatures are 150°C and 35°C, respectively.

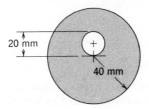

4.6   A supporting strut fabricated from a material with a thermal conductivity of 75 W/m·K has the cross section shown in the following sketch. The end faces are at different temperatures with $T_1 = 100°C$ and $T_2 = 0°C$, while the sides are insulated.

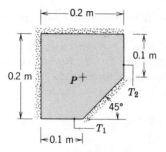

a) Using the flux plot method, estimate the temperature at the location P.

b) Estimate the heat transfer rate through the strut per unit length.

4.7   A hot liquid flows along a V-groove in a solid whose top and side surfaces are well insulated and whose bottom surface is in contact with a coolant.

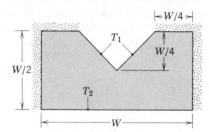

Accordingly, the V-groove surface is at a temperature T_1 which exceeds that of the bottom surface, T_2. Construct an appropriate flux plot and determine the shape factor of the system.

4.8   A plate having a thickness 6 mm and thermal conductivity of 150 W/m·K is subjected to the conditions shown in the sketch. Assuming the side and lower faces of the plate are well insulated, find the shape factor. Determine the heat rate when $T_1 = 50°C$ and $T_2 = 20°C$.

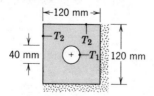

4.9   A long support column of trapezoidal cross section is well insulated on its sides, and temperatures of 100°C and 0°C are maintained at its top and bottom surfaces, respectively. The column is fabricated from AISI 1010 steel, and its width is 0.3 m and 0.6 m at the top and bottom surface, respectively.

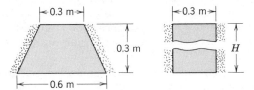

a) Using the flux plot method, determine the heat transfer rate per unit length of the column.

b) If the trapezoidal column is replaced by a bar of rectangular cross section 0.3 m wide and the same material, what is the height H that the bar must have to provide an equivalent thermal resistance?

4.10 The hollow prismatic bars fabricated from plain carbon steel shown in the sketch are 1 m long with both ends insulated. For each bar, find the shape factor and then calculate the heat rate per unit length of the bar when $T_1 = 500$ K and $T_2 = 300$ K. Scale up the figures so the flux plot can be constructed with reasonable detail and take advantage of symmetry wherever possible.

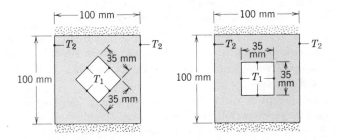

4.11 The two-dimensional rectangular shape shown in the following diagram is subjected to uniform temperature conditions on portions of its upper and lower surfaces. The remaining boundaries are insulated. Use the flux-plot method to estimate the heat transfer rate per unit length normal to the page when the thermal conductivity is 10 W/m·K.

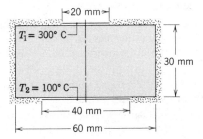

4.12 Radioactive wastes are temporarily stored in a spherical container, the center of which is buried a distance of 10 m below the earth's surface. The outside diameter of the container is 2 m, and 500 W of heat are released due to the radioactive decay process. If the soil surface temperature is 20°C, what is the outside surface temperature of the container under steady-state conditions? On a sketch of the soil-container system drawn to scale, show representative isotherms and heat flow lines in the soil.

4.13 A pipeline, used for the transport of crude oil, is buried in the earth such that its centerline is a distance of 1.5 m below the surface. The pipe has an outer diameter of 0.5 m and is insulated with a layer of cellular glass 10 cm thick. What is the heat loss per unit length of pipe under conditions for which heated oil at 120°C flows through the pipe and the surface of the earth is at a temperature of 0°C?

4.14 An electrical heater 100 mm long and 5 mm in diameter is inserted into a hole drilled normal to the surface of a large block of material having a thermal conductivity of 5 W/m·K. Estimate the temperature reached by the heater when dissipating 50 W with the surface of the block at a temperature of 25°C.

4.15 Two parallel pipelines spaced 0.5 m apart are buried in soil having a thermal conductivity of 0.5 W/m·K. The pipes have outer diameters of 100 mm and 75 mm with surface temperatures of 175°C and 5°C, respectively. Estimate the heat transfer rate per unit length between the two pipelines.

4.16 A tube of diameter 50 mm having a surface temperature of 85°C is imbedded in the center plane of a concrete slab 0.1 m thick with upper and lower surfaces at 20°C.

   a) Using the appropriate tabulated relation for this configuration, find the shape factor. Determine the heat transfer rate per unit length of the tube.

   b) Using the flux-plot method, estimate the shape factor and compare with the result of part (a).

4.17 Pressurized steam at 450 K flows through a long, thin-walled pipe of 0.5 m diameter. The pipe is enclosed in a concrete casing that is of square cross section and 1.5 m on a side. The axis of the pipe is centered in the casing, and the outer surfaces of the casing are maintained at 300 K. What is the heat loss per unit length of pipe?

4.18 An electronic device, in the form of a disc 20 mm in diameter, dissipates 100 W when mounted flush on a large aluminum alloy (2024) block whose temperature is maintained at 27°C. The mounting arrangement is such that a contact resistance of 5×10^{-5} m²·K/W occurs at the interface between the device and the block. Calculate the temperature the device will reach, assuming all the power generated by the device must be transferred by conduction to the block.

4.19 A furnace of cubical shape, with external dimensions of 0.35 m, is constructed from a refractory brick (fire clay). If the wall thickness is 50 mm, the inner surface temperature is 600°C, and the outer surface temperature is 75°C, calculate the heat loss from the furnace.

4.20 Models, as illustrated in the following diagrams, suitable for representing two-dimensional systems with uniform temperature or adiabatic boundaries, were cut from conductive paper and their electrical resistances measured. Model A has a rectangular shape with dimensions $w_A = 50.0$ mm and $L_A = 0.15$ m and has an electrical resistance, $R_A = 6.20$ kΩ. Model B has an irregular shape with dimensions

$y_1 = 0.15$ m, $y_2 = 25$ mm, and $x_1 = 100$ mm and has an electrical resistance, R_B $= 1.10$ kΩ.

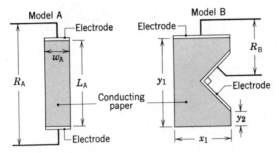

a) What is the conduction shape factor for the one-dimensional system represented by model A?

b) If the conducting paper has an electrical resistivity, ρ_e, and thickness l_e, write an expression for the electrical resistance of model A, R_A, in terms of ρ_e, l_e, and S_A, where S_A is the conduction shape factor for model A.

c) Using the result of part (b) and the electrical resistance measurements of the models, determine the conduction shape factor for the two-dimensional system represented by model B.

d) What is the conduction shape factor for a two-dimensional system that is geometrically similar to model B but is four times larger? That is, $y_1' = 4y_1$, $y_2' = 4y_2$ and $x_1' = 4x_1$.

e) Verify your result of part (c) by performing a flux plot on model B to estimate the conduction shape factor.

4.21 A two-dimensional system has been represented by a piece of conductive paper having the shape shown in the diagram on the following page. This shape, referred to as model B, has the dimensions $y_1 = 0.15$ m, $y_2 = 25$ mm and $x_1 = 100$ mm and has an electrical resistance $R_B = 1.03$ kΩ. Model A, as described in Problem 4.20, is fabricated from the same conductive paper. Also shown on the figure are plots of the constant voltage lines for $E = 0.2$ V, 0.4 V, 0.6 V, and 0.8 V. The electrodes, representing the uniform temperature boundaries, are maintained at 0 and 1.0 V, as shown.

a) What is the conduction shape factor for the one-dimensional system represented by model A?

b) If the conducting paper has an electrical resistivity, ρ_e, and thickness, l_e, write an expression for the electrical resistance of model A, R_A, in terms of ρ_e, l_e, and S_A, where S_A is the conduction shape factor for model A.

c) Using the result of part (b) and the electrical resistance measurements of the models, determine the shape factor for the two-dimensional system represented by model B.

d) Determine the temperatures corresponding to the constant voltage lines on the two-dimensional system if the boundaries are maintained at $T_1 = 80°C$ and $T_2 = 30°C$.

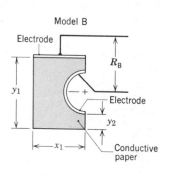

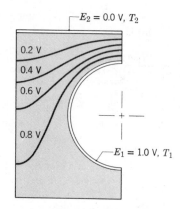

4.22 Consider the nodal configuration (2) in Table 4.2. Derive the finite-difference equations under steady-state conditions for these situations.

    a) The horizontal boundary of the internal corner is perfectly insulated and the vertical boundary is subjected to the convection process (h, T_∞).

    b) Both boundaries of the internal corner are perfectly insulated. How does this result compare with Equation 4.28?

4.23 Consider the nodal configuration (3) on Table 4.2. Derive the finite-difference equation under steady-state conditions when the boundary is insulated. Explain how Equation 4.29 can be modified to agree with your result.

4.24 Consider the nodal configuration (4) on Table 4.2. Derive the finite-difference equations under steady-state conditions for these situations:

    a) The upper boundary of the external corner is perfectly insulated and the side boundary is subjected to the convection process (h, T_∞).

    b) Both boundaries of the external corner are perfectly insulated. How does this result compare with Equation 4.30?

4.25 Apply the general form of the finite-difference equation for a node near the curved surface maintained at a nonuniform temperature, Table 4.2 (5), to nodes 0 and 4 of the configuration shown as follows.

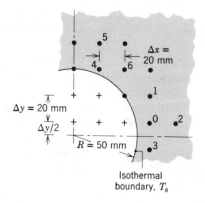

4.26 Beginning with a properly defined control volume for a node near a curved surface maintained at a nonuniform temperature, Table 4.2 (5), derive Equation 4.31.

4.27 Consider the nodal point arrangement in the vicinity of the isothermal surface shown as follows.

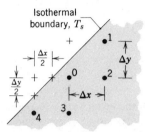

a) Apply Equation 4.31 to the above configuration to obtain the finite-difference equation for node 0.

b) Beginning with the properly defined control volume for node 0, derive the finite-difference equation for that node.

4.28 Consider a one-dimensional fin of uniform cross-sectional area which is insulated at the end, $x = L$. (See Table 3.3, case B.) The temperature at the base of the fin, T_b, and of the adjoining fluid, T_∞, as well as the heat transfer coefficient, h, and the thermal conductivity, k, are known.

a) Derive the finite-difference equation for any interior node, m.

b) Derive the finite-difference equation for the node, n, located at the insulated end where $x = L$.

4.29 The steady-state temperature distribution in a fin of nonuniform cross-sectional area is to be determined by the finite-difference method. The temperatures at the base of the fin, T_b, and of the adjoining fluid, T_∞, as well as the thermal conductivity, k, are known. Furthermore, the convection heat transfer coefficient, h, the perimeter of the fin, P, and the cross-sectional area of the fin, A, are known functions of x, that is, $h(x)$, $P(x)$, and $A(x)$, respectively.

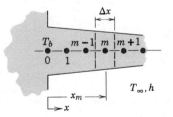

a) Derive the nodal finite-difference equation for point m located at x_m.

b) Describe a procedure for obtaining the temperature distribution. Indicate how you would use this distribution to determine the fin heat transfer rate, q_f.

4.30 Derive the nodal finite-difference equations for the following configurations.

a) Node on a diagonal boundary as shown in the sketch subjected to convection with a fluid at T_∞ and a heat transfer coefficient h. Assume $\Delta x = \Delta y$.

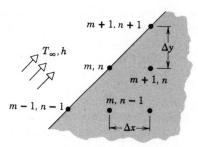

b) Node at the tip of a cutting tool with the upper surface exposed to a constant heat flux, q_o'', and the diagonal surface exposed to a convection cooling process with the fluid at T_∞ and a heat transfer coefficient h. Assume $\Delta x = \Delta y$.

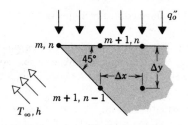

4.31 Consider the nodal point 0 located on the boundary between materials of thermal conductivity k_A and k_B as shown in the following sketch.

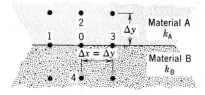

Derive the finite-difference equation under steady-state conditions assuming no internal generation.

4.32 In the two-dimensional cylindrical configuration shown as follows, the radial (Δr) and angular ($\Delta \phi$) spacings of the nodes are uniform. The boundary at $r = r_i$ is of uniform surface temperature T_i. The boundaries in the radial direction are adiabatic (insulated) and surface convection (T_∞, h) with a fluid as illustrated. Derive the finite-difference equations for (a) node 2, (b) node 3, and (c) node 1.

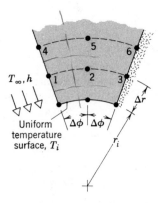

4.33 The temperatures (K) at the nodal points of a two-dimensional object are shown on the following sketch. If surface B is held at a uniform temperature of 500 K, while surface A is subjected to a convection boundary condition, where $h = 10$ W/m$^2 \cdot$ K and the fluid is at $T_\infty = 300$ K, calculate the heat rate leaving surface A per unit thickness normal to the page. Estimate the thermal conductivity of the material.

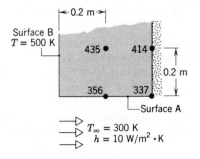

4.34 The steady-state temperatures ($^\circ$C) associated with selected nodal points of a two-dimensional system having a thermal conductivity of 1.5 W/m $\cdot$ K are shown on the following grid network.

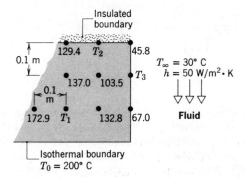

a) Determine the temperatures at nodes 1, 2, and 3.

b) Calculate the heat transfer rate per unit thickness normal to the page from the system to the fluid.

4.35 Assuming two-dimensional, steady-state conduction, determine the temperature at nodes 1, 2, 3, and 4 in the square shape subjected to the uniform surface temperature conditions shown in the figure.

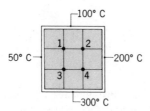

4.36 Consider a long bar of square cross section 0.8 m to the side. Three of these sides are maintained at a uniform temperature of 300°C. The fourth side is exposed to a fluid at 100°C for which the convection heat transfer coefficient is 10 W/m² · K. Using an appropriate numerical technique with a grid spacing of 0.2 m, determine the midpoint temperature and the heat transfer rate between the bar and the fluid per unit length of the bar.

4.37 A long, conducting rod of rectangular cross section (20 mm × 30 mm) and thermal conductivity $k = 20$ W/m · K experiences uniform heat generation at a rate $\dot{q} = 5 \times 10^4$ W/m³, while its surfaces are maintained at 27°C. Using a finite-difference method with a grid spacing of 5 mm, determine the temperature distribution in the rod.

4.38 A solid slab is impregnated with uniformly spaced heating pipes, and the relative dimensions are as shown in the following diagram. Under steady-state conditions fluid passing through the pipes maintains a surface temperature $T_1 = 400$ K, while the top and bottom surfaces of the slab are maintained at $T_2 = 300$ K. Using an appropriate numerical technique with a grid spacing of $\Delta x = \Delta y = D/5$, where $D = 50$ mm, determine the temperature distribution in the slab. If the slab is constructed from granite, what is the heat loss per pipe per unit length?

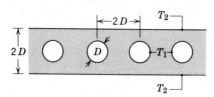

4.39 A long bar of rectangular cross section, 0.4×0.6 m on a side and thermal conductivity 1.5 W/m · K, is subjected to the boundary conditions shown on the sketch. Two of the sides are maintained at a uniform temperature of 200°C. One of the sides is adiabatic; and the remaining side is subjected to a convection process with a fluid at $T_\infty = 30$°C and a heat transfer coefficient of $h = 50$ W/m² · K. Using an appropriate numerical technique with a grid spacing of 0.1 m, determine the temperature distribution in the bar and the heat transfer rate between the bar and the fluid per unit length of the bar.

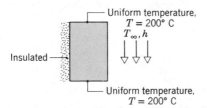

4.40 A straight fin of uniform cross section fabricated from a material with a thermal conductivity of 50 W/m·K has a thickness, $w = 6$ mm, length, $L = 48$ mm, and is very long in the direction normal to the page. The convection heat transfer coefficient is 500 W/m²·K with an ambient air temperature $T_\infty = 30°$C. The base of the fin is maintained at $T_b = 100°$C while the end of the fin is insulated.

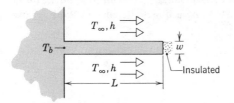

a) Using a finite-difference method with a space increment of 4 mm, estimate the temperature distribution within the fin. Would the assumption of one-dimensional heat transfer be reasonable for this fin?

b) Estimate the fin heat transfer rate per unit length normal to the page. Compare your result with the one-dimensional, analytical solution, Equation 3.75.

4.41 The straight fin shown below has a triangular profile given by the function $y = 0.2$ $(L - x)$ and is very long in the direction normal to the page. Both top and bottom surfaces are subjected to a convection process with $T_\infty = 15°$C and $h = 100$ W/m²·K. The base of the fin is maintained at $T_b = 115°$C and the fin material has a thermal conductivity of 25 W/m·K. Assuming one-dimensional heat transfer and using a finite-difference method with a space increment of 10 mm, determine the fin heat rate and efficiency.

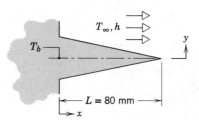

4.42 The surface of a plate and the grooves are maintained at a uniform temperature of $T_1 = 200°C$. Calculate the heat transfer rate per width of groove spacing (w) when the lower surface is at $T_2 = 20°C$, the thermal conductivity is 15 W/m·K, and the groove spacing is 0.16 m. Use a finite-difference method with a space increment of 40 mm.

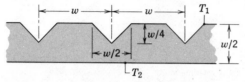

4.43 Refer to the two-dimensional rectangular plate subjected to uniform temperature boundaries as described in Problem 4.2. Using an appropriate numerical method, determine the temperature at the midpoint (1, 0.5). Use a space increment of 0.25 m for a hand-calculator scheme or 0.125 m if you employ digital computer routines.

4.44 The long bar with a trapezoidal shape shown in the following diagram is subjected to uniform temperatures on two surfaces while the remaining ones are well insulated. If the thermal conductivity of the material is 20 W/m·K, estimate the heat transfer rate per unit length of the bar using a finite-difference method. A space increment of 10 mm is suggested.

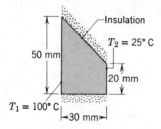

4.45 A long bar of rectangular cross section as shown in the figure is fabricated from two materials with thermal conductivities $k_A = 15$ W/m·K and $k_B = 1$ W/m·K and with thickness $L_A = 50$ mm and $L_B = 100$ mm, respectively. Its width is $w = 300$ mm. Three sides of the bar are subjected to convection conditions with $T_\infty = 100°C$ and $h = 30$ W/m²·K. Calculate the heat transfer rate per unit length of the bar that must be removed by the cooling coil in order to maintain the bottom surface of the bar at $T_0 = 0°C$.

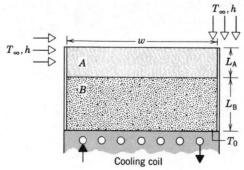

5 Transient Conduction

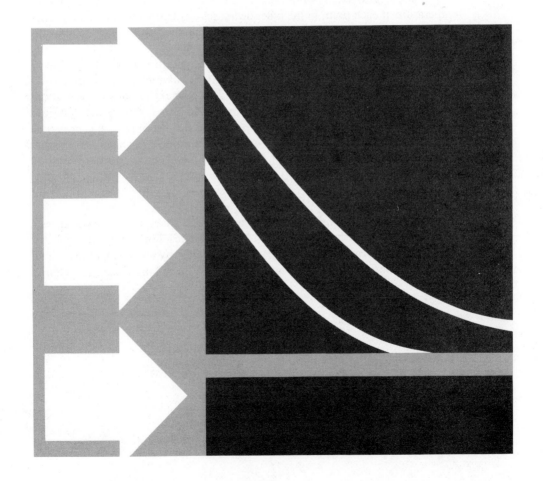

In our treatment of conduction we have gradually considered more complicated conditions. We began with the simple case of one-dimensional, steady-state conduction with no internal generation, and we subsequently considered complications due to multidimensional and generation effects. However, we have not yet considered situations for which conditions change with time.

We now recognize that there are many heat transfer problems that are time dependent. Such *unsteady*, or *transient*, problems typically arise when the boundary conditions of a system are changed. For example, if the surface temperature of a system is altered, the temperature at each point in the system will also begin to change. The changes will continue to occur until a *steady-state* temperature distribution is reached. Consider a hot metal billet that is removed from a furnace and exposed to a cool air stream. Energy is transferred by convection and radiation from its surface to the surroundings. Energy transfer by conduction also occurs from the interior of the metal to the surface, and the temperature at each point in the billet decreases until a steady-state condition is reached. Such time-dependent effects occur in many industrial heating, cooling, and drying processes.

To determine the time dependence of the temperature distribution within a solid during a transient process, we could begin by solving the appropriate form of the heat equation, for example, Equation 2.13. Some cases for which solutions have been obtained are discussed in Sections 5.3 to 5.6. However, such solutions are often difficult to obtain, and where possible a simpler approach is preferred. One such approach may be used under conditions for which temperature gradients within the solid are small. It is termed the *lumped capacitance method*.

5.1   THE LUMPED CAPACITANCE METHOD

A common transient conduction problem is one in which a solid experiences a sudden change in its thermal environment. Consider a hot metal forging that is initially at a uniform temperature T_i and is quenched by immersing it in a liquid of lower temperature $T_\infty < T_i$ (Figure 5.1). If the quenching is said to begin at

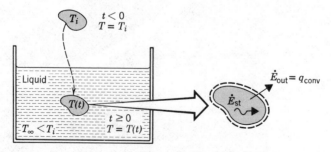

Figure 5.1   Cooling of a hot metal forging.

time $t = 0$, the temperature of the solid will decrease for time $t > 0$, until it eventually reaches T_∞. This reduction is due to convection heat transfer occurring at the solid–liquid interface. The essence of the lumped capacitance method is the assumption that the temperature of the solid is *spatially uniform* at any instant during the transient process. This assumption implies that temperature gradients within the solid are negligible.

From Fourier's law, heat conduction in the absence of a temperature gradient implies the existence of infinite thermal conductivity. Such a condition is clearly impossible. However, although the condition is never satisfied exactly, it is closely approximated if the resistance to conduction within the solid is small compared with the resistance to heat transfer between the solid and its surroundings. For now we assume that this is, in fact, the case.

In neglecting temperature gradients within the solid, we can no longer consider the problem from within the framework of the heat equation. Instead, the transient temperature response is determined by formulating an overall energy balance on the solid. This balance must relate the rate of heat loss at the surface to the rate of change of the internal energy. Applying Equation 1.10 to the control volume of Figure 5.1, this requirement takes the form

$$-\dot{E}_{out} = \dot{E}_{st} \tag{5.1}$$

or

$$-hA_s(T - T_\infty) = \rho Vc \frac{dT}{dt} \tag{5.2}$$

Introducing the temperature difference

$$\theta \equiv T - T_\infty \tag{5.3}$$

and recognizing that $(d\theta/dt) = (dT/dt)$, it follows that

$$\frac{\rho Vc}{hA_s} \frac{d\theta}{dt} = -\theta$$

Separating variables and integrating from the initial condition, for which $t = 0$ and $T(0) = T_i$, we then obtain

$$\frac{\rho Vc}{hA_s} \int_{\theta_i}^{\theta} \frac{d\theta}{\theta} = -\int_0^t dt$$

where

$$\theta_i \equiv T_i - T_\infty \tag{5.4}$$

Evaluating the integrals it follows that

$$\frac{\rho Vc}{hA_s} \ln \frac{\theta_i}{\theta} = t \tag{5.5}$$

or

$$\frac{\theta}{\theta_i} = \frac{T - T_\infty}{T_i - T_\infty} = \exp[-(hA_s/\rho Vc)t] \tag{5.6}$$

Equation 5.5 may be used to determine the time required for the solid to reach some temperature T, or, conversely, Equation 5.6 may be used to compute the temperature reached by the solid at some time t.

The foregoing results indicate that the difference between the solid and fluid temperatures must decay exponentially to zero as t approaches infinity. This behavior is shown in Figure 5.2. From Equation 5.6 it is also evident that the quantity $(\rho Vc/hA_s)$ may be interpreted as a *thermal time constant*. This time constant may be expressed as

$$\tau_t = (1/hA_s)(\rho Vc) = R_t C_t \tag{5.7}$$

where R_t is the resistance to convection heat transfer and C_t is the *lumped thermal capacitance* of the solid. Any increase in R_t or C_t will cause a solid to respond more slowly to changes in its thermal environment and will increase the time required to reach thermal equilibrium ($\theta = 0$).

It is useful to note that the foregoing behavior is analogous to the voltage decay that occurs when a capacitor is discharged through a resistor in an electrical *RC* circuit. Accordingly, the process may be represented by an *equivalent thermal circuit* which is shown in Figure 5.3. With the switch closed the solid is charged to the temperature θ_i. When the switch is opened, the energy that is stored in the solid is discharged through the thermal resistance and the

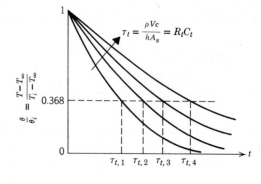

Figure 5.2 Transient temperature response of lumped capacitance solids corresponding to different thermal time constants, τ_t.

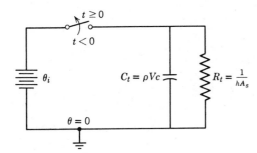

Figure 5.3 Equivalent thermal circuit for a lumped capacitance solid.

temperature of the solid decays with time. This analogy suggests that RC electrical circuits may be used to determine the transient behavior of thermal systems. Implementation of this analogy is discussed in Section 5.8.

To determine the total energy transfer Q occurring up to some time t, we simply write

$$Q = \int_0^t q\,dt = hA_s \int_0^t \theta\,dt$$

Substituting for θ from Equation 5.6 and integrating, we obtain

$$Q = (\rho Vc)\,\theta_i[1 - \exp(-t/\tau_t)] \tag{5.8}$$

The quantity Q is, of course, related to the change in the internal energy of the solid, where

$$-Q = \Delta E_{st}$$

For quenching Q is positive and the solid experiences a decrease in energy. Equations 5.5, 5.6, and 5.8 also apply to situations where the solid is heated ($\theta < 0$), in which case Q is negative and the internal energy of the solid increases.

5.2 VALIDITY OF THE LUMPED CAPACITANCE METHOD

From the foregoing results it is easy to see why there is a strong preference for using the lumped capacitance method. It is certainly the simplest and most convenient method that can be used to solve transient conduction problems. Hence it is important to determine under what conditions it may be used with reasonable accuracy.

To develop a suitable criterion consider steady-state conduction through the plane wall of area A (Figure 5.4). Although we are assuming steady-state conditions, this criterion is readily extended to transient processes. One surface is maintained at a temperature $T_{s,1}$ and the other surface is exposed to a fluid of temperature $T_\infty < T_{s,1}$. The temperature of this surface will be some intermediate value, $T_{s,2}$, for which $T_\infty < T_{s,2} < T_{s,1}$. Hence under steady-state

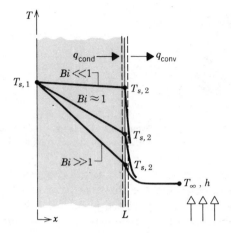

Figure 5.4 Effect of Biot number on steady-state temperature distribution in a plane wall with surface convection.

conditions the surface energy balance, Equation 1.11, reduces to

$$\frac{kA}{L}(T_{s,1} - T_{s,2}) = hA(T_{s,2} - T_{\infty})$$

where k is the thermal conductivity of the solid. Rearranging, we then obtain

$$\frac{T_{s,1} - T_{s,2}}{T_{s,2} - T_{\infty}} = \frac{(L/kA)}{(1/hA)} = \frac{R_{cond}}{R_{conv}} = \frac{hL}{k} \equiv Bi \tag{5.9}$$

The quantity (hL/k) appearing in Equation 5.9 is a *dimensionless parameter*. It is termed the *Biot number*, and it plays a fundamental role in conduction problems that involve surface convection effects. According to Equation 5.9 and as illustrated in Figure 5.4, the Biot number provides a measure of the temperature drop in the solid relative to the temperature difference between the surface and the fluid. Note especially the conditions corresponding to $Bi \ll 1$. The results suggest that, for these conditions, it is reasonable to *assume* a uniform temperature distribution across a solid at any time during a transient process. This result may also be associated with interpretation of the Biot number as a ratio of thermal resistances, Equation 5.9. *If* Bi $\ll$ 1, *the resistance to conduction within the solid is much less than the resistance to convection across the fluid boundary layer.* Hence the assumption of a uniform temperature distribution is reasonable.

We have introduced the Biot number because of its significance to transient conduction problems. Consider the plane wall of Figure 5.5, which is initially at a uniform temperature T_i and experiences convection cooling when it is immersed in a fluid of $T_{\infty} < T_i$. The problem may be treated as one dimensional in x, and we are interested in the temperature variation with position and time,

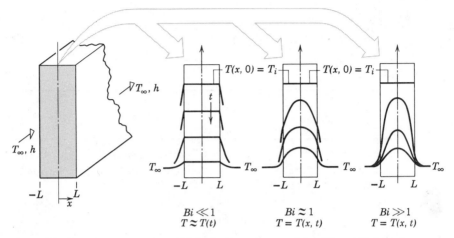

Figure 5.5   Transient temperature distribution in a plane wall for different Biot numbers.

$T(x,t)$. This variation is a strong function of the Biot number, and three conditions are shown in the figure. For $Bi \ll 1$ the temperature gradient in the solid is small and $T(x,t) \approx T(t)$. Virtually all of the temperature difference is between the solid and the fluid, and the solid temperature remains nearly uniform as it decreases to T_∞. For moderate to large values of the Biot number, however, the temperature gradients within the solid are significant. Hence $T = T(x,t)$. Note that for $Bi \gg 1$, the temperature difference across the solid is now much larger than that between the surface and the fluid.

We conclude this section by emphasizing the importance of the lumped capacitance method. Its inherent simplicity renders it the preferred method for solving transient conduction problems. Hence, when confronted with such a problem, *the very first thing that one should do is calculate the Biot number*. If the following condition is satisfied

$$Bi = \frac{hL_c}{k} \leq 0.1 \qquad\qquad (5.10)$$

the error associated with using the lumped capacitance method is small. The *characteristic length*, $L_c \equiv V/A_s$, reduces to the half-thickness, L, for a plane wall of thickness $2L$ (Figure 5.5), to $r_o/2$ for a long cylinder, and to $r_o/3$ for a sphere.

Finally, we note that, with $L_c \equiv V/A_s$, the exponent of Equation 5.6 may be expressed as

$$\frac{hA_st}{\rho Vc} = \frac{ht}{\rho cL_c} = \frac{hL_c}{k}\frac{k}{\rho c}\frac{t}{L_c^2} = \frac{hL_c}{k}\frac{\alpha t}{L_c^2}$$

or

$$\frac{hA_s t}{\rho V c} = Bi \cdot Fo \tag{5.11}$$

where

$$Fo \equiv \frac{\alpha t}{L_c^2} \tag{5.12}$$

is termed the Fourier number. It is a *dimensionless time*, which, with the Biot number, characterizes transient conduction problems. Substituting Equation 5.11 into 5.6, we obtain

$$\frac{\theta}{\theta_i} = \frac{T - T_\infty}{T_i - T_\infty} = \exp(-Bi \cdot Fo) \tag{5.13}$$

EXAMPLE 5.1

A thermocouple junction, which may be approximated as a sphere, is to be used for temperature measurement in a gas stream. The convection coefficient between the junction surface and the gas is known to be $h = 400$ $W/m^2 \cdot K$, and the junction thermophysical properties are $k = 20$ W/m·K, $c = 400$ J/kg·K, and $\rho = 8500$ kg/m^3.

1. What is the junction diameter needed for the thermocouple to have a time constant of 1 s?
2. If the junction is at 25°C and is placed in a gas stream which is at 200°C, how long will it take for the junction to reach 199°C?

SOLUTION

KNOWN:

Thermophysical properties of thermocouple junction used to measure temperature of a gas stream.

FIND:

1. Junction diameter needed for a time constant of 1 s.
2. Time required to reach 199°C in gas stream at 200°C.

SCHEMATIC:

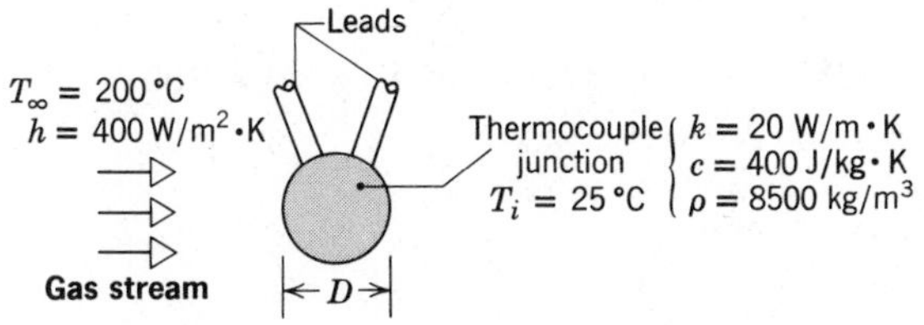

$T_\infty = 200\,°C$
$h = 400\,W/m^2 \cdot K$

Gas stream

Leads

Thermocouple junction
$T_i = 25\,°C$

$k = 20\,W/m \cdot K$
$c = 400\,J/kg \cdot K$
$\rho = 8500\,kg/m^3$

ASSUMPTIONS:

1. Temperature of junction is uniform at any instant.
2. Radiation exchange with the surroundings is negligible.
3. Negligible losses by conduction through the leads.
4. Constant properties.

ANALYSIS:

1. Because the junction diameter is unknown, it is not possible to begin the solution by determining whether the criterion for using the lumped capacitance method, Equation 5.10, is satisfied. However, a reasonable approach is to use the method to find the diameter and to then determine whether the criterion is satisfied. From Equation 5.7 and the fact that $A_s = \pi D^2$ and $V = \pi D^3/6$ for a sphere, it follows that

$$\tau_t = \frac{1}{h\pi D^2} \times \frac{\rho \pi D^3}{6}\,c$$

Rearranging and substituting numerical values,

$$D = \frac{6h\tau_t}{\rho c} = \frac{6 \times 400\,W/m^2 \cdot K \times 1\,s}{8500\,kg/m^3 \times 400\,J/kg \cdot K}$$

$$D = 7.06 \times 10^{-4}\,m = 0.71\,mm \qquad \triangleleft$$

With $L_c = r_o/3$ it then follows from Equation 5.10 that

$$Bi = \frac{h(r_o/3)}{k} = \frac{400\,W/m^2 \cdot K \times 3.53 \times 10^{-4}\,m}{3 \times 20\,W/m \cdot K} = 2.35 \times 10^{-4}$$

Accordingly, the lumped capacitance method may be used to an excellent approximation.

2. From Equation 5.5 the time required for the junction to reach $T = 199\,°C$ is

$$t = \frac{\rho(\pi D^3/6)c}{h(\pi D^2)} \ln \frac{T_i - T_\infty}{T - T_\infty} = \frac{\rho D c}{6h} \ln \frac{T_i - T_\infty}{T - T_\infty}$$

$$t = \frac{8500 \text{ kg/m}^3 \times 7.06 \times 10^{-4} \text{ m} \times 400 \text{ J/kg} \cdot \text{K}}{6 \times 400 \text{ W/m}^2 \cdot \text{K}} \ln \frac{25 - 200}{199 - 200}$$

$$t = 5.2 \text{ s} \qquad \triangleleft$$

COMMENTS:

Recognize that heat losses due to radiation exchange between the junction and the surroundings and conduction through the leads would necessitate using a smaller junction diameter to achieve the desired time response.

5.3 GENERAL SOLUTION METHODS

Situations frequently arise for which the lumped capacitance method is inappropriate, and alternative methods must be used. Regardless of the particular form of the method, we must now cope with the fact that gradients within the medium are no longer negligible.

In their most general form, transient conduction problems are described by the heat equation, Equation 2.13 for rectangular coordinates or Equations 2.18 and 2.19, respectively, for cylindrical and spherical coordinates. The solution to these partial differential equations provides the variation of temperature with both time and the spatial coordinates. However, in many problems, such as the plane wall of Figure 5.5, only one spatial coordinate is needed to describe the internal temperature distribution. With no internal generation and the assumption of constant thermal conductivity, Equation 2.13 then reduces to

$$\frac{\partial^2 T}{\partial x^2} = \frac{1}{\alpha} \frac{\partial T}{\partial t} \tag{5.14}$$

To solve Equation 5.14 for the temperature distribution, $T(x,t)$, it is necessary to specify an *initial* condition and two *boundary conditions*. For the typical transient conduction problem of Figure 5.5, the initial condition is

$$T(x,0) = T_i \tag{5.15}$$

and the boundary conditions are

$$\frac{\partial T}{\partial x}\bigg|_{x=0} = 0 \tag{5.16}$$

and

$$-k\frac{\partial T}{\partial x}\bigg|_{x=L} = h[T(L,t) - T_\infty] \tag{5.17}$$

Equation 5.15 presumes a uniform temperature distribution at time $t = 0$; Equation 5.16 reflects the *symmetry requirement* for the midplane of the wall; and Equation 5.17 describes the surface condition experienced for time $t > 0$. From Equations 5.14 to 5.17, it is evident that, in addition to depending on x and t, temperatures in the wall also depend on a number of physical parameters. In particular

$$T = T(x,t,T_i,T_\infty,L,k,\alpha,h) \tag{5.18}$$

The foregoing problem may be solved analytically, numerically, graphically, or through use of an electrical analog. These methods will be considered in subsequent sections, but before this is done it is important to note the advantages which may be derived by *nondimensionalizing* the governing equations. This may be done by arranging the relevant variables into suitable *groups*. Consider the dependent variable T. If the temperature difference, $\theta \equiv T - T_\infty$, is divided by the *maximum possible temperature difference*, $\theta_i \equiv T_i - T_\infty$, a dimensionless form of the dependent variable may be defined as

$$\theta^* \equiv \frac{\theta}{\theta_i} = \frac{T - T_\infty}{T_i - T_\infty} \tag{5.19}$$

Accordingly, θ^* must lie in the range $0 \le \theta^* \le 1$. Similarly, a dimensionless spatial coordinate may be defined as

$$x^* \equiv \frac{x}{L} \tag{5.20}$$

where L is the half-thickness of the plane wall. Similarly, a dimensionless time may be defined as

$$t^* \equiv \frac{\alpha t}{L^2} \equiv Fo \tag{5.21}$$

where t^* is equivalent to the dimensionless *Fourier number*, Equation 5.12.

Substituting the definitions (5.19) to (5.21) into Equations 5.14 through 5.17, the governing equation becomes

$$\frac{\partial^2 \theta^*}{\partial x^{*2}} = \frac{\partial \theta^*}{\partial Fo} \tag{5.22}$$

and the initial and boundary conditions become

$$\theta^*(x^*,0) = 1 \tag{5.23}$$

$$\left.\frac{\partial \theta^*}{\partial x^*}\right|_{x^*=0} = 0 \tag{5.24}$$

and

$$\left.\frac{\partial \theta^*}{\partial x^*}\right|_{x^*=1} = Bi\,\theta^*(1,t^*) \tag{5.25}$$

where the Biot number is $Bi \equiv hL/k$. In dimensionless form the functional dependence may now be expressed as

$$\theta^* = f(x^*, Fo, Bi) \tag{5.26}$$

Recall that this functional dependence, without the x^* variation, was obtained for the lumped capacitance method, as shown in Equation 5.13.

Comparing Equations 5.18 and 5.26 reveals the considerable advantage associated with casting the problem in dimensionless form. Equation 5.26 implies that, *for a prescribed geometry, the transient temperature distribution is a universal function of x^*, Fo, and Bi*. That is, the *dimensionless solution* assumes a prescribed form that does not depend on the particular value of T_i, T_∞, L, k, α, or h. Since this generalization greatly simplifies the presentation and utilization of transient solutions, the dimensionless variables will be used extensively in subsequent sections.

5.4 ONE-DIMENSIONAL SYSTEMS WITH CONVECTIVE SURFACE CONDITIONS

Analytical solutions to transient conduction problems have been developed for many simplified geometries and boundary conditions [1–4]. These solutions are obtained in much the same way as the solution for two-dimensional, steady-state conditions (Appendix C), and they generally involve lengthy mathematical series. Fortunately, calculations based on these solutions have been performed, and the results have been presented in a convenient graphical form for three important geometries.

The plane wall or *infinite slab* is an approximation to a rectangular solid for which one dimension (its thickness) is small relative to the other two dimensions. Such is the case for the plane wall of thickness $2L$ in Figure 5.5. Conduction is principally in the x direction, and if constant properties are assumed, Equation 5.14 becomes the appropriate form of the heat equation. If the solid is initially at a uniform temperature, $T(x,0) = T_i$, and is suddenly immersed in a fluid of $T_\infty \neq T_i$, the resulting temperatures may be obtained by solving Equation 5.22 subject to the conditions of Equations 5.23 to 5.25. Since the convection conditions for the surfaces at $x^* = \pm 1$ are the same, the temperature distribution at any instant must be symmetrical about the midplane ($x^* = 0$).

The foregoing problem has been solved, and graphical results have been presented by Heisler [5] and Gröber et al. [6]. The results are summarized in Figures 5.6 to 5.8. Note that, as anticipated from the discussion of the preceding section, the results are presented in terms of the Biot and Fourier numbers. Figure 5.6 relates the temperature *at the midplane* of the wall, T_o, to time t during the transient process. Figure 5.7 relates the temperature at any point *off the midplane* to the value of T_o. The absence of Fo from Figure 5.7 implies that the time dependence of the temperature for any point off the midplane is uniquely related to the time dependence of the midplane temperature. Hence Figure 5.7 must be used in conjunction with Figure 5.6. For example, if one wishes to determine the surface temperature ($x^* = \pm 1$) at some time t, Figure 5.6 would first be used to determine T_o at t. Figure 5.7 would then be used to determine the surface temperature from knowledge of T_o. Note that the procedure could be inverted if the problem were one of determining the time required for the surface to reach a prescribed temperature.

The variation of the surface temperature gradient with time may also be obtained from the solution. This variation may, in turn, be used to obtain the total amount of energy, Q, which has either entered or left the wall up to any time t during the transient process. This quantity is equal to the change in internal energy for the wall. The results are presented in Figure 5.8. Note that the quantity Q_o is defined as

$$Q_o \equiv \rho c V (T_i - T_\infty) = \rho c V \theta_i \tag{5.27}$$

and may be interpreted as the initial internal energy of the wall relative to the fluid temperature.

Note that, because the mathematical problem is precisely the same, the foregoing results may also be applied to a plane wall of thickness L, which is *insulated on one side* ($x^* = 0$) and experiences convective transport on the other side ($x^* = +1$). This equivalence is a consequence of the fact that, regardless of whether a symmetrical or an adiabatic requirement is prescribed at $x^* = 0$, the boundary condition is of the form $\partial \theta^* / \partial x^* = 0$.

Results that are similar to those of Figures 5.6 to 5.8 may also be obtained for an infinite cylinder and a sphere. The infinite cylinder is an idealization for real cylinders having $(L/r_o) \gtrsim 10$, which permits the assumption of one-dimensional conduction in the radial direction. Graphical results for the infinite cylinder are presented in Figures 5.9 to 5.11, and those for the sphere are presented in Figures 5.12 to 5.14. Note that, with respect to the use of these figures, the Biot number is defined in terms of r_o. In contrast recall that, for the lumped capacitance method, the characteristic length in the Biot number is $r_o/2$ for the cylinder and $r_o/3$ for the sphere.

In closing it should be noted that the Heisler charts may also be used to determine the transient response of a plane wall, an infinite cylinder or sphere

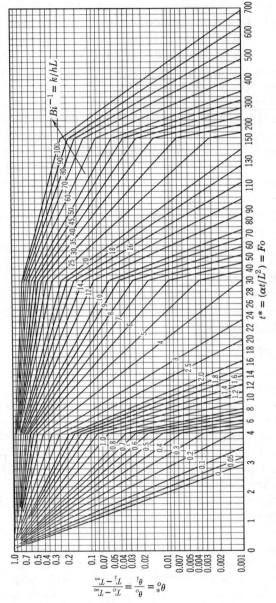

Figure 5.6   Midplane temperature as a function of time for a plane wall of thickness $2L$ [5]. Used with permission.

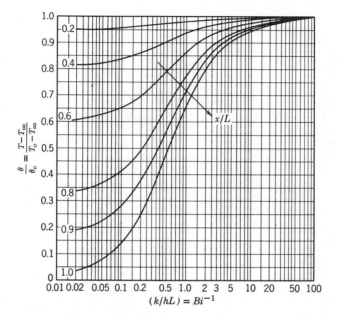

Figure 5.7 Temperature distribution in a plane wall of thickness 2L [5]. Used with permission.

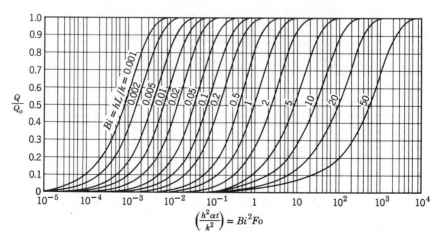

Figure 5.8 Internal energy change as function of time for a plane wall of thickness 2L [6]. Adapted with permission.

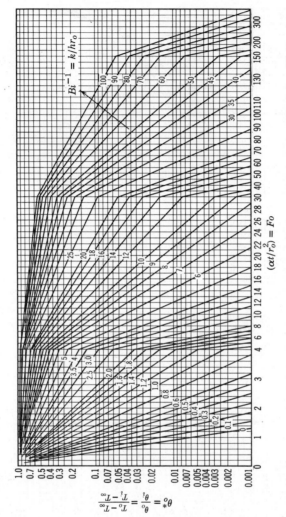

Figure 5.9 Centerline temperature as a function of time for an infinite cylinder of radius r_o [5]. Used with permission.

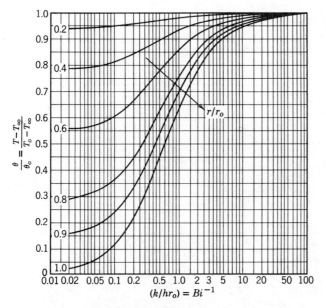

Figure 5.10 Temperature distribution in an infinite cylinder of radius r_o [5]. Used with permission.

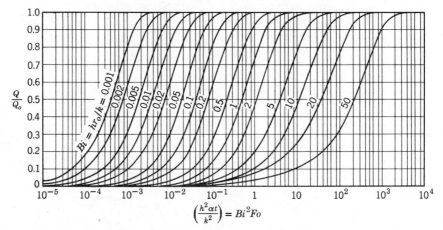

Figure 5.11 Internal energy change as a function of time for an infinite cylinder of radius r_o [6]. Adapted with permission.

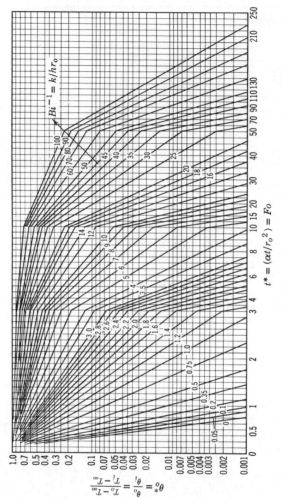

Figure 5.12 Center temperature as a function of time in a sphere of radius r_o [5]. Used with permission.

To - center line TEMP

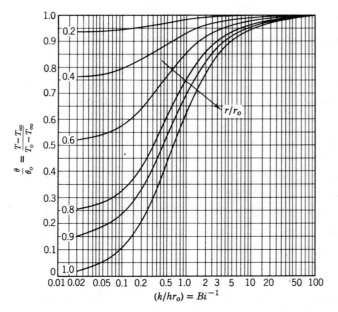

Figure 5.13 Temperature distribution in a sphere of radius r_o [5]. Used with permission.

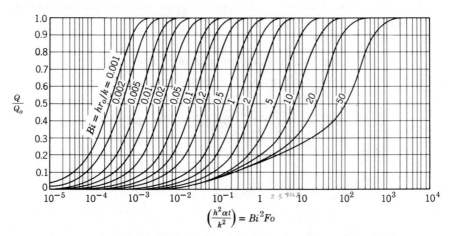

Figure 5.14 Internal energy change as a function of time for a sphere of radius r_o [6]. Adapted with permission.

subjected to a *sudden change in surface temperature*. For such a condition it is only necessary to replace T_∞ by the prescribed surface temperature, T_s, and to set Bi^{-1} equal to zero. In so doing the convection coefficient is tacitly assumed to be infinite, in which case $T_\infty = T_s$.

EXAMPLE 5.2

Consider a steel pipeline, AISI 1010, which is 1 m in diameter and has a wall thickness of 40 mm. The pipe is heavily insulated on the outside, and before the initiation of flow, the walls of the pipe are at a uniform temperature of $-20°C$. With the initiation of flow, hot oil at 60°C is pumped through the pipe creating a convective surface condition corresponding to $h = 500 \text{ W/m}^2 \cdot \text{K}$ at the inner surface of the pipe.

1. What are the appropriate Biot and Fourier numbers 8 min after the initiation of flow?
2. At $t = 8$ min, what is the temperature of the exterior pipe surface covered by the insulation?
3. What is the heat flux, q'' (W/m^2), to the pipe from the oil at $t = 8$ min?
4. How much energy per meter of pipe length has been transferred from the oil to the pipe at $t = 8$ min?

SOLUTION

KNOWN:

Pipeline wall subjected to sudden change in convective surface condition.

FIND:

1. Biot and Fourier numbers after 8 min.
2. Temperature of exterior pipe surface after 8 min.
3. Heat flux to the wall at 8 min.
4. Energy transferred to pipe per unit length after 8 min.

SCHEMATIC:

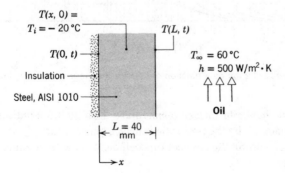

ASSUMPTIONS:

1. Pipe wall can be approximated as plane wall since thickness is much less than diameter.
2. Constant properties.
3. Outer surface of pipe is adiabatic.

PROPERTIES:

Table A.1, steel type AISI 1010 ($T = \dfrac{-20 + 60}{2}$ °C ≈ 300 K): $\rho = 7823$ kg/m^3, $c = 434$ J/kg·K, $k = 63.9$ W/m·K, $\alpha = 18.8 \times 10^{-6}$ m^2/s.

ANALYSIS:

1. At $t = 8$ min, the Biot and Fourier numbers are computed from Equations 5.10 and 5.12, respectively, with $L_c = L$. Hence

$$Bi = \frac{hL}{k} = \frac{500 \text{ W/m}^2 \cdot \text{K} \times 0.04 \text{ m}}{63.9 \text{ W/m} \cdot \text{K}} = 0.313 \qquad \lhd$$

$$Fo = \frac{\alpha t}{L^2} = \frac{18.8 \times 10^{-6} \text{ m}^2/\text{s} \times 8 \text{ min} \times 60 \text{ s/min}}{(0.04 \text{ m})^2} = 5.64 \qquad \lhd$$

2. With $Bi = 0.313$, use of the lumped capacitance method is inappropriate. However, since transient conditions in the insulated pipe wall of thickness L correspond to those in a plane wall of thickness $2L$ experiencing the same surface condition, the desired results may be obtained from the Heisler and Gröber charts for the plane wall. Using the Heisler chart, Figure 5.6, with $Bi^{-1} = 3.2$,

$$\frac{\theta_o}{\theta_i} = \frac{T(0,t) - T_\infty}{T_i - T_\infty} \approx 0.24$$

Hence after 8 min, the temperature of the exterior pipe surface, which corresponds to the midplane temperature of a plane wall, is

$$T_o = T(0, 480 \text{ s}) \approx T_\infty + 0.24(T_i - T_\infty)$$

$$T_o = 60°\text{C} + 0.24 \ (-20 - 60)°\text{C}$$

$$T_o \approx 41°\text{C} \qquad \lhd$$

3. Heat transfer to the inner surface at $x = L$ is by convection, and at any time t the heat flux may be obtained from Newton's law of cooling. Hence at $t = 480$ s,

$$q_L'' = h[T \ (L, \ 480 \text{ s}) - T_\infty]$$

The surface temperature, $T(L, 480\,s)$, may be obtained from the Heisler chart of Figure 5.7. For the prescribed conditions

$$\frac{x}{L} = 1 \qquad Bi^{-1} = 3.2$$

it follows that

$$\frac{\theta(L, 480\,s)}{\theta_o(480\,s)} = \frac{T(L, 480\,s) - T_\infty}{T_o(480\,s) - T_\infty} \approx 0.86$$

Hence

$$T(L, 480\,s) \approx T_\infty + 0.86\,[T_o(480\,s) - T_\infty]$$

$$T(L, 480\,s) \approx 60^\circ C + 0.86\,[41^\circ C - 60^\circ C] \approx 44^\circ C$$

The heat flux at $t = 6$ min is then

$$q_L'' = 500\ W/m^2 \cdot K\ (44 - 60)^\circ C$$

$$q_L'' = -8000\ W/m^2 \qquad \qquad \triangleleft$$

4. The energy transfer to the pipe wall over the 8-min interval may be obtained from the Gröber chart, Figure 5.8, and Equation 5.27. With

$$Bi = 0.313 \qquad Bi^2 Fo = 0.55$$

it follows that

$$\frac{Q}{Q_o} \approx 0.77$$

Hence

$$Q \approx 0.77\,\rho c\,V\,(T_i - T_\infty)$$

or with a volume per unit pipe length of $V' = \pi DL$,

$$Q' \approx 0.77\,\rho c\,\pi DL\,(T_i - T_\infty)$$

$$Q' \approx 0.77 \times 7823\ kg/m^3 \times 434\ J/kg \cdot K \times \pi \times 1\ m \times 0.04\ m\,(-20 - 60)^\circ C$$

$$Q' \approx -2.6 \times 10^7\ J/m \qquad \qquad \triangleleft$$

COMMENTS:

The minus sign associated with q'' and Q' simply implies heat transfer from the oil to the pipe.

EXAMPLE 5.3

A new process for treatment of a special material is to be evaluated. The material, a sphere of radius $r_o = 5$ mm, is initially in equilibrium at 400°C in a furnace. It is suddenly removed from the furnace and subjected to a two-step cooling process.

Step 1 Cooling in air at 20°C for a period of time, t_a, until the center temperature reaches a critical value, $T_a(0, t_a) = 335$°C. For this situation, the convective heat transfer coefficient is $h_a = 10$ W/m$^2 \cdot$K.

After the sphere reaches this critical temperature, the second step is initiated.

Step 2 Cooling in a well-stirred water bath at 20°C wherein the convective heat transfer coefficient is $h_w = 6000$ W/m$^2 \cdot$K.

The thermophysical properties of the material are

$\rho = 3000$ kg/m^3 $\qquad k = 20$ W/m$\cdot$K

$c = 1000$ J/kg$\cdot$K $\qquad \alpha = 6.66 \times 10^{-6}$ m^2/s

1. Calculate the time, t_a, required for step 1 of the cooling process to be completed.

2. Calculate the time, t_w, required during step 2 of the process for the center of the sphere to cool from 335°C (the condition at the completion of step 1) to 50°C.

SOLUTION

KNOWN:

Temperature requirements for a cooling process of a sphere.

FIND:

1. Time, t_a, required to accomplish desired cooling in air.
2. Time, t_w, required to complete cooling in water bath.

SCHEMATIC:

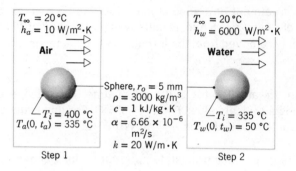

Step 1 — $T_\infty = 20\,°C$, $h_a = 10\ \text{W/m}^2\cdot\text{K}$, Air, $T_i = 400\,°C$, $T_a(0, t_a) = 335\,°C$

Sphere, $r_o = 5$ mm, $\rho = 3000\ \text{kg/m}^3$, $c = 1\ \text{kJ/kg}\cdot\text{K}$, $\alpha = 6.66 \times 10^{-6}\ \text{m}^2/\text{s}$, $k = 20\ \text{W/m}\cdot\text{K}$

Step 2 — $T_\infty = 20\,°C$, $h_w = 6000\ \text{W/m}^2\cdot\text{K}$, Water, $T_i = 335\,°C$, $T_w(0, t_w) = 50\,°C$

ASSUMPTIONS:

1. One-dimensional conduction in r.
2. Constant properties.

ANALYSIS:

1. To determine whether the lumped capacitance method can be used, the Biot number is calculated. From Equation 5.10, with $L_c = r_o/3$,

$$Bi = \frac{h_a r_o}{3k} = \frac{10\ \text{W/m}^2\cdot\text{K} \times 0.005\ \text{m}}{3 \times 20\ \text{W/m}\cdot\text{K}} = 8.33 \times 10^{-4}$$

Accordingly, the lumped capacitance method may be used, and the temperature is nearly uniform throughout the sphere. From Equation 5.5 it follows that

$$t_a = \frac{\rho V c}{h_a A_s} \ln\frac{\theta_i}{\theta_a} = \frac{\rho r_o c}{3 h_a} \ln\frac{T_i - T_\infty}{T_a - T_\infty}$$

where $V = (4/3)\pi r_o^3$ and $A_s = 4\pi r_o^2$. Hence

$$t_a = \frac{3000\ \text{kg/m}^3 \times 0.005\ \text{m} \times 1000\ \text{J/kg}\cdot\text{K}}{3 \times 10\ \text{W/m}^2\cdot\text{K}} \ln\frac{400 - 20}{335 - 20}$$

$$t_a = 94\ \text{s} \qquad \triangleleft$$

2. To determine whether the lumped capacitance method may also be used for the second step of the cooling process, the Biot number is again calculated. In this case

$$Bi = \frac{h_w r_o}{3k} = \frac{6000 \text{ W/m}^2 \cdot \text{K} \times 0.005 \text{ m}}{3 \times 20 \text{ W/m} \cdot \text{K}} = 0.50$$

and the lumped capacitance method is not appropriate. However, to an excellent approximation, the temperature of the sphere is uniform at $t = t_a$ and the Heisler charts may be used for the calculations from $t = t_a$ to $t = t_a + t_w$. Using Figure 5.12 with

$$Bi^{-1} = \frac{k}{h_w r_o} = \frac{20 \text{ W/m} \cdot \text{K}}{6000 \text{ W/m}^2 \cdot \text{K} \times 0.005 \text{ m}} = 0.67$$

$$\frac{\theta_o}{\theta_i} = \frac{T_o - T_\infty}{T_i - T_\infty} = \frac{50 - 20}{335 - 20} = 0.095$$

it follows that

$$Fo \approx 0.75$$

and

$$t_w = Fo \frac{r_o^2}{\alpha} \approx 0.75 \frac{(0.005 \text{ m})^2}{6.66 \times 10^{-6} \text{ m}^2/\text{s}}$$

$$t_w \approx 2.8 \text{ s} \qquad \triangleleft$$

COMMENTS:

1. Note that, if the temperature distribution in the sphere at the conclusion of step 1 were not uniform, the Heisler chart could not be used for the calculations of step 2.

2. The surface temperature of the sphere at the conclusion of step 2 may be obtained from Figure 5.13. With

$$Bi^{-1} = 0.67 \qquad \frac{r}{r_o} = 1$$

$$\frac{\theta(r_o)}{\theta_o} = \frac{T(r_o) - T_\infty}{T_o - T_\infty} \approx 0.52$$

Hence

$$T(r_o) \approx 20°\text{C} + 0.52(50 - 20)°\text{C} \approx 36°\text{C}$$

The variation of the center and surface temperature with time is then as follows.

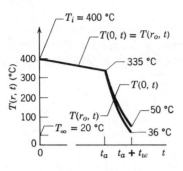

5.5 THE SEMIINFINITE SOLID

Another simple geometry for which analytical solutions may be obtained is the *semiinfinite solid*. Since such a solid extends to infinity in all but one direction, it is characterized by a single identifiable surface (Figure 5.15). If a sudden change of conditions is imposed at this surface, transient, one-dimensional conduction

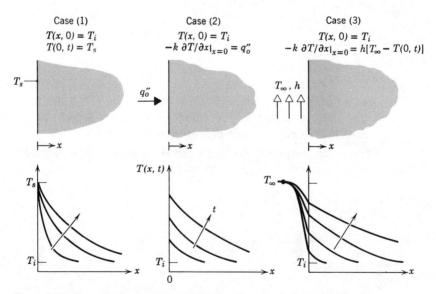

Figure 5.15 Transient temperature distributions in a semiinfinite solid for three surface conditions: constant surface temperature, constant surface heat flux, and surface convection.

will occur within the solid. The semiinfinite solid provides a *useful idealization* for many practical problems. It may be used to determine transient heat transfer effects near the surface of the earth or to approximate the transient response of a finite solid, such as a thick slab. For this second situation the approximation would be reasonable for the early portion of the transient, during which temperatures in the slab interior (well removed from the surface) are uninfluenced by the change in surface conditions.

The heat equation for transient conduction in a semiinfinite solid is given by Equation 5.14. The initial condition is prescribed by Equation 5.15, and the interior boundary condition is of the form

$$T(\infty,t) = T_i \tag{5.28}$$

Closed-form solutions have been obtained for three important surface conditions, instantaneously applied at $t = 0$ [1,2]. These conditions are shown in Figure 5.15. They include application of a constant surface temperature, $T_s \neq T_i$; application of a constant surface heat flux, q_o''; and exposure of the surface to a fluid characterized by $T_\infty \neq T_i$ and the convection coefficient h. The solutions are summarized as follows:

Case 1   Constant Surface Temperature

$$T(0,t) = T_s \tag{5.29}$$

$$\frac{T(x,t) - T_s}{T_i - T_s} = \operatorname{erf}\left(\frac{x}{2\sqrt{\alpha t}}\right) \tag{5.30}$$

$$q_s''(t) = -k\left.\frac{\partial T}{\partial x}\right|_{x=0} = \frac{k(T_s - T_i)}{\sqrt{\pi \alpha t}} \tag{5.31}$$

Case 2   Constant Surface Heat Flux

$$q_s'' = q_o'' \tag{5.32}$$

$$T(x,t) - T_i = \frac{2q_o''(\alpha t/\pi)^{1/2}}{k}\exp\left(\frac{-x^2}{4\alpha t}\right) - \frac{q_o''x}{k}\operatorname{erfc}\left(\frac{x}{2\sqrt{\alpha t}}\right)$$

$$\tag{5.33}$$

Case 3   Surface Convection

$$-k\,\partial T/\partial x|_{x=0} = h[T_\infty - T(0,t)] \tag{5.34}$$

$$\frac{T(x,t) - T_i}{T_\infty - T_i} = \operatorname{erfc}\left(\frac{x}{2\sqrt{\alpha t}}\right)$$

$$-\left[\exp\left(\frac{hx}{k} + \frac{h^2\alpha t}{k^2}\right)\right]\left[\operatorname{erfc}\left(\frac{x}{2\sqrt{\alpha t}} + \frac{h\sqrt{\alpha t}}{k}\right)\right] \tag{5.35}$$

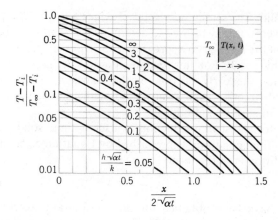

Figure 5.16 Temperature histories in a semiinfinite solid with surface convection [2]. Adapted with permission.

The quantity erf w appearing in Equation 5.30 is the *Gaussian error function*, which is tabulated in Section B.2 of Appendix B. The *complementary error function*, erfc w, is defined as

$$\text{erfc } w \equiv 1 - \text{erf } w$$

Temperature histories for the three cases are also shown in Figure 5.15. Carefully note the distinguishing features of these histories. For Case 3 the specific temperature histories computed from Equation 5.35 are plotted in Figure 5.16. Note that the curve corresponding to $h = \infty$ is equivalent to the result that would be obtained for a sudden change in the *surface temperature* to $T_s = T_\infty$. That is, for $h = \infty$ the second term on the right-hand side of Equation 5.35 goes to zero, and the result is equivalent to Equation 5.30.[1]

EXAMPLE 5.4

In laying water mains, utilities must be concerned with the possibility of freezing during cold periods of the year. Although the problem of determining the temperature in soil as a function of time is complicated by changing surface conditions, it is possible to make some reasonable estimates on the basis of assuming a constant surface temperature over a prolonged period of cold weather.

What is the minimum burial depth, x_m, which you would recommend to avoid freezing under conditions for which soil, initially at a uniform temperature of 20°C, is subjected to a constant surface temperature of -15°C for a period of 60 days?

---

[1] Note that $\dfrac{T - T_i}{T_\infty - T_i} = 1 - \dfrac{T - T_\infty}{T_i - T_\infty} = 1 - \theta^*$.   (5.36)

SOLUTION

KNOWN:

Temperature below freezing imposed at the surface of soil that is initially at 20°C.

FIND:

The depth, x_m, to which the soil has frozen after 60 days.

SCHEMATIC:

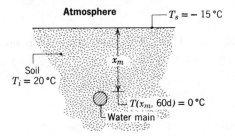

ASSUMPTIONS:

1. One-dimensional conduction in x.
2. Soil is a semiinfinite medium.
3. Constant properties.

PROPERTIES:

Table A.3, soil (300 K): $\rho = 2050$ kg/m^3, $k = 0.52$ W/m·K, $c = 1840$ J/kg·K, $\alpha = (k/\rho c) = 0.138 \times 10^{-6}$ m^2/s.

ANALYSIS:

The prescribed conditions correspond to those of Case 1 of Figure 5.15, and the transient temperature response of the soil is governed by Equation 5.30. Hence at the time $t = 60$ d following the surface temperature change,

$$\frac{T(x_m, t) - T_s}{T_i - T_s} = \text{erf}\left(\frac{x_m}{2\sqrt{\alpha t}}\right)$$

or

$$\frac{0 - (-15)}{20 - (-15)} = 0.429 = \mathrm{erf}\left(\frac{x_m}{2\sqrt{\alpha t}}\right)$$

Hence from Table B.2:

$$\frac{x_m}{2\sqrt{\alpha t}} = 0.40$$

and

$$x_m = 0.80\sqrt{\alpha t} = 0.80(0.138 \times 10^{-6} \text{ m}^2/\text{s} \times 60 \text{ d} \times 24 \text{ h/d} \times 3600 \text{ s/h})^{1/2}$$

$$x_m = 0.68 \text{ m} \qquad \triangleleft$$

COMMENTS:

Note that the properties of soil are highly variable, depending on the nature of the soil and its moisture content.

5.6 MULTIDIMENSIONAL EFFECTS

Transient problems are frequently encountered for which two- and even three-dimensional effects are significant. Solution to a class of such problems can be obtained from the one-dimensional results of Sections 5.4 and 5.5.

Consider immersing the *short* cylinder of Figure 5.17, which is initially at a uniform temperature T_i, in a fluid of temperature $T_\infty \neq T_i$. Because the length and diameter are comparable, the subsequent transfer of energy by conduction will be significant for both the r and x coordinate directions. The temperature within the cylinder will therefore depend on r, x, and t.

Assuming constant properties and no generation, the appropriate form of

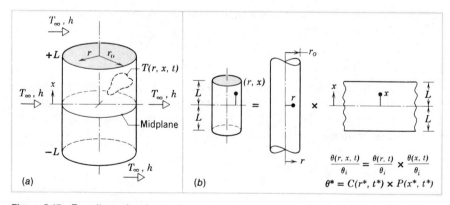

Figure 5.17 Two-dimensional, transient conduction in a short cylinder. (a) Geometry. (b) Form of the product solution.

the heat equation is, from Equation 2.18,

$$\frac{1}{r}\frac{\partial}{\partial r}\left(r\frac{\partial T}{\partial r}\right) + \frac{\partial^2 T}{\partial x^2} = \frac{1}{\alpha}\frac{\partial T}{\partial t}$$

where x has been used in place of z to designate the axial coordinate. A closed form solution to this equation may be obtained by the separation of variables method. Although we will not consider the details of this solution, it is important to note that the end result may be expressed in the following form.

$$\frac{T(r,x,t) - T_\infty}{T_i - T_\infty} = \left.\frac{T(x,t) - T_\infty}{T_i - T_\infty}\right|_{\substack{\text{Plane}\\\text{wall}}} \cdot \left.\frac{T(r,t) - T_\infty}{T_i - T_\infty}\right|_{\text{Infinite cylinder}}$$

That is, the two-dimensional solution may be expressed as a *product* of one-dimensional solutions that correspond to those for a plane wall of thickness $2L$ and an infinite cylinder of radius r_o. These solutions are available from Figures 5.6 and 5.7 for the plane wall and Figures 5.9 and 5.10 for the infinite cylinder.

Results for other multidimensional geometries are summarized in Figure 5.18. In each case the multidimensional solution is prescribed in terms of a product involving one or more of the following one-dimensional solutions.

$$S(x,t) \equiv \left.\frac{T(x,t) - T_\infty}{T_i - T_\infty}\right|_{\substack{\text{Semiinfinite}\\\text{solid}}} \tag{5.37}$$

$$P(x,t) \equiv \left.\frac{T(x,t) - T_\infty}{T_i - T_\infty}\right|_{\substack{\text{Plane}\\\text{wall}}} \tag{5.38}$$

$$C(r,t) \equiv \left.\frac{T(r,t) - T_\infty}{T_i - T_\infty}\right|_{\substack{\text{Infinite}\\\text{cylinder}}} \tag{5.39}$$

The x coordinate for the semiinfinite solid is measured from the surface, whereas for the plane wall it is measured from the midplane. In using Figure 5.18 the coordinate origins should be carefully noted. The transient, three-dimensional temperature distribution in a rectangular parallelepiped, Figure 5.18h, is then, for example, the product of three one-dimensional solutions for plane walls of thicknesses $2L_1$, $2L_2$, and $2L_3$. That is,

$$\frac{T(x_1,x_2,x_3,t) - T_\infty}{T_i - T_\infty} = P(x_1,t) \cdot P(x_2,t) \cdot P(x_3,t)$$

The distances x_1, x_2, and x_3 are all measured with respect to a rectangular coordinate system whose origin is at the center of the parallelepiped.

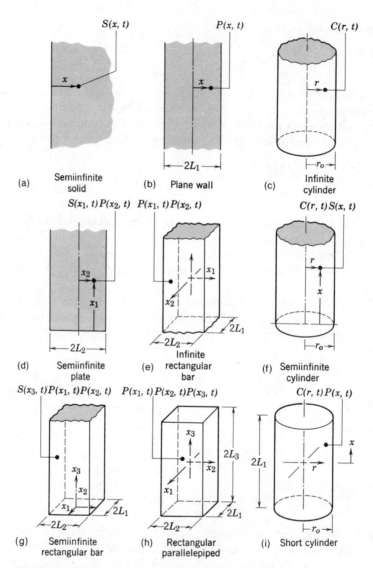

Figure 5.18 Solutions for multidimensional systems expressed as products of one-dimensional results.

EXAMPLE 5.5

In a manufacturing process stainless steel cylinders (AISI 304) initially at 600 K are quenched by submerging them in an oil bath maintained at 300 K with $h = 500$ W/m$^2 \cdot$K. Each cylinder is of length $2L = 60$ mm and diameter $D = 80$ mm. Consider a time 3 min into the cooling process and determine temperatures at the center of the cylinder, at the center of a circular face, and at the midplane of the side.

SOLUTION

KNOWN:

Initial temperature and dimensions of cylinder and temperature and convection conditions of an oil bath.

FIND:

Temperatures, $T(r,x,t)$, after 3 min at the cylinder center, $T(0,0,3\,\text{min})$, at the center of a circular face, $T(0,L,3\,\text{min})$, and at the midplane of the side, $T(r_o,0,3\,\text{min})$.

SCHEMATIC:

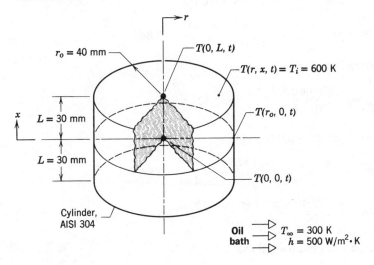

$r_o = 40\ \text{mm}$

$T(0, L, t)$

$T(r, x, t) = T_i = 600\ \text{K}$

x

$L = 30\ \text{mm}$

$T(r_o, 0, t)$

$L = 30\ \text{mm}$

$T(0, 0, t)$

Cylinder, AISI 304

Oil bath

$T_\infty = 300\ \text{K}$
$h = 500\ \text{W/m}^2 \cdot \text{K}$

ASSUMPTIONS:

1. Two-dimensional conduction in r and x.
2. Constant properties.

PROPERTIES:

Table A.1, stainless steel, AISI 304 [$T = (600 + 300)/2 = 450\,\text{K}$]: $\rho = 7900$ kg/m^3, $c = 526$ J/kg·K, $k = 17.4$ W/m·K, $\alpha = k/\rho c = 4.19 \times 10^{-6}\,\text{m}^2/\text{s}$.

ANALYSIS:

The solid steel cylinder corresponds to Case (i) of Figure 5.18, and the temperature at any point in the cylinder may be expressed as the following product of one-dimensional solutions.

$$\frac{T(r,x,t) - T_\infty}{T_i - T_\infty} = P(x,t)C(r,t)$$

where $P(x,t)$ and $C(r,t)$ are defined by Equations 5.38 and 5.39, respectively. Accordingly, for the center of the cylinder,

$$\frac{T(0,0,3\,\text{min}) - T_\infty}{T_i - T_\infty} = \frac{T(0,3\,\text{min}) - T_\infty}{T_i - T_\infty}\bigg|_{\substack{\text{Plane} \\ \text{wall}}} \cdot \frac{T(0,3\,\text{min}) - T_\infty}{T_i - T_\infty}\bigg|_{\substack{\text{Infinite} \\ \text{cylinder}}}$$

Hence, for the plane wall, with

$$Bi^{-1} = \frac{k}{hL} = \frac{17.4\,\text{W/m}\cdot\text{K}}{500\,\text{W/m}^2\cdot\text{K} \times 0.03\,\text{m}} = 1.16$$

$$Fo = \frac{\alpha t}{L^2} = \frac{4.19 \times 10^{-6}\,\text{m}^2/\text{s} \times 180\,\text{s}}{(0.03\,\text{m})^2} = 0.84$$

it follows from Figure 5.6 that

$$\frac{\theta_o}{\theta_i} = \frac{T(0,3\,\text{min}) - T_\infty}{T_i - T_\infty}\bigg|_{\substack{\text{Plane} \\ \text{wall}}} \approx 0.60$$

Similarly, for the infinite cylinder, with

$$Bi^{-1} = \frac{k}{hr_o} = \frac{17.4\,\text{W/m}\cdot\text{K}}{500\,\text{W/m}^2\cdot\text{K} \times 0.04\,\text{m}} = 0.87$$

$$Fo = \frac{\alpha t}{r_o^2} = \frac{4.19 \times 10^{-6}\,\text{m}^2/\text{s} \times 180\,\text{s}}{(0.04\,\text{m})^2} = 0.47$$

it follows from Figure 5.9 that

$$\frac{\theta_o}{\theta_i} = \frac{T(0,3\,\text{min}) - T_\infty}{T_i - T_\infty}\bigg|_{\substack{\text{Infinite} \\ \text{cylinder}}} \approx 0.48$$

Hence, for the center of the cylinder,

$$\frac{T(0,0,3\,\text{min}) - T_\infty}{T_i - T_\infty} \approx 0.60 \times 0.48 \approx 0.29$$

$$T(0,0,3\,\text{min}) \approx 300\,\text{K} + 0.29(600 - 300)\,\text{K}$$

$T(0, 0, 3 \text{ min}) \approx 386 \text{ K}$ ◁

The temperature at the center of a circular face may be obtained from the requirement that

$$\frac{T(0, L, 3 \text{ min}) - T_{\infty}}{T_i - T_{\infty}} = \left.\frac{T(L, 3 \text{ min}) - T_{\infty}}{T_i - T_{\infty}}\right|_{\substack{\text{Plane} \\ \text{wall}}} \cdot \left.\frac{T(0, 3 \text{ min}) - T_{\infty}}{T_i - T_{\infty}}\right|_{\substack{\text{Infinite} \\ \text{cylinder}}}$$

where, from Figure 5.7 with $(x/L) = 1$ and $Bi^{-1} = 1.16$,

$$\frac{\theta(L)}{\theta_o} = \left.\frac{T(L, 3 \text{ min}) - T_{\infty}}{T(0, 3 \text{ min}) - T_{\infty}}\right|_{\substack{\text{Plane} \\ \text{wall}}} \approx 0.66$$

Hence

$$\left.\frac{T(L, 3 \text{ min}) - T_{\infty}}{T_i - T_{\infty}}\right|_{\substack{\text{Plane} \\ \text{wall}}} = \left.\frac{T(L, 3 \text{ min}) - T_{\infty}}{T(0, 3 \text{ min}) - T_{\infty}}\right|_{\substack{\text{Plane} \\ \text{wall}}} \cdot \left.\frac{T(0, 3 \text{ min}) - T_{\infty}}{T_i - T_{\infty}}\right|_{\substack{\text{Plane} \\ \text{wall}}}$$

$$\left.\frac{T(L, 3 \text{ min}) - T_{\infty}}{T_i - T_{\infty}}\right|_{\substack{\text{Plane} \\ \text{wall}}} \approx 0.66 \times 0.60 \approx 0.40$$

Hence

$$\frac{T(0, L, 3 \text{ min}) - T_{\infty}}{T_i - T_{\infty}} \approx 0.40 \times 0.48 \approx 0.19$$

$$T(0, L, 3 \text{ min}) \approx 300 \text{ K} + 0.19(600 - 300) \text{ K}$$

$$T(0, L, 3 \text{ min}) \approx 357 \text{ K} \qquad ◁$$

The temperature at the midplane of the side may be obtained from the requirement that

$$\frac{T(r_o, 0, 3 \text{ min}) - T_{\infty}}{T_i - T_{\infty}} = \left.\frac{T(0, 3 \text{ min}) - T_{\infty}}{T_i - T_{\infty}}\right|_{\substack{\text{Plane} \\ \text{wall}}} \cdot \left.\frac{T(r_o, 3 \text{ min}) - T_{\infty}}{T_i - T_{\infty}}\right|_{\substack{\text{Infinite} \\ \text{cylinder}}}$$

where, from Figure 5.10 with $(r/r_o) = 1$ and $Bi^{-1} = 0.87$,

$$\frac{\theta(r_o)}{\theta_o} = \left.\frac{T(r_o, 3 \text{ min}) - T_{\infty}}{T(0, 3 \text{ min}) - T_{\infty}}\right|_{\substack{\text{Infinite} \\ \text{cylinder}}} \approx 0.61$$

Hence

$$\left.\frac{T(r_o, 3 \text{ min}) - T_{\infty}}{T_i - T_{\infty}}\right|_{\substack{\text{Infinite} \\ \text{cylinder}}} = \left.\frac{T(r_o, 3 \text{ min}) - T_{\infty}}{T(0, 3 \text{ min}) - T_{\infty}}\right|_{\substack{\text{Infinite} \\ \text{cylinder}}} \cdot \left.\frac{T(0, 3 \text{ min}) - T_{\infty}}{T_i - T_{\infty}}\right|_{\substack{\text{Infinite} \\ \text{cylinder}}}$$

$$\left.\frac{T(r_o, 3\,\mathrm{min}) - T_\infty}{T_i - T_\infty}\right|_{\substack{\text{Infinite}\\ \text{cylinder}}} \approx 0.61 \times 0.48 \approx 0.29$$

Hence

$$\frac{T(r_o, 0, 3\,\mathrm{min}) - T_\infty}{T_i - T_\infty} \approx 0.60 \times 0.29 \approx 0.17$$

$$T(r_o, 0, 3\,\mathrm{min}) \approx 300\,\mathrm{K} + 0.17\,(600 - 300)\,\mathrm{K}$$

$$T(r_o, 0, 3\,\mathrm{min}) \approx 351\,\mathrm{K} \qquad\qquad \triangleleft$$

COMMENTS:

Verify that the temperature at the edge of the cylinder is $T(r_o, L, 3\,\mathrm{min}) \approx 335\,\mathrm{K}$.

5.7 FINITE-DIFFERENCE METHODS

Analytical solutions to transient problems are restricted to simple geometries and boundary conditions, such as those considered in the previous sections. Extensive coverage of these and other solutions is treated in the literature [1–4]. However, in many cases the geometry and/or boundary conditions preclude the use of analytical methods, and recourse must be made to *finite-difference* methods. Such methods, introduced in Section 4.4 for steady-state conditions, are readily extended to transient problems.

5.7.1 Discretization of the Heat Equation

Once again consider the two-dimensional system of Figure 4.4. Under transient conditions with constant properties and no internal generation, the appropriate form of the heat equation, Equation 2.15, is

$$\frac{\partial^2 T}{\partial x^2} + \frac{\partial^2 T}{\partial y^2} = \frac{1}{\alpha}\frac{\partial T}{\partial t} \tag{5.40}$$

Using the m and n subscripts, respectively, to designate the x and y positions of an *interior* node, the finite-difference forms of the second derivatives in Equation 5.40 may be expressed as

$$\left.\frac{\partial^2 T}{\partial x^2}\right|_{m,n} \approx \frac{T^p_{m+1,n} + T^p_{m-1,n} - 2T^p_{m,n}}{(\Delta x)^2} \tag{5.41}$$

$$\left.\frac{\partial^2 T}{\partial y^2}\right|_{m,n} \approx \frac{T^p_{m,n+1} + T^p_{m,n-1} - 2T^p_{m,n}}{(\Delta y)^2} \tag{5.42}$$

These expressions are forms of Equations 4.14 and 4.15, with the superscript p added to denote the time dependence of T. It is an integer that is used to designate a *discrete* point in time, where

$$t = p \, \Delta t \tag{5.43}$$

Hence just as a finite-difference solution restricts temperature determination to discrete points in space, it also restricts this determination to discrete points in time. Calculations are made at successive times, separated by the *time interval* Δt.

A finite-difference form of the time derivative in Equation 5.40 may be expressed as

$$\left. \frac{\partial T}{\partial t} \right|_{m,n} \approx \frac{T_{m,n}^{p+1} - T_{m,n}^{p}}{\Delta t} \tag{5.44}$$

where the numerator is a *forward difference* for the temperature change occurring at m,n from time $t(p)$ to $t + \Delta t(p + 1)$. Substituting Equations 5.41, 5.42, and 5.44 into 5.40, assuming a mesh for which $\Delta x = \Delta y$, and solving for $T_{m,n}^{p+1}$, we obtain

$$T_{m,n}^{p+1} = \frac{\alpha \Delta t}{(\Delta x)^2} (T_{m+1,n}^{p} + T_{m-1,n}^{p} + T_{m,n+1}^{p} + T_{m,n-1}^{p})$$

$$+ \left[1 - \frac{4\alpha \Delta t}{(\Delta x)^2} \right] T_{m,n}^{p} \tag{5.45}$$

If the system is one-dimensional in x, the appropriate form of the finite-difference equation is, from Equations 5.40, 5.41, and 5.44,

$$T_m^{p+1} = \frac{\alpha \Delta t}{(\Delta x)^2} (T_{m+1}^{p} + T_{m-1}^{p}) + \left[1 - \frac{2\alpha \Delta t}{(\Delta x)^2} \right] T_m^{p} \tag{5.46}$$

where the subscript n ceases to be relevant.

The foregoing expressions are particularly convenient. They enable determination of the temperature at a future time $t + \Delta t$ for any *interior node* $(m,n$ or $m)$ in terms of *known* temperatures at the preceding time t for $(m,n$ or $m)$ and its adjoining nodes. The temperature of every node in the mesh will be known at $t = 0 (p = 0)$ from prescribed initial conditions. Hence Equation 5.45 or 5.46 could be applied to each interior node to determine its temperature at $t = \Delta t$ $(p = 1)$. With temperatures known for $t = \Delta t$, the appropriate finite-difference equation could then be applied at each interior node to determine its temperature at $t = 2\Delta t$ $(p = 2)$. In this way, the transient temperature distribution may be obtained by *marching out in time*, using intervals of Δt. Note that Equations 5.45 and 5.46 reduce to simpler forms if particular time intervals and spatial increments are selected. It is customary to select the spatial increments on the basis of geometrical considerations and desired accuracy. Once this selection is made, however, a value of Δt may be chosen such that

$$M_{1-D} \equiv \frac{(\Delta x)^2}{\alpha \Delta t} = 2 \qquad (5.47)$$

for the one-dimensional problem and

$$M_{2-D} \equiv \frac{(\Delta x)^2}{\alpha \Delta t} = 4 \qquad (5.48)$$

for the two-dimensional problem. In both cases M is a finite-difference version of the reciprocal of the Fourier number ($M = Fo^{-1}$). The finite-difference forms of the heat equation then reduce to

$$T_m^{p+1} = \frac{1}{2}(T_{m+1}^p + T_{m-1}^p) \qquad (5.49)$$

for one-dimensional conduction and

$$T_{m,n}^{p+1} = \frac{1}{4}(T_{m+1,n}^p + T_{m-1,n}^p + T_{m,n+1}^p + T_{m,n-1}^p) \qquad (5.50)$$

for two-dimensional conduction. According to Equations 5.49 and 5.50, the temperature of an interior node at $t + \Delta t$ is simply an arithmetic average of the temperatures of the neighboring nodes at the preceding time t. These equations are particularly convenient if the transient solution must be obtained by hand calculations.

If the solution is obtained on a digital computer, there is no special advantage to using Equation 5.47 or 5.48. The selection of Δx and Δt would be motivated primarily by the desired accuracy. However, this selection cannot be made arbitrarily. For certain values of M, the finite-difference solution is characterized by a behavior mode known as *instability*. In this mode the value of the temperature oscillates as the solution progresses from one time to the next. The magnitude of the oscillation increases, and the results do not converge to the actual solution. However, it may be shown [7], that a stable solution is insured if the following conditions are satisfied.

$$\text{Stability criteria} \quad \begin{bmatrix} M_{1-D} \geq 2 \\ M_{2-D} \geq 4 \end{bmatrix} \qquad (5.51)$$

5.7.2 The Energy Balance Method

It is important to note that Equations 5.45 and 5.46 apply only to interior nodes. The finite-difference equation for any node may be obtained by formulating an energy balance for the node.

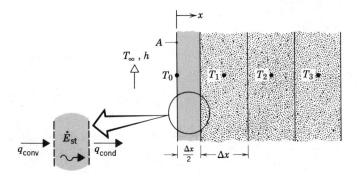

Figure 5.19   Surface node with convection and one-dimensional transient conduction.

Consider the surface node for the one-dimensional system of Figure 5.19. Note that this node has been assigned a thickness that is one-half that of the interior nodes. Although such an assignment is not necessary, it does allow for a more accurate determination of thermal conditions near the surface. Assuming convection transfer from an adjoining fluid to the surface node, the energy conservation requirement, Equation 1.10, may be expressed as

$$hA(T_\infty - T_0^p) - \frac{kA}{\Delta x}(T_0^p - T_1^p) = \rho c A \frac{\Delta x}{2} \frac{(T_0^{p+1} - T_0^p)}{\Delta t}$$

That is, for the time t the rate at which energy is transferred to the surface node by convection minus the rate at which energy leaves this node by conduction must equal the rate of change of internal energy for the node. Solving for T_0^{p+1}, the surface temperature at $t + \Delta t$, we then obtain

$$T_0^{p+1} = \frac{2}{M}(T_1^p + N T_\infty) + \left(1 - \frac{2}{M} - \frac{2N}{M}\right) T_0^p \qquad (5.52)$$

where

$$M \equiv \frac{(\Delta x)^2}{\alpha \Delta t}$$

and

$$N \equiv \frac{h\Delta x}{k}$$

Note that N is a finite-difference version of the Biot number. To insure the stability of a numerical solution involving Equation 5.52, it is necessary that

$$M \geq 2(N + 1)$$

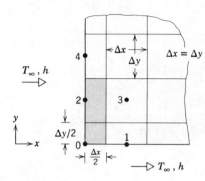

Figure 5.20 Surface and corner nodes with convection and two-dimensional transient conduction.

In a similar fashion, finite-difference equations may be obtained for the surface nodes of a two-dimensional system (Figure 5.20). For the corner node it is easily verified that

$$T_0^{p+1} = \frac{2}{M}(T_1^p + T_2^p) + \frac{4N}{M} T_\infty + \left(1 - \frac{4}{M} - \frac{4N}{M}\right) T_0^p \tag{5.53}$$

The stability criterion for this expression is

$$M \geq 4(N + 1)$$

For the surface node at 2, it may also be shown that the appropriate finite-difference equation is

$$T_2^{p+1} = \frac{1}{M}(T_0^p + T_4^p + 2T_3^p) + \frac{2N}{M} T_\infty + \left(1 - \frac{2N}{M} - \frac{4}{M}\right) T_2^p \tag{5.54}$$

where stability is insured if

$$M \geq 2(N + 2)$$

5.7.3 Additional Considerations

It is appropriate to note that all of the foregoing results are part of an *explicit* finite-difference scheme. In such a scheme the temperature of a node at $t + \Delta t$ may be calculated explicitly in terms of temperatures at the same and neighboring nodes for the *preceding time t*. Hence, determination of a nodal temperature at some time is *independent* of temperatures at other nodes for the *same time*. This facility to solve for a temperature at one time in terms of temperatures at preceding times results from use of a *forward difference* representation for the time derivative. The method offers computational convenience, yet as we noted in previous paragraphs, it suffers from limitations on the selection of Δx and Δt. For a given space increment, the time interval must be compatible with stability requirements. Frequently, this dictates the use of extremely small values of Δt, and an inordinate number of time intervals may be necessary to obtain a solution.

A reduction in the amount of computational time may generally be realized by employing an *implicit*, rather than explicit, finite-difference scheme. Implicit finite-difference equations are obtained by using a *backward difference* for the time derivative and differ from the explicit equations developed in the preceding paragraphs.

Consider the interior point 1 of the one-dimensional system of Figure 5.19. Using the energy balance method with a backward-difference time scheme, the corresponding finite-difference equation has the form

$$k\frac{(T_0^{p+1} - T_1^{p+1})}{\Delta x} - k\frac{(T_1^{p+1} - T_2^{p+1})}{\Delta x} = \rho c \, \Delta x \frac{(T_1^{p+1} - T_1^p)}{\Delta t} \tag{5.55}$$

Note how all the temperatures correspond to the time $p + 1$, except for the temperature of the node at time p appearing in the storage term. Introducing the parameter $M \equiv (\Delta x)^2/\alpha \Delta t$, the finite-difference equation may be expressed as

$$-T_0^{p+1} + (2 + M) T_1^{p+1} - T_2^{p+1} = MT_1^p \tag{5.56}$$

It is apparent that this equation, which applies for node 1 and provides the basis for determining T_1^{p+1}, is no longer explicit. That is, it cannot be used to calculate the temperature of node 1 at $t + \Delta t$ in terms of temperatures at t. Equations of the foregoing form may be used for all interior nodes, and comparable equations may be developed for surface nodes by using the energy balance method.

The implicit equations are of a form that necessitates simultaneously solving (at each time) a set of algebraic equations equal in number to the total number of nodes. Now the situation is one in which the temperature of a node depends on the unknown temperatures of neighboring nodes for the same time. Hence, if a system is comprised of 100 nodes, a solution to 100 equations would have to be simultaneously obtained by using matrix inversion, Gaussian elimination, or some other solution scheme.

Although the implicit scheme is somewhat more complicated than the explicit method, it is not subject to the same instability problems, and hence does not have the foregoing restrictions on Δx and Δt. Larger values of Δt may generally be used with an implicit method, and for this reason the computation time is generally smaller. It is the method that is frequently preferred for solving involved transient diffusion problems, and detailed descriptions may be found in the literature [3, 7, 8, 9].

We conclude by noting that the explicit finite-difference technique may be implemented *graphically*. Such methods were introduced by Schmidt [10] and are discussed in some detail by Razelos [11] and Jacob [12]. With the advent of the digital computer, however, they are rarely used in modern heat transfer practice.

EXAMPLE 5.6

The fuel element of a nuclear reactor, which is in the shape of a plane wall of thickness 2L and is uniformly cooled on both surfaces, experiences a transient temperature variation whenever there is a change in the operating power (heat generation). This temperature variation may be predicted by using finite-difference procedures.

1. Develop a finite-difference equation that could be used to determine the transient response of the midplane temperature, T_0, to a sudden change in the power level. Uniform heat generation may be assumed.

2. If $L = 10$ mm, the coolant is characterized by $T_\infty = 250°C$ and $h = 1100$ W/m^2·K, and the fuel element has the following thermophysical properties,

$$k = 30 \text{ W/m·K} \qquad \alpha = 10^{-5} \text{ m}^2/\text{s}$$

what is the midplane temperature two seconds after the fuel element has gone from operation under steady-state conditions with a uniform generation of $\dot{q}_1 = 10^7$ W/m^3 to operation with $\dot{q}_2 = 10^8$ W/m^3? Use a time increment of $\Delta t = 2s$ for your calculations.

SOLUTION

KNOWN:

Conditions associated with heat generation in a rectangular fuel element with surface cooling.

FIND:

1. Finite-difference equation that governs transient behavior of mid-plane temperature.

2. Midplane temperature reached 2 s after a sudden change in operating power.

SCHEMATIC:

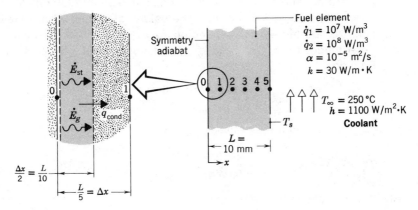

ASSUMPTIONS:

1. One-dimensional conduction in x.
2. Uniform generation.
3. Constant properties.

ANALYSIS:

1. Choosing the nodal network shown in the schematic, as a first approximation, and applying an energy balance, Equation 1.10, to the control volume associated with the midplane nodal region, it follows that

    $$\dot{E}_g - q_{cond} = \dot{E}_{st}$$

or

$$\dot{q}\left(A \times \frac{\Delta x}{2}\right) - kA\frac{(T_0^p - T_1^p)}{\Delta x} = \rho\left(A \times \frac{\Delta x}{2}\right)c\frac{(T_0^{p+1} - T_0^p)}{\Delta t}$$

Hence

$$T_0^{p+1} = T_0^p + \frac{\dot{q}\Delta t}{\rho c} - \frac{2\Delta t k}{(\Delta x)^2 \rho c}(T_0^p - T_1^p)$$

or

$$T_0^{p+1} = \frac{\alpha\Delta t}{(\Delta x)^2}\left[2T_1^p + \dot{q}\frac{(\Delta x)^2}{k}\right] + T_0^p\left[1 - 2\frac{\alpha\Delta t}{(\Delta x)^2}\right]$$

◁

2. The foregoing result may be used to determine the midplane temperature shortly following a change in the generation rate from $\dot{q}_1 = 10^7$ W/m^3 to $\dot{q}_2 = 10^8$ W/m^3. The values of T_0^p and T_1^p are steady-state temperatures corresponding to $\dot{q}_1 = 10^7$ W/m^3 and may be determined from the results of Section 3.4. In particular from Equation 3.45,

$$T_s = T_\infty + \frac{\dot{q}L}{h} = 250°\text{C} + \frac{10^7 \text{ W/m}^3 \times 0.01 \text{ m}}{1100 \text{ W/m}^2 \cdot \text{K}} = 340.9°\text{C}$$

From Equation 3.42 it then follows that

$$T_0^p = \frac{\dot{q}_1 L^2}{2k} + T_s = \frac{10^7 \text{ W/m}^3 (0.01 \text{ m})^2}{2 \times 30 \text{ W/m} \cdot \text{K}} + 340.9°\text{C} = 357.6°\text{C}$$

Also from Equation 3.43,

$$\frac{T_1^p - T_0^p}{T_s - T_0^p} = \left[\frac{(L/5)}{L}\right]^2$$

or

$$T_1^p = T_0^p + \frac{T_s - T_0^p}{25} = 357.6°\text{C} - 0.7°\text{C} = 356.9°\text{C}$$

With

$$\frac{\alpha \Delta t}{(\Delta x)^2} = \frac{10^{-5} \text{ m}^2/\text{s} \times 2 \text{ s}}{(0.002 \text{ m})^2} = 5.0$$

$$\frac{\dot{q}_2 (\Delta x)^2}{k} = \frac{10^8 \text{ W/m}^3 (0.002 \text{ m})^2}{30 \text{ W/m} \cdot \text{K}} = 13.33°\text{C}$$

it then follows that

$$T_0^{p+1} = 5[2 \times 356.9°\text{C} + 13.33°\text{C}] + 357.6°\text{C}[1 - 2 \times 5]$$

or

$$T_0^{p+1} = 417.3°\text{C} \qquad \triangleleft$$

COMMENTS:

The above calculations were performed without any regard for accuracy or the numerical stability criterion. In the interest of satisfying these requirements, a much smaller time increment and a much finer mesh should be used.

EXAMPLE 5.7

A thick slab of copper initially at a uniform temperature of 20°C is suddenly exposed to radiation at one surface such that the net heat flux is maintained at a constant value of 3×10^5 W/m^2 at the surface.

1. Using the finite-difference technique with a space increment of Δx = 75 mm, what is the temperature at the irradiated surface and at an interior point that is 150 mm from the surface after 2 min have elapsed?

2. Compare the foregoing numerical results with those obtained from an appropriate analytical solution.

SOLUTION

KNOWN:

Thick slab of copper, initially at a uniform temperature, is subjected to a constant net heat flux at one surface.

FIND:

1. Using the finite-difference method, determine temperatures at surface and 150 mm from surface after an elapsed time of 2 min.

2. Determine the same temperatures analytically.

SCHEMATIC:

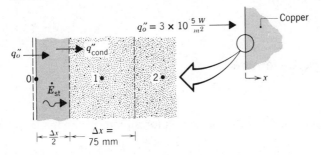

ASSUMPTIONS:

1. One-dimensional conduction in x.

2. Thick slab may be approximated as a semiinfinite medium with constant surface heat flux.

3. Constant properties.

PROPERTIES:

Table A.1, pure copper (300 K): $k = 401$ W/m·K, $\alpha = 117 \times 10^{-6}$ m^2/s.

ANALYSIS:

1. The finite-difference equation for the surface node may be determined by applying an energy balance to a control volume about this node. From Equation 1.10 it follows that

$$q_o'' A - q_{cond}'' A = \dot{E}_{st}$$

or

$$q_o'' - k \frac{T_0^p - T_1^p}{\Delta x} = \rho \frac{\Delta x}{2} c \frac{T_0^{p+1} - T_0^p}{\Delta t}$$

Rearranging

$$q_o'' \frac{2k}{\rho c} \frac{\Delta t}{(\Delta x)^2} \cdot \frac{\Delta x}{k} - \frac{k}{\rho c} \frac{\Delta t}{(\Delta x)^2} 2(T_0^p - T_1^p) = T_0^{p+1} - T_0^p$$

and choosing a time increment Δt, such that

$$M = \frac{(\Delta x)^2}{\alpha \Delta t} = 2$$

it follows that

$$T_0^{p+1} = T_1^p + q_o'' \frac{\Delta x}{k}$$

or

$$T_0^{p+1} = T_1^p + 3 \times 10^5 \text{ W/m}^2 \times \frac{0.075 \text{ m}}{401 \text{ W/m·K}}$$

$$T_0^{p+1} = T_1^p + 56.1°C \tag{1}$$

With $M = 2$ the finite-difference equation for all interior nodes is given by Equation 5.49.

$$T_m^{p+1} = \frac{1}{2}(T_{m+1}^p + T_{m-1}^p) \tag{2}$$

The value of Δt, which is appropriate for use with the above equations, is then

$$\Delta t = \frac{(\Delta x)^2}{M\alpha} = \frac{(0.075 \text{ m})^2}{2 \times 117 \times 10^{-6} \text{ m}^2/s} = 24 \text{ s}$$

With $t = p\Delta t$ the finite-difference calculations may now be performed by using Equation 1 for the surface node and Equation 2 for all interior nodes. The results are tabulated as follows.

p	$t(s)$	T_0	T_1	T_2	T_3	T_4
0	0	20	20	20	20	20
1	24	76.1	20	20	20	20
2	48	76.1	48.1	20	20	20
3	72	104.2	48.1	34.1	20	20
4	96	104.2	69.1	34.1	27.1	20
5	120	125.3	69.1	48.1	27.1	20

After 2 min the surface temperature and the desired interior temperature are therefore

$$T_0 \approx 125°C \qquad \lhd$$

$$T_2 \approx 48°C \qquad \lhd$$

2. Approximating the slab as a semiinfinite medium, the appropriate analytical expression is given by Equation 5.33, which may be applied to any point in the slab.

$$T(x,t) - T_i = \frac{2q_o''(\alpha t/\pi)^{1/2}}{k} \exp\left(-\frac{x^2}{4\alpha t}\right) - \frac{q_o''x}{k} \operatorname{erfc}\left(\frac{x}{2\sqrt{\alpha t}}\right)$$

At the surface, this expression yields

$$T(0, 120\,s) - 20°C = \frac{2 \times 3 \times 10^5 \, W/m^2}{401 \, W/m \cdot K} (117 \times 10^{-6} \, m^2/s \times 120 \, s/\pi)^{1/2}$$

or

$$T(0, 120\,s) - 20°C = 100.0°C$$

$$T(0, 120\,s) = 120.0°C \qquad \lhd$$

At the interior point ($x = 0.15\,m$)

$$T(0.15\,m, \ 120\,s) - 20°C = \frac{2 \times 3 \times 10^5 \, W/m^2}{401 \, W/m \cdot K}$$

$$\times (117 \times 10^{-6} \, m^2/s \times 120 \, s/\pi)^{1/2}$$

$$\times \exp\left[-\frac{(0.15\,\text{m})^2}{4 \times 117 \times 10^{-6}\,\text{m}^2/\text{s} \times 120\,\text{s}}\right] - \frac{3 \times 10^5\,\text{W/m}^2 \times 0.15\,\text{m}}{401\,\text{W/m}\cdot\text{K}}$$

$$\times \left[1 - \text{erf}\left(\frac{0.15\,\text{m}}{2\sqrt{117 \times 10^{-6}\,\text{m}^2/\text{s} \times 120\,\text{s}}}\right)\right]$$

or

$$T(0.15\,\text{m},\ 120\,\text{s}) - 20°\text{C} = 67.0°\text{C} - 41.6°\text{C}$$

$$T(0.15\,\text{m},\ 120\,\text{s}) = 45.4°\text{C} \qquad \triangleleft$$

COMMENTS:

1. The agreement between the finite-difference and analytical results is good, particularly since a coarse grid and a large time increment have been used for the finite-difference solution. The accuracy of this solution may be improved by reducing the values of Δx and Δt.

2. Calculation of identical temperatures at successive times for the same node is an idiosyncrasy of the explicit finite-difference technique. The actual physical condition is, of course, one in which the temperature changes continuously with time.

EXAMPLE 5.8

Consider the copper slab initially at a uniform temperature and suddenly subjected to a net heat flux as described in Example 5.7. Using the same grid size and time increment, find the required temperatures using the backward-difference, or implicit, finite-difference scheme.

SOLUTION

FIND:

Required temperatures using the implicit, finite-difference scheme.

ANALYSIS:

Referring to the schematic of Example 5.7 and using the energy balance method with a backward-difference time scheme, the finite-difference equation for node 0 has the form

$$q_o'' - k\frac{T_0^{p+1} - T_1^{p+1}}{\Delta x} = \rho\frac{\Delta x}{2}c\frac{T_0^{p+1} - T_0^p}{\Delta t}$$

Recognizing that $M = (\Delta x)^2/\alpha\Delta t$ and regrouping with the forward temperatures $(p + 1)$ on the left-hand side, the result is

$$\left(1 + \frac{M}{2}\right)T_0^{p+1} - T_1^{p+1} = q_o'' \frac{\Delta x}{k} + \frac{M}{2} T_0^p$$

Using the appropriate numerical values for M, q_o'', Δx, and k, the equation has the form

$$2T_0^{p+1} - T_1^{p+1} = 56.1 + T_0^p \tag{1}$$

The finite-difference equations for the interior nodes $(1,2,\ldots)$ are given by Equation 5.56 and comprise a set of equations having the form

$$-T_0^{p+1} + 4T_1^{p+1} - T_2^{p+1} \ldots\ldots\ldots\ldots\ldots\ldots\ldots = 2T_1^p \tag{2}$$

$$-T_1^{p+1} + 4T_2^{p+1} - T_3^{p+1} \ldots\ldots\ldots\ldots\ldots = 2T_2^p \tag{3}$$

$$-T_2^{p+1} + 4T_3^{p+1} - T_4^{p+1} \ldots\ldots\ldots = 2T_3^p \tag{4}$$

$$\vdots \qquad \vdots \qquad \vdots$$

$$-T_{i-1}^{p+1} + 4T_i^{p+1} - T_{i+1}^{p+1} \quad = 2T_i^p \tag{5}$$

Since we are dealing with a semiinfinite solid, the number of nodes is infinite. However, we can limit the set by prescribing the number of nodes N that are likely to be affected by the change in the boundary conditions within a time period of interest. Let us assume that for $N = 8$, the temperature at node 8 will not be affected by the sudden heat flux at the boundary after 120 s has elapsed. This means that $T_9^{p+1} = 20°C$ and the finite-difference equation for node 8 $(i = 8)$ is written as

$$-T_7^{p+1} + 4T_8^{p+1} = 2T_8^p + T_9^{p+1} \tag{6}$$

We now have a set of $N = 8$ equations that must be solved simultaneously for each time increment p. We chose to use the matrix inversion method and using the notation of Equation 4.33 we note the above finite-difference equations are of the form $[A][T] = [C]$ where

$$[A] = \begin{bmatrix} 2 & -1 & 0 & 0 & 0 & 0 & 0 & 0 \\ -1 & 4 & -1 & 0 & 0 & 0 & 0 & 0 \\ 0 & -1 & 4 & -1 & 0 & 0 & 0 & 0 \\ 0 & 0 & -1 & 4 & -1 & 0 & 0 & 0 \\ 0 & 0 & 0 & -1 & 4 & -1 & 0 & 0 \\ 0 & 0 & 0 & 0 & -1 & 4 & -1 & 0 \\ 0 & 0 & 0 & 0 & 0 & -1 & 4 & -1 \\ 0 & 0 & 0 & 0 & 0 & 0 & -1 & 4 \end{bmatrix}$$

$$[C] = \begin{bmatrix} 56.1 + T_0^p \\ 2T_1^p \\ 2T_2^p \\ 2T_3^p \\ 2T_4^p \\ 2T_5^p \\ 2T_6^p \\ 2T_7^p \\ 2T_8^p + T_9^{p+1} \end{bmatrix}$$

Note that the numerical values for the components of $[C]$ are determined from the backward time, p, values of the nodal temperatures. Note also how the eighth finite-difference equation, Equation 6, appears in the matrices $[A]$ and $[C]$. A temperature distribution table, as that used in Example 5.7, can be established with the rows for $p = 0$ corresponding to the initial uniform temperature condition. To obtain the temperature distribution for $p = 1$, the second row, the matrix inversion for $[A]$ and $[C]$ must be performed, where the values for $[C]$, determined with $T_i^0 = 20°C$ ($p = 0$), are

$$[C]_{p=0} = \begin{bmatrix} 76.1 \\ 40 \\ 40 \\ 40 \\ 40 \\ 40 \\ 40 \\ 60 \end{bmatrix}$$

Using the above numerical values for $[C]$ corresponding to $p = 0$ and the matrix $[A]$, the matrix inversion method provides a simultaneous solution for T_i^1 ($p = 1$), which appears in the second row of the table. This procedure is then repeated; that is, values of $[C]_p$ are determined from the previous time step in order to obtain T_i^{p+1}. The table below shows the results for five time steps giving the distribution for $t = p\Delta t = 5 \times 24\,s = 120\,s$.

Tabulated Temperature Distributions

p	t(s)	T_0	T_1	T_2	T_3	T_4	T_5	T_6	T_7	T_8
0	0	20.0	20.0	20.0	20.0	20.0	20.0	20.0	20.0	20.0
1	24	52.4	28.7	22.3	20.6	20.2	20.0	20.0	20.0	20.0
2	48	74.0	39.5	26.6	22.1	20.7	20.2	20.1	20.0	20.0
3	72	90.2	50.3	32.0	24.4	21.6	20.6	20.2	20.1	20.0
4	96	103.4	60.5	38.0	27.4	22.9	21.1	20.4	20.2	20.1
5	120	114.7	70.0	44.2	30.9	24.7	21.9	20.8	20.3	20.1	◁

COMMENTS:

1. Note that our choice of $N = 8$ nodes was sufficient since $T_8 = 20.1°C$ after 120 s. If T_8 had increased significantly, it would have been necessary to enlarge the number of finite-difference equations that need to be solved simultaneously.

2. Comparing the above results with those of the forward time-difference method, we see that the forward method gives higher values for $T(0, 120 \text{ s})$ and $T(150 \text{ mm}, 120 \text{ s})$. The present method will give more accurate results than the forward time-difference method for the same space-time increment values.

5.8 THE ELECTRICAL ANALOG

The notion of an analogy between the transfer of energy and electrical charge was introduced in Section 3.1.2 and was used to infer expressions for one-dimensional heat transfer resistances. In Section 4.3 it was considered as a means for solving two-dimensional, steady-state problems. The analogy may also be applied to transient problems.

To develop the specific form of the analogy, let us begin by substituting the expression for dimensionless time, Equation 5.21, into the dimensionless form of the heat equation, Equation 5.22. We obtain

$$\frac{\partial^2 \theta^*}{\partial x^{*2}} = \frac{\partial \theta^*}{\partial(\alpha t/L^2)} = \frac{\partial \theta^*}{\partial\left[t \bigg/ \left(\dfrac{L}{kA} \cdot \rho c\, LA\right)\right]} = \frac{\partial \theta^*}{\partial(t/R_t C_t)}$$

where the *thermal resistance* and *capacitance* are defined as

$$R_t = \frac{L}{kA} \tag{5.57}$$

and

$$C_t = \rho c A L \tag{5.58}$$

The heat equation may therefore be expressed as

$$\frac{\partial^2 \theta^*}{\partial x^{*2}} = R_t C_t \frac{\partial \theta^*}{\partial t} \tag{5.59}$$

For the transient conduction of electrical charge in a one-dimensional system with negligible induction, it is well known that the charge conservation equation may be expressed as

$$\frac{\partial^2 E}{\partial x^2} = \frac{R_e C_e}{L^2} \frac{\partial E}{\partial t} \tag{5.60}$$

where E is the electric potential and R_e (Ω) and C_e (F) are the electrical resistance and capacitance, respectively. Note that the product $R_e C_e$ has the units of time (s). Introducing a normalized potential of the form

$$\varepsilon^* \equiv \frac{E - E_r}{E_i - E_r} \tag{5.61}$$

where E_r and E_i are reference and initial potentials, respectively, and substituting Equations 5.20 and 5.61 into Equation 5.60, we obtain

$$\frac{\partial^2 \varepsilon^*}{\partial x^{*2}} = R_e C_e \frac{\partial \varepsilon^*}{\partial t} \tag{5.62}$$

The analogy that exists between Equations 5.59 and 5.62 is apparent. Hence, if the initial and boundary conditions are of equivalent forms, a solution to Equation 5.59 may be inferred from a solution to Equation 5.62.

From the foregoing results it is evident that a transient heat transfer problem may be solved by establishing an equivalent electrical circuit and measuring $\varepsilon^*(x^*,t)$. However, in so doing, it is important that the *time scaling* be properly effected. That is, for ε^* to equal θ^*, Equations 5.59 and 5.62 require that

$$\frac{t_t}{R_t C_t} = \frac{t_e}{R_e C_e} \tag{5.63}$$

where t_t and t_e are times measured in the thermal and electrical systems, respectively.

Hence, if the value of ε^* is measured at some time t_e for a particular x^* in the electrical system, it will correspond to the value of θ^* for the same x^* in the thermal system but for the time

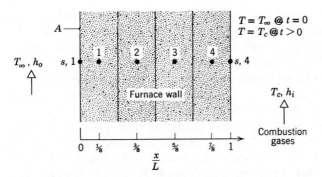

Figure 5.21 Transient heating of a furnace wall.

$$t_t = \frac{R_t C_t}{R_e C_e} t_e \tag{5.64}$$

To illustrate the foregoing concepts, consider a furnace wall of area A that is initially at uniform temperature T_i equal to the air temperature T_∞ on each side (Figure 5.21). At time $t > 0$ the furnace is fired, and the combustion gases instantaneously reach an elevated temperature $T_c > T_\infty$. Energy is transferred by convection from the combustion gases to the inner surface, by conduction through the wall, and by convection to the ambient air, which remains at T_∞.

Using the electrical analog to solve the above problem is facilitated by dividing the wall into a finite number of nodes. We have elected to use four nodes, each of thermal capacitance, $C_t/4 = (\rho c\, A)(L/4)$. In addition, *surface nodes of zero capacitance* are introduced to obtain a better estimate of surface thermal conditions. An *equivalent thermal circuit* for this system is represented in Figure 5.22. Eight *conduction resistances* of magnitude $R_t/8 = (L/8)/kA$ are

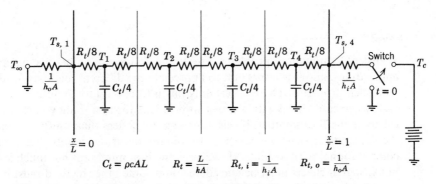

Figure 5.22 Equivalent thermal circuit for transient heating of a furnace wall.

placed in series with inner and outer *convection resistances* of $R_{t,i} = 1/h_iA$ and $R_{t,o} = 1/h_oA$, respectively. Initially, each node is at a temperature $T_i = T_\infty$, but at time $t = 0$ the switch is thrown and heat transfer from the combustion gases at T_c to the wall is initiated. During the transient process, the internal thermal energy of each node of finite capacitance increases, as does the temperature of all nodes. Eventually, a new steady-state is reached, which (for constant k) is characterized by a linear temperature distribution through the solid.

It would be a simple matter to construct an electrical circuit that models the foregoing thermal system. Eight equivalent electrical resistors, $R_e/8$, may be strung in series to represent the total conduction resistance, and four electrical capacitors, $C_e/4$, may be arranged as shown in Figure 5.22. Specific values of R_e and C_e are generally chosen on the basis of establishing a convenient time scale factor in Equation 5.64. That is, with R_t and C_t specified by the nature of the thermal problem, R_e and C_e may be chosen to provide a convenient value of (R_tC_t/R_eC_e). Electrical resistors, $R_{e,i}$ and $R_{e,o}$, corresponding to the inner and outer surface convection resistances, $R_{t,i}$ and $R_{t,o}$, respectively, may then be selected subject to the requirements that $(R_{e,i}/R_e) = [(1/h_iA)/R_t]$ and $(R_{e,o}/R_e) = [(1/h_oA)/R_t]$. For time $t_e > 0$ the electric potential may be measured at any node in the electrical system, and for the thermal time given by Equation 5.64 the corresponding temperature in the thermal system may be inferred from the relation

$$\frac{T - T_c}{T_\infty - T_c} = \frac{E - E_o}{E_\infty - E_o} \tag{5.65}$$

If E_o is the open-circuit voltage of the power supply used in the electrical system and E_∞ is selected as ground ($E_\infty = 0$), Equation 5.65 reduces to

$$\frac{T - T_c}{T_\infty - T_c} = 1 - \frac{E}{E_o} \tag{5.66}$$

5.9   SUMMARY

Transient conduction problems are to be found in numerous engineering applications, and it is important to appreciate the different methods for dealing with such problems. There is certainly much to be said for simplicity, in which case, when confronted with a transient problem, the first thing you should do is calculate the Biot number. If this number is much less than unity, you may use the lumped capacitance method to obtain accurate results with minimal computational requirements. However, if the Biot number is not much less than unity, spatial effects must be considered, and some other method must be used. Analytical results are available in convenient graphical form for the plane wall, the infinite cylinder and the sphere, as well as in graphical and equation form for the semiinfinite solid. You should know when and how to use these results. If geometrical complexities and/or the form of the boundary conditions preclude their use, recourse must be made to finite-difference methods. With the digital computer, such methods may be used to solve any conduction problem, regardless of complexity.

REFERENCES

1. Carslaw, H. S. and J. C. Jaeger, *Conduction of Heat in Solids*, 2nd Ed., Oxford University Press, London, 1959.
2. Schneider, P. J., *Conduction Heat Transfer*, Addison-Wesley, Cambridge, Mass., 1955.
3. Arpaci, V. S., *Conduction Heat Transfer*, Addison-Wesley, Reading, Mass., 1966.
4. Crank, J., *The Mathematics of Diffusion*, Oxford University Press, London, 1955.
5. Heisler, M. P., "Temperature Charts for Induction and Constant Temperature Heating," *Trans. ASME, 69,* 227–236, 1947.
6. Gröber, H., S. Erk, and U. Grigull, *Fundamentals of Heat Transfer*, McGraw-Hill, New York, 1961.
7. Richtmyer, R. D. and K. W. Morton, *Difference Methods for Initial Value Problems*, Wiley-Interscience, New York, 1967.
8. Dusinberre, G. M., *Heat Transfer Calculations by Finite Differences*, International Textbook Co., Scranton, Pa., 1961.
9. Myers, G. E., *Analytical Methods in Conduction Heat Transfer*, McGraw-Hill, New York, 1971.
10. Schmidt, E., "Über die Auswendung der Differenzenrechnung auf technische Anheiz- und Abkühlungs-Probleme," *Beitr. tech. Mech. und tech. Phys.*, August Föppl Festschrift, Berlin, 1924.
11. Razelos, P., "Methods of Obtaining Approximate Solutions" in W. M. Rohsenow and J. P. Hartnett, Eds., *Handbook of Heat Transfer*, McGraw-Hill, New York, 1973.
12. Jakob, M., *Heat Transfer*, Vol. 1, Wiley, New York, 1949.

PROBLEMS

5.1 Consider a thin electrical heater attached to a plate and backed by insulation as shown in the sketch. Initially, the heater and plate are at the temperature of the

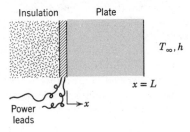

ambient air, T_∞. Suddenly, the power to the heater is switched on giving rise to a constant heat flux, q_o'' (W/m^2), at the inner surface of the plate.

a) Sketch and label, on $T - x$ coordinates, the temperature distributions: initial, steady-state, and two intermediate curves.

b) Sketch the heat flux at the outer surface, $q_x''(L,t)$, as a function of time.

5.2 Steel balls 12 mm in diameter are annealed by heating to 1150 K and then slowly cooling to 400 K in an air environment for which $T_\infty = 325$ K and $h = 20$ $W/m^2 \cdot K$. Assuming the properties of the steel to be $k = 40$ $W/m \cdot K$, $\rho = 7800$ kg/m^3, and $c = 600$ $J/kg \cdot K$, estimate the time required for the cooling process.

5.3 The heat transfer coefficient for air flowing over a sphere is to be determined by observing the temperature-time history of a sphere fabricated from pure copper. The temperature of the sphere, which is 12.7 mm in diameter, is 66°C before it is inserted into an air stream having a temperature of 27°C. A thermocouple on the outer surface of the sphere indicates a temperature of 55°C 69 s after the sphere is inserted in the air stream. Assume, and then justify, that the sphere behaves as a spacewise isothermal object and calculate the heat transfer coefficient.

5.4 In an industrial process requiring high d-c currents, water-jacketed copper rods, 20 mm in diameter, are used to carry the current. The water, which flows continuously between the jacket and the rod, maintains the rod temperature at 75°C during normal operation at 1000 A. The electrical resistance of the rod is known to be 0.15 Ω/m. A matter of concern relates to problems that would arise if the coolant water ceased to be available due, for example, to a valve malfunction. In such a situation heat transfer from the rod surface would diminish greatly, and the rod would eventually melt. Estimate the time required for melting to occur.

5.5 A cube of dimension L on a side is at a uniform temperature T_i and is insulated on all surfaces except for one. This surface is suddenly brought into contact with a larger, cooler surface at T_o fabricated from a material with very high thermal conductivity. Using the lumped capacitance method, develop an expression for estimating the temperature of the cube as a function of time in terms of the thermophysical properties of the cube material, L, T_i, and T_o. Assume that there is negligible contact resistance between the cube and the cool surface and that the cool surface remains at T_o during the process. Hint: write an energy balance for the cube approximating it as a lumped capacitance, and assume the resistance to heat transfer to be that due to conduction over a distance $L/2$ of the cube.

5.6 A microwave oven operates on the principle that application of a high-frequency field causes electrically polarized molecules in food to oscillate. The net effect is a *uniform generation* of thermal energy within the food, which enables it to be heated from refrigeration temperatures to 90°C in as short a time span as 30 s.

Consider the process of cooking a slab of beef of thickness $2L$ in a microwave oven and compare it with cooking in a conventional oven, where *each side* of the slab is *heated by radiation* for a period of approximately 30 min. In each case the meat is to be heated from 0°C to a *minimum* temperature of 90°C. Base your comparison on a sketch of the temperature distribution at selected times for each of the cooking processes. In particular consider the time t_0 at which heating is initiated, a time t_1 during the heating process, the time t_2 corresponding to the conclusion of heating, and a time t_3 well into the subsequent cooling process.

5.7   Spheres of varying diameter are made out of glass with a thermal conductivity k and are heated to a high temperature. When the spheres are suddenly cooled by a fluid at T_∞ with a heat transfer coefficient h, it is observed that spheres of large diameter crack during the cooling process. Develop a model that could be used to predict how small the diameter should be to avoid cracking.

5.8   Consider the system of Problem 5.1 where the temperature of the plate is spacewise isothermal during the transient process.

   a) Develop an expression for the temperature of the plate as a function of time, $T(t)$, in terms of q_o'', T_∞, h, and the plate properties ρ and c.

   b) Determine the thermal time constant and the steady-state temperature for a 12 mm thick plate of pure copper when $T_\infty = 27°C$, $h = 50$ W/m$^2 \cdot$K, and $q_o'' = 5000$ W/m^2. Estimate the time required to reach steady-state conditions.

5.9   Construction of a gas sampling bottle is shown in the following diagram. The inner glass flask of cylindrical shape has fittings on its hemispherical ends to connect with vacuum or chemical analysis equipment. With the glass bottle evacuated, the system is punctured with a fine needle and the gas sample (such as the exhaust fumes from a spray painting hood) is drawn into the bottle. The electrical heater maintains the system at an elevated temperature to eliminate condensation and assure transport of the sample back to the laboratory for analysis. The total mass of the system under design is estimated to be 0.4 kg and the effective specific heat to be 800 J/kg$\cdot$K.

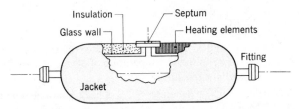

   a) Assuming there are no heat losses from the system, estimate the power required by the electrical heater to bring the system from 25°C to operating temperature, 150°C, in 5 min.

   b) It is estimated that the effective area of the fittings and system assemblies is 5×10^{-2} m^2. Assuming a convection heat transfer coefficient of 10 W/m$^2 \cdot$K and ambient air at 25°C, estimate the electrical power required to accomplish the same rate of temperature rise as specified in part (a).

5.10 It is desired to develop a thermal model of a product, such as a flat, single layer cake, which will predict the time required to bring the product to its "done stage" in a baking oven. Experience has shown that the done stage is reached when the product temperature is $T_d = 93°C$.

   Assume that the cake can be modeled as a homogeneous solid of uniform temperature with mass, M (kg), and surface area, A_s. A consequence of the baking process is to cause the evaporation of moisture at a nearly constant rate, $\dot{m}$ (kg/s).

However, although this rate is sufficient to influence the thermal condition of the cake, its effect on the mass M may be neglected.

a) Using a lumped capacitance approach, develop the differential equation for the temperature variation of the cake with time under conditions for which the product is taken from its initial state at T_i and suddenly placed inside the oven where the combined convection-radiation mode can be expressed in terms of a heat transfer coefficient, h, and a single temperature, T_∞, which represents both the oven cavity walls and the air temperature.

b) Determine the time required to bake a typical cake, initially at $T_i = 20°C$, in a 350°F (177°C) oven having these characteristics:

$$h = 15 \text{ W/m}^2 \cdot \text{K} \qquad A_s = 0.082 \text{ m}^2$$
$$M = 0.460 \text{ kg} \qquad \dot{m} = 1.8 \times 10^{-5} \text{ kg/s}$$

where all necessary thermophysical properties of the cake are considered to be those of water.

5.11 An electronic device, such as a power transistor mounted on a finned heat sink, can be modeled as a spatially isothermal object with internal heat generation and an external convection resistance.

a) Consider such a system of mass, M, specific heat, c, and surface area, A_s, which is initially in equilibrium with the environment at T_∞. Suddenly, the electronic device is energized such that a constant heat generation, $\dot{E}_g$ (W), occurs. Beginning with an energy balance on the system show that the temperature response of the device is

$$\frac{\theta}{\theta_i} = \exp(-t/RC)$$

where $\theta \equiv T - T(\infty)$ and $T(\infty)$ is the steady-state temperature corresponding to $t \to \infty$; $\theta_i = T_i - T(\infty)$; T_i = initial temperature of device; R = thermal resistance, $1/\bar{h}A_s$; and C = thermal capacitance, Mc.

b) An electronic device, which generates 60 W of heat, is mounted on an aluminum heat sink weighing 0.31 kg and reaches a temperature of 100°C in ambient air at 20°C under steady-state conditions. If the device is initially at 20°C, what temperature will it reach 5 min after the power is switched on?

5.12 Prior to being injected into a furnace, pulverized coal is preheated by passing it through a cylindrical tube whose surface is maintained at $T_{sur} = 1000°C$. The coal pellets are suspended in an airflow and are known to move with a speed of 3 m/s. If the pellets may be approximated as spheres of 1 mm diameter and it may be assumed that they are heated by radiation transfer from the tube surface, how long must the tube be to heat coal entering at 25°C to a temperature of 600°C? Is the use of the lumped capacitance method justified? Note: the integral that needs to be evaluated is given in Appendix B.1.

5.13 In our treatment of the lumped capacitance method, Section 5.1, the solid was exposed to convective surface conditions. Frequently applications involve surface heat transfer that is dominated by radiation, and it is interesting to determine the transient response of a lumped element to such conditions.

    a) Derive an expression for the time rate of change of the element temperature when it is suddenly exposed to a radiation exchange process characterized by the rate equation, Equation 1.6. Hint: the method of analysis will be similar to that used in Section 5.1 and the integral which needs evaluation is given in Appendix B.1.

    b) Develop a criterion that could be used to determine the suitability of the lumped capacitance method for the above conditions. Hint: consider using Equation 1.8, which defines a radiation heat transfer coefficient.

5.14 A metal sphere of diameter D, which is at a uniform temperature T_i, is suddenly removed from a furnace and suspended from a fine wire in a large room with air at a uniform temperature T_∞ and the surrounding walls at a temperature T_{sur}.

    a) Assuming temperature gradients within the sphere to be negligible, develop a differential equation that governs the temperature change with time in the sphere.

    b) Neglecting heat transfer by radiation, determine the time it takes for the sphere to cool to some temperature T.

    c) Neglecting heat transfer by convection, determine the time it takes for the sphere to cool to the temperature T. Hint: the integral that needs evaluation is given in Appendix B.1.

    d) How would you go about determining the time required for the sphere to cool to the temperature T if both convection and radiation are of the same order of magnitude?

    e) Consider an aluminum sphere 50 mm in diameter, which is at an initial temperature of $T_i = 800$ K. Both the air and the surroundings are at 300 K. Calculate and compare the time it will take for the sphere to cool to 400 K using the results of parts (b), (c), and (d).

5.15 A large aluminum (2024 alloy) plate of thickness 0.15 m, initially at a uniform temperature of 300 K, is placed in a furnace having an ambient temperature of 800 K for which the convection heat transfer coefficient is estimated to be 500 W/m$^2 \cdot$K.

    a) Determine the time required for the plate centerplane to reach 700 K.

    b) What is the surface temperature of the plate for this condition?

    c) Repeat the calculations if the material were stainless steel (type 304).

5.16 After a long, hard week on the books, you and your friend are ready to relax. You take a steak 50 mm thick from the freezer. How long do you have to let the good times roll before the steak has thawed? Assume that the steak is initially at $-6°$C, thaws when the midplane temperature reaches 4°C, and that the room temperature is 23°C with a convection heat transfer coefficient of 10 W/m$^2 \cdot$K. Treat the steak as a slab having the properties of liquid water at 0°C. Neglect the heat of fusion associated with the melting phase change.

5.17 A one-dimensional, plane wall with a thickness of 0.1 m initially at a uniform temperature of 250°C is suddenly immersed in an oil bath at a temperature of 30°C. Assuming the convection heat transfer coefficient for the wall in the bath is

500 W/m²·K, calculate the surface temperature of the wall 9 min after immersion. The properties of the wall are $k = 50$ W/m·K, $\rho = 7835$ kg/m³, and $c = 465$ J/kg·K.

5.18 Cylindrical steel rods (AISI 1010), 50 mm in diameter, are heat treated by drawing them through an oven 5 m long in which air is maintained at 750°C. The rods enter at 50°C and achieve a centerline temperature of 600°C before leaving. For a convection coefficient between the air and the rod of 125 W/m²·K, use the Heisler charts to estimate the speed at which the rods must be drawn through the oven.

5.19 Estimate the time required to cook a hot dog in boiling water. Assume that the hot dog is initially at 6°C, that the convection heat transfer coefficient is 100 W/m²·K, and that the final temperature is 80°C at the centerline. Treat the hot dog as a long cylinder of 20 mm diameter having these properties: $\rho = 880$ kg/m³, $c = 3350$ J/kg·K, and $k = 0.52$ W/m·K.

5.20 A long rod 40 mm in diameter, fabricated from sapphire (aluminum oxide) and initially at a uniform temperature of 800 K, is suddenly exposed to a cooling process with a fluid at 300 K having a heat transfer coefficient of 1600 W/m²·K. After 35 s of exposure to the cooling process, the rod is wrapped in insulation and experiences no heat losses. What will be the temperature of the rod after a long period of time?

5.21 The cold air chamber shown as follows is proposed for quenching steel ball bearings of diameter $D = 0.2$ m and initial temperature $T_i = 400°C$. Air in the chamber is maintained at $-15°C$ by a refrigeration system, and the steel balls pass through the chamber on a conveyor belt. Optimum bearing production requires that 70 percent of the initial thermal energy content of the ball above $-15°C$ be removed. Radiation effects may be neglected, and the convection heat transfer coefficient within the chamber is 1000 W/m²·K. Estimate the residence time of the balls within the chamber, and recommend a drive velocity of the conveyor. The following properties may be used for the steel: $k = 50$ W/m·K, $\alpha = 2 \times 10^{-5}$ m²/s, and $c = 450$ J/kg·K.

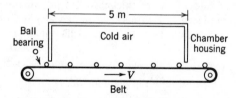

5.22 Stainless steel (AISI 304) ball bearings, which have been uniformly heated to 850°C, are hardened by quenching them in an oil bath that is maintained at 40°C. The ball diameter is 20 mm, and the convection coefficient associated with the oil bath is 1000 W/m²·K.

a) If quenching is to occur until the surface temperature of the balls reaches 100°C, how long must the balls be kept in the oil? What is the center temperature at the conclusion of the cooling period?

b) If 10,000 balls are to be quenched per hour, what is the rate at which energy must be removed by the oil bath cooling system in order to maintain its temperature at 40°C?

5.23 A spherical hailstone that is 5 mm in diameter is formed in a high altitude cloud at $-30°C$. If the stone begins to fall through warmer air at $5°C$, how long will it take before the outer surface begins to melt? What is the temperature of the stone's center at this point in time, and how much energy (J) has been transferred to the stone? A convection heat transfer coefficient of $250 \text{ W/m}^2 \cdot \text{K}$ may be assumed, and the properties of the hailstone may be taken to be those of ice.

5.24 A sphere 30 mm in diameter initially at 800 K is quenched in a large bath having a constant temperature of 320 K with a convection heat transfer coefficient of 75 $\text{W/m}^2 \cdot \text{K}$. The thermophysical properties of the sphere material are: $\rho = 400 \text{ kg/m}^3$, $c = 1600 \text{ J/kg} \cdot \text{K}$, and $k = 1.7 \text{ W/m} \cdot \text{K}$.

a) Show, in a qualitative manner on $T - t$ coordinates, the temperatures at the center and at the surface of the sphere as a function of time.

b) Calculate the time required for the surface of the sphere to reach 415 K.

c) Determine the heat flux (W/m^2) at the outer surface of the sphere at the time determined in part (b).

d) Determine the energy (J) that has been lost by the sphere during the process of cooling to the surface temperature of 415 K.

e) At the time determined by part (b), the sphere is quickly removed from the bath and covered with perfect insulation, such that there is no heat loss from the surface of the sphere. What will be the temperature of the sphere after a long period of time elapses?

5.25 Two spheres, A and B, are initially at 800 K, and they are simultaneously quenched in large constant temperature baths, each having a temperature of 320 K. The following parameters are associated with each of the spheres and their cooling processes.

	SPHERE A	SPHERE B
Diameter, mm	300	30
Density, kg/m^3	1600	400
Specific heat, kJ/kg$\cdot$K	0.400	1.60
Thermal conductivity, W/m$\cdot$K	170	1.70
Convection coefficient, W/m$^2\cdot$K	5	50

a) Show in a qualitative manner, on T versus t coordinates, the temperatures at the center and at the surface for each sphere as a function of time. Briefly explain the reasoning by which you determine the relative positions of the curves.

b) Calculate the time required for the surface of each sphere to reach 415 K.

c) Determine the energy that has been gained by each of the baths during the process of the spheres cooling to 415 K.

5.26 Two large blocks of different materials, such as copper and concrete, have been sitting in a room ($23°C$) for a very long time. Which of the two blocks, if either, will feel colder to the touch? Begin your analysis by assuming the blocks to be semiinfinite solids and your hand to be at a temperature of $37°C$.

5.27 Asphalt pavement may achieve temperatures as high as 50°C on a hot summer day. Assume that such a temperature exists throughout the pavement, when suddenly a rainstorm reduces the surface temperature to 20°C. Calculate the total amount of energy (J/m^2) that will be transferred from the asphalt over a 30-min period in which the surface is maintained at 20°C.

5.28 An insurance company has hired you as a consultant to improve their understanding of burn injuries. They are especially interested in injuries induced when some portion of a worker's body comes into contact with machinery that is at elevated temperatures in the range from 50 to 100°C. Their medical consultant informs them that irreversible thermal injury (cell death) will occur in any living tissue that is maintained at $T \geq 48°C$ for a duration $\Delta t \geq 10$ s. They are interested in acquiring information concerning the extent of irreversible tissue damage (as measured by distance from the skin surface) as a function of the machinery temperature and the time during which contact is made between the skin and the machinery. Can you help them?

Assume living tissue to have a normal temperature of 37°C, to be isotropic, and to have constant properties equivalent to those of liquid water.

5.29 An electric heater in the form of a sheet is placed in good contact with the surface of a thick slab of bakelite having a uniform temperature of 300 K. Determine the temperature of the slab at the surface and at a depth of 25 mm, 10 min after the heater is energized and provides a constant heat flux to the surface of 2500 W/m^2.

5.30 A very thick slab with thermal diffusivity 5.6×10^{-6} m^2/s and thermal conductivity 20 $W/m \cdot K$ is initially at a uniform temperature of 325°C. Suddenly, the surface is exposed to a coolant at 15°C for which the convection heat transfer coefficient is 100 $W/m^2 \cdot K$. Determine the temperatures at the surface and at a depth of 45 mm after 3 min have elapsed.

5.31 A thick wooden wall constructed from oak, initially at 25°C, is suddenly exposed to combustion products at 800°C. Determine the time of exposure required for the surface to reach the ignition temperature of 400°C, assuming the convection heat transfer coefficient between the wall and products to be 20 $W/m^2 \cdot K$.

5.32 A steel (plain carbon) billet of square cross section 0.3 × 0.3 m, initially at a uniform temperature of 30°C, is placed in a soaking oven having a temperature of 750°C. If the convection heat transfer coefficient for the heating process is 100 $W/m^2 \cdot K$, how long must the billet remain in the oven before its center temperature reaches 600°C?

5.33 Fire clay brick of dimensions 0.06 m × 0.09 m × 0.20 m is removed from a kiln at 1600 K and cooled in air at 40°C with $h = 50$ $W/m^2 \cdot K$. What is the temperature at the center and at the corners of the brick after 50 min of cooling?

5.34 A cylindrical copper pin 100 mm long and 50 mm in diameter is initially at a uniform temperature of 20°C. The end faces are suddenly subjected to an intense heating rate that suddenly raises them to a temperature of 500°C. At the same time, the cylindrical surface is subjected to heating by gas flow with a temperature 500°C and a heat transfer coefficient 100 $W/m^2 \cdot K$.

a) Determine the temperature at the center point of the cylinder after a time of 8 s from sudden application of the heating.

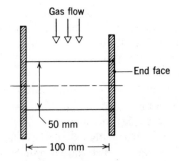

Gas flow

End face

50 mm

|←——— 100 mm ———→|

b) Considering the parameters governing the temperature distribution in transient heat diffusion problems, are there any simplifying assumptions that could be justified in analyzing this particular problem? Explain briefly.

5.35 Recalling that your mother once said that meat should be cooked until every portion has attained a temperature of 80°C, how long will it take to cook a 2.25-kg roast? Assume that the meat is initially at 6°C and that the oven temperature is 175°C with a convection heat transfer coefficient of $15 \text{ W/m}^2 \cdot \text{K}$. Treat the roast as a cylinder having a diameter equal to its length with properties of liquid water.

5.36 A thin rod of diameter D is initially in equilibrium with its surroundings, a large vacuum enclosure at temperature, T_{sur}. Suddenly an electrical current, I (A), is passed through the rod having an electrical resistivity, ρ_e, and emissivity, ε. Other pertinent thermophysical properties are identified on the figure. Derive the transient, finite-difference equation for node m.

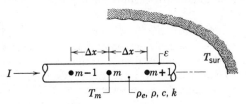

5.37 A tantalum rod of diameter 3 mm and length 120 mm is supported by two electrodes within a large vacuum enclosure. Initially the rod is in equilibrium with

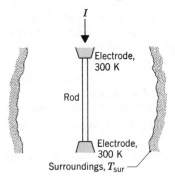

the electrodes and its surroundings, which are maintained at 300 K. Suddenly, an electrical current, $I = 80$ A, is passed through the rod. Assume the emissivity of the rod is 0.1 and the electrical resistivity is 95×10^{-8} $\Omega \cdot$m. Use Table A.1 to obtain the other thermophysical properties required in your solution. Use a finite-difference method with a space increment of 10 mm.

a) Estimate the time required for the midlength of the rod to reach 1000 K.

b) Determine the steady-state temperature distribution and estimate approximately how long it will take to reach this condition.

5.38 A one-dimensional slab of thickness $2L$ is initially at a uniform temperature T_i. Suddenly electric current is passed through the slab causing a uniform volumetric heating, $\dot{q}$ (W/m³), to be present. At the same time, both outer surfaces ($x = \pm L$) are subjected to a convection process at T_∞ with a heat transfer coefficient, h.

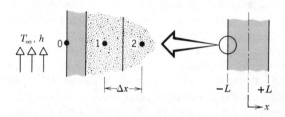

Write the finite-difference equation expressing conservation of energy for node 0 located on the outer surface at $x = -L$ as shown in the accompanying figure. Rearrange your equation and clearly identify any important dimensionless coefficients.

5.39 A wall 0.12 m thick having a thermal diffusivity of 1.5×10^{-6} m²/s is initially at a uniform temperature of 85°C. Suddenly one face is lowered to a temperature of 20°C while the other face is perfectly insulated. Using a numerical method with space and time increments of 30 mm and 300 s, respectively, determine the temperature distribution within the wall after 45 min have elapsed.

5.40 A large plastic casting with thermal diffusivity 6.0×10^{-7} m²/s is removed from its mold at a uniform temperature of 150°C. The casting is then exposed to a high-velocity airstream such that the surface experiences a sudden change in temperature to 20°C. Assuming the casting approximates a semiinfinite medium and using a finite-difference method with a space increment of 6 mm, estimate the temperature at a distance 18 mm from the surface after 3 min have elapsed. Verify your result by comparison with the appropriate analytical solution.

5.41 A thick slab of copper initially at a uniform temperature of 40°C is suddenly exposed to a constant heat flux at the surface of 3×10^5 W/m². Assume that surface convective effects and radiative exchange with the surroundings are negligible.

a) Use a finite-difference method to estimate the temperature at the surface and at a depth of 0.15 m after 5 min have elapsed. A grid spacing of 75 mm is suggested.

b) Compare your value with the analytical solution for the semiinfinite solid with the appropriate surface condition.

c) Explain what effect the use of a finer grid spacing would have on the finite-difference solution.

5.42 A very thick plate with thermal diffusivity 5.6×10^{-6} m²/s and thermal conductivity 20 W/m·K is initially at a uniform temperature of 325°C. Suddenly, the surface is exposed to a coolant at 15°C for which the convection heat transfer coefficient is 100 W/m²·K. Using the finite-difference method with a space increment of $\Delta x = 15$ mm and a time increment of 18 s, determine temperatures at the surface and at a depth of 45 mm after 3 min have elapsed.

5.43 A plane wall of thickness 100 mm with a uniform volumetric heat generation of $\dot{q}$ $= 1.5 \times 10^6$ W/m³ is exposed to convection conditions of $T_\infty = 30°C$ and $h = 1000$ W/m²·K on both surfaces. The wall is maintained under steady-state conditions when suddenly the heat generation level ($\dot{q}$) is reduced to zero. The thermal diffusivity and thermal conductivity of the wall material are 1.6×10^{-6} m²/s and 75 W/m·K, respectively. A space increment of 10 mm is suggested with your finite-difference method.

a) Estimate the center plane temperature of the wall 3 min after the heat generation level is switched off.

b) Plot on $T - x$ coordinates the temperature distribution obtained in part (a). Show also the initial and steady-state temperature distributions for the wall.

5.44 For the conditions described in Example 5.6, use the finite-difference method to estimate the temperature at the center plane ($x = 0$) 20 s after the power level is changed from $\dot{q}_1$ to $\dot{q}_2$.

5.45 One end of a stainless steel (AISI 316) rod of diameter 10 mm and length 0.16 m is inserted into a fixture maintained at 200°C. The rod, covered with an insulating sleeve, reaches a uniform temperature throughout its length. When the sleeve is removed, the rod is subjected to ambient air at 25°C such that the convection heat transfer coefficient is 30 W/m²·K. Using a numerical technique, estimate the time required for the midlength of the rod to reach 100°C.

5.46 The cross section of an oven wall is comprised of 30 mm thick insulation sandwiched between two thin (1.5 mm thickness) stainless steel sheets. Under steady-state conditions, the oven is operating with an inside air temperature of $T_{\infty,i}$

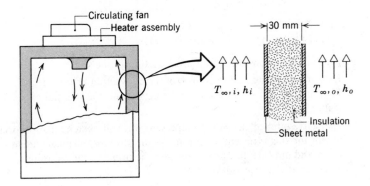

$= 150°C$ and an outside ambient air temperature of $T_{\infty,o} = 20°C$ with convection heat transfer coefficients $h_i = 100$ W/m$^2 \cdot$K and $h_o = 10$ W/m$^2 \cdot$K, respectively. When the oven heater level is changed and the fan speed changed to substantially increase air circulation within the oven, the inside surface of the oven experiences a sudden temperature change to $100°C$. The insulation has a thermal conductivity of 0.03 W/m$\cdot$K and a thermal diffusivity of 7.5×10^{-7} m^2/s. In your finite-difference solution, use a space increment of 6 mm. Assume that the effect of the stainless steel sheets is negligible and that the outside convection heat transfer coefficient, h_o, remains unchanged. Estimate the time required for the oven wall to approximate steady-state conditions after the inner wall temperature is changed to $100°C$.

5.47 The rectangular stainless steel (AISI 304) bar shown as follows is initially at a uniform temperature of $25°C$. Suddenly the upper surface is exposed to convection conditions with an air stream of $200°C$ and a heat transfer coefficient of 15 W/m$^2 \cdot$K. Using a numerical method, and assuming two-dimensional conduction, estimate the time required for the midpoint to reach $100°C$.

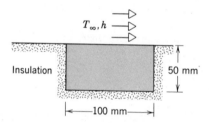

5.48 An electrical analog is to be used to determine the transient heating of the wall of a space vehicle during reentry into the earth's atmosphere. Frictional heating (viscous dissipation) in the atmosphere near the vehicle surface suddenly heats the air in the boundary layer to an effective temperature of $T_{\infty} = 1200°C$. The outside convection heat transfer coefficient is $h_o = 285$ W/m$^2 \cdot$K. Just before the sudden heating, the wall is at a uniform temperature $T_i = 30°C$.

The wall thickness is small compared with its length and width so that one-dimensional diffusion occurs. The inner surface of the wall is well insulated and the wall has the following properties:

$L = 30$ mm $c = 855$ J/kg$\cdot$K
$\rho = 2400$ kg/m^3 $k = 0.855$ W/m$\cdot$K

In your analysis, consider a unit cross section of the wall.

a) Determine the thermal resistance of the convection process at the outer surface, $R_{t,o}$, the thermal resistance of the wall, R_t, and the thermal capacitance of the wall, C_t.

b) Using five lumped elements to represent the wall, sketch an appropriate *thermal* circuit for this transient situation. Be sure to label elements, identify temperatures, and indicate distances. Consider the origin of the coordinate system $x = 0$ to occur at the *inner* surface.

c) The electrical resistance and electrical capacitance values selected to represent the *wall* are $R_e = 2000 \text{ k}\Omega$ and $C_e = 3.0 \text{ } \mu\text{F}$, respectively. What electrical resistance value, $R_{e,o}$, should be chosen to represent the convection thermal resistance, $R_{t,o}$?

d) Sketch and label the *electrical* circuit that is analogous to the thermal circuit.

e) Sketch, on $T - x$ coordinates, the temperature distributions for these time conditions: initial, steady state, and two intermediate time conditions.

f) What is the relationship between time for the electrical system (t_e) and for the thermal system (t_t)? That is, determine the time scale factor.

g) For the point $x = L/2$, the observation of the voltage–time relationship is shown in the following diagram.

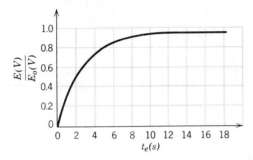

From this information, determine the time required for this point to reach $615°\text{C}$.

h) Verify your result of part (g) by using a Heisler chart solution method.

5.49 Consider the transient heating of a furnace wall (0.15 m thick) as described in Figure 5.21 with the corresponding electrical analog as shown in Figure 5.22. Initially the wall is in equilibrium with the ambient air at a temperature $T_\infty = 300 \text{ K}$, when suddenly the furnace is fired and the combustion gases at the inner surface are $T_c = 800 \text{ K}$. The convection heat transfer coefficients on the inner and outer surfaces are $h_i = 100 \text{ W/m}^2 \cdot \text{K}$ and $h_o = 20 \text{ W/m}^2 \cdot \text{K}$, respectively. The thermophysical properties of the wall material are $\rho = 2000 \text{ kg/m}^3$, $c = 970 \text{ J/kg} \cdot \text{K}$, and $k = 0.75 \text{ W/m} \cdot \text{K}$. The electrical resistance, R_e, and electrical capacitance, C_e, values selected to represent the wall are $3600 \text{ k}\Omega$ and $4.0 \text{ } \mu\text{F}$, respectively.

a) Determine the thermal resistance, R_t, and thermal capacitance, C_t, of the wall for a unit cross-sectional area.

b) Determine the thermal resistance of the convection heat transfer processes occurring on the inner and outer surfaces, $R_{t,i}$ and $R_{t,o}$, respectively.

c) What values of electrical resistance (Ω) must be used to represent the thermal resistances corresponding to the convection processes of part (b)?

d) What is the relationship between time for the electrical system (t_e) and for the thermal system (t_t)? That is, determine the time scale factor.

e) Sketch on $T - x$ coordinates, the temperature distributions for these time conditions: initial, steady-state, and two intermediate time conditions. For the steady-state distribution, be sure to determine and label the temperatures for the inner and outer surfaces, $T(0, \infty)$ and $T(L, \infty)$, respectively.

f) For the point $x = 3L/8$, the observation of the voltage–time relationship is shown in the following diagram.

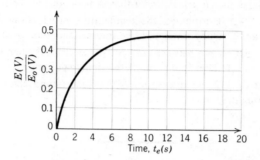

From this information, determine $T(3L/8, 1\ h)$, the temperature at the point $x = 3L/8$ one hour after the combustion gases have been introduced into the furnace.

6 Introduction to Convection

Thus far we have focused on heat transfer by diffusion and have considered convection only to the extent that it provides a possible boundary condition for diffusion problems. Recall that in Section 1.2.2 we used the term convection to describe energy transfer between a surface and a fluid moving over the surface. Although the mechanism of diffusion (random motion of fluid molecules) contributes to this transfer, the dominant contribution is generally made by the bulk or gross motion of fluid particles.

In our treatment of convection, we have two major objectives. In addition to obtaining a firm understanding of the physical mechanisms that underlie convection transfer, we wish to develop the means to perform convection transfer calculations. The present chapter is devoted primarily to achieving the former objective. In particular an effort has been made to concentrate many of the fundamentals in one place. Physical origins are carefully discussed, and the relevant dimensionless parameters, as well as important analogies, are developed.

A unique feature of this chapter is the manner in which convection mass transfer effects are introduced by analogy to those of convection heat transfer. In mass transfer by convection, gross fluid motion combines with diffusion to promote the transport of a species for which there exists a concentration gradient. In this text, we focus on convection mass transfer that occurs at the surface of a volatile solid or a liquid due to motion of a gas over the surface.

With conceptual underpinnings well established, subsequent chapters are used to develop useful tools for quantifying convection effects. In Chapters 7 and 8 methods are presented for computing the coefficients associated with *forced convection* in *external* and *internal flow* configurations, respectively. In Chapter 9 methods are described for determining these coefficients in *free convection*, and in Chapter 10 the problem of *convection with phase change* (*boiling* and *condensation*) is considered. In Chapter 11 methods are developed for designing and evaluating the performance of *heat exchangers*, devices that are widely used in engineering practice to effect heat transfer between fluids.

Accordingly, we begin by developing our understanding of the nature of convection.

6.1 THE CONVECTION TRANSFER PROBLEM

Consider the flow condition of Figure 6.1*a*. A fluid of velocity V and temperature T_∞ flows over a surface of arbitrary shape and of area A_s. If the surface conditions are such that $T_s \neq T_\infty$, we know that convection heat transfer will occur. The *local heat flux*, q'', may be expressed as

$$q'' = h(T_s - T_\infty) \tag{6.1}$$

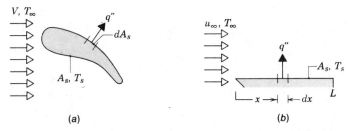

Figure 6.1   Local and total convection heat transfer effects. (a) Surface of arbitrary shape. (b) Flat plate.

where h is the *local convection coefficient*. Because flow conditions vary from point to point on the surface, both q'' and h also vary along the surface. The *total heat transfer rate, q*, may be obtained by integrating the local flux over the entire surface. That is,

$$q = \int_{A_s} q'' \, dA_s \tag{6.2}$$

or, from Equation 6.1, assuming a uniform surface temperature,

$$q = (T_s - T_\infty) \int_{A_s} h \, dA_s \tag{6.3}$$

Defining an *average convection coefficient*, $\bar{h}$, for the entire surface, the total heat transfer rate may also be expressed as

$$q = \bar{h} A_s (T_s - T_\infty) \tag{6.4}$$

Equating Equations 6.3 and 6.4, it follows that the average and local convection coefficients are related by an expression of the form

COMBINING 6.2 & 6.4

$$\bar{h} = \frac{1}{A_s} \int_{A_s} h \, dA_s \tag{6.5}$$

Note that for the special case of flow over a flat plate, Figure 6.1b, h varies with the distance x from the leading edge and Equation 6.5 reduces to

$$\bar{h} = \frac{1}{L} \int_0^L h \, dx \tag{6.6}$$

Similar results may be obtained for convection mass transfer. If a fluid of species molar concentration $C_{A,\infty}$ flows over a surface at which the species concentration is maintained at some value $C_{A,s} \neq C_{A,\infty}$, Figure 6.2a, transfer of the species by convection will occur. Species A is typically a vapor that is

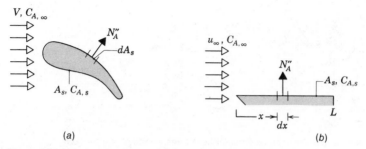

Figure 6.2 Local and total convection species transfer effects. (a) Surface of arbitrary shape. (b) Flate plate.

transferred into a gas stream due to evaporation or sublimation at a liquid or solid surface, respectively, and we are interested in determining the rate at which this transfer occurs. As for the case of heat transfer, such a calculation may be based on the use of a convection coefficient. In particular we may relate the molar flux of species A to the product of a transfer coefficient and a concentration difference. Referring to Figure 6.2, the molar flux of species A, N_A'' (kmol/s·m²), may be expressed as

$$N_A'' = h_m(C_{A,s} - C_{A,\infty}) \tag{6.7}$$

where h_m (m/s) is the *convection mass transfer coefficient*. The molar concentrations, $C_{A,s}$ and $C_{A,\infty}$, have units of kmol/m³. The total molar transfer rate for an entire surface, N_A (kmol/s), may then be expressed as

$$N_A = \bar{h}_m A_s(C_{A,s} - C_{A,\infty}) \tag{6.8}$$

where the average and local mass transfer convection coefficients are related by an equation of the form

$$\bar{h}_m = \frac{1}{A_s} \int_{A_s} h_m \, dA_s \tag{6.9}$$

For the flat plate of Figure 6.2b, it follows that

$$\bar{h}_m = \frac{1}{L} \int_0^L h_m dx \tag{6.10}$$

Note that species transfer may also be expressed as a mass flux, n_A'' (kg/s·m²), or as a mass transfer rate, n_A (kg/s), simply by multiplying both sides of Equations 6.7 and 6.8, respectively, by the molecular weight, $\mathcal{M}_A$ (kg/kmol), of species A. Accordingly,

$$n_A'' = h_m(\rho_{A,s} - \rho_{A,\infty}) \tag{6.11}$$

and

$$n_A = \bar{h}_m A_s (\rho_{A,s} - \rho_{A,\infty}) \tag{6.12}$$

where ρ_A (kg/m^3) is the mass density of species A.

To use the foregoing equations, it is necessary to determine the value of $C_{A,s}$ or $\rho_{A,s}$. Such a determination may be readily made by first noting that thermodynamic equilibrium will exist at the interface between the gas and the liquid or solid phase. One implication of this equilibrium condition is that the temperature of the vapor at the interface is equal to the surface temperature, T_s. A second implication is that the vapor is in a *saturated state*, in which case thermodynamic tables, such as Table A.6 for water, may be used to obtain its density from knowledge of T_s. To a first approximation, the molar concentration of the vapor at the surface may also be determined from the vapor pressure through application of the perfect gas law

$$C_{A,s} = \frac{p_{sat}(T_s)}{\mathscr{R} T_s} \tag{6.13}$$

where $\mathscr{R}$ is the universal gas constant and $p_{sat}(T_s)$ is the vapor pressure corresponding to saturation at T_s. Note that the vapor mass density at the surface may be obtained from the requirement that $\rho_{A,s} = \mathscr{M}_A C_{A,s}$.

The local flux and/or the total transfer rate are of paramount importance in any convection problem. These quantities may be determined from the rate equations, Equations 6.1, 6.4, 6.7, and 6.8, which depend on knowledge of the local and average convection coefficients. It is for this reason that determination of these coefficients is viewed as *the problem of convection*. However, the problem is not a simple one, for in addition to depending on numerous *fluid properties* such as density, viscosity, thermal conductivity, specific heat, and mass diffusivity, the coefficients depend on the *surface geometry* and the *flow conditions*. This multiplicity of independent variables results from the fact that convection transfer is determined by the *boundary layers* that develop on the surface.

EXAMPLE 6.1

Experimental results for the local heat transfer coefficient, h_x, for flow over a flat plate with an extremely rough surface were found to fit the relation

$$h_x(x) = ax^{-0.1}$$

where a is a coefficient (W/m$^{1.9}$·K) and $x(m)$ is the distance from the leading edge of the plate.

1. Develop an expression for the ratio of the average heat transfer coefficient, $\bar{h}_x$, for a plate of length x, to the local heat transfer coefficient, h_x, at x.

2. Show in a qualitative manner, the variation of h_x and $\bar{h}_x$ as a function of x.

SOLUTION

KNOWN:

Variation of the local heat transfer coefficient, $h_x(x)$.

FIND:

1. The ratio of the average heat transfer coefficient, $\bar{h}(x)$, to the local value, $h_x(x)$.
2. Sketch of the variation of h_x and $\bar{h}_x$ with x.

SCHEMATIC:

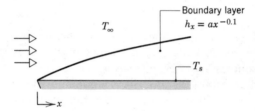

ANALYSIS:

1. From Equation 6.6 the average value of the convection heat transfer coefficient over the region from 0 to x is

$$\bar{h}_x = \bar{h}_x(x) = \frac{1}{x} \int_0^x h_x(x)\,dx$$

Substituting the expression for the local heat transfer coefficient

$$h_x(x) = ax^{-0.1}$$

and integrating, we obtain

$$\bar{h}_x = \frac{1}{x} \int_0^x ax^{-0.1}\,dx = \frac{a}{x} \int_0^x x^{-0.1}\,dx$$

$$\bar{h}_x = \frac{a}{x} \left(\frac{x^{+0.9}}{0.9} \right) = 1.11ax^{-0.1}$$

or

$$\bar{h}_x = 1.11 h_x \qquad \triangleleft$$

2. The variation of h_x and $\bar{h}_x$ with x is shown as follows.

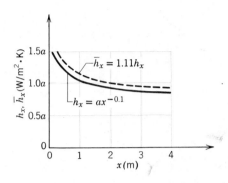

COMMENTS:

Note that, due to boundary layer development, both the local and average coefficients decrease with increasing distance from the leading edge. The average coefficient up to x must therefore exceed the local value at x.

EXAMPLE 6.2

A long circular cylinder 20 mm in diameter is fabricated from solid naphthalene, a common moth repellant, and is exposed to an airstream that provides for a convection mass transfer coefficient of $\bar{h}_m = 0.05$ m/s. The saturated vapor concentration of naphthalene is 5×10^{-6} kmol/m^3, and its molecular weight is 128 kg/kmol. What is the mass sublimation rate per unit length of cylinder?

SOLUTION

KNOWN:

Saturated vapor density of naphthalene sublimating into an airstream.

FIND:

Sublimation rate per unit length, n'_A (kg/s·m).

SCHEMATIC:

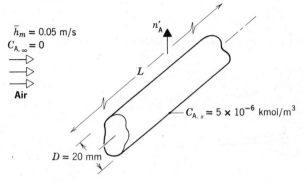

ASSUMPTIONS:

1. Steady-state conditions.
2. Negligible concentration of naphthalene in freestream of air.

ANALYSIS:

Naphthalene is transported to the air by convection, and from Equation 6.8, the molar transfer rate for the cylinder is

$$N_A = \bar{h}_m \pi D L (C_{A.s} - C_{A.\infty})$$

With $C_{A.\infty} = 0$ and $N'_A = N_A/L$, it follows that

$$N'_A = (\pi D)\bar{h}_m C_{A.s} = \pi \times 0.02 \text{ m} \times 0.05 \text{ m/s} \times 5 \times 10^{-6} \text{ kmol/m}^3$$

$$N'_A = 1.57 \times 10^{-8} \text{ kmol/s} \cdot \text{m}$$

The mass sublimation rate is then

$$n'_A = \mathscr{M}_A N'_A = 128 \text{ kg/kmol} \times 1.57 \times 10^{-8} \text{ kmol/s} \cdot \text{m}$$

$$n'_A = 2.01 \times 10^{-6} \text{ kg/s} \cdot \text{m} \qquad \triangleleft$$

6.2   THE CONVECTION BOUNDARY LAYERS

6.2.1   The Velocity Boundary Layer

To introduce the concept of a boundary layer, consider flow over the flat plate of Figure 6.3. When fluid particles make contact with the surface of the plate, they must assume zero velocity. These particles then act to retard the motion of particles in the adjoining fluid layer, which act to retard the motion of particles

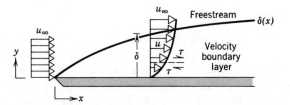

Figure 6.3 Velocity boundary layer on a flat plate.

in the next layer and so on until, at a distance $y = \delta$ from the surface, the effect becomes negligible. This retardation of fluid motion is associated with *shear stresses* τ acting in planes which are parallel to the fluid velocity (Figure 6.3). With increasing distance y from the surface, the x velocity component of the fluid, u, must then increase until it approaches the freestream value, u_∞. The subscript ∞ is used to designate conditions in the *free stream* outside the boundary layer.

The quantity δ is termed the *boundary layer thickness*, and it is formally defined as the value of y for which $u = 0.99u_\infty$. The *boundary layer velocity profile* refers to the manner in which u varies with y through the boundary layer. Accordingly, the fluid flow is characterized by two distinct regions, a thin fluid layer (the boundary layer) in which velocity gradients and shear stresses are large and a region outside the boundary layer in which velocity gradients and shear stresses are negligible. With increasing distance from the leading edge, the effects of viscosity penetrate further into the free stream and the boundary layer grows (δ increases with x).

Because it pertains to the fluid velocity, the foregoing boundary layer may be referred to more specifically as the *velocity boundary layer*. It develops whenever there is fluid flow over a surface, and it is of fundamental importance to problems involving convection transport. In fluid mechanics its significance to the engineer stems from its relation to the surface shear stress, τ_s, and hence to surface frictional effects. For external flows it provides the basis for determining the *friction coefficient*

$$C_f \equiv \frac{\tau_s}{\rho u_\infty^2 / 2} \tag{6.14}$$

a key dimensionless parameter from which the surface frictional drag may be determined. The surface shear stress may be evaluated from knowledge of the velocity gradient at the surface. That is,

$$\tau_s = \mu \left. \frac{\partial u}{\partial y} \right|_{y=0} \tag{6.15}$$

6.2.2   The Thermal Boundary Layer

Just as a velocity boundary layer develops when there is fluid flow over a surface, a *thermal boundary layer* must develop if the fluid freestream and surface temperatures differ. Again consider flow over a flat plate (Figure 6.4). At the leading edge the *temperature profile* is uniform, with $T = T_\infty$. However, fluid particles that come into contact with the plate achieve thermal equilibrium at the plate's surface temperature. In turn these particles exchange energy with those in the adjoining fluid layer, and temperature gradients develop in the fluid. The region of the fluid in which these temperature gradients exist is the thermal boundary layer, and its thickness δ_t is formally defined as the value of y for which the ratio $[(T_s - T)/(T_s - T_\infty)] = 0.99$. Note that, with increasing distance from the leading edge, the effects of heat transfer penetrate further into the free stream and the thermal boundary layer grows.

    The relationship between conditions in this boundary layer and the convection heat transfer coefficient may readily be demonstrated. At any distance from the leading edge, the *local* heat flux may be obtained by applying Fourier's law to the fluid at $y = 0$. That is,

$$q_s'' = -k_f \frac{\partial T}{\partial y}\bigg|_{y=0} \qquad \text{AT Plate surface} \tag{6.16}$$

This expression is appropriate because *at the surface, there is no fluid motion and energy transfer can occur only by conduction.* By combining Equation 6.16 with Newton's law of cooling, Equation 6.1, we then obtain

$$h = \frac{-k_f \,\partial T/\partial y|_{y=0}}{T_s - T_\infty} \qquad \begin{array}{l}\text{combine } q_s'' = h(T_s - T_\infty) \\[4pt] \text{\& } q_s'' = -k_f \frac{\partial T}{\partial y}|_{y=0}\end{array} \tag{6.17}$$

Hence conditions in the thermal boundary layer, which strongly influence the wall temperature gradient, $\partial T/\partial y|_{y=0}$, will determine the rate of heat transfer across the boundary layer. Note that, since $(T_s - T_\infty)$ is a constant, independent of x, while δ_t increases with increasing x, temperature gradients in the boundary

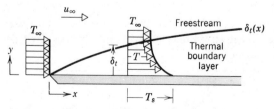

Figure 6.4   Thermal boundary layer on a flat plate.

layer must decrease with increasing x. Accordingly, the magnitude of $\partial T/\partial y|_{y=0}$ decreases with increasing δ_t, and it follows that q_s'' and h decrease with increasing x.

EXAMPLE 6.3

The temperature profile at a particular x location in a thermal boundary layer is given by an expression of the form

$$T(y) = A - By + Cy^2$$

where A, B, and C are constants. Obtain an expression for the corresponding local heat transfer coefficient h_x.

SOLUTION

KNOWN:

Temperature profile, $T(y)$, in a thermal boundary layer at a particular location.

FIND:

An expression for the local convection heat transfer coefficient, h_x.

SCHEMATIC:

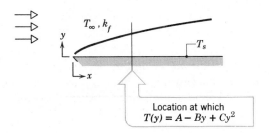

Location at which
$T(y) = A - By + Cy^2$

ANALYSIS:

From Equation 6.17, the heat transfer coefficient is calculated as

$$h_x = \frac{-k_f \, \partial T/\partial y|_{y=0}}{T_s - T_\infty}$$

or, substituting for the boundary layer temperature profile,

$$h_x = \frac{-k_f(-B + 2Cy)|_{y=0}}{T_s - T_\infty}$$

$$h_x = \frac{k_f B}{T_s - T_\infty} \qquad\qquad \lhd$$

COMMENTS:

If the boundary layer temperature distribution is known, it is a simple matter to obtain the convection heat transfer coefficient.

6.2.3 The Concentration Boundary Layer

Just as the velocity and thermal boundary layers determine wall friction and convection heat transfer, the *concentration boundary layer* determines convection mass transfer. If the concentration of some species A in the fluid at the surface, $C_{A,s}$, differs from that in the free stream, $C_{A,\infty}$ (Figure 6.5), a concentration boundary layer will develop. It is the region of the fluid in which concentration gradients exist, and its thickness δ_c is defined as the value of y for which $[(C_{A,s} - C_A)/(C_{A,s} - C_{A,\infty})] = 0.99$. Species transfer by convection between the surface and the free stream fluid is determined by conditions in this boundary layer.

The relationship between species transfer by convection and the concentration boundary layer may be demonstrated by first recognizing that the molar flux associated with species transfer by diffusion is determined by an expression that is analogous to Fourier's law. The expression, which is termed *Fick's law*, may be approximated as

$$N_A'' = -D_{AB}\frac{\partial C_A}{\partial y} \qquad\qquad (6.18)$$

where D_{AB} is a property known as the *binary diffusion coefficient*. At any point corresponding to $y > 0$ in the concentration boundary layer of Figure 6.5, species transfer is due to both bulk fluid motion and diffusion. However, at $y = 0$

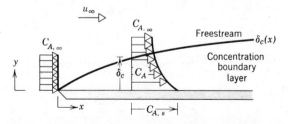

Figure 6.5 Species concentration boundary layer on a flat plate.

there is no fluid motion and species transfer is only due to diffusion. Since the species flux does not vary across the boundary layer, it follows that the species flux at any distance from the leading edge may be expressed as

$$N''_A = -D_{AB} \frac{\partial C_A}{\partial y}\bigg|_{y=0} \tag{6.19}$$

Combining Equations 6.7 and 6.19, it follows that

$$h_m = \frac{-D_{AB} \partial C_A/\partial y|_{y=0}}{C_{A,s} - C_{A,\infty}} \tag{6.20}$$

Hence conditions in the concentration boundary layer, which strongly influence the surface concentration gradient, $\partial C_A/\partial y|_{y=0}$, will also influence the convection mass transfer coefficient and hence the rate of species transfer across the boundary layer.

The foregoing results may also be expressed on a mass, rather than a molar, basis. Multiplying both sides of Equation 6.18 by the species molecular weight $\mathcal{M}_A$, the species mass flux due to diffusion may be approximated as

$$n''_A = -D_{AB} \frac{\partial \rho_A}{\partial y} \tag{6.21}$$

Applying this equation at $y = 0$ and combining with Equation 6.11, it follows that

$$h_m = \frac{-D_{AB} \partial \rho_A/\partial y|_{y=0}}{\rho_{A,s} - \rho_{A,\infty}} \tag{6.22}$$

EXAMPLE 6.4

At some location on the surface of a pan of water, measurements of the partial pressure of water vapor, p_A (atm), are made as a function of the distance, y, from the surface, and the results are shown as follows.

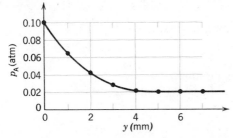

Determine the convection mass transfer coefficient, $h_{m,x}$, at this location.

SOLUTION

KNOWN:

Partial pressure, p_A, of water vapor as a function of distance, y, at a particular location on the surface of a water layer.

FIND:

Convection mass transfer coefficient at the prescribed location.

SCHEMATIC:

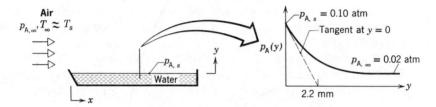

ASSUMPTIONS:

1. Water vapor may be approximated as a perfect gas.
2. Isothermal conditions.

PROPERTIES:

Table A.8, water vapor–air (298 K): $D_{AB} = 0.26 \times 10^{-4}$ m²/s.

ANALYSIS:

From Equation 6.22 the local convection mass transfer coefficient is

$$h_{m,x} = \frac{-D_{AB}\, \partial \rho_A / \partial y |_{y=0}}{\rho_{A,s} - \rho_{A,\infty}}$$

or, approximating the vapor as a perfect gas

$$p_A = \rho_A R T$$

with constant T (isothermal conditions),

$$h_{m,x} = \frac{-D_{AB}\, \partial p_A / \partial y |_{y=0}}{p_{A,s} - p_{A,\infty}}$$

From the measured vapor pressure distribution

$$\left. \frac{\partial p_A}{\partial y} \right|_{y=0} = \frac{(0 - 0.1) \text{ atm}}{(0.0022 - 0) \text{ m}} = -45.5 \text{ atm/m}$$

Hence

$$h_{m,x} = \frac{-0.26 \times 10^{-4} \text{ m}^2/\text{s} \, (-45.5 \text{ atm/m})}{(0.1 - 0.02) \text{ atm}}$$

$$h_{m,x} = 0.0148 \text{ m/s} \qquad \qquad \triangleleft$$

COMMENTS:

Note that, from the existence of thermodynamic equilibrium at the liquid–vapor interface, the temperature at this interface may be determined from Table A.6. With $p_{A,s} = 0.1$ atm $= 0.101$ bar, it follows that $T_s = 319$ K. Note also that evaporative cooling effects may cause the value of T_s to be less than T_∞.

6.2.4 Significance of the Boundary Layers

In summary we note that the velocity boundary layer is of extent $\delta(x)$ and is characterized by the presence of velocity gradients and shear stresses. The thermal boundary layer is of extent $\delta_t(x)$ and is characterized by temperature gradients and heat transfer. Finally, the concentration boundary layer is of extent $\delta_c(x)$ and is characterized by concentration gradients and species transfer. For the engineer the principal manifestations of the three boundary layers are, respectively, *surface friction, convection heat transfer*, and *convection mass transfer*. The key boundary layer parameters are then the *friction coefficient, C_f*, and the *heat* and *mass transfer convection coefficients, h* and *h_m*, respectively.

For flow over any surface, there will always exist a velocity boundary layer and hence surface friction. However, a thermal boundary layer, and hence convection heat transfer, only exists if the surface and freestream temperatures differ. Similarly, a concentration boundary layer and convection mass transfer only exist if the surface concentration of a species differs from its freestream concentration. Situations can arise in which all three boundary layers are present. In such cases, the boundary layers rarely grow at the same rate, and the values of δ, δ_t, and δ_c at a given x location are not the same.

6.3 LAMINAR AND TURBULENT FLOW

An essential first step in the treatment of any convection problem is to determine whether the boundary layer is *laminar* or *turbulent*. Surface friction and the convection transfer rates depend strongly on which of these conditions exists.

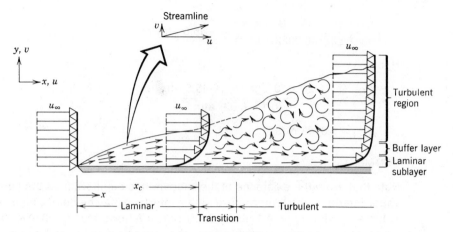

Figure 6.6 Velocity boundary layer development on a flat plate.

As shown in Figure 6.6, there are sharp differences between laminar and turbulent flow conditions. In the laminar boundary layer, fluid motion is highly ordered and it is possible to identify streamlines along which particles move. Fluid motion along a streamline is characterized by velocity components in both the x and y directions. Since the velocity component v is in the direction normal to the surface, it can contribute significantly to the transfer of momentum, energy or species through the boundary layer. Fluid motion normal to the surface is necessitated by boundary layer growth in the x direction.

In contrast, fluid motion in the turbulent boundary layer is highly irregular and is characterized by velocity fluctuations. These fluctuations enhance the transfer of momentum, energy, and species and hence increase surface friction, as well as convection transfer rates. Due to fluid mixing resulting from the fluctuations, turbulent boundary layer thicknesses are larger and boundary layer profiles (velocity, temperature, and concentration) are flatter than in laminar flow.

The foregoing conditions are shown schematically in Figure 6.6 for velocity boundary layer development on a flat plate. The boundary layer is initially laminar, but at some distance from the leading edge, transition to turbulent flow begins to occur. Fluid fluctuations begin to develop in the *transition region*, and the boundary layer eventually becomes completely turbulent. The transition to turbulence is accompanied by significant increases in the boundary layer thicknesses, the wall shear stress, and the convection coefficients. These effects are illustrated in Figure 6.7 for the velocity boundary layer thickness δ and the local convection heat transfer coefficient h. In the turbulent boundary layer, three different regions may be delineated. We may speak of a *laminar sublayer* in which transport is dominated by diffusion and the velocity profile is nearly

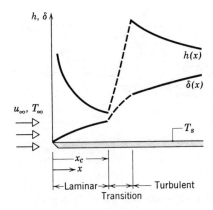

Figure 6.7 Variation of velocity boundary layer thickness, δ, and the local heat transfer coefficient, h, for flow over a flat plate.

linear. There is an adjoining *buffer layer* in which diffusion and turbulent mixing are comparable, and there is a *turbulent zone* in which transport is dominated by turbulent mixing.

In calculating boundary layer behavior it is frequently reasonable to assume that transition occurs at some location x_c. This location is determined by a dimensionless grouping of variables called the *Reynolds number*,

$$Re_x \equiv \frac{\rho u_\infty x}{\mu} \tag{6.23}$$

where the characteristic length, x, is the distance from the leading edge. The *critical Reynolds number* is the value of Re_x for which transition occurs, and for external flow it is known to vary from 10^5 to 3×10^6, depending on surface roughness, the turbulence level of the free stream and the nature of the pressure variation along the surface. A representative value of

$$Re_{x,c} = \frac{\rho u_\infty x_c}{\mu} = 5 \times 10^5 \tag{6.24}$$

is generally assumed for boundary layer calculations and, unless otherwise noted, will be used for the calculations of this text.

6.4 THE CONVECTION TRANSFER EQUATIONS

We can improve our understanding of the physical effects that determine boundary layer behavior and further illustrate its relevance to convection transport by developing the equations which govern boundary layer conditions. Consider the simultaneous development of velocity, thermal and concentration boundary layers over the surface of Figure 6.8. The fluid is considered to be a

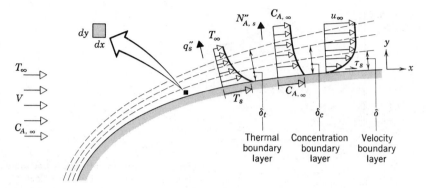

Figure 6.8 Development of the velocity, thermal, and concentration boundary layers for an arbitrary surface.

binary mixture of species A and B, and the species A concentration boundary layer originates from a difference between the freestream and surface concentrations $(C_{A,\infty} \neq C_{A,s})$. Selection of the relative thicknesses $(\delta_t > \delta_c > \delta)$ is arbitrary, for the moment, and the factors that influence relative boundary layer development will be discussed later in this chapter. To simplify the development we assume two-dimensional flow conditions for which x is in the direction along the surface and y is normal to the surface. Extension of this development to three-dimensional flows may be readily made [1–3].

For each of the boundary layers we will identify the relevant physical effects and apply the appropriate conservation laws to control volumes of infinitesimal size. Although it is not essential for you to follow in detail the derivation of the conservation equations, you should attempt to develop an understanding of the underlying physical effects.

6.4.1 The Velocity Boundary Layer

One conservation law that is pertinent to the velocity boundary layer is that matter may neither be created nor destroyed. Stated in the context of the differential control volume of Figure 6.9, this law requires that *the net rate at which mass enters the control volume* (inflow–outflow) *must equal the rate of increase of mass stored within the control volume*. Mass enters and leaves the control volume exclusively through gross fluid motion. Transport due to such motion is often referred to as *convective transport*. If one corner of the control volume, $dxdy$, is located at (x,y), the rate at which mass enters the control volume through the surface perpendicular to x may be expressed as $(\rho u)dy$, where ρ is the total mass density $(\rho = \rho_A + \rho_B)$ and u is the x component of the *mass average velocity*. Note that the control volume is of unit depth in the z direction. Since ρ and u may

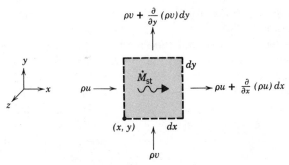

Figure 6.9 Differential control volume ($dx \cdot dy \cdot 1$) for mass conservation in the two-dimensional velocity boundary layer.

vary with x, the rate at which mass leaves the surface at $x + dx$ may be expressed by an expansion of the form

$$\left[(\rho u) + \frac{\partial(\rho u)}{\partial x} dx \right] dy$$

A similar result applies for the y direction. Moreover, at any instant the mass *stored* in the control volume may be changing, and its rate of change, $\dot{M}_{st}$, is of the form

$$\dot{M}_{st} = \frac{\partial}{\partial t} (\rho \, dxdy) = \frac{\partial \rho}{\partial t} dxdy$$

The conservation of mass requirement may now be expressed as

$$(\rho u)dy + (\rho v)dx - \left[\rho u + \frac{\partial(\rho u)}{\partial x} dx \right]dy - \left[\rho v + \frac{\partial(\rho v)}{\partial y} dy \right]dx = \left(\frac{\partial \rho}{\partial t} \right)dxdy$$

Canceling terms and dividing by $dxdy$, we obtain

$$\frac{\partial \rho}{\partial t} + \frac{\partial(\rho u)}{\partial x} + \frac{\partial(\rho v)}{\partial y} = 0 \qquad \text{Sum mass flows in } \uparrow \text{ out} \qquad (6.25)$$

Equation 6.25, the *continuity equation*, is a general expression of the *overall* mass conservation requirement, and it must be satisfied at every point in the velocity boundary layer. The equation applies for a single species fluid, as well as for mixtures in which species diffusion and chemical reactions may be occurring.

The second conservation law which is pertinent to the velocity boundary layer is *Newton's second law of motion*. For a differential control volume in the velocity boundary layer, this requirement states that the sum of all forces acting on the control volume must equal the rate of increase of the fluid momentum within the control volume plus the net rate at which momentum leaves the control volume (outflow–inflow).

Two kinds of forces may act on the fluid in the boundary layer: *body forces*, which are proportional to the volume, and *surface forces*, which are proportional to area. Gravitational, centrifugal, magnetic, and/or electric fields may contribute to the total body force, and we designate the x and y components of this force per unit volume of fluid as X and Y, respectively. The surface forces, F_s, are due to the fluid static pressure, as well as to *viscous stresses*. At any point in the boundary layer, the viscous stress (a force per unit area) may be resolved into two perpendicular components, which include a *normal stress*, σ_{ii}, and a *shear stress*, τ_{ij} (Figure 6.10).

A double subscript notation is used to specify the stress components. The first subscript indicates the surface orientation by providing the direction of its outward normal, and the second subscript indicates the direction of the force component. Accordingly, for the x surface of Figure 6.10, the normal stress σ_{xx} corresponds to a force component normal to the surface, and the shear stress τ_{xy} corresponds to a force in the y direction along the surface. All the stress components shown on the figure are positive in the sense that *both* the surface normal and the force component are in the same direction. That is, they are both in either the positive coordinate direction or in the negative coordinate direction. By this convention the normal viscous stresses are *tensile* stresses. In contrast the static pressure originates from an external force acting on the fluid in the control volume and is therefore a *compressive* stress.

Several features of the viscous stress should be noted. The associated force is between adjoining fluid elements and is a natural consequence of the fluid motion and viscosity. The surface forces of Figure 6.10 are therefore presumed to act on the fluid within the control volume and to be due to its interaction with the surrounding fluid. These stresses would vanish if the fluid velocity, or more specifically the velocity gradient, went to zero. In this respect the normal viscous

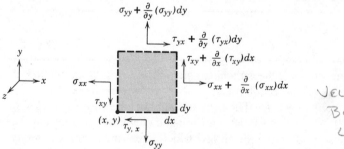

Figure 6.10 Normal and shear viscous stresses for a differential control volume ($dx \cdot dy \cdot 1$) in the two-dimensional velocity boundary layer.

stresses (σ_{xx} and σ_{yy}) must not be confused with the static pressure, which does not vanish for zero velocity.

Each of the stresses may change continuously in each of the coordinate directions. Using the customary expansion for the stresses, the *net* surface force for each of the two directions may then be expressed as

$$F_{s,x} = \left(\frac{\partial \sigma_{xx}}{\partial x} - \frac{\partial p}{\partial x} + \frac{\partial \tau_{yx}}{\partial y} \right) dx dy \tag{6.26}$$

$$F_{s,y} = \left(\frac{\partial \tau_{xy}}{\partial x} + \frac{\partial \sigma_{yy}}{\partial y} - \frac{\partial p}{\partial y} \right) dx dy \tag{6.27}$$

To use Newton's second law, the fluid momentum fluxes for the control volume must also be evaluated. If we focus on the x direction, the relevant fluxes are as shown in Figure 6.11. A contribution to the total x-momentum flux is made by the mass flow in each of the two directions. For example, the mass flux through the x surface (in the y-z plane) is (ρu), and the corresponding x-momentum flux is $(\rho u)u$. Similarly, the x-momentum flux due to mass flow through the y surface (in the x-z plane) is $(\rho v)u$. These fluxes may change in each of the coordinate directions, and the *net* rate at which x-momentum leaves the control volume is

$$\frac{\partial [(\rho u)u]}{\partial x} dx(dy) + \frac{\partial [(\rho v)u]}{\partial y} dy(dx)$$

In addition, there may be a time rate of change of the x momentum of the fluid within the control volume, which is given by

$$\frac{\partial}{\partial t} (\rho u) dx dy$$

Equating the total rate of change in the x momentum of the fluid to the sum of the forces in the x direction, we then obtain

$$\frac{\partial (\rho u)}{\partial t} + \frac{\partial [(\rho u)u]}{\partial x} + \frac{\partial [(\rho v)u]}{\partial y} = \frac{\partial \sigma_{xx}}{\partial x} - \frac{\partial p}{\partial x} + \frac{\partial \tau_{yx}}{\partial y} + X \tag{6.28}$$

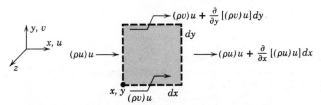

Figure 6.11 Momentum fluxes for a differential control volume $(dx \cdot dy \cdot 1)$ in the two-dimensional velocity boundary layer.

This expression may be put in a more convenient form by expanding the derivatives on the left-hand side and substituting from the continuity equation, Equation 6.25, in which case we obtain

$$\rho\left(\frac{\partial u}{\partial t} + u\frac{\partial u}{\partial x} + v\frac{\partial u}{\partial y}\right) = \frac{\partial}{\partial x}(\sigma_{xx} - p) + \frac{\partial \tau_{yx}}{\partial y} + X \qquad (6.29)$$

A similar expression may be obtained for the y direction and is of the form

$$\rho\left(\frac{\partial v}{\partial t} + u\frac{\partial v}{\partial x} + v\frac{\partial v}{\partial y}\right) = \frac{\partial \tau_{xy}}{\partial x} + \frac{\partial}{\partial y}(\sigma_{yy} - p) + Y \qquad (6.30)$$

Although Equations 6.29 and 6.30 are somewhat complicated, we should not lose sight of the physics that they represent. The first term on the left-hand side of the equations represents the increase in momentum of the fluid in the control volume, and the remaining terms represent the net rate of momentum flow from the control volume. The terms on the right-hand side of the equations account for net viscous and pressure forces, as well as the body force. These equations must be satisfied at each point in the boundary layer, and with Equation 6.25 they may be solved for the velocity field.

Before a solution to the foregoing equations can be obtained, it is necessary to relate the viscous stresses to other flow variables. These stresses are related to the deformation of the fluid and are a function of the fluid viscosity and velocity gradients. From Figure 6.12 it is evident that a *normal stress* must produce a *linear deformation* of the fluid, whereas a *shear stress* produces an *angular deformation*. Moreover, the magnitude of a stress is proportional to the *rate* at which the deformation occurs. The deformation rate is, in turn, related to the fluid viscosity and to the velocity gradients in the flow. For a *Newtonian fluid*[1] the stresses are proportional to the velocity gradients, where the proportionality constant is the fluid viscosity. Because of its complexity, however, development of

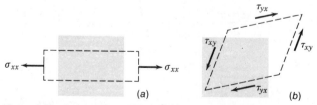

Figure 6.12 Deformations of a fluid element due to viscous stresses. (a) Linear deformation due to a normal stress. (b) Angular deformation due to shear stresses.

---

[1] A Newtonian fluid is one for which the shear stress is linearly proportional to the rate of angular deformation. All fluids of interest in this text are Newtonian.

the specific relations is left to the literature [1], and we limit ourselves to a presentation of the results. In particular it has been shown that

$$\sigma_{xx} = 2\mu \frac{\partial u}{\partial x} - \frac{2}{3}\mu(\nabla \cdot \mathbf{v}) \tag{6.31}$$

$$\sigma_{yy} = 2\mu \frac{\partial v}{\partial y} - \frac{2}{3}\mu(\nabla \cdot \mathbf{v}) \tag{6.32}$$

$$\tau_{xy} = \tau_{yx} = \mu\left(\frac{\partial u}{\partial y} + \frac{\partial v}{\partial x}\right) \tag{6.33}$$

where

$$\nabla \cdot \mathbf{v} = \frac{\partial u}{\partial x} + \frac{\partial v}{\partial y} \tag{6.34}$$

Equation 6.25 and Equations 6.29 to 6.33 provide a complete representation of conditions in a two-dimensional velocity boundary layer, and the velocity field in the boundary layer may be determined by solving these equations. Once the velocity field is known, it is a simple matter to obtain the wall shear stress, τ_s, from Equation 6.15.

6.4.2 The Thermal Boundary Layer

To apply the energy conservation requirement, Equation 1.10, to a differential control volume in the thermal boundary layer, Figure 6.13, it is necessary to first delineate the relevant physical processes. Neglecting potential energy effects, the energy per unit mass of the fluid includes the thermal internal energy, e, and the kinetic energy, $V^2/2$, where $V^2 \equiv u^2 + v^2$. Accordingly the rate of increase of

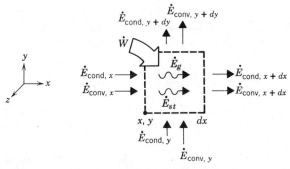

Figure 6.13 Differential control volume ($dx \cdot dy \cdot 1$) for energy conservation in the two-dimensional thermal boundary layer.

energy stored within the control volume may be expressed as

$$\dot{E}_{st} = \frac{\partial}{\partial t}\left[\rho\left(e + \frac{V^2}{2}\right)\right]dx\,dy \tag{6.35}$$

Thermal and kinetic energy is also *convected* with the *bulk fluid* motion across the control surfaces. For the x direction, the *net* rate at which this energy *enters* the control volume is

$$\dot{E}_{conv,x} - \dot{E}_{conv,x+dx} \equiv \rho u\left(e + \frac{V^2}{2}\right)dy$$

$$\qquad - \left\{\rho u\left(e + \frac{V^2}{2}\right) + \frac{\partial}{\partial x}\left[\rho u\left(e + \frac{V^2}{2}\right)\right]dx\right\}dy$$

$$= -\frac{\partial}{\partial x}\left[\rho u\left(e + \frac{V^2}{2}\right)\right]dx\,dy \tag{6.36}$$

In addition to convection, energy is also transferred across the control surface by *molecular processes*. There may be two contributions: that due to *conduction* and energy transfer due to the *diffusion* of *species* A and B. However, generally it is only in chemically reacting boundary layers that species diffusion will strongly influence thermal conditions. Hence the effect is neglected in the following development. For the conduction process, the *net* transfer of energy into the control volume is

$$\dot{E}_{cond,x} - \dot{E}_{cond,x+dx} = -\left(k\frac{\partial T}{\partial x}\right)dy - \left[-k\frac{\partial T}{\partial x} - \frac{\partial}{\partial x}\left(k\frac{\partial T}{\partial x}\right)dx\right]dy$$

$$= \frac{\partial}{\partial x}\left(k\frac{\partial T}{\partial x}\right)dx\,dy \tag{6.37}$$

Energy may also be transferred to and from the fluid in the control volume by *work* interactions involving the *body* and *surface forces*. The *net* rate at which work is done *on* the fluid by forces in the x direction may be expressed as

$$\dot{W}_{net,x} = (X\,u)dx\,dy + \frac{\partial}{\partial x}[(\sigma_{xx} - p)u]dx\,dy + \frac{\partial}{\partial y}(\tau_{yx}u)dx\,dy \tag{6.38}$$

The first term on the right-hand side of Equation 6.38 represents the work done by the body force, and the remaining terms account for the *net* work done by the pressure and viscous forces.

Using Equation 6.35 and Equations 6.36 to 6.38, as well as analogous equations for the y direction, the energy conservation requirement, Equation 1.10, may be expressed as

$$\frac{\partial}{\partial t}\left[\rho\left(e + \frac{V^2}{2}\right)\right] = -\frac{\partial}{\partial x}\left[\rho u\left(e + \frac{V^2}{2}\right)\right] - \frac{\partial}{\partial y}\left[\rho u\left(e + \frac{V^2}{2}\right)\right]$$

$$+ \frac{\partial}{\partial x}\left(k\frac{\partial T}{\partial x}\right) + \frac{\partial}{\partial y}\left(k\frac{\partial T}{\partial y}\right) + (Xu + Yv) - \frac{\partial}{\partial x}(pu) - \frac{\partial}{\partial y}(pv)$$

$$+ \frac{\partial}{\partial x}\left(\sigma_{xx}u + \tau_{xy}v\right) + \frac{\partial}{\partial y}(\tau_{yx}u + \sigma_{yy}v) + \dot{q} \tag{6.39}$$

where $\dot{q}$ is the rate at which energy is generated per unit volume. This expression provides a general form of the energy conservation requirement for the thermal boundary layer.

The foregoing equation may be cast in several alternative forms [2]. A particularly convenient form is obtained by substituting from modified versions of the momentum equations, Equations 6.29 and 6.30, and from the continuity equation, Equation 6.25. The desired form is

$$\rho\frac{\partial i}{\partial t} + \rho u\frac{\partial i}{\partial x} + \rho v\frac{\partial i}{\partial y} = \frac{\partial}{\partial x}\left(k\frac{\partial T}{\partial x}\right) + \frac{\partial}{\partial y}\left(k\frac{\partial T}{\partial y}\right)$$

$$+ \left(\frac{\partial p}{\partial t} + u\frac{\partial p}{\partial x} + v\frac{\partial p}{\partial y}\right) + \mu\Phi + \dot{q} \tag{6.40}$$

where i is the enthalpy per unit mass of mixture

$$i = e + \frac{p}{\rho} \tag{6.41}$$

and $\mu\Phi$ is the *viscous dissipation*, defined as

$$\mu\Phi \equiv \mu\left\{\left(\frac{\partial u}{\partial y} + \frac{\partial v}{\partial x}\right)^2 + 2\left[\left(\frac{\partial u}{\partial x}\right)^2 + \left(\frac{\partial v}{\partial y}\right)^2\right] - \frac{2}{3}\left(\frac{\partial u}{\partial x} + \frac{\partial v}{\partial y}\right)^2\right\} \tag{6.42}$$

The first term on the right-hand side of Equation 6.42 originates from the viscous shear stresses, and the remaining terms arise from the viscous normal stresses. Collectively, the terms account for the rate at which *mechanical energy is converted to thermal energy due to viscous effects in the fluid.*

6.4.3 The Concentration Boundary Layer

Since we are considering a binary mixture in which there are species concentration gradients (Figure 6.8), there will be *relative* transport of the species and it is necessary that *species conservation* be satisfied at each point in the concentration boundary layer. The pertinent form of the conservation equation may be obtained by identifying the processes that affect the *transport, generation* and *storage* of species A for a differential control volume in the boundary layer.

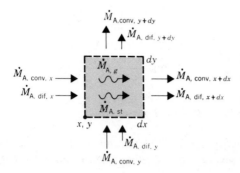

Figure 6.14 Differential control volume ($dx \cdot dy \cdot 1$) for species conservation in the two-dimensional boundary layer.

Consider the control volume of Figure 6.14. Species A may be transported by *convection* (with the mean velocity of the mixture) and by *diffusion* (relative to the mean motion) in each of the coordinate directions. The concentration may also be affected by chemical reactions, and we designate the rate at which the mass of species A is generated per unit volume due to such reactions as $\dot{n}_A$. Of course, the amount of A stored in the volume may change with time, and we designate this rate of change as $\dot{M}_{A, \text{st}}$.

The *net* rate at which species A *enters* the control volume due to *convection* in the x direction is

$$\dot{M}_{A, \text{conv}, x} - \dot{M}_{A, \text{conv}, x+dx} = (\rho_A u)dy \ - \left[(\rho_A u) + \frac{\partial(\rho_A u)}{\partial x}\,dx\right]dy$$

$$= -\frac{\partial(\rho_A u)}{\partial x}\,dx\,dy \tag{6.43}$$

Similarly, using Fick's law, Equation 6.21, to evaluate the diffusion flux in the x direction, $-D_{AB}(\partial\rho_A/\partial x)$, the rate at which species A enters the control volume is

$$\dot{M}_{A, \text{dif}, x} = \left(-D_{AB}\,\frac{\partial\rho_A}{\partial x}\right)dy$$

Hence, the *net* rate at which species A *enters* the control volume due to *diffusion* in the x direction is

$$\dot{M}_{A, \text{dif}, x} - \dot{M}_{A, \text{dif}, x+dx} = \left(-D_{AB}\,\frac{\partial\rho_A}{\partial x}\right)dy - \left[\left(-D_{AB}\,\frac{\partial\rho_A}{\partial x}\right)\right.$$

$$\left.+ \frac{\partial}{\partial x}\left(-D_{AB}\,\frac{\partial\rho_A}{\partial x}\right)dx\right]dy = \frac{\partial}{\partial x}\left(D_{AB}\,\frac{\partial\rho_A}{\partial x}\right)dx\,dy \tag{6.44}$$

Expressions similar to Equations 6.43 and 6.44 may be formulated for the y direction, and the storage effect may be expressed as

$$\dot{M}_{\text{A, st}} = \left(\frac{\partial \rho_{\text{A}}}{\partial t}\right) dx dy \tag{6.45}$$

Referring to Figure 6.14, the species conservation requirement may be expressed as

$$\dot{M}_{\text{A, conv}, x} - \dot{M}_{\text{A, conv}, x+dx} + \dot{M}_{\text{A, conv}, y} - \dot{M}_{\text{A, conv}, y+dy}$$

$$+ \dot{M}_{\text{A, dif}, x} - \dot{M}_{\text{A, dif}, x+dx} + \dot{M}_{\text{A, dif}, y} - \dot{M}_{\text{A, dif}, y+dy}$$

$$+ \dot{M}_{\text{A}, g} = \dot{M}_{\text{A, st}} \tag{6.46}$$

Substituting from Equations 6.43 and 6.44, as well as from similar forms for the y direction, and from Equation 6.45, it follows that

$$\frac{\partial \rho_{\text{A}}}{\partial t} + \frac{\partial (\rho_{\text{A}} u)}{\partial x} + \frac{\partial (\rho_{\text{A}} v)}{\partial y} = \frac{\partial}{\partial x}\left(D_{AB}\frac{\partial \rho_{\text{A}}}{\partial x}\right) + \frac{\partial}{\partial y}\left(D_{AB}\frac{\partial \rho_{\text{A}}}{\partial y}\right) + \dot{n}_{\text{A}} \tag{6.47}$$

A more useful form of this equation may be obtained by expanding the second and third terms on the left-hand side and substituting from the overall continuity equation, Equation 6.25. If the total mass density ρ is assumed to be constant, which will be reasonable for all applications of this text, Equation 6.47 then reduces to

$$\frac{\partial \rho_{\text{A}}}{\partial t} + u\frac{\partial \rho_{\text{A}}}{\partial x} + v\frac{\partial \rho_{\text{A}}}{\partial y} = \frac{\partial}{\partial x}\left(D_{AB}\frac{\partial \rho_{\text{A}}}{\partial x}\right) + \frac{\partial}{\partial y}\left(D_{AB}\frac{\partial \rho_{\text{A}}}{\partial y}\right) + \dot{n}_{\text{A}} \tag{6.48}$$

or in molar form

$$\frac{\partial C_{\text{A}}}{\partial t} + u\frac{\partial C_{\text{A}}}{\partial x} + v\frac{\partial C_{\text{A}}}{\partial y} = \frac{\partial}{\partial x}\left(D_{AB}\frac{\partial C_{\text{A}}}{\partial x}\right) + \frac{\partial}{\partial y}\left(D_{AB}\frac{\partial C_{\text{A}}}{\partial y}\right) + \dot{N}_{\text{A}} \tag{6.49}$$

EXAMPLE 6.5

Because of the inherent complexity, there are few situations for which *exact* solutions to the convection transfer equations may be obtained. One such situation, however, involves what is termed *parallel flow*. In this case gross (convective) fluid motion is only in one direction. Consider a special case of parallel flow involving stationary and moving plates of infinite extent separated by a distance L, with the intervening space filled by an incompressible fluid. This situation is referred to as Couette flow with heat transfer and occurs, for example, in a journal bearing. In this case gross fluid motion is in only the x direction.

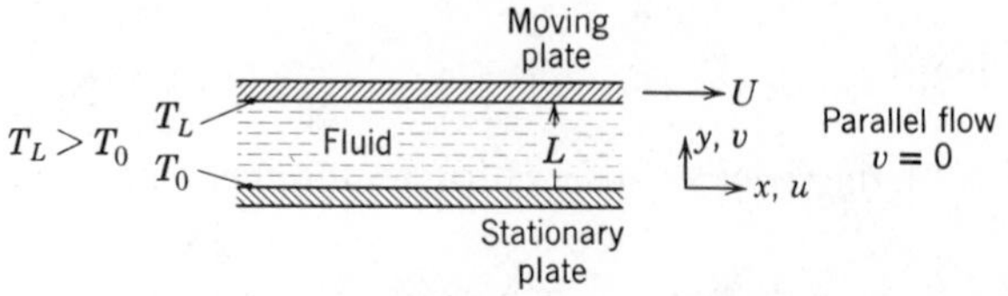

1. What is the appropriate form of the continuity equation, Equation 6.25?

2. Beginning with the momentum equation, Equation 6.29, determine the velocity distribution between the plates.

3. Beginning with the energy equation, Equation 6.40, determine the temperature distribution between the plates.

4. Consider conditions for which the fluid is engine oil with $L = 3$ mm, $U = 10$ m/s, $T_0 = 10°C$ and $T_L = 30°C$. Calculate the heat flux to each of the plates and determine the maximum temperature in the oil.

SOLUTION

KNOWN:

Couette flow with heat transfer.

FIND:

1. Form of the continuity equation.

2. Velocity distribution.

3. Temperature distribution.

4. Surface heat fluxes and maximum temperature for prescribed conditions.

SCHEMATIC:

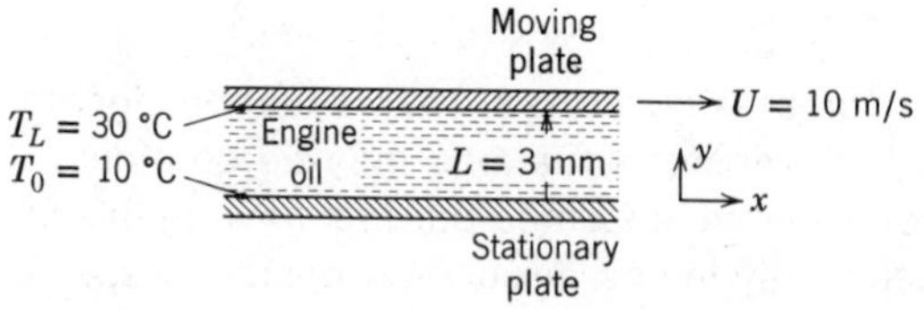

ASSUMPTIONS:

1. Steady-state conditions.

2. Two-dimensional flow (no variations in z).

3. Incompressible fluid with constant properties.

4. No body forces.

5. No internal energy generation.

PROPERTIES:

Table A.8, engine oil (20°C): $\rho = 888.2 \text{ kg/m}^3$, $k = 0.145 \text{ W/m·K}$, $v = 900 \times 10^{-6} \text{ m}^2/\text{s}$, $\mu = v\rho = 0.799 \text{ N·s/m}^2$.

ANALYSIS:

1. For two-dimensional, steady-state conditions, Equation 6.25 reduces to

$$\frac{\partial(\rho u)}{\partial x} + \frac{\partial(\rho v)}{\partial y} = 0$$

Hence for an incompressible fluid (constant ρ) and parallel flow ($v = 0$),

$$\frac{\partial u}{\partial x} = 0 \qquad \qquad \lhd$$

The important implication of this result is that, while depending on y, the x velocity component u is independent of x. It may then be said that the velocity field is *fully developed*.

2. For two-dimensional, steady-state conditions with $v = 0$, $(\partial u/\partial x) = 0$ and $X = 0$, Equation 6.29 reduces to

$$0 = \frac{\partial}{\partial x}(\sigma_{xx} - p) + \frac{\partial \tau_{yx}}{\partial y}$$

However, in Couette flow, motion of the fluid is sustained not by the pressure gradient, dp/dx, but by an external force that provides for motion of the top plate relative to the bottom plate. Hence

$$\frac{\partial p}{\partial x} = 0$$

From Equation 6.31 it also follows that, with $v = (\partial u/\partial x) = 0$,

$$\sigma_{xx} = 0$$

and from Equation 6.33, with $v = 0$,

$$\tau_{xy} = \mu \frac{\partial u}{\partial y}$$

Accordingly, the x-momentum equation reduces to

$$\frac{\partial^2 u}{\partial y^2} = 0$$

The desired velocity distribution may be obtained by solving this equation. Integrating twice, we obtain

$$u(y) = C_1 y + C_2$$

where C_1 and C_2 are the constants of integration. Applying the boundary conditions

$$u(0) = 0 \qquad u(L) = U$$

it follows that $C_2 = 0$ and $C_1 = U/L$. The velocity distribution is then

$$u(y) = \frac{y}{L} U \qquad \qquad \triangleleft$$

3. The energy equation, Equation 6.40, may be simplified for the prescribed conditions. In particular for two-dimensional, steady-state conditions with $v = 0$, $(\partial u/\partial x) = 0$, $(dp/dx) = 0$ and $\dot{q} = 0$, it follows that

$$\rho u \frac{\partial i}{\partial x} = \frac{\partial}{\partial x}\left(k \frac{\partial T}{\partial x}\right) + \frac{\partial}{\partial y}\left(k \frac{\partial T}{\partial y}\right) + \mu \left(\frac{\partial u}{\partial y}\right)^2$$

However, because the top and bottom plates are at uniform temperatures, the temperature field must also be fully developed, in which case $(\partial T/\partial x) = 0$. With $\rho u(\partial i/\partial x) = \rho u c_p (\partial T/\partial x)$ and constant properties, it then follows that the appropriate form of the energy equation is

$$0 = k \frac{\partial^2 T}{\partial y^2} + \mu \left(\frac{\partial u}{\partial y}\right)^2$$

The desired temperature distribution may be obtained by solving this equation. Rearranging and substituting for the velocity distribution,

$$k \frac{d^2 T}{dy^2} = -\mu \left(\frac{du}{dy}\right)^2 = -\mu \left(\frac{U}{L}\right)^2$$

Integrating twice, we obtain

$$T(y) = -\frac{\mu}{2k} \left(\frac{U}{L}\right)^2 y^2 + C_3 y + C_4$$

The constants of integration may be obtained from the boundary conditions

$$T(0) = T_0 \qquad T(L) = T_L$$

in which case

$$C_4 = T_0$$

$$C_3 = \frac{T_L - T_0}{L} + \frac{\mu}{2k}\frac{U^2}{L}$$

and

$$T(y) = T_0 + \frac{\mu}{2k} U^2 \left[\frac{y}{L} - \left(\frac{y}{L}\right)^2\right] + (T_L - T_0)\frac{y}{L} \qquad \triangleleft$$

4. Knowing the temperature distribution, the surface heat fluxes may be obtained by applying Fourier's law. Hence

$$q_y'' = -k\frac{dT}{dy} = -k\left[\frac{\mu}{2k}U^2\left(\frac{1}{L} - \frac{2y}{L^2}\right) + \frac{T_L - T_0}{L}\right]$$

At the bottom and top surfaces, respectively, it follows that

$$q_0'' = -\frac{\mu U^2}{2L} - \frac{k}{L}(T_L - T_0)$$

$$q_L'' = +\frac{\mu U^2}{2L} - \frac{k}{L}(T_L - T_0)$$

Hence, for the prescribed numerical values,

$$q_0'' = -\frac{0.799 \text{ N·s/m}^2 \times 100 \text{ m}^2/\text{s}^2}{2 \times 3 \times 10^{-3} \text{ m}} - \frac{0.145 \text{ W/m·K}}{3 \times 10^{-3} \text{ m}}(30 - 10)°\text{C}$$

$$q_0'' = -13{,}315 \text{ W/m}^2 - 967 \text{ W/m}^2 = -14.3 \text{ kW/m}^2 \qquad \triangleleft$$

$$q_L'' = +13{,}315 \text{ W/m}^2 - 967 \text{ W/m}^2 = 12.3 \text{ kW/m}^2 \qquad \triangleleft$$

The location of the maximum temperature in the oil may be found from the requirement that

$$\frac{dT}{dy} = \frac{\mu}{2k}U^2\left(\frac{1}{L} - \frac{2y}{L^2}\right) + \frac{T_L - T_0}{L} = 0$$

Solving for y it follows that

$$y_{max} = \left[\frac{k}{\mu U^2}(T_L - T_0) + \frac{1}{2}\right]L$$

or for the prescribed conditions

$$y_{max} = \left[\frac{0.145 \text{ W/m} \cdot \text{K}}{0.799 \text{ N} \cdot \text{s/m}^2 \times 100 \text{ m}^2/\text{s}^2} (30 - 10)^\circ\text{C} + \frac{1}{2} \right] L$$

$$y_{max} = 0.536 L$$

Substituting the value of y_{max} into the expression for $T(y)$, it follows that

$$T_{max} = 89.3^\circ\text{C} \qquad \triangleleft$$

COMMENTS:

1. Due to the strong effect of viscous dissipation for the prescribed conditions, the maximum temperature occurs in the oil and there is heat transfer to the hot, as well as to the cold, plate. The temperature distribution is a function of the velocity of the moving plate, and the effect is shown schematically as follows.

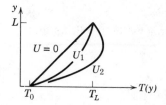

For velocities less than U_1 the maximum temperature corresponds to that of the hot plate. For $U = 0$ there is no viscous dissipation, and the temperature distribution is linear.

2. Recognize that the properties were evaluated at $\bar{T} = (T_L + T_0)/2 = 20^\circ\text{C}$, which is *not* a good measure of the average oil temperature. For more precise calculations, the properties should be evaluated at a more appropriate value of the average temperature (e.g., $\bar{T} \approx 55^\circ\text{C}$), and the calculations should be repeated.

6.5 APPROXIMATIONS AND SPECIAL CONDITIONS

The equations of the preceding section provide a complete account of the physical processes that may influence conditions in the velocity, thermal and concentration boundary layers. However, it is a rare situation when all of the associated terms need to be considered, and it is customary to work with simplified forms of the equations. The usual situation is one in which the two-dimensional boundary layer can be characterized as: *steady* (time-independent), of *constant fluid properties* (k, μ, D_{AB}, etc.), *incompressible* (ρ is constant), with *negligible body forces* $(X = Y = 0)$, *nonreacting* $(\dot{n}_A = 0)$, and *without energy generation* $(\dot{q} = 0)$.

Additional simplifications may be made by invoking what are known as the boundary layer approximations. Because the boundary layer thicknesses are typically very small, the following inequalities are known to apply.

$$\left.\begin{array}{l} u \gg v \\ \dfrac{\partial u}{\partial y} \gg \dfrac{\partial u}{\partial x}, \dfrac{\partial v}{\partial y}, \dfrac{\partial v}{\partial x} \end{array}\right] \begin{array}{l} \text{Velocity} \\ \text{boundary layer} \end{array}$$

$$\left.\dfrac{\partial T}{\partial y} \gg \dfrac{\partial T}{\partial x}\right] \begin{array}{l} \text{Thermal} \\ \text{boundary layer} \end{array}$$

$$\left.\dfrac{\partial C_A}{\partial y} \gg \dfrac{\partial C_A}{\partial x}\right] \begin{array}{l} \text{Concentration} \\ \text{boundary layer} \end{array}$$

That is, the velocity component in the direction along the surface is much larger than that normal to the surface, and gradients normal to the surface are much larger than those along the surface. The normal stresses given by Equations 6.31 and 6.32 are then negligible, and the single relevant shear stress component of Equation 6.33 reduces to

$$\tau_{xy} = \tau_{yx} = \mu\left(\frac{\partial u}{\partial y}\right) \tag{6.50}$$

Moreover, conduction and species diffusion rates for the y direction are much larger than those for the x direction.

Special attention needs to be given to the effect of species transfer on the velocity boundary layer. Recall that velocity boundary layer development is generally characterized by the existence of zero fluid velocity *at the surface*. This condition pertains to the velocity component v normal to the surface, as well as to the velocity component u along the surface. However, if there is sumultaneous mass transfer to or from the surface, it is evident that v can no longer be zero at the surface. Nevertheless, for the mass transfer problems of interest in this text, it will be reasonable to assume that $v = 0$, which is equivalent to assuming that mass transfer has a negligible effect on the velocity boundary layer. The assumption is reasonable for problems involving evaporation or sublimation from gas–liquid or gas–solid interfaces, respectively. It is not reasonable, however, for *mass transfer cooling* problems which involve large surface mass transfer rates. The treatment of such problems is left to the literature [4]. In addition we note that, with mass transfer, the boundary layer fluid is a binary mixture of species A and B, and its properties should be those of the mixture. However, in all problems of interest, $C_A \ll C_B$ and it is reasonable to assume that the boundary layer properties (such as k, μ, c_p, etc.) are those associated with species B.

With the foregoing simplifications and approximations, the overall continuity equation, Equation 6.25, and the x-momentum equation, Equation 6.29, reduce to

$$\frac{\partial u}{\partial x} + \frac{\partial v}{\partial y} = 0 \qquad \text{CONTINUITY SIMPLIFIED} \tag{6.51}$$

$$u\frac{\partial u}{\partial x} + v\frac{\partial u}{\partial y} = -\frac{1}{\rho}\frac{\partial p}{\partial x} + v\frac{\partial^2 u}{\partial y^2} \qquad \text{MOMENTUM SIMPLIFIED} \tag{6.52}$$

In addition, from an order-of-magnitude analysis that uses the velocity boundary layer approximations [1,3], it may be shown that the y-momentum equation, Equation 6.30, reduces to

$$\frac{\partial p}{\partial y} = 0 \tag{6.53}$$

That is, the *pressure does not vary in the direction normal to the surface.* Hence, the pressure in the boundary layer depends only on x and is equal to the pressure in the freestream outside the boundary layer. The form of $p(x)$, which depends on the surface geometry, may then be obtained from a separate consideration of flow conditions in the freestream. Hence, as far as Equation 6.52 is concerned, $(\partial p/\partial x) = (dp/dx)$, and the pressure gradient may be treated as a known quantity.

PRESSURE
GRADIENT
IN X-DIR.
ONLY

With the foregoing simplifications the energy equation, Equation 6.40, reduces to

$$u\frac{\partial T}{\partial x} + v\frac{\partial T}{\partial y} = \alpha\frac{\partial^2 T}{\partial y^2} + \frac{v}{c_p}\left(\frac{\partial u}{\partial y}\right)^2 \qquad \text{ENERGY SIMPLIFIED} \tag{6.54}$$

and the species continuity equation, Equation 6.49, becomes

$$u\frac{\partial C_A}{\partial x} + v\frac{\partial C_A}{\partial y} = D_{AB}\frac{\partial^2 C_A}{\partial y^2} \tag{6.55}$$

Note that the last term on the right-hand side of Equation 6.54 is what remains of the viscous dissipation, Equation 6.42. In most situations this term may be neglected relative to those which account for convection (the left-hand side of the equation) and conduction (the first term on the right-hand side). In fact it is only for supersonic flows or the high-speed motion of lubricating oils that viscous dissipation may not be neglected.

Despite the considerable simplification that has been achieved, the resulting conservation equations, Equations 6.51, 6.52, 6.54, and 6.55 remain difficult to solve. However, it is evident that from such solutions, conditions in the different boundary layers may be determined. For the velocity boundary layer the solution to Equations 6.51 and 6.52 yields the velocity profiles, $u(y)$ and $v(y)$, as a function of x. From knowledge of $u(y)$ the velocity gradient $(\partial u/\partial y)_{y=0}$ could be evaluated, and hence the wall shear stress could be obtained from Equation 6.15. With $u(x, y)$ and $v(x, y)$ known, Equations 6.54 and 6.55 could then be solved for the temperature and species concentration profiles, $T(y)$ and $C_A(y)$, as a function of x. From knowledge of these profiles, the convection heat and mass transfer coefficients could then be determined from Equations 6.17 and 6.20, respectively.

The major objective of a boundary layer analysis is to determine the velocity, temperature, and concentration profiles by solving the foregoing conservation equations. Such solutions are complicated, involving mathematics generally beyond the scope of this text. However, it is not for the purpose of obtaining solutions that the equations have been developed. Instead, a major motivation for this development has been to cultivate an appreciation for the different physical processes which occur in the boundary layers. These processes will, of course, affect wall friction, as well as energy and species transfer across the boundary layers. A second motivation comes from the fact that the equations may be used to suggest key *boundary layer similarity parameters*, as well as important *analogies* between convective *momentum*, *heat*, and *mass* transfer.

6.6 BOUNDARY LAYER SIMILARITY: THE NORMALIZED CONVECTION TRANSFER EQUATIONS

If we examine Equations 6.52, 6.54, and 6.55 more carefully, we note a strong similarity. In fact, if the pressure gradient appearing in Equation 6.52 and the viscous dissipation term of Equation 6.54 are negligible, the three equations are of the same form. *Each equation is characterized by convection terms on the left-hand side and a diffusion term on the right-hand side.* This situation describes *low-speed, forced convection flows*, which are found in many engineering applications and which will occupy much of our attention in this text. Implications of this similarity may be developed in a rational manner by first *nondimensionalizing* the governing equations.

6.6.1 Boundary Layer Similarity Parameters

The boundary layer equations are normalized by first defining dimensionless independent variables of the form

$$x^* \equiv \frac{x}{L} \qquad y^* \equiv \frac{y}{L} \tag{6.56}$$

where L is some *characteristic length* for the surface of interest (e.g., the length of a flat plate). Moreover, dependent dimensionless variables may also be defined as

$$u^* \equiv \frac{u}{u_\infty} \qquad v^* \equiv \frac{v}{u_\infty} \tag{6.57}$$

$$T^* \equiv \frac{T - T_s}{T_\infty - T_s} \tag{6.58}$$

$$C_A^* \equiv \frac{C_A - C_{A,s}}{C_{A,\infty} - C_{A,s}} \tag{6.59}$$

Equations 6.56 to 6.59 may be substituted into Equations 6.52, 6.54, and 6.55 to obtain the dimensionless forms of the conservation equations shown in Table 6.1. Note that viscous dissipation has been neglected and that $p^* \equiv (p/\rho u_\infty^2)$ is a dimensionless pressure. The boundary conditions required to solve the equations are also shown in the table.

From the form of Equations 6.60 to 6.62 three *similarity parameters* may be inferred. Similarity parameters are important because they allow us to apply results obtained for a surface experiencing one set of conditions to *geometrically similar* surfaces experiencing entirely different conditions. These conditions may vary, for example, with the nature of the fluid, the fluid velocity, and/or with the size of the surface (as determined by L).

Beginning with Equation 6.60, we note that the quantity $v/u_\infty L$ is a *dimensionless group* whose reciprocal is termed the Reynolds number.

Table 6.1 The convection transfer equations and their boundary conditions in nondimensional form

BOUNDARY LAYER	CONSERVATION EQUATION	BOUNDARY CONDITIONS		SIMILARITY PARA- METER(S)
		WALL	FREESTREAM	
Velocity	$u^* \dfrac{\partial u^*}{\partial x^*} + v^* \dfrac{\partial u^*}{\partial y^*} = -\dfrac{dp^*}{dx^*} + \dfrac{v}{u_\infty L} \dfrac{\partial^2 u^*}{\partial y^{*2}}$ (6.60)	$u^*(x^*,0) = 0$ $v^*(x^*,0) = 0$	$u^*(x^*,\infty) = 1$ (6.63)	Re_L
Thermal	$u^* \dfrac{\partial T^*}{\partial x^*} + v^* \dfrac{\partial T^*}{\partial y^*} = \dfrac{\alpha}{u_\infty L} \dfrac{\partial^2 T^*}{\partial y^{*2}}$ (6.61)	$T^*(x^*,0) = 0$	$T^*(x^*,\infty) = 1$ (6.64)	Re_L, Pr
Concentration	$u^* \dfrac{\partial C_A^*}{\partial x^*} + v^* \dfrac{\partial C_A^*}{\partial y^*} = \dfrac{D_{AB}}{u_\infty L} \dfrac{\partial^2 C_A^*}{\partial y^{*2}}$ (6.62)	$C_A^*(x^*,0) = 0$	$C_A^*(x^*,\infty) = 1$ (6.65)	Re_L, Sc

Reynolds Number

$$\ast \quad Re_L \equiv \frac{u_\infty L}{v} \tag{6.66}$$

From Equation 6.61 we also note that the term $\alpha/u_\infty L$ is a dimensionless group that may be expressed as $(v/u_\infty L)(\alpha/v) = (Re_L)^{-1}(\alpha/v)$. The ratio of properties, α/v, is also dimensionless and its reciprocal is referred to as the Prandtl number.

Prandtl Number

$$\ast \quad Pr \equiv \frac{v}{\alpha} \tag{6.67}$$

Finally, from the species continuity equation, Equation 6.62, we note that the term $D_{AB}/u_\infty L$ is equivalent to $(v/u_\infty L)(D_{AB}/v) = (Re_L)^{-1}(D_{AB}/v)$. The ratio D_{AB}/v is dimensionless and its reciprocal is termed the Schmidt number.

Schmidt Number

$$Sc \equiv \frac{v}{D_{AB}} \tag{6.68}$$

Using Equations 6.66 through 6.68 with the boundary layer equations, Equations 6.60 to 6.62, and including the dimensionless form of the continuity equation, Equation 6.51, the complete set of boundary layer equations becomes

$$\frac{\partial u^*}{\partial x^*} + \frac{\partial v^*}{\partial y^*} = 0 \tag{6.69}$$

$$u^* \frac{\partial u^*}{\partial x^*} + v^* \frac{\partial u^*}{\partial y^*} = -\frac{dp^*}{dx^*} + \frac{1}{Re_L} \frac{\partial^2 u^*}{\partial y^{*2}} \tag{6.70}$$

$$u^* \frac{\partial T^*}{\partial x^*} + v^* \frac{\partial T^*}{\partial y^*} = \frac{1}{Re_L Pr} \frac{\partial^2 T^*}{\partial y^{*2}} \tag{6.71}$$

$$u^* \frac{\partial C_A^*}{\partial x^*} + v^* \frac{\partial C_A^*}{\partial y^*} = \frac{1}{Re_L Sc} \frac{\partial^2 C_A^*}{\partial y^{*2}} \tag{6.72}$$

6.6.2 Functional Form of the Solutions

The preceding equations are extremely useful from the standpoint of suggesting how important boundary layer results may be simplified and generalized. The momentum equation, Equation 6.70, suggests that, although conditions in the velocity boundary layer depend on the fluid properties, ρ and μ, the velocity u_∞ and the length scale L, this dependence may be simplified by grouping these variables in the form of the Reynolds number. We therefore anticipate that the solution to Equation 6.70 will be of the functional form

$$u^* = f_1\left(x^*, y^*, Re_L, \frac{dp^*}{dx^*}\right) \tag{6.73}$$

Note that the pressure distribution, $p^*(x^*)$, depends on the surface geometry and may be obtained independently by considering flow conditions in the freestream. Hence the appearance of dp^*/dx^* in Equation 6.73 represents the influence of geometry on the velocity distribution.

From Equation 6.15, the shear stress at the surface, $y^* = 0$, may be expressed as

$$\tau_s = \mu \left.\frac{\partial u}{\partial y}\right|_{y=0} = (\mu u_\infty / L) \left.\partial u^* / \partial y^*\right|_{y^*=0}$$

and from Equations 6.14 and 6.66 it follows that the friction coefficient is

$$C_f = \frac{\tau_s}{\rho u_\infty^2 / 2} = \frac{2}{Re_L} \left.\partial u^* / \partial y^*\right|_{y^*=0} \tag{6.74}$$

From Equation 6.73 we also know that

$$\left.\frac{\partial u^*}{\partial y^*}\right|_{y^*=0} = f_2\left(x^*, Re_L, \frac{dp^*}{dx^*}\right)$$

Hence *for a prescribed geometry* Equation 6.74 may be expressed as

$$C_f = \frac{2}{Re_L} f_2(x^*, Re_L) \tag{6.75}$$

The significance of this result should not be overlooked. Equation 6.75 states that the friction coefficient, a dimensionless parameter of considerable importance to the engineer, may be expressed exclusively in terms of a dimensionless space coordinate and the Reynolds number. Hence for a prescribed geometry we expect the function that relates C_f to x^* and Re_L to be *universally* applicable. That is, we expect it to apply to different fluids and over a wide range of values for u_∞ and L.

Similar results may be obtained for the convection coefficients of heat and mass transfer. Intuitively, we might anticipate that h depends on the fluid properties (k, c_p, μ, and ρ), the fluid velocity u_∞, the length scale L, and the surface geometry. However, Equation 6.71 suggests the manner in which this dependence may be simplified. In particular the solution to this equation may be expressed in the form

$$T^* = f_3\left(x^*, y^*, Re_L, Pr, \frac{dp^*}{dx^*}\right) \tag{6.76}$$

where the dependence on dp^*/dx^* originates from the influence of the fluid motion (u^* and v^*) on the thermal conditions. Once again the term dp^*/dx^* represents the effect of surface geometry. From the definition of the convection coefficient, Equation 6.17, and the dimensionless variables, Equations 6.56 and 6.58, we also obtain

$$h = -\frac{k_f}{L}\frac{(T_\infty - T_s)}{(T_s - T_\infty)}\frac{\partial T^*}{\partial y^*}\bigg|_{y^*=0} = +\frac{k_f}{L}\frac{\partial T^*}{\partial y^*}\bigg|_{y^*=0}$$

This expression suggests defining a dependent dimensionless parameter of the form

$$Nu \equiv \frac{hL}{k_f} = +\frac{\partial T^*}{\partial y^*}\bigg|_{y^*=0} \tag{6.77}$$

This parameter, termed the *Nusselt number*, is equal to the dimensionless temperature gradient at the surface, and it provides a measure of the convection heat transfer occurring at the surface. From Equation 6.76 it follows that, *for a prescribed geometry*,

$$Nu = f_4(x^*, Re_L, Pr) \tag{6.78}$$

The Nusselt number is to the thermal boundary layer what the friction coefficient is to the velocity boundary layer. Equation 6.78 implies that for a given geometry, the Nusselt number must be some *universal function* of x^*, Re_L, and Pr. If this function were known, it could then be used to compute the value of Nu for different fluids and for different values of u_∞ and L. From knowledge of Nu, the local convection coefficient h may be found and the *local* heat flux may then be computed from Equation 6.1. Moreover, since the *average* heat transfer coefficient is obtained by integrating over the surface of the body, it must be independent of the spatial variable x^*. Hence the functional dependence of the *average* Nusselt number is

$$\overline{Nu} = \frac{\bar{h}L}{k_f} = f_5(Re_L, Pr) \tag{6.79}$$

Similarly, it may be argued that, for mass transfer in a gas flow over an evaporating liquid or a sublimating solid, the convection mass transfer coefficient h_m depends on the properties D_{AB}, ρ and μ, the velocity u_∞, and the characteristic length L. However, Equation 6.72 suggests that this dependence may be simplified. The solution to this equation must be of the form

$$C_A^* = f_6\left(x^*, y^*, Re_L, Sc, \frac{dp^*}{dx^*}\right) \tag{6.80}$$

where the dependence on dp^*/dx^* again originates from the influence of the fluid motion. Moreover, from the definition of the convection coefficient,

Equation 6.20, and the dimensionless variables, Equations 6.56 and 6.59, we know that

$$h_m = -\frac{D_{AB}}{L}\frac{(C_{A,\infty}-C_{A,s})}{(C_{A,s}-C_{A,\infty})}\frac{\partial C_A^*}{\partial y^*}\Big|_{y^*=0} = +\frac{D_{AB}}{L}\frac{\partial C_A^*}{\partial y^*}\Big|_{y^*=0}$$

Hence we may define a dependent dimensionless parameter of the form

$$Sh \equiv \frac{h_m L}{D_{AB}} = +\frac{\partial C_A^*}{\partial y^*}\Big|_{y^*=0} \tag{6.81}$$

This parameter, termed the *Sherwood number*, is equal to the dimensionless concentration gradient at the surface, and it provides a measure of the convection mass transfer occurring at the surface. From Equation 6.80 it follows that *for a prescribed geometry*

$$Sh = f_7(x^*,\ Re_L,\ Sc) \tag{6.82}$$

The Sherwood number is to the concentration boundary layer what the Nusselt number is to the thermal boundary layer, and Equation 6.82 implies that it must be a universal function of x^*, Re_L, and Sc. As for the Nusselt number, it is also possible to work with an *average* Sherwood number that depends on only Re_L and Sc.

$$\overline{Sh} = \frac{\overline{h}_m L}{D_{AB}} = f_8(Re_L, Sc) \tag{6.83}$$

From the foregoing development we have obtained the relevant dimensionless parameters for low-speed, forced convection boundary layers. We have done so by nondimensionalizing the differential equations that describe the physical processes occurring within the boundary layers. An alternative approach could have involved the use of dimensional analysis in the form of the Buckingham Pi theorem [5]. However, the success of this method depends on one's ability to select, largely from intuition, the various parameters that influence a problem. For example, knowing beforehand that $\overline{h} = f(k, c_p, \rho, \mu, u_\infty, L)$, one could use the Buckingham Pi theorem to obtain Equation 6.79. However, having begun with the differential form of the conservation equations, we have eliminated the guesswork and have established the similarity parameters in a rigorous fashion.

The value of an expression such as Equation 6.79 should be fully appreciated. It states that convection heat transfer results, whether obtained theoretically or experimentally, can be represented in terms of three dimensionless groups, instead of the original seven parameters. The convenience afforded by such simplifications is evident. Moreover, once the form of the functional dependence of Equation 6.79 has been obtained for a particular surface geometry, let us say from laboratory measurements, it is known to be *universally* applicable. By this we mean that it may be applied for different fluids, velocities, and length scales, as long as the assumptions implicit in the originating boundary layer equations remain valid (e.g., negligible viscous dissipation and body forces).

EXAMPLE 6.6

Experimental tests on a portion of the turbine blade shown below indicate a heat flux to the blade of $q'' = 95{,}000 \ \text{W/m}^2$. To maintain a steady-state surface temperature of 800°C, heat transferred to the blade is removed by circulating a coolant inside the blade.

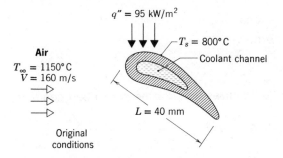

1. Determine the heat flux to the blade if the blade temperature is reduced to 700°C by increasing the coolant flow.
2. Determine the heat flux at the same dimensionless location for a similar turbine blade having a chord length of $L = 80$ mm, when the blade operates in an air flow at $T_\infty = 1150$°C and $V = 80$ m/s, with $T_s = 800$°C.

SOLUTION

KNOWN:

Operating conditions of an internally cooled turbine blade.

FIND:

1. Heat flux to the blade when the surface temperature is reduced.
2. Heat flux to a larger turbine blade of the same shape with reduced air velocity.

SCHEMATIC:

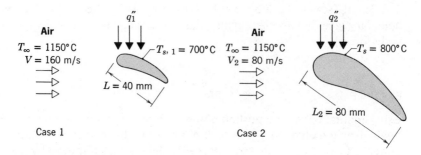

Case 1

Case 2

ASSUMPTIONS:

1. Steady-state conditions.
2. Constant air properties.

ANALYSIS:

1. From Equation 6.78 it follows that, for the prescribed geometry,

$$Nu = \frac{hL}{k} = f_4(x^*, Re_L, Pr)$$

Hence, since there is no change in x^*, Re_L, or Pr associated with a change in T_s for constant properties, the local Nusselt number is unchanged. Moreover, since L and k are also unchanged, the local convection coefficient remains the same. The desired heat flux for Case 1 may then be obtained from Newton's law of cooling

$$q_1'' = h_1(T_\infty - T_s)_1$$

where

$$h_1 = h = \frac{q''}{(T_\infty - T_s)}$$

Hence

$$q_1'' = q'' \frac{(T_\infty - T_s)_1}{(T_\infty - T_s)} = 95{,}000 \text{ W/m}^2 \frac{(1150 - 700)°C}{(1150 - 800)°C}$$

$$q_1'' = 122{,}000 \text{ W/m}^2 \qquad \triangleleft$$

2. To determine the heat flux associated with the larger blade and the reduced air flow (Case 2), we first note that, although L has increased

by a factor of 2, the velocity has decreased by the same factor and the Reynolds number has not changed. That is,

$$Re_{L,2} = \frac{V_2 L_2}{\nu} = \frac{80 \text{ m/s} \times 0.08 \text{ m}}{\nu \text{ (m}^2/\text{s)}} = \frac{6.4}{\nu}$$

$$Re_L = \frac{VL}{\nu} = \frac{160 \text{ m/s} \times 0.04 \text{ m}}{\nu \text{ (m}^2/\text{s)}} = \frac{6.4}{\nu}$$

Accordingly, since x^* and Pr are also unchanged, the local Nusselt number remains the same.

$$Nu_2 = Nu$$

Because the characteristic length is different, however, the convection coefficient changes, where

$$\frac{h_2 L_2}{k} = \frac{hL}{k}$$

$$h_2 = h \frac{L}{L_2} = \frac{q''}{(T_\infty - T_s)} \frac{L}{L_2}$$

The heat flux is then

$$q_2'' = h_2 (T_\infty - T_s) = q'' \frac{(T_\infty - T_s)}{(T_\infty - T_s)} \frac{L}{L_2}$$

$$q_2'' = 95{,}000 \text{ W/m}^2 \times \frac{0.04 \text{ m}}{0.08 \text{ m}}$$

$$q_2'' = 47{,}500 \text{ W/m}^2 \qquad \qquad \triangleleft$$

COMMENTS:

Note that, if the Reynolds number for the two situations of part 2 differed, that is, $Re_{L,2} \neq Re_L$, the heat flux, q_2'', could only be obtained if the particular form of the function f_4 were known. Such forms are provided for many different shapes in subsequent chapters.

6.7 PHYSICAL SIGNIFICANCE OF THE DIMENSIONLESS PARAMETERS

All of the foregoing dimensionless parameters have physical interpretations that relate to conditions in the boundary layers. Consider the *Reynolds number, Re*, Equation 6.66, which may be interpreted as the *ratio of inertia to viscous forces* in the velocity boundary layer. For a differential control volume in this boundary layer, inertia forces are associated with an increase in the momentum flux of fluid moving through the control volume. From Equation 6.28, it is evident that these forces are of the form $\partial[(\rho u)u]/\partial x$, in which case an order of magnitude approximation gives $F_I \approx \rho u_\infty^2/L$. Similarly, the net shear force is of the form $\partial \tau_{yx}/\partial y = \partial[\mu(\partial u/\partial y)]/\partial y$ and may be approximated as $F_s \approx \mu u_\infty/L^2$. Therefore the ratio of forces is

$$\frac{F_I}{F_s} \approx \frac{\rho u_\infty^2/L}{\mu u_\infty/L^2} = \frac{\rho u_\infty L}{\mu} = Re_L$$

We therefore expect inertia forces to dominate for large values of Re and viscous forces to dominate for small Re.

There are several important implications of the above result. Recall that the Reynolds number determines the existence of laminar or turbulent flow. In any flow there exist small disturbances that can be amplified to produce turbulent conditions. However for small Re, viscous forces are sufficiently large relative to inertia forces to prevent this amplification. Hence laminar flow is maintained. However, with increasing Re viscous effects become progressively less important relative to inertia effects, and small disturbances may be amplified to a point where transition occurs. We should also expect the magnitude of the Reynolds number to influence the velocity boundary layer thickness, δ. With increasing Re at a fixed location on a surface, we expect viscous forces to become less influential relative to inertia forces. Hence the effects of viscosity do not penetrate as far into the freestream, and the value of δ diminishes.

The physical interpretation of the *Prandtl number* follows from its definition as a ratio of the momentum diffusivity, v, to the thermal diffusivity, α, Equation 6.67. The Prandtl number provides a *measure of the relative effectiveness of momentum and energy transport by diffusion in the velocity and thermal boundary layers*, respectively. From Table A.4 we see that the Prandtl number of gases is near unity, in which case energy and momentum transfer by diffusion are comparable. In a liquid metal (Table A.7), $Pr \ll 1$ and the energy diffusion rate greatly exceeds the momentum diffusion rate. The opposite is true for oils (Table A.5), for which $Pr \gg 1$. From this interpretation it then follows that the value of Pr strongly influences the relative growth of the velocity and thermal boundary layers. In fact for laminar boundary layers (in which transport by diffusion is *not* overshadowed by turbulent mixing) it is reasonable to expect that

$$\frac{\delta}{\delta_t} \approx Pr^n \tag{6.84}$$

where n is a positive exponent. Hence for a gas $\delta_t \approx \delta$; for a liquid metal $\delta_t \gg \delta$; for an oil $\delta_t \ll \delta$.

Similarly, the *Schmidt number*, which is defined by Equation 6.68, provides *a measure of the relative effectiveness of momentum and mass transport by diffusion in the velocity and concentration boundary layers*, respectively. For convection mass transfer in laminar flows, it therefore determines the relative velocity and concentration boundary layer thicknesses, where

$$\frac{\delta}{\delta_c} \approx Sc^n \tag{6.85}$$

Another parameter, which is related to *Pr* and *Sc*, is the *Lewis number*. It is defined as

$$Le = \frac{\alpha}{D_{AB}} = \frac{Sc}{Pr} \tag{6.86}$$

and is relevant to any situation involving simultaneous heat and mass transfer by convection. From Equations 6.84 to 6.86 it then follows that

$$\frac{\delta_t}{\delta_c} \approx Le^n \tag{6.87}$$

The Lewis number is therefore a measure of the relative thermal and concentration boundary layer thicknesses.

In Table 6.2 those dimensionless groups that appear frequently in the heat and mass transfer literature have been listed. The list includes groups already considered, as well as those yet to be introduced for special convection problems. As a new group is confronted, its definition and interpretation should be commited to memory. Note that the *Grashof number* provides a measure of the ratio of buoyancy forces to viscous forces in the velocity boundary layer. Its role in free convection (Chapter 9) is much the same as that of the Reynolds number in forced convection. The *Eckert number* provides a measure of the kinetic energy of the flow relative to the enthalpy difference across the thermal boundary layer. It plays an important role in high-speed flows for which viscous dissipation is significant [3]. Note also that, although similar in form, the Nusselt and Biot numbers differ in both definition and interpretation. Whereas the Nusselt number is defined in terms of the thermal conductivity of the fluid, the Biot number is based on the solid thermal conductivity, Equation 5.9. Moreover, from the expression

$$Nu = \frac{hL}{k_f} = \frac{h(T_s - T_\infty)}{(k_f/L)(T_s - T_\infty)}$$

the Nusselt number may be interpreted as the ratio of the actual convection heat flux to the conduction heat flux that would occur through a fluid slab of thickness *L*.

Table 6.2   Selected dimensionless groups of heat and mass transfer

GROUP	DEFINITION	INTERPRETATION
Biot number (Bi)	$\dfrac{hL}{k_s}$	Ratio of the internal thermal resistance of a solid to the boundary layer thermal resistance.
Mass transfer Biot number (Bi_m)	$\dfrac{h_m L}{D_{AB}}$	Ratio of the internal species transfer resistance to the boundary layer species transfer resistance.
Coefficient of friction (C_f)	$\dfrac{\tau_s}{\rho u_\infty^2 / 2}$	Dimensionless surface shear stress.
Eckert number (Ec)	$\dfrac{u_\infty^2}{c_p(T_s - T_\infty)}$	Kinetic energy of the flow relative to the boundary layer enthalpy difference.
Fourier number (Fo)	$\dfrac{\alpha t}{L^2}$	Ratio of the heat conduction rate to the rate of thermal energy storage in a solid. Dimensionless time.
Mass transfer Fourier number (Fo_m)	$\dfrac{D_{AB} t}{L^2}$	Ratio of the species diffusion rate to the rate of species storage. Dimensionless time.
Friction factor (f)	$\dfrac{\Delta p}{(L/D)(\rho u_m^2 / 2)}$	Dimensionless pressure drop for internal flow.
Grashof number (Gr_L)	$\dfrac{g\beta(T_s - T_\infty)L^3}{\nu^2}$	Ratio of buoyancy to viscous forces.
Colburn j factor (j_H)	$St Pr^{2/3}$	Dimensionless heat transfer coefficient.
Colburn j factor (j_m)	$St_m Sc^{2/3}$	Dimensionless mass transfer coefficient.
Lewis number (Le)	$\dfrac{\alpha}{D_{AB}}$	Ratio of the molecular thermal and mass diffusivities.
Nusselt number (Nu_L)	$\dfrac{hL}{k_f}$	Ratio of convection heat transfer to conduction in a fluid slab of thickness L.
Peclet number (Pe_L)	$\dfrac{u_\infty L}{\alpha} = Re_L Pr$	Dimensionless independent heat transfer parameter.
Mass transfer Peclet number ($Pe_{L,m}$)	$\dfrac{u_\infty L}{D_{AB}} = Re_L Sc$	Dimensionless independent mass transfer parameter.

Table 6.2 Continued

GROUP	DEFINITION	INTERPRETATION
Prandtl number (Pr)	$\dfrac{c_p\mu}{k} = \dfrac{\nu}{\alpha}$	Ratio of the molecular momentum and thermal diffusivities.
Reynolds number (Re_L)	$\dfrac{u_\infty L}{\nu}$	Ratio of the inertia and viscous forces.
Schmidt number (Sc)	$\dfrac{\nu}{D_{AB}}$	Ratio of the molecular momentum and mass diffusivities.
Sherwood number (Sh_L)	$\dfrac{h_m L}{D_{AB}}$	Ratio of convection mass transfer to diffusion in a slab of thickness L.
Stanton number (St)	$\dfrac{h}{\rho u_\infty c_p} = \dfrac{Nu_L}{Re_L Pr}$	Dimensionless heat transfer coefficient.
Mass transfer Stanton number (St_m)	$\dfrac{h_m}{u_\infty} = \dfrac{Sh_L}{Re_L Sc}$	Dimensionless mass transfer coefficient.

6.8 BOUNDARY LAYER ANALOGIES

As engineers, our interest in boundary layer behavior is directed principally toward the dimensionless parameters C_f, Nu, and Sh. It is from knowledge of these parameters that we may then compute the wall shear stress and the convection heat and mass transfer rates. It is therefore understandable that expressions that relate C_f, Nu, and Sh to each other can be useful tools in convection analysis. Such expressions are available in the form of *boundary layer analogies*.

6.8.1 The Heat and Mass Transfer Analogy

If two or more processes are governed by dimensionless equations of the same form, the processes are said to be *analogous*. Clearly then, from Equations 6.61 and 6.62 and the boundary conditions, Equations 6.64 and 6.65, of Table 6.1, convection heat and mass transfer are similar. Each of the differential equations is comprised of convection and diffusion terms of the same form. Moreover, as shown in Equations 6.71 and 6.72 each equation is related to the velocity field through Re_L, and the parameters Pr and Sc assume analogous roles. The major implication of this similarity is that dimensionless relations that govern thermal boundary layer behavior must be the same as those that govern the concentration boundary layer. Hence the boundary layer temperature and concentration profiles must be of the same functional form.

Recalling the discussion of Section 6.6.2, key features of which are summarized in Table 6.3, the most important result of the heat and mass transfer analogy may be obtained. From the foregoing paragraph, it follows that f_3 of Equation 6.76 must be of the same form as f_6 of Equation 6.80. From Equations 6.77 and 6.81 it then follows that the dimensionless temperature and concentration gradients evaluated at the surface, and therefore the values of Nu and Sh, are similar. That is, f_4 of Equation 6.78 is of the same form as f_7 of Equation 6.82. Similarly, expressions for the average Nusselt and Sherwood numbers, which involve the functions f_5 and f_8 of Equations 6.79 and 6.83, respectively, are also of the same form. *Accordingly, heat and mass transfer relations for a particular geometry are interchangeable.* If, for example, one has performed a set of heat transfer experiments to determine the form of f_4 for a particular surface geometry, the results may be used for convection mass transfer involving the same geometry, simply by replacing Nu with Sh and Pr with Sc.

The analogy may also be used to directly relate the two convection coefficients. In subsequent chapters we will find that Nu and Sh are generally proportional to Pr^n and Sc^n, respectively, where n is a positive exponent less than 1. Anticipating this dependence, we may then use Equations 6.78 and 6.82 to obtain

$$Nu = f'_4(x^*, Re_L) Pr^n$$

$$Sh = f'_7(x^*, Re_L) Sc^n$$

Table 6.3   Functional relations pertinent to the boundary layer analogies

FLUID FLOW	HEAT TRANSFER	MASS TRANSFER
$u^* = f_1\left(x^*, y^*, Re_L, \dfrac{dp^*}{dx^*}\right)$ (6.73)	$T^* = f_3\left(x^*, y^*, Re_L, Pr, \dfrac{dp^*}{dx^*}\right)$ (6.76)	$C_A^* = f_6\left(x^*, y^*, Re_L, Sc, \dfrac{dp^*}{dx^*}\right)$ (6.80)
$C_f = \dfrac{2}{Re_L} \left.\dfrac{\partial u^*}{\partial y^*}\right|_{y^*=0}$ (6.74)	$Nu = \dfrac{hL}{k} = + \left.\dfrac{\partial T^*}{\partial y^*}\right|_{y^*=0}$ (6.77)	$Sh = \dfrac{h_m L}{D_{AB}} = + \left.\dfrac{\partial C_A^*}{\partial y^*}\right|_{y^*=0}$ (6.81)
$C_f = \dfrac{2}{Re_L} f_2(x^*, Re_L)$ (6.75)	$Nu = f_4(x^*, Re_L, Pr)$ (6.78)	$Sh = f_7(x^*, Re_L, Sc)$ (6.82)
– – – – – – –	$\overline{Nu} = f_5(Re_L, Pr)$ (6.79)	$\overline{Sh} = f_8(Re_L, Sc)$ (6.83)

in which case

$$\frac{Nu}{Pr^n} = f'_4(x^*, Re_L) = f'_7(x^*, Re_L) = \frac{Sh}{Sc^n} \tag{6.88}$$

Substituting from Equations 6.77 and 6.81 we then obtain

$$\frac{hL/k}{Pr^n} = \frac{h_m L/D_{AB}}{Sc^n}$$

or, from Equation 6.86,

$$\frac{h}{h_m} = \frac{k}{D_{AB} Le^n} = \rho c_p Le^{1-n} \tag{6.89}$$

The above result may often be put to good use in determining one convection coefficient, for example, h_m, from knowledge of the other coefficient. The same relation may be applied to the average coefficients $\bar{h}$ and $\bar{h}_m$, and it may be used in turbulent, as well as laminar, flow.

EXAMPLE 6.7

A solid of arbitrary shape is suspended in atmospheric air having a free stream temperature and velocity of 20°C and 100 m/s, respectively. The solid has a characteristic length of 1 m, and its surface is maintained at a temperature of 80°C. Under these conditions measurements of the heat flux at a particular point (x^*) on the surface and of the temperature in the boundary layer above this point (x^*, y^*) reveal values of 10^4 W/m² and 60°C, respectively. A mass transfer operation is to be effected for a second solid having the same shape but a characteristic length of 2 m. In particular a thin film of water on the solid is to be evaporated in dry atmospheric air having a free stream velocity of 50 m/s, with the air and the solid both at a temperature of 50°C. What are the molar concentration and the species molar flux of the water vapor at a location (x^*, y^*) corresponding to the point at which the temperature and heat flux measurements were made in the first case?

SOLUTION

KNOWN:

A boundary layer temperature and heat flux at a location on a solid in an air stream of prescribed temperature and velocity.

FIND:

Water vapor concentration and flux associated with the same location on a larger surface of the same shape.

SCHEMATIC:

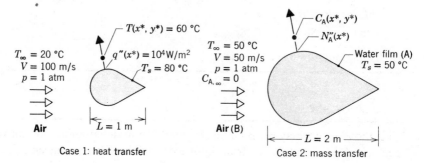

$T_\infty = 20\,°C$
$V = 100\,m/s$
$p = 1\,atm$

Air

$T(x^*, y^*) = 60\,°C$
$q''(x^*) = 10^4\,W/m^2$
$T_s = 80\,°C$

$L = 1\,m$

Case 1: heat transfer

$T_\infty = 50\,°C$
$V = 50\,m/s$
$p = 1\,atm$
$C_{A,\infty} = 0$

Air (B)

$C_A(x^*, y^*)$
$N_A''(x^*)$

Water film (A)
$T_s = 50\,°C$

$L = 2\,m$

Case 2: mass transfer

ASSUMPTIONS:

1. Steady-state, two-dimensional, incompressible boundary layer behavior. Constant properties.
2. Boundary layer approximations are valid.
3. Negligible viscous dissipation.
4. Mole fraction of water vapor in concentration boundary layer is much less than unity.

PROPERTIES:

Table A.4, air (50°C): $\nu = 18.2 \times 10^{-6}\,m^2/s$, $k = 28 \times 10^{-3}\,W/m\cdot K$, $Pr = 0.70$.
Table A.6, saturated water vapor (50°C): $\rho_{A,sat} = v_g^{-1} = 0.082\,kg/m^3$.
Table A.8, water vapor–air (50°C): $D_{AB} \approx 0.26 \times 10^{-4}\,m^2/s$.

ANALYSIS:

The desired molar concentration and flux may be determined by invoking the analogy between heat and mass transfer. From Equations 6.76 and 6.80, we know that

$$T^* \equiv \frac{T - T_s}{T_\infty - T_s} = f_3\left(x^*, y^*, Re_L, Pr, \frac{dp^*}{dx^*}\right)$$

and

$$C_A^* \equiv \frac{C_A - C_{A,s}}{C_{A,\infty} - C_{A,s}} = f_6\left(x^*, y^*, Re_L, Sc, \frac{dp^*}{dx^*}\right)$$

However, for Case 1

$$Re_{L,1} = \frac{V_1 L_1}{\nu} = \frac{100 \text{ m/s} \times 1 \text{ m}}{18.2 \times 10^{-6} \text{ m}^2/\text{s}} = 5.5 \times 10^6, \ Pr = 0.70$$

while for Case 2

$$Re_{L,2} = \frac{V_2 L_2}{\nu} = \frac{50 \text{ m/s} \times 2 \text{ m}}{18.2 \times 10^{-6} \text{ m}^2/\text{s}} = 5.5 \times 10^6,$$

$$Sc = \frac{\nu}{D_{AB}} = \frac{18.2 \times 10^{-6} \text{ m}^2/\text{s}}{26 \times 10^{-6} \text{ m}^2/\text{s}} = 0.70$$

Since $Re_{L,1} = Re_{L,2}$, $Pr = Sc$, $x_1^* = x_2^*$, $y_1^* = y_2^*$, and the surface geometries are the same, it follows that

$$f_3 = f_6$$

Hence

$$\frac{C_A(x^*, y^*) - C_{A,s}}{C_{A,\infty} - C_{A,s}} = \frac{T(x^*, y^*) - T_s}{T_\infty - T_s} = \frac{60 - 80}{20 - 80} = 0.33$$

or, with $C_{A,\infty} = 0$,

$$C_A(x^*, y^*) = C_{A,s}(1 - 0.33) = 0.67 C_{A,s}$$

With

$$C_{A,s} = C_{A,\text{sat}}(50°C) = \frac{\rho_{A,\text{sat}}}{\mathcal{M}_A} = \frac{0.082 \text{ kg/m}^3}{18 \text{ kg/kmol}} = 0.0046 \text{ kmol/m}^3$$

it follows that

$$C_A(x^*, y^*) = 0.67 \ (0.0046 \text{ kmol/m}^3)$$

$$C_A(x^*, y^*) = 0.0031 \text{ kmol/m}^3 \qquad \triangleleft$$

The molar flux may be obtained from Equation 6.7

$$N_A''(x^*) = h_m(C_{A,s} - C_{A,\infty})$$

with h_m evaluated from the analogy. From Equations 6.78 and 6.82 we know that, since $x_1^* = x_2^*$, $Re_{L,1} = Re_{L,2}$, $Pr = Sc$,

$$f_4 = f_7$$

Hence

$$Sh = \frac{h_m L_2}{D_{AB}} = Nu = \frac{hL_1}{k}$$

With $h = q''/(T_s - T_\infty)$ from Newton's law of cooling,

$$h_m = \frac{L_1}{L_2} \times \frac{D_{AB}}{k} \times \frac{q''}{(T_s - T_\infty)} = \frac{1}{2} \times \frac{0.26 \times 10^{-4} \text{ m}^2/\text{s}}{0.028 \text{ W/m} \cdot \text{K}} \times \frac{10^4 \text{ W/m}^2}{(80 - 20)^\circ\text{C}}$$

$$h_m = 0.077 \text{ m/s}$$

Hence

$$N_A''(x^*) = 0.077 \text{ m/s } (0.0046 - 0.0) \text{ kmol/m}^3$$

or

$$N_A''(x^*) = 3.54 \times 10^{-4} \text{ kmol/s} \cdot \text{m}^2 \qquad \triangleleft$$

COMMENTS:

Recognize that the kinematic viscosity of air (ν_B) may be used to evaluate $Re_{L,2}$ since the mole fraction of water vapor in the concentration boundary layer is small.

6.8.2 Evaporative Cooling

An important application of the heat and mass transfer analogy is to the process of evaporative cooling, which occurs whenever a gas flows over a liquid (Figure 6.15). Evaporation must occur from the liquid surface, and the energy associated with the phase change is the latent heat of vaporization of the liquid.[2] Evaporation occurs when liquid molecules near the surface experience collisions

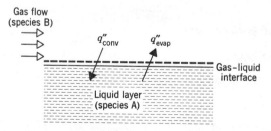

Figure 6.15   Latent and sensible heat exchange at a gas–liquid interface.

---

[2] The latent heat of vaporization, h_{fg}(J/kg), represents the amount of energy required to convert a unit mass of liquid to vapor.

that increase their energy above that needed to overcome the surface binding energy. The energy required to sustain the evaporation must come from the internal energy of the water, which then must experience a reduction in temperature (the cooling effect). However, assuming steady-state conditions, the latent energy lost by the liquid through evaporation is replenished by the transfer of energy to the liquid from its surroundings. Assuming radiative effects to be negligible, this transfer must be by the convection of sensible energy from the gas.

Under steady-state conditions, application of the energy balance, Equation 1.11, to a control surface at the interface then yields

$$q''_{conv} - q''_{evap} = 0 \tag{6.90}$$

where q''_{evap} may be expressed as a product of the mass evaporation rate and the latent heat of vaporization

$$q''_{evap} = \mathcal{M}_A N''_A h_{fg} \tag{6.91}$$

Substituting from Equations 6.1, 6.7, and 6.91, Equation 6.90 becomes

$$h(T_\infty - T_s) = h_{fg} \mathcal{M}_A h_m [C_{A,\,sat}(T_s) - C_{A,\,\infty}] \tag{6.92}$$

where the vapor concentration at the surface is that associated with saturated conditions at T_s. Hence the magnitude of the cooling effect may be expressed as

$$T_\infty - T_s = h_{fg} \mathcal{M}_A (h_m/h)[C_{A,\,sat}(T_s) - C_{A,\,\infty}]$$

Substituting for (h_m/h) from Equation 6.89 and for the molar concentrations from the perfect gas law, Equation 6.13, we then obtain

$$(T_\infty - T_s) = \frac{\mathcal{M}_A h_{fg}}{\mathcal{R} \rho c_p Le^{2/3}} \left[\frac{p_{A,\,sat}(T_s)}{T_s} - \frac{p_{A,\,\infty}}{T_\infty} \right] \tag{6.93}$$

where, in the interest of accuracy, the gas (species B) properties ρ, c_p, and Le should be evaluated at the arithmetic mean temperature of the thermal boundary layer, $T_{am} = (T_s + T_\infty)/2$.

Equation 6.93 may generally be applied to a good approximation. A somewhat less accurate, but more convenient, form may be obtained by assuming that T_s and T_∞ are approximately equal to T_{am}. Equation 6.93 may then be expressed as

$$(T_\infty - T_s) \approx \frac{\mathcal{M}_A h_{fg}}{\mathcal{R} c_p Le^{2/3} \rho T_{am}} [p_{A,\,sat}(T_s) - p_{A,\,\infty}]$$

or, recognizing that $m_A \ll m_B$, we may introduce the expression $(\rho T_{am}) = p/(\mathcal{R}/\mathcal{M}_B)$ from the perfect gas law to obtain

$$(T_\infty - T_s) \approx \frac{(\mathcal{M}_A/\mathcal{M}_B)h_{fg}}{c_p Le^{2/3}} \left[\frac{p_{A,\,sat}(T_s)}{p} - \frac{p_{A,\infty}}{p} \right] \tag{6.94}$$

Numerous environmental and industrial applications of the foregoing results arise for situations in which the gas is *air* and the liquid is *water*.

EXAMPLE 6.8

A container, which is wrapped in a fabric that is continually moistened with a highly volatile liquid, may be used to keep beverages cool in hot arid regions. Consider a situation for which the container is placed in dry ambient air at 40°C, with heat and mass transfer between the wetting agent and the air occurring by forced convection. The wetting agent is known to have a molecular weight of 200 kg/kmol and a latent heat of vaporization of 100 kJ/kg. Its saturated vapor pressure for the prescribed conditions is approximately 5000 N/m^2, and the diffusion coefficient of the vapor in air is 0.2×10^{-4} m^2/s. What is the steady-state temperature of the beverage?

SOLUTION

KNOWN:

Properties of wetting agent used to evaporatively cool a beverage container.

FIND:

Steady-state temperature of beverage.

SCHEMATIC:

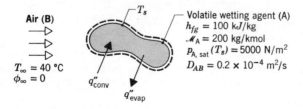

ASSUMPTIONS:

1. Heat and mass transfer analogy is applicable.
2. Perfect gas behavior for vapor.
3. Radiation effects are negligible.

4. Air properties may be evaluated at a mean boundary layer temperature assumed to be 300 K.

PROPERTIES:

Table A.4, air (300 K): $\rho = 1.16$ kg/m^3, $c_p = 1.007$ kJ/kg·K, $\alpha = 22.5 \times 10^{-6}$ m^2/s.

ANALYSIS:

Subject to the foregoing assumptions, the evaporative cooling effect is given by Equation 6.93.

$$(T_\infty - T_s) = \frac{\mathscr{M}_A h_{fg}}{\mathscr{R}\rho c_p Le^{2/3}}\left[\frac{p_{A,sat}(T_s)}{T_s} - \frac{p_{A,\infty}}{T_\infty}\right]$$

Setting $p_{A,\infty} = 0$ and rearranging, it follows that

$$T_s^2 - T_\infty T_s + B = 0$$

where the coefficient B is

$$B = \frac{\mathscr{M}_A h_{fg} p_{A,sat}}{\mathscr{R}\rho c_p Le^{2/3}}$$

or

$$B = [200 \text{ kg/kmol} \times 100 \text{ kJ/kg} \times 5000 \text{ N/m}^2 \times 10^{-3} \text{ kJ/N·m}]/$$
$$\left[8.315 \text{ kJ/kmol·K} \times 1.16 \text{ kg/m}^3 \times 1.007 \text{ kJ/kg·K}\right.$$
$$\left.\times \left(\frac{22.5 \times 10^{-6} \text{ m}^2/\text{s}}{20 \times 10^{-6} \text{ m}^2/\text{s}}\right)^{2/3}\right] = 9514 \text{ K}^2$$

Hence

$$T_s = \frac{T_\infty \pm \sqrt{T_\infty^2 - 4B}}{2} = \frac{313 \text{ K} \pm \sqrt{(313)^2 - 4(9514)} \text{ K}}{2}$$

Rejecting the minus sign on physical grounds (T_s must equal T_∞ if there is no evaporation, in which case $p_{A,sat} = 0$ and $B = 0$), it follows that

$$T_s = 278.9 \text{ K} = 5.9°\text{C} \qquad \triangleleft$$

COMMENTS:

The above result is independent of the shape of the container as long as the heat and mass transfer analogy may be used.

6.8.3 The Reynolds Analogy

A second boundary layer analogy may be obtained by noting from Table 6.1 that, for $dp^*/dx^* = 0$ and $Pr = Sc = 1$, the conservation equations, Equations 6.60 to 6.62, are of precisely the same form. Accordingly, since the boundary conditions, Equations 6.63 to 6.65, also have the same form, it follows that the solutions for u^*, T^*, and C_A^* are equivalent. That is, from Equations 6.73, 6.76, and 6.80 of Table 6.3, $f_1 = f_3 = f_6$. Moreover, the friction coefficient, Nusselt number, and Sherwood number are related by the requirement that $f_2 = f_4 = f_7$, and , from Equations 6.75, 6.78, and 6.82, we conclude that

$$C_f \frac{Re_L}{2} = Nu = Sh \tag{6.95}$$

Replacing Nu and Sh by alternative dependent dimensionless parameters, the *Stanton number*, *St*, and the *mass transfer Stanton number*, St_m, respectively, where

$$St \equiv \frac{h}{\rho u_\infty c_p} = \frac{Nu}{Re\ Pr} \tag{6.96}$$

$$St_m \equiv \frac{h_m}{u_\infty} = \frac{Sh}{Re\ Sc} \tag{6.97}$$

Equation 6.95 may also be expressed in the form

$$\frac{C_f}{2} = St = St_m \tag{6.98}$$

Equation 6.98 is known as the *Reynolds analogy*, and it relates the key engineering parameters of the velocity, thermal, and concentration boundary layers. If the velocity parameter is known, the analogy may then be used to obtain the other parameters and vice versa. However, there are obviously numerous restrictions associated with using this result. In addition to relying on the validity of the assumptions inherent in the boundary layer conservation equations, Equations 6.69 to 6.72, the accuracy of Equation 6.98 depends on having Pr and $Sc \approx 1$ and $dp^*/dx^* \approx 0$. However, it has been shown that the analogy may be applied over a wide range of Pr and Sc, if certain corrections are added. In particular the *modified Reynolds*, or *Chilton-Colburn*, analogies [6,7] have the form

$$\frac{C_f}{2} = St Pr^{2/3} \equiv j_H \qquad\qquad 0.6 < Pr < 60 \tag{6.99}$$

$$\frac{C_f}{2} = St_m Sc^{2/3} \equiv j_m \qquad\qquad 0.6 < Sc < 3000 \qquad\qquad (6.100)$$

where j_H and j_m are the *Colburn j factors* for heat and mass transfer, respectively. For laminar flow Equations 6.99 and 6.100 are only appropriate when $dp^*/dx^* \approx 0$, but in turbulent flow, conditions are less sensitive to the effect of pressure gradients and these equations remain approximately valid. Note that, if the analogy is applicable at every point on a surface, it may then be applied to the surface average coefficients.

6.9 THE EFFECTS OF TURBULENCE

At this point we recognize that turbulent conditions characterize numerous flows of practical interest. In fact the practicing engineer deals more often with turbulent flows than with laminar flows. It is well known that small disturbances associated with distortions in the fluid streamlines of a laminar flow can eventually lead to turbulent conditions. These disturbances may originate from the free stream or be induced by surface roughness. The onset of turbulence depends on whether these disturbances are amplified or attenuated in the direction of fluid flow, which in turn depends on the ratio of the inertia to viscous forces (the Reynolds number). Recall that if the Reynolds number is small, inertia forces are small relative to viscous forces. The naturally occurring disturbances are then dissipated, and the flow remains laminar. For a large Reynolds number, however, the inertia forces are sufficiently large to amplify the disturbances, and a transition to turbulence occurs. In Section 6.3 we observed that the critical Reynolds number, Re_c, required for transition is approximately 5×10^5 for flow over a flat plate.

The onset of turbulence is associated with the existence of *random fluctuations* in the fluid, and at least on a small scale, the flow is inherently *unsteady*. This behavior is shown in Figure 6.16, where the variation of an arbitrary flow property, P, is plotted as a function of time at some location in a turbulent boundary layer. The property P could be a velocity component, the

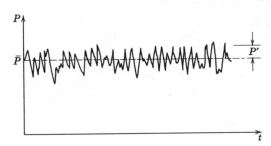

Figure 6.16 Property variation with time at some point in a turbulent boundary layer.

fluid temperature or a species concentration, and at any instant it may be represented as the sum of a *time-mean* value $\bar{P}$ and a fluctuating component P'. The average is taken over a time that is large compared with the period of a typical fluctuation, and if $\bar{P}$ is independent of time, the time-mean flow is said to be *steady*.

The existence of turbulent flow can be advantageous in the sense of providing increased heat and mass transfer rates. However the motion is extremely complicated and difficult to describe theoretically. Although the boundary layer conservation equations developed in previous sections remain applicable, the dependent variables (u, v, T, C_A) must be interpreted as *instantaneous values* and it is impossible to predict their exact variation with time. In the practical sense, however, this inability to determine the variation of the instantaneous properties P with time is not a serious restriction, since the engineer is generally concerned only with the time-mean properties $\bar{P}$. Equations of the form $P = \bar{P} + P'$ may be substituted for each of the flow variables into the boundary layer equations. For steady, incompressible, constant-property flow, using well established time-averaging procedures [1,8], the following forms of the x momentum, energy, and species conservation equations may be obtained

$$\rho\left(\bar{u}\frac{\partial\bar{u}}{\partial x} + \bar{v}\frac{\partial\bar{u}}{\partial y}\right) = -\frac{dp}{dx} + \frac{\partial}{\partial y}\left(\mu\frac{\partial\bar{u}}{\partial y} - \rho\overline{u'v'}\right) \tag{6.101}$$

$$\rho c_p\left(\bar{u}\frac{\partial\bar{T}}{\partial x} + \bar{v}\frac{\partial\bar{T}}{\partial y}\right) = \frac{\partial}{\partial y}\left(k\frac{\partial\bar{T}}{\partial y} - \rho c_p\overline{v'T'}\right) \tag{6.102}$$

$$\left(\bar{u}\frac{\partial\bar{C}_A}{\partial x} + \bar{v}\frac{\partial\bar{C}_A}{\partial y}\right) = \frac{\partial}{\partial y}\left(D_{AB}\frac{\partial\bar{C}_A}{\partial y} - \overline{v'C'_A}\right) \tag{6.103}$$

The equations are like those for the laminar boundary layer, except for the presence of additional terms of the form $\overline{a'b'}$. These terms account for the effect of the turbulent fluctuations on momentum, energy, and species transport.

On the basis of the foregoing results it is customary to speak of a *total* shear stress and *total* fluxes, which are defined as

$$\tau_{\text{tot}} = \left(\mu\frac{\partial\bar{u}}{\partial y} - \rho\overline{u'v'}\right) \tag{6.104}$$

$$q''_{\text{tot}} = -\left(k\frac{\partial\bar{T}}{\partial y} - \rho c_p\overline{v'T'}\right) \tag{6.105}$$

$$N''_{A,\text{tot}} = -\left(D_{AB}\frac{\partial\bar{C}_A}{\partial y} - \overline{v'C'_A}\right) \tag{6.106}$$

and which consist of contributions due to molecular diffusion and turbulent mixing. From the form of these equations we see how momentum, energy, and species transfer rates are enhanced by the existence of turbulence. The term $\overline{\rho u'v'}$ appearing in Equation 6.104 represents the momentum flux due to the turbulent fluctuations, and it is often termed the *Reynolds stress*.

A simple conceptual model, which is widely used, attributes the transport of momentum, heat, and mass in a turbulent boundary layer to the motion of *eddies*, small portions of fluid in the boundary layer which move about for a short time before losing their identity. Due to this motion, the transport of momentum, energy, and species is greatly enhanced. The notion of transport by eddies has prompted the introduction of a transport coefficient which is defined as the *eddy diffusivity for momentum transfer, ε_M*, and which has the form

$$\rho \, \varepsilon_M \frac{\partial \bar{u}}{\partial y} \equiv -\overline{\rho u'v'} \tag{6.107}$$

Hence the total shear stress may be expressed as

$$\tau_{\text{tot}} = \rho \left(v + \varepsilon_M \right) \frac{\partial \bar{u}}{\partial y} \tag{6.108}$$

Similarly, *eddy diffusivities for heat and mass transfer, ε_H and ε_m*, may be defined by the relations

$$\varepsilon_H \frac{\partial \bar{T}}{\partial y} \equiv -\overline{v'T'} \tag{6.109}$$

$$\varepsilon_m \frac{\partial \bar{C}_A}{\partial y} \equiv -\overline{v'C_A'} \tag{6.110}$$

in which case

$$q_{\text{tot}}'' = \rho c_p (\alpha + \varepsilon_H) \frac{\partial \bar{T}}{\partial y} \tag{6.111}$$

$$N_{A,\text{tot}}'' = -(D_{AB} + \varepsilon_m) \frac{\partial \bar{C}_A}{\partial y} \tag{6.112}$$

It is important to note that, in the region of a turbulent boundary layer removed from the surface (the core region), the eddy diffusivities are much larger than the molecular diffusivities. The enhanced mixing associated with this condition has the effect of making velocity, temperature, and concentration profiles more uniform in the core. This behavior is shown in Figure 6.17 for laminar and turbulent velocity boundary layers corresponding to the same free-stream velocity. Accordingly, the velocity gradient at the surface, and therefore

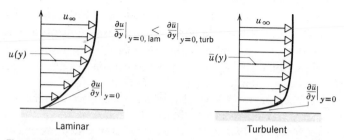

Figure 6.17 Comparison of laminar and turbulent velocity boundary layer profiles for the same freestream velocity.

the surface shear stress, is much larger for the turbulent boundary layer than for the laminar boundary layer. In like manner, it may be argued that the surface temperature or concentration gradient, and therefore the heat or mass transfer rates, are much larger for turbulent than for laminar flow. Because of this enhancement of convection heat and mass transfer rates, it is desirable to have turbulent flow conditions in many engineering applications. However, the increase in wall shear stress will always have the adverse effect of increasing pump or fan power requirements. A fundamental problem in performing a turbulent boundary layer analysis involves determining the eddy diffusivities as a function of the mean properties of the flow. Unlike the molecular diffusivities, which are strictly fluid properties, the eddy diffusivities depend strongly on the nature of the flow, and they can vary from point to point in a boundary layer. Although the problem of determining the eddy diffusivities has received considerable attention [8], there is to date no formulation that holds for all flow conditions.

6.10 THE CONVECTION COEFFICIENTS

In this chapter much material has been presented for the purpose of providing a fundamental understanding of convection transport phenomena. In the process, however, it is hoped that you have not lost sight of what remains *the problem of convection.* Our primary objective is still one of developing the means of determining the convection coefficients, h and h_m. Although these coefficients may be obtained by solving the boundary layer equations, it is only for simple flow situations that such solutions are readily effected. The more practical approach frequently involves calculating h and h_m from empirical equations of the form given by Equations 6.78 and 6.82. The particular form of these equations is obtained by *correlating* measured convection heat and mass transfer results in terms of the appropriate dimensionless groups. It is this approach that is emphasized in the following chapters on convection.

6.11 SUMMARY

This chapter has considered much material, and you may well be fatigued, not to mention overwhelmed, from your efforts to assimilate it. However, it is an important chapter. Attempts have been made to develop, in a logical fashion, the mathematical and physical basis of convection transport, and the ideas are essential to anyone wishing to deal rationally with convection processes.

To test your comprehension of the material, you should challenge yourself with appropriate questions. What are the velocity, thermal, and concentration boundary layers; under what conditions do they develop; and why are they of interest to the engineer? How do laminar and turbulent boundary layers differ, and how does one determine whether a particular boundary layer is laminar or turbulent? There are numerous processes that affect momentum, energy, and species transfer in a boundary layer. What are they? How are they represented mathematically? What are the boundary layer approximations, and in what way do they alter the conservation equations? What are the relevant dimensionless groups for the various boundary layers? How may they be physically interpreted? How will the use of these groups facilitate convection calculations? How is velocity, thermal, and concentration boundary layer behavior analogous? How may the effects of turbulence be treated in a boundary layer analysis? Finally, what is *the* convection transfer problem?

If you can answer questions such as the above, you should feel confident in your understanding of the fundamentals of convection transfer.

REFERENCES

1. Schlichting, H., *Boundary Layer Theory*, 6th Ed., McGraw-Hill, New York, 1968.
2. Bird, R. B., W. E. Stewart, and E. N. Lightfoot, *Transport Phenomena*, Chapters 10 and 18, Wiley, New York, 1966.
3. Eckert, E. R. G. and R. M. Drake, Jr., *Analysis of Heat and Mass Transfer*, McGraw-Hill, New York, 1973, pp. 270–275.
4. Hartnett, J. P., "Mass Transfer Cooling" in W. M. Rohsenow and J. P. Hartnett, Eds., *Handbook of Heat Transfer*, McGraw-Hill, New York, 1973.
5. Kreith, F., *Principles of Heat Transfer*, Chapter 6, 2nd Ed., International Text Book Co., Scranton, Pa., 1965.
6. Colburn, A. P., *Trans. Amer. Inst. Chem. Engrs.*, *29*, 174, 1933.
7. Chilton, T. H. and A. P. Colburn, *Ind. Eng. Chem.*, *26*, 1183, 1934.
8. Hinze, J. O., *Turbulence*, 2nd Ed., McGraw-Hill, New York, 1975.

PROBLEMS

6.1 For flow over a flat plate, the local heat transfer coefficient, h_x, is known to vary as $x^{-1/2}$, where x is the distance from the leading edge ($x = 0$) of the plate. What is the ratio of the average coefficient between the leading edge and some location x on the plate to the local coefficient at x?

6.2 For laminar free convection from a heated vertical surface, the local convection coefficient may be expressed as

$$h_x = C x^{-1/4}$$

where h_x is the coefficient at a distance x from the leading edge of the surface and the quantity C, which depends on the fluid properties, is independent of x. Obtain an expression for the ratio $\bar{h}_x/h_x$, where $\bar{h}_x$ is the average coefficient between the leading edge ($x = 0$) and the x location. Sketch the variation of h_x and $\bar{h}_x$ with x.

6.3 On a summer day the air temperature is $27°C$ and the relative humidity is 30 percent. Water evaporates from the surface of a lake at a rate of $0.10\ \text{kg/h}$ per square meter of water surface area. The temperature of the water is also $27°C$. Determine the value of the convection mass transfer coefficient (m/h).

6.4 It is observed that a 230-mm-diameter pan of water at $23°C$ has a mass loss rate of 1.5×10^{-5} kg/s when the ambient air is dry and at $23°C$.

a) Determine the convection mass transfer coefficient for this situation.

b) Estimate the evaporation mass loss rate when the ambient air has a relative humidity of 50 percent.

c) Estimate the evaporation mass loss rate when the water and ambient air temperatures are $47°C$ assuming the convection mass transfer coefficient remains unchanged and the ambient air is dry.

6.5 The rate at which water is lost due to evaporation from the surface of a waterbody may be determined by measuring the surface recession rate. Consider a summer day for which the temperature of the water and the ambient air are both 305 K and the relative humidity of the air is 40 percent. If the surface recession rate is known to be 0.1 mm/h, what is the rate at which mass is lost due to evaporation per unit surface area? What is the convection mass transfer coefficient?

6.6 The process of photosynthesis, as it occurs in the leaves of a green plant, involves transport of CO_2 from the atmosphere to the chloroplasts of the leaves, and the rate of photosynthesis may be quantified in terms of the rate of CO_2 assimilation by the chloroplasts. This assimilation is strongly influenced by CO_2 transfer through the atmospheric boundary layer which develops on the leaf surface. Under conditions for which the density of CO_2 is 6×10^{-4} kg/m³ in the air and 5×10^{-4} kg/m³ at the leaf surface and the convection mass transfer coefficient is 10^{-2} m/s, what is the rate of photosynthesis in terms of kilograms of CO_2 assimilated per unit time and area of leaf surface?

6.7 In flow over a surface the velocity and temperature profiles are known to be of the form

$$u(y) = Ay + By^2 - Cy^3$$

and

$$T(y) = D + Ey + Fy^2 - Gy^3$$

where the coefficients A through G can be considered as constants at any location on the surface. Obtain expressions for the friction coefficient, C_f, and the convection coefficient, h, in terms of u_∞, T_∞, and appropriate profile coefficients and fluid properties.

6.8) Water at a temperature of $T_\infty = 25°C$ flows over one of the surfaces of a steel wall (AISI 1010) whose temperature is $T_{s,1} = 40°C$. The wall is 0.35 m thick, and its other surface temperature is $T_{s,2} = 100°C$.

a) For steady-state conditions what is the convection coefficient associated with the water flow?

b) What is the temperature gradient in the wall and in the water which is in contact with the wall? Sketch the temperature distribution in the wall and in the adjoining water.

6.9   Species A is evaporating from a flat surface into species B. Assume the concentration profile for species A in the concentration boundary layer is of the form

$$C_A(y) = Dy^2 + Ey + F$$

where D, E, and F are constants at any x location and y is measured along a normal from the surface.

a) Develop an expression for the mass transfer convection coefficient, h_m, in terms of the above constants, the concentration of A in the freestream, $C_{A,\infty}$, and the mass diffusivity, D_{AB}.

b) Using the result from part (a), write an expression for the molar flux of mass transfer by convection for species A.

6.10 A fan, which can provide air speeds up to 50 m/s, is to be used in a low-speed wind tunnel with atmospheric air at 25°C. If one wishes to use the wind tunnel to study flat plate boundary layer behavior up to Reynolds numbers of $Re_x = 10^8$, what is the minimum plate length that should be used? At what distance from the leading edge would transition occur, if the critical Reynolds number is $Re_{x,c} = 5 \times 10^5$?

6.11) Assuming a transition Reynolds number of 5×10^5, determine the distance from the leading edge of a flat plate at which transition will occur for each of the following fluids when $u_\infty = 1$ m/s: atmospheric air, water, engine oil, and mercury. In each case the fluid temperature is 27°C.

6.12 Consider the system with stresses acting on it as shown below for the special case of steady-state conditions with $v = 0$, $T = T(y)$, and $\rho = \text{const}$.

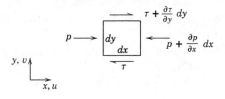

a) Prove that $u = u(y)$ if $v = 0$ everywhere.

b) Derive the x-momentum equation and simplify it as much as possible.

c) Derive the energy equation and simplify it as much as possible.

6.13 A wire manufacturing process involves extrusion of wire of diameter D through platinum dies. The wire is extruded at a constant speed V and is cooled by forced air convection and radiation, as it passes some distance between fixed points at $x = 0$ and $x = L$. The convection is characterized by the coefficient h, and the temperature of the ambient air is T_∞. Radiation is exchanged with surroundings having an effective temperature T_{sur}. After it is cooled, the wire is rolled onto large spools for subsequent shipment. Develop an equation that governs the temperature distribution along the wire between $x = 0$ and $x = L$.

6.14 Consider a lightly loaded journal bearing using oil having the constant properties $\mu = 10^{-2}$ kg/s·m and $k = 0.15$ W/m·K. If the journal and the bearing are each maintained at a temperature of 40°C, what is the maximum temperature in the oil when the journal is rotating at 10 m/s?

6.15 Consider a lightly loaded journal bearing using oil having the constant properties $\rho = 800$ kg/m³, $v = 10^{-5}$ m²/s, and $k = 0.13$ W/m·K. The journal diameter is 75 mm; the clearance is 0.25 mm; and the bearing operates at 3600 rpm.

a) Determine the temperature distribution in the oil film assuming that there is no heat transfer into the journal and that the bearing surface is maintained at 75°C.

b) What is the rate of heat transfer from the bearing, and how much power is needed to rotate the journal?

6.16 Consider two large (infinite) parallel plates separated by a distance of 5 mm. One plate is stationary, while the other plate is moving at a speed of 200 m/s. Both plates are maintained at a temperature of 27°C. Consider two cases, one for which the plates are separated by water and the other for which the plates are separated by air.

a) For each of the two fluids, what is the force per unit surface area required to maintain the above condition? What is the corresponding power requirement?

b) What is the viscous dissipation associated with each of the two fluids?

c) What is the maximum temperature in each of the two fluids?

6.17 Consider Couette flow with heat transfer as described in Example 6.5.

a) Rearrange the temperature distribution to obtain the dimensionless form

$$\theta(\eta) = \eta\left[1 + \frac{1}{2} PrEc(1 - \eta)\right]$$

where $\theta \equiv [T(y) - T_0]/[T_L - T_0]$ and $\eta = y/L$. The dimensionless groups are the Prandtl number, $Pr = \mu c_p/k$, and the Eckert number, $Ec = U^2/c_p(T_L - T_0)$.

b) Derive an expression that prescribes the conditions under which there will be no heat transfer to the upper plate.

c) Derive an expression for the heat transfer rate to the lower plate for the conditions identified in part (b).

6.18 Consider the problem of steady, incompressible laminar flow between two stationary, infinite parallel plates maintained at different temperatures. The fluid is a binary mixture of species A and B, with the molar concentration of A maintained at different values on the two plates.

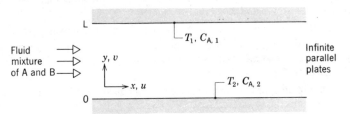

Referred to as *Poiseuille flow* with heat and mass transfer, this special case of parallel flow is one for which the x velocity component is finite, but the y and z components (v and w) are zero.

a) What is the form of the continuity equation for this case? In what way is the flow *fully developed*?

b) What forms do the x and y momentum equations take? What is the form of the velocity profile? Note that, unlike Couette flow, fluid motion between the plates is now sustained by a finite pressure gradient. How is this pressure gradient related to the maximum fluid velocity?

c) Assuming viscous dissipation to be significant and recognizing that conditions must be thermally fully developed, what is the appropriate form of the energy equation? Solve this equation for the temperature distribution. What is the heat flux at the upper ($y = L$) surface?

d) What is the appropriate form of the species A continuity equation? Solve this equation for the species concentration distribution. What is the species flux at the upper surface?

e) Comment on the applicability of an analogy between heat and mass transfer in this situation.

6.19 A simple scheme for desalination involves maintaining a thin film of salt water on the lower surface of two large (infinite) parallel plates which are slightly inclined and separated by a distance L.

A slow, incompressible, laminar air flow exists between the plates, such that the x velocity component is finite while the y and z components are zero. Evaporation occurs from the liquid film on the lower surface, which is maintained at an elevated temperature, T_0, while condensation occurs at the upper surface, which is maintained at a reduced temperature T_L. The corresponding molar concentrations of water vapor at the lower and upper surfaces are designated as $C_{A,0}$ and $C_{A,L}$, respectively. The species concentration and temperature may be assumed to be independent of x and z.

a) Obtain an expression for the distribution of the water vapor molar concentration, $C_A(y)$, in the air. What is the mass rate of pure water production per unit surface area? Express your results in terms of $C_{A,0}$, $C_{A,L}$, L, and the vapor–air diffusion coefficient, D_{AB}.

b) Obtain an expression for the rate at which heat must be supplied per unit area to maintain the lower surface at T_0. Express your result in terms of $C_{A,0}$, $C_{A,L}$, T_0, T_L, L, D_{AB}, h_{fg} (the latent heat vaporization of water), and the thermal conductivity k.

6.20 Consider the convection transfer conservation equations relating to heat transfer, Equations 6.25, 6.29, and 6.40.

a) Identify each of these conservation equations and describe, briefly and clearly, the physical significance of each term in these equations.

b) Identify the approximations and special conditions that are made to reduce these general expressions to the boundary layer equations, Equations 6.51, 6.52, and 6.54.

c) Based on comparison of Equations 6.52 and 6.54, identify under what conditions these equations have the same form and hence the heat and momentum transfer analogy may be applied.

6.21 Consider the conservation equations, Equations 6.40 and 6.49.

a) Describe, briefly and clearly, the physical significance of each of the terms in these equations.

b) Identify the approximations and special conditions that are made to reduce these general expressions to the boundary layer equations, Equations 6.54 and 6.55. Again, describe the physical significance of each term in these equations.

c) Based on a comparison of Equations 6.54 and 6.55, identify under what conditions the equations have the same form and hence the heat and mass transfer analogy may be applied.

6.22 The *falling film* is widely used in chemical processing for the removal of gaseous species. It involves the flow of a liquid along a surface that may be inclined at some $\phi \geq 0$.

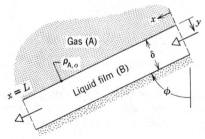

The flow is sustained by gravity, and the gas species A outside the film is absorbed at the liquid–gas interface. The film may be assumed to be in fully developed laminar flow over the entire plate, such that its velocity components in the y and z directions are zero. The mass density of A at $y = 0$ in the liquid is a constant, $\rho_{A,o}$, independent of x.

a) Write the appropriate form of the x-momentum equation for the film. Solve this equation for the distribution of the x velocity component, $u(y)$, in the film. Express your result in terms of δ, g, ϕ, and the liquid properties μ and ρ. Write an expression for the maximum velocity, u_{max}.

b) Obtain an appropriate form of the A species conservation equation for conditions within the film. If it is further assumed that the transport of species A across the gas–liquid interface does not penetrate very far into the film, the position $y = \delta$ may, for all practical purposes, be viewed as $y = \infty$. Moreover, this condition implies that, to a good approximation, $u = u_{max}$ in the region of penetration. Subject to these assumptions, determine an expression for $\rho_A(x,y)$ that applies in the film. (Hint: this problem is analogous to heat transfer in a semiinfinite medium with a sudden change in surface temperature.)

c) If a local mass transfer convection coefficient is defined as

$$ h_{m,x} \equiv \frac{n''_{A,x}}{\rho_{A,o}} $$

where $n''_{A,x}$ is the local mass flux at the gas–liquid interface, develop a suitable correlation for Sh_x as a function of Re_x and Sc.

d) Develop an expression for the total gas absorption rate per unit width for a film of length L (kg/s·m).

e) A water film which is 1 mm thick runs down the inside surface of a vertical tube that is 2 m in length and has an inside diameter of 50 mm. An airstream containing NH_3 moves through the tube, such that the mass density of NH_3 at the gas–liquid interface (but in the liquid) is 25 kg/m³. A dilute solution of ammonia in water is formed, and the diffusion coefficient is 2×10^{-9} m²/s. What is the mass rate of NH_3 removal by absorption?

6.23 An object of irregular shape has a characteristic length of $L = 1$ m and is maintained at a uniform surface temperature of $T_s = 400$ K. When placed in atmospheric air at a temperature of $T_\infty = 300$ K and moving with a velocity of $V = 100$ m/s, the average heat flux from the surface to the air is known to be 20,000 W/m². If a second object of the same shape, but with a characteristic length of $L = 5$ m, is maintained at a surface temperature of $T_s = 400$ K and is placed in atmospheric air at $T_\infty = 300$ K, what will the value of the average convection coefficient be if the air velocity is $V = 20$ m/s?

6.24 Experimental tests have shown that, for air flow conditions of $T_\infty = 35°C$ and $V_1 = 100$ m/s, the rate of heat transfer from a turbine blade of characteristic length $L_1 = 0.15$ m and surface temperature $T_{s,1} = 300°C$ is $q_1 = 1500$ W. What would be the heat transfer rate from a second turbine blade of characteristic length $L_2 = 0.3$ m operating at $T_{s,2} = 400°C$ in air flow of $T_\infty = 35°C$ and $V_2 = 50$ m/s? The surface area of the blade may be assumed to be directly proportional to its characteristic length.

6.25 Experimental measurements of the convection heat transfer coefficient for a square bar in crossflow, as shown below, resulted in these values:

$$\bar{h}_1 = 50 \text{ W/m}^2 \cdot \text{K} \quad \text{when} \quad V_1 = 20 \text{ m/s}$$
$$\bar{h}_2 = 40 \text{ W/m}^2 \cdot \text{K} \quad \text{when} \quad V_2 = 15 \text{ m/s}$$

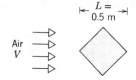

Assume that the functional form of the Nusselt number is $\overline{Nu} = C Re^m Pr^n$ where C, m, and n are constants.

a) What will be the convection heat transfer coefficient for a similar bar with $L = 1$ m when $V = 15$ m/s?

b) What will be the convection heat transfer coefficient for a similar bar with $L = 1$ m when $V = 30$ m/s?

c) Would your results be the same if the side of the bar, rather than its diagonal, were used as the characteristic length?

6.26 An object of irregular shape has a characteristic length of $L = 1$ m and is maintained at a uniform surface temperature of $T_s = 325$ K. It is suspended in an airstream that is at atmospheric pressure ($p = 1$ atm) and has a velocity of $V = 100$ m/s and a temperature of $T_\infty = 275$ K. The average heat flux from the surface to the air is known to be 12,000 W/m^2. Referring to the foregoing situation as Case 1, consider the following cases and determine whether conditions are analogous (similar) to those of Case 1. Each of these cases involves an object of the same shape, which is suspended in an airstream in the same manner. Where similar behavior does exist, determine the corresponding value of the average convection coefficient.

a) The values of T_s, T_∞, and p remain the same, but $L = 2$ m and $V = 50$ m/s.

b) The values of T_s and T_∞ remain the same, but $L = 2$ m, $V = 50$ m/s and $p = 0.2$ atm.

c) The surface is coated with a liquid film that evaporates into the air. The entire system is at 300 K, and the binary diffusion coefficient associated with the air–vapor mixture is $D_{AB} = 1.12 \times 10^{-4}$ m^2/s. Also, $L = 2$ m, $V = 50$ m/s, and $p = 1$ atm.

d) The surface is coated with another liquid film for which $D_{AB} = 1.12 \times 10^{-4}$ m^2/s, and the system is again at 300 K. In this case $L = 2$ m, $V = 250$ m/s, and $p = 0.2$ atm.

6.27 On a cool day in April a scantily clothed runner is known to lose heat at a rate of 2000 W due to convection to the surrounding air at $T_\infty = 10°$C. The runner's skin remains dry and at a temperature of $T_s = 30°$C. Three months later, the runner is moving at the same speed but the day is warm and humid with a temperature of T_∞ = 30°C and a relative humidity of ϕ_∞ = 60 percent. The runner is now drenched in sweat and has a uniform surface temperature of 35°C. Under both conditions constant air properties may be assumed with $v = 1.6 \times 10^{-5}$ m²/s, $k = 0.026$ W/m·K, $Pr = 0.70$, and D_{AB} (water vapor–air) = 2.3×10^{-5} m²/s.

a) What is the rate of water loss due to evaporation on the summer day?

b) What is the total convective heat loss on the summer day?

6.28 Experimental results for heat transfer over a flat plate with an extremely rough surface were found to be correlated by an expression of the form

$$Nu_x = 0.04 Re_x^{0.9} Pr^{1/3}$$

where Nu_x is the local value of the Nusselt number at a position x measured from the leading edge of the plate. Obtain an expression for the ratio of the average heat transfer coefficient $\bar{h}_x$ to the local coefficient h_x.

6.29 Consider conditions for which a fluid with a freestream velocity of $V = 1$ m/s flows over an evaporating or subliming surface with a characteristic length of $L = 1$ m, providing an average mass transfer convection coefficient of $\bar{h}_m = 10^{-2}$ m/s. Calculate the dimensionless parameters $\overline{Sh}_L$, Re_L, Sc, and $\bar{j}_m$ for the following combinations: (a) air flow over water, (b) air flow over naphthelene and (c) warm glycerol over ice. Assume the fluids to be at a temperature of 300 K.

6.30 Consider conditions for which a fluid with a freestream velocity of $V = 1$ m/s flows over a surface with a characteristic length of $L = 1$ m, providing an average convection heat transfer coefficient of $\bar{h} = 100$ W/m²·K. Calculate the dimensionless parameters $\overline{Nu}_L$, Re_L, Pr, and $\bar{j}_H$ for the following fluids: (a) air, (b) engine oil, (c) mercury, and (d) water. Assume the fluids to be at a temperature of 300 K.

6.31 For flow over a flat plate of length L, the local heat transfer coefficient, h_x, is known to vary as $x^{-1/2}$, where x is the distance from the leading edge of the plate. What is the ratio of the average Nusselt number for the entire plate, $\overline{Nu}_L$, to the local Nusselt number at $x = L$, Nu_L?

6.32 For laminar boundary layer flow of air at 20°C and 1 atm, the thermal boundary layer thickness, δ_t, is approximately 13 percent larger than the velocity boundary layer thickness, δ. Determine the ratio δ/δ_t if the fluid is ethylene glycol under the same flow conditions.

6.33 Sketch the variation of the velocity and concentration boundary layer thicknesses with distance from the leading edge of a flat plate for the following laminar flow conditions: (a) air flow over a water film, (b) air flow over a layer of dry ice, (c) air flow over a layer of naphthalene, and (d) the flow of warm glycerol over a layer of ice, which melts and dissolves in the glycerol. For each case assume a mean fluid temperature of 300 K.

6.34 Sketch the variation of the velocity and thermal boundary layer thicknesses with distance from the leading edge of a flat plate for the laminar flow of (a) air, (b) water, (c) engine oil, and (d) mercury. For each case assume a mean fluid temperature of 300 K.

6.35 An object of irregular shape 1 m long maintained at a constant temperature of 100°C is suspended in an airstream having a free stream temperature of 0°C, a pressure of 1 atm, and a velocity of 120 m/s. The air temperature measured at a point near the object in the airstream is found to be 80°C.

A second object having exactly the same shape, but which is 2 m long, is suspended in an airstream in the same manner. The air freestream velocity is 60 m/s. Both the air and the object are at 50°C, and the total pressure is 1 atm. A plastic coating on the surface of the object is being dried by this process. The molecular weight of the vapor is 82 and the saturation pressure at 50°C for the plastic material is 0.0323 atm. The mass diffusivity for the vapor in air at 50°C is 2.60×10^{-5} m²/s. The freestream air can be considered to contain a negligible amount of the vapor.

a) For the second object, at a location corresponding to the point of measurement in the first object, determine the concentration and the partial pressure for the vapor.

b) If the average heat flux, q'', is 2000 W/m² for the first object, determine the average mass flux, n''_A (kg/s·m²), for the second object.

6.36 An industrial process involves the evaporation of water from a liquid film which forms on a contoured surface. Dry air is passed over the surface, and it is known from laboratory measurements that the convection heat transfer correlation for the surface is given by an expression of the form

$$\overline{Nu}_L = 0.43 Re_L^{0.58} Pr^{0.4}$$

a) For an air temperature and velocity of 27°C and 10 m/s, respectively, what is the rate of evaporation of water from a surface of 1 m² area and of characteristic length $L = 1$ m? Approximate the density of saturated vapor corresponding to the liquid temperature as $\rho_{A, sat} = 0.0077$ kg/m³.

b) What is the steady-state temperature of the liquid film?

6.37 A streamlined strut supporting a bearing housing is exposed to a hot air flow from an engine exhaust as shown in the sketch. It is necessary to run experiments to determine the average convection heat transfer coefficient, $\bar{h}$, from the air to the strut in order to be able to cool the strut to the desired surface temperature, T_s. The decision is made to run mass transfer experiments on an object of the same shape and to obtain the desired heat transfer results by using the heat and mass transfer analogy.

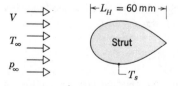

The mass transfer experiments were conducted using a half-size model strut constructed from naphthalene exposed to an airstream at 27°C. Mass transfer measurements yielded these results:

$\overline{Sh}_L$	Re_L
282	60,000
491	120,000
568	144,000
989	288,000

The Nusselt and Sherwood numbers may be assumed to be proportional to $Pr^{1/3}$ and $Sc^{1/3}$, respectively.

a) Using the mass transfer experimental results, determine the coefficients C and m for a correlation of the form $\overline{Sh}_L = C\,Re_L^m\,Sc^{1/3}$.

b) Determine the average convection heat transfer coefficient, $\bar{h}$, for the full-sized strut, $L_H = 60$ mm, when exposed to a freestream air flow with $V = 60$ m/s, $T_\infty = 184°C$, and $p_\infty = 1$ atm when $T_s = 70°C$.

c) The surface area of the strut can be expressed as $A_s = 2.2\,L_H \cdot l$ where l is the length normal to the page. For the conditions of part (b), what would be the change in the total rate of heat transfer to the strut if the characteristic length, L_H, were doubled?

6.38 Warm glycerol flows over a layer of ice whose shape is such that convection heat transfer effects can be correlated by an equation of the form

$$\overline{Nu}_L = 0.25 Re_L^{0.7}\,Pr^{1/3}$$

The ice melts and dissolves in the glycerol. Under conditions for which the average convection heat transfer coefficient is $\bar{h}_L = 100$ W/m²·K, what is the average convection mass transfer coefficient?

6.39 It is known that on clear nights the air temperature need not drop below 0°C before a thin layer of water on the ground will freeze. Consider such a layer of water on a clear night for which the effective sky temperature is $-30°C$ and the convection heat transfer coefficient due to wind motion is $h = 25$ W/m²·K. The water may be assumed to have an emissivity of 1.0 and to be insulated from the ground as far as conduction is concerned.

a) Neglecting evaporation, determine the lowest temperature which the air can have without the water freezing.

b) For the conditions given, estimate the mass transfer coefficient for water evaporation, h_m, in m/s.

c) Accounting now for the effect of evaporation, what is the lowest temperature which the air can have without the water freezing? Assume the air to be dry.

6.40 An expression for the actual water-vapor partial pressure in terms of wet-bulb and dry-bulb temperatures, referred to as the Carrier equation, is given as

$$p_v = p_{gw} - \frac{(p - p_{gw})(T_{DB} - T_{WB})}{1810 - T_{WB}}$$

where

p_v = actual partial pressure, bar

p_{gw} = saturation pressure corresponding to the
wet-bulb temperature, bar

p = total mixture pressure, bar

T_{DB} = dry-bulb temperature, K

T_{WB} = wet-bulb temperature, K

Using this relation, and others to be identified, the subsequent questions relate to air at 1 atm and 37.8°C flowing over a wet-bulb thermometer that indicates 21.1°C.

a) Using Carrier's equation, calculate the partial pressure of the water vapor in the freestream. What is the relative humidity?

b) Refer to a pyschrometric chart and give the relative humidity directly for the conditions indicated. Compare the result with part (a).

c) Use the value of vapor pressure obtained in part (a) with the evaporative cooling relation, Equation 6.94, to calculate a temperature difference between the bulk air and the wet bulb. Compare this calculated temperature difference with the actual temperature difference. What is the percentage difference between the two values?

6.41) A thin flat plate that is 0.2 m by 0.2 m on a side is oriented parallel to an atmospheric airstream having a velocity of 40 m/s. The air is at a temperature of $T_\infty = 20°C$, while the plate is maintained at $T_s = 120°C$. The air flows over the top and bottom surfaces of the plate, and measurement of the resulting drag force reveals a value of 0.075 N. What is the rate of heat transfer from both sides of the plate to the air?

6.42) For flow over a flat plate with an extremely rough surface, convection heat transfer effects are known to be correlated by an expression of the form

$$Nu_x = 0.04Re_x^{0.9} Pr^{1/3}$$

For air flow at a velocity of $u_\infty = 50$ m/s over the plate, what is the surface shear stress at a distance of $x = 1$ m from the leading edge of the plate? Assume the air to be at a temperature of 300 K.

6.43 Many thermal systems involve mass transfer, as well as heat transfer, and the mass transfer effects can significantly influence system performance. For the following systems, identify the pertinent processes, designating them by an appropriately labeled arrow on a sketch of the system.

a) The air–water interface is an important part of the natural environment, since it separates bodies of water from the atmosphere. For a stagnant body of water during the daylight hours, identify all processes that contribute to energy transfer at this interface.

b) Hot coffee is poured into a cup that is sitting on a table in a room. Identify all heat transfer processes relevant to cooling of the coffee. Obtain an equation that could be solved for the variation of the coffee temperature with time (do not attempt to solve).

c) Identify all heat transfer processes occurring at the surface of a sunbather's skin on a day with moderate winds. What additional process must be considered if the sunbather has just climbed from a swimming pool and is covered with a thin film of water? Will the presence of the film make the sunbather feel warmer or cooler? Why?

d) You have perhaps heard it said that, when placed in a freezer, warm water will solidify faster than cold water. A plausible explanation for whether or not such an event is possible may be based on heat and mass transfer considerations. Consider a layer of warm water in a pan that is placed in a freezer compartment. Identify all processes relevant to heat and mass exchange between the free surface of the water layer and the air and walls of the freezer compartment. Describe clearly and concisely the sequence of events that lead to complete freezing of water and explain how it could be possible for warm water to freeze faster than cold water.

e) Solar distillation units are being considered for the desalination of sea water. A simple system is shown as follows.

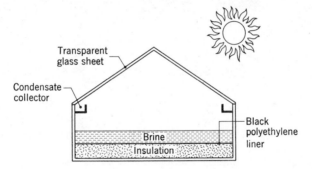

Describe the manner in which the system operates, and identify parameters which influence its performance. Two important variables are the temperature of the water and the temperature of the glass cover plate. Identify the transport processes which determine these temperatures.

f) Paper is dried by passing it over a steam-heated drum. Identify the processes that provide for heat transfer to and from the paper.

7 External Flow

In this chapter we focus on the problem of computing heat and mass transfer rates to or from a surface in *external flow*. In such a flow boundary layers develop freely, without constraints imposed by adjacent surfaces. Accordingly, there will always exist a region of the flow outside the boundary layer in which velocity, temperature, and/or concentration gradients are negligible. Examples include fluid motion over a flat plate (inclined or parallel to the free stream velocity) and flow over curved surfaces such as a sphere, cylinder, airfoil, or turbine blade.

For the moment we confine our attention to problems of *low-speed, forced convection*, with *no phase change* occurring within the fluid. *Forced* convection refers to situations in which the relative motion between the fluid and the surface is maintained by external means, such as a fan or a pump, and not by buoyancy forces due to temperature gradients in the fluid (*natural* convection). *Internal flows, natural convection,* and *convection with phase change* are treated in Chapters 8, 9, and 10, respectively.

Our primary objective is to determine convection coefficients for different flow geometries. In particular we wish to obtain specific forms of the functions which represent these coefficients. By nondimensionalizing the boundary layer conservation equations in Chapter 6, we found that the local and average convection coefficients may be correlated by equations of the form

Heat transfer:

$$Nu_x = f_4(x^*, Re_x, Pr) \tag{6.78}$$

$$\overline{Nu}_x = f_5(Re_x, Pr) \tag{6.79}$$

Mass transfer:

$$Sh_x = f_7(x^*, Re_x, Sc) \tag{6.82}$$

$$\overline{Sh}_x = f_8(Re_x, Sc) \tag{6.83}$$

Note that the subscript x has been added to emphasize our interest in conditions at a particular location on the surface. The overbar then indicates an average from $x^* = 0$, where the boundary layer begins to develop, to the location of interest. Recall that *the problem of convection* is one of obtaining these functions. There are two approaches that we could take, one theoretical and the other experimental.

The *theoretical approach* involves solving the boundary layer equations for a particular geometry. For example, obtaining the temperature profile T^* from such a solution, Equation 6.77 could then be used to evaluate the local Nusselt number, Nu_x, and therefore the local convection coefficient, h_x. With knowledge of how h_x varies over the surface, Equation 6.5 may then be used to determine the average convection coefficient, $\overline{h}_x$, and therefore the Nusselt number, $\overline{Nu}_x$.

Unfortunately, the first step in this procedure, that of solving the boundary layer equations, is often extremely difficult, and even for simple geometries the mathematics are generally cumbersome and beyond the scope of this text.

In Appendices D and E methods for obtaining boundary layer solutions are developed for the simple case of a flat plate in parallel, laminar flow. In Appendix D an exact solution is obtained by using a similarity method; in Appendix E an approximate solution is obtained by using an integral method. Since we will use some of the results in subsequent sections, you should read this material to obtain an appreciation for how dimensionless momentum, heat, and mass transfer correlations may be obtained from boundary layer solutions. However, for the most part we will rely on the *experimental approach* to obtain these correlations.

7.1 THE NATURE OF EMPIRICAL CORRELATIONS

The manner in which a convection heat transfer correlation may be obtained experimentally is illustrated in Figure 7.1. If a prescribed geometry, such as the flat plate in parallel flow, is heated electrically to maintain $T_s > T_\infty$, convection heat transfer occurs from the surface to the fluid. It would be a simple matter to measure T_s and T_∞, as well as the electrical power, $E \cdot I$, which is equal to the total heat transfer rate, q. The convection coefficient, $\bar{h}_L$, which is an average associated with the entire plate, could then be computed from Newton's law of cooling, Equation 6.4. Moreover, from knowledge of the characteristic length L and the fluid properties, the Nusselt, Reynolds, and Prandtl numbers could be computed from their definitions, Equations 6.79, 6.66, and 6.67, respectively.

The foregoing procedure could be repeated for a variety of test conditions. We could vary the velocity u_∞ and the plate length L, as well as the nature of the fluid, using, for example, fluids such as air, water, and engine oil, which have substantially different Prandtl numbers. We would then be left with many different values of the Nusselt number corresponding to a wide range of Reynolds and Prandtl numbers, and the results could be plotted on a *log-log* scale, as shown in Figure 7.2a. Each symbol represents a unique set of test

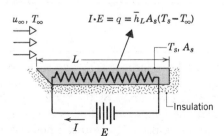

Figure 7.1 Experiment for measuring the average convection heat transfer coefficient, $\bar{h}_L$.

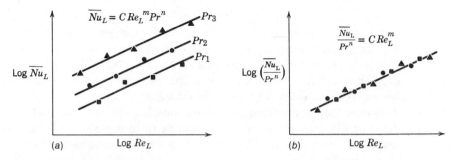

Figure 7.2 Dimensionless representation of convection heat transfer measurements.

conditions. As is often the case, the results associated with a given fluid, and hence a fixed Prandtl number, fall close to a straight line, which may be represented by an algebraic expression of the form

$$\overline{Nu}_L = C\,Re_L^m\,Pr^n \tag{7.1}$$

Since the values of C, m, and n are independent of the nature of the fluid, the family of straight lines corresponding to different Prandtl numbers can be collapsed to a single line by plotting the results in terms of the ratio, $\overline{Nu}_L/Pr^n$, as shown in Figure 7.2b.

Because Equation 7.1 is inferred from experimental measurements, it is termed an *empirical correlation*. Note, however, that the specific values of the coefficient C and the exponents m and n vary with the nature of the surface geometry and the type of flow (laminar or turbulent).

We will use expressions of the form given by Equation 7.1 for many special cases, and it is important to note that the assumption of *constant fluid properties* will often be implicit in the results. However, we know that the fluid properties vary with temperature across the boundary layer and that this variation can certainly influence the heat transfer rate. This influence may be handled in one of two ways. In one method, Equation 7.1 is used with all properties evaluated at a mean boundary layer temperature, T_f, termed the *film temperature*.

$$T_f \equiv \frac{T_s + T_\infty}{2} \tag{7.2}$$

The alternate method is to evaluate all properties at T_∞ and to multiply the right-hand side of Equation 7.1 by an additional parameter to account for the property variations. The parameter is commonly of the form $(Pr_\infty/Pr_s)^r$ or $(\mu_\infty/\mu_s)^r$, where the subscripts ∞ and s designate evaluation of the properties at the freestream and surface temperatures, respectively. Both methods are used in the results that follow.

Finally, we note that experiments may also be performed to obtain convection mass transfer correlations. However, under conditions for which the heat and mass transfer analogy may be applied (Section 6.8.1), the mass transfer correlation must be of precisely the same form as the corresponding heat transfer correlation. Accordingly, we anticipate correlations of the form

$$\overline{Sh}_L = C\,Re_L^m\,Sc^n \tag{7.3}$$

where, for a given geometry and flow condition, the values of C, m, and n are the same as those appearing in Equation 7.1.

7.2 THE FLAT PLATE IN PARALLEL FLOW

Despite its simplicity, parallel flow over a flat plate (Figure 7.3) occurs in numerous engineering applications. Moreover, this geometry is often a reasonable approximation for flow over slightly contoured surfaces, such as an airfoil or turbine blade.

7.2.1 Laminar Flow

The pertinent convection parameters for laminar flow have been obtained from theory (Appendix D), and the accuracy of the results has been confirmed experimentally. They are summarized as follows. The *local* friction coefficient is given by an expression of the form

$$C_{f,x} \equiv \frac{\tau_{s,x}}{\rho u_\infty^2/2} = 0.664 Re_x^{-1/2} \tag{7.4}$$

and the local boundary layer thickness is given by

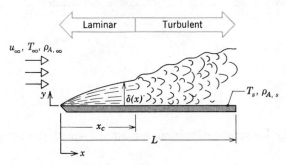

Figure 7.3 The flat plate in parallel flow.

$$\frac{\delta}{x} = 5Re_x^{-1/2} \tag{7.5}$$

where $Re_x \equiv (u_\infty x/v)$. These expressions may be used for any $0 < x < x_c$, where x_c is the distance from the leading edge at which transition occurs. Generally, this distance may be computed from the transition criterion given by Equation 6.24. Moreover, for convection heat and mass transfer at a surface of uniform temperature or species concentration, respectively, the *local* Nusselt and Sherwood numbers are of the form

$$Nu_x \equiv \frac{h_x x}{k} = 0.332 Re_x^{1/2} Pr^{1/3} \qquad (Pr \gtrsim 0.6) \tag{7.6}$$

$$Sh_x \equiv \frac{h_{m,x} x}{D_{AB}} = 0.332 Re_x^{1/2} Sc^{1/3} \qquad (Sc \gtrsim 0.6) \tag{7.7}$$

and the thermal and concentration boundary layer thicknesses are

$$\frac{\delta_t}{\delta} = Pr^{-1/3} \tag{7.8}$$

$$\frac{\delta_c}{\delta} = Sc^{-1/3} \tag{7.9}$$

where δ is given by Equation 7.5. Note that Equations 7.4. 7.6, and 7.7 imply that $\tau_{s,x}$, h_x and $h_{m,x}$ are, in principle, infinite at the leading edge and decrease as $x^{-1/2}$ in the flow direction. Equations 7.8 and 7.9 also imply that for values of Pr and Sc close to unity, which is the case for most gases, the three boundary layers experience nearly identical growth.

From the foregoing local results, the average boundary layer parameters may be determined. With the average friction coefficient defined as

$$\bar{C}_{f,x} \equiv \frac{\bar{\tau}_{s,x}}{\rho u_\infty^2/2} \tag{7.10}$$

where

$$\bar{\tau}_{s,x} \equiv \frac{1}{x} \int_0^x \tau_{s,x}\, dx$$

the form of $\tau_{s,x}$ may be substituted from Equation 7.4 and the integration performed to obtain

✳ $$\bar{C}_{f,x} = 1.328 Re_x^{-1/2} \qquad\qquad (7.11)$$

Moreover, from Equations 6.6 and 7.6 the *average* heat transfer coefficient for laminar flow is

$$\bar{h}_x = \frac{1}{x} \int_0^x h_x \, dx = 0.332(k/x)Pr^{1/3}(u_\infty/\nu)^{1/2} \int_0^x \frac{dx}{x^{1/2}}$$

Integrating and substituting from Equation 7.6, it follows that

✳ $$\bar{h}_x = 2h_x$$

Hence

✳ $$\overline{Nu}_x \equiv \frac{\bar{h}_x x}{k} = 0.664 Re_x^{1/2} Pr^{1/3} \qquad (Pr \gtrsim 0.6) \qquad\qquad (7.12)$$

In a similar fashion, it may be shown that

$$\overline{Sh}_x \equiv \frac{\bar{h}_{m,x} x}{D_{AB}} = 0.664 Re_x^{1/2} Sc^{1/3} \qquad (Sc \gtrsim 0.6) \qquad\qquad (7.13)$$

Note that, if the flow is laminar over the entire surface, the subscript x may be replaced by L, and Equations 7.11, 7.12, and 7.13 may be used to predict average conditions for the entire surface.

From the foregoing expressions we see that, for laminar flow over a flat plate, the *average* friction and convection coefficients from the leading edge up to a point x on the surface are *twice* the *local* coefficients at that point. We also note that, in using these expressions, the effect of variable properties can be treated by evaluating all properties at the *film temperature*, Equation 7.2.

For fluids of small Prandtl number, namely *liquid metals*, Equation 7.6 does not apply. However, for this case the thermal boundary layer development is much more rapid than that of the velocity boundary layer ($\delta_t \gg \delta$), and it is reasonable to assume uniform velocity ($u = u_\infty$) throughout the thermal boundary layer. From a solution to the thermal boundary layer equation based on this assumption [1], it may then be shown that

$$Nu_x \approx 0.565 Pe_x^{1/2} \qquad (Pr \lesssim 0.05) \qquad\qquad (7.14)$$

where $Pe_x \equiv Re_x Pr$ is the *Peclet number* (Table 6.2). Heat transfer in liquid metals is of considerable current interest, particularly as it relates to nuclear reactor cooling. Despite the corrosive and reactive nature of liquid metals, their unique properties (low melting point and vapor pressure, as well as high thermal capacity and conductivity) render them attractive as coolants in applications requiring high heat transfer rates.

7.2.2 Turbulent Flow

From experiment [2] it is known that the *local* friction coefficient for Reynolds numbers between 5×10^5 and 10^7 is well correlated by an expression of the form

$$C_{f,x} = 0.0592 Re_x^{-1/5} \qquad (5 \times 10^5 < Re_x < 10^7) \tag{7.15}$$

The expression may also be used to within 15 percent accuracy for values of Re_x up to 10^8. Moreover, it is known that, to a reasonable approximation, the velocity boundary layer thickness may be expressed as

$$\delta = 0.37x \, Re_x^{-1/5} \tag{7.16}$$

Comparing these results with those for the laminar boundary layer, Equations 7.4 and 7.5, we see that the turbulent boundary layer growth is much more rapid (δ varies as $x^{4/5}$ in contrast to $x^{1/2}$ for laminar flow) and that the decay in the friction coefficient is more gradual ($x^{-1/5}$ versus $x^{-1/2}$). Note, that for turbulent flow, boundary layer development is influenced strongly by random fluctuations in the fluid and not by molecular diffusion. Hence relative boundary layer growth does not depend upon the value of Pr or Sc, and Equation 7.16 may be used to obtain the thermal and concentration, as well as the velocity, boundary layer thicknesses. That is, for turbulent flow, $\delta \approx \delta_t \approx \delta_c$.

Using Equation 7.15 with the modified Reynolds, or Chilton-Colburn, analogy, Equations 6.99 and 6.100, the *local* Nusselt number for turbulent flow is

$$Nu_x = StRe_xPr = 0.0296Re_x^{4/5}Pr^{1/3} \qquad (0.6 < Pr < 60) \tag{7.17}$$

and the *local* Sherwood number is

$$Sh_x = St_mRe_xSc = 0.0296Re_x^{4/5}Sc^{1/3} \qquad (0.6 < Sc < 3000) \tag{7.18}$$

Note that, due to enhanced mixing, the turbulent boundary layer grows more rapidly than the laminar boundary layer and is characterized by larger friction and convection coefficients. This result is consistent with the discussion of laminar and turbulent boundary layer effects provided in Section 6.9.

Expressions for the average coefficients could now be determined using the procedures of the preceding section. Since the turbulent boundary layer is generally preceded by a laminar boundary layer, however, it is convenient to first consider *mixed* boundary layer conditions.

7.2.3 Mixed Boundary Layer Conditions

For laminar flow over the entire plate, Equations 7.11 to 7.13 may be used to compute the average coefficients. Moreover, if transition occurs toward the rear of the plate, for example in the range $0.95 \lesssim (x_c/L) \leq 1$, these equations may be used to compute the average coefficients to a reasonable approximation. However, when transition occurs sufficiently upstream of the rear edge, $(x_c/L) \lesssim 0.95$, the surface average coefficients will be influenced by conditions in both the laminar and turbulent boundary layers.

In the mixed boundary layer situation (Figure 7.3), Equation 6.6 may be used to obtain the average convection heat transfer coefficient for the entire plate. Integrating over the laminar region $(0 \leq x \leq x_c)$ and then over the turbulent region $(x_c < x \leq L)$, this equation may be expressed as

$$\bar{h}_L = \frac{1}{L} \left(\int_0^{x_c} h_{\text{lam}}\, dx + \int_{x_c}^{L} h_{\text{turb}}\, dx \right)$$

where it is assumed that transition occurs abruptly at $x = x_c$. Substituting from Equations 7.6 and 7.17, for h_{lam} and h_{turb}, respectively, we obtain

$$\bar{h}_L = \left(\frac{k}{L} \right) \left[0.332 \left(\frac{u_\infty}{\nu} \right)^{1/2} \int_0^{x_c} \frac{dx}{x^{1/2}} + 0.0296 \left(\frac{u_\infty}{\nu} \right)^{4/5} \int_{x_c}^{L} \frac{dx}{x^{1/5}} \right] Pr^{1/3}$$

Integrating, we then obtain

$$\overline{Nu}_L = [0.664 Re_{x,c}^{1/2} + 0.037(Re_L^{4/5} - Re_{x,c}^{4/5})] Pr^{1/3}$$

or,

$$\overline{Nu}_L = (0.037 Re_L^{4/5} - A) Pr^{1/3} \tag{7.19}$$

where the constant A is determined by the value of the critical Reynolds number, $Re_{x,c}$. That is,

$$A = 0.037 Re_{x,c}^{4/5} - 0.664 Re_{x,c}^{1/2} \tag{7.20}$$

If, from Equation 6.24, the typical transition Reynolds number of $Re_{x,c} = 5 \times 10^5$ is assumed, Equation 7.19 reduces to

$$\overline{Nu}_L = (0.037 Re_L^{4/5} - 871) Pr^{1/3} \tag{7.21}$$

$$\begin{bmatrix} 0.6 < Pr < 60 \\ 5 \times 10^5 < Re_L \lesssim 10^8 \\ Re_{x,c} = 5 \times 10^5 \end{bmatrix}$$

where the bracketed relations indicate the range of applicability of Equation 7.21.

In a similar manner it may be shown that for convection mass transfer

$$\overline{Sh}_L = (0.037Re_L^{4/5} - 871)Sc^{1/3} \tag{7.22}$$

$$\begin{bmatrix} 0.6 < Sc < 3000 \\ 5 \times 10^5 < Re_L \lesssim 10^8 \\ Re_{x,c} = 5 \times 10^5 \end{bmatrix}$$

and for the friction coefficient

$$\overline{C}_{f,L} = \frac{0.074}{Re_L^{1/5}} - \frac{1742}{Re_L} \tag{7.23}$$

$$\begin{bmatrix} 5 \times 10^5 < Re_L \lesssim 10^8 \\ Re_{x,c} = 5 \times 10^5 \end{bmatrix}$$

It is useful to note that in situations for which $L \gg x_c$ $(Re_L \gg Re_{x,c})$, $A \ll 0.037Re_L^{4/5}$ and to a reasonable approximation Equations 7.21 and 7.22 reduce to

$$\overline{Nu}_L = 0.037Re_L^{4/5} Pr^{1/3} \tag{7.24}$$

$$\overline{Sh}_L = 0.037Re_L^{4/5} Sc^{1/3} \tag{7.25}$$

Similarly, it follows that

$$\overline{C}_{f,L} = 0.074Re_L^{-1/5} \tag{7.26}$$

Use of the above results is also appropriate when a turbulent boundary layer exists over the entire plate. Such a condition may be realized by *tripping* the boundary layer at the leading edge, using a fine wire or some other turbulence promoter.

All of the foregoing correlations require evaluation of the fluid properties at the film temperature, Equation 7.2. An alternative approach for treating the temperature dependence of the fluid properties has been suggested by Whitaker [3] on the basis of measurements by Zhukauskas [4]. The heat transfer correlation is of the form

$$\overline{Nu}_L = 0.036(Re_L^{4/5} - 9200)Pr^{0.43}(\mu_\infty/\mu_s)^{1/4} \begin{bmatrix} 0.7 < Pr < 380 \\ 10^5 < Re_L < 5.5 \times 10^6 \\ 0.26 < (\mu_\infty/\mu_s) < 3.5 \end{bmatrix} \tag{7.27}$$

where Pr and all fluid properties appearing in $\overline{Nu}_L$ and Re_L are evaluated at T_∞, while μ_∞ and μ_s are evaluated at T_∞ and T_s, respectively.More recently, Churchill [5] has critically reviewed theoretical and experimental results available in the literature and has suggested a single correlating equation that applies over the complete range of Re and Pr. Different constants are suggested for use with the equation, where specific values depend on the existence of uniform surface temperature or heat flux and whether one is interested in local or average conditions.

The foregoing correlations are suitable for most engineering calculations. However, they should not be viewed as providing exact determinations of the coefficients. Conditions may vary according to the freestream turbulence and surface roughness, and it would be unreasonable to expect the correlations to predict results to better than 25 percent accuracy. This statement may be made for virtually all of the correlations to be considered.

7.3 METHODOLOGY FOR A CONVECTION CALCULATION

Although we have only discussed correlations for parallel flow over a flat plate, it is useful to note that the selection and application of a convection correlation for *any flow situation* is facilitated by following a few simple rules.

1. *Become immediately cognizant of the flow geometry.* Does the problem involve flow over a flat plate, a sphere, a cylinder, etc.? The specific form of the convection correlation depends, of course, on the geometry.

2. *Specify the appropriate reference temperature and then evaluate the pertinent fluid properties at that temperature.* For moderate boundary layer temperature differences, it has been found that the film temperature, Equation 7.2, may be used for this purpose. However, there are correlations, such as Equation 7.27, which require property evaluation at the freestream temperature and include a property ratio to account for the nonconstant property effect.

3. *In mass transfer problems the pertinent fluid properties are those of species B.* In our treatment of convection mass transfer, we are only concerned with *dilute, binary mixtures.* That is, the problems will involve transport of some species A for which $x_A \ll 1$. To a good approximation the properties of the mixture may then be assumed to be the properties of species B. The Schmidt number, for example, would then be $Sc = \nu_B/D_{AB}$ and the Reynolds number would be $Re_L = (u_\infty L/\nu_B)$.

4. *Determine whether the flow is laminar or turbulent.* This determination is made by calculating the Reynolds number and comparing the value with the appropriate transition criterion. For example, if a problem involves

parallel flow over a flat plate for which the Reynolds number is $Re_L = 10^6$ and the transition criterion is $Re_{x,c} = 5 \times 10^5$, it is obvious that a mixed boundary layer condition exists.

5. *Decide whether a local or surface average coefficient is required.* Recall that the local coefficient is used to determine the flux at a particular point on the surface; whereas the average coefficient determines the transfer rate for the entire surface.

Having complied with the foregoing rules, you will then have sufficient information to select the appropriate correlation for the problem.

EXAMPLE 7.1

Air at a pressure of 6 kN/m² and a temperature of 300°C flows with a velocity of 10 m/s over a flat plate of length 0.5 m. Estimate the cooling rate per unit width of the plate needed to maintain it at a surface temperature of 27°C.

SOLUTION

KNOWN:

Air flow over an isothermal flat plate.

FIND:

Cooling rate per unit width of the plate, q'(W/m).

SCHEMATIC:

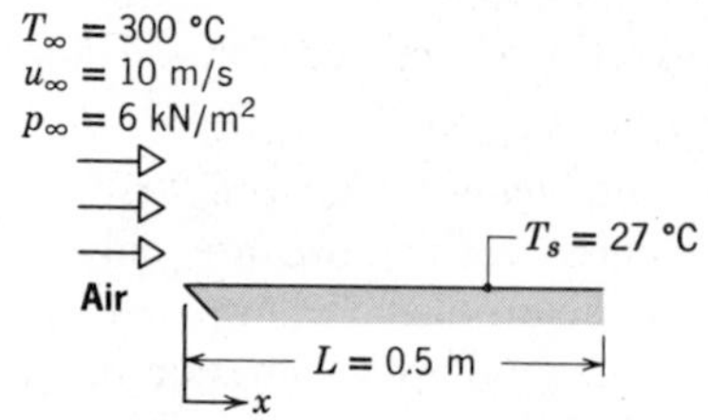

$T_\infty = 300 \ °C$
$u_\infty = 10 \ m/s$
$p_\infty = 6 \ kN/m^2$

Air

$T_s = 27 \ °C$

$L = 0.5 \ m$

x

ASSUMPTIONS:

1. Steady-state conditions.

2. Negligible radiation effects.

PROPERTIES:

Table A.4, air ($T_f = 437$ K, $p = 1$ atm): $v = 30.84 \times 10^{-6}$ m^2/s, $k = 36.4 \times 10^{-3}$ W/m·K, $Pr = 0.687$. Properties such as k, Pr, and μ may be assumed to be independent of pressure to an excellent approximation. However, for a gas the kinematic viscosity, $v = \mu/\rho$, will vary with temperature through its dependence on density. From the perfect gas law, $\rho = p/RT$, it follows that the ratio of kinematic viscosities for a gas at the same temperature but at different pressures, p_1 and p_2, is

$$\frac{v_1}{v_2} = \frac{p_2}{p_1}$$

Hence the kinematic viscosity of air at 437 K and $p_\infty = 6 \times 10^3$ N/m^2 is

$$v = 30.84 \times 10^{-6} \text{ m}^2/\text{s} \times \frac{1.0133 \times 10^5 \text{ N/m}^2}{6 \times 10^3 \text{ N/m}^2}$$

$$v = 5.21 \times 10^{-4} \text{ m}^2/\text{s}$$

ANALYSIS:

For a plate of unit width, it follows from Newton's law of cooling that the rate of convection heat transfer *to* the plate is

$$q' = \bar{h}L(T_\infty - T_s)$$

To determine the appropriate convection correlation for computing $\bar{h}$, the Reynolds number must first be computed

$$Re_L = \frac{u_\infty L}{v} = \frac{10 \text{ m/s} \times 0.5 \text{ m}}{5.21 \times 10^{-4} \text{ m}^2/\text{s}} = 9597$$

Hence the flow is laminar over the entire plate, and the appropriate correlation is given by Equation 7.12.

$$\overline{Nu}_L = 0.664 Re_L^{1/2} Pr^{1/3}$$

$$\overline{Nu}_L = 0.664 (9597)^{1/2} (0.687)^{1/3} = 57.4$$

The average convection coefficient is then

$$\bar{h} = \frac{\overline{Nu}_L k}{L} = \frac{57.4 \times 0.0364 \text{ W/m·K}}{0.5 \text{ m}} = 4.18 \text{ W/m}^2\text{·K}$$

and the required cooling rate per unit width of plate is

$$q' = 4.18 \text{ W/m}^2\text{·K} \times 0.5 \text{ m} (300°C - 27°C)$$

$$q' = 570 \text{ W/m} \qquad \triangleleft$$

COMMENTS:

Recognize that the results of Table A.4 apply to gases at atmospheric pressure. Except for the kinematic viscosity, they may generally be used at other pressures without correction. The kinematic viscosity for a pressure other than one atmosphere may be obtained by dividing the tabulated value by the pressure in atmospheres.

EXAMPLE 7.2

A flat plate of width $w = 1$ m is maintained at a uniform surface temperature, $T_s = 230°C$, by using independently controlled, electrical, strip heaters, each of which is 50 mm long.

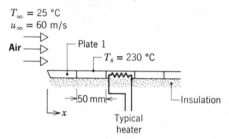

If atmospheric air at 25°C flows over the plate at a velocity of 60 m/s, at what heater is the electrical input a maximum? What is the value of this input?

SOLUTION

KNOWN:

Air flowing over a flat plate with imbedded, segmented heaters.

FIND:

Maximum heater power requirement.

SCHEMATIC:

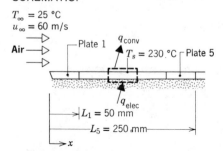

ASSUMPTIONS:

1. Steady-state conditions.
2. Negligible radiation effects.
3. Bottom surface of plate is adiabatic.

PROPERTIES:

Table A.4, air ($T_f = 400$ K, $p = 1$ atm): $v = 26.41 \times 10^{-6}$ m²/s, $k = 0.0338$ W/m·K, $Pr = 0.690$.

ANALYSIS:

The location of the heater requiring the maximum electrical power may be determined by first finding the point of boundary layer transition. The Reynolds number based on the length, L_1, of the first heater is

$$Re_1 = \frac{u_\infty L_1}{v} = \frac{60 \text{ m/s} \times 0.05 \text{ m}}{26.41 \times 10^{-6} \text{ m}^2/\text{s}} = 1.14 \times 10^5$$

If the transition Reynolds number is assumed to be $Re_{x,c} = 5 \times 10^5$, it follows that transition will occur on the fifth heater, or more specifically at

$$x_c = \frac{v}{u_\infty} Re_{x,c} = \frac{26.41 \times 10^{-6} \text{ m}^2/\text{s}}{60 \text{ m/s}} 5 \times 10^5 = 0.22 \text{ m}$$

The heater requiring the maximum electrical power is that for which the average convection coefficient is largest. From knowledge of how the local convection coefficient varies with distance from the leading edge, we conclude that there are three possibilities:

1. heater 1, since it corresponds to the largest local, laminar convection coefficient,
2. heater 5, since it corresponds to the largest local, turbulent convection coefficient, and
3. heater 6, since turbulent conditions exist over the entire heater.

For each of these heaters, conservation of energy requires that

$$q_{\text{elec}} = q_{\text{conv}}$$

For the first heater,

$$q_{\text{conv}, 1} = \bar{h}_1 L_1 w (T_s - T_\infty)$$

where $\bar{h}_1$ is determined from Equation 7.12.

$$\overline{Nu}_1 = 0.664 Re_1^{1/2} Pr^{1/3}$$

$$\overline{Nu}_1 = 0.664(1.14 \times 10^5)^{1/2}(0.69)^{1/3} = 198$$

Hence

$$\bar{h}_1 = \frac{\overline{Nu}_1 k}{L_1} = \frac{198 \times 0.0338 \text{ W/m} \cdot \text{K}}{0.05 \text{ m}} = 134 \text{ W/m}^2 \cdot \text{K}$$

and

$$q_{\text{conv},1} = 134 \text{ W/m}^2 \cdot \text{K} \ (0.05 \times 1)\text{m}^2 (230 - 25)^{\circ}\text{C}$$

$$q_{\text{conv},1} = 1370 \text{ W}$$

The power requirement for the fifth heater may be obtained by subtracting the total heat loss associated with the first four heaters from that associated with the first five heaters. Accordingly,

$$q_{\text{conv},5} = \bar{h}_{1-5}L_5 w(T_s - T_{\infty}) - \bar{h}_{1-4}L_4 w(T_s - T_{\infty})$$

$$q_{\text{conv},5} = (\bar{h}_{1-5}L_5 - \bar{h}_{1-4}L_4)w(T_s - T_{\infty})$$

The value of $\bar{h}_{1-4}$ may be obtained from Equation 7.12, where

$$\overline{Nu}_4 = 0.664 Re_4^{1/2} Pr^{1/3}$$

With $Re_4 = 4Re_1 = 4.56 \times 10^5$,

$$\overline{Nu}_4 = 0.664(4.56 \times 10^5)^{1/2}(0.69)^{1/3} = 396$$

Hence

$$\bar{h}_{1-4} = \frac{\overline{Nu}_4 k}{L_4} = \frac{396 \times 0.0338 \text{ W/m} \cdot \text{K}}{0.2 \text{ m}} = 67 \text{ W/m}^2 \cdot \text{K}$$

In contrast, the fifth heater is characterized by mixed boundary layer conditions, and $\bar{h}_{1-5}$ must be obtained from Equation 7.21. With $Re_5 = 5Re_1 = 5.70 \times 10^5$,

$$\overline{Nu}_5 = (0.037 Re_5^{4/5} - 871)Pr^{1/3}$$

$$\overline{Nu}_5 = [0.037(5.70 \times 10^5)^{4/5} - 871](0.69)^{1/3} = 546$$

Hence

$$\bar{h}_{1-5} = \frac{\overline{Nu}_5 k}{L_5} = \frac{546 \times 0.0338 \text{ W/m} \cdot \text{K}}{0.25 \text{ m}} = 74 \text{ W/m}^2 \cdot \text{K}$$

The rate of heat transfer from the fifth heater is then

$$q_{\text{conv},5} = (74 \text{ W/m}^2 \cdot \text{K} \times 0.25 \text{ m} - 67 \text{ W/m}^2 \cdot \text{K} \times 0.20 \text{ m}) \, 1 \text{ m} (230 - 25)^{\circ}\text{C}$$

$$q_{conv,5} = 1050 \text{ W}$$

Similarly, the power requirement for the sixth heater may be obtained by subtracting the total heat loss associated with the first five heaters from that associated with the first six heaters. Hence

$$q_{conv,6} = (\bar{h}_{1-6}L_6 - \bar{h}_{1-5}L_5)w(T_s - T_\infty)$$

where $\bar{h}_{1-6}$ may be obtained from Equation 7.21. With $Re_6 = 6Re_1 = 6.84 \times 10^5$,

$$\overline{Nu}_6 = [0.037(6.84 \times 10^5)^{4/5} - 871](0.69)^{1/3} = 753$$

Hence

$$\bar{h}_{1-6} = \frac{\overline{Nu}_6 k}{L_6} = \frac{753 \times 0.0338 \text{ W/m}\cdot\text{K}}{0.30 \text{ m}} = 85 \text{ W/m}^2\cdot\text{K}$$

and

$$q_{conv,6} = (85 \text{ W/m}^2\cdot\text{K} \times 0.30 \text{ m} - 74 \text{ W/m}^2\cdot\text{K} \times 0.25 \text{ m}) \, 1 \text{ m} \, (230 - 25)^\circ\text{C}$$

$$q_{conv,6} = 1440 \text{ W} \qquad \triangleleft$$

Hence $q_{conv,6} > q_{conv,1} > q_{conv,5}$, and the sixth plate has the largest power requirement.

COMMENTS:

An alternative method of finding the convection heat transfer rate from a particular plate involves estimating an average local convection coefficient for the surface. For example, Equation 7.17 could be used to evaluate the local convection coefficient at the midpoint of the sixth plate. With $x = 0.275$ m, $Re_x = 6.27 \times 10^5$, $Nu_x = 1136$, and $h_x = 140 \text{ W/m}^2\cdot\text{K}$. The convection heat transfer rate from the sixth plate is then

$$q_{conv,6} = h_x(L_6 - L_5)w(T_s - T_\infty)$$

$$q_{conv,6} = 140 \text{ W/m}^2\cdot\text{K} \, (0.30 - 0.25) \text{ m} \times 1 \text{ m} \, (230 - 25)^\circ\text{C}$$

$$q_{conv,6} = 1440 \text{ W}$$

Recognize that this procedure may only be used when the variation of the local convection coefficient with distance is gradual, such as in turbulent flow. It could lead to significant error when used for a surface that experiences transition.

EXAMPLE 7.3

Drought conditions in the southwestern region of the country have prompted officials to question whether the operation of residential swimming pools should be permitted. As the chief engineer of a city that has a high density of such pools, you have been asked to estimate the *daily* water loss due to pool evaporation. As representative average conditions you may assume water and ambient air temperatures of 25°C, an ambient air relative humidity of 50 percent, pool surface dimensions of 6 m × 12 m, and a wind speed of 2 m/s in the direction of the long (12 m) side of the pool. You may assume the freestream turbulence of the air to be negligible and the surface of the water to be smooth and at a level of the pool deck. What is the water loss for the pool in kilograms per day?

SOLUTION

KNOWN:

Ambient air conditions above a swimming pool.

FIND:

Daily evaporative water loss.

SCHEMATIC:

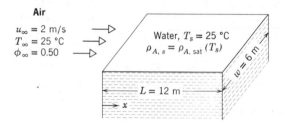

ASSUMPTIONS:

1. Steady-state conditions.
2. Smooth water surface and negligible freestream turbulence.
3. Heat and mass transfer analogy is applicable.
4. Water and air are at the same temperature.

PROPERTIES:

Table A.4, air (25°C): $\nu = 15.7 \times 10^{-6}$ m²/s.
Table A.8, water vapor–air (25°C): $D_{AB} = 0.26 \times 10^{-4}$ m²/s, $Sc = \nu/D_{AB} = 0.60$.
Table A.6, saturated water vapor (25°C): $\rho_{A,\,sat} = v_g^{-1} = 0.0226$ kg/m³.

ANALYSIS:

To determine the nature of boundary layer conditions at the pool surface, the Reynolds number is computed.

$$Re_L = \frac{u_\infty L}{\nu} = \frac{2\text{ m/s} \times 12\text{ m}}{15.7 \times 10^{-6}\text{ m}^2/\text{s}} = 1.53 \times 10^6$$

Hence transition occurs at $x_c = (5 \times 10^5/1.53 \times 10^6)\,12 = 3.9$ m. A mixed boundary layer condition exists, and Equation 7.22 gives

$$\overline{Sh}_L = (0.037\,Re_L^{4/5} - 871)Sc^{1/3}$$

or

$$\overline{Sh}_L = [0.037(1.53 \times 10^6)^{4/5} - 871](0.60)^{1/3} = 2032$$

Hence

$$\bar{h}_{m,\,L} = \overline{Sh}_L\left(\frac{D_{AB}}{L}\right) = 2032\,\frac{0.26 \times 10^{-4}\text{ m}^2/\text{s}}{12\text{ m}} = 4.4 \times 10^{-3}\text{ m/s}$$

The evaporation rate for the pool is

$$n_A = \bar{h}_m A(\rho_{A,\,s} - \rho_{A,\,\infty})$$

or, with the relative humidity defined as,

$$\phi_\infty = \frac{\rho_{A,\,\infty}}{\rho_{A,\,sat}(T_\infty)}$$

and $\rho_{A,\,s} = \rho_{A,\,sat}(T_s)$,

$$n_A = \bar{h}_m A\,[\rho_{A,\,sat}(T_s) - \phi_\infty\,\rho_{A,\,sat}(T_\infty)]$$

Since $T_s = T_\infty = 25°C$, it follows that

$$n_A = \bar{h}_m A\,\rho_{A,\,sat}(25°C)(1 - \phi_\infty)$$

Hence

$$n_A = 4.4 \times 10^{-3}\text{ m/s} \times 72\text{ m}^2 \times 0.0226\text{ kg/m}^3\ (1 - 0.5)3600\text{ s/h} \times 24\text{ h/day}$$

or

$$n_A = 309 \text{ kg/day} \qquad \triangleleft$$

COMMENTS:

The water surface temperature is likely to be slightly less than the air temperature due to the evaporative cooling effect.

7.4 THE CYLINDER IN CROSS FLOW

7.4.1 Flow Considerations

Another common external flow configuration involves the circular cylinder in *cross flow*, where flow is normal to the axis of the cylinder. Consider the fluid mechanics of the situation in Figure 7.4. Boundary layer formation is initiated at the *forward stagnation point*, where the fluid is brought to rest with an accompanying rise in pressure. The pressure is a maximum at this point, and it decreases with increasing x, the streamline coordinate, and θ, the angular coordinate. The boundary layer is then said to be developing under the influence of a *favorable pressure gradient* $(dp/dx < 0)$. However, the pressure must eventually reach a minimum, and towards the rear of the cylinder further boundary layer development occurs in the presence of an *adverse pressure gradient* $(dp/dx > 0)$.

From Euler's equation for an inviscid flow [6], it is also known the velocity distribution, $u_\infty(x)$, must exhibit the opposite behavior. That is, $u_\infty = 0$ at the stagnation point and the fluid must accelerate in the presence of a favorable pressure gradient $(du_\infty/dx > 0$ when $dp/dx < 0)$. However, u_∞ will reach a maximum when $dp/dx = 0$, and beyond this point the fluid must decelerate $(du_\infty/dx < 0$ when $dp/dx > 0)$. From Figure 7.4 note the distinction that has been made between the fluid velocity upstream of the cylinder, V, and the velocity

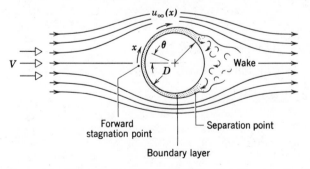

Figure 7.4 Boundary layer formation and separation on a circular cylinder in cross flow.

outside the boundary layer, $u_\infty(x)$. Unlike conditions for the flat plate in parallel flow, these velocities differ, with u_∞ now depending on the distance x from the stagnation point.

The presence of an adverse pressure gradient complicates flow conditions. Its effect is to decelerate the fluid, particularly the slower moving particles near the surface, and a condition is eventually reached for which the velocity gradient at the surface becomes zero (Figure 7.5). At this location, fluid near the surface lacks sufficient momentum to overcome the pressure gradient, and continued downstream movement is impossible. Since the oncoming fluid also precludes flow back upstream, *boundary layer separation* must occur. This is a condition for which the boundary layer detaches from the surface, and a *wake* is formed in the downstream region. Flow in this region is characterized by vortex formation and is highly irregular. The *separation point* is the location for which $(\partial u/\partial y)_s = 0$.

It is important to note that the occurrence of *boundary layer transition*, which depends on the Reynolds number, strongly influences the position of the separation point. For the circular cylinder the characteristic length is the diameter, and the Reynolds number is defined as

$$Re_D \equiv \frac{\rho V D}{\mu} = \frac{V D}{\nu}$$

Since the momentum of fluid in a turbulent boundary layer is larger than in the laminar boundary layer, it is reasonable to expect transition to delay the occurrence of separation. If $Re_D \lesssim 2 \times 10^5$, the boundary layer remains laminar, and separation occurs at $\theta \approx 80°$ (Figure 7.6). However, if $Re_D \gtrsim 2 \times 10^5$, boundary layer transition occurs, and separation is delayed to $\theta \approx 140°$.

The foregoing processes strongly influence the drag force, F_D, acting on the

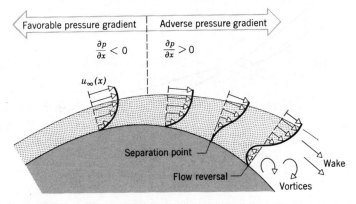

Figure 7.5 Velocity profile associated with separation on a circular cylinder in cross flow.

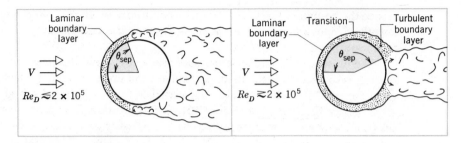

Figure 7.6 The effect of turbulence on separation.

cylinder. This force has two components, one of which is due to the boundary layer surface shear stress (*friction drag*). The other component is due to a pressure differential in the flow direction resulting from formation of the wake (*form*, or *pressure*, *drag*). A dimensionless *drag coefficient*, C_D, may be defined as

$$C_D \equiv \frac{F_D}{A_f(\rho V^2/2)} \tag{7.28}$$

where A_f is the cylinder frontal area (the area projected perpendicular to the free stream velocity). The drag coefficient is a function of Reynolds number and results are presented in Figure 7.7. For $Re_D < 2$ separation effects are negligible,

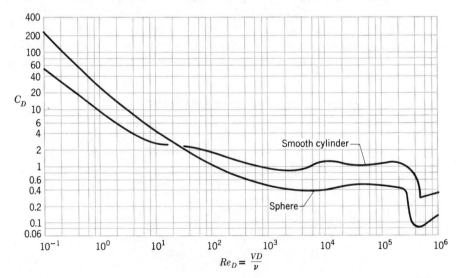

Figure 7.7 Drag coefficient for smooth circular cylinder in cross flow and for a sphere [2]. Adapted with permission.

and conditions are dominated by friction drag. However, with increasing Reynolds number, the effect of separation, and therefore form drag, becomes more important. The large reduction in C_D that occurs for $Re_D > 2 \times 10^5$, is due to boundary layer transition, which delays separation, thereby reducing the extent of the wake region and the magnitude of the form drag.

7.4.2 Convection Heat and Mass Transfer

Because of the complexities associated with flow over a cylinder, emphasis is placed on the use of experimental methods to determine heat and mass transfer effects. Experimental results for the variation of the local Nusselt number with θ are shown in Figure 7.8 for the cylinder in a cross flow of air. Consider the results for $Re_D \lesssim 10^5$. Starting at the stagnation point, Nu_θ decreases with increasing θ due to the laminar boundary layer development. However, a minimum is reached at $\theta \approx 80°$. At this point separation occurs, and Nu_θ

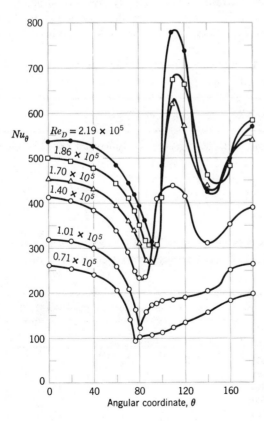

Figure 7.8 Local Nusselt number for airflow normal to a circular cylinder. Adapted with permission from W. H. Giedt, *Trans. ASME*, 71, 375, 1949.

increases with θ due to the mixing associated with vortex formation in the wake. In contrast for $Re_D \gtrsim 10^5$ the variation of Nu_θ with θ is characterized by two minima. The decline in Nu_θ from the value at the stagnation point is again due to laminar boundary layer development, but the sharp increase which occurs between 80 and 100° is now due to boundary layer transition to turbulence. With further development of the turbulent boundary layer, Nu_θ must again begin to decline. However, eventually separation occurs ($\theta \approx 140°$), and Nu_θ must again increase due to the considerable mixing associated with the wake region.

From the standpoint of engineering calculations, we are more interested in overall average conditions. For this purpose an empirical correlation of the form

$$\overline{Nu}_D \equiv \frac{\bar{h}D}{k} = C\,Re_D^m\,Pr^{1/3} \tag{7.29}$$

is widely used, where the constants, C and m, are listed in Table 7.1. Equation 7.29 may also be used for *gas flow* over cylinders of noncircular cross section, with the characteristic length, D, and the constants obtained from Table 7.2. In working with Equation 7.29 all properties are evaluated at the film temperature.

More recently Whitaker [3] has recommended an improved correlation for the circular cylinder in cross flow

$$\overline{Nu}_D = (0.4Re_D^{1/2} + 0.06Re_D^{2/3})Pr^{0.4}(\mu_\infty/\mu_s)^{1/4} \tag{7.30}$$

which is accurate to within ± 25 percent for the following range of conditions

$$\begin{bmatrix} 0.67 < Pr < 300 \\ 10 < Re_D < 10^5 \\ 0.25 < (\mu_\infty/\mu_s) < 5.2 \end{bmatrix}$$

In Equation 7.30 all properties are evaluated at T_∞, except μ_s, which is evaluated at T_s. In addition Zhukauskas [10] has suggested a correlation of the form

Table 7.1 Constants of Equation 7.29 for the circular cylinder in cross flow [7, 8]

Re_D	C	m
0.4–4	0.989	0.330
4–40	0.911	0.385
40–4000	0.683	0.466
4000–40,000	0.193	0.618
40,000–400,000	0.027	0.805

Table 7.2   Constants of Equation 7.29 for noncircular cylinders in cross flow of a gas [9]

GEOMETRY		Re_D	C	m
Square				
$V \rightarrow \diamondsuit$	$\overset{T}{\underset{\perp}{D}}$	$5 \times 10^3 - 10^5$	0.246	0.588
$V \rightarrow \square$	$\overset{T}{\underset{\perp}{D}}$	$5 \times 10^3 - 10^5$	0.102	0.675
Hexagon				
$V \rightarrow \hexagon$	$\overset{T}{\underset{\perp}{D}}$	$5 \times 10^3 - 1.95 \times 10^4$ $1.95 \times 10^4 - 10^5$	0.160 0.0385	0.638 0.782
$V \rightarrow \hexagon$	$\overset{T}{\underset{\perp}{D}}$	$5 \times 10^3 - 10^5$	0.153	0.638
Vertical Plate				
$V \rightarrow \vert$	$\overset{T}{\underset{\perp}{D}}$	$4 \times 10^3 - 1.5 \times 10^4$	0.228	0.731

$$\overline{Nu}_D = C Re_D^m Pr^n (Pr_\infty / Pr_s)^{1/4} \tag{7.31}$$

$$\begin{bmatrix} 0.7 < Pr < 500 \\ 1 < Re_D < 10^6 \end{bmatrix}$$

where all properties are evaluated at T_∞, except Pr_s, which is evaluated at T_s. Values of C and m are listed in Table 7.3. If $Pr \leq 10$, $n = 0.37$; if $Pr > 10$, $n = 0.36$. Most recently, Churchill and Bernstein [11] have proposed a single comprehensive equation which covers the entire range of Re_D for which data are available, as well as a wide range of Pr. The equation is recommended for all $Re_D Pr > 0.2$ and has the form

$$\overline{Nu}_D = 0.3 + \frac{0.62 Re_D^{1/2} Pr^{1/3}}{[1 + (0.4/Pr)^{2/3}]^{1/4}} \left[1 + \left(\frac{Re_D}{28200} \right)^{5/8} \right]^{4/5} \tag{7.32}$$

where all properties are evaluated at the film temperature.

Table 7.3   Constants of Equation 7.31 for the circular cylinder in cross flow [10]

Re_D	C	m
1–40	0.75	0.4
40 – 1000	0.51	0.5
$10^3 - 2 \times 10^5$	0.26	0.6
$2 \times 10^5 - 10^6$	0.076	0.7

Again we are prompted to caution the reader not to view any of the foregoing correlations as sacrosanct. Each correlation is reasonable over a certain range of conditions, but for most engineering calculations one should not expect accuracy to much better than 25 percent. Because they are based on more recent results encompassing a wide range of conditions, Equations 7.31 and 7.32 are used for the calculations of this text. A detailed review of the many correlations that have been developed for the circular cylinder is provided by Morgan [12].

Finally, we note that by invoking the heat and mass transfer analogy, Equations 7.29 to 7.32 may be applied to problems involving convection mass transfer from a cylinder in cross flow. It is simply a matter of replacing $\overline{Nu}_D$ by $\overline{Sh}_D$ and Pr by Sc. Note that, in mass transfer problems, boundary layer property variations are typically small. Hence, when using the mass transfer analog of Equations 7.30 and 7.31, the property ratio, which accounts for nonconstant property effects, may be neglected.

EXAMPLE 7.4

Experiments have been conducted by students in our laboratory on a metallic cylinder of 12.7 mm diameter and 94 mm length. The cylinder is heated internally by an electrical heater and is subjected to a cross flow of air in a low-speed wind tunnel. Under a specific set of operating conditions for which the free stream air velocity and temperature were maintained at $V = 10$ m/s and 26.2°C, respectively, the heater power dissipation was measured to be $P = 46$ W, while the average cylinder surface temperature was determined to be $T_s = 128.4$°C. It is estimated that 15 percent of the power dissipation is lost through the insulated end pieces.

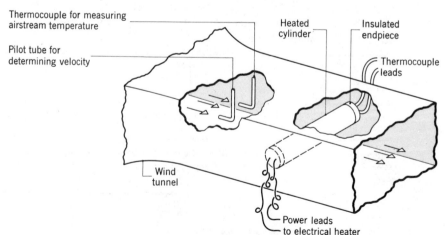

1. Determine the convection heat transfer coefficient from the experimental observations.
2. Compare the foregoing result with convection coefficients obtained by using appropriate convection correlations for the prescribed conditions.

SOLUTION

KNOWN:

Operating conditions associated with a heated cylinder in a wind tunnel.

FIND:

1. Convection coefficient associated with the operating conditions.
2. Convection coefficients obtained from appropriate correlations.

SCHEMATIC:

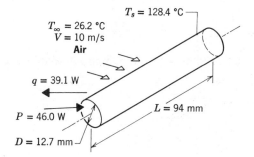

ASSUMPTIONS:

1. Steady-state conditions.
2. Uniform cylinder surface temperature.
3. Negligible radiation effects.

PROPERTIES:

Table A.4, air $(T_\infty = 26.2°C \approx 300$ K): $\mu = 184.6 \times 10^{-7}$ N·s/m², $\nu = 15.89 \times 10^{-6}$ m²/s, $k = 26.3 \times 10^{-3}$ W/m·K, $Pr = 0.707$.
Table A.4, air $(T_f \approx 350$ K): $\nu = 20.92 \times 10^{-6}$ m²/s, $k = 30 \times 10^{-3}$ W/m·K, $Pr = 0.700$.
Table A.4, air $(T_s = 128.4°C = 401$ K): $\mu = 230.5 \times 10^{-7}$ N·s/m², $Pr = 0.690$.

ANALYSIS:

1. The convection heat transfer coefficient may be determined from the data by using Newton's law of cooling. That is,

$$\bar{h} = \frac{q}{A(T_s - T_\infty)}$$

With $q = 0.85P$ and $A = \pi DL$, it follows that

$$\bar{h} = \frac{0.85 \times 46 \text{ W}}{\pi \times 0.0127 \text{ m} \times 0.094 \text{ m } (128.4 - 26.2)°C}$$

$$\bar{h} = 102 \text{ W/m}^2 \cdot \text{K} \qquad \lhd$$

2. There are four standard correlations that may be used to predict the convection coefficient. From one of the recommended correlations, Equation 7.31,

$$\overline{Nu}_D = CRe_D^m Pr^n (Pr_\infty/Pr_s)^{1/4}$$

where all properties, except Pr_s, are evaluated at T_∞. Accordingly,

$$Re_D = \frac{VD}{\nu} = \frac{10 \text{ m/s} \times 0.0127 \text{ m}}{15.89 \times 10^{-6} \text{ m}^2/\text{s}} = 7992$$

Hence, from Table 7.3, $C = 0.26$ and $m = 0.6$. Also, since $Pr < 10$, $n = 0.37$. It follows that

$$\overline{Nu}_D = 0.26(7992)^{0.6}(0.707)^{0.37}\left(\frac{0.707}{0.690}\right)^{0.25}$$

$$\overline{Nu}_D = 50.5$$

Hence

$$\bar{h}_1 = \overline{Nu}_D \frac{k}{D} = 50.5 \frac{0.0263 \text{ W/m} \cdot \text{K}}{0.0127 \text{ m}}$$

$$\bar{h}_1 = 105 \text{ W/m}^2 \cdot \text{K} \qquad \lhd$$

From the other recommended correlation, Equation 7.32,

$$\overline{Nu}_D = 0.3 + \frac{0.62 Re_D^{1/2} Pr^{1/3}}{[1 + (0.4/Pr)^{2/3}]^{1/4}}\left[1 + \left(\frac{Re_D}{28200}\right)^{5/8}\right]^{4/5}$$

where all properties are evaluated at T_f. Hence $Pr = 0.70$ and

$$Re_D = \frac{VD}{\nu} = \frac{10 \text{ m/s} \times 0.0127 \text{ m}}{20.92 \times 10^{-6} \text{ m}^2/\text{s}} = 6071$$

Hence

$$\overline{Nu}_D = 0.3 + \frac{0.62(6071)^{1/2}(0.70)^{1/3}}{[1+(0.4/0.70)^{2/3}]^{1/4}} \left[1 + \left(\frac{6071}{28200}\right)^{5/8}\right]^{4/5}$$

$$\overline{Nu}_D = 49.1$$

and

$$\overline{h}_2 = \overline{Nu}_D \frac{k}{D} = 49.1 \frac{0.030 \text{ W/m} \cdot \text{K}}{0.0127 \text{ m}}$$

$$\overline{h}_2 = 115.9 \text{ W/m}^2 \cdot \text{K} \qquad \triangleleft$$

Similarly, from Equation 7.30

$$\overline{Nu}_D = (0.4 Re_D^{1/2} + 0.06\, Re_D^{2/3}) Pr^{0.4} \left(\frac{\mu_\infty}{\mu_s}\right)^{1/4}$$

where all properties, except μ_s, are again evaluated at T_∞. Hence

$$\overline{Nu}_D = [0.4(7992)^{1/2} + 0.06(7992)^{2/3}](0.707)^{0.4}$$

$$\times \left(\frac{184.6 \times 10^{-7} \text{ N} \cdot \text{s/m}^2}{230.5 \times 10^{-7} \text{ N} \cdot \text{s/m}^2}\right)^{0.25}$$

$$\overline{Nu}_D = 49.1$$

and

$$\overline{h}_3 = \overline{Nu}_D \frac{k}{D} = 49.1 \frac{0.0263 \text{ W/m} \cdot \text{K}}{0.0127 \text{ m}}$$

$$\overline{h}_3 = 102 \text{ W/m}^2 \cdot \text{K} \qquad \triangleleft$$

Finally, from Equation 7.29

$$\overline{Nu}_D = C Re_D^m Pr^{1/3}$$

where all properties are now evaluated at the film temperature. Accordingly, $Re_D = 6071$ and $Pr = 0.700$. Hence, from Table 7.1, $C = 0.193$ and $m = 0.618$. The Nusselt number is then

$$\overline{Nu}_D = 0.193(6071)^{0.618}(0.700)^{0.333}$$

$$\overline{Nu}_D = 37.3$$

and

$$\overline{h}_4 = \overline{Nu}_D \frac{k}{D} = 37.3 \frac{0.030 \text{ W/m} \cdot \text{K}}{0.0127 \text{ m}}$$

$$\overline{h}_4 = 88 \text{ W/m}^2 \cdot \text{K} \qquad \triangleleft$$

COMMENTS:

1. In view of the experimental uncertainty associated with the data, the excellent agreement of the Zhukauskas and the Whitaker results, $\bar{h}_1$ and $\bar{h}_3$, respectively, with the measured result, $\bar{h}$, is fortuitous. Uncertainties associated with measuring the air velocity, estimating the heat loss from cylinder ends, and averaging the cylinder surface temperature, which varies axially and circumferentially, render the experimental result accurate to no better than 15 percent. Accordingly, even the Hilpert result, $\bar{h}_4$, is within the experimental uncertainty of the measured result.

2. Recognize the importance of using the proper temperature when evaluating fluid properties.

7.5 THE SPHERE

Boundary layer effects associated with flow over a sphere are much like those for the circular cylinder, with transition and separation both playing prominent roles. Results for the drag coefficient, which is defined by Equation 7.28, are presented in Figure 7.7.

Numerous heat transfer correlations have been proposed, and Whitaker [3] recommends an expression of the form

$$\overline{Nu}_D = 2 + (0.4Re_D^{1/2} + 0.06Re_D^{2/3})Pr^{0.4}(\mu_\infty/\mu_s)^{1/4} \tag{7.33}$$

$$\left[\begin{array}{l} 0.71 < Pr < 380 \\ 3.5 < Re_D < 7.6 \times 10^4 \\ 1.0 < (\mu_\infty/\mu_s) < 3.2 \end{array} \right]$$

Except for the additive factor of 2, whose presence may be justified theoretically [13], this expression is equivalent to Equation 7.30 for the circular cylinder. The correlation is accurate to within ± 30 percent for the range of parameter values listed. Note that all properties except μ_s should be evaluated at T_∞, and the result may be applied to mass transfer problems simply by replacing $\overline{Nu}_D$ and Pr with $\overline{Sh}_D$ and Sc, respectively.

A special case of convection heat and mass transfer from spheres relates to transport from freely falling liquid drops. Although the correlation of Ranz and Marshall [14] is often used

$$\overline{Nu}_D = 2 + 0.6Re_D^{1/2}Pr^{1/3} \tag{7.34}$$

it has recently been suggested [15] that the expression be modified to account for the effects of droplet oscillations and distortion. In particular the recommended correlation is

$$\overline{Nu}_D = 2 + 0.6Re_D^{1/2} Pr^{1/3} [25(x/D)^{-0.7}] \tag{7.35}$$

where x is the falling distance measured from rest. The properties are evaluated at T_∞, and the corresponding mass transfer correlation may be obtained by replacing $\overline{Nu}_D$ and Pr by $\overline{Sh}_D$ and Sc, respectively.

EXAMPLE 7.5

The decorative plastic film on a copper sphere of 10 mm diameter is cured in an oven at 75°C. Upon removal from the oven, the sphere is subjected to an airstream at 1 atm and 23°C having a velocity of 10 m/s. Estimate how long it will take to cool the sphere to 35°C.

SOLUTION

KNOWN:

Sphere cooling in an airstream.

FIND:

Time, t, required to cool from $T_i = 75$°C to $T(t) = 35$°C.

SCHEMATIC:

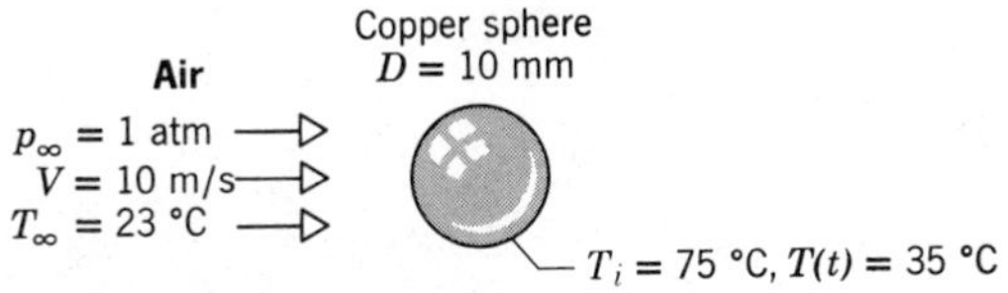

ASSUMPTIONS:

1. Negligible thermal resistance and capacitance for the plastic film.
2. The sphere is spatially isothermal.
3. Radiation effects are negligible.

PROPERTIES:

Table A.1, copper ($T \approx 328$ K): $\rho = 8933$ kg/m^3, $k = 399$ W/m·K, $c_p = 387$ J/kg·K.

Table A.4, air ($T_\infty = 296$ K): $\mu = 181.6 \times 10^{-7}$ N·s/m², $\nu = 15.36 \times 10^{-6}$ m²/s, $k = 0.0258$ W/m·K, $Pr = 0.709$.
Table A.4, air ($T_s \approx 328$ K): $\mu = 197.8 \times 10^{-7}$ N·s/m².

ANALYSIS:

The time required to complete the cooling process may be obtained from results for a lumped capacitance. In particular, from Equations 5.4 and 5.5

$$t = \frac{\rho V c_p}{\bar{h} A_s} \ln \frac{T_i - T_\infty}{T - T_\infty}$$

or, with $V = \pi D^3/6$ and $A_s = \pi D^2$

$$t = \frac{\rho c_p D}{6\bar{h}} \ln \frac{T_i - T_\infty}{T - T_\infty}$$

The convection heat transfer coefficient may be evaluated from Equation 7.33

$$\overline{Nu}_D = 2 + (0.4\, Re_D^{1/2} + 0.06 Re_D^{2/3}) Pr^{0.4} (\mu_\infty/\mu_s)^{1/4}$$

where

$$Re_D = \frac{VD}{\nu} = \frac{10 \text{ m/s} \times 0.01 \text{ m}}{15.36 \times 10^{-6} \text{ m}^2/\text{s}} = 6510$$

Hence

$$\overline{Nu}_D = 2 + [0.4(6510)^{1/2} + 0.06(6510)^{2/3}](0.709)^{0.4}$$

$$\times \left(\frac{181.6 \times 10^{-7} \text{ N·s/m}^2}{197.8 \times 10^{-7} \text{ N·s/m}^2} \right)^{1/4}$$

$$\overline{Nu}_D = 47.4$$

and

$$\bar{h} = \overline{Nu}_D \frac{k}{D} = 47.4 \frac{0.0258 \text{ W/m·K}}{0.01 \text{ m}} = 122 \text{ W/m}^2\cdot\text{K}$$

The time required for cooling is then

$$t = \frac{8933 \text{ kg/m}^3 \times 387 \text{ J/kg·K} \times 0.01 \text{ m}}{6 \times 122 \text{ W/m}^2\cdot\text{K}} \ln \left(\frac{75 - 23}{35 - 23} \right)$$

$$t = 69.2 \text{ s} \qquad \triangleleft$$

COMMENTS:

1. The validity of the lumped capacitance method may be determined by calculating the Biot number. From Equation 5.10

$$Bi = \frac{\bar{h}L_c}{k_s} = \frac{\bar{h}(r_o/3)}{k_s} = \frac{122 \text{ W/m}^2 \cdot \text{K} \times 0.005 \text{ m/3}}{399 \text{ W/m} \cdot \text{K}} = 5.1 \times 10^{-4}$$

    and the criterion is satisfied.

2. Remember that, although having similar definitions, the Nusselt number is defined in terms of the thermal conductivity of the fluid, whereas the Biot number is defined in terms of the thermal conductivity of the solid.

7.6 FLOW ACROSS BANKS OF TUBES

Heat transfer to or from a bank (or bundle) of tubes in cross flow is relevant to numerous industrial applications, such as steam generation in a boiler or air cooling in the coil of an air conditioner. The geometric arrangement is shown schematically in Figure 7.9. Typically, one fluid moves over the tubes, while a

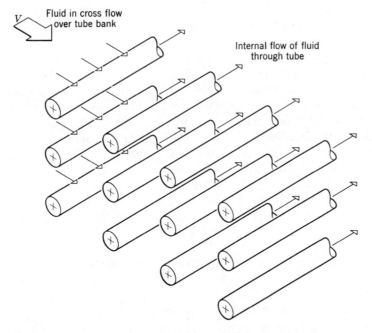

Figure 7.9 Schematic of a tube bank in cross flow.

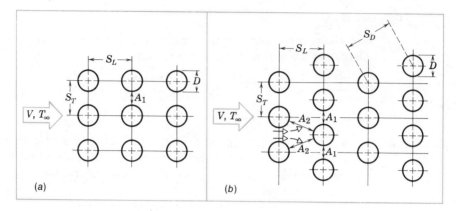

Figure 7.10 Tube arrangements in a bank. (a) Aligned. (b) Staggered.

second fluid at a different temperature passes through the tubes. In this section we are specifically interested in the convection heat transfer associated with cross flow over the tubes.

The tube rows of a bank are either *staggered* or *aligned* in the direction of the fluid velocity V (Figure 7.10). The configuration is characterized by the tube diameter, D, and by the *transverse pitch*, S_T, and *longitudinal pitch*, S_L, measured between tube centers. Flow conditions within the bank are dominated by boundary layer separation effects and by wake interactions, which in turn influence convection heat transfer.

The heat transfer coefficient associated with a tube is determined by its position in the bank [9]. The coefficient for a tube in the first row is approximately equal to that for a single tube in cross flow, whereas larger heat transfer coefficients are associated with tubes of the inner rows. The tubes of the first few rows act as turbulence grids, which increase the heat transfer coefficient for tubes in the following rows. In most configurations, however, heat transfer conditions stabilize, such that little change occurs in the convection coefficient for a tube beyond the fourth or fifth row.

Generally, we wish to know the *average* heat transfer coefficient for the *entire* tube bundle. For air flow across tube bundles comprised of *10 or more rows* ($N \geq 10$), Grimison [16] has obtained a correlation of the form

$$\overline{Nu}_D = C_1 Re_{D,\max}^m \quad \left[\begin{array}{l} N \geq 10 \\ 2000 < Re'_{D,\max} < 40{,}000 \\ Pr = 0.7 \end{array} \right] \qquad (7.36)$$

where C_1 and m are listed in Table 7.4 and

$$Re_{D,\max} \equiv \frac{\rho V_{\max} D}{\mu} \qquad (7.37)$$

Table 7.4 Constants of Equations 7.36 and 7.38 for a tube bank of 10 or more rows [16]

	S_T/D							
	1.25		1.5		2.0		3.0	
S_L/D	C_1	m	C_1	m	C_1	m	C_1	m
Aligned								
1.25	0.348	0.592	0.275	0.608	0.100	0.704	0.0633	0.752
1.50	0.367	0.586	0.250	0.620	0.101	0.702	0.0678	0.744
2.00	0.418	0.570	0.299	0.602	0.229	0.632	0.198	0.648
3.00	0.290	0.601	0.357	0.584	0.374	0.581	0.286	0.608
Staggered								
0.600	—	—	—	—	—	—	0.213	0.636
0.900	—	—	—	—	0.446	0.571	0.401	0.581
1.000	—	—	0.497	0.558	—	—	—	—
1.125	—	—	—	—	0.478	0.565	0.518	0.560
1.250	0.518	0.556	0.505	0.554	0.519	0.556	0.522	0.562
1.500	0.451	0.568	0.460	0.562	0.452	0.568	0.488	0.568
2.000	0.404	0.572	0.416	0.568	0.482	0.556	0.449	0.570
3.000	0.310	0.592	0.356	0.580	0.440	0.562	0.428	0.574

It has become common practice to extend this result to other fluids through insertion of the factor $1.13Pr^{1/3}$, in which case

$$\overline{Nu}_D = 1.13 C_1\, Re_{D,\max}^m\, Pr^{1/3} \qquad (7.38)$$

$$\begin{bmatrix} N \geq 10 \\ 2000 < Re_{D,\max} < 40{,}000 \\ Pr \geq 0.7 \end{bmatrix}$$

All properties appearing in the above equations are evaluated at the film temperature. If $N < 10$ a correction factor may be applied such that

$$\overline{Nu}_D|_{(N<10)} = C_2\, \overline{Nu}_D|_{(N \geq 10)} \qquad (7.39)$$

where C_2 is given in Table 7.5.

Table 7.5 Correction factor C_2 of Equation 7.39 for $N < 10$ [17]

N	1	2	3	4	5	6	7	8	9
Aligned	0.64	0.80	0.87	0.90	0.92	0.94	0.96	0.98	0.99
Staggered	0.68	0.75	0.83	0.89	0.92	0.95	0.97	0.98	0.99

Note that the Reynolds number, $Re_{D,\,max}$, for the foregoing correlations is based on the *maximum fluid velocity* occurring within the tube bank. For the aligned arrangement V_{max} occurs at the transverse plane A_1 of Figure 7.10*a*, and from the mass conservation requirement for an incompressible fluid

$$V_{max} = \frac{S_T}{S_T - D}\, V \tag{7.40}$$

For the staggered configuration, the maximum velocity may occur at either the transverse plane A_1 or the diagonal plane A_2 of Figure 7.10*b*. It will occur at A_2 if the rows are spaced such that

$$2(S_D - D) < (S_T - D)$$

The factor of 2 results from the bifurcation experienced by the fluid moving from the A_1 to the A_2 planes. Hence V_{max} occurs at A_2 if

$$S_D = [S_L^2 + (S_T/2)^2]^{1/2} < \frac{S_T + D}{2}$$

in which case it is given by

$$V_{max} = \frac{S_T}{2(S_D - D)}\, V \tag{7.41}$$

If V_{max} occurs at A_1 for the staggered configuration, it may again be computed from Equation 7.40.

More recent results have been obtained [3,10], and Zhukauskas [10] has proposed a correlation of the form

$$\overline{Nu}_D = C\, Re_{D,\,max}^m\, Pr^{0.36}(Pr_\infty / Pr_s)^{1/4} \tag{7.42}$$

$$\begin{bmatrix} N \geq 20 \\ 0.7 < Pr < 500 \\ 1000 < Re_{D,\,max} < 2 \times 10^6 \end{bmatrix}$$

where all properties, except Pr_s, are evaluated at T_∞ and the constants, C and m, are listed in Table 7.6. This result correlates data very well and may be used for tube banks having $N < 20$. In fact accuracy to within 25 percent is retained for N as low as 4.

Note that neither Equation 7.38 nor 7.42 may be used in the low Reynolds number range, $Re_{D,\,max} \lesssim 1000$. Heat transfer results for this range are provided by Zhukauskas [10] and Bergelin et al. [18]. Note also that the foregoing results may be used to determine mass transfer rates associated with evaporation or sublimation from the surfaces of a bank of cylinders in cross flow. Once again it is only necessary to replace $\overline{Nu}_D$ and Pr by $\overline{Sh}_D$ and Sc, respectively.

Table 7.6 Constants of Equation 7.42 for the tube bank in cross flow [10]

CONFIGURATION	$Re_{D,max}$	C	m
Aligned	$10^3 - 2 \times 10^5$	0.27	0.63
Staggerred $(S_T/S_L < 2)$	$10^3 - 2 \times 10^5$	$0.35(S_T/S_L)^{1/5}$	0.60
Staggered $(S_T/S_L > 2)$	$10^3 - 2 \times 10^5$	0.40	0.60
Aligned	$2 \times 10^5 - 2 \times 10^6$	0.021	0.84
Staggered	$2 \times 10^5 - 2 \times 10^6$	0.022	0.84

We close by recognizing that we will generally have as much interest in the pressure drop associated with flow across a tube bank as we will have in the overall heat transfer rate. The power required to move the fluid across the bank is often a major operating expense for the system and is directly proportional to the pressure drop.

The most recent and reliable results are those of Zhukauskas [10] where the pressure drop is given by

$$\Delta p = N\chi\left(\frac{\rho V_{max}^2}{2}\right)f \tag{7.43}$$

The friction factor, f, and the correction factor, χ, are plotted in Figures 7.11 and 7.12. Figure 7.11 pertains to a square, in-line tube arrangement for which the

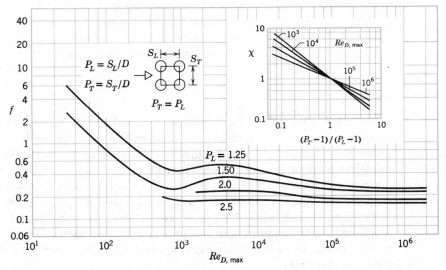

Figure 7.11 Friction factor, f, and correction factor, χ, for Equation 7.43; in-line tube bundle arrangement [10]. Used with permission.

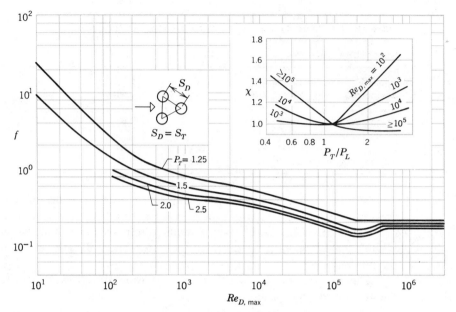

Figure 7.12 Friction factor, f, and correction factor, χ, for Equation 7.43. Staggered tube bundle arrangement [10]. Used with permission.

dimensionless longitudinal and transverse pitches, $P_L \equiv S_L/D$ and $P_T \equiv S_T/D$, respectively, are equal. The correction factor, χ, plotted in the insert, may then be used to apply the results to other in-line arrangements. Similarly, Figure 7.12 applies to a staggered arrangement of tubes in the form of an equilateral triangle ($S_T = S_D$), and the correction factor enables extension of the results to other staggered arrangements. Note that the Reynolds number appearing in Figures 7.11 and 7.12 is based on the maximum fluid velocity, $V_{\max}$.

EXAMPLE 7.6

Pressurized water is often available at elevated temperatures and may be used to heat air for space heating or industrial process applications. In such cases it is customary to use a tube bundle in which the water is passed through the tubes, while air is passed in cross flow over the tubes. Consider a staggered tube arrangement for which the tube outside diameter is 16.4 mm and the longitudinal and transverse pitches are $S_L = 34.3$ mm and $S_T = 31.3$ mm, respectively. There are seven rows of tubes in the air flow direction. Under typical operating conditions the cylinder surface temperature is at 70°C, while the air upstream temperature and velocity are 15°C and 6 m/s, respectively.

1. What is the air side convection coefficient?
2. What is the pressure drop across the tube bundle?

SOLUTION

KNOWN:

Geometry and operating conditions of a tube bank.

FIND:

1. Air side convection coefficient.
2. Pressure drop.

SCHEMATIC:

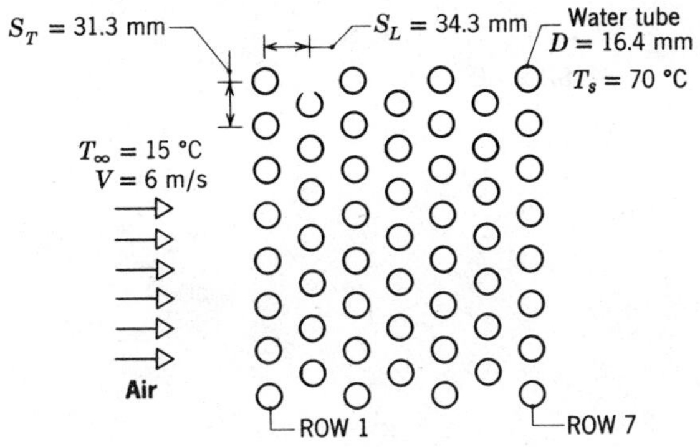

ASSUMPTIONS:

1. Steady-state conditions.
2. Negligible radiation effects.

PROPERTIES:

Table A.4, air ($T_\infty = 15°C$): $\rho = 1.217$ kg/m^3, $\nu = 14.82 \times 10^{-6}$ m^2/s, $k = 0.0253$ W/m·K, $Pr = 0.710$.
Table A.4, air ($T_s = 70°C$): $Pr = 0.701$.
Table A.4, air ($T_f = 43°C$): $\nu = 17.4 \times 10^{-6}$ m^2/s, $k = 0.0274$ W/m·K, $Pr = 0.705$.

ANALYSIS:

1. To a first approximation, the air side convection coefficient may be

estimated from Equation 7.42.

$$\overline{Nu}_D = CRe^m_{D,\,max}Pr^{0.36}(Pr_\infty/Pr_s)^{1/4}$$

Since $S_D = [S_L^2 + (S_T/2)^2]^{1/2} = 37.7$ mm is greater than $(S_T + D)/2$, the maximum velocity occurs on the transverse plane, A_1, of Figure 7.10. Hence from Equation 7.40

$$V_{max} = \frac{S_T}{S_T - D}V$$

$$V_{max} = \frac{31.3 \text{ mm}}{(31.3 - 16.4) \text{ mm}} 6 \text{ m/s}$$

$$V_{max} = 12.6 \text{ m/s}$$

With

$$Re_{D,\,max} = \frac{V_{max}D}{\nu} = \frac{12.6 \text{ m/s} \times 0.0164 \text{ m}}{14.82 \times 10^{-6} \text{ m}^2/\text{s}} = 13,943$$

and

$$\frac{S_T}{S_L} = \frac{31.3 \text{ mm}}{34.3 \text{ mm}} = 0.91 < 2$$

it follows from Table 7.6 that

$$C = 0.35 \left(\frac{S_T}{S_L}\right)^{1/5} = 0.34$$

$$m = 0.60$$

Hence

$$\overline{Nu}_D = 0.34(13,943)^{0.60}(0.71)^{0.36}\left(\frac{0.710}{0.701}\right)^{0.25}$$

$$\overline{Nu}_D = 92.5$$

and

$$\overline{h} = \overline{Nu}_D \frac{k}{D} = 92.5 \times \frac{0.0253 \text{ W/m·K}}{0.0164 \text{ m}}$$

$$\overline{h} = 142.7 \text{ W/m}^2\text{·K} \qquad \triangleleft$$

Alternatively, with properties based on T_f,

$$Re_{D,max} = \frac{12.6 \text{ m/s } (0.0164 \text{ m})}{17.4 \times 10^{-6} \text{ m}^2/\text{s}} = 11,876$$

and, from Tables 7.4 and 7.5, with $S_T/D \approx 2$ and $S_L/D \approx 2$,

$$C_1 = 0.482 \qquad m = 0.556 \qquad C_2 = 0.97$$

From Equation 7.38 it follows that

$$\overline{Nu}_D = [1.13(0.482)(11,876)^{0.556}(0.705)^{1/3}]0.97$$

$$\overline{Nu}_D = 91.7$$

Hence

$$\bar{h} = \overline{Nu}_D \frac{k}{D} = 91.7 \frac{0.0274 \text{ W/m·K}}{0.0164 \text{ m}}$$

$$\bar{h} = 153.7 \text{ W/m}^2\text{·K} \qquad \triangleleft$$

2. The pressure drop may be obtained from Equation 7.43.

$$\Delta p = N \chi \left(\frac{\rho V_{max}^2}{2} \right) f$$

With $Re_{D,max} = 13,943$, $P_T = (S_T/D) = 1.91$ and $(P_T/P_L) = 0.91$, it follows from Figure 7.12 that

$$\chi \approx 1.04$$

$$f \approx 0.35$$

Hence with $N = 7$

$$\Delta p = 7 \times 1.04 \left[\frac{1.217 \text{ kg/m}^3 (12.6 \text{ m/s})^2}{2} \right] 0.35$$

$$\Delta p = 246 \text{ N/m}^2 = 2.46 \times 10^{-3} \text{ bars} \qquad \triangleleft$$

COMMENTS:

The values of $\bar{h}$ obtained from Equations 7.38 and 7.42 agree to within 8 percent, which is well within the uncertainties associated with the two correlations.

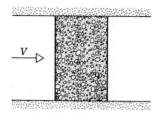

Figure 7.13   Gas flow through a packed bed of solid particles.

7.7 PACKED BEDS

Gas flow through a *packed bed* of solid particles (Figure 7.13) is relevant to many industrial processes, which include the transfer and storage of thermal energy, heterogeneous catalytic reactions and drying. The term *packed bed* refers to a condition for which the position of the particles is *fixed*. In contrast a *fluidized bed* is one in which the particles are in motion due to convection with the fluid.

For a packed bed a large amount of heat or mass transfer surface area can be contained in a small volume, and the irregular flow which exists in the voids of the bed enhances transport through turbulent mixing. Many correlations, which have been developed for different particle shapes, sizes and packing densities, are described in the literature [13, 19–21]. One such correlation, which has been recommended for gas flow in a bed of spheres, is of the form

$$\varepsilon \bar{j}_H = \varepsilon \bar{j}_m = 2.06 Re_D^{-0.575} \quad \left[\begin{array}{c} Pr \approx 0.7 \\ 90 \leq Re_D \leq 4000 \end{array} \right] \qquad (7.44)$$

where $\bar{j}_H$ and $\bar{j}_m$ are the Colburn j factors defined by Equations 6.99 and 6.100. The Reynolds number, $Re_D = VD/v$, is defined in terms of the sphere diameter and the upstream velocity V, which would exist in the empty channel without the packing. The quantity ε is the *porosity*, or *void fraction*, of the bed (volume of void space per unit volume of bed), and its value typically ranges from 0.30 to 0.50. The correlation may be applied to packing materials other than spheres by multiplying the right-hand side by an appropriate correction factor. For a bed of uniformly sized cylinders, with length-to-diameter ratio of 1, the factor is 0.79; for a bed of cubes it is 0.71.

7.8 SUMMARY

In this chapter we have compiled convection correlations for a variety of external flow conditions. For simple surface geometries these results may be derived from a boundary layer analysis, but in most cases they can only be obtained from generalizations based on experiment. The correlations of this chapter may be used to estimate convection transfer rates in many practical problems. You should know when and how to use the various expressions, and you should be familiar with the general methodology of a convection calculation.

The content of this chapter may give the appearance of being a collection of *recipes*. This is perhaps because, in many respects, it is just that. Proper use of these recipes is essential to solving practical problems, and to facilitate their use, they are summarized in Table 7.7.

Table 7.7 Summary of convection heat transfer correlations for extermal flow[a]

CORRELATION		GEOMETRY	CONDITIONS
$C_{f,x} = 0.664 Re_x^{-1/2}$	(7.4)	Flat plate	Laminar, local, T_f
$\delta = 5x Re_x^{-1/2}$	(7.5)	Flat plate	Laminar, T_f
$Nu_x = 0.332 Re_x^{1/2} Pr^{1/3}$	(7.6)	Flat plate	Laminar, local, T_f, $Pr \gtrsim 0.6$
$\delta_t = \delta Pr^{-1/3}$	(7.8)	Flat plate	Laminar, T_f
$\bar{C}_{f,x} = 1.328 Re_x^{-1/2}$	(7.11)	Flat plate	Laminar, average, T_f
$\overline{Nu}_x = 0.664 Re_x^{1/2} Pr^{1/3}$	(7.12)	Flat plate	Laminar, average, T_f, $Pr \gtrsim 0.6$
$Nu_x = 0.565 Pe_x^{1/2}$	(7.14)	Flat plate	Laminar, local, T_f, $Pr \lesssim 0.05$
$C_{f,x} = 0.0592 Re_x^{-1/5}$	(7.15)	Flat plate	Turbulent, local, T_f, $Re_x \lesssim 10^8$
$\delta = 0.37x Re_x^{-1/5}$	(7.16)	Flat plate	Turbulent, local, T_f, $Re_x \lesssim 10^8$
$Nu_x = 0.0296 Re_x^{4/5} Pr^{1/3}$	(7.17)	Flat plate	Turbulent, local, T_f, $Re_x \lesssim 10^8$, $0.6 \lesssim Pr \lesssim 60$
$\bar{C}_{f,L} = 0.074 Re_L^{-1/5} - 1742 Re_L^{-1}$	(7.23)	Flat plate	Mixed, average, T_f, $Re_{x,c} = 5 \times 10^5$, $Re_L \lesssim 10^8$
$\overline{Nu}_L = (0.037 Re_L^{4/5} - 871) Pr^{1/3}$	(7.21)	Flat plate	Mixed, average, T_f, $Re_{x,c} = 5 \times 10^5$, $Re_L \lesssim 10^8$, $0.6 < Pr < 60$
or,			
$\overline{Nu}_L = 0.036(Re_L^{4/5} - 9200) Pr^{0.43}$ $\cdot (\mu_\infty / \mu_s)^{1/4}$	(7.27)	Flat plate	Mixed, average, T_∞, $10^5 < Re_L < 5.5$ $\times 10^6$, $0.7 < Pr < 380$, $0.26 < (\mu_\infty / \mu_s) < 3.5$
$\overline{Nu}_D = C Re_D^m Pr^{1/3}$ (Tables 7.1 and 7.2)	(7.29)	Cylinder	Average, T_f, $0.4 < Re_D < 4 \times 10^5$, $Pr \gtrsim 0.7$
$\overline{Nu}_D = (0.4 Re_D^{1/2} + 0.06 Re_D^{2/3}) Pr^{0.4}$ $\cdot (\mu_\infty / \mu_s)^{1/4}$	(7.30)	Cylinder	Average, T_∞, $10 < Re_D < 10^5$, $0.67 < Pr < 300$, $0.25 < (\mu_\infty / \mu_s) < 5.2$

Table 7.7 Continued

CORRELATION		GEOMETRY	CONDITIONS
$\overline{Nu}_D = C\,Re_D^m Pr^n (Pr_\infty/Pr_s)^{1/4}$ (7.31) (Table 7.3)		Cylinder	Average, T_∞, $1 < Re_D < 10^6$, $0.7 < Pr < 500$
$\overline{Nu}_D = 0.3 + [0.62Re_D^{1/2}Pr^{1/3}/[1 + (0.4/Pr)^{2/3}]^{1/4}] \cdot [1 + (Re_D/28200)^{5/8}]^{4/5}$ (7.32)		Cylinder	Average, T_f, $Re_D Pr > 0.2$
$\overline{Nu}_D = 2 + (0.4Re_D^{1/2} + 0.06Re_D^{2/3})Pr^{0.4}$ $\cdot(\mu_\infty/\mu_s)^{1/4}$ (7.33)		Sphere	Average, T_∞, $3.5 < Re_D < 7.6 \times 10^4$, $0.71 < Pr < 380$, $1.0 < (\mu_\infty/\mu_s) < 3.2$
$\overline{Nu}_D = 2 + 0.6Re_D^{1/2}Pr^{1/3}$ $\cdot[25(x/D)^{-0.7}]$ (7.35)		Falling drop	Average, T_∞
$\overline{Nu}_D = 1.13C_1 Re_{D,max}^m Pr^{1/3}$ (7.38) (Tables 7.4, 7.5)		Tube bank	Average, T_f, $2 \times 10^3 < Re_{D,max}$ $< 4 \times 10^4$, $Pr \geq 0.7$
$\overline{Nu}_D = C\,Re_{D,max}^m Pr^{0.36}(Pr_\infty/Pr_s)^{1/4}$ (Table 7.6) (7.42)		Tube bank	Average, T_∞, $1000 < Re_D < 2 \times 10^6$, $0.7 < Pr < 500$
$\varepsilon\overline{j}_H = \varepsilon\overline{j}_m = 2.06Re_D^{-0.575}$ (7.44)		Packed bed of spheres	Average, T_∞, $90 \leq Re_D \leq 4000$, $Pr \approx 0.7$

[a] When the heat and mass transfer analogy is applicable, the corresponding mass transfer correlations may be obtained by replacing Nu and Pr by Sh and Sc, respectively.

REFERENCES

1. Kays, W. M., *Convective Heat and Mass Transfer*, McGraw-Hill, New York, 1966.
2. Schlichting, H., *Boundary Layer Theory*, 6th Ed., McGraw-Hill, New York, 1968.
3. Whitaker, S., *AIChE J.*, *18*, 361, 1972.
4. Zhukauskas, A. and A. B. Ambrazyavichyus, *Int. J. Heat Mass Transfer*, *3*, 305, 1961.
5. Churchill, S. W., *AIChE J.*, *22*, 264, 1976.
6. Fox, R. W. and A. T. McDonald, *Introduction to Fluid Mechanics*, 2nd Ed., Wiley, New York, 1978.
7. Hilpert, R., *Forsch. Gebiete Ingenieurwes.*, *4*, 215, 1933.
8. Knudsen, J. D. and D. L. Katz, *Fluid Dynamics and Heat Transfer*, McGraw-Hill, New York, 1958.
9. Jakob, M., *Heat Transfer*, Vol. 1, Wiley, New York, 1949.
10. Zhukauskas, A., "Heat Transfer from Tubes in Cross Flow" in J. P. Hartnett and T. F. Irvine, Jr., Eds., *Advances in Heat Transfer*, Vol. 8, Academic Press, New York, 1972.
11. Churchill, S. W. and M. Bernstein, *J. Heat Transfer*, *99*, 300, 1977.
12. Morgan, V. T., "The Overall Convective Heat Transfer from Smooth Circular

Cylinders" in T. F. Irvine, Jr. and J. P. Hartnett, Eds., *Advances in Heat Transfer*, Vol. 11, Academic Press, New York, 1975.

13. Bird, R. B., W. E. Stewart, and E. N. Lightfoot, *Transport Phenomena*, Wiley, New York, 1966.

14. Ranz, W. and W. Marshall, *Chem. Eng. Progr.*, *48*, 141, 1952.

15. Yao, S-C and V. E. Schrock, ASME Publication 75-WA/HT-37, Winter Annual Meeting, Houston, Texas, November 30–December 4, 1975.

16. Grimison, E. D., *Trans. ASME*, *59*, 583, 1937.

17. Kays, W. M. and R. K. Lo, Stanford University Technical Report No. 15, 1952.

18. Bergelin, O. P., G. A. Brown, and S. C. Doberstein, *Trans. ASME*, *74*, 953, 1952.

19. Jakob, M., *Heat Transfer*, Vol. 2, Wiley, New York, 1957.

20. Geankopplis, C. J., *Mass Transport Phenomena*, Holt, Rinehart and Winston, New York, 1972.

21. Sherwood, T. K., R. L. Pigford, and C. R. Wilkie, *Mass Transfer*, McGraw-Hill, New York, 1975.

PROBLEMS

7.1 Consider the following fluids at a film temperature of 300 K in parallel flow over a flat plate with velocity of 1 m/s: atmospheric air, water, engine oil, and mercury. For each fluid determine the velocity and thermal boundary layer thicknesses at a distance of 40 mm from the leading edge.

7.2 Consider atmospheric air at 25°C in parallel flow at 5 m/s over a 1-m-long flat plate maintained at 75°C.

   a) Determine the boundary layer thickness, the surface shear stress, and the heat flux at the trailing edge.

   b) Determine the drag force on the plate and the total heat transfer from the plate, each per unit width of the plate.

7.3 Engine oil at 100°C and a velocity of 0.1 m/s flows over a 1-m-long flat plate whose surface temperature is maintained at 20°C. Determine:

   a) The velocity and thermal boundary layer thicknesses at the trailing edge of the plate.

   b) The local heat flux and surface shear stress at the trailing edge.

   c) The total drag force and heat transfer per unit width of the plate.

7.4 A common flow geometry, which includes the flat plate in parallel flow as a special case, is the *wedge*.

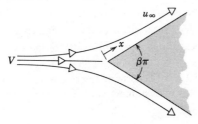

From a potential flow solution, it is known that the velocity at the edge of the boundary layer increases with x according to the relation $u_\infty = Vx^m$, where $m = \beta/(2 - \beta)$ and $\beta\pi$ is the wedge angle. From a laminar boundary layer solution, it is also known that the local Nusselt number may be expressed as

$$Nu_x = C_1 Re_x^{1/2}$$

where $Re_x \equiv (u_\infty x/\nu)$ and C_1 is a known function of Pr and β, as tabulated below.

		Pr				
		0.7	0.8	1.0	5.0	10.0
β	m	C_1				
0	0	0.292	0.307	0.332	0.585	0.730
0.2	0.111	0.331	0.348	0.378	0.669	0.851
0.5	0.333	0.384	0.403	0.440	0.792	1.013
1.0	1.000	0.496	0.523	0.570	1.043	1.344

a) Comment on the nature of flow conditions at $x = 0$ for $\beta > 0$. How does u_∞ vary with x for $\beta = 1$?

b) Obtain the ratio of the average convection coefficient, $\bar{h}_x$, to the local coefficient, h_x, for $\beta = 0.5$ and for $\beta = 1.0$.

c) It is common practice to approximate wedge flow heat transfer effects by using convection correlations associated with the flat plate in parallel flow. Comment on the accuracy of such an approximation by computing the ratio $(\bar{h}_{x,\beta>0}/\bar{h}_{x,\beta=0})$ at $x = 1$ m for air flow over wedges corresponding to $\beta = 0.5$ and to $\beta = 1.0$.

7.5 An electric air heater consists of a horizontal array of thin metal strips that are 10 mm wide and oriented normal to an airstream that is in parallel flow over the top of the strips. Each strip is 0.2 m long, and 25 strips are arranged side by side, forming a continuous and smooth surface over which the air flows at 2 m/s. During operation each strip is maintained at 500°C and the air is at 25°C.

a) What is the rate of convection heat transfer from the first strip? The fifth strip? The tenth strip?

b) What is the rate of convection heat transfer from all of the strips?

7.6 Atmospheric air at 25°C flows along a flat plate at 25 m/s. The plate is 1 m long and is maintained at a temperature of 125°C. Determine:

a) The velocity and thermal boundary layer thicknesses at the trailing edge.

b) The local heat flux and surface shear stress at the trailing edge.

c) The total drag force and heat transfer per unit width of the plate.

7.7 Consider atmospheric air at 25°C and a velocity of 25 m/s flowing over a 1-m-long flat plate that is maintained at 125°C. Determine the rate of heat transfer from the plate for values of the critical Reynolds number corresponding to 10^5, 5×10^5, and 10^6.

7.8   Consider air at 27°C and 1 atm in parallel flow over a 1-m-long flat plate with a velocity of 10 m/s. Plot the variation of the local convection heat transfer coefficient with distance along the plate. What is the value of the average convection coefficient?

7.9   Consider water at 27°C in parallel flow over a 1-m-long flat plate with a velocity of 2 m/s. Plot the variation of the local convection heat transfer coefficient with distance along the plate. What is the value of the average convection coefficient?

7.10 Air at a pressure and temperature of 1 atm and 15°C, respectively, is in parallel flow at a velocity of 10 m/s over a 3-m-long flat plate that is heated to a uniform temperature of 140°C.

a) What is the average heat transfer coefficient for the plate?

b) What is the local heat transfer coefficient at the midpoint of the plate?

c) Sketch qualitatively the manner in which the heat flux varies along the length of the plate.

7.11 Explain under what conditions the total rate of heat transfer from a flat plate of dimensions L by $2L$ would be the same, independent of whether parallel flow over the plate is directed along the side of length L or $2L$. With a critical Reynolds number of 5×10^5, for what values of Re_L would the total heat transfer be independent of orientation?

7.12 The surface of a 1.5-m-long flat plate is maintained at 40°C, and water at a temperature of 4°C and a velocity of 0.6 m/s flows over the surface.

a) Using the film temperature, T_f, for evaluation of the properties, calculate the heat transfer rate per unit width of the plate, q' (W/m).

b) Calculate the error in q' that would be incurred in (a) if the thermophysical properties of the water were evaluated at the freestream temperature and the same empirical correlation were used.

c) In part (a), if a wire were placed near the leading edge of the plate to induce turbulence over its entire length, what would be the heat transfer rate to the water?

7.13 Air at a pressure and temperature of 1 atm and 50°C, respectively, is in parallel flow over the top surface of a flat plate that is heated to a uniform temperature of 100°C. The plate has a length of 0.20 m (in the flow direction) and a width of 0.10 m. The Reynolds number based on the plate length is 40,000.

a) What is the rate of heat transfer from the plate to the air?

b) If the freestream velocity of the air is doubled and the pressure is increased to $p = 10$ atm, what is the rate of heat transfer from the plate to the air?

7.14 A thin, flat plate of length $L = 1$ m separates two airstreams that are in parallel flow over opposite surfaces of the plate. One airstream has a temperature of $T_{\infty,1} = 200$°C and a velocity of $u_{\infty,1} = 60$ m/s, while the other airstream has a temperature of $T_{\infty,2} = 25$°C and a velocity of $u_{\infty,2} = 10$ m/s. What is the heat flux between the two streams at the midpoint of the plate?

7.15 Consider a rectangular fin that is used to cool a motorcycle engine. The fin is 0.15 m long and at a temperature of 250°C, while the motorcycle is moving at 80 km/h in air at 27°C. The air is in parallel flow over both surfaces of the fin, and turbulent flow conditions may be assumed to exist throughout. What is the rate of heat removal per unit width of fin?

7.16 Consider the wing of an aircraft as a flat plate of length $L = 2.5$ m in the flow direction. The plane is moving at a speed of 100 m/s in air that is at a pressure of $p_\infty = 0.7$ bar and a temperature of $-10°C$. If the top surface of the wing absorbs solar radiation at a rate of 800 W/m², estimate its steady-state temperature. Assume the wing to be of solid construction and to have a single, uniform temperature.

7.17 The top surface of a heated compartment consists of very smooth (A) and highly roughened (B) portions, and the surface is placed in an atmospheric air stream as shown below.

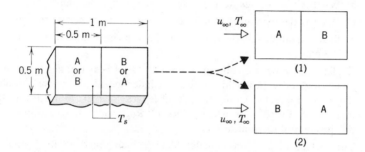

In the interest of minimizing total convection heat transfer from the surface, which orientation, (1) or (2), is preferred? If $T_s = 100°C$, $T_\infty = 20°C$, and $u_\infty = 20$ m/s, what is the convection heat transfer from the entire surface for this orientation?

7.18 Two rooms are arranged on the side of a building, shown schematically as follows. The wall of room A is 13 m long and 3.5 m high, while that of room B is 7 m long and 3.5 m high. Both walls are 0.25 m thick and are characterized by an effective thermal conductivity of 1 W/m·K.

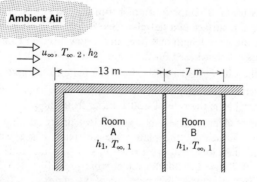

Consider conditions for which the ambient air is in parallel flow over the outer surface of the wall with $u_\infty = 7$ m/s and $T_{\infty,2} = -20°C$. The inside air temperature is to be maintained at $T_{\infty,1} = 20°C$, and the inside convection coefficient may be approximated as $h_1 = 5$ W/m²·K. Estimate the rate at which heat is lost through the 13 m long and 7 m long walls of rooms A and B, respectively. Turbulent boundary layer conditions may be assumed for the ambient air flow over the entire wall.

7.19 As a means of supplying fresh water to arid regions of the world, it has been advocated that icebergs be towed from Antarctica or Arctica. Icebergs that are considered to be best suited for towing are those that are relatively broad and flat. Consider an iceberg that is 1 km long by 0.5 km wide and of depth $D = 0.25$ km. It is proposed that this iceberg be towed at 1 km/h in the direction of its length for 6000 km through water whose average temperature (over the trip) is 10°C. As a first approximation, the interaction of the iceberg with its surroundings may be assumed to be dominated by conditions at the bottom (1 km × 0.5 km) surface. The latent heat of fusion of ice is 3.34×10^5 J/kg.

a) What is the average recession (melting) rate, dD/dt, at the bottom surface?

b) What is the power required to move the iceberg at the designated speed?

c) If towing costs amount to \$1/kW·h of power requirement, what is the minimum cost of fresh water at the destination?

7.20 Consider mass loss from a smooth wet flat plate due to forced convection at atmospheric pressure. The plate is 0.5 m long and 3 m wide. Dry air at 300 K and a freestream velocity of 35 m/s flows over the surface, which is also at a temperature of 300 K.

a) Estimate the average mass transfer coefficient, $\bar{h}_m$, for the plate.

b) Determine the water vapor mass loss rate (kg/s) from the plate.

7.21 Consider dry, atmospheric air in parallel flow over a 0.5-m-long plate whose surface is wetted. The air velocity is 35 m/s, and the air and water are each at a temperature of 300 K. What is the heat loss per unit width of the plate?

7.22 Dry air at atmospheric pressure and 350 K, with a freestream velocity of 25 m/s, flows over a smooth, porous plate of 1 m length. Assuming the plate to be saturated with liquid water at 350 K, estimate the mass rate of evaporation per unit width of the plate (kg/s·m).

7.23 A Volkswagen van traveling 90 km/h has just passed through a thunderstorm that left a film of water 0.1 mm thick on the top of the van. The top of the van can be assumed to be a flat plate 6 m long. Assume isothermal conditions at 27°C, an ambient air relative humidity of 80 percent, and turbulent flow over the entire surface.

a) What location on the van top will be the last to dry?

b) What is the water evaporation rate per unit area (kg/s·m²) at the trailing edge of the van top, that is, at $x = 6$ m?

7.24 Benzene, a known carcinogen, has been spilled on the floor of a laboratory and has spread to a length of 2 m. If a film 1 mm deep is formed, how long will it take for the benzene to completely evaporate? Ventilation in the laboratory provides for air flow parallel to the surface at 1 m/s, and the benzene and air are both at 25°C. The mass densities of benzene in the saturated vapor and liquid states are known to be 0.417 and 900 kg/m³, respectively.

7.25 Condenser cooling water for a power plant is stored in a cooling pond that is 1000 m long by 500 m wide. However, due to evaporative losses, it is necessary to periodically add "make-up" water to the pond in order to maintain a suitable water level. Assuming isothermal conditions at 27°C for the water and the air, that the freestream air is dry and moving at a velocity of 2 m/s in the direction of the 1000-m pond length, and that the boundary layer on the water surface is everywhere turbulent, determine the amount of "make-up" water that should be added to the pond daily.

7.26 A channel of triangular cross section, which is 25 m long and 1 m deep, is used for the storage of water.

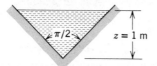

The water and the surrounding air are each at a temperature of 25°C, and the relative humidity of the air is 50 percent.

a) If the air moves at a velocity of 5 m/s along the length of the channel, what is the rate of water loss due to evaporation from the surface?

b) Obtain an expression for the rate at which the water depth would decrease with time due to evaporation. For the above conditions, how long would it take for all of the water to evaporate?

7.27 A stream of atmospheric air is used to dry a series of photographic plates, which are each of length $L_i = 0.25$ m in the direction of the airflow. The air is dry and at a temperature which is equal to that of the plates $(T_\infty = T_s = 50°C)$. The air speed is $u_\infty = 9.1$ m/s.

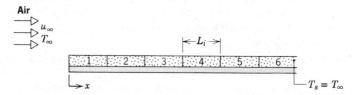

a) Sketch the variation of the local convection mass transfer coefficient, $h_{m,x}$, with distance x from the leading edge. Indicate the specific nature of the x dependence.

b) Which of the plates will dry the fastest? Calculate the drying rate per meter of width for this plate (kg/s·m).

c) What is the rate at which heat would have to be supplied to the fastest drying plate in order to maintain it at $T_s = 50°C$ during the drying process?

7.28 Calculate the rate of heat transfer per unit length for a 10-mm diameter cylinder maintained at 50°C with the following fluids in cross flow over the cylinder at $T_\infty = 20°C$ and $V = 5$ m/s:

a) Atmospheric air.

b) Saturated water.

c) Engine oil.

7.29 A circular pipe of 25 mm outside diameter is placed in an airstream at 25°C and 1 atmosphere pressure. The air moves in cross flow over the pipe with velocity of 15 m/s, while the outer surface of the pipe is maintained at 100°C.

a) What is the drag force exerted on the pipe per unit length?

b) What is the rate of heat transfer from the pipe per unit length?

7.30 A circular cylinder of 30 mm diameter is heated to a surface temperature of 90°C, while water at 25°C is in cross flow over the cylinder with a velocity of 3 m/s.

a) What is the drag force exerted on the cylinder per unit length?

b) What is the rate of heat transfer from the cylinder per unit length?

7.31 A circular cylinder of 25 mm diameter is initially at 150°C and is quenched by immersion in a 80°C oil bath, which moves at a velocity of 2 m/s in cross flow over the cylinder. What is the initial rate of heat loss per unit length of the cylinder?

7.32 Consider atmospheric air at 25°C and flowing at a velocity of 15 m/s. What is the rate of heat transfer per unit length from the following surfaces, each at 75°C, when the air is in cross flow over the surface?

a) A circular cylinder of 10 mm diameter.

b) A square cylinder of 10 mm length on a side.

c) A vertical plate of 10 mm height.

7.33 A fine wire, of diameter D, is positioned across a flow passage to determine flow velocity from heat transfer characteristics. Current is passed through the wire to heat it, and the heat is dissipated to the flowing fluid by convection. The resistance of the wire is determined from electrical measurements, and the temperature is known from the resistance.

a) For a fluid of arbitrary Prandtl number, develop an expression for its velocity in terms of the difference between the temperature of the wire and the freestream temperature of the fluid.

b) What is the velocity of an airstream at one atmosphere and 25°C, if a wire of 0.5 mm diameter achieves a temperature of 40°C while dissipating 35 W/m?

7.34 A horizontal copper rod of 10 mm diameter and 100 mm length is inserted in the airspace between surfaces of an electronic device in order to enhance heat dissipation. The ends of the rod are at 90°C, while air at 25°C is in cross flow over the cylinder with a velocity of 25 m/s.

a) What is the temperature at the midplane of the rod?

b) What is the rate of heat transfer from the rod?

7.35 An uninsulated steam pipe is used to transport high temperature steam from one building to another. The pipe is of 0.5 m diameter, has a surface temperature of 150°C, and is exposed to ambient air at −10°C. The air moves in cross flow over the pipe with a velocity of 5 m/s. What is the heat loss per unit length of pipe?

7.36 Assume that a person can be approximated as a cylinder of 0.3 m diameter and

1.8 m height with a surface temperature 24°C. Calculate the body heat loss while this person is subjected to a 15 m/s wind whose temperature is −5°C.

7.37 Dry air at 1 atm pressure and a velocity of 15 m/s is to be humidified by passing it in crossflow over a porous, cylinder of diameter $D = 40$ mm which is saturated with water.

a) Assuming the water and air to be at 300 K, calculate the mass rate of water evaporated under steady-state conditions from the cylindrical medium per unit length.

b) The evaporation rate will increase, remain the same, or decrease if the temperature of the air and water are maintained at 325 K rather than 300 K? Briefly explain.

7.38 Mass transfer experiments have been conducted on a naphthalene cylinder of 18.4 mm diameter and 88.9 mm length subjected to a cross flow of air in a low-speed wind tunnel. After exposure for 39 minutes to the airstream at a temperature of 26°C and a velocity of 12 m/s, it was determined that the cylinder mass decreased by 0.35 g. The barometric pressure was recorded at 750.6 mm Hg. The saturation pressure, p_{sat}, of naphthalene vapor in equilibrium with solid naphthalene is given by the relation

$$p_{sat} = p \times 10^E$$

where $E = 8.67 - (3766/T)$, with $T(K)$ and $p(bar)$ being the temperature and pressure of air. Naphthalene has a molecular weight of 128.16 kg/kmol.

a) Determine the convection mass transfer coefficient from the experimental observations.

b) Compare this result with estimates from appropriate correlations for the prescribed flow conditions.

7.39 Approximate the human form as an unclothed vertical cylinder of 0.3 m diameter and 1.75 m length with a surface temperature of 30°C.

a) Calculate the heat loss in a 10 m/s wind at 20°C.

b) What is the heat loss if the skin is covered with a thin layer of water at 30°C and the relative humidity of the air is 60 percent?

7.40 Water at 20°C flows over a 20-mm diameter sphere with a velocity of 5 m/s. The surface of the sphere is at 60°C.

a) What is the drag force on the sphere?

b) What is the rate of heat transfer from the sphere?

7.41 Air at 25°C flows over a 10-mm diameter sphere with a velocity of 25 m/s, while the surface of the sphere is maintained at 75°C.

a) What is the drag force on the sphere?

b) What is the rate of heat transfer from the sphere?

7.42 Atmospheric air at 25°C and a velocity of 0.5 m/s flows over a 50-W incandescent bulb whose surface temperature is at 140°C. The bulb may be approximated as a sphere of 50 mm diameter. What is the rate of heat loss by convection to the air?

7.43 In manufacturing lead shot, molten lead droplets are allowed to fall from a tower through cool air into a tank of water. The intent is to have the lead solidify, as a result of heat transfer to the air, before hitting the water. Consider each pellet to be a sphere of 3 mm diameter and to be moving at the terminal velocity (for which the drag force equals the gravitational force) throughout the descent. If pellets at the melting point are to be converted from the molten to the solid state in air at 15°C, what is the necessary height of the tower above the tank? The latent heat of fusion of the lead is 2.45×10^4 J/kg.

7.44 The humidity of air may be controlled by spraying water droplets into an airflow. Consider droplets of diameter $D = 3$ mm in an airflow for which the relative velocity is 5 m/s. If the droplet and air temperatures are 25°C and 35°C, respectively, what is the rate of evaporation from a single drop?

7.45 Repeat Example 7.6 for a more compact tube bank in which the longitudinal and transverse pitches are $S_L = S_T = 20.5$ mm. All other conditions remain the same.

7.46 A preheater involves the use of condensing steam at 100°C on the inside of a bank of tubes to heat air which enters at one atmosphere and 25°C. The air moves at 5 m/s in cross flow over the tubes. Each tube is one meter long and has an outside diameter of 10 mm, and the bank consists of 196 tubes in a square, aligned array for which $S_T = S_L = 15$ mm.

   a) What is the total rate of heat transfer to the air?

   b) What is the pressure drop associated with the air flow?

7.47 A tube bank uses an aligned arrangement of 10 mm diameter tubes with $S_T = S_L = 20$ mm. There are 10 rows of tubes with 50 tubes in each row. Consider an application for which cold water flows through the tubes, maintaining the outer surface temperature at 27°C, while flue gases at 427°C and a velocity of 5 m/s are in cross flow over the tubes. The properties of the flue gas may be approximated as those of atmospheric air at 427°C. What is the total rate of heat transfer per unit length of the tubes in the bank?

7.48 Consider an air conditioning system comprised of a bank of tubes arranged normal to air flowing in a duct at a mass rate of $\dot{m}_a$ (kg/s). A coolant flowing through the tubes is able to maintain the surface temperature of the tubes at a constant value of $T_s < T_{a,i}$, where $T_{a,i}$ is the inlet air temperature (upstream of the tube bank). It has been suggested that air cooling may be enhanced in this situation if a thin, uniform film of water is maintained on the outer surface of each of the tubes.

   a) Assuming the water film to be at the temperature T_s, develop an expression for the amount of cooling which occurs with the water film relative to the amount of cooling which occurs without the film. The amount of cooling may be defined as $(T_{a,i} - T_{a,o})$, where $T_{a,o}$ is the outlet air temperature (downstream of the tube bank). The upstream air may be assumed to be dry. Note: the total rate of heat loss from the air may be expressed as $q = \dot{m}_a c_{p,a}(T_{a,i} - T_{a,o})$. Estimate the value of this ratio under conditions for which $T_{a,i} = 35°C$ and $T_s = 10°C$.

   b) Consider a tube bank that is five rows deep, with 12 tubes in a row. Each tube is 0.5 m long, with an outside diameter of 8 mm, and a staggered arrangement is used for which $S_T = S_L = 24$ mm. Under conditions for which $\dot{m}_a = 0.5$ kg/s, V

$= 3$ m/s, where V is the upstream air velocity, $T_{a,i} = 35°C$ and $T_s = 10°C$, what is the value of $T_{a,o}$ if the tubes are wetted? What is the specific humidity of the air leaving the tube bank?

7.49 The use of rock pile thermal energy storage systems has been considered for solar energy and industrial process heat applications. A particular system involves a cylindrical container, 2 m long by 1 m in diameter, in which nearly spherical rocks of 0.03 m diameter are packed. The bed has a void space of 0.42, and the density and specific heat of the rock are $\rho = 2300$ kg/m³ and $c_p = 879$ J/kg·K, respectively. Consider conditions for which atmospheric air is supplied to the rock pile at a steady flow rate of 1 kg/s and a temperature of 90°C. The air flows in the axial direction through the container. If the rock is at a temperature of 25°C, what is the total rate of heat transfer from the air to the rock pile?

8 Internal Flow

Having acquired the means to compute convection transfer rates for external flow, we now consider the convection transfer problem for *internal flow*. Recall that an external flow is one for which boundary layer development on a surface is allowed to continue without external constraints, as for the flat plate of Figure 6.6. In contrast an internal flow, such as flow in a pipe, is one for which the fluid is *confined* by a surface. Hence the boundary layer is unable to develop without eventually being constrained. The internal flow configuration represents a convenient geometry for the heating and cooling of fluids used in the chemical processing, environmental control, and energy conversion areas.

We begin by considering velocity (hydrodynamic) effects pertinent to internal flows, focusing on certain unique features of boundary layer development. Thermal boundary layer effects will then be considered, and an overall energy balance will be applied to determine fluid temperature variations in the flow direction. Empirical correlations for estimating the convection heat transfer coefficient will then be presented for a variety of internal flow conditions.

8.1   HYDRODYNAMIC CONSIDERATIONS

When considering external flow, it is only necessary to ask whether this flow is laminar or turbulent. However, for an internal flow we must also be concerned with the existence of *entrance* and *fully developed* regions.

8.1.1   Flow Conditions

Consider laminar flow in a circular tube of radius r_o (Figure 8.1), where fluid enters the tube with a uniform velocity. We know that, when the fluid makes contact with the surface, viscous effects become important and a boundary layer develops with increasing x. This development occurs at the expense of a shrinking inviscid flow region and concludes with boundary layer merger at the centerline. The distance from the entrance at which this merger occurs is termed

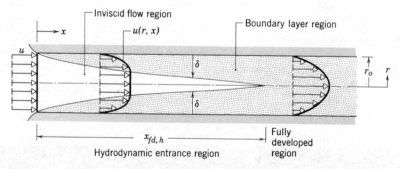

Figure 8.1   Laminar, hydrodynamic boundary layer development in a circular tube.

the *hydrodynamic entry length*, $x_{fd,h}$. For $x > x_{fd,h}$ the boundary layer, and therefore the effect of viscosity, extends over the entire cross section, and the flow is said to be *fully developed*.

When dealing with internal flows, it is important to be cognizant of the extent of the entry region, which depends on whether the flow is laminar or turbulent. The Reynolds number for flow in a circular tube is defined as

$$Re_D \equiv \frac{\rho u_m D}{\mu} \qquad (8.1)$$

where u_m is the mean fluid velocity over the tube cross section and D is the tube diameter. A representative value of the critical Reynolds number is

$$Re_{D,c} = 2300 \qquad (8.2)$$

For laminar flow in a circular tube ($Re_D \lesssim 2300$), the hydrodynamic entry length may be obtained from an expression of the form [1]

$$(x_{fd,h}/D)_{lam} \approx 0.05 Re_D \qquad (8.3)$$

Although there is no satisfactory general expression for the entry length in turbulent flow ($Re_D \gtrsim 2300$), we know that it is independent of Reynolds number and that, as a first approximation,

$$10 \lesssim (x_{fd,h}/D)_{turb} \lesssim 60 \qquad (8.4)$$

8.1.2 The Mean Velocity

Because the velocity varies over the cross section and there is no well-defined freestream, it is necessary to work with a mean velocity, u_m, when dealing with internal flows. Calculation of the Reynolds number from Equation 8.1 dictates that this velocity be known. It is defined such that, when multiplied by the fluid density, ρ, and the cross-sectional area of the tube, A_c, it provides the rate of mass flow through the tube. Hence

$$\dot{m} = \rho u_m A_c \qquad (8.5)$$

For steady, incompressible flow in a tube of uniform cross-sectional area, $\dot{m}$ and u_m are constants independent of x. From Equations 8.1 and 8.5 it is evident that for flow in a circular tube ($A_c = \pi D^2/4$), the Reynolds number may be expressed as

$$Re_D = \frac{4\dot{m}}{\pi D \mu} \qquad (8.6)$$

Since the mass flow rate may also be expressed as the integral of the mass flux, (ρu), over the cross section

$$\dot{m} = \int_{A_c} \rho u(r,x)dA_c \tag{8.7}$$

it follows that, for *incompressible* flow in a *circular* tube,

$$u_m = \frac{\int_{A_c} \rho u(r,x)dA_c}{\rho A_c} = \frac{2\pi\rho}{\rho\pi r_o^2} \int_0^{r_o} u(r,x)rdr = \frac{2}{r_o^2} \int_0^{r_o} u(r,x)rdr \tag{8.8}$$

The foregoing expression may be used to determine u_m at any axial location, x, from knowledge of the velocity profile, $u(r)$, at that location.

8.1.3   Velocity Profile in the Fully Developed Region

The form of the velocity profile may readily be determined for the *laminar flow* of an *incompressible, constant-property fluid* in the *fully developed region* of a *circular tube*. An important feature of hydrodynamic conditions in the fully developed region is the fact that both the radial velocity component, v, and the gradient of the axial velocity component, $(\partial u/\partial x)$, are everywhere zero.

Fully Developed Hydrodynamic Conditions

$$v = 0 \qquad \left(\frac{\partial u}{\partial x}\right) = 0 \qquad \text{CONST. AXIAL VELOCITY} \tag{8.9}$$

Hence for these conditions the axial velocity component depends only on r, $u(x,r) = u(r)$.

The radial dependence of the axial velocity may be obtained by solving the appropriate form of the x-momentum equation for the flow. This form is determined by first recognizing that, for the conditions of Equation 8.9, the net momentum flux is everywhere zero in the fully developed region. Hence the momentum conservation requirement reduces to a simple balance between shear and pressure forces in the flow. For the annular differential element of Figure 8.2, this force balance may be expressed as

$$-\tau_r(2\pi rdx) + \left\{\tau_r(2\pi rdx) + \frac{d}{dr}[\tau_r(2\pi rdx)]dr\right\}$$

$$+ p(2\pi rdr) - \left\{p(2\pi rdr) + \frac{d}{dx}[p(2\pi rdr)]dx\right\} = 0$$

Hence

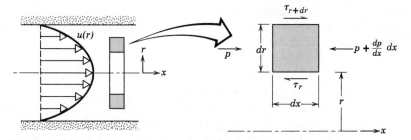

Figure 8.2 Force balance on a differential element for laminar, fully developed flow in a circular tube.

$$+ 2\pi dr dx \frac{d}{dr}(r\tau_r) - 2\pi r dr dx \frac{dp}{dx} = 0$$

or

$$+ \frac{d}{dr}(r\tau_r) = r\frac{dp}{dx} \tag{8.10}$$

Substituting from Newton's law of viscosity for simple shear flow, Equation 6.50, which may be expressed as

$$\tau_r = \mu \frac{du}{dr} \tag{8.11}$$

Equation 8.10 then becomes

$$\frac{\mu}{r} \frac{d}{dr}\left(r\frac{du}{dr}\right) = \frac{dp}{dx} \tag{8.12}$$

Since the axial pressure gradient is independent of r, Equation 8.12 may be solved by integrating twice to obtain

$$r\frac{du}{dr} = \frac{1}{\mu}\left(\frac{dp}{dx}\right)\frac{r^2}{2} + C_1$$

and

$$u(r) = \frac{1}{\mu}\left(\frac{dp}{dx}\right)\frac{r^2}{4} + C_1 \ln r + C_2$$

The integration constants may be determined by invoking the boundary conditions

$$u(r_o) = 0 \qquad \left.\frac{\partial u}{\partial r}\right|_{r=0} = 0$$

which reflect the requirement of zero slip at the tube surface and radial symmetry about the centerline, respectively. It is a simple matter to evaluate the constants, and it follows that

$$u(r) = -\frac{1}{4\mu}\left(\frac{dp}{dx}\right)r_o^2\left[1 - \left(\frac{r}{r_o}\right)^2\right]$$

Pressure decreases with x (8.13)

From this expression we conclude that the velocity profile is *parabolic*. Note that the pressure gradient must always be negative.

The foregoing result may be used to determine the mean velocity of the flow. Substituting Equation 8.13 into Equation 8.8 and integrating, we obtain

$$u_m = -\frac{r_o^2}{8\mu}\frac{dp}{dx}$$

Mean velocity (8.14)

Substituting this result back into Equation 8.13, the velocity profile may then be expressed as

$$\frac{u(r)}{u_m} = 2\left[1 - \left(\frac{r}{r_o}\right)^2\right]$$

Velocity profile in terms of mean velocity AND r (8.15)

Remember that u_m can also be computed from knowledge of the mass flow rate, Equation 8.5, in which case, Equation 8.14 could be used to determine the pressure gradient.

8.1.4 Pressure Gradient and Friction Factor in Fully Developed Flow

The engineer is frequently interested in knowing the pressure drop required to sustain an internal flow, since it determines pump or fan power requirements. To determine the pressure drop it is convenient to work with the (*Moody* or *Darcy*) *friction factor*. It is a dimensionless parameter defined as

$$f \equiv \frac{-(dp/dx)D}{\rho u_m^2/2}$$

(8.16)[1]

---

[1] This quantity is not to be confused with the *friction coefficient*, sometimes called the Fanning friction factor, which is defined as

$$C_f \equiv \frac{\tau_s}{\rho u_m^2/2}$$

(8.17)

Since the shear force of the fluid on the surface is equal and opposite to the force of the surface on the fluid, the surface shear stress may be expressed as $\tau_s = -\mu(du/dr)_{r=r_o}$. It follows from Equation 8.13 that

$$C_f = \frac{f}{4}$$

(8.18)

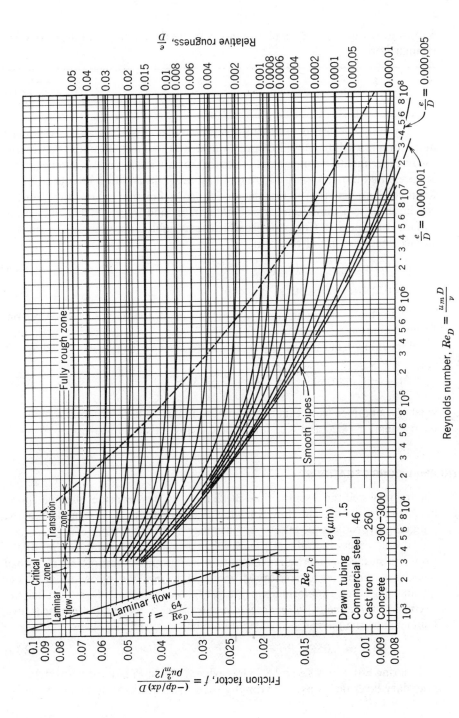

Figure 8.3 Friction factor for fully developed flow in a circular tube [5]. Used with permission.

Substituting Equations 8.1 and 8.14 into 8.16, we then obtain for fully developed laminar flow

$$f = \frac{64}{Re_D} \tag{8.19}$$

For fully developed turbulent flow in circular tubes, the analysis is much more complicated than that for laminar flow, and we must ultimately rely on the use of experimental results. Friction factors for a wide Reynolds number range are presented in the *Moody diagram* of Figure 8.3. In addition to depending on Reynolds number, the friction factor is also a function of the tube surface condition. It is a minimum for *smooth* surfaces and increases with increasing surface roughness, e. Correlations which reasonably approximate the smooth surface condition are of the form

$$f = 0.316Re_D^{-1/4} \qquad (Re_D \lesssim 2 \times 10^4) \tag{8.20}$$

$$f = 0.184Re_D^{-1/5} \qquad (Re_D \gtrsim 2 \times 10^4) \tag{8.21}$$

Note that f, and hence dp/dx, is a constant in the fully developed region. From Equation 8.16 the pressure drop, $\Delta p = p_1 - p_2$, associated with fully developed flow from the axial position x_1 to x_2 may then be expressed as

$$\Delta p = -\int_{p_1}^{p_2} dp = f\frac{\rho u_m^2}{2D}\int_{x_1}^{x_2} dx = f\frac{\rho u_m^2}{2D}(x_2 - x_1) \tag{8.22}$$

where f is obtained from Figure 8.3 or from Equation 8.19 for laminar flow and from Equations 8.20 and 8.21 for turbulent flow in smooth tubes.

8.2 THERMAL CONSIDERATIONS

Having reviewed the essential fluid mechanics, we now consider thermal effects associated with internal flow. In what way are these effects similar to the foregoing hydrodynamic effects; in what way do they differ?

If fluid enters the tube of Figure 8.4 at a uniform temperature that differs from the surface temperature, convection heat transfer occurs and a *thermal boundary layer* begins to develop. Moreover, if the tube *surface* condition is fixed by imposing either a uniform temperature ($T_s = $ const.) or a uniform heat flux ($q_s'' = $ const.), a *thermally fully developed condition* is eventually reached. For laminar flow the *thermal entry length* may be expressed as [2]

$$(x_{fd,t}/D)_{lam} \approx 0.05Re_DPr \qquad \text{THermal entry length} \tag{8.23}$$

Comparing Equations 8.3 and 8.23 it is evident that, if $Pr > 1$, the hydrodynamic boundary layer develops more rapidly than the thermal boundary layer ($x_{fd,h}$

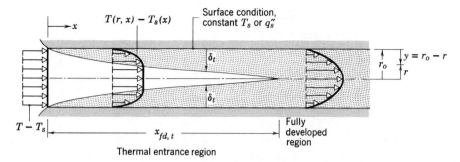

Figure 8.4 Thermal boundary layer development in a circular tube.

$< x_{\mathrm{fd},t}$), while the inverse is true for $Pr < 1$. Moreover, for extremely high Prandtl number fluids, such as oils ($Pr \gtrsim 100$), $x_{\mathrm{fd},h}$ is very much smaller than $x_{\mathrm{fd},t}$ and it is reasonable to assume a fully developed velocity profile throughout the thermal entry region. However, for turbulent flow, conditions are nearly independent of Prandtl number, and Equation 8.4 may be used to obtain $x_{\mathrm{fd},t}$ to a first approximation.

Thermal conditions in the fully developed region are characterized by several interesting and useful features. Before we can consider these features (Section 8.2.3), however, it is necessary to first introduce the notion of a mean temperature and the appropriate form of Newton's law of cooling.

8.2.1 The Mean Temperature

Just as the absence of a freestream velocity requires use of a mean velocity to describe an internal flow, the absence of a fixed freestream temperature necessitates using a *mean temperature*. The *mean temperature* of the fluid at a given cross section can be defined in terms of the *thermal* energy transported with the bulk motion of the fluid as it moves past the cross section. The rate at which this transport occurs, $\dot{E}_t$, may be obtained by integrating the product of the mass flux (ρu) and the internal energy per unit mass ($c_v T$) over the cross section. That is,

$$\dot{E}_t = \int_{A_c} \rho u c_v T \, dA_c \tag{8.24}$$

Hence if a mean temperature is defined such that

$$\dot{E}_t \equiv \dot{m} c_v T_m \tag{8.25}$$

we obtain

$$T_m = \frac{\int_{A_c} \rho u c_v T \, dA_c}{\dot{m} c_v} \tag{8.26}$$

For *incompressible flow* in a *circular tube* with constant c_v, it follows from Equations 8.5 and 8.26 that

$$T_m = \frac{2}{u_m r_o^2} \int_0^{r_o} u T r \, dr \tag{8.27}$$

Note that, in any internal flow for which there is convection heat transfer to or from the fluid, T_m must change with the distance x along the flow.

It is important to note that, when multiplied by the mass flow rate and the specific heat, T_m provides the rate at which thermal energy is transported with the fluid as it moves along the tube.

8.2.2 Newton's Law of Cooling

The mean temperature, T_m, is a convenient reference temperature for internal flows, playing much the same role as the freestream temperature, T_∞, for external flows. Accordingly, Newton's law of cooling may be expressed as

$$q_s'' = h(T_s - T_m) \tag{8.28}$$

where h is the *local* convection heat transfer coefficient. However, there is an essential difference between T_m and T_∞. Whereas T_∞ is a constant in the flow direction, T_m must vary in this direction. That is, dT_m/dx is never equal to zero if heat transfer is occurring. The value of T_m increases with x if heat transfer is from the surface to the fluid ($T_s > T_m$); it decreases with x if the opposite is true ($T_s < T_m$).

8.2.3 Fully Developed Conditions

Since the existence of convection heat transfer between the surface and the fluid dictates that the fluid temperature must continue to change with x, one might legitimately question whether fully developed thermal conditions could ever be reached. The situation is certainly different than the hydrodynamic case, for which $(\partial u/\partial x) = 0$ in the fully developed region. In contrast, if there is heat transfer, not only is (dT_m/dx) not equal to zero but $(\partial T/\partial x)$ at any radius r is also not equal to zero. Accordingly, the temperature profile, $T(r)$, is continuously changing with x, and it would seem that a fully developed condition could never be reached. This apparent contradiction may be reconciled by working with a dimensionless form of the temperature.

Recall that analyses have often been simplified by working with dimensionless temperature differences, as for transient conduction (Chapter 5) and the energy conservation equation (Chapter 6). A dimensionless temperature difference for internal flow may be inferred by first recognizing, from Equation 8.28, that the driving potential for convection heat transfer is $(T_s - T_m)$. Since the local temperature $T(r)$ must appear in the resulting term, it follows that an appropriate form of the dimensionless temperature difference is

$$\frac{T_s - T(r)}{T_s - T_m}$$

When the temperature profile, plotted in this manner, becomes independent of x, the flow is said to be thermally fully developed. Accordingly, just as fully developed hydrodynamic conditions imply that the velocity profile is independent of x, fully developed thermal conditions simply imply that the *dimensionless temperature profile* is independent of x. Hence for *fully developed thermal conditions*:

$$\frac{\partial}{\partial x}\left[\frac{T_s(x) - T(r,x)}{T_s(x) - T_m(x)}\right]_{\text{fd},t} = 0 \tag{8.29}$$

where T_s is the tube surface temperature, T is the local fluid temperature, and T_m is the mean temperature of the fluid over the cross section of the tube.

The condition given by Equation 8.29 is eventually reached in a tube for which there is either a *uniform surface heat flux* (q_s'' is constant) *or* a *uniform surface temperature* (T_s is constant). These surface conditions arise in many different engineering applications. For example, a constant surface heat flux would exist if the tube wall were heated electrically or if the outer surface were being uniformly irradiated. In contrast a constant surface temperature would exist if a phase change (due to boiling or condensation) were occurring at the outer surface. Note that it is impossible to *simultaneously* impose the conditions of constant surface heat flux and constant surface temperature. If q_s'' is constant, T_s must vary with x; conversely, if T_s is constant, q_s'' must vary with x.

It is now possible to obtain several important features of fully developed flow. If Equation 8.29 is satisfied, the temperature ratio $(T_s - T)/(T_s - T_m)$ must be independent of x and therefore can depend only on r. Hence for any value of r the derivative of this ratio with respect to r must also be independent of x. Evaluating this derivative at the tube surface (note that T_s and T_m are constants in so far as differentiation with respect to r is concerned), we then obtain

$$\frac{\partial}{\partial r}\left(\frac{T_s - T}{T_s - T_m}\right)\bigg|_{r=r_o} = \frac{-\partial T/\partial r|_{r=r_o}}{T_s - T_m} \neq f(x)$$

Substituting for $\partial T/\partial r$ from Fourier's law, which, from Figure 8.4, must be of the form,

$$q_s'' = -k \frac{\partial T}{\partial y}\bigg|_{y=0} = k \frac{\partial T}{\partial r}\bigg|_{r=r_o}$$

and for q_s'' from Newton's law of cooling, Equation 8.28, we then obtain

$$\frac{h}{k} = \text{constant} \tag{8.30}$$

Hence *in the thermally fully developed flow* of a fluid *with constant properties*, the *local convection coefficient is a constant, independent of* x.

When Equation 8.29 is not satisfied, the convection coefficient must vary with x. For example, h must vary with x in the entry region, as shown in Figure 8.5. Because the thermal boundary layer thickness is zero at the tube entrance, the convection coefficient is extremely large at x = 0. However, h decays rapidly as the thermal boundary layer begins to develop, until the constant value associated with fully developed conditions is reached.

Additional simplifications are associated with the special case of *uniform surface heat flux*. Since both h and q_s'' are constant in the fully developed region, it follows from Equation 8.28 that

$$\frac{dT_s}{dx}\bigg|_{\text{fd},t} = \frac{dT_m}{dx}\bigg|_{\text{fd},t} \qquad (q_s'' = \text{const.}) \tag{8.31}$$

If we expand Equation 8.29 and solve for $\partial T/\partial x$, it also follows that in the fully developed region

$$\frac{\partial T}{\partial x}\bigg|_{\text{fd},t} = \frac{dT_s}{dx}\bigg|_{\text{fd},t} - \frac{(T_s - T)}{(T_s - T_m)}\frac{dT_s}{dx}\bigg|_{\text{fd},t} + \frac{(T_s - T)}{(T_s - T_m)}\frac{dT_m}{dx}\bigg|_{\text{fd},t} \tag{8.32}$$

Substituting from Equation 8.31, we then obtain

$$\frac{\partial T}{\partial x}\bigg|_{\text{fd},t} = \frac{dT_m}{dx}\bigg|_{\text{fd},t} \qquad (q_s'' = \text{const.}) \tag{8.33}$$

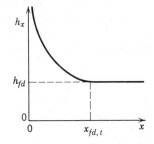

Figure 8.5 Axial variation of the convection heat transfer coefficient for flow in a tube.

Hence the axial temperature gradient is independent of the radial location. For the case of *constant surface temperature* $(dT_s/dx = 0)$, it also follows from Equation 8.32 that

$$\left.\frac{\partial T}{\partial x}\right|_{\text{fd},t} = \frac{(T_s - T)}{(T_s - T_m)} \left.\frac{dT_m}{dx}\right|_{\text{fd},t} \qquad (T_s = \text{const.}) \qquad (8.34)$$

in which case the value of $\partial T/\partial x$ depends on the radial coordinate.

From all of the foregoing results, it is evident that the mean temperature is a very important variable for internal flows. To describe such flows, it is essential that its variation with x be known. This variation may be obtained by applying an *overall energy balance* to the flow.

EXAMPLE 8.1

For flow of a liquid metal through a circular tube, the velocity and temperature profiles at a particular axial location may be approximated as being uniform and parabolic, respectively. That is, $u(r) = C_1$ and $T(r) - T_s = C_2[1 - (r/r_o)^2]$, where C_1 and C_2 are constants. What is the value of the Nusselt number, Nu_D, at this location?

SOLUTION

KNOWN:

Form of the velocity and temperature profiles at a particular axial location for flow in a circular tube.

FIND:

Nusselt number at the prescribed location.

SCHEMATIC:

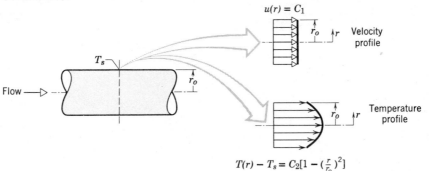

ASSUMPTIONS:

Incompressible, constant property flow.

ANALYSIS:

The Nusselt number may be obtained by first determining the convection coefficient, which, from Equation 8.28, is given as

$$h = \frac{q_s''}{(T_s - T_m)}$$

From Equation 8.27, the mean temperature is then

$$T_m = \frac{2}{u_m r_o^2} \int_0^{r_o} u T r\, dr = \frac{2C_1}{u_m r_o^2} \int_0^{r_o} \left\{ T_s + C_2 \left[1 - \left(\frac{r}{r_o}\right)^2 \right] \right\} r\, dr$$

or, since $u_m = C_1$ from Equation 8.8,

$$T_m = \frac{2}{r_o^2} \int_0^{r_o} \left\{ T_s + C_2 \left[1 - \left(\frac{r}{r_o}\right)^2 \right] \right\} r\, dr$$

$$T_m = \frac{2}{r_o^2} \left[T_s \frac{r^2}{2} + C_2 \frac{r^2}{2} - \frac{C_2}{4} \frac{r^4}{r_o^2} \right]\Big|_0^{r_o}$$

$$T_m = \frac{2}{r_o^2} \left(T_s \frac{r_o^2}{2} + \frac{C_2}{2} r_o^2 - \frac{C_2}{4} r_o^2 \right)$$

$$T_m = T_s + C_2$$

The heat flux may be obtained from Fourier's law, in which case,

$$q_s'' = k \frac{\partial T}{\partial r}\Big|_{r=r_o} = -kC_2 2 \frac{r}{r_o^2}\Big|_{r=r_o}$$

$$q_s'' = -2C_2 \frac{k}{r_o}$$

Hence

$$h = \frac{q_s''}{T_s - T_m} = \frac{-2C_2(k/r_o)}{-C_2} = \frac{2k}{r_o}$$

and

$$Nu_D = \frac{hD}{k} = \frac{(2k/r_o) \times 2r_o}{k}$$

$$Nu_D = 4 \qquad\qquad \triangleleft$$

8.3 THE ENERGY BALANCE

8.3.1 General Considerations

Because the flow in a tube is completely enclosed, it is a simple matter to apply an energy balance. From such a balance it is possible to determine how the mean temperature, $T_m(x)$, varies with position along the tube and how the total convection heat transfer, q_{conv}, is related to the difference in temperatures at the tube inlet and outlet.

Consider the tube flow shown in Figure 8.6. Fluid moves at a constant flowrate $\dot{m}$, and convection heat transfer occurs at the inner surface. In the usual situation fluid kinetic and potential energy changes, as well as energy transfer by conduction in the axial direction, are negligible. Hence if no shaft work is done by the fluid as it moves through the tube, the only significant effects will be those associated with *thermal energy changes* and with *flow work*. Flow work is associated with a control-volume analysis, and it is related to the work which must be done to move fluid through a control surface [3]. The amount of work that is done per unit mass of fluid may be expressed as the product of the fluid pressure p and specific volume v ($v = 1/\rho$).

Applying the energy balance equation, Equation 1.10, to the differential control volume of Figure 8.6, and recalling the definition of the mean temperature, Equation 8.25, we then obtain

$$dq_{conv} + \dot{m}(c_v T_m + pv) - \left[\dot{m}(c_v T_m + pv) + \dot{m}\frac{d(c_v T_m + pv)dx}{dx} \right] = 0$$

or

$$dq_{conv} = \dot{m}d(c_v T_m + pv) \qquad (8.35)$$

That is, the rate of convection heat transfer to the fluid must equal the rate at which the fluid thermal energy increases plus the net rate at which work is done in moving the fluid through the control volume. A simplified version of

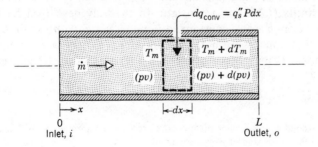

Figure 8.6 Control volume for internal flow in a tube.

Equation 8.35 may be applied to most heat transfer problems. For example, if the fluid is assumed to be a *perfect gas*, $pv = RT_m$ and $c_p = c_v + R$. Hence, if c_p is assumed to be constant, Equation 8.35 reduces to

$$dq_{conv} = \dot{m}c_p dT_m \qquad (8.36)$$

Moreover, this expression may generally be used to a good approximation for *incompressible liquids*. In this case $c_v = c_p$, and since v is extremely small, $d(pv)$ is generally much less than $d(c_v T_m)$.[2] Accordingly, Equation 8.36 again follows from Equation 8.35.

A special form of Equation 8.36 relates to conditions for the *entire* tube. In particular, integrating from the tube inlet, i, to the outlet, o, it follows that

$$q_{conv} = \dot{m}c_p(T_{m,o} - T_{m,i}) \qquad (8.37)$$

where q_{conv} is the total tube heat transfer rate. Note that this simple energy balance relates three important thermal variables (q_{conv}, $T_{m,o}$, $T_{m,i}$) and allows for determination of one of the variables from knowledge of the other two. *It is a general expression that applies irrespective of the nature of the surface thermal condition or tube flow conditions.*

Equation 8.36 may be cast in a convenient form by first noting that the rate of convection heat transfer to the differential element may be expressed as $dq_{conv} = q''_s \, Pdx$, where P is the surface perimeter ($P = \pi D$ for a circular tube). Substituting from Equation 8.28, it follows that

$$\frac{dT_m}{dx} = \frac{q''_s P}{\dot{m}c_p} = \frac{P}{\dot{m}c_p}h(T_s - T_m) \qquad (8.38)$$

CHANGE IN mean TEMP. with x.

This expression is an extremely useful result, from which the axial variation of T_m may be determined. If $T_s > T_m$, heat is transferred to the fluid and T_m increases with x; if $T_s < T_m$, the opposite is true.

It is important to note how the terms on the right-hand side of Equation 8.38 depend on x. Although P may vary with x, the most common situation is one in which it is a constant (a tube of constant cross-sectional area). Hence, the quantity $(P/\dot{m}c_p)$ is a constant. In the fully developed region, the convection coefficient h is also constant, although it varies with x in the entrance region (Figure 8.5). Finally, we note that, although T_s may be constant under certain conditions, T_m must always vary with x (except for the trivial case of no heat transfer, $T_s = T_m$).

---

[2] The only exception arises when the pressure gradient is extremely large. This situation occurs when $\dot{m}$ is very large and/or A_c is very small.

The solution to Equation 8.38 for $T_m(x)$ depends on the nature of the surface thermal conditions. Recall that the two special cases of interest include *constant surface heat flux* and *constant surface temperature*. It is common to find one of these conditions existing to a reasonable approximation.

EXAMPLE 8.2

In a particular application involving fluid flow at a rate $\dot{m}$ through a circular tube of length L and diameter D, the surface heat flux is known to have the following sinusoidal variation with x

$$q_s''(x) = q_{s,m}'' \frac{\pi x}{L}$$

The maximum flux, $q_{s,m}''$, is a known constant, and the fluid enters the tube at a known temperature, $T_{m,i}$. Assuming the convection coefficient to be constant, how do the mean temperature of the fluid and the surface temperature vary with x?

SOLUTION

KNOWN:

Axial variation of surface heat flux for flow through a tube.

FIND:

Axial variation of mean and surface temperatures.

SCHEMATIC:

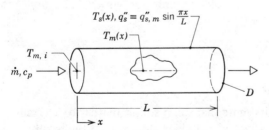

ASSUMPTIONS:

1. Local convection coefficient is independent of x.
2. Kinetic and potential energy changes and axial conduction are negligible.

3. Fluid is an ideal gas or a liquid for which $d(pv) \ll (c_v T_m)$.

ANALYSIS:

The energy balance given by Equation 8.38 is applicable. Hence

$$\frac{dT_m}{dx} = \frac{q_s'' P}{\dot{m} c_p} = \frac{(\pi D) q_{s,m}'' \sin \dfrac{\pi x}{L}}{\dot{m} c_p}$$

Integrating from $x = 0$

$$\int_{T_{m,i}}^{T_m(x)} dT_m = \frac{\pi D q_{s,m}''}{\dot{m} c_p} \int_0^x \sin \frac{\pi x}{L} dx$$

Hence

$$T_m(x) - T_{m,i} = - \frac{L D q_{s,m}''}{\dot{m} c_p} \cos \frac{\pi x}{L} \Bigg|_0^x$$

or

$$T_m(x) = T_{m,i} + \frac{L D q_{s,m}''}{\dot{m} c_p} \left(1 - \cos \frac{\pi x}{L} \right) \qquad \triangleleft$$

From Equation 8.28 it also follows that $T_s(x) = (q_s''/h) + T_m$. Hence

$$T_s(x) = \frac{q_{s,m}''}{h} \sin \frac{\pi x}{L} + T_{m,i} + \frac{L D q_{s,m}''}{\dot{m} c_p} \left(1 - \cos \frac{\pi x}{L} \right) \qquad \triangleleft$$

COMMENTS:

For the prescribed surface condition the flow is not fully developed. The assumption of constant h should therefore be viewed as a first approximation.

8.3.2 Constant Surface Heat Flux

For this condition we first note that it is a simple matter to determine the total heat transfer rate q_{conv}. Since q_s'' is independent of x, it follows that

$$q_{\text{conv}} = q_s'' (P \cdot L) \qquad (8.39)$$

This expression could be used with Equation 8.37 to determine the tube inlet and outlet temperature difference.

For constant q_s'' it also follows that the right-hand side of Equation 8.38 is a constant independent of x. Hence

$$\frac{dT_m}{dx} = \frac{q_s'' P}{\dot{m}c_p} \neq f(x) \tag{8.40}$$

Integrating from $x = 0$, it then follows that

$$T_m(x) = T_{m,i} + \frac{q_s'' P}{\dot{m}c_p} x \qquad (q_s'' = \text{const.}) \tag{8.41}$$

Accordingly, the mean temperature must vary *linearly* with x along the tube. This variation is shown in Figure 8.7a. Moreover, from Equation 8.28 and Figure 8.5 we also expect the temperature difference $(T_s - T_m)$ to vary with x as shown in Figure 8.7a. This difference is initially small (due to the large value of h at the entrance) but increases with increasing x due to the decrease in h that occurs as the boundary layer develops. However, in the fully developed region we know from Equation 8.30 that h is independent of x. Hence from Equation 8.28 it follows that $(T_s - T_m)$ must also be independent of x in this region. Note also from Equations 8.31 and 8.33 that the axial temperature gradient assumes a fixed value, regardless of radial location in the fully developed region.

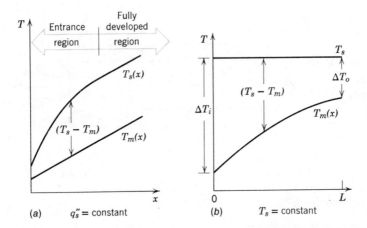

(a) $q_s'' = \text{constant}$ (b) $T_s = \text{constant}$

Figure 8.7   Axial temperature variations for heat transfer in a tube. (a) Constant surface heat flux. (b) Constant surface temperature.

EXAMPLE 8.3

A system for heating water from an inlet temperature of $T_{m,i} = 20°C$ to an outlet temperature of $T_{m,o} = 60°C$ involves passing the water through a thick-walled tube of inner and outer diameters equal to 20 mm and 40 mm, respectively. The outer surface of the tube is well insulated, and electrical heating within the wall provides for a uniform generation rate of $\dot{q} = 10^6$ W/m^3.

1. For a water mass flowrate of $\dot{m} = 0.1$ kg/s, how long must the tube be to achieve the desired outlet temperature?

2. If the inner surface temperature of the tube is measured to be $T_s = 70°C$ at the outlet, what is the local convection heat transfer coefficient at the outlet?

SOLUTION

KNOWN:

Internal flow through a thick-walled tube having uniform heat generation.

FIND:

1. Length of tube needed to achieve the desired outlet temperature.

2. Local convection coefficient at the outlet.

SCHEMATIC:

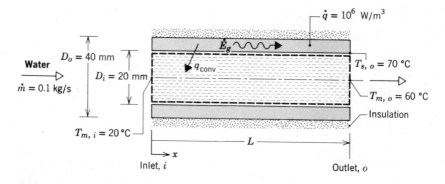

ASSUMPTIONS:

1. Steady-state conditions.

2. Uniform heat generation.

3. Negligible potential energy, kinetic energy, and flow work changes.

4. Constant properties.

5. Adiabatic outer tube surface.

PROPERTIES:

Table A.6, water ($\bar{T}_m = 313$ K): $c_p = 4.179$ kJ/kg·K.

ANALYSIS:

1. Since the outer surface of the tube is adiabatic, the rate at which energy is generated within the tube wall must equal the rate at which it is convected to the water.

$$\dot{E}_g = q_{conv}$$

With

$$\dot{E}_g = \dot{q}\,\frac{\pi}{4}\,(D_o^2 - D_i^2)L$$

it follows from Equation 8.37 that

$$\dot{q}\,\frac{\pi}{4}\,(D_o^2 - D_i^2)L = \dot{m}c_p\,(T_{m,o} - T_{m,i})$$

or

$$L = \frac{4\dot{m}c_p}{\pi(D_o^2 - D_i^2)\dot{q}}\,(T_{m,o} - T_{m,i})$$

$$L = \frac{4 \times 0.1 \text{ kg/s} \times 4179 \text{ J/kg·K}}{\pi(0.04^2 - 0.02^2)\text{m}^2 \times 10^6 \text{ W/m}^3}\,(60 - 20)°\text{C}$$

$$L = 17.7 \text{ m} \qquad\qquad\qquad \triangleleft$$

2. From Newton's law of cooling, Equation 8.28, the local convection coefficient at the tube exit is

$$h_o = \frac{q_s''}{(T_{s,o} - T_{m,o})}$$

Uniform heat generation in the wall provides for a constant surface heat flux, with

$$q_s'' = \frac{\dot{E}_g}{\pi D_i L} = \frac{\dot{q}}{4}\,\frac{(D_o^2 - D_i^2)}{D_i}$$

$$q_s'' = \frac{10^6 \text{ W/m}^3}{4} \frac{(0.04^2 - 0.02^2)\text{m}^2}{0.02 \text{ m}}$$

$$q_s'' = 1.5 \times 10^4 \text{ W/m}^2$$

Hence

$$h_o = \frac{1.5 \times 10^4 \text{ W/m}^2}{(70 - 60)°\text{C}}$$

$$h_o = 1500 \text{ W/m}^2 \cdot \text{K} \qquad \triangleleft$$

COMMENTS:

1. Note that, if conditions are fully developed over the entire tube, the local convection coefficient and the temperature difference, $T_s - T_m$, are independent of x. Hence $h = 1500$ W/m$^2 \cdot$K and $(T_s - T_m) = 10°$C over the entire tube, and the inner surface temperature at the tube inlet is $T_{s,i} = 30°$C.

2. Note also that the required tube length, L, could have been computed by applying the expression for $T_m(x)$, Equation 8.41, at $x = L$.

8.3.3   Constant Surface Temperature

Results for the total heat transfer rate and the axial distribution of the mean temperature are entirely different for the *constant surface temperature* condition. Defining ΔT as $T_s - T_m$, Equation 8.38 may be expressed as

$$\frac{dT_m}{dx} = -\frac{d(\Delta T)}{dx} = \frac{P}{\dot{m}c_p} h\Delta T$$

Separating variables and integrating from the tube inlet to the outlet, we then obtain

$$\int_{\Delta T_i}^{\Delta T_o} \frac{d(\Delta T)}{\Delta T} = -\frac{P}{\dot{m}c_p} \int_0^L h\, dx$$

or,

$$\ln \frac{\Delta T_o}{\Delta T_i} = -\frac{PL}{\dot{m}c_p} \left(\frac{1}{L} \int_0^L h\, dx\right)$$

From the definition of the average convection heat transfer coefficient given by Equation 6.5, it follows that

$$\ln \frac{\Delta T_o}{\Delta T_i} = -\frac{PL}{\dot{m}c_p}\bar{h}_L \qquad (T_s = \text{const.}) \tag{8.42a}$$

where $\bar{h}_L$, or simply $\bar{h}$, is the average value of h for the entire tube. Rearranging, we therefore obtain

$$\frac{\Delta T_o}{\Delta T_i} = \frac{T_s - T_{m,o}}{T_s - T_{m,i}} = \exp\left(-\frac{PL}{\dot{m}c_p}\bar{h}\right) \qquad (T_s = \text{const.}) \tag{8.42b}$$

Had we integrated from the tube inlet to some axial position x within the tube, we would have obtained the similar, but more general, result that

$$\frac{T_s - T_m(x)}{T_s - T_{m,i}} = \exp\left(-\frac{Px}{\dot{m}c_p}\bar{h}\right) \qquad (T_s = \text{const.}) \tag{8.43}$$

where $\bar{h}$ is now the average value of h from the tube inlet to x. This result suggests that the temperature difference $(T_s - T_m)$ *decays exponentially* with distance along the tube axis. The axial surface and mean temperature distributions are therefore as shown in Figure 8.7b.

Determination of an expression for the total heat transfer rate, q_{conv}, is complicated by the exponential nature of the temperature decay. Expressing Equation 8.37 in the form

$$q_{conv} = \dot{m}c_p[(T_s - T_{m,i}) - (T_s - T_{m,o})] = \dot{m}c_p(\Delta T_i - \Delta T_o)$$

and substituting for $\dot{m}c_p$ from Equation 8.42a, we obtain

$$q_{conv} = \bar{h} A_s \Delta T_{lm} \qquad (T_s = \text{const.}) \tag{8.44}$$

where A_s is the tube surface area ($A_s = P \cdot L$) and ΔT_{lm} is the *log mean temperature difference*

$$\Delta T_{lm} \equiv \frac{\Delta T_o - \Delta T_i}{\ln(\Delta T_o / \Delta T_i)} \tag{8.45}$$

Equation 8.44 is a form of Newton's law of cooling for the entire tube, and ΔT_{lm} is the appropriate *average* of the temperature difference over the tube length. The logarithmic nature of this average temperature difference (in contrast, for example, to an *arithmetic mean temperature* difference of the form $\Delta T_{am} = (\Delta T_i + \Delta T_o)/2$) is a consequence of the exponential nature of the temperature decay.

EXAMPLE 8.4

Steam condensing on the outer surface of a thin-walled circular tube of diameter $D = 50$ mm and length $L = 6$ m maintains a uniform surface temperature of 100°C. Water flows through the tube at a rate of $\dot{m} = 0.25$ kg/s, and its inlet and outlet temperatures are known to be $T_{m,i} = 15°C$ and $T_{m,o} = 45°C$, respectively. What is the average convection coefficient associated with the water flow?

SOLUTION

KNOWN:

Flowrate and inlet and outlet temperatures of water flowing through a tube of prescribed dimensions and surface temperature.

FIND:

Average convection heat transfer coefficient.

SCHEMATIC:

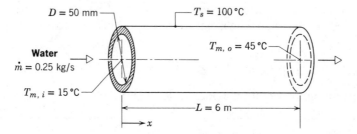

ASSUMPTIONS:

1. Constant surface temperature, T_s.
2. Negligible kinetic energy, potential energy, and flow work changes.
3. Constant properties.

PROPERTIES:

Table A.6, water (30°C): $c_p = 4178$ J/kg·K.

ANALYSIS:

Combining the energy balance, Equation 8.37, with the rate equation, Equation 8.44, the average convection coefficient is given by

$$\bar{h} = \frac{\dot{m}c_p}{\pi DL} \frac{(T_{m,o} - T_{m,i})}{\Delta T_{lm}}$$

From Equation 8.45

$$\Delta T_{lm} = \frac{(T_s - T_{m,o}) - (T_s - T_{m,i})}{\ln [(T_s - T_{m,o})/(T_s - T_{m,i})]}$$

$$\Delta T_{lm} = \frac{(100 - 45) - (100 - 15)}{\ln [(100 - 45)/(100 - 15)]} = 68.9°C$$

Hence

$$\bar{h} = \frac{0.25 \text{ kg/s} \times 4178 \text{ J/kg} \cdot \text{K}}{\pi \times 0.05 \text{ m} \times 6 \text{ m}} \frac{(45 - 15)°C}{68.9°C}$$

or

$$\bar{h} = 483 \text{ W/m}^2 \cdot \text{K} \qquad \triangleleft$$

COMMENTS:

1. In this case use of an arithmetic mean temperature difference, ΔT_{am} $= T_s - (T_{m,i} + T_{m,o})/2 = 70°C$, in place of the log mean temperature difference, $\Delta T_{lm} = 68.9°C$, would have been a reasonable approximation.

2. If conditions were fully developed over the entire tube, the local convection coefficient would be everywhere equal to 483 W/m$^2 \cdot$K.

8.4 CONVECTION CORRELATIONS: LAMINAR FLOW IN CIRCULAR TUBES

To use many of the foregoing results, the convection coefficients must be known. In this section we outline the manner in which such coefficients may be obtained theoretically for laminar flow in a circular tube. In subsequent sections we will consider empirical correlations pertinent to turbulent flow in a circular tube, as well as to flows in tubes of noncircular cross section.

8.4.1 The Fully Developed Region

The problem of laminar flow in a circular tube has been treated theoretically, and the results may be used to determine the convection coefficients. At any point in the tube the boundary layer approximations (Section 6.5) may be applied, and the appropriate form of the energy equation is

$$u \frac{\partial T}{\partial x} + v \frac{\partial T}{\partial r} = \frac{\alpha}{r} \frac{\partial}{\partial r} \left(r \frac{\partial T}{\partial r} \right) \tag{8.46}$$

This equation, which applies for cylindrical coordinates, is of the same form as the boundary layer equation, Equation 6.54, which was developed in rectangular coordinates, except that viscous dissipation has been neglected. The terms on the left-hand side of Equation 8.46 account for the net transfer of energy by gross fluid motion, and the term on the right side accounts for the net transfer of energy by conduction in the radial direction. Note that constant properties are assumed.

The solution to Equation 8.46 is readily obtained for the *fully developed region*. In this region the velocity boundary layer approximations are satisfied exactly. That is, $v = 0$ and $(\partial u / \partial x) = 0$, in which case the axial velocity component is given by the parabolic profile of Equation 8.15. Moreover, for the case of *constant surface heat flux*, the thermal boundary layer approximation is also satisfied exactly. That is, $(\partial^2 T / \partial x^2) = 0$. Substituting for the axial temperature gradient from Equation 8.33 and for the axial velocity component from Equation 8.15, the energy equation, Equation 8.46, reduces to

$$\frac{1}{r} \frac{d}{dr} \left(r \frac{dT}{dr} \right) = \frac{2u_m}{\alpha} \left(\frac{dT_m}{dx} \right) [1 - (r/r_o)^2] \qquad (q_s'' = \text{const.}) \tag{8.47}$$

where the term $(2u_m/\alpha)(dT_m/dx)$ is a constant. Separating variables and integrating twice, we then obtain the following expression for the radial temperature distribution

$$T(r) = \frac{2u_m}{\alpha} \left(\frac{dT_m}{dx} \right) \left[\frac{r^2}{4} - \frac{r^4}{16r_o^2} \right] + C_1 \ln r + C_2$$

The constants of integration may be evaluated by applying appropriate boundary conditions. From the requirement that the temperature remain finite at $r = 0$, it follows that $C_1 = 0$. From the requirement that $T(r_o) = T_s$, where T_s varies with x, it also follows that

$$C_2 = T_s - \frac{2u_m}{\alpha} \left(\frac{dT_m}{dx} \right) \left(\frac{3r_o^2}{16} \right)$$

Accordingly, for the fully developed region with constant surface heat flux, the temperature profile is of the form

$$T(r) = T_s - \frac{2u_m r_o^2}{\alpha} \left(\frac{dT_m}{dx} \right) \left[\frac{3}{16} + \frac{1}{16} \left(\frac{r}{r_o} \right)^4 - \frac{1}{4} \left(\frac{r}{r_o} \right)^2 \right] \tag{8.48}$$

From knowledge of the temperature profile, all other thermal parameters may be determined. For example, if the velocity and temperature profiles,

Equations 8.15 and 8.48, respectively, are substituted into Equation 8.27 and the integration over r is performed, the mean temperature is found to be

$$T_m = T_s - \frac{11}{48}\left(\frac{u_m r_o^2}{\alpha}\right)\left(\frac{dT_m}{dx}\right) \tag{8.49}$$

From Equation 8.40, where $P = \pi D$ and $\dot{m} = \rho u_m(\pi D^2/4)$, we then obtain

$$T_m - T_s = -\frac{11}{48}\frac{q_s'' D}{k} \tag{8.50}$$

Combining Newton's law of cooling, Equation 8.28, and 8.50, it follows that

$$h = \frac{48}{11}\left(\frac{k}{D}\right)$$

or

$$Nu_D \equiv \frac{hD}{k} = 4.36 \qquad (q_s'' = \text{const.}) \tag{8.51}$$

Hence in a *circular tube* characterized by *uniform surface heat flux* and *laminar, fully developed conditions*, the *Nusselt number is a constant*, independent of Re_D, Pr, and axial location.

For *laminar, fully developed conditions* with a *constant surface temperature*, the velocity boundary layer approximations are again satisfied exactly, and the thermal boundary layer approximation, $(\partial^2 T/\partial x^2) \ll (\partial^2 T/\partial r^2)$, is often reasonable. Substituting for the velocity profile from Equation 8.15 and for the axial temperature gradient from Equation 8.34, the energy equation, Equation 8.46, then becomes

$$\frac{1}{r}\frac{d}{dr}\left(r\frac{dT}{dr}\right) = \frac{2u_m}{\alpha}\left(\frac{dT_m}{dx}\right)[1-(r/r_o)^2]\frac{T_s-T}{T_s-T_m} \qquad (T_s = \text{const.}) \tag{8.52}$$

A solution to this equation may be obtained by an iterative procedure which involves making successive approximations to the temperature profile. The resulting profile is not described by a simple algebraic expression, but the resulting Nusselt number may be shown to be of the form [2]

$$Nu_D = 3.66 \qquad (T_s = \text{const.}) \tag{8.53}$$

Note that in using Equation 8.51 or 8.53 to determine h, the thermal conductivity should be evaluated at T_m.

EXAMPLE 8.5

One concept being considered for solar energy collection involves placing a tube at the focal point of a parabolic reflector and passing a fluid through the tube.

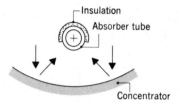

The net effect of this arrangement *may be approximated* as one of creating a condition of uniform heating at the surface of the tube. That is, the resulting heat flux to the fluid, q_s'', may be assumed to be a constant along the axis of the tube. Consider operation with a tube of diameter $D = 60$ mm on a sunny day for which $q_s'' = 2000$ W/m^2.

1. If water enters the tube at $\dot{m} = 0.01$ kg/s and $T_{m,i} = 20°$C, what tube length L is required to obtain an exit temperature of 80°C?

2. What is the tube surface temperature at the outlet of the tube, where fully developed conditions may be assumed to exist?

SOLUTION

KNOWN:

Internal flow with uniform surface heat flux.

FIND:

1. Length of tube, L, to achieve required heating.
2. Tube wall temperature, $T_s(L)$, at the outlet section, $x = L$.

SCHEMATIC:

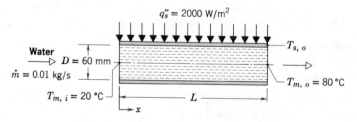

ASSUMPTIONS:

1. Steady-state conditions.
2. Incompressible flow with constant properties.
3. Negligible kinetic and potential energy and flow work changes.
4. Fully developed conditions at tube outlet.

PROPERTIES:

Table A.6, water ($\bar{T}_m = 323$ K): $c_p = 4181$ J/kg·K.
Table A.6, water ($T_{m,o} = 353$ K): $k = 0.670$ W/m·K, $\mu = 352 \times 10^{-6}$ N·s/m²,
$Pr = 2.2$.

ANALYSIS:

1. For constant surface heat flux, Equation 8.39 may be used with the energy balance, Equation 8.37, to obtain

$$A_s = \pi DL = \frac{\dot{m}c_p(T_{m,o} - T_{m,i})}{q_s''}$$

$$L = \frac{\dot{m}c_p}{\pi Dq_s''}(T_{m,o} - T_{m,i})$$

Hence

$$L = \frac{0.01 \text{ kg/s} \times 4181 \text{ J/kg·K}}{\pi \times 0.060 \text{ m} \times 2000 \text{ W/m}^2}(80 - 20)°\text{C} = 6.65 \text{ m} \qquad \lhd$$

2. The surface temperature at the outlet may be obtained from Newton's law of cooling, Equation 8.28, where

$$T_{s,o} = \frac{q_s''}{h} + T_{m,o}$$

To find the local convection coefficient at the tube outlet, the nature of the flow condition must first be established. From Equation 8.6

$$Re_D = \frac{4\dot{m}}{\pi D\mu} = \frac{4 \times 0.01 \text{ kg/s}}{\pi \times 0.060 \text{ m} \times 352 \times 10^{-6} \text{ N·s/m}^2} = 603$$

Hence the flow is laminar. With the assumption of fully developed conditions, the appropriate heat transfer correlation is then

$$Nu_D = \frac{hD}{k} = 4.36$$

and

$$h = 4.36 \frac{k}{D} = 4.36 \frac{0.670 \text{ W/m} \cdot \text{K}}{0.06 \text{ m}} = 48.7 \text{ W/m}^2 \cdot \text{K}$$

The surface temperature at the tube outlet is then

$$T_{s,o} = \frac{2000 \text{ W/m}^2}{48.7 \text{ W/m}^2 \cdot \text{K}} + 80°\text{C} = 121°\text{C} \qquad\qquad \triangleleft$$

COMMENTS:

Note that, for the conditions of this problem, $(x_{fd}/D) = 0.05 Re_D Pr = 66.3$, while $L/D = 110$. Hence the assumption of fully developed conditions is justified.

8.4.2   The Entry Region

The solution to the energy equation, Equation 8.46, for the entry region is considerably more difficult to obtain, since velocity and temperature now depend on x as well as r. Even if the radial convection term is neglected, as is often done, the axial temperature gradient, $\partial T/\partial x$, may no longer be simplified through Equations 8.33 or 8.34. However, two different types of entry length solutions have been obtained. The simplest solution is for the *thermal entry length problem*, and it is based on the assumption that thermal conditions develop in the presence of a *fully developed velocity profile*. Such a situation would exist if the point at which heat transfer begins is preceded by an *unheated starting length*. It could also be assumed to a reasonable approximation for large Prandtl number fluids, such as oils. Even in the absence of an unheated starting length, velocity boundary layer development would occur far more rapidly than the thermal boundary layer development, and a thermal entry length approximation could be made. In contrast the *combined* (thermal and velocity) *entry length problem* corresponds to the case for which the temperature and velocity profiles develop simultaneously.

Solutions have been obtained for both entry length conditions [2], and selected results are presented in Figure 8.8. The Nusselt numbers are, in principle, infinite at $x = 0$ and decay to their asymptotic (fully developed) values with increasing x. Note that the results for the combined entry length problem depend on the Prandtl number and have been presented only for $Pr = 0.7$. However, these results are representative of conditions for most gases. Note also that fully developed conditions are reached for $[(x/D)/Re_D Pr] \approx 0.05$.

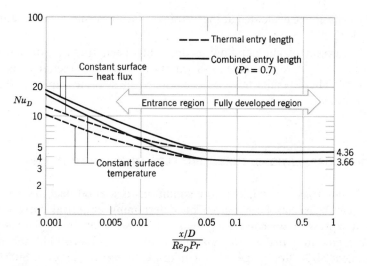

Figure 8.8 Local Nusselt number obtained from entry length solutions for laminar flow in a circular tube [2]. Adapted with permission.

For the *constant surface temperature* condition, it is desirable to know the *average* convection coefficient for use with Equation 8.44. Kays [6] presents a correlation attributed to Hausen [7], which is of the form

$$\overline{Nu}_D = 3.66 + \frac{0.0668(D/L)Re_D Pr}{1 + 0.04[(D/L)Re_D Pr]^{2/3}} \tag{8.54}$$

where $\overline{Nu}_D \equiv \bar{h}D/k$. However, because this result presumes a thermal entry length, it is not generally applicable. A preferred correlation for the combined entry length is the Sieder-Tate equation [8], which is of the form

$$\overline{Nu}_D = 1.86 \left(\frac{Re_D Pr}{L/D}\right)^{1/3} \left(\frac{\mu}{\mu_s}\right)^{0.14} \tag{8.55}$$

NON-FULLY DEVELOPED

$$\begin{bmatrix} T_s = \text{const.} \\ 0.48 < Pr < 16{,}700 \\ 0.0044 < (\mu/\mu_s) < 9.75 \end{bmatrix}$$

The correlation has been recommended by Whitaker [9] for values of $[(Re_D Pr/L/D)^{1/3} (\mu/\mu_s)^{0.14}] \gtrsim 2$. Below this limit fully developed conditions encompass much of the tube, and Equation 8.53 may be used to a good approximation. Note that all properties appearing in Equations 8.54 and 8.55, except μ_s, should be evaluated at the average value of the mean temperature, $\bar{T}_m$ $\equiv (T_{m,i} + T_{m,o})/2$.

The subject of laminar flow in ducts has been studied extensively, and numerous results are available for a variety of duct cross sections and surface conditions. These results have been compiled in a monograph by Shah and London [10].

8.5 CONVECTION CORRELATIONS: TURBULENT FLOW IN CIRCULAR TUBES

Since the analysis of turbulent flow conditions is a good deal more involved, greater emphasis has been placed on determining empirical correlations. A classical expression for computing the *local* Nusselt number for *fully developed turbulent flow* in a *smooth circular tube* is due to Colburn [11] and may be determined by applying the Chilton-Colburn analogy. Substituting Equation 6.99 into Equation 8.18, the analogy may be expressed in the following form for internal flow.

$$\frac{C_f}{2} = \frac{f}{8} = St \, Pr^{2/3} = \frac{Nu_D}{Re_D Pr} \, Pr^{2/3} \tag{8.56}$$

Hence substituting for the friction factor from Equation 8.21, we obtain the *Colburn equation*

$$Nu_D = 0.023 Re_D^{4/5} \, Pr^{1/3} \tag{8.57}$$

The *Dittus-Boelter equation* [12] is a slightly different and preferred version of the above result and is of the form

$$Nu_D = 0.023 Re_D^{4/5} \, Pr^n \tag{8.58}$$

where $n = 0.4$ for heating $(T_s > T_m)$ and 0.3 for cooling $(T_s < T_m)$. The above equations have been confirmed experimentally for the range of conditions

$$\begin{bmatrix} 0.7 \le Pr \le 160 \\ Re_D \gtrsim 10{,}000 \\ L/D \gtrsim 60 \end{bmatrix}$$

However, the equations should only be used for small to moderate temperature differences, $T_s - T_m$, with all of the properties evaluated at T_m. For flows characterized by large property variations it is recommended that the following equation due to Sieder and Tate [8] be used.

$$Nu_D = 0.027 Re_D^{4/5} Pr^{1/3} \left(\frac{\mu}{\mu_s}\right)^{0.14} \tag{8.59}$$

$$\begin{bmatrix} 0.7 \leq Pr \leq 16{,}700 \\ Re_D \gtrsim 10{,}000 \\ L/D \gtrsim 60 \end{bmatrix}$$

where all properties except μ_s are evaluated at T_m. Note that to a good approximation, the foregoing correlations may be applied for both the constant surface temperature and heat flux conditions.

Since the average Nusselt number, $\overline{Nu}_D$, for the entire tube depends on the nature of the inlet conditions, as well as on the Reynolds and Prandtl numbers, generally applicable correlations are not available. The manner in which the Nusselt number varies in the entry region is described by Notter and Sleicher [13] and Kays and Perkins [14], but as a first approximation the foregoing equations may be used to obtain a reasonable estimate of $\overline{Nu}_D$ if $(L/D) \gtrsim 60$. Note that, when determining $\overline{Nu}_D$, all fluid properties should be evaluated at the arithmetic average of the mean temperature, $\overline{T}_m \equiv (T_{m,i} + T_{m,o})/2$.

It should also be noted that the foregoing correlations apply for smooth surface conditions. The heat transfer rate is larger for rough surfaces, and the Nusselt number may be estimated by applying the Chilton-Colburn analogy, Equation 8.56, to the friction factor data of Figure 8.3. However, more accurate correlations are available in the literature [14].

Finally, note that the foregoing correlations do not apply to liquid metals (3 $\times 10^{-3} \lesssim Pr \lesssim 5 \times 10^{-2}$). For fully developed turbulent flow $(L/D \gtrsim 60)$ in smooth circular tubes with constant surface heat flux, Skupinski et al. [15] recommend a correlation of the form

$$Nu_D = 4.82 + 0.0185 Pe_D^{0.827} \qquad (q_s'' = \text{const.}) \tag{8.60}$$

$$\begin{bmatrix} 3.6 \times 10^3 < Re_D < 9.05 \times 10^5 \\ 10^2 < Pe_D < 10^4 \end{bmatrix}$$

Similarly, for constant surface temperature Seban and Shimazaki [16] recommend the following correlation for $Pe_D > 100$

$$Nu_D = 5.0 + 0.025 Pe_D^{0.8} \qquad (T_s = \text{const.}) \tag{8.61}$$

Extensive data and additional correlations are available in the literature [17].

EXAMPLE 8.6

Hot air flows with a mass rate of $\dot{m} = 0.050$ kg/s through an uninsulated sheet-metal duct of diameter $D = 0.15$ m, which passes through the crawl-space of a house. The hot air enters the duct at a temperature of 103°C and, after a distance of $L = 5$ m, cools to a temperature of 77°C. The heat transfer coefficient between the duct outer surface and the cold ambient air at $T_\infty = 0$°C is assumed to have a constant value of $h_o = 6$ W/m$^2\cdot$K.

1. Calculate the heat loss (W) from the duct over the length L.
2. Determine the heat flux and the duct surface temperature at $x = L$.

SOLUTION

KNOWN:

Hot air flowing in a duct passing through crawlspace of house.

FIND:

1. Heat loss from the duct over the length L, $q(W)$.
2. Heat flux and surface temperature at $x = L$.

SCHEMATIC:

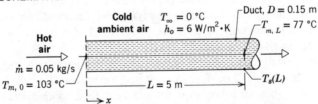

ASSUMPTIONS:

1. Steady-state conditions.
2. Constant properties.
3. Perfect gas behavior.
4. Negligible kinetic and potential energy changes.
5. Negligible duct wall thermal resistance.
6. Uniform convection coefficient at outer surface of duct.

PROPERTIES:

Table A.4, air ($\overline{T}_m = 363$ K): $c_p = 1010$ J/kg$\cdot$K.

Table A.4, air $(T_{m,L} = 350 \text{ K})$: $k = 0.030 \text{ W/m·K}$, $\mu = 208 \times 10^{-7} \text{ N·s/m}^2$, $Pr = 0.70$.

ANALYSIS:

1. From the energy balance for the entire tube, Equation 8.37,

$$q = \dot{m}c_p(T_{m,L} - T_{m,0})$$

$$q = 0.05 \text{ kg/s} \times 1010 \text{ J/kg·K} (77 - 103)°C$$

$$q = -1313 \text{ W} \qquad \triangleleft$$

2. An expression for the heat flux at $x = L$ may be inferred from the resistance network

where $h_x(L)$ is the inside convection heat transfer coefficient at $x = L$. Hence

$$q''_s(L) = \frac{T_{m,L} - T_\infty}{[1/h_x(L)] + (1/h_o)}$$

The inside convection coefficient may be obtained from knowledge of the Reynolds number. From Equation 8.6,

$$Re_D = \frac{4\dot{m}}{\pi D \mu} = \frac{4 \times 0.05 \text{ kg/s}}{\pi \times 0.15 \text{ m} \times 208 \times 10^{-7} \text{ N·s/m}^2} = 20{,}404$$

Hence the flow is turbulent. Moreover, with $(L/D) = (5/0.15) = 33.3$, it is reasonable to assume fully developed conditions at $x = L$. Hence from Equation 8.58, with $n = 0.3$,

$$Nu_D = \frac{h_x(L)D}{k} = 0.023Re_D^{4/5}Pr^{0.3}$$

$$Nu_D = 0.023(20{,}404)^{4/5}(0.70)^{0.3} = 57.9$$

$$h_x(L) = Nu_D\frac{k}{D} = 57.9\frac{0.030 \text{ W/m·K}}{0.15 \text{ m}} = 11.6 \text{ W/m}^2\text{·K}$$

Hence

$$q''_s(L) = \frac{(77 - 0)°C}{[(1/11.6) + (1/6.0)] \text{ m}^2\text{·K/W}} = 304.5 \text{ W/m}^2$$

Referring back to the network, it also follows that

$$q_s''(L) = \frac{T_{m,L} - T_{s,L}}{1/h_x(L)}$$

in which case

$$T_{s,L} = T_{m,L} - \frac{q_s''(L)}{h_x(L)}$$

$$T_{s,L} = 77°C - \frac{304.5 \text{ W/m}^2}{11.6 \text{ W/m}^2 \cdot \text{K}}$$

$$T_{s,L} = 50.7°C \qquad\qquad\qquad\qquad \triangleleft$$

COMMENTS:

1. Note that in using the energy balance of part (1) for the entire tube, properties (in this case, only c_p) are evaluated at $\bar{T}_m = (T_{m,0} + T_{m,L})/2$. However, in using the correlation, Equation 8.58, properties are evaluated at the local mean temperature, $T_{m,L} = 77°C$; this follows because we are using the correlation for a local heat transfer coefficient.

2. Note that this problem is characterized neither by constant surface temperature nor constant surface heat flux. It would therefore be erroneous to presume that the total heat loss from the tube is given by $q_s''(L)\pi DL = 717$ W. This result is substantially less than the actual heat loss of 1313 W because $q_s''(x)$ decreases with increasing x. This decrease in $q_s''(x)$ is due to reductions in both $h_x(x)$ and $[T_m(x) - T_\infty]$ with increasing x.

8.6 CONVECTION CORRELATIONS: NONCIRCULAR TUBES

Although we have thus far restricted our consideration to internal flows of circular cross section, many engineering applications involve convection transport in *noncircular tubes*. At least to a first approximation, however, many of the circular tube results may be applied by using *an effective diameter* as the characteristic length. It is termed the *hydraulic diameter* and is defined as

$$D_h \equiv \frac{4A_c}{P} \qquad\qquad\qquad\qquad (8.62)$$

where A_c and P are the *flow* cross-sectional area and the *wetted perimeter*, respectively [18]. It is this diameter that should be used in calculating parameters such as Re_D and Nu_D.

For turbulent flow, which still occurs if $Re_D \gtrsim 2300$, it is reasonable to use the correlations of Section 8.5 for $Pr \gtrsim 0.7$. However, we should note that in a noncircular tube the convection coefficients will vary around the periphery, approaching zero in the corners. Hence in using a circular tube correlation, the coefficient is presumed to be an average over the perimeter.

For laminar flow, the use of circular tube correlations is less accurate, particularly with cross sections characterized by sharp corners. For such cases the Nusselt number corresponding to fully developed conditions may be obtained from Table 8.1.

Although the foregoing procedures are generally satisfactory, exceptions do

Table 8.1 Nusselt numbers for fully developed laminar flow in tubes of different cross section

CROSS SECTION	$\dfrac{b}{a}$	$Nu_D \equiv \dfrac{hD_h}{k}$ (Constant q_s'')	(Constant T_s)
circle	—	4.36	3.66
square	1.0	3.63	2.98
rectangle	1.4	3.78	—
rectangle	2.0	4.11	3.39
rectangle	3.0	4.77	—
rectangle	4.0	5.35	4.44
rectangle	8.0	6.60	5.95
rectangle	∞	8.23	7.54
triangle	—	3.00	2.35

exist. Detailed discussions of heat transfer in noncircular tubes are provided by
Kays and Perkins [14] and Shah and London [10, 19].

8.7   THE CONCENTRIC TUBE ANNULUS

Many internal flow problems involve heat transfer in a *concentric tube annulus*
(Figure 8.9). Fluid passes through the space (annulus) formed by the concentric
tubes, and convection heat transfer may occur to or from both the inner and
outer tube surfaces. It is possible to independently specify the heat flux or
temperature, that is, the thermal condition, at each of these surfaces. In any case
the heat flux at each surface may be computed from expressions of the form

$$q_i'' = h_i(T_{s,i} - T_m) \tag{8.63}$$

$$q_o'' = h_o(T_{s,o} - T_m) \tag{8.64}$$

Note that separate convection coefficients are associated with the inner and
outer surfaces. The corresponding Nusselt numbers are of the form

$$Nu_i \equiv \frac{h_i D_h}{k} \tag{8.65}$$

$$Nu_o \equiv \frac{h_o D_h}{k} \tag{8.66}$$

where, from Equation 8.62, the hydraulic diameter, D_h, is

$$D_h = \frac{4(\pi/4)(D_o^2 - D_i^2)}{\pi D_o + \pi D_i} = D_o - D_i \tag{8.67}$$

For the case of fully developed laminar flow with one surface insulated and
the other surface at a constant temperature, Nu_i or Nu_o may be obtained from

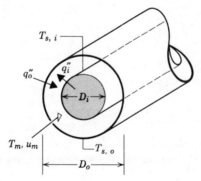

Figure 8.9   The concentric tube annulus.

Table 8.2 Nusselt number for fully developed laminar flow in a circular tube annulus with one surface insulated and the other at constant temperature

D_i/D_o	Nu_i	Nu_o
0	—	3.66
0.05	17.46	4.06
0.10	11.56	4.11
0.25	7.37	4.23
0.50	5.74	4.43
1.00	4.86	4.86

Table 8.2. Note that in such cases we would only be interested in the convection coefficient associated with the isothermal (nonadiabatic) surface.

If uniform heat flux conditions exist at both surfaces, the Nusselt numbers may be computed from expressions of the form

$$Nu_i = \frac{Nu_{ii}}{1 - (q_o''/q_i'')\theta_i^*} \tag{8.68}$$

$$Nu_o = \frac{Nu_{oo}}{1 - (q_i''/q_o'')\theta_o^*} \tag{8.69}$$

The influence coefficients (Nu_{ii}, Nu_{oo}, θ_i^* and θ_o^*) appearing in these equations may be obtained from Table 8.3. Note that q_i'' and q_o'' may be positive or negative, depending on whether heat transfer is to or from the fluid, respectively.

Table 8.3 Influence coefficients for fully developed laminar flow in a circular tube annulus with uniform heat flux maintained at both surfaces

D_i/D_o	Nu_{ii}	Nu_{oo}	θ_i^*	θ_o^*
0	—	4.364	∞	0
0.05	17.81	4.792	2.18	0.0294
0.10	11.91	4.834	1.383	0.0562
0.20	8.499	4.833	0.905	0.1041
0.40	6.583	4.979	0.603	0.1823
0.60	5.912	5.099	0.473	0.2455
0.80	5.58	5.24	0.401	0.299
1.00	5.385	5.385	0.346	0.346

Moreover, situations may arise for which the values of h_i and h_o are negative. Such results, when used with the sign convention implicit in Equations 8.63 and 8.64, reveal the relative magnitudes of T_s and T_m.

For fully developed turbulent flow the influence coefficients are a function of the Reynolds and Prandtl numbers [14]. However, to a first approximation the inner and outer convection coefficients may be assumed to be equal, and they may be evaluated by using the hydraulic diameter, Equation 8.67, with the Dittus-Boelter equation, Equation 8.58.

8.8 CONVECTION MASS TRANSFER

Mass transfer by convection may also occur for internal flows. For example, a gas may flow through a tube whose surface has been wetted or is sublimable. Evaporation or sublimation will then occur, and a concentration boundary layer will develop. Just as the mean temperature is the appropriate reference temperature for heat transfer considerations, the mean species concentration, $\rho_{A,m}$, plays an analogous role for mass transfer. By analogy to Equation 8.27 it follows that, for incompressible flow in a circular tube,

$$\rho_{A,m} = \frac{2}{u_m r_o^2} \int_0^{r_o} u\,\rho_A\,dr \qquad (8.70)$$

The concentration boundary layer development is characterized by entrance and fully developed regions, and Equation 8.23 may be used (with Pr replaced by Sc) to determine the *concentration entry length*, $x_{\mathrm{fd},c}$, for laminar flow. Equation 8.4 may again be used as a first approximation for turbulent flow. Moreover, by analogy to Equation 8.29, for both laminar and turbulent flows, fully developed conditions exist when

$$\frac{\partial}{\partial x}\left[\frac{\rho_{A,s}(x) - \rho_A(r,x)}{\rho_{A,s}(x) - \rho_{A,m}(x)}\right]_{\mathrm{fd},c} = 0 \qquad (8.71)$$

The mass flux of species A may be computed from an expression of the form

$$n''_{A,s} = h_m(\rho_{A,s} - \rho_{A,m}) \qquad (8.72)$$

where the convection mass transfer coefficient, h_m, may be obtained from appropriate correlations involving the Sherwood number, Sh_D, defined as

$$Sh_D \equiv \frac{h_m D}{D_{AB}} \qquad (8.73)$$

Invoking the heat and mass transfer analogy, the specific form of the correlation may be inferred from the foregoing heat transfer results simply by replacing Nu_D with Sh_D and Pr with Sc.

EXAMPLE 8.7

A thin liquid film of ammonia, which has formed on the inner surface of a tube of diameter $D = 10$ mm and length $L = 1$ m, is removed by passing dry air through the tube at a flow rate of 3×10^{-4} kg/s. The tube and the air are at 25°C. What is the average mass transfer convection coefficient for the tube?

SOLUTION

KNOWN:

Liquid ammonia on the inner surface of a tube is removed by evaporation into an airstream.

FIND:

Average mass transfer convection coefficient for the tube.

SCHEMATIC:

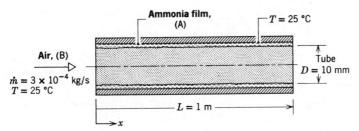

ASSUMPTIONS:

1. Thin ammonia film with smooth surface.
2. Evaporative cooling effects are negligible.

PROPERTIES:

Table A.4, air (25°C): $v = 15.7 \times 10^{-6}$ m²/s, $\mu = 183.6 \times 10^{-7}$ N·s/m².
Table A.8, ammonia–air (25°C): $D_{AB} = 0.28 \times 10^{-4}$ m²/s, $Sc = (v/D_{AB}) = 0.56$.

ANALYSIS:

From Equation 8.6

$$Re_D = \frac{4 \times 3 \times 10^{-4} \text{ kg/s}}{\pi \times 0.01 \text{ m} \times 183.6 \times 10^{-7} \text{ N·s/m}^2} = 2080$$

in which case the flow is laminar. Hence, since a constant ammonia vapor concentration is maintained at the surface of the film, which is analogous to a constant surface temperature condition, and since

$$\left(\frac{Re_D Sc}{L/D}\right)^{1/3} = \left[\frac{(2080)(0.56)}{1/0.01}\right]^{1/3} = 2.27 > 2$$

the mass transfer analog to Equation 8.55 may be used to determine the average mass transfer convection coefficient. It follows that

$$\overline{Sh}_D = 1.86 \left(\frac{Re_D Sc}{L/D}\right)^{1/3} = 1.86 \times 2.27 = 4.22$$

Hence

$$\bar{h}_m = \overline{Sh}_D \frac{D_{AB}}{D} = \frac{4.22 \times 0.28 \times 10^{-4} \text{ m}^2/\text{s}}{0.01 \text{ m}}$$

or

$$\bar{h}_m = 0.012 \text{ m/s} \qquad \triangleleft$$

COMMENTS:

From Equation 8.23 $x_{fd,c} \approx (0.05 Re_D Sc)D = 0.58$ m, and fully developed conditions exist over approximately 40 percent of the tube length. An assumption of fully developed conditions over the entire tube would provide a value of $\overline{Sh}_D = 3.66$, which is 13 percent less than the above result.

8.9 SUMMARY

Internal flow is encountered in numerous applications, and it is important to appreciate its unique features. What is the nature of fully developed flow, and how does it differ from flow in the entry region? How does the Prandtl number influence boundary layer development in the entry region? How do thermal conditions in the fluid depend on the nature of the surface condition? For example, how do the surface and mean temperatures vary with x for the case of uniform surface heat flux? Or, how do the mean temperature and the surface heat flux vary for the case of uniform surface temperature?

You must be able to perform engineering calculations that involve an energy balance and appropriate convection correlations. The methodology is one which involves first determining whether the flow is laminar or turbulent

Table 8.4 Summary of convection correlations for internal flow.[a, b, c]

CORRELATION		CONDITIONS
$f = 64/Re_D$	(8.19)	Laminar, fully developed
$Nu_D = 4.36$	(8.51)	Laminar, fully developed, constant q_s'', $Pr \gtrsim 0.6$
$Nu_D = 3.66$	(8.53)	Laminar, fully developed, constant T_s, $Pr \gtrsim 0.6$
$\overline{Nu}_D = 3.66 + \dfrac{0.0668(D/L)Re_D Pr}{1 + 0.04[(D/L)Re_D Pr]^{2/3}}$ or,	(8.54)	Laminar, thermal entry length ($Pr \gg 1$ or an unheated starting length), constant T_s
$\overline{Nu}_D = 1.86 \left(\dfrac{Re_D Pr}{L/D}\right)^{1/3} (\mu/\mu_s)^{0.14}$	(8.55)	Laminar, $[(Re_D Pr/L/D)^{1/3}(\mu/\mu_s)^{0.14}] \gtrsim 2$, constant T_s, $0.48 < Pr < 16{,}700$, $0.0044 < (\mu/\mu_s) < 9.75$
$f = 0.0316Re_D^{-1/4}$	(8.20)[d]	Turbulent, fully developed, $Re_D \lesssim 2 \times 10^4$
$f = 0.184Re_D^{-1/5}$	(8.21)[d]	Turbulent, fully developed, $Re_D \gtrsim 2 \times 10^4$
$Nu_D = 0.023Re_D^{4/5}Pr^{1/3}$ or,	(8.57)[e]	Turbulent, fully developed, $0.6 \le Pr \le 160$, $Re_D \gtrsim 10{,}000$, $L/D \gtrsim 60$
$Nu_D = 0.023Re_D^{4/5}Pr^n$ or,	(8.58)[e]	Turbulent, fully developed, $0.6 \le Pr \le 160$, $Re_D \ge 10{,}000$, $L/D \gtrsim 60$, $n = 0.4$ for $T_s > T_m$ and $n = 0.3$ for $T_s < T_m$
$Nu_D = 0.027Re_D^{4/5}Pr^{1/3}(\mu/\mu_s)^{0.14}$	(8.59)[e]	Turbulent, fully developed, $0.7 \le Pr \le 16{,}700$, $Re_D \gtrsim 10{,}000$, $L/D \gtrsim 60$
$Nu_D = 4.82 + 0.0185(Re_D Pr)^{0.827}$	(8.60)	Liquid metals, turbulent, fully developed, constant q_s'', $3.6 \times 10^3 < Re_D < 9.05 \times 10^5$, $10^2 < Pe_D < 10^4$
$Nu_D = 5.0 + 0.025(Re_D Pr)^{0.8}$	(8.61)	Liquid metals, turbulent, fully developed, constant T_s, $Pe_D > 100$

[a]The mass transfer correlations may be obtained by replacing Nu_D and Pr by Sh_D and Sc, respectively.

[b]Properties in Equations 8.51, 8.53, 8.57, 8.58, 8.59, 8.60, and 8.61 are based on T_m; properties in Equations 8.19, 8.20, and 8.21 are based on $T_f \equiv (T_s + T_m)/2$; properties in Equations 8.54 and 8.55 are based on $\overline{T}_m \equiv (T_{m,i} + T_{m,o})/2$.

[c]$Re_D \equiv D_h u_m/\nu$; $D_h \equiv 4A_c/P$; $u_m \equiv \dot{m}/\rho A_c$.

[d]Equations 8.20 and 8.21 pertain to smooth tubes. For rough tubes the Chilton-Colburn analogy, Equation 8.56, should be used with the results of Figure 8.3.

[e]As a first approximation, Equations 8.57, 8.58, or 8.59 may be used to evaluate the average Nusselt number $\overline{Nu}_D$ over the entire tube length, if $(L/D) \gtrsim 60$. The properties should then be evaluated at the average of the mean temperature, $\overline{T}_m \equiv (T_{m,i} + T_{m,o})/2$.

and then establishing the length of the entry region. After deciding whether you are interested in local conditions (at a particular axial location) or in average conditions (for the entire tube), the convection correlation may be selected and used with the appropriate form of the energy balance to solve the problem. A summary of the correlations is provided in Table 8.4.

You should note several features that complicate internal flows and that have not been considered in this chapter. For example, a situation may exist for which there is a prescribed axial variation in T_s or q_s'', rather than uniform surface conditions. Among other things, such a variation would preclude the existence of a fully developed region. There may also exist surface roughness effects, circumferential heat flux or temperature variations, widely varying fluid properties, or transition flow conditions. A complete discussion of these effects is provided by Kays and Perkins [14].

REFERENCES

1. Langhaar, H. L., *J. Appl. Mech.*, *64*, A-55, 1942.
2. Kays, W. M., *Convective Heat and Mass Transfer*, McGraw-Hill, New York, 1966.
3. Wark, K. *Thermodynamics*, 2nd Ed., McGraw-Hill, New York, 1971.
4. Bird, R. B., W. E. Stewart, and E. N. Lightfoot, *Transport Phenomena*, Wiley, New York, 1966.
5. Moody, L. F., *Trans. ASME*, *66*, 671, 1944.
6. Kays, W. M., *Trans. ASME*, *77*, 1265, 1955.
7. Hausen, H., *Z. VDI. Beih. Verfahrenstech*, *4*, 91, 1943.
8. Sieder, E. N. and G. E. Tate, *Ind. Eng. Chem.*, *28*, 1429, 1936.
9. Whitaker, S., *AIChE J.*, *18*, 361, 1972.
10. Shah, R. K. and A. L. London, *Laminar Flow Forced Convection in Ducts*, Academic Press, New York, 1978.
11. Colburn, A. P., *Trans. AIChE*, *29*, 174, 1933.
12. Dittus, F. W. and L. M. K. Boelter, Univ. Calif., Berkeley, Publ. Eng., Vol. 2, 1930, p. 443.
13. Notter, R. H. and C. A. Sleicher, *Chem. Eng. Sci.*, *27*, 2073, 1972.
14. Kays, W. M. and H. C. Perkins, In *Handbook of Heat Transfer*, Chapter 7, W. M. Rohsenow and J. P. Hartnett, Eds., McGraw-Hill, New York, 1972.
15. Skupinski, E. S., J. Tortel, and L. Vautrey, *Int. J. Heat Mass Transfer*, *8*, 937, 1965.
16. Seban, R. A. and T. T. Shimazaki, *Trans. ASME*, *73*, 803, 1951.
17. Stein, R. In J. P. Hartnett and T. F. Irvine, Eds., *Advances in Heat Transfer*, Vol. 3, Academic Press, New York, 1966.
18. Fox, R. W. and A. T. McDonald, *Introduction to Fluid Mechanics*, 2nd Ed., Wiley, New York, 1978.
19. Shah, R. K. and A. L. London, *J. Heat Trans.*, *96*, 159, 1974.

PROBLEMS

8.1 Fully developed conditions are known to exist for water flowing through a 25-mm-diameter tube at a rate of 0.01 kg/s and a temperature of 27°C. What is the maximum velocity of the water in the tube? What is the pressure gradient associated with the flow?

8.2 What is the pressure drop associated with water at 27°C flowing with a mean velocity of 0.2 m/s through a 600-m-long cast iron pipe of 0.15 m inside diameter?

8.3 Water at 27°C flows with a mean velocity of 1 m/s through a 1-km-long cast iron pipe of 0.25 m inside diameter.

a) Determine the pressure drop over the pipe length and the corresponding pump power requirement, if the pipe surface is clean.

b) If the pipe surface roughness is increased by 25 percent due to contamination, what is the new pressure drop and pump power requirement?

8.4 Consider a 25-mm-diameter circular tube through which liquid mercury, water, or engine oil at 27°C may flow at a rate of 0.03 kg/s. Determine the velocity, the hydrodynamic entry length, and the thermal entry length for each of the fluids.

8.5 Compare the thermal and velocity entry lengths for oil, water, and mercury flowing through a 25-mm-diameter tube with a mean velocity and temperature of $u_m = 5$ mm/s and $T_m = 27$°C, respectively.

8.6 Water at a flow rate of 2 kg/s enters a long section of pipe with a temperature of 25°C and a pressure of 100 bars. The pipe wall is heated such that 10^5 W are transferred to the water as it flows through the pipe.

a) If the water leaves the pipe with a pressure of 2 bars, what is its outlet temperature?

b) What value of the outlet temperature would be obtained if Equation 8.37 were used for the calculation?

8.7 A thin plastic rod of density ρ and diameter D, which is extruded from a die at T_0, moves with a velocity V to a rolling station where the rod temperature is T_L. Cool air at T_∞ is blown across the rod, and the local heat transfer coefficient h_x is constant over the length of the rod. It can be assumed that radial temperature gradients and axial conduction in the rod are negligible and that steady state conditions exist.

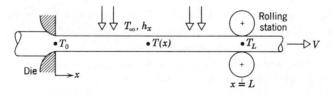

a) Use the conservation of energy requirement on a properly defined control volume to find a differential equation that governs the variation of the rod temperature with x.

b) Solve the foregoing equation to determine the temperature distribution along the rod.

c) Write an expression for the total heat loss by the rod in terms of the heat transfer coefficient h_x and other appropriate variables.

8.8 A process for thermal curing of a thin plastic film of thickness, l, employs a radiation source that provides a uniform heat flux, q_o'', on the film traveling at a lineal speed V. As shown in the sketch, the film is at a temperature T_o as it enters under the radiant heater section, and it is desired to have the film reach the curing temperature, T_c, before it exits the section. Energy exchange by convection occurs between the film and the ambient air at T_∞, but radiative exchange between the film and its surroundings may be neglected.

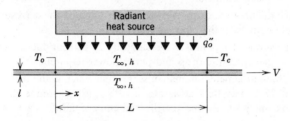

a) Beginning with a properly defined control volume or control mass, perform an energy balance to obtain a differential equation for the temperature distribution in the film. List any assumptions that you make.

b) Indicate how you would solve this equation. Be specific but do not attempt to solve it.

c) In order to *estimate* the length required to reach T_c, neglect internal conduction and surface convection effects to obtain a simplified model. Calculate the length of the heating section, L, using these parameters: $c_p = 1$ kJ/kg·K, $l = 0.10$ mm, $q_o'' = 2500$ W/m^2, $\rho = 1000$ kg/m^3, $V = 0.5$ m/s, $T_o = 30°$C, and $T_c = 175°$C.

8.9 Water enters a tube at 27°C with a flow rate of 450 kg/h. The heat transfer from the tube wall to the fluid is given as

$$q'(W/m) = ax$$

where the coefficient a is equal to 20 W/m^2 and $x(m)$ is the axial distance from the tube entrance.

a) Beginning with a properly defined differential control volume in the tube, derive an expression for the temperature distribution, $T(x)$, of the water.

b) What is the outlet temperature of the water for a heated section 30 m long?

8.10 Consider the flow in a circular tube as shown in the following sketch. Within the test section length (between 1 and 2) a constant heat flux, q_s'', is maintained.

a) For the two cases identified, show a qualitative sketch of the surface temperature, $T_s(x)$, and the fluid mean temperature, $T_m(x)$, as a function of distance along the test section, x.
Case A: flow is hydrodynamically and thermally fully developed.
Case B: flow is not developed.

b) Assuming that the surface flux, q_s'', and the inlet mean temperature, $T_{m,1}$, are identical for both cases, will the exit mean temperature, $T_{m,2}$, for case A be greater, equal to, or less than $T_{m,2}$ for case B? Briefly explain why.

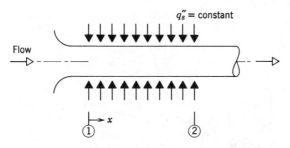

$q_s'' = \text{constant}$

8.11 Consider a cylindrical nuclear fuel rod of length L and diameter D that is encased within a concentric tube. Pressurized water flows through the annular region between the rod and the tube at a rate $\dot{m}$, and the outer surface of the tube is well insulated. Heat generation occurs within the fuel rod, and the volumetric generation rate is known to vary sinusoidally with distance along the rod. That is, $\dot{q}(x) = \dot{q}_o \sin(\pi x/L)$, where $\dot{q}_o$ (W/m³) is a constant. A uniform convection coefficient h may be assumed to exist between the surface of the rod and the water.

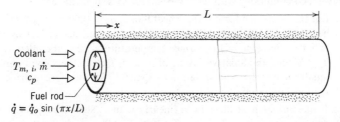

Coolant
$T_{m,i}, \dot{m}$
c_p
Fuel rod
$\dot{q} = \dot{q}_o \sin(\pi x/L)$

a) Obtain expressions for the local heat flux, $q''(x)$, and the total heat transfer, q, from the fuel rod to the water.

b) Obtain an expression for the variation of the mean temperature, $T_m(x)$, of the water with distance x along the tube.

c) Obtain an expression for the variation of the rod surface temperature, $T_s(x)$, with distance x along the tube. Develop an expression for the x location at which this temperature is maximized.

8.12 A flat plate solar collector is used to heat atmospheric air flowing through a rectangular channel, as shown in the following diagram. The bottom surface of the channel is well insulated, while the top surface is subjected to a uniform heat flux q_o'', which is due to the net effect of solar radiation absorption and heat exchange between the absorber and cover plates.

a) Beginning with an appropriate differential control volume, obtain an equation that could be used to determine the mean air temperature, $T_m(x)$, as a function of distance along the channel. Solve this equation to obtain an expression for the mean temperature of the air leaving the collector.

b) With air inlet conditions of $\dot{m} = 0.1$ kg/s and $T_{m,i} = 40°C$, what is the air outlet temperature if $L = 3$ m, $w = 1$ m, and $q_o'' = 700$ W/m²? The specific heat of air is $c_p = 1008$ J/kg·K.

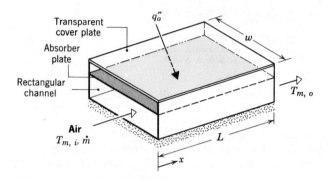

8.13 Atmospheric air enters the heated section of a circular tube at a flowrate of 0.005 kg/s and a temperature of 20°C. The tube is of diameter $D = 50$ mm, and fully developed conditions with $h = 25$ W/m²·K exist over the entire length of $L = 3$ m.

a) For the case of uniform surface heat flux at $q_s'' = 1000$ W/m², determine the total heat transfer rate, q, and the mean temperature of the air leaving the tube, $T_{m,o}$. What is the value of the surface temperature at the tube inlet, $T_{s,i}$, and outlet, $T_{s,o}$? Sketch the axial variation of T_s and T_m. On the same figure, also sketch (qualitatively) the axial variation of T_s and T_m for the more realistic case in which the local convection coefficient varies with x.

b) If the surface heat flux varies linearly with x, such that q_s'' (W/m²) $= 500x(m)$, what are the values of q, $T_{m,o}$, $T_{s,i}$, and $T_{s,o}$? Sketch the axial variation of T_s and T_m. On the same figure, also sketch (qualitatively) the axial variation of T_s and T_m for the more realistic case in which the local convection coefficient varies with x.

8.14 Slug flow is an idealized tube flow condition for which the velocity is assumed to be uniform over the entire tube cross section. For the case of laminar slug flow with a uniform surface heat flux, determine the form of the fully developed temperature distribution, $T(r)$, and the Nusselt number, Nu_D.

8.15 Engine oil at a rate of 0.02 kg/s flows through a 3-mm-diameter tube of length 30 m. The oil has an inlet temperature of 60°C while the tube wall temperature is maintained at 100°C by steam condensing on its outer surface.

a) Estimate the average heat transfer coefficient for internal flow of the oil.

b) Determine the outlet temperature of the oil.

8.16 Oil is heated by flowing through a circular tube of diameter $D = 50$ mm and length $L = 25$ m and whose outer surface is maintained at 150°C. If the flow rate and inlet temperature of the oil are 0.5 kg/s and 20°C, respectively, what is the oil outlet temperature? What is the total heat transfer rate for the tube?

8.17 Engine oil flows through a 25 mm diameter, 10 m long tube at a rate of 0.5 kg/s. The oil enters the tube at a temperature of 25°C, while the tube surface temperature is maintained at 100°C.

a) Determine the total heat transfer to the oil and the oil outlet temperature.

b) Repeat part (a), but subject to the assumption of fully developed conditions throughout the tube.

8.18 Engine oil flows through a 25-mm-diameter tube at a rate of 0.5 kg/s. The oil enters the tube at a temperature of 25°C, while the tube surface temperature is maintained at 100°C. Determine the oil outlet temperature for a 5-m-long tube and for a 100-m-long tube. For each case compare the log mean temperature difference to the arithmetic mean temperature difference.

8.19 A 25-mm-diameter circular tube, whose outer surface is maintained at 100°C, is used to heat water flowing at 1 kg/s from 30°C to 70°C. How long must the tube be?

8.20 Water flowing at 2 kg/s through a 40-mm-diameter tube is to be heated from 25°C to 75°C by maintaining the tube surface temperature at 100°C. What is the required tube length?

8.21 Water flows at 2 kg/s through a 40-mm-diameter tube 4 m long. If the water enters the tube at 25°C and the surface temperature is maintained at 90°C, what is the temperature of the water as it leaves the tube? What is the rate of heat transfer to the water?

8.22 A thick-walled, stainless steel (AISI 316) pipe of inside and outside diameters $D_i = 20$ mm and $D_o = 40$ mm, respectively, is heated electrically to provide a uniform heat generation rate of $\dot{q} = 10^6$ W/m^3. The outer surface of the pipe is insulated, while water flows through the pipe at a rate of $\dot{m} = 0.1$ kg/s.

a) If the water inlet temperature is $T_{m,i} = 20°C$ and the desired outlet temperature is $T_{m,o} = 40°C$, what is the required pipe length?

b) What is the location of the maximum pipe temperature and what is the value of this temperature?

8.23 Atmospheric air enters a 10-m long, 150-mm diameter uninsulated heating duct at 60°C and 0.04 kg/s. If the duct surface temperature is approximately constant at $T_s = 15°C$, what is the outlet air temperature and what is the duct heat loss?

8.24 Liquid mercury at 0.5 kg/s is to be heated from 300 to 400 K by passing it through a 50-mm-diameter tube whose surface is maintained at 450 K. Calculate the required tube length by using an appropriate liquid metal convection heat transfer correlation. Compare your result with that which would have been obtained by using a correlation appropriate for $Pr \gtrsim 0.7$.

8.25 The surface of a 50-mm-diameter, thin-walled tube is maintained at 100°C. In one case air is in cross flow over the tube with a temperature and velocity of 25°C and 30 m/s, respectively. In another case air is in fully developed flow through the tube with a temperature of 25°C and a mean velocity of 30 m/s. Compare the heat flux from the tube to the air for the two cases.

8.26 Freon is being transported at 0.1 kg/s through a Teflon tube of inside diameter $D_i = 25$ mm and outside diameter $D_o = 28$ mm, while atmospheric air at $V = 25$ m/s and 300 K is in cross flow over the tube. What is the heat transfer per unit length of tube to Freon at 240 K?

8.27 Consider a thin-walled, metallic tube of length $L = 1$ m and inside diameter $D_i = 3$ mm. Water enters the tube at $\dot{m} = 0.015$ kg/s and $T_{m,i} = 97°C$.

a) What is the outlet temperature of the water if the tube surface temperature is maintained at 27°C?

b) If a 0.5-mm-thick layer of insulation of $k = 0.05$ W/m·K is applied to the tube and its outer surface is maintained at 27°C, what is the outlet temperature of the water?

c) If the outer surface of the insulation is no longer maintained at 27°C but is allowed to exchange heat by free convection with ambient air at 27°C, what is the outlet temperature of the water? The free convection heat transfer coefficient is 5 W/m²·K.

8.28 In recent years the problem of heat losses from a fluid moving through a buried pipeline has received considerable attention. Practical applications include the Alaskan pipeline, as well as powerplant steam and water distribution lines. Consider a steel pipe of diameter D that is used to transport oil flowing at a rate $\dot{m}_o$ through a cold region. The pipe is covered with a layer of insulation of thickness t and thermal conductivity k_i and is buried in soil to a depth z (distance from the soil surface to the pipe centerline). Each section of pipe is of length L and extends between pumping stations in which the oil is heated to insure low viscosity and hence low pump power requirements. The temperature of the oil entering the pipe from a pumping station and the temperature of the ground above the pipe are designated as $T_{m,i}$ and T_s, respectively, and are presumed to be known.

Consider conditions for which the oil (o) properties may be approximated as $\rho_o = 900$ kg/m³, $c_{p,o} = 2000$ J/kg·K, $\nu_o = 8.5 \times 10^{-4}$ m²/s, $k_o = 0.140$ W/m·K, $Pr_o = 10^4$; the oil flowrate is $\dot{m}_o = 500$ kg/s; and the pipe diameter is 1.2 m.

a) Expressing your results in terms of D, L, z, t, $\dot{m}_o$, $T_{m,i}$ and T_s, as well as the appropriate oil (o), insulation (i) and soil (s) properties, obtain all of the expressions needed to estimate the temperature, $T_{m,o}$, of the oil leaving the pipe.

b) If $T_s = -40°C$, $T_{m,i} = 120°C$, $t = 0.15$ m, $k_i = 0.05$ W/m·K, $k_s = 0.5$ W/m·K, $z = 3$ m, and $L = 100$ km, what is the value of $T_{m,o}$? What is the total rate of heat transfer from a section of the pipeline?

8.29 You are designing an operating room heat exchange device that cools blood (bypassed from a human) from 40°C to 30°C by passing the blood through a coiled tube sitting in a vat of water–ice mixture. The volumetric flow rate (V) is 10^{-4} m³/min, the inside tube diameter (D) is 2.5 mm, and $T_{m,i}$ and $T_{m,o}$ represent the inlet and outlet temperatures of the blood, respectively.

a) At what temperatures would you evaluate the fluid properties in determining $\bar{h}$ for the entire tube length?

b) If the properties of blood evaluated at the temperature for part (a) are: $\rho = 1000$ kg/m³, $\nu = 7 \times 10^{-7}$ m²/s, $k = 0.5$ W/m·K, and $c_p = 4.0$ kJ/kg·K, what is the Prandtl number for the blood?

c) Is the blood flow laminar or turbulent?

d) Neglecting all entrance effects and assuming fully developed conditions, calculate the value of $\bar{h}$ for heat transfer from the blood.

e) What is the total heat rate lost from the blood as it passes through the tube?

f) When free convection effects on the outside of the tube are included, the average overall heat transfer coefficient, $\bar{U}$, between the blood and the ice–water mixture can be approximated as 300 W/m$^2 \cdot$K. Determine the tube length, L, required to obtain the outlet temperature, $T_{m,o}$.

8.30 Pressurized water at an elevated temperature of $T_{m,i} = 200°$C is pumped at a flowrate of $\dot{m} = 2$ kg/s from a power plant to a nearby industrial user through a thin-walled, round pipe of inside diameter $D = 1$ m. The pipe is covered with a layer of insulation of thickness $t = 0.15$ m and thermal conductivity $k = 0.05$ W/m$\cdot$K. The pipe, which is of length $L = 500$ m, is exposed to a crossflow of air at $T_\infty = -10°$C and $V = 4$ m/s.

a) Obtain a differential equation that could be used to solve for the variation of the mixed-mean temperature of the water, $T_m(x)$, with the axial coordinate. As a first approximation, the internal flow may be assumed to be fully developed throughout the pipe. Express your results in terms of $\dot{m}$, V, T_∞, D, t, k and appropriate water (w) and air (a) properties.

b) Evaluate the heat loss per unit length of the pipe at the inlet. What is the mean temperature of the water at the outlet?

8.31 Air at $\dot{m} = 3 \times 10^{-4}$ kg/s and $T_{m,i} = 27°$C enters a rectangular duct which is 1 m long and 4 mm by 16 mm on a side. A uniform heat flux of $q_s'' = 600$ W/m^2 is imposed on the duct surface. What is the temperature of the air and of the duct surface at the outlet?

8.32 Air at $\dot{m} = 4 \times 10^{-4}$ kg/s and $T_{m,i} = 27°$C enters a triangular duct that is 20 mm on a side and 2 m long. The duct surface is maintained at 100°C. Assuming fully developed flow conditions throughout the duct, determine the air outlet temperature.

8.33 Air at 1 atm and 285 K enters a 2-m-long rectangular duct with cross section 75 mm $\times$ 150 mm. The duct is maintained at a constant surface temperature of 400 K, and the air mass flow rate is 0.10 kg/s. Determine the heat transfer rate from the duct to the air and the air outlet temperature.

8.34 Referring to Figure 8.9, consider conditions in an annulus whose outer surface is insulated ($q_o'' = 0$) and which has a uniform heat flux q_i'' at the inner surface. Fully developed, laminar flow may be assumed to exist.

a) Determine the velocity profile, $u(r)$, in the annular region.

b) Determine the temperature profile, $T(r)$, and obtain an expression for the Nusselt number, Nu_i, associated with the inner surface.

8.35 Consider a concentric tube annulus for which the inner and outer diameters are $D_i = 25$ mm and $D_o = 50$ mm, respectively. Water enters the annular region at a rate of $\dot{m} = 0.04$ kg/s and a temperature of 25°C. If the inner tube wall is heated electrically at a rate (per unit length) of $q' = 4000$ W/m, while the outer tube wall is insulated, how long must the tubes be in order for the water to achieve an outlet temperature of 85°C? What is the inner tube surface temperature at the outlet, where fully developed flow conditions may be assumed.

8.36 An industrial process heat application involves passing ethylene glycol through a concentric tube annulus, which is itself enclosed in a transparent glass tube. Under sunny skies, reflectors are situated such that the blackened outer surface of the annulus is uniformly irradiated to heat the ethylene glycol as it passes through the system. The space between the glass cover and the absorbing surface is evacuated, and the fluid flow is fully developed throughout the annulus. An electrically heated rod is used to heat the fluid during the evening hours, as well as during cloudy days.

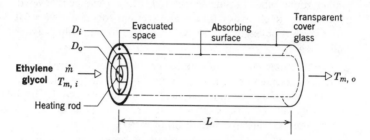

Consider a system for which $D_i = 10$ mm, $D_o = 100$ mm and $L = 20$ m. The ethylene glycol enters at a rate of $\dot{m} = 0.25$ kg/s and a temperature of $T_{m,i} = 300$ K.

a) During sunny skies, there is a net radiation heat flux of $q_o'' = 3000$ W/m^2 uniformly distributed on the outer surface of the annulus, and the heating rod is not operational. What is the outlet fluid temperature, $T_{m,o}$? What is the temperature of the absorbing surface at the inlet and outlet?

b) What is the surface heat flux q_i'' that must be provided by the heating rod during the evening to maintain the fluid outlet temperature predicted in part (a)? What would be the surface temperature of the rod at the inlet and outlet?

8.37 Water at $\dot{m} = 0.02$ kg/s and $T_{m,i} = 20°$C enters an annular region formed by an inner tube of diameter $D_i = 25$ mm and an outer tube of diameter $D_o = 100$ mm. Saturated steam flows through the inner tube, maintaining its surface at a uniform temperature of $T_{s,i} = 100°$C, while the outer surface of the outer tube is well insulated. If fully developed conditions may be assumed throughout the annulus, how long must the system be to provide an outlet water temperature of $75°$C? What is the heat flux from the inner tube at the outlet?

8.38 For the conditions of the preceding problem, how long must the annulus be if the water flowrate is 0.30 kg/s instead of 0.02 kg/s?

8.39 In the processing of very long plastic tubes of 2 mm inside diameter, air flows inside the tubing with a Reynolds number of 1000. The interior layer of the plastic material evaporates into the air under fully developed conditions. Both the plastic and air are at a temperature of 400 K, and the Schmidt number for the mixture of the plastic vapor and the air is 2.0. Determine the convection mass transfer coefficient.

8.40 What is the convection mass transfer coefficient associated with fully developed atmospheric air flow at $27°$C and 0.04 kg/s through a 50-mm-diameter tube whose surface has been coated with a thin layer of naphthalene? Determine the velocity and concentration entry lengths.

9 Free Convection

In previous chapters we focused on convection transfer in fluid flows that originate from an *external forcing* condition. For example, fluid motion could be induced by a fan or a pump, or it could result from the propulsion of a solid through the fluid. In the presence of a temperature or concentration gradient, *forced convection* heat or mass transfer will then occur.

It is now time to consider situations for which there is no *forced* velocity, and yet convection currents exist within the fluid. Such situations are referred to as *free* or *natural convection*, and they originate when a *body force* acts on a fluid in which there are *density gradients*. The net effect is a *buoyancy force*, which induces free convection currents. The most common case is one in which the density gradient is due to a temperature gradient, and the body force is due to the gravitational field.

Since free convection flow velocities are generally much smaller than those associated with forced convection, the corresponding convection transfer rates are also smaller. It is perhaps tempting to therefore attach less significance to free convection processes. This temptation should be resisted. In many systems involving multimode heat transfer effects, free convection provides the largest resistance to heat transfer and therefore plays an important role in the design or performance of the system. Moreover, when it is desirable to minimize heat transfer rates or to minimize operating costs, free convection is often preferred to forced convection.

There are, of course, many applications. Free convection strongly influences heat transfer from pipes and transmission lines, as well as from various electronic devices. It is important in transferring heat from electric baseboard heaters or steam radiators to room air and in dissipating heat from the coil of a refrigeration unit to the surrounding air. It is also relevant to the environmental sciences, where it is responsible for oceanic and atmospheric motions, as well as related heat and species transfer processes.

9.1 PHYSICAL CONSIDERATIONS

In free convection fluid motion is due to buoyancy effects, whereas in forced convection it is externally imposed. *Buoyancy is due to the combined presence of a density gradient within the fluid* and *a body force that is proportional to the fluid density.* In practice the relevant body force is usually *gravitational*, although it may be a centrifugal force in problems involving rotating fluid machinery or a Coriolis force in problems involving atmospheric and oceanic rotational motions. There are also several ways in which a mass density gradient may arise in a fluid, but the most common situation is one in which it is due to the presence of a temperature gradient. We know that the density of gases and liquids depends upon temperature, generally decreasing (due to fluid expansion) with increasing temperature ($\partial \rho / \partial T < 0$). Accordingly, there must be a density gradient associated with any temperature gradient in a fluid.

In this text we focus on free convection problems in which the density gradient is due to a temperature gradient and the body force is gravitational. Note, however, that the presence of a fluid density gradient in a gravitational field does not insure the existence of free convection currents. Consider the conditions of Figure 9.1. A fluid is enclosed by two large, horizontal plates of different temperature $(T_1 \neq T_2)$. In case (a) the temperature of the lower plate exceeds that of the upper plate, and the density decreases in the direction of the gravitational force. This condition is *unstable*, and free convection currents must exist. The gravitational force on the denser fluid in the upper layers exceeds that acting on the lighter fluid in the lower layers, and the designated circulation pattern will exist. The heavier fluid will descend, being warmed in the process, while the lighter fluid will rise, cooling as it moves. However, this condition does not characterize case (b), for which $T_1 > T_2$ and the density no longer decreases in the direction of the gravitational force. Conditions are now *stable*, and there is no bulk fluid motion. Note that in case (a) heat transfer occurs from the bottom to the top surface by free convection. In contrast for case (b) heat transfer (from top to bottom) occurs by conduction.

Free convection flows may be classified according to whether or not the flow is bounded by a surface. In the absence of an adjoining surface, *free boundary flows* may occur in the form of a *plume* or a *buoyant jet* (Figure 9.2). A plume is associated with fluid rising from a submerged heated object. Consider the heated wire of Figure 9.2a, which is immersed in an *extensive, quiescent* fluid.[1] Fluid that is heated by the wire rises due to buoyancy forces, entraining fluid from the quiescent region. Although the width of the plume increases with distance from the wire, the plume itself will eventually dissipate due to viscous effects and a reduction in the buoyancy force caused by cooling of the fluid in the plume. The distinction between a plume and a buoyant jet is generally made on the basis of the *initial* fluid velocity. This velocity is zero for the plume, but finite

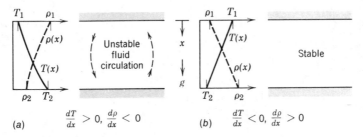

Figure 9.1 Conditions in a fluid between large horizontal plates at different temperatures. (a) Unstable temperature gradient. (b) Stable temperature gradient.

---

[1] An extensive medium is, in principle, an infinite medium. Since a quiescent fluid is one which is otherwise at rest, the velocity of fluid far from the heated wire is zero.

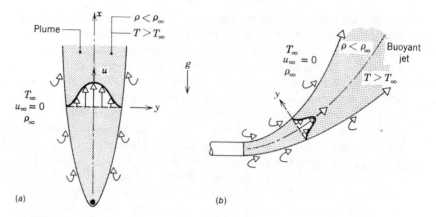

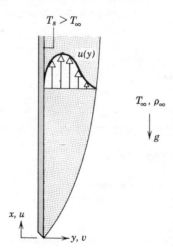

Figure 9.2 Buoyancy driven free boundary layer flows in an extensive, quiescent medium. (*a*) Plume formation above a heated wire. (*b*) Buoyant jet associated with a heated discharge.

for the buoyant jet. Figure 9.2*b* shows a heated fluid being discharged as a horizontal jet into a quiescent medium of lower temperature. The vertical motion that the jet begins to assume is due to the buoyancy force. Such a condition occurs when warm water from the condenser of a central power station is discharged into a reservoir of cooler water.

In this text we focus on free convection flows bounded by a surface, and a classical example relates to boundary layer development on a heated vertical plate (Figure 9.3). The plate is immersed in an extensive, quiescent fluid, and with $T_s > T_\infty$ the density of fluid close to the plate is less than that of fluid that is

Figure 9.3 Boundary layer development on a heated vertical plate.

further removed. Buoyancy forces therefore induce a free convection boundary layer in which the heated fluid rises vertically, entraining fluid from the quiescent region. Note that the resulting velocity distribution is unlike that associated with forced convection boundary layers. In particular the velocity must be zero as $y \to \infty$, as well as at $y = 0$. A free convection boundary layer will also develop if $T_s < T_\infty$. In this case, however, the boundary layer development is vertically downward.

9.2 THE GOVERNING EQUATIONS

As for forced convection, the equations that describe momentum, energy, and species transfer in free convection flows must originate from the related conservation principles. Moreover, the specific processes related to free convection are much like those which dominate in forced convection. Inertia and viscous forces remain important, as do energy and species transfer by convection and diffusion. The major difference between the two kinds of flows is that, in free convection, a major role is played by buoyancy forces. It is such forces that, in fact, sustain the flow.

Consider a laminar boundary layer flow (Figure 9.3) that is driven by buoyancy forces. Assume steady-state, two-dimensional conditions in which the gravity force acts in the negative x direction. Assume that temperature differences are small to moderate, in which case the fluid may be treated as having constant properties. Also, with one exception, assume the fluid to be incompressible. The exception involves accounting for the effect of variable density in the buoyancy force, since it is this variation that induces fluid motion. Finally, assume that the boundary layer approximations are valid.

With the foregoing simplifications the x-momentum equation, Equation 6.29, reduces to the boundary layer equation, Equation 6.52, except that the body force term X is retained. If we further assume that the only contribution to this force is made by gravity, the body force per unit volume is $X = -\rho g$, where g is the local acceleration due to gravity. The appropriate form of the x-momentum equation is then

$$u \frac{\partial u}{\partial x} + v \frac{\partial u}{\partial y} = -\frac{1}{\rho} \frac{\partial p}{\partial x} - g + \nu \frac{\partial^2 u}{\partial y^2} \tag{9.1}$$

The above equation may be couched in a more convenient form by first noting that, if there is no body force in the y direction, $(\partial p/\partial y) = 0$ from the y-momentum equation, Equation 6.53. Hence the x-pressure gradient at any point *in* the boundary layer must equal the pressure gradient in the quiescent region outside the boundary layer. However, in this region $u = v = 0$ and Equation 9.1 reduces to

$$\frac{\partial p}{\partial x} = -\rho_\infty g \tag{9.2}$$

Substituting Equation 9.2 into 9.1, we then obtain the following equation

$$u\frac{\partial u}{\partial x} + v\frac{\partial u}{\partial y} = \frac{g}{\rho}(\rho_\infty - \rho) + v\frac{\partial^2 u}{\partial y^2} \tag{9.3}$$

which must apply at every point in the free convection boundary layer.

The first term on the right-hand side of Equation 9.3 is the buoyancy force, and flow originates because the density ρ is a variable. The nature of this variation may be treated by introducing the *volumetric thermal expansion coefficient*

$$\beta = -\frac{1}{\rho}\left(\frac{\partial \rho}{\partial T}\right)_p \tag{9.4}$$

This *thermodynamic* property of the fluid provides a measure of the amount by which the density changes in response to a change in temperature at constant pressure. If it is expressed in the following approximate form,

$$\beta \approx -\frac{1}{\rho}\frac{\rho_\infty - \rho}{T_\infty - T}$$

it follows that

$$(\rho_\infty - \rho) \approx \rho\beta(T - T_\infty)$$

Substituting into Equation 9.3, the x-momentum equation then becomes

$$u\frac{\partial u}{\partial x} + v\frac{\partial u}{\partial y} = g\beta (T - T_\infty) + v\frac{\partial^2 u}{\partial y^2} \tag{9.5}$$

where it is now apparent how the buoyancy force is related to temperature difference.

Since the overall mass, energy, and species conservation equations are unchanged from forced convection,[2] Equations 6.51, 6.54, and 6.55 may be used to complete the problem formulation. The set of governing equations may then be expressed as

$$\frac{\partial u}{\partial x} + \frac{\partial v}{\partial y} = 0 \tag{9.6}$$

$$u\frac{\partial u}{\partial x} + v\frac{\partial u}{\partial y} = g\beta(T - T_\infty) + v\frac{\partial^2 u}{\partial y^2} \tag{9.7}$$

[2] In free convection, buoyancy effects are confined to the momentum equation.

$$u \frac{\partial T}{\partial x} + v \frac{\partial T}{\partial y} = \alpha \frac{\partial^2 T}{\partial y^2} \tag{9.8}$$

$$u \frac{\partial C_A}{\partial x} + v \frac{\partial C_A}{\partial y} = D_{AB} \frac{\partial^2 C_A}{\partial y^2} \tag{9.9}$$

Note that viscous dissipation has been neglected in the energy equation, Equation 9.8, an assumption that is certainly reasonable for the small velocities associated with free convection. In the mathematical sense the appearance of the buoyancy term in Equation 9.7 complicates matters. No longer may the hydrodynamic problem, given by Equations 9.6 and 9.7, be uncoupled from and solved to the exclusion of the thermal problem, given by Equation 9.8. The solution to the momentum equation depends on knowledge of T and hence on the solution to the energy equation. Equations 9.6 to 9.8 are therefore strongly coupled, and their solutions must be obtained simultaneously.

Free convection effects obviously depend on the expansion coefficient β. The manner in which β is obtained depends upon the nature of the fluid. For a perfect gas, $\rho = p/RT$ and

$$\beta = -\frac{1}{\rho} \left(\frac{\partial \rho}{\partial T} \right)_p = \frac{1}{\rho} \frac{p}{RT^2} = \frac{1}{T} \tag{9.10}$$

where T is the *absolute* temperature. For liquids and nonideal gases, β must be obtained from appropriate property tables (Appendix A).

9.3 SIMILARITY CONSIDERATIONS

Let us now consider the nature of the dimensionless parameters that govern free convective flow and heat transfer. As for forced convection (Chapter 6), the parameters may be obtained by nondimensionalizing the governing equations. Introducing

$$x^* \equiv \frac{x}{L} \qquad y^* \equiv \frac{y}{L}$$

$$u^* \equiv \frac{u}{u_0} \qquad v^* \equiv \frac{v}{u_0} \qquad T^* \equiv \frac{T - T_\infty}{T_s - T_\infty}$$

where L is a characteristic length and u_0 is an *arbitrary* reference velocity, the x-momentum and energy equations, Equations 9.7 and 9.8, reduce to

$$u^* \frac{\partial u^*}{\partial x^*} + v^* \frac{\partial u^*}{\partial y^*} = \frac{g\beta (T_s - T_\infty)L}{u_0^2} T^* + \frac{1}{Re_L} \frac{\partial^2 u^*}{\partial y^{*2}} \tag{9.11}$$

$$u^* \frac{\partial T^*}{\partial x^*} + v^* \frac{\partial T^*}{\partial y^*} = \frac{1}{Re_L Pr} \frac{\partial^2 T^*}{\partial y^{*2}} \tag{9.12}$$

The dimensionless parameter in the first term on the right-hand side of Equation 9.11 is a direct consequence of the buoyancy force. However, because it is expressed in terms of the unknown reference velocity u_0, it is inconvenient in its present form. It is therefore customary to work with an alternative form that is obtained from multiplying by $Re_L^2 = (u_0 L/v)^2$. The result is termed the *Grashof number*, Gr_L.

$$Gr_L \equiv \frac{g\beta (T_s - T_\infty)L}{u_0^2} \left(\frac{u_0 L}{v}\right)^2 = \frac{g\beta (T_s - T_\infty)L^3}{v^2} \tag{9.13}$$

The Grashof number plays the same role in free convection that the Reynolds number plays in forced convection. Recall that the *Reynolds number* provides a measure of the *ratio of* the *inertial to viscous forces* acting on a fluid element. In contrast the *Grashof number* indicates the *ratio of* the *buoyancy force relative to* the *viscous force* acting on the fluid.

Although Equations 9.11 to 9.13 prompt us to expect heat transfer correlations of the form, $Nu_L = f(Re_L, Gr_L, Pr)$, it is important to note that such correlations are only pertinent when forced and free convection effects are of comparative importance. For such cases an external flow is superposed on the buoyancy-driven flow, and there exists a well-defined forced convection velocity. Generally, the combined effects of free and forced convection must be considered whenever $(Gr_L/Re_L^2) \approx 1$. If the inequality $(Gr_L/Re_L^2) \ll 1$ is satisfied, free convection effects may be neglected and $Nu_L = f(Re_L, Pr)$. Conversely, if $(Gr_L/Re_L^2) \gg 1$, forced convection effects may be neglected and $Nu_L = f(Gr_L, Pr)$. In the strict sense, a free convection flow is one that is induced solely by buoyancy forces, in which case there is no well-defined forced convection velocity and $(Gr_L/Re_L^2) = \infty$.

9.4 LAMINAR FREE CONVECTION ON A VERTICAL SURFACE

Numerous solutions to the laminar free convection boundary layer equations have been obtained, and a special case that has received considerable attention is that of free convection from an isothermal vertical surface in an extensive quiescent medium (Figure 9.3). For this geometry Equations 9.6 to 9.8 must be solved subject to boundary conditions of the form[3]

---

[3] Note that the boundary layer approximations are assumed in using Equations 9.6 to 9.8. However, the approximations are only valid for $(Gr_x Pr) \gtrsim 10^4$. Below this value (close to the leading edge), the boundary layer thickness is too large relative to the characteristic length, x, to insure the validity of the approximations.

$$y = 0: \qquad u = v = 0 \qquad T = T_s$$

$$y \to \infty: \qquad u \to 0 \qquad T \to T_\infty$$

A similarity solution to the foregoing problem has been obtained by Ostrach [1]. The solution involves transforming variables by introducing a *similarity parameter* of the form

$$\eta \equiv \frac{y}{x}\left(\frac{Gr_x}{4}\right)^{1/4} \tag{9.14}$$

and representing the velocity components in terms of a stream function defined as

$$\psi(x,y) \equiv f(\eta)\left[4v\left(\frac{Gr_x}{4}\right)^{1/4}\right] \tag{9.15}$$

With the above definition of the stream function, the x-velocity component may be expressed as

$$u = \frac{\partial\psi}{\partial y} = \frac{\partial\psi}{\partial\eta}\frac{\partial\eta}{\partial y} = 4v\left(\frac{Gr_x}{4}\right)^{1/4}f'(\eta)\frac{1}{x}\left(\frac{Gr_x}{4}\right)^{1/4}$$

$$= \frac{2v}{x}Gr_x^{1/2}f'(\eta) \tag{9.16}$$

where primed quantities indicate diffemtion with respect to η. Hence $f'(\eta) \equiv df/d\eta$. Introducing the dimensionless temperature

$$T^* \equiv \frac{T - T_\infty}{T_s - T_\infty} \tag{9.17}$$

the three original partial differential equations, Equations 9.6 to 9.8, may then be reduced to two ordinary differential equations of the form

$$f''' + 3ff'' - 2(f')^2 + T^* = 0 \tag{9.18}$$

$$T^{*''} + 3Pr\,fT^{*'} = 0 \tag{9.19}$$

where f and T^* are functions of only η and the double and triple primes, respectively, refer to second and third derivatives with respect to η. Note that f assumes the role of the key dependent variable for the velocity boundary layer. Note also that the continuity equation, Equation 9.6, is automatically satisfied through introduction of the stream function.

The transformed boundary conditions required to solve the momentum and energy equations, Equations 9.18 and 9.19, are of the form

$$\eta = 0: \qquad f = f' = 0 \qquad T^* = 1$$

$$\eta \to \infty: \qquad f' \to 0 \qquad T^* \to 0$$

A numerical solution has been obtained by Ostrach [1], and selected results are shown in Figure 9.4. Note that the x velocity component, u, may readily be obtained from Figure 9.4a through the use of Equation 9.16. Note also that, through the definition of the similarity parameter η, Figure 9.4 may be used to obtain values of u and T for any value of x and y.

Figure 9.4b may also be used to infer the appropriate form of the heat transfer correlation. Using Newton's law of cooling for the local convection coefficient, h, the local Nusselt number may be expressed as

$$Nu_x = \frac{hx}{k} = \frac{[q_s''/(T_s - T_\infty)]x}{k}$$

Using Fourier's law to obtain q_s'' and expressing the surface temperature gradient in terms of η, Equation 9.14, and T^*, Equation 9.17, it follows that

$$q_s'' = -k \left.\frac{\partial T}{\partial y}\right|_{y=0} = -\frac{k}{x}(T_s - T_\infty)\left(\frac{Gr_x}{4}\right)^{1/4} \left.\frac{dT^*}{d\eta}\right|_{\eta=0}$$

Hence

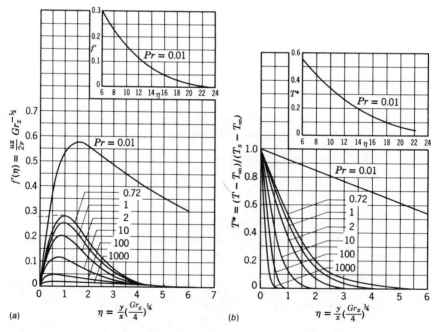

Figure 9.4 Laminar, free convection boundary layer conditions on an isothermal, vertical surface. (a) Velocity profiles. (b) Temperature profiles [1].

Table 9.1 Coefficients of Equations 9.20 and 9.21 for free convection on a vertical plate

Pr	0.01	0.72	1	2	10	100	1000
g(Pr)	0.081	0.505	0.567	0.716	1.169	2.191	3.966

$$Nu_x = \frac{hx}{k} = -\left(\frac{Gr_x}{4}\right)^{1/4} \frac{dT^*}{d\eta}\bigg|_{\eta=0} = \left(\frac{Gr_x}{4}\right)^{1/4} g(Pr) \tag{9.20}$$

which acknowledges that the dimensionless temperature gradient at the surface is a function of the Prandtl number, $g(Pr)$. This dependence is evident from Figure 9.4b and may be determined from the results of the numerical solution presented in Table 9.1.

Using Equation 9.20 for the local convection coefficient and substituting for the local Grashof number, which is of the form

$$Gr_x = \frac{g\beta (T_s - T_\infty)x^3}{\nu^2}$$

the average convection coefficient for a surface of length L is then

$$\bar{h} = \frac{1}{L}\int_0^L h\,dx = \frac{k}{L}\left[\frac{g\beta(T_s - T_\infty)}{4\nu^2}\right]^{1/4} g(Pr) \int_0^L \frac{dx}{x^{1/4}}$$

Integrating, it follows that

$$\overline{Nu_L} = \frac{\bar{h}L}{k} = \frac{4}{3}\left(\frac{Gr_L}{4}\right)^{1/4} g(Pr) \tag{9.21}$$

or substituting from Equation 9.20, with $x = L$,

$$\overline{Nu_L} = \frac{4}{3} Nu_L \tag{9.22}$$

It should be noted that the foregoing results apply irrespective of whether $T_s > T_\infty$ or $T_s < T_\infty$. If $T_s < T_\infty$, conditions are inverted from those of Figure 9.3. The leading edge is at the top of the plate, and positive x is defined in the direction of the gravity force. Note also that Sparrow and Gregg [2] have considered free convection for a vertical surface with *constant heat flux*, q_s'', and obtained results that are within 5 percent of the foregoing results. Accordingly, Equations 9.20 to 9.22 may be used with reasonable accuracy for constant surface heat flux, as well as constant surface temperature.

9.5 THE EFFECTS OF TURBULENCE

It is important to note that free convection boundary layers are not restricted to laminar flow conditions. All free convection flows of interest in this chapter originate from a *thermal instability*. That is, warmer, lighter fluid moves vertically upward relative to cooler, heavier fluid. However, as with forced convection, *hydrodynamic instabilities* may also arise. That is, disturbances in the flow may be amplified, leading to transition from laminar to turbulent flow conditions. This transition process is shown schematically in Figure 9.5 for a heated vertical plate.

Transition in a free convection boundary layer depends on the relative magnitude of the buoyancy and viscous forces in the fluid. It is customary to correlate its occurrence in terms of the *Rayleigh number*, which is simply the product of the Grashof and Prandtl numbers. For vertical plates it is known that

$$Ra_{x,c} = Gr_{x,c}\, Pr = \frac{g\beta(T_s - T_\infty)x^3}{\nu\alpha} \approx 10^9 \qquad (9.23)$$

An extensive discussion of stability and transition effects is given by Gebhart [3].

As in forced convection, transition to turbulence will have an important effect on heat transfer rates. Hence the results of the foregoing section only apply if $Ra_L \lesssim 10^9$. To obtain the appropriate correlations for turbulent flow, heavy reliance must be placed on experimental results.

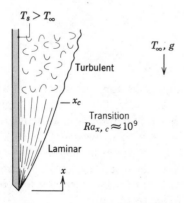

Figure 9.5 Free convection boundary layer transition on a vertical plate.

EXAMPLE 9.1

Consider a vertical plate that is 0.25 m long and is maintained at a temperature of 70°C. The plate is suspended in air which is at a temperature of 25°C. Estimate the boundary layer thickness at the trailing edge of the plate if the air is quiescent. How does this thickness compare with that which would exist if the air were flowing over the plate at a freestream velocity of 5 m/s?

SOLUTION

KNOWN:

Vertical plate is in quiescent air at a lower temperature.

FIND:

Boundary layer thickness at trailing edge. Compare with thickness corresponding to an air speed of 5 m/s.

SCHEMATIC:

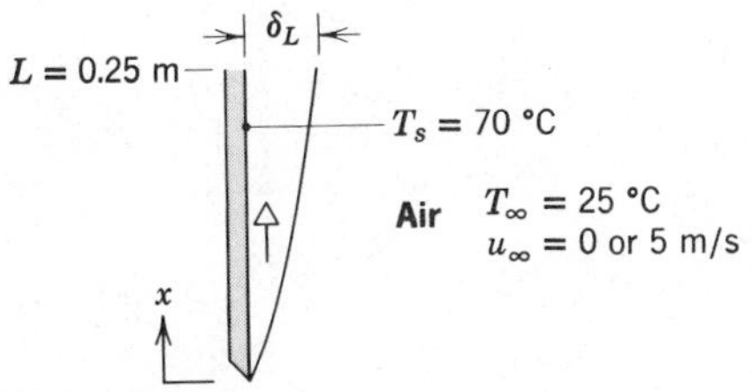

$L = 0.25$ m

δ_L

$T_s = 70$ °C

Air $\quad T_\infty = 25$ °C
$\quad u_\infty = 0$ or 5 m/s

x

ASSUMPTIONS:

1. Constant properties.
2. Buoyancy effects are negligible when $u_\infty = 5$ m/s.

PROPERTIES:

Table A.4, air ($T_f = 320.5$ K): $\nu = 17.95 \times 10^{-6}$ m²/s, $Pr = 0.7$, $\beta = T_f^{-1} = 3.12 \times 10^{-3}$ K^{-1}.

ANALYSIS:

For the quiescent air, Equation 9.13 gives

$$Gr_L = \frac{g\beta(T_s - T_\infty)L^3}{\nu^2}$$

$$= \frac{9.8 \text{ m/s}^2 \times (3.12 \times 10^{-3} \text{ K}^{-1})(70 - 25)°\text{C}(0.25 \text{ m})^3}{(17.95 \times 10^{-6} \text{ m}^2\text{/s})^2}$$

$$= 6.69 \times 10^7$$

Hence from Equation 9.23 the free convection boundary layer is laminar, and the analysis of Section 9.4 is applicable. The results of Figure 9.4 indicate that for $Pr \approx 0.7$, $\eta \approx 6.0$ at the edge of the boundary layer, that is, at $y = \delta$. Hence

$$\delta_L \approx \frac{6L}{(Gr_L/4)^{1/4}} = \frac{6(0.25 \text{ m})}{(1.67 \times 10^7)^{1/4}}$$

or

$$\delta_L \approx 0.024 \text{ m} \qquad \triangleleft$$

For air flow at $u_\infty = 5$ m/s

$$Re_L = \frac{u_\infty L}{v} = \frac{(5 \text{ m/s}) \times 0.25 \text{ m}}{17.95 \times 10^{-6} \text{ m}^2\text{/s}} = 6.97 \times 10^4$$

and the boundary layer is laminar. Hence from Equation 7.5

$$\delta_L \approx \frac{5L}{Re_L^{1/2}} = \frac{5(0.25 \text{ m})}{(6.97 \times 10^4)^{1/2}}$$

or

$$\delta_L \approx 0.0047 \text{ m} \qquad \triangleleft$$

COMMENTS:

1. Boundary layer thicknesses are typically larger for free convection than for forced convection.

2. $(Gr_L/Re_L^2) = 0.014 \ll 1$ and the assumption of negligible buoyancy effects for $u_\infty = 5$ m/s is justified.

9.6   EMPIRICAL CORRELATIONS: EXTERNAL FREE CONVECTION FLOWS

In this section we summarize appropriate empirical correlations that have been developed for common external flow geometries. The correlations are suitable for most engineering calculations and are generally of the form

$$\overline{Nu}_L = \frac{\bar{h}L}{k} = CRa_L^n \qquad (9.24)$$

where the Rayleigh number,

$$Ra_L = Gr_L Pr = \frac{g\beta(T_s - T_\infty)L^3}{v\alpha} \tag{9.25}$$

is based on the characteristic length L of the geometry. Typically, $n = 1/4$ and $1/3$ for laminar and turbulent flows, respectively. For turbulent flow it then follows that $\overline{Nu_L}$ is independent of L. Note that all properties are evaluated at the film temperature, $T_f \equiv (T_s + T_\infty)/2$. The specific correlations that follow are summarized in Table 9.3.

9.6.1 The Vertical Plate

Equations of the form given by (9.24) have been developed for the vertical plate [4–6] and are plotted in Figure 9.6. The coefficient C and the exponent n depend upon the Rayleigh number range; for Rayleigh numbers less than 10^4, the Nusselt number should be obtained directly from the figure.

More recently, Churchill and Chu [7] have recommended a correlation which may be applied over the *entire* range of Ra_L. It is of the form

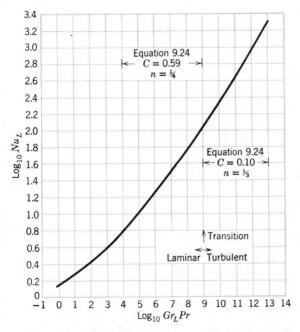

Figure 9.6 Nusselt number for free convection heat transfer from a vertical plate [4–6].

✳ $$\overline{Nu}_L = \left\{ 0.825 + \frac{0.387 Ra_L^{1/6}}{[1 + (0.492/Pr)^{9/16}]^{8/27}} \right\}^2 \qquad (9.26)$$

Although Equation 9.26 is suitable for most engineering calculations, slightly better accuracy may be obtained for laminar flow by using [7]

$$\overline{Nu}_L = 0.68 + \frac{0.670 Ra_L^{1/4}}{[1 + (0.492/Pr)^{9/16}]^{4/9}} \qquad (0 < Ra_L < 10^9) \qquad (9.27)$$

The foregoing results may be applied for constant heat flux, as well as for constant surface temperature. They may also be applied to *vertical* cylinders of height L, if the boundary layer thickness δ is much less than the cylinder diameter D. This condition is known to be satisfied [3,8] when

$$\frac{D}{L} \gtrsim \frac{35}{Gr_L^{1/4}} \qquad (9.28)$$

Cebeci [9] and Minkowycz and Sparrow [10] present results for slender, vertical cylinders not meeting the above condition, where transverse curvature does influence boundary layer development and heat transfer rates.

EXAMPLE 9.2

A glass-door firescreen, used to reduce exfiltration of room air through a chimney, has a height of 0.71 m and a width of 1.02 m and reaches a temperature of 232°C. If the room temperature is 23°C, estimate the convection heat rate from the fireplace to the room.

SOLUTION

KNOWN:

Glass screen situated in fireplace opening.

FIND:

Heat transfer by free convection between screen and room air.

SCHEMATIC:

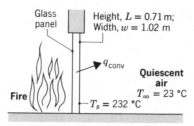

Glass panel

Height, $L = 0.71$ m; Width, $w = 1.02$ m

q_{conv}

Fire

Quiescent air
$T_\infty = 23$ °C

$T_s = 232$ °C

ASSUMPTIONS:

1. Screen is at a uniform temperature, T_s.
2. Room air is quiescent.

PROPERTIES:

Table A.4, air ($T_f = 400$ K): $k = 33.8 \times 10^{-3}$ W/m·K, $v = 26.4 \times 10^{-6}$ m²/s, $\alpha = 38.3 \times 10^{-6}$ m²/s, $Pr = 0.690$, $\beta = (1/T_f) = 0.0025$ K^{-1}.

ANALYSIS:

The rate of heat transfer by free convection from the panel to the room is given by Newton's law of cooling

$$q = \bar{h}A_s(T_s - T_\infty)$$

where $\bar{h}$ may be obtained from knowledge of the Rayleigh number. Using Equation 9.25:

$$Ra_L = \frac{g\beta(T_s - T_\infty)L^3}{\alpha v}$$

$$= \frac{9.8 \text{ m/s}^2 \times 1/400 \text{ K } (232 - 23)°\text{C} \times (0.71 \text{ m})^3}{38.3 \times 10^{-6} \text{ m/s}^2 \times 26.4 \times 10^{-6} \text{ m/s}^2} = 1.813 \times 10^9$$

and from Equation 9.23 it follows that transition to turbulence will occur on the panel. The appropriate heat transfer correlation is then given by Equation 9.26

$$\overline{Nu}_L = \left\{0.825 + \frac{0.387Ra_L^{1/6}}{[1 + (0.492/Pr)^{9/16}]^{8/27}}\right\}^2$$

$$\overline{Nu}_L = \left\{0.825 + \frac{0.387(1.813 \times 10^9)^{1/6}}{[1 + (0.492/0.690)^{9/16}]^{8/27}}\right\}^2$$

$$\overline{Nu}_L = 147$$

Hence

$$h = \frac{\overline{Nu}_L \cdot k}{L} = \frac{147 \times 33.8 \times 10^{-3} \text{ W/m} \cdot \text{K}}{0.71 \text{ m}} = 7.00 \text{ W/m}^2 \cdot \text{K}$$

and

$$q = 7.0 \text{ W/m}^2 \cdot \text{K} \ (1.02 \times 0.71) \text{m}^2 (232 - 23)°\text{C}$$

$$q = 1060 \text{ W} \qquad \qquad \lhd$$

COMMENTS:

1. If $\bar{h}$ were computed from Equation 9.24, with $C = 0.10$ and $n = 1/3$, we would obtain $\bar{h} = 5.8 \text{ W/m}^2 \cdot \text{K}$ and the heat transfer prediction would be approximately 20 percent lower than the foregoing result. This difference is within the uncertainty normally associated with using such correlations.

2. Radiation heat transfer effects are often significant relative to free convection.

Using Equation 1.6 with an assumed emissivity of $\varepsilon = 1.0$ for the glass surface and a temperature of $T_{\text{sur}} = 23°\text{C}$, it follows that the net rate of radiation heat transfer between the glass and the surroundings is

$$q_{\text{rad}} = \varepsilon A_s \sigma (T_s^4 - T_\infty^4)$$

$$q_{\text{rad}} = 1(1.02 \times 0.71) \text{m}^2 \times 5.67 \times 10^{-8} \text{ W/m}^2 \cdot \text{K}^4 \ (505^4 - 296^4) \text{K}^4$$

or

$$q_{\text{rad}} = 2355 \text{ W}$$

Hence in this case radiation heat transfer exceeds free convection heat transfer by more than a factor of 2.

9.6.2 The Horizontal Plate

For this geometry the specific form of the correlation depends on whether the plate is *warmer* or *cooler* than the surrounding fluid and on whether it is *facing upward* or *downward*. Although correlations suggested by McAdams [4] are widely used, improved accuracy may be obtained by altering the form of the characteristic length on which the correlations are based [11,12]. In particular with the characteristic length defined as

$$L \equiv \frac{A_s}{P} \qquad \qquad (9.29)$$

where A_s and P are the plate surface area and perimeter, respectively, the recommended correlations are

Upper Surface Heated or Lower Surface Cooled

$$\overline{Nu}_L = 0.54 Ra_L^{1/4} \qquad (10^5 \lesssim Ra_L \lesssim 10^7) \tag{9.30}$$

$$\overline{Nu}_L = 0.15 Ra_L^{1/3} \qquad (10^7 \lesssim Ra_L \lesssim 10^{10}) \tag{9.31}$$

Lower Surface Heated or Upper Surface Cooled

$$\overline{Nu}_L = 0.27 Ra_L^{1/4} \qquad (10^5 \lesssim Ra_L \lesssim 10^{10}) \tag{9.32}$$

EXAMPLE 9.3

Airflow through a long rectangular heating duct, whose width and height are 0.75 m and 0.30 m, respectively, maintains the outer duct surface temperature at 45°C. If the duct is uninsulated and exposed to air at 15°C in the crawlspace beneath a home, what is the heat loss from the duct per meter of length?

SOLUTION

KNOWN:

Surface temperature of a long rectangular duct.

FIND:

Heat loss from duct per meter of length.

SCHEMATIC:

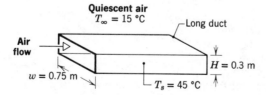

ASSUMPTIONS:

1. Ambient air is quiescent.
2. Surface radiation effects are negligible.

PROPERTIES:

Table A.4, air ($T_f = 303$ K): $v = 16.2 \times 10^{-6}$ m^2/s, $\alpha = 22.9 \times 10^{-6}$ m^2/s, $k = 0.0265$ W/m·K, $\beta = 0.0033$ K^{-1}, $Pr = 0.71$

ANALYSIS:

Surface heat loss is by free convection from the vertical sides and the horizontal top and bottom. From Equation 9.25

$$Ra_L = \frac{g\beta(T_s - T_\infty)L^3}{v\alpha} = \frac{(9.8 \text{ m/s}^2)(0.0033 \text{ K}^{-1})(30 \text{ K})L^3(\text{m}^3)}{(16.2 \times 10^{-6} \text{ m/s}^2)(22.9 \times 10^{-6} \text{ m/s}^2)}$$

or

$$Ra_L = 2.62 \times 10^9 L^3$$

For the two sides, $L = H = 0.3$ m. Hence $Ra_L = 7.07 \times 10^7$. The free convection boundary layer is therefore laminar, and from Equation 9.27

$$\overline{Nu}_L = 0.68 + \frac{0.670 \, Ra_L^{1/4}}{[1 + (0.492/Pr)^{9/16}]^{4/9}}$$

The convection coefficient associated with the sides is then

$$\bar{h}_s = \frac{k}{H} \overline{Nu}_L$$

$$\bar{h}_s = \frac{0.0265 \text{ W/m·K}}{0.3 \text{ m}} \left\{ 0.68 + \frac{0.670(7.07 \times 10^7)^{1/4}}{[1 + (0.492/0.71)^{9/16}]^{4/9}} \right\} = 4.23 \text{ W/m}^2\text{·K}$$

For the top and bottom, $L = (A_s/P) \approx (w/2) = 0.375$ m. Hence $Ra_L = 1.38 \times 10^8$, and from Equations 9.31 and 9.32, respectively,

$$\bar{h}_t = k/(w/2) \times 0.15 \, Ra_L^{1/3} = \frac{0.0265 \text{ W/m·K}}{0.375 \text{ m}}$$

$$\times 0.15(1.38 \times 10^8)^{1/3} = 5.47 \text{ W/m}^2\text{·K}$$

$$\bar{h}_b = k/(w/2) \times 0.27 \, Ra_L^{1/4} = \frac{0.0265 \text{ W/m·K}}{0.375 \text{ m}}$$

$$\times 0.27(1.38 \times 10^8)^{1/4} = 2.07 \text{ W/m}^2\text{·K}$$

The rate of heat loss per unit length of duct is then

$$q' = 2q'_s + q'_t + q'_b$$

or

$$q' = (2\bar{h}_s \cdot H + \bar{h}_t \cdot w + \bar{h}_b \cdot w)(T_s - T_\infty)$$

$$q' = (2 \times 4.23 \times 0.3 + 5.47 \times 0.75 + 2.07 \times 0.75)(45 - 15) \text{ W/m}$$

$$q' = 246 \text{ W/m} \qquad \qquad \triangleleft$$

COMMENTS:

Note that, even in this situation, radiation effects are significant. From Equation 1.6 with ε assumed to be unity and $T_{\text{sur}} = 288$ K, it follows that $q'_{\text{rad}} = 398$ W/m.

9.6.3 The Inclined Plate

As a first approximation, Churchill and Chu [7] recommend using Equations 9.26 and 9.27 for surfaces that are inclined up to $\theta = 60°$ from the vertical. For laminar flow Equation 9.27 may be used if g is replaced by $g \cos \theta$ in evaluating Ra_L. Equation 9.26 may be used for turbulent flow without this modification. More accurate representations may be obtained from the literature [13,14].

9.6.4 The Long Horizontal Cylinder

This important geometry has been studied extensively, and the many existing correlations have been reviewed by Morgan [15]. Based on this review Morgan suggests using an expression of the form

$$\overline{Nu}_D = \frac{\overline{h}D}{k} = C\, Ra_D^n \tag{9.33}$$

where C and n are given in Table 9.2. Churchill and Chu [16] have recommended a single correlation for a wide Rayleigh number range.

$$\overline{Nu}_D = \left\{ 0.60 + \frac{0.387 Ra_D^{1/6}}{[1 + (0.559/Pr)^{9/16}]^{8/27}} \right\}^2 \qquad (10^{-5} < Ra_D < 10^{12}) \tag{9.34}$$

Table 9.2 Constants of Equation 9.33 for free convection on a horizontal circular cylinder [15]

Ra_D	C	n
$10^{-10} - 10^{-2}$	0.675	0.058
$10^{-2} - 10^2$	1.02	0.148
$10^2 - 10^4$	0.850	0.188
$10^4 - 10^7$	0.480	0.250
$10^7 - 10^{12}$	0.125	0.333

EXAMPLE 9.4

A horizontal, high-pressure steam pipe of 0.1 m outside diameter passes through a large room whose walls and air temperature are 23°C. The pipe has an outside surface temperature of 165°C and an emissivity of $\varepsilon = 0.85$. Estimate the heat loss from the pipe per unit length.

SOLUTION

KNOWN:

Surface temperature of a horizontal steam pipe passing through a room.

FIND:

Heat loss from the pipe per unit length, q'(W/m).

SCHEMATIC:

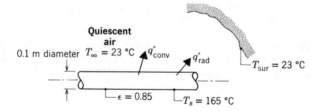

ASSUMPTIONS:

1. Pipe surface area is small compared to surroundings.
2. Room air is quiescent.

PROPERTIES:

Table A.4, air ($T_f = 367$ K): $k = 0.0313$ W/m·K, $\nu = 22.8 \times 10^{-6}$ m²/s, $\alpha = 32.8 \times 10^{-6}$ m²/s, $Pr = 0.697$, $\beta = 2.725 \times 10^{-3}$ K^{-1}.

ANALYSIS:

The total heat loss per unit length of pipe is

$$q' = q'_{conv} + q'_{rad}$$

or

$$q' = \bar{h}\pi D(T_s - T_\infty) + \varepsilon \pi D\sigma(T_s^4 - T_{sur}^4)$$

The convection coefficient may be obtained from Equation 9.34

$$\overline{Nu}_D = \left\{ 0.60 + \frac{0.387 Ra_D^{1/6}}{[1 + (0.559/Pr)^{9/16}]^{8/27}} \right\}^2$$

where

$$Ra_D = \frac{g\beta(T_s - T_\infty)D^3}{\nu\alpha}$$

$$Ra_D = \frac{9.8 \text{ m/s}^2 \times 2.725 \times 10^{-3} \text{ 1/K}(165 - 23)^\circ\text{C}(0.1 \text{ m})^3}{22.8 \times 10^{-6} \text{ m}^2/\text{s} \times 32.8 \times 10^{-6} \text{ m}^2/\text{s}} = 5.073 \times 10^6$$

Hence

$$\overline{Nu}_D = \left\{ 0.60 + \frac{0.387(5.073 \times 10^6)^{1/6}}{[1 + (0.559/0.697)^{9/16}]^{8/27}} \right\}^2 = 23.3$$

and

$$\bar{h} = \frac{k}{D} \overline{Nu}_D = \frac{0.0313 \text{ W/m·K}}{0.1 \text{ m}} \times 23.3 = 7.29 \text{ W/m}^2\text{·K}$$

The total heat loss is then

$$q' = 7.29 \text{ W/m}^2\text{·K} (\pi \times 0.1 \text{ m})(165 - 23)^\circ\text{C}$$

$$+ 0.85(\pi \times 0.1 \text{ m})5.67 \times 10^{-8} \text{ W/m}^2\text{·K}^4(438^4 - 296^4)\text{K}^4$$

$$q' = (325 + 441) \text{ W/m}$$

$$q' = 766 \text{ W/m} \qquad \triangleleft$$

COMMENTS:

Equation 9.33 could also be used to estimate the Nusselt number with the result that $\overline{Nu}_D = 22.8$.

9.6.5 Spheres

The following correlation by Yuge [17] is recommended for spheres in fluids of Pr close to unity.

$$\overline{Nu}_D = 2 + 0.43 Ra_D^{1/4} \qquad (9.35)$$

$$\begin{bmatrix} 1 < Ra_D < 10^5 \\ Pr \approx 1 \end{bmatrix}$$

The foregoing correlations are summarized in Table 9.3.

Table 9.3 Summary of free convection empirical correlations for external flow geometries

GEOMETRY	RECOMMENDED CORRELATION	RESTRICTIONS
1. Vertical plates[a]	Equation 9.26	None
2. Horizontal plates (a) Hot surface up or cold surface down	Equation 9.30 Equation 9.31	$10^5 \lesssim Ra_L \lesssim 10^7$ $10^7 \lesssim Ra_L \lesssim 10^{10}$
(b) Cold surface up or hot surface down	Equation 9.32	$10^5 \lesssim Ra_L \lesssim 10^{10}$
3. Inclined plates	Equation 9.26 Equation 9.27	$Ra_L \gtrsim 10^9$ $\theta < 60°$ $Ra_L \lesssim 10^9$ $\theta < 60°$ $g \to g\cos\theta$
4. Horizontal cylinder	Equation 9.34	$10^{-5} < Ra_D < 10^{12}$
5. Sphere	Equation 9.35	$1 < Ra_D < 10^5$ $Pr \approx 1$

[a]The correlation may be applied to a vertical cylinder if $(D/L) \gtrsim (35/Gr_L^{1/4})$.

9.7 EMPIRICAL CORRELATIONS: ENCLOSURES

The correlations of the preceding section pertain to free convection between a surface and an extensive fluid medium. However, engineering applications frequently involve heat transfer between surfaces that are at different temperatures and are separated by an *enclosed* fluid. In this section we present correlations that are pertinent to the most common geometries.

9.7.1 Rectangular Cavities

The rectangular cavity, shown schematically in Figure 9.7, has been studied extensively and comprehensive reviews of both experimental and theoretical results are available in the literature [18, 19]. The tilt angle τ between the heated and cooled surfaces of the cavity and the horizontal plane strongly influences the heat flux across the cavity. This flux may be expressed as

$$q'' = h(T_1 - T_2) \tag{9.36}$$

The horizontal cavity heated from below ($\tau = 0$) has been considered by many investigators. It is known that when

$$Ra_L \equiv \frac{g\beta(T_1 - T_2)L^3}{\alpha v} > 1700$$

conditions are thermally unstable and there is heat transfer by free convection. As a first approximation, the convection coefficient may be obtained from a correlation proposed by Globe and Dropkin [20]

$$\overline{Nu}_L = \frac{\overline{h}L}{k} = 0.069 Ra_L^{1/3} Pr^{0.074} \qquad (3 \times 10^5 < Ra_L < 7 \times 10^9) \tag{9.37}$$

where all properties are evaluated at the average temperature, $\overline{T} \equiv (T_1 + T_2)/2$. More detailed correlations, which apply over a wider range of Ra_L, have been suggested by Hollands et al. [21]. In each case, however, the height ratios, L/H and L/w, were small and the effects of lateral boundaries on heat transfer were

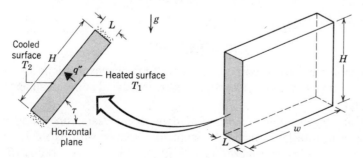

Figure 9.7 Free convection in a rectangular cavity.

not measurable. For $Ra_L < 1700$ or for $\tau = 180°$, heat transfer between the surfaces occurs by conduction and $Nu_L = 1$.

In the vertical rectangular cavity ($\tau = 90°$), it is the vertical surfaces that are heated and cooled, with the horizontal surfaces serving only to complete the cavity. A number of correlations may be found in the literature, but it has recently been recommended that, where appropriate, the following relationships be used [19].

$$* \quad \overline{Nu}_L = 0.22 \left(\frac{Pr}{0.2 + Pr} Ra_L \right)^{0.28} (H/L)^{-1/4} \qquad \text{Vertical cavity} \tag{9.38}$$

$$\begin{bmatrix} 2 < (H/L) < 10 \\ Pr < 10^5 \\ Ra_L < 10^{10} \end{bmatrix}$$

$$* \quad \overline{Nu}_L = 0.18 \left(\frac{Pr}{0.2 + Pr} Ra_L \right)^{0.29} \qquad \text{Vertical cavity} \tag{9.39}$$

$$\begin{bmatrix} 1 < (H/L) < 2 \\ 10^{-3} < Pr < 10^5 \\ 10^3 < (Ra_L Pr)/(0.2 + Pr) \end{bmatrix}$$

For large values of the aspect ratio, (H/L), MacGregor and Emery [22] have proposed the following correlations

$$\overline{Nu}_L = 0.42 Ra_L^{1/4} Pr^{0.012} (H/L)^{-0.3} \qquad \begin{bmatrix} 10 < (H/L) < 40 \\ 1 < Pr < 2 \times 10^4 \\ 10^4 < Ra_L < 10^7 \end{bmatrix} \tag{9.40}$$

$$\overline{Nu}_L = 0.046 Ra_L^{1/3} \qquad \begin{bmatrix} 1 < (H/L) < 40 \\ 1 < Pr < 20 \\ 10^6 < Ra_L < 10^9 \end{bmatrix} \tag{9.41}$$

Convection coefficients computed from the foregoing expressions are to be used with Equation 9.36. Again, all properties are evaluated at the mean temperature, $(T_1 + T_2)/2$.

Many publications have appeared in recent years concerning free convection heat transfer in inclined, rectangular spaces, with particular attention given to solar collector applications [23–28]. For large aspect ratios, $(H/L) \gtrsim 12$,

---

Table 9.4 Critical angle for inclined rectangular cavities

(H/L)	1	3	6	12	> 12
τ^*	25°	53°	60°	67°	70°

---

and tilt angles less than the critical value τ^* given in Table 9.4, the following correlation due to Hollands et al. [28] is in excellent agreement with available data

$$\overline{Nu}_L = 1 + 1.44\left[1 - \frac{1708}{Ra_L\cos\tau}\right]^{\cdot}\left[1 - \frac{1708(\sin 1.8\tau)^{1.6}}{Ra_L\cos\tau}\right]$$
$$+ \left[\left(\frac{Ra_L\cos\tau}{5830}\right)^{1/3} - 1\right]^{\cdot}\qquad \begin{bmatrix}(H/L) \gtrsim 12\\ 0 < \tau \leq \tau^*\end{bmatrix} \qquad (9.42)$$

The notation $[\;\;]^{\cdot}$ implies that, if the quantity in brackets is negative, it must be set equal to zero. For small-aspect ratios Catton [19] suggests that reasonable results may be obtained from a correlation of the form

$$\overline{Nu}_L = \overline{Nu}_L(\tau = 0)\left[\frac{\overline{Nu}_L(\tau = 90)}{\overline{Nu}_L(\tau = 0)}\right]^{\tau/\tau^*}(\sin\tau^*)^{(\tau/4\tau^*)}\qquad \begin{bmatrix}(H/L) \lesssim 12\\ 0 < \tau \leq \tau^*\end{bmatrix}\qquad (9.43)$$

Beyond the critical tilt angle, the following correlations due to Ayyaswamy and Catton [23] and Arnold et al. [26], respectively, have been recommended [19] for all aspect ratios, (H/L).

$$\overline{Nu}_L = \overline{Nu}_L(\tau = 90°)(\sin\tau)^{1/4}\qquad (\tau^* < \tau < 90°)\qquad (9.44)$$

$$\overline{Nu}_L = 1 + [\overline{Nu}_L(\tau = 90°) - 1]\sin\tau\quad (90° < \tau < 180°)\qquad (9.45)$$

9.7.2 Concentric Cylinders

Free convection heat transfer in the annular space between *long*, horizontal concentric cylinders (Figure 9.8) has been considered by Raithby and Hollands

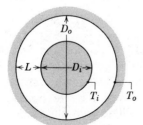

Figure 9.8 Free convection heat transfer in the annular space between long concentric cylinders or concentric spheres.

[29]. The heat transfer rate per unit length of cylinder (W/m) may be expressed as

$$q' = \frac{2\pi k_{eff}}{\ln(D_o/D_i)} (T_i - T_o) \tag{9.46}$$

where the *effective thermal conductivity*, k_{eff}, is the thermal conductivity that a stationary fluid should have to transfer the same amount of heat as the moving fluid. The suggested correlation for k_{eff} is of the form

$$\frac{k_{eff}}{k} = 0.386 \left(\frac{Pr}{0.861 + Pr}\right)^{1/4} (Ra_c^*)^{1/4} \tag{9.47}$$

where

$$Ra_c^* = \frac{[\ln(D_o/D_i)]^4}{L^3(D_i^{-3/5} + D_o^{-3/5})^5} Ra_L \tag{9.48}$$

Equation 9.47 may be used for the range $10^2 \lesssim Ra_c^* \lesssim 10^7$.

9.7.3   Concentric Spheres

Raithby and Hollands [29] have also considered free convection heat transfer between concentric spheres (Figure 9.8) and express the total heat transfer rate as

$$q = k_{eff} \pi (D_i D_o/L)(T_i - T_o) \tag{9.49}$$

The effective thermal conductivity may be expressed as

$$\frac{k_{eff}}{k} = 0.74 \left(\frac{Pr}{0.861 + Pr}\right)^{1/4} (Ra_s^*)^{1/4} \tag{9.50}$$

where

$$Ra_s^* = \left[\frac{L}{(D_o D_i)^4} \frac{Ra_L}{(D_i^{-7/5} + D_o^{-7/5})^5}\right]^{1/4} \tag{9.51}$$

The result may be used to a reasonable approximation for $10^2 \lesssim Ra_s^* \lesssim 10^4$.

EXAMPLE 9.5

A long tube of 0.1 m diameter is maintained at 120°C by the passage of steam through its interior. A radiation shield is installed concentric to the tube with an air gap of 10 mm. If the shield is at 35°C, estimate the heat transfer by free convection from the tube per unit length. What would be the heat loss if the space between the tube and the shield were filled with glass-fiber blanket insulation?

SOLUTION

KNOWN:

Temperatures and diameters of a steam tube and a concentric radiation shield.

FIND:

1. Heat loss per unit length of tube.
2. Heat loss if air space is filled with glass-fiber blanket insulation.

SCHEMATIC:

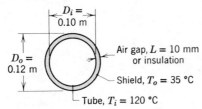

$D_i = 0.10$ m

$D_o = 0.12$ m

Air gap, $L = 10$ mm or insulation

Shield, $T_o = 35$ °C

Tube, $T_i = 120$ °C

ASSUMPTIONS:

1. Radiation heat transfer effects may be neglected.
2. Negligible contact resistance with insulation.

PROPERTIES:

Table A.4, air [$T = (T_i + T_o)/2 = 350$ K]: $k = 0.030$ W/m·K, $\nu = 20.92 \times 10^{-6}$ m^2/s, $\alpha = 29.9 \times 10^{-6}$ m^2/s, $Pr = 0.70$, $\beta = 0.00285$ K^{-1}.
Table A.3, insulation, glass-fiber ($T \approx 300$ K): $k = 0.038$ W/m·K.

ANALYSIS:

1. From Equation 9.46 the heat loss by free convection is

$$q' = \frac{2\pi k_{\text{eff}}}{\ln (D_o/D_i)} (T_i - T_o)$$

where k_{eff} may be obtained from Equations 9.47 and 9.48. With

$$Ra_L = \frac{g\beta(T_i - T_o)L^3}{\nu\alpha}$$

$$Ra_L = \frac{9.8 \text{ m/s}^2 \times 0.00285 \text{ K}^{-1} \times (120 - 35)°\text{C}(0.01)^3}{20.92 \times 10^{-6} \text{ m}^2/\text{s} \times 29.9 \times 10^{-6} \text{ m}^2/\text{s}} = 3795$$

it follows that

$$Ra_c^* = \frac{[\ln(D_o/D_i)]^4 Ra_L}{L^3(D_i^{-3/5} + D_o^{-3/5})^5}$$

$$Ra_c^* = \frac{[\ln(0.12\ m/0.10\ m)]^4 3795}{(0.01\ m)^3[(0.10\ m)^{-3/5} + (0.12\ m)^{-3/5}]^5} = 171$$

Hence

$$\frac{k_{eff}}{k} = 0.386\left(\frac{Pr}{0.861 + Pr}\right)^{1/4}(Ra_c^*)^{1/4}$$

$$\frac{k_{eff}}{k} = 0.386\left(\frac{0.70}{0.861 + 0.70}\right)^{1/4}(171)^{1/4}$$

$$\frac{k_{eff}}{k} = 1.14$$

The effective thermal conductivity is then

$$k_{eff} = 1.14(0.030\ W/m \cdot K)$$

$$k_{eff} = 0.0343\ W/m \cdot K$$

and the heat loss is

$$q' = \frac{2\pi(0.0343\ W/m \cdot K)}{\ln(0.12\ m/0.10\ m)}(120 - 35)°C$$

$$q' = 100\ W/m \qquad \triangleleft$$

2. With insulation in the space between the tube and the shield, heat loss is by conduction and from Equation 3.26

$$q' = \frac{2\pi k(T_i - T_o)}{\ln(D_o/D_i)}$$

$$q' = \frac{2\pi(0.038\ W/m \cdot K)}{\ln(0.12\ m/0.10\ m)}(120 - 35)°C$$

$$q' = 111\ W/m \qquad \triangleleft$$

COMMENTS:

Although there is slightly more heat loss by conduction through the insulation than by free convection across the air space, the total heat loss across the air space may exceed that through the insulation due to the effects of radiation. The heat loss due to radiation may be minimized by using a radiation shield of low emissivity, and the means for calculating the loss will be developed in Chapter 13.

9.8 COMBINED FREE AND FORCED CONVECTION

In dealing with forced convection (Chapters 6–8), we ignored the effects of free convection. This was, of course, an assumption; for, as we now know, free convection will always exist when there is an unstable temperature gradient. Similarly, in the preceding sections of this chapter, we assumed that forced convection effects were negligible. It is now time to acknowledge that situations may arise for which free and forced convection effects are of comparable significance, in which case it is inappropriate to neglect either process. It is therefore useful to know under what conditions free convection is dominant, when the combined effects of free and forced convection must be considered, or when forced convection dominates.

On intuitive grounds we recognize that, when the velocities associated with the forced flow are large compared to those which are generated by buoyancy forces, free convection may be neglected. However, criteria concerning the relative significance of free and forced convection may be more firmly based by resorting to the relevant dimensionless parameters. From Section 9.3 we know that the dimensionless group, (Gr/Re^2), provides a measure of the ratio of buoyancy forces to inertial forces in the flow. This parameter may then be used to delineate the various flow regimes, with the following criteria known to be generally applicable.

$(Gr/Re^2) \gg 1$: "Pure" free convection

$(Gr/Re^2) \approx 1$: Mixed convection

$(Gr/Re^2) \ll 1$: "Pure" forced convection

9.8.1 Vertical and Horizontal Plates

Lloyd and Sparrow [30] have considered the problem of mixed convection for laminar parallel flow on an isothermal vertical plate $(3 \times 10^{-3} \le Pr \le 10^2)$. Their results suggest that the local Nusselt number, Nu_x, may be predicted to within approximately 20 percent accuracy, if the pertinent correlation for pure forced convection, Equation 7.6, is used for $(Gr_x/Re_x^2) \le C_0$ and if the correlation for pure free convection, Equation 9.20, is used for $(Gr_x/Re_x^2) > C_0$. The value of C_0 is approximately 1.0 for $Pr = 100$ and approximately 0.6 for $Pr \lesssim 10$.

Mori [31] considered mixed convection for laminar parallel flow over the lower and upper surfaces of an isothermal horizontal plate. His results indicate that for air the local Nusselt number may be predicted to within 10 percent accuracy with the appropriate forced convection correlation, Equation 7.6, if

$$(Gr_x/Re_x^{2.5}) \le 0.083$$

9.8.2 Horizontal Cylinders and Spheres

Mixed convection heat transfer to or from cylinders and spheres in external flow depends strongly on the relative orientation of the forced convection flow velocity and the gravitational field. Fand and Keswani [32] have reviewed the pertinent literature for horizontal cylinders and have recommended a variety of correlations for the different orientations and flow regimes. Heat transfer results for spheres have been obtained by Yuge [17].

9.8.3 Tube Flow

The various regimes that characterize flow in a horizontal circular tube have been delineated by Metais and Eckert [33]. Convection is mixed when $(Gr_D/Re_D^2) \approx 1$, where Gr_D is defined in terms of $\Delta T = T_s - T_m$ and T_m is the fluid mean temperature. The flow may be assumed to be turbulent if $Re_D \gtrsim 2000$ when $Ra_D(D/L) \lesssim 2 \times 10^4$, or if $Re_D \gtrsim 800$, when $Ra_D \ (D/L) \gtrsim 2 \times 10^4$. For mixed convection in laminar flow, Brown and Gauvin [34] recommend a correlation of the form

$$\overline{Nu}_D = 1.75 \left(\frac{\mu_m}{\mu_s}\right)^{0.14} [Gz + 0.012(GzGr_D^{1/3})^{4/3}]^{1/3} \tag{9.52}$$

where Gz is the *Graetz number*, $Gz \equiv Re_D \ Pr \ (D/L)$. For mixed convection in turbulent flow, Metais and Eckert suggest

$$\overline{Nu}_D = 4.69 Re_D^{0.27} Pr^{0.21} Gr_D^{0.07}(D/L)^{0.36} \tag{9.53}$$

Metais and Eckert also review results for flow through vertical tubes. Turbulent flow will exist if $Re_D \gtrsim 2000$ when $Ra_D(D/L) \lesssim 5000$ or if $Re_D \gtrsim 400$ when $Ra_D(D/L) \gtrsim 5000$.

9.9 CONVECTION MASS TRANSFER

Mass density gradients may exist in a fluid as a consequence of spatial variations in the fluid composition. Such variations exist when species transfer is occurring due to a concentration gradient. Accordingly, under the influence of the gravitational field, free convection flows can be induced in species transfer processes. However, such flows are complicated, and their treatment is beyond the scope of this text. In fact, consideration in the literature is limited.

Another aspect of free convection mass transfer relates to the fact that thermally driven free convection flows can enhance evaporation or sublimation occurring at a surface. For example, evaporation from a horizontal water layer is augmented by the free convection flow that is induced when the temperature of the water exceeds the temperature of the quiescent air above the water. Few empirical correlations have been developed to treat such situations. However, an approximate solution to problems of this kind may be obtained by invoking the heat and mass transfer analogy (Section 6.8.1). Assuming that the species transport process has a negligible influence on the free convection flow, heat transfer correlations, such as those described in the preceding sections, could be used to determine the heat transfer coefficient. From knowledge of this coefficient and use of the analogy given by Equation 6.89, the convection transfer coefficient can then be estimated. It must be emphasized that this solution method is often a gross approximation. In the absence of more definitive results, however, there may be no acceptable alternative to its use.

9.10 SUMMARY

In this chapter we have considered convective flows that originate in part, or exclusively, from buoyancy forces, and we have introduced the dimensionless parameters needed to characterize such flows. You should be able to discern when free convection effects are important and should be able to quantify the associated heat transfer rates. An assortment of empirical correlations has been provided for this purpose.

REFERENCES

1. Ostrach, S., "An Analysis of Laminar Free Convection Flow and Heat Transfer about a Flat Plate Parallel to the Direction of the Generating Body Force," NACA Report 1111, 1953.

2. Sparrow, E. M. and J. L. Gregg, "Laminar Free Convection from a Vertical Flat Plate," *Trans. ASME, 78*, 435, 1956.

3. Gebhart, B., *Heat Transfer*, Chapter 8, 2nd Ed., McGraw-Hill, New York, 1971.

4. McAdams, W. H., *Heat Transmission*, Chapter 7, 3rd Ed., McGraw-Hill, New York, 1954.

5. Warner, C. Y. and V. S. Arpaci, "An Experimental Investigation of Turbulent Natural Convection in Air at Low Pressure for a Vertical Heated Flat Plate," *Int. J. Heat Mass Transfer, 11*, 397, 1968.

6. Bayley, F. J., "An Analysis of Turbulent Free Convection Heat Transfer," *Proc. Inst. Mech. Eng., 169*, 361, 1955.

7. Churchill, S. W. and H. H. S. Chu, "Correlating Equations for Laminar and Turbulent Free Convection from a Vertical Plate," *Int. J. Heat Mass Transfer, 18*, 1323, 1975.

8. Sparrow, E. M. and J. L. Gregg, "Laminar Free Convection Heat Transfer from the Outer Surface of a Vertical Circular Cylinder," *Trans. ASME, 78*, 1823, 1956.

9. Cebeci, T., "Laminar-Free-Convective-Heat Transfer from the Outer Surface of a Vertical Slender Circular Cylinder," *Proc. Fifth Intl. Heat Transfer Conference*, Paper NCl.4, pp. 15–19, 1974.

10. Minkowycz, W. J. and E. M. Sparrow, "Local Nonsimilar Solutions for Natural Convection on a Vertical Cylinder," *J. Heat Transfer, 96*, 178, 1974.

11. Goldstein, R. J., E. M. Sparrow, and D. C. Jones, "Natural Convection Mass Transfer Adjacent to Horizontal Plates," *Int. J. Heat Mass Transfer, 16*, 1025, 1973.

12. Lloyd, J. R. and W. R. Moran, "Natural Convection Adjacent to Horizontal Surfaces of Various Planforms," ASME Paper 74-WA/HT-66, 1974.

13. Vliet, G. C., "Natural Convection Local Heat Transfer on Constant-Heat-Flux Inclined Surfaces," *Trans. ASME, 91C*, 511, 1969.

14. Fujii, T. and H. Imura, "Natural Convection Heat Transfer from a Plate with Arbitrary Inclination," *Int. J. Heat Mass Transfer, 15*, 755, 1972.

15. Morgan, V. T., "The Overall Convective Heat Transfer from Smooth Circular Cylinders" in T. F. Irvine and J. P. Hartnett, Eds., *Advances in Heat Transfer*, Vol. 11, Academic Press, New York, 1975, pp. 199–264.

16. Churchill, S. W. and H. H. S. Chu, "Correlating Equations for Laminar and Turbulent Free Convection from a Horizontal Cylinder," *Int. J. Heat Mass Transfer, 18*, 1049, 1975.

17. Yuge, T., "Experiments on Heat Transfer from Spheres Including Combined Natural and Forced Convection," *J. Heat Transfer, 82*, 214, 1960.

18. Ostrach, S., "Natural Convection in Enclosures" in J. P. Hartnett and T. F. Irvine, Eds., *Advances in Heat Transfer*, Vol. 8, Academic Press, New York, 1972, pp. 161–227.

19. Catton, I., "Natural Convection in Enclosures," *Proc. 6th International Heat Transfer Conference, 6*, 13, Toronto, Canada, 1978.

20. Globe, S. and D. Dropkin, "Natural Convection Heat Transfer in Liquids Confined between Two Horizontal Plates," *J. Heat Transfer, 81C*, 24, 1959.

21. Hollands, K. G. T., G. D. Raithby, and L. Konicek, "Correlation Equations for Free Convection Heat Transfer in Horizontal Layers of Air and Water," *Int. J. Heat Mass Transfer, 18*, 879, 1975.

22. MacGregor, R. K. and A. P. Emery, "Free Convection through Vertical Plane Layers: Moderate and High Prandtl Number Fluids," *J. Heat Transfer, 91*, 391, 1969.

23. Ayyaswamy, P. S. and I. Catton, "The Boundary-Layer Regime for Natural Convection in a Differentially Heated, Tilted Rectangular Cavity," *J. Heat Transfer, 95*, 543, 1973.

24. Catton, I., P. S. Ayyaswamy, and R. M. Clever, "Natural Convection Flow in a Finite, Rectangular Slot Arbitrarily Oriented with Respect to the Gravity Vector," *Int. J. Heat Mass Transfer, 17*, 173, 1974.

25. Clever, R. M., "Finite Amplitude Longitudinal Convection Rolls in an Inclined Layer," *J. Heat Transfer, 95*, 407, 1973.

26. Arnold, J. N., I. Catton, and D. K. Edwards, "Experimental Investigation of Natural Convection in Inclined Rectangular Regions of Differing Aspect Ratios," ASME Paper 75-HT-62, 1975.

27. Buchberg, H., I. Catton, and D. K. Edwards, "Natural Convection in Enclosed Spaces — A Review of Application to Solar Energy," *J. Heat Transfer, 98*, 182, 1976.

28. Hollands, K. G. T., S. E. Unny, G. D. Raithby, and L. Konicek, "Free Convective Heat Transfer across Inclined Air Layers," *J. Heat Transfer, 98,* 189, 1976.

29. Raithby, G. D. and K. G. T. Hollands, "A General Method of Obtaining Approximate Solutions to Laminar and Turbulent Free Convection Problems" in T. F. Irvine and J. P. Hartnett, Eds., *Advances in Heat Transfer,* Vol. 11, Academic Press, New York, 1975, pp. 265–315.

30. Lloyd, J. R. and E. M. Sparrow, "Combined Forced and Free Convection on Vertical Surfaces," *Int. J. Heat Mass Transfer, 13,* 434, 1970.

31. Mori, Y., "Buoyancy Effects in Forced Laminar Convection Flow Over a Horizontal Flat Plate," *J. Heat Transfer, 83,* 479, 1961.

32. Fand, R. M. and K. K. Keswani, "Combined Natural and Forced Convection Heat Transfer from Horizontal Cylinders to Water," *Int. J. Heat Mass Transfer, 16,* 1175, 1973.

33. Metais, B. and E. R. G. Eckert, "Forced, Mixed and Free Convection Regimes," *J. Heat Transfer, 86,* 295, 1964.

34. Brown, C. K. and W. H. Gauvin, "Combined Free and Forced Convection, I and II," *Can. J. Chem. Eng., 43,* 306, 1965.

PROBLEMS

9.1 Using the tabulated values of density for water in Table A.6, calculate the volumetric thermal expansion coefficient at 300 K from its definition, Equation 9.4, and compare your result with the tabulated value.

9.2 The heat transfer rate due to free convection from a vertical surface, 1 m high and 0.6 m wide, to quiescent air that is 20 K colder than the surface is known. What is the ratio of the heat transfer rate for that situation to the rate corresponding to a vertical surface, 0.6 m high and 1 m wide, when the quiescent air is 20 K warmer than the surface. Neglect heat transfer by radiation and any influence of temperature on the relevant thermophysical properties of air.

9.3 The heat transfer rate per unit length due to free convection from a horizontal tube is 200 W/m when its surface is maintained at 65°C and the ambient air is 25°C. Estimate the heat transfer rate per unit length when the tube surface is maintained at 145°C. Neglect heat transfer by radiation and any influence of temperature on the relevant thermophysical properties of air.

9.4 Consider a large vertical plate with a uniform surface temperature of 130°C suspended in quiescent air at 25°C and atmospheric pressure.

a) Estimate the boundary layer thickness at a location 0.25 m measured from the lower edge.

b) What is the maximum velocity in the boundary layer at this location and at what position in the boundary layer does the maximum occur?

c) Using the similarity solution result, Equation 9.20, determine the heat transfer coefficient for the location 0.25 m measured from the lower edge.

d) At what location on the plate measured from the lower edge will the boundary layer become turbulent?

9.5   A number of thin plates are to be cooled by vertically suspending them in a water bath at a temperature of 20°C. If the plates are initially at 54°C and have a length of 0.15 m, what is the minimum spacing permitted such that no interference will occur between their free convection boundary layers?

9.6   Determine the average convection heat transfer coefficient for the 2.5-m-high vertical walls of a home having interior air and wall surface temperatures of (a) 20°C and 10°C, respectively, and (b) 27°C and 37°C, respectively.

9.7   Beginning with the free convection correlation of the form given by Equation 9.24, show that for air at atmospheric pressure and a film temperature of 400 K, the average heat transfer coefficient for a vertical plate can be expressed as

$$\bar{h}_L = 1.40 \left(\frac{\Delta T}{L}\right)^{1/4} \qquad 10^4 < Ra_L < 10^9$$

$$\bar{h}_L = 0.98 \Delta T^{1/3} \qquad 10^9 < Ra_L < 10^{13}$$

9.8   A household oven door of 0.5 m height and 0.7 m width reaches an average surface temperature of 32°C during operation. Estimate the heat loss to the room with ambient air and surroundings at 22°C.

9.9   An aluminum alloy (2024) plate, heated to a uniform temperature of 227°C, is allowed to cool while vertically suspended in a room where the ambient air and surroundings are at 27°C. The plate is 0.3 m square with a thickness of 15 mm and has an emissivity of 0.25.

a) Develop an expression for the time rate of change of the plate temperature assuming the temperature to be uniform at any time.

b) Determine the initial rate of cooling (K/s) when the plate temperature is 227°C.

c) Justify the uniform plate temperature assumption.

9.10   Air flow through a long, 0.2 m square air conditioning duct maintains the outer duct surface temperature at 10°C. If the duct is uninsulated and exposed to air at 35°C in the crawl space beneath a home, what is the heat gain to the duct per unit length of the duct?

9.11   Air at 3 atm and 100°C is discharged from a compressor into a vertical receiver of height 2.5 m and diameter 0.75 m. Estimate the receiver wall temperature and the heat transfer to the ambient air at 25°C. Neglect radiation effects and heat losses from the ends. Assume that the receiver walls are at a uniform surface temperature.

9.12   An electrical heater in the form of a horizontal disc of 400 mm diameter is used to heat a tank filled with engine oil at a temperature of 5°C. Calculate the power required to maintain the heater surface temperature at 70°C.

9.13   Consider a horizontal straight fin fabricated from plain carbon steel with a thickness 6 mm and length 100 mm. Estimate the fin heat rate per unit width when the base is maintained at 150°C and the surrounding atmospheric air is quiescent and at 25°C.

9.14   A circular grill of diameter 0.25 m and emissivity 0.9 is maintained at a constant surface temperature of 130°C. What electrical power is required when the room air and surroundings are at 24°C?

9.15 A plate 1 m × 1 m, inclined at an angle of 45° with the horizontal, is exposed to the environment which provides a net radiation heat flux of 300 W/m². If the backside of the plate is well insulated, estimate the temperature the plate reaches when the ambient air is still and at a temperature of 0°C.

9.16 A horizontal rod 5 mm in diameter is immersed in water maintained at 18°C. If the rod surface temperature is 56°C, estimate the free convection heat transfer rate per unit length of the rod.

9.17 A horizontal uninsulated steam pipe passes through a large room whose walls and ambient air are at 300 K. The pipe of 150 mm diameter has an emissivity of 0.85 and an outer surface temperature of 400 K. Calculate the heat loss per unit length from the pipe.

9.18 Beer in cans of length 150 mm and diameter 60 mm is initially at 27°C and is to be cooled by placement in a refrigerator compartment at 4°C. In the interest of maximizing the cooling rate, should the cans be laid horizontally or vertically in the compartment? As a first approximation, neglect heat transfer from the ends.

9.19 A horizontal pipe of 70 mm outside diameter passes through a room in which the temperature of the air and the walls is 25°C. The outer surface has an emissivity of 0.8 and is maintained at 150°C by steam passing through the pipe. What is the heat loss per unit length of the pipe?

9.20 A horizontal tube of 12.5 mm diameter with an outer surface temperature of 240°C is located in a room with an air temperature of 20°C. Estimate the heat transfer rate per unit length of the tube due to free convection.

9.21 Saturated steam at 3 bars with a mean velocity of 3 m/s flows through a tube whose inner and outer diameters are 55 mm and 60 mm, respectively. The heat transfer coefficient for the steam flow is known to be 11,000 W/m²·K. The tube is covered with a 25-mm-thick layer of 85 percent magnesia insulation and is exposed to atmospheric air at 25°C.

   a) Estimate the heat transfer rate by free convection to the room per unit length of the tube.

   b) If the steam is saturated at the inlet of the tube, estimate its quality at the outlet of a tube 30 m long.

9.22 A sphere of 25 mm diameter contains an imbedded electrical heater. Calculate the power required to maintain the surface temperature at 94°C when the sphere is exposed to a quiescent medium at 20°C for these fluids: (a) air at atmospheric pressure, (b) water, and (c) ethylene glycol.

9.23 Under steady-state operation the surface temperature of a small 20 W incandescent light bulb is 125°C when the temperature of the room air and walls is 25°C. Approximating the bulb as a sphere 40 mm in diameter with a surface emissivity of 0.8, what is the rate of heat transfer from the surface of the bulb to the surroundings?

9.24 A building window pane that has a height of 1.2 m and a width of 0.8 m is separated from the ambient air by a storm window of the same height and width. The thickness of the air space between the two windows is 0.06 m. If the building and storm windows are at temperatures of 20°C and −10°C, respectively, what is the rate of heat loss by free convection across the air space?

9.25 The absorber plate and the adjoining cover plate of a flat plate solar collector are at temperatures of 70°C and 35°C, respectively, and are separated by an air space of 0.05 m. What is the rate of free convection heat transfer per unit surface area between the two plates if they are inclined at an angle of 60° from the horizontal?

9.26 In a solar collector of 1 m width × 3 m length, the glass cover plate at an average temperature of 29°C is spaced 20 mm from the absorber plate at an average temperature of 50°C.

a) Estimate the convection heat loss from the absorber plate to the glass when the collector is positioned horizontally.

b) What is the convection heat loss if the spacing is reduced to 10 mm?

9.27 A rectangular cavity consists of two parallel, 0.5 m square plates separated by a distance of 50 mm, with the lateral boundaries insulated. The heated plate is maintained at 325 K and the cooled plate at 275 K. Estimate the heat flux between the surfaces for these three orientations of the cavity using the notation of Figure 9.7: (a) vertical, $\tau = 90°$, (b) horizontal with $\tau = 0°$, and (c) horizontal with $\tau = 180°$.

9.28 Consider the rectangular cavity of Problem 9.27. Determine the heat flux between the two surfaces for tilt angles of (a) 45° and (b) 75°.

9.29 A solar collector design consists of an inner tube enclosed concentrically within an outer tube which is transparent to solar radiation. The tubes are thin walled with inner and outer diameters of 0.10 m and 0.15 m, respectively. The annular space between the tubes is completely enclosed and filled with air at atmospheric pressure. Under operating conditions for which the inner and outer tube surface temperatures are 70°C and 30°C, respectively, what is the convective heat loss per meter of tube length across the air space?

9.30 The annulus formed by two concentric, horizontal tubes with inner and outer diameters of 50 mm and 75 mm, respectively, is filled with water. If the inner and outer surfaces are maintained at 300 K and 350 K, respectively, estimate the convection heat transfer rate per unit length of the tubes.

9.31 The surfaces of two long, horizontal, concentric thin-walled tubes having radii of 100 mm and 125 mm are maintained at 300 K and 400 K, respectively. If the annular space is pressurized with nitrogen at 5 atm, estimate the convection heat transfer rate per unit length of the tubes.

9.32 The surfaces of two concentric spheres having radii of 75 mm and 100 mm are maintained at 325 K and 275 K, respectively. If the space between the spheres is filled with air at 3 atm, estimate the convection heat transfer rate.

9.33 Water at a temperature 17°C and a mass rate of 0.012 kg/s is to be used for maintaining a small plate (on which a special sensor is to be mounted) at a fixed temperature. The plate is situated within a hot air environment at a temperature of 235°C. The tube is horizontal and 1 m in length. Fabricated from a plastic with a thermal conductivity of 0.05 W/m·K, the tube has an inner diameter, $D_i = 1.4$ mm, and an outer diameter, $D_o = 3.2$ mm.

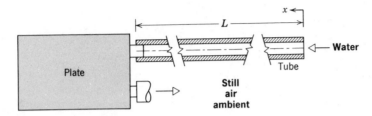

a) Assuming that the average outer surface temperature of the tube is 120°C, estimate the heat transfer coefficient for free convection between the tube and the ambient air.

b) Assuming the flow and thermal conditions within the tube are fully developed, estimate the heat transfer coefficient between the tube and the water.

c) Determine the overall heat transfer coefficient based upon the outer tube area, U_o.

d) Estimate the temperature of the water at the end of the tube ($x = 1$ m) where it is connected to the plate.

9.34 Water at 35°C with a velocity of 0.05 m/s flows over a horizontal, 50 mm diameter cylinder maintained at a uniform surface temperature of 20°C. Do you anticipate that heat transfer by free convection will be significant? What would be the situation if the fluid were air at atmospheric pressure?

9.35 According to experimental results for parallel air flow over a uniform temperature, heated vertical plate, the effect of free convection on the heat transfer convection coefficient will be 5 percent when $Gr_L/Re_L^2 = 0.08$. Consider a heated vertical plate of length 0.3 m maintained at a surface temperature of 60°C in atmospheric air at 25°C. What is the minimum vertical velocity required of the air flow such that free convection effects will be less than 5 percent of the heat transfer rate?

9.36 A wet garment, soaked through with water, is hung up to dry in a warm room for which the still air is dry and at a temperature of 40°C. The garment may be assumed to have a temperature of 25°C and a characteristic length of 1 m in the vertical direction. Estimate the drying rate per unit width of the garment.

10 Boiling and Condensation

In this chapter we focus on convection processes associated with the change in phase of a fluid. We are already equipped to handle such processes when they occur as evaporation at a liquid–gas interface or as sublimation at a solid–gas interface. In particular they are treated by using the methods of convection mass transfer. However, we have not yet considered important processes that can occur at a solid–liquid interface, that is, *boiling* and *condensation*. For these cases *latent* heat effects associated with the phase change are significant. The change from the liquid to the vapor state due to boiling is sustained by heat transfer from the solid surface; conversely, condensation of a vapor to the liquid state results in heat transfer to the solid surface.

Since they involve fluid motion, boiling and condensation are classified as forms of the convection mode of heat transfer. However, they are characterized by unique features. Because there is a phase change, heat transfer to or from the fluid can occur without influencing the fluid temperature. Moreover, heat transfer coefficients and rates are generally much larger than those characteristic of convection heat transfer without phase change. In fact, through boiling or condensation, high heat transfer rates may be achieved with small temperature differences.

There are many engineering problems that involve boiling and condensation. For example, both processes are essential to all closed-loop power and refrigeration cycles. In a power cycle, pressurized liquid is converted to vapor in a *boiler*. After expansion in a turbine, the vapor is restored to its liquid state in a *condenser*, whereupon it is pumped to the boiler to repeat the cycle. Evaporators, in which the boiling process occurs, and condensers are also essential components in vapor-compression refrigeration cycles. The rational design of such components dictates that the associated phase change processes be well understood.

10.1 BOILING: PHYSICAL MECHANISMS

When evaporation occurs at a solid–liquid interface, it is termed *boiling*. The process occurs when the temperature of the surface, T_s, exceeds the saturation temperature corresponding to the liquid pressure, T_{sat}. Heat is transferred from the solid surface to the liquid, and the appropriate form of Newton's law of cooling is

$$q_s'' = h(T_s - T_{sat}) = h\Delta T_e \tag{10.1}$$

where $\Delta T_e \equiv T_s - T_{sat}$ is termed the *excess temperature*. The process is characterized by the formation of vapor bubbles, which grow and subsequently detach from the surface. Vapor bubble growth and dynamics depend, in a complicated manner, on the excess temperature, the nature of the surface and on thermophysical properties of the fluid, such as its surface tension. In turn, the dynamics of vapor bubble formation affect fluid motion near the surface and therefore strongly influence the heat transfer coefficient. Detailed description of these effects is beyond the scope of this text and is left to the literature [1, 2].

Boiling may occur under various conditions. For example, the term *pool boiling* refers to a situation in which the liquid is quiescent and its motion near the surface is due to free convection and to mixing induced by bubble growth and detachment. In contrast, for *forced-convection boiling*, fluid motion is induced by external means, as well as by free convection and bubble-induced mixing. Boiling may also be classified according to whether it is *subcooled* or *saturated*. In subcooled (or local) boiling, the temperature of the liquid is below the saturation temperature and bubbles formed at the surface eventually condense in the liquid. In contrast the temperature of the liquid exceeds the saturation temperature in *saturated boiling*. Bubbles formed at the surface are then propelled through the liquid by buoyancy forces, eventually escaping from a free surface.

Saturated pool boiling conditions have been studied extensively (Figure 10.1). Although there is a sharp decline in the liquid temperature close to the solid surface, the temperature through most of the liquid remains slightly above saturation. Bubbles generated at the liquid–solid interface therefore rise to and are transported across the liquid–vapor interface. An appreciation for the underlying physical mechanisms may be obtained by examining the *boiling curve* of Figure 10.2. The specific curve pertains to water, although similar trends characterize the behavior of other fluids. The curve may be generated by performing an experiment in which the surface temperature, T_s, is varied and the resulting heat flux, q_s'', is measured. From Equation 10.1 we note that q_s'' depends on the convection coefficient, h, as well as on the excess temperature, ΔT_e. Different boiling regimes may be delineated according to the value of ΔT_e.

Free convection boiling is said to exist if $\Delta T_e \leq \Delta T_{e,A}$, where $\Delta T_{e,A} \approx 5°C$. In

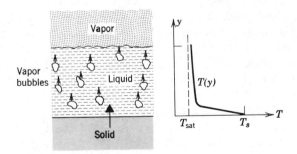

Figure 10.1 Temperature distribution in saturated pool boiling with a liquid–vapor interface.

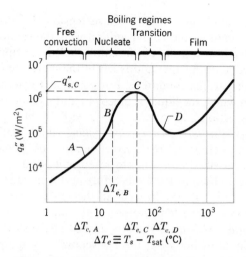

Figure 10.2 The boiling curve for water: surface heat flux, q_s'', as a function of excess temperature, $\Delta T_e \equiv T_s - T_{sat}$.

this regime bubble formation is restricted to isolated spots on the surface, and fluid motion is determined principally by free convection effects. According to whether laminar or turbulent flow conditions exist, h therefore varies as ΔT_e to the $\frac{1}{4}$ or $\frac{1}{3}$ power, respectively, in which case q_s'' varies as ΔT_e to the $\frac{5}{4}$ or $\frac{4}{3}$ power. Note, however, that the magnitude of the convection coefficient is much larger than that associated with free convection without phase change.

Nucleate boiling exists in the range, $\Delta T_{e,A} \leq \Delta T_e \leq \Delta T_{e,C}$, where $\Delta T_{e,C} \approx 50°C$. In this range the surface becomes densely populated with bubbles, and bubble separation induces considerable fluid mixing near the surface, substantially increasing h and q_s''. In this situation most of the heat exchange is through direct transfer from the surface to liquid in motion at the surface and not through the vapor bubbles rising from the surface. Point B corresponds to an inflection point in the boiling curve. For $\Delta T_e < \Delta T_{e,B}$, h varies as ΔT_e to the second or third power, with q_s'' varying as ΔT_e to the third or fourth power. However, for $\Delta T_e > \Delta T_{e,B}$, increased bubble formation causes coalescence and interference between bubbles, which inhibits the motion of liquid near the surface. At this point h begins to decrease with increasing ΔT_e, although q_s'', which is the product of h and ΔT_e, continues to increase. This trend results from the fact that, for $\Delta T_e > \Delta T_{e,B}$, the relative increase in ΔT_e exceeds the relative reduction in h. At point C, however, further increase in ΔT_e is exactly balanced by the reduction in h. The maximum heat flux, $q_{s,C}''$, is usually termed the *critical heat flux* and is generally in excess of 1 MW/m². '

Because high heat transfer rates and convection coefficients are associated

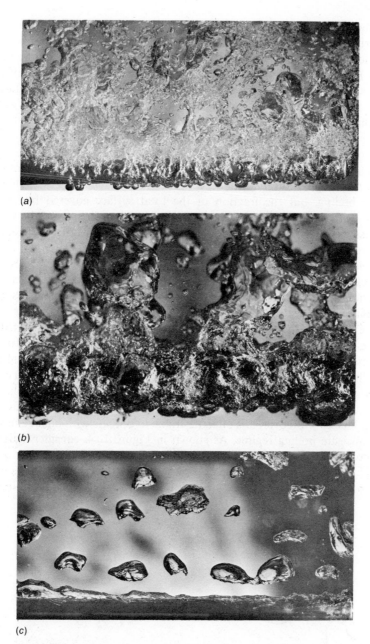

Figure 10.3 Boiling of methanol on a horizontal tube. (*a*) Nucleate boiling. (*b*) Transition boiling. (*c*) Film boiling. Photographs courtesy of Professor J. W. Westwater, University of Illinois at Champaign-Urbana.

with small values of the excess temperature, it is desirable to operate many engineering devices in the nucleate boiling regime. The magnitude of the convection coefficient may be inferred by using Equation 10.1 with the boiling curve of Figure 10.2. Dividing q_s'' by ΔT_e, it is evident that convection coefficients in excess of 10^4 W/m²·K are characteristic of this regime. These values are considerably larger than those normally corresponding to convection with no phase change.

The region corresponding to $\Delta T_{e,C} \leq \Delta T_e \leq \Delta T_{e,D}$, where $\Delta T_{e,D} \approx 150°C$, is termed *transition boiling*, *unstable film boiling*, or *partial film boiling*. Bubble formation is now so rapid that a vapor film or blanket begins to form on the surface. At any point on the surface, conditions may oscillate between film and nucleate boiling, but the fraction of the total surface covered by the film increases with increasing ΔT_e. Because the thermal conductivity of the vapor is much less than that of the liquid, the value of h (and q_s'') must then decrease with increasing ΔT_e.

Film boiling exists for $\Delta T_e \geq \Delta T_{e,D}$. The surface is completely covered by a vapor blanket, and heat transfer from the surface to the liquid occurs by conduction and radiation through the vapor. From the minimum associated with point D, the heat flux, q_s'', increases with increasing ΔT_e.

The nature of the vapor formation and bubble dynamics associated with nucleate boiling, transition boiling, and film boiling is shown in Figure 10.3. The photographs were obtained for the boiling of methanol on a horizontal tube.

Although the foregoing discussion of the boiling curve assumes that control may be maintained over T_s, it is important to note that many applications involve controlling q_s'' (e.g., in a nuclear reactor or in an electric resistance heating device). Moreover, it is often desirable to operate with high values of q_s'' in the nucleate boiling regime. As shown in Figure 10.4, consider starting at some point P in the nucleate boiling regime and gradually increasing q_s''. The value of ΔT_e, and hence the value of T_s, will also increase, following the boiling curve to point C. However, any increase in q_s'' beyond this point will induce a

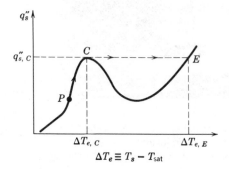

Figure 10.4   Onset of the boiling crisis.

departure from the boiling curve in which surface conditions change abruptly from $\Delta T_{e,C}$ to $\Delta T_{e,E} \equiv T_{s,E} - T_{sat}$. Because $T_{s,E}$ generally exceeds the melting point of the solid, destruction or failure of the system may occur. For this reason point C is often termed the *burnout point* or the *boiling crisis*, and accurate knowledge of the critical flux, $q''_{s,C}$, is important. We may want to operate a heat transfer surface close to this value, but rarely would we want to exceed it.

10.2 BOILING CORRELATIONS

From the shape of the boiling curve and the fact that various physical mechanisms characterize the different regimes, it is no surprise that a multiplicity of heat transfer correlations exist for the boiling process. In this section we review some of the more widely used correlations for nucleate and film boiling.

10.2.1 Nucleate Pool Boiling

The nucleate pool boiling regime characterizes many applications, and the recommended correlation is of the form [3]

$$q''_s = \mu_l h_{fg} \left[\frac{g(\rho_l - \rho_v)}{g_c \sigma} \right]^{1/2} \left(\frac{c_{p,l} \Delta T_e}{C_{s,f} h_{fg} Pr_l^{1.7}} \right)^3 \tag{10.2}$$

where the subscripts l and v, respectively, denote the saturated liquid and vapor states. The latent heat of vaporization is h_{fg} (J/kg), and the surface tension of the liquid is σ(N/m). The appearance of the surface tension follows from the significant effect which this fluid property has on bubble formation and development. The quantity g_c is the proportionality constant discussed in Section 1.6 (1 kg·m/N·s²), and g is the local gravitational acceleration (m/s²). The coefficient $C_{s,f}$ depends on the surface-liquid combination and typically may vary from 0.0027 to 0.014. Values for representative surfaces and water are presented in Table 10.1, and values for other liquid-surface combinations may be obtained from the literature [4–6]. Values of the surface tension, as well as the latent heat of vaporization, for water are presented in Table A.6. Values for other liquids may be obtained from any recent edition of the *Handbook of Chemistry and Physics*. For clean surfaces submerged in single-component liquids, Equation 10.2 may be used to compute the heat flux from knowledge of the temperature excess, or vice versa. Note that q''_s is proportional to the cube of ΔT_e.

A simple expression has been recommended for the convection coefficient associated with free convection and nucleate boiling in *water* [8]. Accounting for the effects of pressure, it is of the form

Table 10.1   Values of $C_{s,f}$ for various water surface combinations [5–7]

SURFACE	$C_{s,f}$
Copper	
Scored	0.0068
Polished	0.0130
Stainless Steel	
Chemically etched or	
mechanically polished	0.0130
Ground and polished	0.0080
Brass	0.0060
Platinum	0.0130

Table 10.2   Values of C and n for use with Equation 10.3 [8]

SURFACE	$q_s''(\text{kW/m}^2)$	C	n
Horizontal	$q_s'' < 15.8$	1040	1/3
	$15.8 < q_s'' < 236$	5.56	3
Vertical	$q_s'' < 3.15$	539	1/7
	$3.15 < q_s'' < 63.1$	7.95	3

$$h = C(\Delta T_e)^n \left(\frac{p}{p_a}\right)^{0.4} \tag{10.3}$$

where p and p_a are the system pressure and standard atmospheric pressure, respectively. Values of the coefficient C and the exponent n are given in Table 10.2 for h in units of W/m²·K and ΔT_e in units of K.

10.2.2   Critical Heat Flux for Nucleate Pool Boiling

We recognize that the critical heat flux, $q_{s,C}''$, represents an important point on the boiling curve. We may wish to operate a boiling process close to this point, but we appreciate the danger of dissipating heat in excess of this amount. Zuber [9] has theoretically obtained an expression of the form

$$q_{s,C}'' = \frac{\pi}{24} h_{fg}\rho_v \left[\frac{\sigma g g_c(\rho_l - \rho_v)}{\rho_v^2}\right]^{1/4}\left(\frac{\rho_l + \rho_v}{\rho_l}\right)^{1/2} \tag{10.4}$$

which, as a first approximation, is independent of surface material and geometry. Replacing the constant $(\pi/24) = 0.131$ by 0.18 to obtain better agreement with experiment and approximating the last term in parentheses by unity, Equation 10.4 may then be expressed as

$$q''_{s,c} = 0.18 h_{fg} \rho_v \left[\frac{\sigma g g_c (\rho_l - \rho_v)}{\rho_v^2} \right]^{1/4} \tag{10.5}$$

EXAMPLE 10.1

The bottom of a copper pan, 0.3 m in diameter, is maintained at 118°C by an electric heater.

1. Estimate the power required to boil water in this pan.
2. What is the rate at which water will evaporate from the pan due to the boiling process?
3. Estimate the critical heat flux for these conditions.

SOLUTION

KNOWN:

Water boiling in a copper pan with known bottom surface temperature.

FIND:

1. Power required by electric heater to cause boiling.
2. Rate of water evaporation due to boiling.
3. Critical heat flux corresponding to the burnout point.

SCHEMATIC:

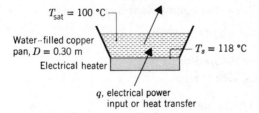

$T_{sat} = 100$ °C

Water-filled copper pan, $D = 0.30$ m

$T_s = 118$ °C

Electrical heater

q, electrical power input or heat transfer

ASSUMPTIONS:

1. Steady-state conditions.
2. Water is exposed to standard atmospheric pressure, 1.01 bar.

3. Water is at uniform temperature $T_{sat} = 100°C$.
4. Pan bottom surface is polished copper.

PROPERTIES:

Table A.6, saturated water, liquid (100°C): $\rho_l = 1/v_f = 957.9$ kg/m³, $c_{p,l}$
$= c_{p,f} = 4.217$ kJ/kg·K, $\mu_l = \mu_f = 279 \times 10^{-6}$ N·s/m², $Pr_l = Pr_f = 1.76$, h_{fg}
$= 2257$ kJ/kg, $\sigma = 58.9 \times 10^{-3}$ N/m.
Table A.6, saturated water, vapor (100°C): $\rho_v = 1/v_g = 0.5955$ kg/m³.

ANALYSIS:

1. From knowledge of the saturation temperature, T_{sat}, of water boiling
   at 1 atm and the temperature of the heated copper surface, T_s, the
   excess temperature, ΔT_e, is

$$\Delta T_e \equiv T_s - T_{sat} = 118°C - 100°C = 18°C$$

According to the boiling curve of Figure 10.2, nucleate pool boiling will
occur and the recommended correlation for estimating the heat transfer
rate per unit area of plate surface is given by Equation 10.2.

$$q_s'' = \mu_l h_{fg} \left[\frac{g(\rho_l - \rho_v)}{g_c \sigma} \right]^{1/2} \left(\frac{c_{p,l} \Delta T_e}{C_{s,f} h_{fg} Pr_l^{1.7}} \right)^3$$

The coefficient, $C_{s,f}$, corresponding to the polished copper surface-water
combination is determined from the experimental results of Table 10.1,
where

$$C_{s,f} = 0.0130$$

Substituting numerical values, the boiling heat flux rate is

$$q_s'' = 279 \times 10^{-6} \text{ N·s/m}^2 \times 2257 \times 10^3 \text{ J/kg}$$

$$\times \left[\frac{9.8 \text{ m/s}^2 \, (957.9 - 0.5955) \text{kg/m}^3}{1 \text{ kg·m/N·s}^2 \times 58.9 \times 10^{-3} \text{ N/m}} \right]^{1/2}$$

$$\times \left(\frac{4.217 \times 10^3 \text{ J/kg·K} \times 18°C}{0.0130 \times 2257 \times 10^3 \text{ J/kg} \times 1.76^{1.7}} \right)^3$$

$$q_s'' = 243.5 \text{ kW/m}^2$$

Hence the boiling heat transfer rate for the given pan area is

$$q_s = q_s'' \times A = q_s'' \times \frac{\pi D^2}{4}$$

$$q_s = 2.435 \times 10^5 \text{ W/m}^2 \times \frac{\pi(0.30 \text{ m})^2}{4}$$

$$q_s = 17.2 \text{ kW} \qquad \triangleleft$$

2. Under steady-state conditions all heat addition to the pan will result in water evaporation from the pan. Hence

$$q_s = \dot{m}_b h_{fg}$$

where $\dot{m}_b$ is the rate at which water vapor evaporates from the free surface to the room. It follows that

$$\dot{m}_b = \frac{q_s}{h_{fg}} = \frac{1.72 \times 10^4 \text{ W}}{2257 \times 10^3 \text{ J/kg}}$$

$$\dot{m}_b = 7.621 \times 10^{-3} \text{ kg/s} = 27.4 \text{ kg/h} \qquad \triangleleft$$

3. The critical heat flux for nucleate pool boiling can be estimated from the correlation, Equation 10.5.

$$q_{s,C}'' = 0.18 h_{fg} \rho_v \left[\frac{\sigma g g_c (\rho_l - \rho_v)}{\rho_v^2} \right]^{1/4}$$

Substituting the appropriate numerical values

$$q_{s,C}'' = 0.18 \times 2257 \times 10^3 \text{ J/kg} \times 0.5955 \text{ kg/m}^3$$

$$\times \left[\frac{58.9 \times 10^{-3} \text{ N/m} \times 9.8 \text{ m/s}^2 \times 1 \text{ kg} \cdot \text{m/N} \cdot \text{s}^2 (957.9 - 0.5955) \text{ kg/m}^3}{(0.5955)^2 (\text{kg/m}^3)^2} \right]^{1/4}$$

$$q_{s,C}'' = 1.52 \text{ MW/m}^2 \qquad \triangleleft$$

COMMENTS:

1. Note that the critical heat flux, $q_{s,C}'' = 1.52 \text{ MW/m}^2$, represents the maximum heat flux for boiling water at normal atmospheric pressure. Operation of the heater at $q_s'' = 0.2435 \text{ MW/m}^2$ is therefore well below the critical condition.

2. Note that, since surface shape does not have a significant influence on bubble formation and development, Equations 10.2 and 10.5 for nucleate pool boiling may be applied to different geometries, such as plates, wires, cylinders, etc.

10.2.3 Film Pool Boiling

Heat transfer in stable film boiling is due to both convection and radiation. Bromley [10] investigated film boiling from the outer surface of horizontal tubes and suggested calculating the total heat transfer coefficient from a transcendental equation of the form

$$h^{4/3} = h_{conv}^{4/3} + h_{rad} h^{1/3} \tag{10.6}$$

The convection coefficient, h_{conv}, may be obtained from

$$h_{conv} = 0.62 \left[\frac{k_v^3 \rho_v (\rho_l - \rho_v) g (h_{fg} + 0.4 c_{p,v} \Delta T_e)}{\mu_v D \Delta T_e} \right]^{1/4} \tag{10.7}$$

where D is the tube outer diameter and vapor properties are evaluated at the film temperature, $T_f = (T_s + T_{sat})/2$. The effective radiation coefficient, h_{rad}, may be approximated as

$$h_{rad} = \frac{5.67 \times 10^{-8} \varepsilon (T_s^4 - T_{sat}^4)}{T_s - T_{sat}} \tag{10.8}$$

where ε is the emissivity of the solid (Table A.11) and h_{rad} is in $W/m^2 \cdot K$. Although the above expressions correlate much of the existing data, they should be used with caution, particularly for small tube diameters.

EXAMPLE 10.2

A metal-clad heating element of 6 mm diameter and emissivity $\varepsilon = 1$ is horizontally immersed in a water bath. The surface temperature of the metal is 255°C under steady-state boiling conditions. Estimate the power dissipation per unit length for the heater.

SOLUTION

KNOWN:

Boiling from the outer surface of a horizontal cylindrical heater in a water bath.

FIND:

Power dissipation per unit length for the heater, q_s'.

SCHEMATIC:

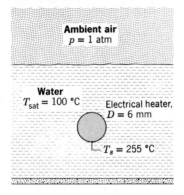

ASSUMPTIONS:

1. Steady-state conditions.
2. Water is exposed to standard atmospheric pressure and is at a uniform temperature, T_{sat}.

PROPERTIES:

Table A.6, saturated water, liquid (100°C): $\rho_l = 1/v_f = 957.9$ kg/m^3, $h_{fg} = 2257$ kJ/kg.
Table A.6, saturated water, vapor (450 K): $\rho_v = 1/v_g = 4.808$ kg/m^3, $c_{p,v} = c_{p,g} = 2.56$ kJ/kg·K, $k_v = k_g = 0.0331$ W/m·K, $\mu_v = \mu_g = 14.85 \times 10^{-6}$ N·s/m^2.

ANALYSIS:

The excess temperature for this boiling situation is given as

$$\Delta T_e = T_s - T_{sat}$$

$$\Delta T_e = 255°C - 100°C = 155°C$$

According to the boiling curve of Figure 10.2, film pool boiling conditions are approached, in which case heat transfer is due to both convection and radiation. The heat transfer rate follows from Equation 10.1, written on a per unit length basis for a cylindrical surface of diameter D.

$$q'_s = q''_s \pi D = h \pi D \, \Delta T_e$$

The heat transfer coefficient, h, is calculated from the transcendental equation, Equation 10.6

$$h^{4/3} = h_{conv}^{4/3} + h_{rad} h^{1/3}$$

where the convection and radiation heat transfer coefficients follow from Equations 10.7 and 10.8, respectively. For the convection coefficient:

$$h_{conv} = 0.62 \left[\frac{k_v^3 \rho_v (\rho_l - \rho_v) g (h_{fg} + 0.4 c_{p,v} \Delta T_e)}{\mu_v D \, \Delta T_e} \right]^{1/4}$$

$$h_{conv} = 0.62$$

$$\times \left[\frac{(0.0331)^3 (W/m \cdot K)^3 \times 4.808 \text{ kg/m}^3 (957.9 - 4.808) \text{ kg/m}^3 \times 9.8 \text{ m/s}^2}{1} \right.$$

$$\times \left. \frac{(2257 \times 10^3 \text{ J/kg} + 0.4 \times 2.56 \times 10^3 \text{ J/kg} \cdot K \times 155°C)}{14.85 \times 10^{-6} \text{ N} \cdot \text{s/m}^2 \times 6 \times 10^{-3} \text{ m} \times 155°C} \right]^{1/4}$$

$$h_{conv} = 453 \text{ W/m}^2 \cdot K$$

For the radiation heat transfer coefficient:

$$h_{rad} = \frac{5.67 \times 10^{-8} \varepsilon (T_s^4 - T_{sat}^4)}{T_s - T_{sat}}$$

$$h_{rad} = \frac{5.67 \times 10^{-8} \text{ W/m}^2 \cdot K^4 \times 1(528^4 - 373^4) K^4}{(528 - 373) K}$$

$$h_{rad} = 21.3 \text{ W/m}^2 \cdot K$$

Solving, by trial and error, the transcendental equation, Equation 10.6, with numerical values for h_{conv} and h_{rad}

$$h^{4/3} = 453^{4/3} + 21.3 h^{1/3}$$

it follows that

$$h = 469 \text{ W/m}^2 \cdot K$$

Hence the heat transfer rate per unit length of heater element is

$$q_s' = 469 \text{ W/m}^2 \cdot K \times \pi \times 6 \times 10^{-3} \text{ m} \times 155°C$$

$$q_s' = 1.37 \text{ kW/m} \qquad \triangleleft$$

COMMENTS:

Due to the low value of ΔT_e for which partial film boiling, rather than film boiling, may exist, considerable uncertainty is associated with the use of Equations 10.6 to 10.8 and therefore with the foregoing result.

10.2.4   Forced-Convection Boiling

Forced-convection boiling is usually associated with boiling from the inner surface of a heated tube through which a liquid is flowing. Bubble growth and separation are strongly influenced by the flow velocity, and hydrodynamic effects are significantly different than those corresponding to pool boiling. The process is complicated by the existence of different two-phase flow patterns that preclude the development of generalized theories.

Consider flow development in the heated tube of Figure 10.5. Heat transfer to the subcooled liquid which enters the tube is initially by forced convection and may be predicted from the correlations of Chapter 8. However, boiling is soon initiated, with bubbles appearing at the surface growing and being carried into the mainstream of the liquid. There is a sharp increase in the convection heat transfer coefficient associated with this *bubbly flow regime.* As the volume fraction of the vapor increases, individual bubbles coalesce to form plugs or slugs of vapor. This *slug-flow regime* is followed by an *annular-flow* regime in which the liquid forms a film. This film moves along the inner surface, while vapor moves at a faster velocity through the core of the tube. The heat transfer coefficient continues to increase through the *bubbly flow* and much of the *annular-flow* regimes. However, dry spots eventually appear on the inner surface,

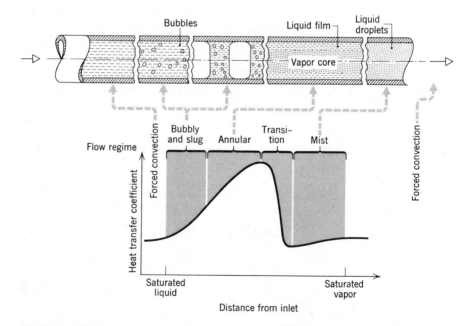

Figure 10.5   Flow regimes for forced-convection boiling inside a tube.

at which point the convection coefficient begins to decrease. The *transition regime* is characterized by the growth of the dry spots, until the surface is completely dry and all remaining liquid is in the form of droplets appearing in the vapor core. The convection coefficient continues to decrease through this regime. There is little change in this coefficient through the *mist-flow* regime, which persists until all of the droplets are converted to vapor. The vapor is then *superheated* by forced convection from the surface.

Discussion of the many correlations which have been developed to quantify the foregoing two-phase flow phenomena is left to the literature [2,11,12].

10.3 CONDENSATION: PHYSICAL MECHANISMS

Condensation occurs when the temperature of a vapor is reduced below its saturation temperature. In engineering practice the process generally occurs when the vapor comes in contact with a cool surface. The latent energy of the vapor is released, and there is heat transfer to the surface.

As shown in Figure 10.6, condensation may occur in one of two possible ways, depending on the condition of the surface. The dominant form of condensation is one in which a liquid film covers the entire condensing surface, and under the action of gravity the film flows continuously from the surface. *Film condensation* is generally characteristic of clean, uncontaminated surfaces. However, if the surface is coated with a substance which inhibits wetting, it is possible to maintain *dropwise condensation*. The drops form in cracks and pits on the surface and may grow and coalesce through condensation. Typically, more than 90 percent of the surface is covered by drops, ranging from a few micrometers in diameter to agglomerations visible to the naked eye. The

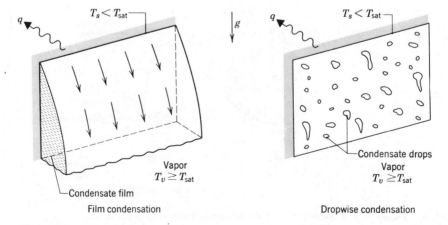

Figure 10.6 Film and dropwise condensation on a vertical surface.

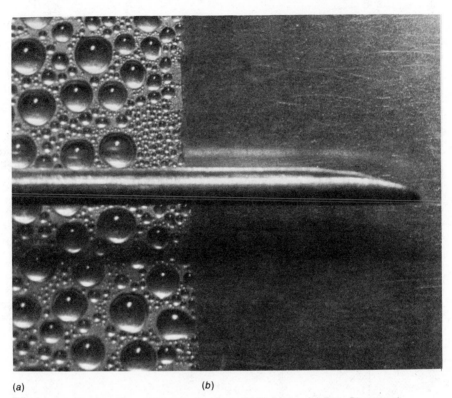

(a)   (b)

Figure 10.7 Condensation on a vertical surface. (a) Dropwise. (b) Film. Photograph courtesy of Professor J. W. Westwater, University of Illinois at Champaign-Urbana.

droplets flow from the surface due to the action of gravity. Film and dropwise condensation of steam on a vertical copper surface are also shown in Figure 10.7. A thin coating of cupric oleate was applied to the left-hand portion of the surface in order to promote the dropwise condensation. A thermocouple probe of 1 mm diameter extends across the photograph.

It is important to note that, irrespective of whether it is in the form of a film or droplets, the condensate provides a resistance to heat transfer between the vapor and the surface. Because this resistance increases with condensate thickness, which increases in the flow direction, it is desirable to use short vertical surfaces or horizontal cylinders in situations involving film condensation. Most condensers therefore consist of horizontal tube bundles through which a liquid coolant flows and around which the vapor to be condensed is circulated. In terms of maintaining high condensation and heat transfer rates, droplet formation is superior to that of film formation. In dropwise condensation most of the heat transfer is through drops of less than 100 μm diameter, and transfer rates that are more than an order of magnitude larger than those associated with film condensation may be achieved. It is therefore common practice to use surface coatings that inhibit wetting and hence stimulate dropwise condensation. Silicones, Teflon, and an assortment of waxes and fatty acids are often used for this purpose. However, such coatings gradually lose their effectiveness due to oxidation, fouling, or outright removal, and film condensation would eventually develop.

Although it is desirable to achieve dropwise condensation in industrial applications, it is often difficult to maintain this condition. For this reason and because the convection coefficients for film condensation are smaller than those for the dropwise case, condenser design calculations are often based on the assumption of film condensation. In the remaining sections of this chapter, we therefore focus on film condensation effects. A review of the limited results available for dropwise condensation is provided by Griffith [13].

10.4 LAMINAR FILM CONDENSATION ON A VERTICAL PLATE

As shown in Figure 10.8, there may be several complicating features associated with film condensation. The film originates at the top of the plate and flows

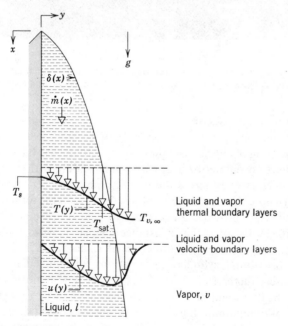

Figure 10.8 Boundary layer effects related to film condensation on a vertical surface.

downward under the influence of gravity. The thickness, δ, and the condensate mass flowrate, $\dot{m}$, increase with increasing x due to continuous condensation at the liquid–vapor interface, which is at T_{sat}. There is then heat transfer from this interface through the film to the surface, which is maintained at $T_s < T_{sat}$. In the most general case the vapor may be superheated ($T_{v,\infty} > T_{sat}$) and may be part of a mixture containing one or more noncondensable gases. Moreover, there exists a finite shear stress at the liquid–vapor interface, contributing to a velocity gradient in the vapor, as well as in the film [14,15].

Despite the complexities associated with film condensation, useful results may be obtained by making assumptions that originated with an analysis by Nusselt [16].

1. Laminar flow and constant properties are assumed for the liquid film.

2. The gas is assumed to be a pure vapor and at a uniform temperature equal to T_{sat}. With no temperature gradient in the vapor, heat transfer to the liquid–vapor interface can only occur by condensation at the interface and not by conduction from the vapor.

3. The shear stress at the liquid–vapor interface is assumed to be negligible, in which case $\partial u/\partial y|_{y=\delta} = 0$. With this assumption and the foregoing assumption of a uniform vapor temperature, there is then no need to consider the vapor velocity or thermal boundary layers shown in Figure 10.8.

4. Momentum and energy transfer by convection in the condensate film are assumed to be negligible. This assumption is reasonable by virtue of the low

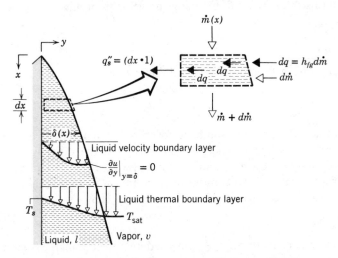

Figure 10.9 Velocity and thermal boundary layer conditions associated with the Nusselt analysis.

velocities associated with the film. It follows that heat transfer across the film occurs only by conduction, in which case the liquid temperature distribution is linear.

Film conditions resulting from the above assumptions are shown in Figure 10.9.

From the fourth approximation, inertial (fluid acceleration) terms may be neglected, and the x-momentum equation, Equation 6.52, may be expressed as

$$\frac{\partial^2 u}{\partial y^2} = \frac{1}{\mu_l}\frac{dp}{dx} - \frac{X}{\mu_l} \tag{10.9}$$

where the body force X, originally introduced in Equation 6.29, has been retained. The body force within the film is equal to $\rho_l g$, and the pressure gradient may be approximated in terms of conditions outside the film. That is, invoking the boundary layer approximation that $(\partial p/\partial y) \approx 0$, it follows that $(dp/dx) \approx \rho_v g$. The momentum equation may therefore be expressed as

$$\frac{\partial^2 u}{\partial y^2} = -\frac{g}{\mu_l}(\rho_l - \rho_v) \tag{10.10}$$

Integrating twice and applying boundary conditions of the form $u(0) = 0$ and $\partial u/\partial y|_{y=\delta} = 0$, the velocity profile in the film becomes

$$u(y) = \frac{g(\rho_l - \rho_v)\delta^2}{\mu_l}\left[\frac{y}{\delta} - \frac{1}{2}\left(\frac{y}{\delta}\right)^2\right] \tag{10.11}$$

From this result the condensate mass flowrate may be obtained. Expressing this flowrate in terms of an integral involving the velocity profile

$$\dot{m}(x) = \int_0^{\delta(x)} \rho_l u(y) dy \tag{10.12}$$

and substituting from Equation 10.11, it follows that

$$\dot{m}(x) = \frac{g\rho_l(\rho_l - \rho_v)\delta^3}{3\mu_l} \tag{10.13}$$

The specific variation of δ, and hence of $\dot{m}$, with x may be obtained by first applying the conservation of energy requirement to the differential element shown in Figure 10.9. At a portion of the liquid–vapor interface of unit width and length dx, the rate of heat transfer into the film, dq, must equal the rate of energy release due to condensation at the interface. Hence

$$dq = h_{fg} d\dot{m} \tag{10.14}$$

Since convective effects are negligible, according to the fourth assumption, it also follows that the rate of heat transfer across the interface must equal the rate of heat transfer to the surface. Hence

$$dq = q_s''(dx \cdot 1) \tag{10.15}$$

Since the liquid temperature distribution is linear, Fourier's law may be used to express the surface heat flux as

$$q_s'' = \frac{k_l(T_{sat} - T_s)}{\delta} \tag{10.16}$$

Combining Equations 10.14 to 10.16, we then obtain

$$\frac{dm}{dx} = \frac{k_l(T_{sat} - T_s)}{\delta h_{fg}} \tag{10.17}$$

Differentiating Equation 10.13 we also obtain

$$\frac{dm}{dx} = \frac{g\rho_l(\rho_l - \rho_v)\delta^2}{\mu_l}\frac{d\delta}{dx} \tag{10.18}$$

Combining Equations 10.17 and 10.18 it follows that

$$\delta^3 d\delta = \frac{k_l\mu_l(T_{sat} - T_s)}{g\rho_l(\rho_l - \rho_v)h_{fg}}dx$$

Integrating from $x = 0$, where $\delta = 0$, to any x location of interest on the surface, it follows that

$$\delta(x) = \left[\frac{4k_l\mu_l(T_{sat} - T_s)x}{g\rho_l(\rho_l - \rho_v)h_{fg}}\right]^{1/4} \tag{10.19}$$

This result may then be substituted into Equation 10.13 to obtain $\dot{m}(x)$.

The surface heat flux may also be expressed as

$$q_s'' = h_x(T_{sat} - T_s) \tag{10.20}$$

Substituting from Equation 10.16, the local convection coefficient is then of the form

$$h_x = \frac{k_l}{\delta} \tag{10.21}$$

or, from Equation 10.19,

$$h_x = \left[\frac{g\rho_l(\rho_l - \rho_v)k_l^3 h_{fg}}{4\mu_l(T_{sat} - T_s)x}\right]^{1/4} \tag{10.22}$$

Since h_x depends on $x^{-1/4}$, it follows that the average convection coefficient for the entire plate is

$$\bar{h}_L = \frac{1}{L}\int_0^L h_x dx = \frac{4}{3}h_L$$

or

$$\bar{h}_L = 0.943 \left[\frac{g\rho_l(\rho_l - \rho_v)k_l^3 h_{fg}}{\mu_l(T_{\text{sat}} - T_s)L} \right]^{1/4}$$

(10.23)

Because this expression has been found to underpredict most experimental results for laminar film condensation by approximately 20 percent [17], it is customary to use a value of 1.13 in place of the coefficient 0.943. Hence

$$\bar{h}_L = 1.13 \left[\frac{g\rho_l(\rho_l - \rho_v)k_l^3 h_{fg}}{\mu_l(T_{\text{sat}} - T_s)L} \right]^{1/4}$$

(10.24)

In using Equation 10.24 all liquid properties should be evaluated at the film temperature, $T_f = (T_{\text{sat}} + T_s)/2$, and h_{fg} should be evaluated at T_{sat}. The correlation may also be used for condensation on the inner or outer surface of a vertical tube of radius R if $R \gg \delta$. It can also be used for inclined surfaces if g is replaced by $g \cdot \cos\theta$, where θ is the angle between the vertical and the surface. However, it must be used with caution for large values of θ and does not apply if $\theta = \pi/2$.

The total heat transfer to the surface may be obtained by using Equation 10.24 with the following form of Newton's law of cooling

$$q = \bar{h}_L A(T_{\text{sat}} - T_s)$$

(10.25)

The total condensation rate may then be determined from the relation

$$\dot{m} = \frac{q}{h_{fg}} = \frac{\bar{h}_L A (T_{\text{sat}} - T_s)}{h_{fg}}$$

(10.26)

Equations 10.25 and 10.26 are generally applicable to any surface geometry, although the form of $\bar{h}_L$ will vary according to geometry and flow conditions.

10.5   TURBULENT FILM CONDENSATION

As for all previous convection phenomena, turbulent flow conditions may exist in film condensation. Consider the vertical surface of Figure 10.10. The transition criterion may be expressed in terms of a Reynolds number defined as

$$Re = \frac{\rho_l u_m D_h}{\mu_l}$$

(10.27)

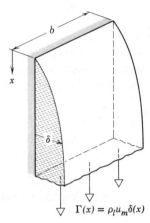

$\Gamma(x) = \rho_l u_m \delta(x)$ **Figure 10.10** Film condensation on a vertical plate of width b.

where u_m is the average velocity in the film and D_h, the hydraulic diameter, is defined as

$$D_h = \frac{4A_c}{P} \tag{8.62}$$

For the vertical surface the flow area A_c is $b \cdot \delta$ and the wetted perimeter P of the solid interface is simply b. The characteristic length is therefore 4δ, and the Reynolds number may be expressed as

$$Re_\delta = \frac{4\delta \rho_l u_m}{\mu_l} = \frac{4\dot{m}}{\mu_l b} \equiv \frac{4\Gamma}{\mu_l} \tag{10.28}$$

where the mass flowrate per unit width ($\dot{m}/b = \rho_l u_m \delta$) is defined as Γ. Note that $\Gamma = 0$ at the top of the plate and increases with increasing x. When the value of Γ corresponding to a critical Reynolds number of $Re_{\delta,c} \approx 1800$ is reached, transition to turbulence occurs. Equation 10.24 is recommended for estimating the average heat transfer coefficient when $Re_\delta \lesssim Re_{\delta,c}$. For $Re_\delta \gtrsim Re_{\delta,c}$ the following expression is recommended [17]:

$$\overline{h}_L = 0.0077 \left[\frac{g \rho_l (\rho_l - \rho_v) k_l^3}{\mu_l^2} \right]^{1/3} Re_\delta^{0.4} \tag{10.29}$$

where Re_δ is given by Equation 10.28. Equations 10.25 and 10.26 may again be used to determine the total heat transfer and condensation rates.

EXAMPLE 10.3

The outer surface of a vertical tube, which is of length $L = 1$ m and outer diameter $D = 80$ mm is exposed to saturated steam at atmospheric pressure and is maintained at 50°C by the flow of cool water through the tube. What is the rate of heat transfer to the coolant, and what is the rate at which steam is condensed at the surface?

SOLUTION

KNOWN:

Dimensions and temperature of a vertical tube experiencing condensation of steam at its outer surface.

FIND:

Heat transfer and condensation rates.

SCHEMATIC:

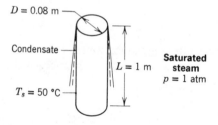

$D = 0.08$ m

Condensate

$L = 1$ m

Saturated steam
$p = 1$ atm

$T_s = 50$ °C

ASSUMPTIONS:

1. Laminar film condensation on a vertical surface.
2. Negligible concentration of noncondensable gases in the steam.

PROPERTIES:

Table A.6, saturated vapor ($p = 1.0133$ bars): $T_{\text{sat}} = 100°C$, $\rho_v = (1/v_g)$ $= 0.596$ kg/m^3, $h_{fg} = 2257$ kJ/kg.
Table A.6, saturated water ($T_f = 75°C$): $\rho_l = (1/v_f) = 975$ kg/m^3, $\mu_l = 375$ $\times 10^{-6}$ N·s/m^2, $k_l = 0.668$ W/m·K.

ANALYSIS:

The heat transfer rate may be determined from Equation 10.25, where $A = \pi DL$. Hence

$$q = \bar{h}_L(\pi DL)(T_{sat} - T_s)$$

Having assumed laminar film condensation on a vertical surface, it follows from Equation 10.24 that

$$\bar{h}_L = 1.13 \left[\frac{g\rho_l(\rho_l - \rho_v)k_l^3 h_{fg}}{\mu_l(T_{sat} - T_s)L} \right]^{1/4}$$

Hence

$$\bar{h}_L = 1.13$$
$$\times [\{9.8 \text{ m/s}^2 \times 975 \text{ kg/m}^3(975 - 0.596)\text{kg/m}^3(0.668 \text{ W/m·K})^3 2.257$$
$$\times 10^6 \text{ J/kg}\}/\{375 \times 10^{-6} \text{ kg/s·m}(100 - 50)\text{K} \times 1 \text{ m}\}]^{1/4}$$

$$\bar{h}_L = 4831 \text{ W/m}^2\text{·K}$$

and

$$q = 4831 \text{ W/m}^2\text{·K} \times \pi \times 0.08 \text{ m} \times 1 \text{ m}(100 - 50)\text{K}$$

$$q = 60{,}708 \text{ W} \qquad \triangleleft$$

From Equation 10.26 the condensation rate is then

$$\dot{m} = \frac{q}{h_{fg}} = \frac{60{,}708 \text{ W}}{2.257 \times 10^6 \text{ J/kg}}$$

$$\dot{m} = 0.0269 \text{ kg/s} \qquad \triangleleft$$

COMMENTS:

The assumption of laminar film conditions may be checked by calculating Re_δ from Equation 10.28.

$$Re_\delta = \frac{4\dot{m}}{\mu_l b} = \frac{4 \times 0.0269 \text{ kg/s}}{375 \times 10^{-6} \text{ kg/s·m } \pi(0.08 \text{ m})} = 1142$$

Hence $Re_\delta < Re_{\delta,c}$ and the film is, in fact, laminar. Moreover, from Equation 10.19

$$\delta(L) = \left[\frac{4k_l\mu_l(T_{sat} - T_s)L}{g\rho_l(\rho_l - \rho_v)h_{fg}} \right]^{1/4}$$

$$\delta(L) = \left[\frac{4 \times 0.668 \text{ W/m·K} \times 375 \times 10^{-6} \text{ kg/s·m}(100 - 50)\text{K} \times 1 \text{ m}}{9.8 \text{ m/s}^2 \times 975 \text{ kg/m}^3(975 - 0.596)\text{kg/m}^3 \times 2.257 \times 10^6 \text{ J/kg}} \right]^{1/4}$$

$$\delta(L) = 2.21 \times 10^{-4} \text{ m} = 0.221 \text{ mm}$$

Hence $\delta(L) \ll (D/2)$ and the use of the vertical plate correlation for a vertical cylinder is justified.

10.6 FILM CONDENSATION ON HORIZONTAL TUBES AND TUBE BANKS

The foregoing (Nusselt) analysis may be extended to film condensation on the outer surface of a horizontal tube, Figure 10.11a, and the result for the average convection coefficient is of the form [18].

$$\bar{h}_D = 0.728 \left[\frac{g\rho_l(\rho_l - \rho_v)k_l^3 h_{fg}}{\mu_l(T_{\text{sat}} - T_s)D} \right]^{1/4} \tag{10.30}$$

For a vertical tier of N horizontal tubes, Figure 10.9b, the average (over the N tubes) may be expressed as

$$\bar{h}_D = 0.728 \left[\frac{g\rho_l(\rho_l - \rho_v)k_l^3 h_{fg}}{N\mu_l(T_{\text{sat}} - T_s)D} \right]^{1/4} \tag{10.31}$$

Such an arrangement is often used in condenser design. The reduction in $\bar{h}$ with increasing N may be attributed to an increase in the average film thickness for each successive tube. Equations 10.30 and 10.31 are generally in agreement with or slightly lower than experimental results for pure vapors. Departures may be attributed to ripples in the liquid surface or, for Equation 10.31, to splashing effects. However, if noncondensable gases are present, the convection coefficient may be lower than results predicted from the foregoing correlations.

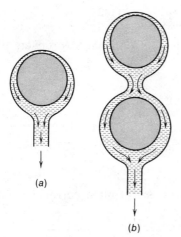

(a)

(b)

Figure 10.11   Film condensation on (a) a single horizontal tube and (b) a vertical tier of horizontal tubes.

EXAMPLE 10.4

A steam condenser consists of a square array of 400 tubes, each 6 mm in diameter. If the tubes are exposed to saturated steam at a pressure of 0.15 bar and the tube surface temperature is maintained at 25°C, what is the rate at which steam is condensed per unit length of the tubes?

SOLUTION

KNOWN:

Configuration and surface temperature of condenser tubes exposed to saturated steam at 0.15 bar.

FIND:

Condensation rate per unit length of tubes.

SCHEMATIC:

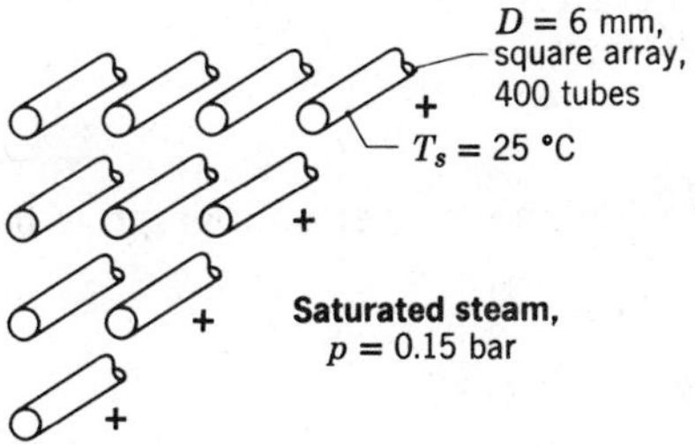

$D = 6$ mm, square array, 400 tubes
$T_s = 25$ °C

Saturated steam, $p = 0.15$ bar

ASSUMPTIONS:

1. Negligible concentration of noncondensable gases in the steam.
2. Film condensation on the tubes.

PROPERTIES:

Table A.6, saturated vapor ($p = 0.15$ bar): $T_{sat} = 327$ K $= 54$°C, $\rho_v = (1/v_g)$ $= 0.098$ kg/m³, $h_{fg} = 2373$ kJ/kg.
Table A.6, saturated water ($T_f = 312.5$ K): $\rho_l = (1/v_f) = 992$ kg/m³, $\mu_l = 663 \times 10^{-6}$ N·s/m², $k_l = 0.631$ W/m·K.

ANALYSIS:

The condensation rate for a single tube of the array may be obtained from

Equation 10.26, where for a unit length of the tube,

$$\dot{m}'_1 = \frac{q'_1}{h_{fg}} = \frac{\bar{h}_D(\pi D)(T_{sat} - T_s)}{h_{fg}}$$

From Equation 10.31

$$\bar{h}_D = 0.728 \left[\frac{g\rho_l(\rho_l - \rho_v)k_l^3 h_{fg}}{N\mu_l(T_{sat} - T_s)D} \right]^{1/4}$$

or with $N = 20$,

$$\bar{h}_D = 0.728$$

$$\times [\{9.8 \text{ m/s}^2 \times 992 \text{ kg/m}^3(992 - 0.098)\text{kg/m}^3(0.631 \text{ W/m·K})^3 2.373$$
$$\times 10^6 \text{ J/kg}\}/\{20 \times 663 \times 10^{-6} \text{ kg/s·m } (54 - 25)\text{K} \times 0.006 \text{ m}\}]^{1/4}$$

$$\bar{h}_D = 5144 \text{ W/m}^2 \cdot \text{K}$$

Hence for a single tube

$$\dot{m}'_1 = \frac{5144 \text{ W/m}^2 \cdot \text{K } (\pi \times 0.006 \text{ m})(54 - 25)\text{K}}{2.373 \times 10^6 \text{ J/kg}}$$

$$\dot{m}'_1 = 1.18 \times 10^{-3} \text{ kg/s·m}$$

For the complete array, the condensation rate per unit length is then

$$\dot{m}' = N^2 \dot{m}'_1 = 400 \times 1.18 \times 10^{-3} \text{ kg/s·m}$$

$$\dot{m} = 0.474 \text{ kg/s·m} \qquad \qquad \triangleleft$$

10.7 FILM CONDENSATION IN HORIZONTAL TUBES

Condensers used for refrigeration and airconditioning systems generally involve vapor condensation inside horizontal or vertical tubes. Unfortunately, conditions within the tube are complicated and depend strongly on the velocity of the vapor flowing through the tube. If this velocity is small, condensation occurs in the manner depicted by Figure 10.12a for a horizontal tube. That is, the condensate flow is from the upper portion of the tube to the bottom, from whence it flows in a longitudinal direction with the vapor. For low-vapor velocities such that

$$Re_{v,i} = \left(\frac{\rho_v u_{m,v} D}{\mu_v} \right)_i < 35,000$$

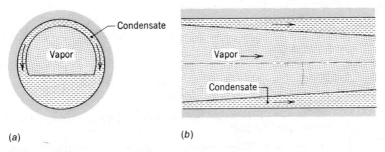

Figure 10.12   Film condensation in a horizontal tube. (*a*) Cross section of condensate flow for low vapor velocities. (*b*) Longitudinal section of condensate flow for large vapor velocities.

where *i* refers to the tube inlet, Chato [19] recommends an expression of the form

$$\bar{h}_D = 0.555 \left[\frac{g\rho_l(\rho_l - \rho_v)k_l^3 h'_{fg}}{\mu_l(T_{\text{sat}} - T_s)D} \right]^{1/4} \tag{10.32}$$

where

$$h'_{fg} \equiv h_{fg} + \tfrac{3}{8} c_{p,l}(T_{\text{sat}} - T_s) \tag{10.33}$$

At higher vapor velocities the two-phase flow regime becomes annular in nature, as shown in Figure 10.12*b*. The vapor occupies the core of the annulus, diminishing in diameter as the thickness of the outer condensate layer increases in the flow direction. Results for this complicated flow condition are provided by Rohsenow [18].

10.8   SUMMARY

It is apparent that boiling and condensation are complicated processes for which the existence of generalized relations is somewhat limited. The purpose of this chapter has been to identify the essential physical features of the processes and to present correlations suitable for approximate engineering calculations. However, a great deal of additional information is available in the literature, and much of it has been summarized in several extensive reviews of the subject [1, 11–13, 18, 20, 21].

REFERENCES

1. Leppert, G. and C. C. Pitts, "Boiling" in T. F. Irvine Jr. and J. P. Hartnett, Eds., *Advances in Heat Transfer*, Vol. 1, Academic Press, New York, 1964, pp. 185–266.

2. Tong, L. S., *Boiling Heat Transfer and Two-Phase Flow*, Wiley, New York, 1965.

3. Rohsenow, W. M., "A Method of Correlating Heat Transfer Data for Surface Boiling Liquids," *Trans. ASME*, *74*, 969, 1952.

4. Cichelli, M. T. and C. F. Bonilla, "Heat Transfer to Liquids Boiling under Pressure," *Trans. AIChE*, *41*, 755, 1945.

5. Piret, E. L. and H. S. Isbin, "Natural Circulation Evaporation Two-Phase Heat Transfer," *Chem. Eng. Prog.*, *50*, 305, 1954.

6. Vachon, R. I., G. H. Nix, and G. E. Tanger, "Evaluation of Constants for the Rohsenow Pool-Boiling Correlation," *J. Heat Transfer*, *90*, 239, 1968.

7. Cryder, D. S. and A. C. Finalbargo, "Heat Transmission from Metal Surfaces to Boiling Liquids: Effect of Temperature of the Liquid on Film Coefficient," *Trans. AIChE*, *33*, 346, 1937.

8. Jakob, M. and G. A. Hawkins, *Elements of Heat Transfer*, 3rd Ed. Wiley, New York, 1957.

9. Zuber, N., "On the Stability of Boiling Heat Transfer," *Trans. ASME*, *80*, 711, 1958.

10. Bromley, L. A., "Heat Transfer in Stable Film Boiling," *Chem. Eng. Prog.*, *46*, 221, 1950.

11. Rohsenow, W. M., "Boiling" in W. M. Rohsenow and J. P. Hartnett, Eds., *Handbook of Heat Transfer*, Chapter 13, McGraw-Hill, New York, 1973.

12. Griffith, P., "Two-Phase Flow" in W. M. Rohsenow and J. P. Hartnett, Eds., *Handbook of Heat Transfer*, Chapter 14, McGraw-Hill, New York, 1973.

13. Griffith, P., "Dropwise Condensation" in W. M. Rohsenow and J. P. Hartnett, Eds., *Handbook of Heat Transfer*, Chapter 12B, McGraw-Hill, New York, 1973.

14. Sparrow, E. M. and J. L. Gregg, "A Boundary Layer Treatment of Laminar Film Condensation," *J. Heat Transfer*, *81*, 13, 1959.

15. Koh, J. C. Y., E. M. Sparrow, and J. P. Hartnett, "The Two-Phase Boundary Layer in Laminar Film Condensation," *Int. J. Heat Mass Transfer*, *2*, 69, 1961.

16. Nusselt, W., "Die Oberflachenkondensation des Wasserdampfes," *Z. Ver. Deut. Ing.*, *60*, 541, 1916.

17. McAdams, W. H., *Heat Transmission*, 3rd Ed., McGraw-Hill, New York, 1954.

18. Rohsenow, W. M., "Film Condensation" in W. M. Rohsenow and J. P. Hartnett, Eds., *Handbook of Heat Transfer*, Chapter 12A, McGraw-Hill, New York, 1973.

19. Chato, J. C., "Laminar Condensation Inside Horizontal and Inclined Tubes," *J. Am. Soc. Heating Refrig. Aircond. Engrs.*, *4*, 52, 1962.

20. Jordan, D. P., "Film and Transition Boiling" in T. F. Irvine Jr. and J. P. Hartnett, Eds., *Advances in Heat Transfer*, Vol. 5, Academic Press, New York, 1968, pp. 55–128.

21. Merte, H., Jr., "Condensation Heat Transfer," in T. F. Irvine Jr. and J. P. Hartnett, Eds., *Advances in Heat Transfer*, Vol. 9, Academic Press, New York, 1973, pp. 181–272.

PROBLEMS

10.1 From experience it is known that the risk of burning foods with silicon-coated pans is less than with ordinary, uncoated utensils. Can you explain why this is so?

10.2 A long, 1 mm diameter wire passes an electrical current dissipating 4085 W/m and reaches a surface temperature of 128°C when submerged in water at one atmosphere.

   a) What is the boiling heat transfer coefficient?

   b) If nucleate pool boiling is occurring, estimate the value of the correlation coefficient $C_{s,f}$.

10.3 Estimate the nucleate pool boiling heat transfer coefficient for water boiling at atmospheric pressure on the outer surface of a vertical, 10 mm diameter tube maintained 7°C above the saturation temperature. What effects occur when the tube is positioned horizontally?

10.4 Boiling occurs at the surface of a 0.5-m square plate maintained at 106°C and vertically suspended in water exposed to atmospheric pressure. Estimate the heat transfer rate.

10.5 Estimate the nucleate pool boiling heat transfer coefficient for water under atmospheric pressure in contact with ground/polished stainless steel when the excess temperature is 15°C.

10.6 Determine the maximum heat flux for nucleate pool boiling of water under normal atmospheric conditions. What will this heat flux be when the pressure is 2 bars?

10.7 Estimate the current at which a 1-mm diameter nickel wire will burn out when submerged horizontally in water at atmospheric pressure. The electrical resistance of the wire is 0.129 Ω/m.

10.8 For forced-convection local boiling of water inside vertical tubes, the heat transfer coefficient can be estimated by the correlation.

$$h = 2.54(\Delta T_e)^3 \exp\left(\frac{p}{15.3}\right)$$

for h in units of $W/m^2 \cdot K$, ΔT_e in units of K, and p in units of bar. Consider water at 4 bars flowing through a vertical tube of 50 mm inner diameter. Local boiling occurs when the tube wall is 15°C above the saturation temperature. Estimate the boiling heat transfer rate per unit length of the tube.

10.9 A steel bar, 20 mm in diameter and 200 mm in length, with an emissivity of 0.9, is removed from a furnace at 455°C and suddenly submerged in a water bath under atmospheric pressure. Estimate the initial heat transfer rate from the bar.

10.10 The passage of electrical current through a horizontal, 2 mm diameter conductor with emissivity 0.5 causes its surface temperature to reach 555°C when immersed in water under atmospheric pressure. Estimate the power dissipation per unit length of the conductor.

10.11 A polished stainless steel bar of 50 mm diameter having an emissivity 0.10 is maintained at a surface temperature of 150°C while horizontally submerged in water at 25°C under atmospheric pressure. Estimate the heat rate per unit length of the bar.

10.12 For forced-convection boiling in smooth tubes, the heat flux can be estimated as the separate effects due to boiling and to forced convection. For nucleate boiling, Equation 10.2 would be appropriate to estimate that effect. To determine the forced convection effect, the correlation Equation 8.58 has been found suitable with the coefficient 0.023 replaced by 0.019. The appropriate temperature difference is based upon the tube wall and liquid mean temperatures. Consider water at 1 atm with a mean velocity of 1.5 m/s and a mean temperature of 95°C flowing through a 15-mm diameter brass tube whose surface is maintained at 110°C. Estimate the heat transfer rate per unit length of the tube.

10.13 Saturated steam at 0.1 bar condenses with a convection coefficient of 6800 W/m²·K on the outside of a brass tube having inner and outer diameters of 16.5 mm and 19 mm, respectively. The convection coefficient for water flowing inside the tube is 5200 W/m²·K. Estimate the steam condensation rate per unit length of the tube when the mean water temperature is 30°C.

10.14 For laminar film condensation on a vertical plate, develop an expression for the Reynolds number in terms of the relevant fluid thermophysical properties, temperature difference, and plate length.

10.15 Saturated steam at one atmosphere is exposed to a vertical plate 1 m in height and 0.5 m in width having a uniform surface temperature of 70°C. Estimate the heat transfer rate to the plate and the steam condensation rate.

10.16 A 0.5-m square vertical plate with a uniform surface temperature of 84°C is exposed to saturated steam at one atmosphere.

a) Estimate the local convection coefficient at the middle and at the bottom of the plate.

b) Estimate the average convection coefficient for the entire plate.

c) Determine the total condensation rate and the total heat transfer rate to the plate.

10.17 Saturated steam at one atmosphere condenses on the outer surface of a vertical, 100 mm diameter pipe of length 1 m and uniform surface temperature of 94°C. Estimate the total condensation rate and the heat transfer rate to the pipe.

10.18 Determine the total condensation rate and the heat transfer rate for the condensation process of Problem 10.17 when the steam is saturated at 1.5 bars.

10.19 Consider the condensation process of Problem 10.17. In order to maintain the pipe wall at the uniform surface temperature of 94°C, cooling water passes through the steel pipe with an inner diameter 92 mm. What water flow rate will result in a 4°C temperature difference of the water between the outlet and inlet of the pipe?

10.20 For turbulent film condensation on a vertical plate, develop an expression for the total condensation rate in terms of the relevant fluid thermophysical properties, temperature difference, and the plate dimensions.

10.21 A vertical plate 2.5 m high, maintained at a uniform temperature of 54°C, is exposed to saturated steam at atmospheric pressure.

a) Estimate the condensation rate and the heat transfer rate per unit width of the plate.

b) If the plate height were halved, would turbulent flow conditions still exist? What effect would reducing the height have on the heat transfer rate?

10.22 Determine the total condensation rate and heat transfer rate for the condensation process of Problem 10.17 when the pipe is horizontal.

10.23 An uninsulated, 25 mm diameter pipe with a surface temperature of 15°C passes through a room having an air temperature of 37°C and relative humidity of 75 percent. Estimate the water condensation rate per unit length of the pipe assuming film rather than dropwise condensation occurs.

10.24 Consider laminar film condensation conditions on the outer surface of a tube of length L and diameter D. Determine the ratio of the condensation heat transfer rate for the horizontal orientation relative to the vertical orientation.

10.25 A horizontal tube of 50 mm diameter, with a surface temperature of 34°C, is exposed to steam at 0.2 bar. Estimate the condensation rate and heat transfer rate per unit length of the tube.

10.26 Saturated steam at a pressure of 0.1 bar is condensed over a square array of 100 tubes each of diameter 8 mm. If the tube surfaces are maintained at 27°C, estimate the condensation rate per unit length of the tubes.

10.27 Water at a mean temperature and velocity of 17°C and 2 m/s, respectively, flows through a horizontal, brass tube with inner and outer diameters of 28.0 mm and 30.5 mm, respectively. Estimate the condensation rate per unit tube length of saturated steam at a pressure of 0.15 bar surrounding the tube.

10.28 Saturated steam at 1.5 bars, condenses inside a horizontal, 75 mm diameter pipe whose surface is maintained at 100°C. Assuming low-vapor velocities and film condensation, estimate the heat transfer coefficient and the condensation rate per unit length of the pipe.

11 Heat Exchangers

The process of heat exchange between two fluids that are at different temperatures and separated by a solid wall occurs in many engineering applications. The device used to implement this exchange is termed a *heat exchanger*, and specific applications may be found in space heating and air-conditioning systems, power production systems, and chemical processing systems. In this chapter we consider the principles of heat transfer needed to design and/or to evaluate the performance of a heat exchanger.

11.1   HEAT EXCHANGER TYPES

There are several different heat exchanger types, which may be classified according to the fluid flow arrangement. The simplest flow configuration is one in which the hot and cold fluids move in either the same or opposite directions. The fluids could be separated by a plane wall, but more commonly they are separated by a *concentric-tube* (or *double-pipe*) arrangement. The latter configuration is shown schematically in Figure 11.1. In the *parallel-flow* arrangement of Figure 11.1*a*, the hot and cold fluids enter at the same end, flow in the same direction, and leave at the same end. In the *counterflow* arrangement of Figure 11.1*b*, the fluids enter at opposite ends, flow in opposite directions, and leave at opposite ends.

An alternative flow configuration is one for which the fluids move in *cross flow* (at right angles to each other) through the heat exchanger. The configuration is normally used in applications where a gas is forced over a tube bundle through which a liquid is pumped. Two possible arrangements are shown in Figure 11.2. For the finned-tube bundle of Figure 11.2*a*, the gas flow is said to be *unmixed* because it cannot move freely in the transverse direction (normal to the flow). Similarly, because it is confined to separate tubes, the liquid passing through the tubes is also unmixed. Such finned-tube, cross-flow heat exchangers are widely used in air conditioning applications. Conversely, if the tubes are without fins, gas motion, and hence mixing, in the transverse direction is possible. Gas flow through the arrangement of Figure 11.2*b* is therefore said to be *mixed*. Note that, when unmixed, a fluid is characterized by a two-dimensional temperature distribution. That is, its temperature varies along and

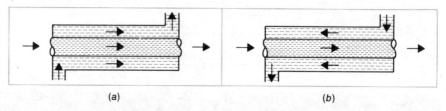

(a)   (b)

Figure 11.1   Concentric-tube heat exchangers. (a) Parallel flow. (b) Counterflow.

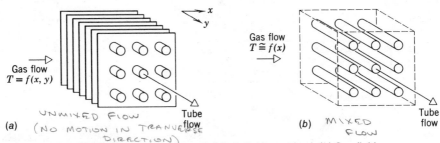

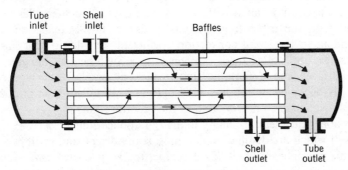

Figure 11.2 Cross-flow heat exchangers. (a) Both fluids unmixed. (b) One fluid mixed and the other unmixed.

normal to the flow direction. However, for the mixed flow condition temperature variations are principally in the flow direction. It is not surprising that the nature of the mixing condition can have a significant influence on the overall heat transfer for the exchanger.

Flow conditions become more complicated for *shell-and-tube* heat exchangers, which are commonly used for liquid-to-liquid heat transfer and consist of a bundle of circular tubes mounted in a cylindrical shell. Many specific forms are available according to the number of tube and/or shell passes. The simplest configuration, which involves a single shell pass and one tube pass, is shown schematically in Figure 11.3. Cross baffles are generally installed in the exchanger in order to generate turbulence in the shell side fluid and to promote a cross-flow component in the velocity of this fluid relative to the tubes. Collectively, these effects contribute to a higher heat transfer coefficient for the outer tube surface. The overall heat transfer may also be increased by using multiple tube and shell passes. Baffled heat exchangers with one shell pass and two tube passes and with two shell passes and four tube passes are shown schematically in Figures 11.4a and 11.4b, respectively.

Figure 11.3 Shell-and-tube heat exchanger with one shell pass and one tube pass.

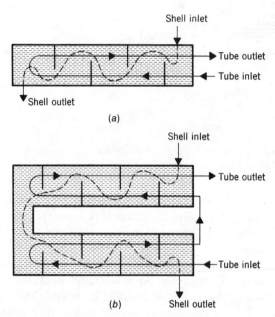

Shell inlet

Tube outlet

Tube inlet

Shell outlet

(a)

Shell inlet

Tube outlet

Tube inlet

(b) Shell outlet

Figure 11.4 Shell-and-tube heat exchangers. (a) One shell pass and two tube passes. (b) Two shell passes and four tube passes.

Although we shall focus on the foregoing heat exchanger configurations in the remaining sections of this chapter, we should note that there are many other special cases which often arise in industrial practice. There are, for example, many different kinds of *compact heat exchangers*. Such heat exchangers are an outgrowth of efforts to achieve a very large heat transfer surface area per unit volume of the exchanger. They are generally used when the convection heat transfer coefficient associated with one of the fluids is much smaller than that associated with the other fluid. Extensive use of fins is made on the side corresponding to the small convection coefficient. In addition for gas-to-gas heat exchange, use is often made of *regenerators*. Such heat exchangers involve rotating surfaces and a periodic flow, which allow the same space to be alternatively occupied by the hot and cold gases. Detailed descriptions of the above heat exchangers are available in the literature [1–5].

Also note that there are many industrial applications of *mass exchangers*, examples of which are distillation columns, spray driers, and scrubbers. There are also many applications of *direct contact* (or *evaporative*) *heat exchangers*. This system is a liquid-to-gas heat exchanger which allows for direct contact between the liquid and the gas (there is no separating wall). Through the latent heat exchange associated with evaporation, large heat transfer rates per unit exchanger volume are possible. Again, discussion of such systems is relegated to the literature [6–8].

11.2 THE OVERALL HEAT TRANSFER COEFFICIENT

An essential, and often the most uncertain, part of any heat exchanger analysis is to determine the *overall heat transfer coefficient*. Recall from Equation 3.18 that this coefficient is defined in terms of the total thermal resistance. If the two fluids of the heat exchanger are separated by a plane wall, as in Figure 11.5*a* or Figure 3.1, the coefficient is of the form

$$U = \frac{1}{(1/h_i) + (L/k) + (1/h_o)} \qquad \frac{1}{R_{TOT}} \tag{11.1}$$

If the fluids are separated by a tube wall, as in Figure 11.5*b* or Figure 3.5, the coefficient is of the form

$$U_o = \frac{1}{\dfrac{1}{h_o} + \dfrac{r_o}{k}\ln\left(\dfrac{r_o}{r_i}\right) + \left(\dfrac{r_o}{r_i}\right)\dfrac{1}{h_i}} \tag{11.2}$$

or,

$$U_i = \frac{1}{\dfrac{1}{h_i} + \dfrac{r_i}{k}\ln\left(\dfrac{r_o}{r_i}\right) + \left(\dfrac{r_i}{r_o}\right)\dfrac{1}{h_o}} \tag{11.3}$$

where $U_i A_i = U_o A_o$.

It should be noted that Equations 11.1 to 11.3 apply only for *clean* surfaces. During normal heat exchanger operation, surfaces are often subject to fouling, due to fluid impurities, rust formation or other reactions between the fluid and the wall material. The subsequent deposition of a film or scale on the surface

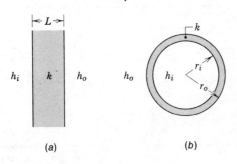

Figure 11.5 Nomenclature for overall heat transfer coefficient associated with (a) a plane wall, (b) a tube wall.

can, in turn, greatly increase the resistance to heat transfer between the fluids. This effect can be treated by introducing an additional thermal resistance, termed the *fouling factor*, R_f. Its value depends on the operating temperature, fluid velocity and length of service of the heat exchanger. With the inclusion of this resistance for the inner and outer surfaces of a tube, the overall heat transfer coefficient may be expressed as

$$U_o = \frac{1}{\frac{1}{h_o} + R_{f,o} + \frac{r_o}{k}\ln\left(\frac{r_o}{r_i}\right) + \left(\frac{r_o}{r_i}\right)R_{f,i} + \left(\frac{r_o}{r_i}\right)\frac{1}{h_i}} \tag{11.4}$$

for the outer surface or

$$U_i = \frac{1}{\frac{1}{h_i} + R_{f,i} + \frac{r_i}{k}\ln\left(\frac{r_o}{r_i}\right) + \left(\frac{r_i}{r_o}\right)R_{f,o} + \left(\frac{r_i}{r_o}\right)\frac{1}{h_o}} \tag{11.5}$$

for the inner surface. Representative values of the fouling factor are given in Table 11.1.

The overall heat transfer coefficient may be determined from knowledge of h_o, $R_{f,o}$, h_i, and $R_{f,i}$, where the convection coefficients may be estimated from appropriate forms of the heat transfer correlations presented in the preceding chapters. Note that the wall conduction term in Equations 11.4 and 11.5 may often be neglected since a thin wall of high thermal conductivity is generally used. Moreover, situations are often encountered for which one of the convection coefficients is much smaller than the other and hence dominates determination of the overall coefficient. Representative values of the overall coefficient are presented in Table 11.2.

Table 11.1   Representative fouling factors [9]

FLUID	$R_f(\text{m}^2\cdot\text{K/W})$
Seawater and treated boiler feedwater (below 50°C)	0.0001
Seawater and treated boiler feedwater (above 50°C)	0.0002
River water (below 50°C)	0.0002–0.0001
Fuel oil	0.0009
Refrigerating liquids	0.0002
Steam (nonoil bearing)	0.0009

Table 11.2   Representative values of the overall heat transfer coefficient.

FLUID COMBINATION	$U(\text{W/m}^2 \cdot \text{K})$
Water to water	850–1700
Water to oil	110–350
Steam condenser (water in tubes)	1000–6000
Ammonia condenser (water in tubes)	800–1400
Alcohol condenser (water in tubes)	250–700
Finned-tube heat exchanger (water in tubes, air in cross flow)	25–50

11.3   HEAT EXCHANGER ANALYSIS: USE OF THE LOG MEAN TEMPERATURE DIFFERENCE

To design or to predict the performance of a heat exchanger, it is essential to have expressions that relate the total heat transfer rate to quantities such as the inlet and outlet fluid temperatures, the overall heat transfer coefficient, and the total surface area for heat transfer. Two such expressions may readily be obtained by applying overall energy balances to the hot and cold fluids, as shown in Figure 11.6. In particular, if q is designated as the total rate of heat transfer between the hot and cold fluids and it is assumed that there is negligible heat transfer between the exchanger and its surroundings, as well as negligible potential and kinetic energy changes, application of an energy balance, Equation 1.10, gives

$$q = \dot{m}_h c_{p,h}(T_{h,i} - T_{h,o}) \tag{11.6}$$

and

$$q = \dot{m}_c c_{p,c}(T_{c,o} - T_{c,i}) \tag{11.7}$$

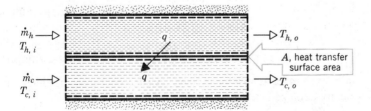

Figure 11.6   Overall energy balances for the hot and cold fluids of a two-fluid heat exchanger.

The subscripts h and c refer to the hot and cold fluids, respectively, whereas the subscripts i and o designate the fluid inlet and outlet conditions, respectively. In Equation 11.6 the heat transfer rate is related to the reduction in enthalpy experienced by the hot fluid, and in Equation 11.7 it is related to the increase in enthalpy experienced by the cold fluid. The temperatures appearing in these expressions refer to the *mean* fluid temperatures at the designated locations. Note that Equations 11.6 and 11.7 are independent of the flow arrangement.

Another useful expression may be obtained by relating the total heat transfer rate, q, to the temperature difference, ΔT, between the hot and cold fluids, where

$$\Delta T \equiv T_h - T_c \tag{11.8}$$

Such an expression would be an extension of Newton's law of cooling, with the overall heat transfer coefficient, U, used in place of the single convection coefficient, h. However, since ΔT varies with position in the heat exchanger, it is necessary to work with a rate equation of the form

$$q = UA\Delta T_m \tag{11.9}$$

where ΔT_m is an appropriate *mean* value of the temperature difference. Equation 11.9 may be used with Equations 11.6 and 11.7 to perform a heat exchanger analysis, but before this can be done, the specific form of ΔT_m must be established. Consider first the parallel-flow heat exchanger.

11.3.1 The Parallel-Flow Heat Exchanger

The hot and cold fluid temperature distributions associated with a parallel-flow heat exchanger are shown in Figure 11.7. The temperature difference, ΔT, is initially large but decays rapidly with increasing x, approaching zero asymptotically. It is important to note that, for such an exchanger, the outlet temperature of the cold fluid never exceeds that of the hot fluid. In Figure 11.7 the subscripts 1 and 2 are used to designate opposite ends of the heat exchanger. This convention will be used for all types of heat exchangers to be considered. For parallel flow, it follows that $T_{h,i} = T_{h,1}$, $T_{h,o} = T_{h,2}$, $T_{c,i} = T_{c,1}$ and $T_{c,o} = T_{c,2}$.

The form of ΔT_m may be determined by applying an energy balance to differential elements in the hot and cold fluids. Each element is of length dx and heat transfer surface area dA, as shown in Figure 11.7. The energy balances and the subsequent analysis are performed subject to the following assumptions.

1. The heat exchanger is insulated from its surroundings, in which case the only heat exchange is between the hot and cold fluids.
2. Axial conduction along the tubes is negligible.
3. Potential and kinetic energy changes are negligible.

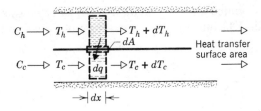

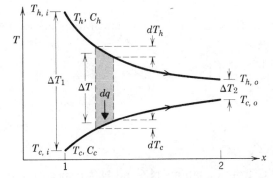

Figure 11.7 Temperature distributions for a parallel-flow heat exchanger.

4. The fluid specific heats are constant.
5. The overall heat transfer coefficient is constant.

The specific heats may of course change due to temperature variations, and the overall heat transfer coefficient may change due to variations in fluid properties and flow conditions. However, in many applications such variations are not significant, and it is reasonable to work with average values of $c_{p,c}$, $c_{p,h}$, and U for the heat exchanger.

Applying an energy balance to each of the differential elements of Figure 11.7, it follows that

$$dq = -\dot{m}_h c_{p,h}\, dT_h \equiv -C_h\, dT_h \qquad\qquad C_h = \dot{m}_h c_{p,h} \qquad (11.10)$$

and

$$dq = \dot{m}_c c_{p,c}\, dT_c \equiv C_c\, dT_c \qquad\qquad C_c = \dot{m}_c c_{p,c} \qquad (11.11)$$

where C_h and C_c are the hot and cold fluid *heat capacity rates*, respectively. Note that these expressions may be integrated across the heat exchanger to obtain the overall energy balances given by Equations 11.6 and 11.7. The heat transfer across the surface area dA may also be expressed as

$$\boxed{dq = U\Delta T\, dA} \qquad (11.12)$$

where $\Delta T = T_h - T_c$ is the *local* temperature difference between the hot and cold fluids.

To determine the integrated form of Equation 11.12, we begin by substituting Equations 11.10 and 11.11 into the differential form of Equation 11.8

$$d(\Delta T) = dT_h - dT_c$$

to obtain

$$d(\Delta T) = -dq\left(\frac{1}{C_h} + \frac{1}{C_c}\right)$$

Substituting for dq from Equation 11.12 and integrating across the heat exchanger, we then obtain

$$\int_1^2 \frac{d(\Delta T)}{\Delta T} = -U\left(\frac{1}{C_h} + \frac{1}{C_c}\right)\int_1^2 dA$$

or

$$\ln\left(\frac{\Delta T_2}{\Delta T_1}\right) = -UA\left(\frac{1}{C_h} + \frac{1}{C_c}\right) \qquad (11.13)$$

Substituting for C_h and C_c from Equations 11.6 and 11.7, respectively, it follows that

$$\ln\left(\frac{\Delta T_2}{\Delta T_1}\right) = -UA\left(\frac{T_{h,i} - T_{h,o}}{q} + \frac{T_{c,o} - T_{c,i}}{q}\right)$$

$$= -\frac{UA}{q}\left[(T_{h,i} - T_{c,i}) - (T_{h,o} - T_{c,o})\right]$$

Recognizing that, for the parallel-flow heat exchanger of Figure 11.7, $\Delta T_1 = (T_{h,i} - T_{c,i})$ and $\Delta T_2 = (T_{h,o} - T_{c,o})$, we then obtain

$$q = UA\frac{\Delta T_2 - \Delta T_1}{\ln(\Delta T_2/\Delta T_1)}$$

Comparing the above expression with Equation 11.9, we conclude that the appropriate average temperature difference is a *log mean temperature difference*, ΔT_{lm}. Accordingly, we may write

$$q = UA\Delta T_{lm} \qquad (11.14)$$

where

$$\Delta T_{lm} = \frac{\Delta T_2 - \Delta T_1}{\ln(\Delta T_2/\Delta T_1)} = \frac{\Delta T_1 - \Delta T_2}{\ln(\Delta T_1/\Delta T_2)} \qquad (11.15)$$

Remember that, for the *parallel-flow exchanger,*

$$\left[\begin{array}{l} \Delta T_1 \equiv T_{h,1} - T_{c,1} = T_{h,i} - T_{c,i} \\ \Delta T_2 \equiv T_{h,2} - T_{c,2} = T_{h,o} - T_{c,o} \end{array}\right] \tag{11.16}$$

11.3.2 The Counterflow Heat Exchanger

The hot and cold fluid temperature distributions associated with a counterflow heat exchanger are shown in Figure 11.8. In contrast to the parallel-flow exchanger, this configuration provides for heat transfer between the hotter portions of the two fluids near the entrance, as well as between the colder portions near the exit. For this reason, the change in the temperature difference, $\Delta T = T_h - T_c$, with respect to x is nowhere as large as it is for the inlet region of the parallel-flow exchanger. Note that the outlet temperature of the cold fluid may now exceed the outlet temperature of the hot fluid.

Equations 11.6 and 11.7 apply to any heat exchanger and hence may be used for the conterflow arrangement. Moreover, from an analysis like that performed in the preceding section, it may be shown that Equations 11.14 and 11.15 also apply. However, for the *counterflow exchanger* the end-point temperature differences must now be defined as

$$\left[\begin{array}{l} \Delta T_1 \equiv T_{h,1} - T_{c,1} = T_{h,i} - T_{c,o} \\ \Delta T_2 \equiv T_{h,2} - T_{c,2} = T_{h,o} - T_{c,i} \end{array}\right] \tag{11.17}$$

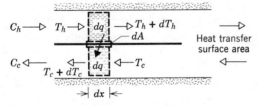

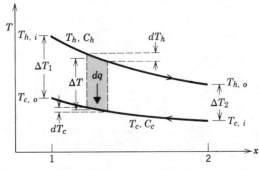

Figure 11.8 Temperature distributions for a counterflow heat exchanger.

Note that, for the same inlet and outlet temperatures, the log mean temperature difference for counterflow exceeds that for parallel flow, $\Delta T_{\text{lm,CF}} > \Delta T_{\text{lm,PF}}$. Hence the surface area required to effect a prescribed heat transfer rate, q, is smaller for the counterflow than for the parallel-flow arrangement, assuming the same value of U.

11.3.3 Special Operating Conditions

It is useful to note certain special conditions under which heat exchangers may be operated. In Figure 11.9a temperature distributions are shown for a heat exchanger in which the hot fluid has a heat capacity rate, $C_h \equiv \dot{m}_h c_{p,h}$, which is much larger than that of the cold fluid, $C_c \equiv \dot{m}_c c_{p,c}$. For this case the temperature of the hot fluid remains approximately constant throughout the heat exchanger, while the temperature of the cold fluid increases. The same condition is achieved if the hot fluid is a condensing vapor. Condensation occurs at constant temperature, and, for all practical purposes, $C_h \rightarrow \infty$. Conversely, in an evaporator or a boiler, Figure 11.9b, it is the cold fluid that experiences a change in phase and remains at a nearly uniform temperature $(C_c \rightarrow \infty)$. The same effect is achieved without phase change if $C_h \ll C_c$. The third special case, Figure 11.9c, involves a counterflow heat exchanger for which the heat capacity rates are equal $(C_h = C_c)$. The temperature difference, ΔT, must then be a constant throughout the exchanger, in which case $\Delta T_1 = \Delta T_2 = \Delta T_{\text{lm}}$.

11.3.4 Multipass and Cross-Flow Heat Exchangers

Although conditions become more complicated in multipass and crossflow heat exchangers, Equations 11.6, 11.7, 11.14 and 11.15 may still be used for the purpose of analysis. The only modification that need be made involves the log

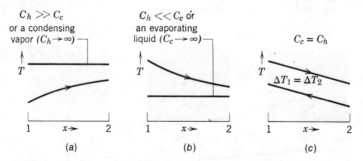

Figure 11.9 Special heat exchanger conditions. (a) $C_h \gg C_c$ or a condensing vapor. (b) An evaporating liquid or $C_h \ll C_c$. (c) A counterflow heat exchanger with equivalent fluid heat capacities $(C_h = C_c)$.

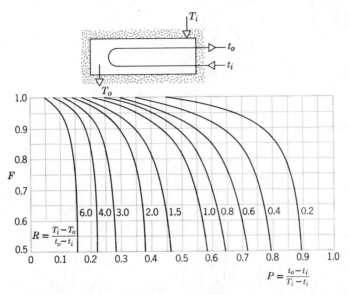

Figure 11.10 Correction factor for a shell-and-tube heat exchanger with one shell and any multiple of two tube passes (two, four, etc. tube passes).

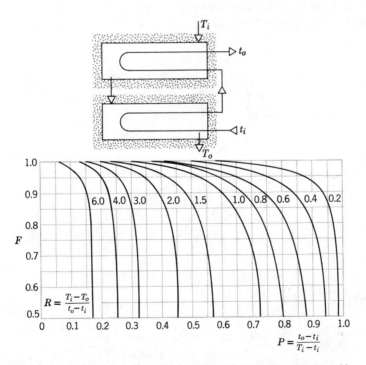

Figure 11.11 Correction factor for a shell-and-tube heat exchanger with two shell passes and any multiple of four tube passes (four, eight, etc. tube passes).

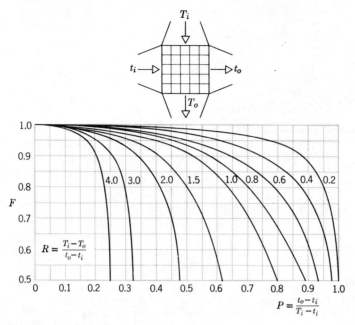

Figure 11.12 Correction factor for a single-pass, cross-flow heat exchanger with both fluids unmixed.

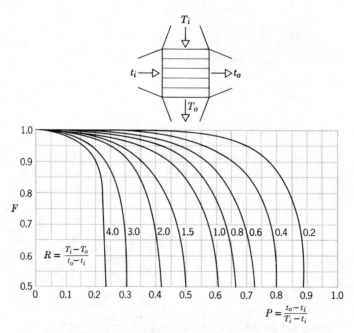

Figure 11.13 Correction factor for a single-pass, cross-flow heat exchanger with one fluid mixed and the other unmixed.

mean temperature difference and is of the form

$$✳ \quad \Delta T_{lm} = F \Delta T_{lm,CF} \tag{11.18}$$

That is, the appropriate form of ΔT_{lm} is obtained by applying a correction factor to the value of ΔT_{lm} that would be computed *under the assumption of counterflow conditions.* Hence from Equation 11.17 $\Delta T_1 = T_{h,i} - T_{c,o}$ and $\Delta T_2 = T_{h,o} - T_{c,i}$.

Algebraic expressions for the correction factor, F, have been developed for various shell-and-tube and cross-flow heat exchanger configurations [5, 9, 10], and the results may be represented graphically. Selected results are shown in Figures 11.10 to 11.13 for common heat exchanger configurations. The notation (T, t) is used to specify the fluid temperatures, with the variable t always assigned to the tube-side fluid. With this convention it does not matter whether the hot fluid or the cold fluid flows through the shell or the tubes.

EXAMPLE 11.1

A counterflow, concentric tube heat exchanger is used to cool the lubricating oil for a large industrial gas turbine engine. The flow rate of cooling water through the inner tube $(D_i = 25$ mm$)$ is $\dot{m}_c = 0.2$ kg/s, while the flow rate of oil through the outer annulus $(D_o = 45$ mm$)$ is $\dot{m}_h = 0.1$ kg/s. The oil and water enter at temperatures of $T_{h,i} = 100°C$ and $T_{c,i} = 30°C$, respectively. How long must the tube be made if the outlet temperature of the oil is to be $T_{h,o} = 60°C$?

SOLUTION

KNOWN:

Fluid flow rates and inlet temperatures for a counterflow, concentric tube heat exchanger of prescribed inner and outer diameter.

FIND:

Tube length required to achieve a desired hot fluid outlet temperature.

SCHEMATIC:

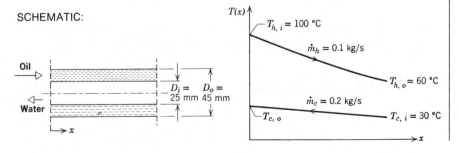

ASSUMPTIONS:

1. Negligible heat loss to the surroundings.
2. Negligible kinetic and potential energy changes.
3. Constant properties.
4. Negligible tube wall thermal resistance and fouling factors.
5. Fully developed conditions for the water and oil (U is independent of x).

PROPERTIES:

Table A.5, unused engine oil ($\bar{T}_{h,m} = 80°C = 353$ K): $c_p = 2131$ J/kg·K, $\mu = 3.25 \times 10^{-2}$ N·s/m², $k = 0.138$ W/m·K.
Table A.6, water ($\bar{T}_{c,m} \approx 35°C$): $c_p = 4178$ J/kg·K, $\mu = 725 \times 10^{-6}$ N·s/m², $k = 0.625$ W/m·K, $Pr = 4.85$. (Note, the value of $T_{c,o}$, and hence $\bar{T}_{c,m} = (T_{c,i} + T_{c,o})/2$, may be obtained from use of the energy balance equations in the early stages of the analysis).

ANALYSIS:

The required heat transfer rate may be obtained from the overall energy balance for the hot fluid, Equation 11.6.

$$q = \dot{m}_h c_{p,h}(T_{h,i} - T_{h,o})$$

$$q = 0.1 \text{ kg/s} \times 2131 \text{ J/kg·K} (100°C - 60°C)$$

$$q = 8524 \text{ W}$$

Applying the overall energy balance for the cold fluid, Equation 11.7, it follows that the water outlet temperature is

$$T_{c,o} = \frac{q}{\dot{m}_c c_{p,c}} + T_{c,i}$$

$$T_{c,o} = \frac{8524 \text{ W}}{0.2 \text{ kg/s} \times 4178 \text{ J/kg·K}} + 30°C$$

$$T_{c,o} = 40.2°C$$

Accordingly, use of $\bar{T}_{c,m} = 35°C$ is justified. The required heat exchanger length may now be obtained from Equation 11.14

$$q = UA\Delta T_{lm}$$

where $A = \pi D_i L$ and from Equations 11.15 and 11.17

$$\Delta T_{lm} = \frac{(T_{h,i} - T_{c,o}) - (T_{h,o} - T_{c,i})}{\ln [(T_{h,i} - T_{c,o})/(T_{h,o} - T_{c,i})]} = \frac{59.8 - 30}{\ln (59.8/30)}$$

$$\Delta T_{lm} = 43.2°C$$

The overall heat transfer coefficient may be obtained from Equation 11.4 or 11.5

$$U = \frac{1}{(1/h_i) + (1/h_o)}$$

For the flow of water through the tube,

$$Re_D = \frac{4\dot{m}_c}{\pi D_i \mu} = \frac{4 \times 0.2 \text{ kg/s}}{\pi(0.025 \text{ m})\, 725 \times 10^{-6} \text{ N·s/m}^2} = 14{,}050$$

Accordingly, the flow is turbulent and the convection coefficient may be computed from Equation 8.58

$$Nu_D = 0.023 Re_D^{4/5}\, Pr^{0.4}$$
$$Nu_D = 0.023(14{,}050)^{4/5}(4.85)^{0.4} = 90$$

Hence

$$h_i = Nu_D \frac{k}{D_i} = \frac{90 \times 0.625 \text{ W/m·K}}{0.025 \text{ m}} = 2250 \text{ W/m}^2\text{·K}$$

For the flow of oil through the annulus, the hydraulic diameter is, from Equation 8.67, $D_h = D_o - D_i = 0.02$ m, and the Reynolds number is

$$Re_D = \frac{\rho u_m D_h}{\mu} = \frac{\rho(D_o - D_i)}{\mu} \times \frac{\dot{m}_h}{\rho \pi(D_o^2 - D_i^2)/4}$$

$$Re_D = \frac{4\dot{m}_h}{\pi(D_o + D_i)\mu} = \frac{4 \times 0.1 \text{ kg/s}}{\pi(0.045 + 0.025)\text{m} \times 3.25 \times 10^{-2} \text{ kg/s·m}}$$

$$Re_D = 56.0$$

The annular flow is therefore laminar. Assuming uniform temperature along the inner surface of the annulus and a perfectly insulated outer surface, the convection coefficient at the inner surface of the annulus may be obtained from Table 8.2. With $(D_i/D_o) = 0.56$, it follows from linear interpolation that

$$Nu_i = \frac{h_o D_h}{k} = 5.56$$

and

$$h_o = 5.56\frac{0.138 \text{ W/m·K}}{0.020 \text{ m}} = 38.4 \text{ W/m}^2\text{·K}$$

The overall convection coefficient is then

$$U = \frac{1}{(1/2250 \text{ W/m}^2 \cdot \text{K}) + (1/38.4 \text{ W/m}^2 \cdot \text{K})} = 37.8 \text{ W/m}^2 \cdot \text{K}$$

From the rate equation it then follows that

$$L = \frac{q}{U \pi D_i \Delta T_{\text{lm}}} = \frac{8524 \text{ W}}{37.8 \text{ W/m}^2 \cdot \text{K} \, \pi (0.025 \text{ m}) (43.2^\circ\text{C})}$$

$$L = 66.5 \text{ m} \qquad\qquad\qquad \triangleleft$$

COMMENTS:

1. Note that the hot side convection coefficient controls the rate of heat transfer between the two fluids, and the low value of h_o is responsible for the large value of L required. A spiral tube arrangement would be needed.

2. Because $h_i \gg h_o$ the tube wall temperature will follow closely that of the coolant water. Accordingly, the assumption of uniform wall temperature used to obtain h_o is reasonable.

EXAMPLE 11.2

A shell-and-tube heat exchanger must be designed to heat 2.5 kg/s of water from 15°C to 85°C. The heating is to be accomplished by passing hot engine oil, which is available at 160°C, through the shell side of the exchanger. The oil is known to provide an average convection coefficient of $h_o = 400 \text{ W/m}^2 \cdot \text{K}$ on the outside of the tubes. Ten tubes are used to pass the water through the shell. Each tube is thin walled, of diameter $D = 25$ mm, and makes eight passes through the shell.

1. If the oil is to leave the exchanger at 100°C, what must its flow rate be?

2. How long must the tubes be to accomplish the desired heating?

SOLUTION

KNOWN:

Fluid inlet and outlet temperatures for a shell-and-tube heat exchanger with 10 tubes making eight passes.

FIND:

1. Oil flow rate required to achieve specified outlet temperature.
2. Tube length required to achieve specified water heating.

SCHEMATIC:

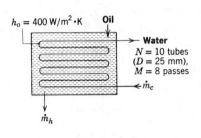

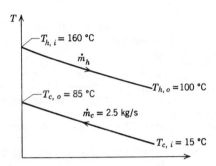

ASSUMPTIONS:

1. Negligible heat loss to the surroundings and kinetic and potential energy changes.
2. Constant properties.
3. Negligible tube wall thermal resistance and fouling effects.
4. Fully developed water flow in tubes.

PROPERTIES:

Table A.5, unused engine oil ($\bar{T}_m = 130°C$): $c_p = 2350$ J/kg·K.
Table A.6, water ($\bar{T}_m = 50°C$): $c_p = 4181$ J/kg·K, $\mu = 548 \times 10^{-6}$ N·s/m^2, $k = 0.643$ W/m·K, $Pr = 3.56$.

ANALYSIS:

1. From the overall energy balance, Equation 11.7, the heat transfer required of the exchanger is

$$q = \dot{m}_c c_{p,c}(T_{c,o} - T_{c,i}) = 2.5 \text{ kg/s} \times 4181 \text{ J/kg·K}(85-15)°C$$
$$= 7.317 \times 10^5 \text{ W}$$

Hence, from Equation 11.6,

$$\dot{m}_h = \frac{q}{c_{p,h}(T_{h,i} - T_{h,o})} = \frac{7.317 \times 10^5 \text{ W}}{2350 \text{ J/kg·K} \times (160-100)°C}$$

$$\dot{m}_h = 5.19 \text{ kg/s} \qquad \triangleleft$$

2. The required tube length may be obtained from Equations 11.14 and 11.18, where

$$q = UAF\Delta T_{lm, CF}$$

From Equation 11.4 or 11.5, $\quad U = \dfrac{1}{(1/h_i) + (1/h_o)}$

where h_i may be obtained by first calculating Re_D. With $\dot{m}_1 \equiv \dot{m}_c/N = 0.25$ kg/s defined as the water flowrate per tube, it follows from Equation 8.6 that

$$Re_D = \frac{4\dot{m}_1}{\pi D \mu} = \frac{4 \times 0.25 \text{ kg/s}}{\pi(0.025 \text{ m})548 \times 10^{-6} \text{ kg/s}\cdot\text{m}} = 23{,}234$$

Hence the water flow is turbulent, and from Equation 8.58

$$Nu_D = 0.023 Re_D^{4/5} \, Pr^{0.4} = 0.023(23234)^{4/5}(3.56)^{0.4} = 119$$

$$h_i = \frac{k}{D} Nu_D = \frac{0.643 \text{ W/m}\cdot\text{K}}{0.025 \text{ m}} 119 = 3061 \text{ W/m}^2\cdot\text{K}$$

Hence

$$U = \frac{1}{(1/400) + (1/3061)} = 354 \text{ W/m}^2\cdot\text{K}$$

The correction factor F may be obtained from Figure 11.10, where

$$R = \frac{160 - 100}{85 - 15} = 0.86 \qquad P = \frac{85 - 15}{160 - 15} = 0.48$$

Hence

$$F \approx 0.87$$

From Equations 11.15 and 11.17, it follows that

$$\Delta T_{lm, CF} = \frac{(T_{h,i} - T_{c,o}) - (T_{h,o} - T_{c,i})}{\ln[(T_{h,i} - T_{c,o})/(T_{h,o} - T_{c,i})]} = \frac{75 - 85}{\ln(75/85)} = 79.9°C$$

Hence, since $A = N\pi DL$, where $N = 10$ is the number of tubes,

$$L = \frac{q}{UN\pi DF\Delta T_{lm, CF}} = \frac{7.317 \times 10^5 \text{ W}}{354 \text{ W/m}^2\cdot\text{K} \times 10 \, \pi(0.025 \text{ m})0.87(79.9°C)}$$

$$L = 37.9 \text{ m} \qquad \qquad \triangleleft$$

COMMENTS:

1. With $(L/D) = 37.9$ m$/0.025$ m $= 1516$, the assumption of fully developed conditions throughout the tube is justified.

2. With eight passes, the shell length is approximately $L/M = 4.7$ m.

11.4 HEAT EXCHANGER ANALYSIS: THE EFFECTIVENESS–NTU METHOD

It is a simple matter to use the log mean temperature difference (LMTD) method of heat exchanger analysis when the fluid inlet temperatures are known and the outlet temperatures are specified or readily determined from the energy balance expressions, Equations 11.6 and 11.7. The value of ΔT_{lm} for the exchanger may then be determined. However, situations often arise for which only the inlet temperatures are known, and use of the LMTD method requires an iterative procedure. In such cases it is preferable to use an alternative approach, which we term the *effectiveness–NTU* method.

11.4.1 Definitions

To define the effectiveness of a heat exchanger, we must first determine the *maximum possible heat transfer rate*, q_{max}, for the exchanger. This heat transfer rate could, in principle, be achieved in a counterflow heat exchanger (Figure 11.8) of infinite length. In such an exchanger, one of the fluids would experience the maximum possible temperature difference, $T_{h,i} - T_{c,i}$. To illustrate this point, consider a situation for which $C_c < C_h$, in which case, from Equations 11.10 and 11.11, $|dT_c| > |dT_h|$. The cold fluid would then experience the larger temperature change, and since $L \rightarrow \infty$, it would be heated to the inlet temperature of the hot fluid ($T_{c,o} = T_{h,i}$). Accordingly, from Equation 11.7

$$C_c < C_h: \quad q_{max} = C_c(T_{h,i} - T_{c,i})$$

Similarly, if $C_h < C_c$, the hot fluid would experience the larger temperature change and would be cooled to the inlet temperature of the cold fluid ($T_{h,o} = T_{c,i}$). From Equation 11.6 we then obtain

$$C_h < C_c: \quad q_{max} = C_h(T_{h,i} - T_{c,i})$$

From the foregoing results we are then prompted to write the general expression

$$q_{max} = C_{min}(T_{h,i} - T_{c,i}) \tag{11.19}$$

where C_{min} is equal to C_c or C_h, whichever is smaller. For prescribed hot and cold fluid inlet temperatures, Equation 11.19 provides the maximum heat transfer rate that could possibly be experienced by the heat exchanger.

It is now logical to define the *effectiveness*, ε, as the ratio of the actual heat transfer rate for a heat exchanger to the maximum possible heat transfer rate. That is,

$$\varepsilon \equiv \frac{q}{q_{max}} \tag{11.20}$$

From Equations 11.6, 11.7, and 11.19, it follows that

$$\varepsilon = \frac{C_h(T_{h,i} - T_{h,o})}{C_{min}(T_{h,i} - T_{c,i})} \tag{11.21}$$

or

$$\varepsilon = \frac{C_c(T_{c,o} - T_{c,i})}{C_{min}(T_{h,i} - T_{c,i})} \tag{11.22}$$

By definition the effectiveness, which is a dimensionless parameter, must be in the range $0 \leq \varepsilon \leq 1$. Its utility rests with the fact that, if ε, $T_{h,i}$, and $T_{c,i}$ are known, the actual heat transfer rate for the exchanger may readily be determined from the expression

$$q = \varepsilon C_{min}(T_{h,i} - T_{c,i}) \qquad \text{CAN FIND H.T. FROM INLET CONDITIONS} \tag{11.23}$$

For any heat exchanger it can be shown that

$$\varepsilon = f(\text{NTU}, C_{min}/C_{max}) \tag{11.24}$$

where C_{min}/C_{max} is equal to C_c/C_h or C_h/C_c, depending on the relative magnitudes of the hot and cold fluid heat capacity rates. The *number of transfer units*, NTU, is a dimensionless parameter that is widely used for heat exchanger analysis [3] and is defined as

$$\text{NTU} \equiv \frac{UA}{C_{min}} \tag{11.25}$$

It is therefore the ratio of the heat transfer rate per degree of average temperature difference between the fluids, Equation 11.14, to the heat transfer rate per degree of temperature change for the fluid of minimum heat capacity rate.

11.4.2 Effectiveness–NTU Relations

To determine a specific form of the effectiveness–NTU relation, Equation 11.24, consider a parallel-flow heat exchanger for which $C_{min} = C_h$. From Equation 11.21 we then obtain

$$\varepsilon = \frac{T_{h,i} - T_{h,o}}{T_{h,i} - T_{c,i}} \tag{11.26}$$

and from Equations 11.6 and 11.7 it follows that

$$\frac{C_{min}}{C_{max}} = \frac{\dot{m}_h c_{p,h}}{\dot{m}_c c_{p,c}} = \frac{T_{c,o} - T_{c,i}}{T_{h,i} - T_{h,o}} \tag{11.27}$$

Now consider Equation 11.13, which may be expressed as

$$\ln\left(\frac{T_{h,o} - T_{c,o}}{T_{h,i} - T_{c,i}}\right) = -\frac{UA}{C_{min}}\left(1 + \frac{C_{min}}{C_{max}}\right)$$

or from Equation 11.25

$$\frac{T_{h,o} - T_{c,o}}{T_{h,i} - T_{c,i}} = \exp\left[-\text{NTU}\left(1 + \frac{C_{min}}{C_{max}}\right)\right] \tag{11.28}$$

Rearranging the left-hand side of this expression as

$$\frac{T_{h,o} - T_{c,o}}{T_{h,i} - T_{c,i}} = \frac{T_{h,o} - T_{h,i} + T_{h,i} - T_{c,o}}{T_{h,i} - T_{c,i}}$$

and substituting for $T_{c,o}$ from Equation 11.27, it follows that

$$\frac{T_{h,o} - T_{c,o}}{T_{h,i} - T_{c,i}} = \frac{(T_{h,o} - T_{h,i}) + (T_{h,i} - T_{c,i}) - (C_{min}/C_{max})(T_{h,i} - T_{h,o})}{T_{h,i} - T_{c,i}}$$

or from Equation 11.26

$$\frac{T_{h,o} - T_{c,o}}{T_{h,i} - T_{c,i}} = -\varepsilon + 1 - (C_{min}/C_{max})\varepsilon = 1 - \varepsilon\left(1 + \frac{C_{min}}{C_{max}}\right)$$

Substituting the above expression into Equation 11.28 and solving for ε, we then obtain for the *parallel-flow heat exchanger*

$$\varepsilon = \frac{1 - \exp\{-\text{NTU}[1 + (C_{min}/C_{max})]\}}{[1 + (C_{min}/C_{max})]} \tag{11.29}$$

Since precisely the same result may be obtained for $C_{min} = C_c$, Equation 11.29 applies for any parallel–flow heat exchanger, irrespective of whether the minimum heat capacity rate is associated with the hot or cold fluid.

Results of similar form have been developed for a variety of heat exchangers

[3]. For the *counterflow heat exchanger*

$$\varepsilon = \frac{1 - \exp\{-NTU[1 - (C_{min}/C_{max})]\}}{1 - (C_{min}/C_{max})\exp\{-NTU[1 - (C_{min}/C_{max})]\}} \tag{11.30}$$

For a *shell-and-tube heat exchanger* with *one shell pass* and any multiple of two tube passes

$$\varepsilon_1 = 2\left\{1 + C_r + (1 + C_r^2)^{1/2}\frac{1 + \exp[-NTU(1 + C_r^2)^{1/2}]}{1 - \exp[-NTU(1 + C_r^2)^{1/2}]}\right\}^{-1} \tag{11.31}$$

where $C_r = (C_{min}/C_{max})$. For a *shell-and-tube heat exchanger* with *two shell passes* and any multiple of four tube passes

$$\varepsilon_2 = \left[\left(\frac{1 - \varepsilon_1 C_r}{1 - \varepsilon_1}\right)^2 - 1\right]\left[\left(\frac{1 - \varepsilon_1 C_r}{1 - \varepsilon_1}\right)^2 - C_r\right]^{-1} \tag{11.32}$$

where ε_1 is given by Equation 11.31. For a single-pass, *cross-flow heat exchanger* with *both fluids unmixed*

$$\varepsilon \approx 1 - \exp[C_r(NTU)^{0.22}\{\exp[-C_r(NTU)^{0.78}] - 1\}] \tag{11.33}$$

Finally, for a single-pass, *cross-flow heat exchanger* with *one fluid mixed* and the *other unmixed*, it is known that

$$\varepsilon = (1/C_r)(1 - \exp\{C_r[1 - \exp(-NTU)]\}) \tag{11.34}$$

if C_{max} and C_{min} are associated with the mixed and unmixed fluids, respectively, or

$$\varepsilon = 1 - \exp(C_r^{-1}\{1 - \exp[-C_r(NTU)]\}) \tag{11.35}$$

if C_{max} and C_{min} are associated with the unmixed and mixed fluids, respectively.

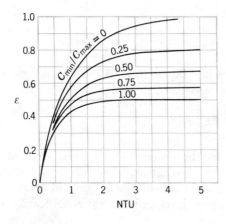

Figure 11.14 Effectiveness of a parallel-flow heat exchanger.

Note that Equation 11.33 may only be used with confidence for $C_r \approx 1$. In general, it is recommended that the graphical results of Figure 11.18 be used, rather than the approximation of Equation 11.33.

The foregoing expressions are represented graphically in Figures 11.14 to 11.19. For Figure 11.19 the solid curves correspond to C_{min} mixed and C_{max} unmixed, while the dashed curves correspond to C_{min} unmixed and C_{max} mixed. Recall from the discussion of Section 11.3.3 that, for an evaporator or a condenser, $C_r = 0$, since the heat capacity rate of the phase change fluid is considered to be infinite.

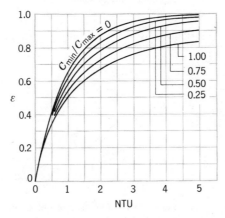

Figure 11.15 Effectiveness of a counterflow heat exchanger.

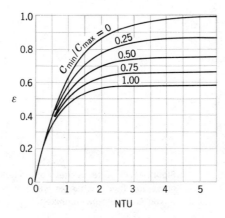

Figure 11.16 Effectiveness of a shell-and-tube heat exchanger with one shell and any multiple of two tube passes (two, four, etc. tube passes).

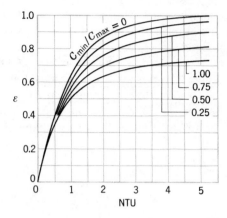

Figure 11.17 Effectiveness of a shell-and-tube heat exchanger with two shell passes and any multiple of four tube passes (four, eight, etc. tube passes).

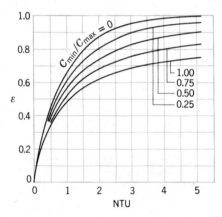

Figure 11.18 Effectiveness of a single-pass, cross-flow heat exchanger with both fluids unmixed.

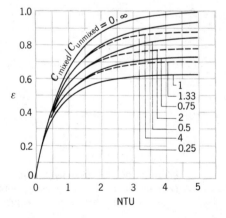

Figure 11.19 Effectiveness of a single-pass, cross-flow heat exchanger with one fluid mixed and the other unmixed.

EXAMPLE 11.3

Hot exhaust gases, which enter a finned-tube, cross-flow heat exchanger at 300°C and leave at 100°C, are used to heat pressurized water at a flowrate of 1 kg/s from 35°C to 125°C. The exhaust gas specific heat is approximately 1000 J/kg·K, and the overall heat transfer coefficient based on the gas side surface area is $U_h = 100$ W/m²·K. Determine the required gas side surface area, A_h, using the NTU method.

SOLUTION

KNOWN:

Inlet and outlet temperatures of hot gases and water used in a finned-tube, cross-flow heat exchanger. Water flowrate and gas side overall heat transfer coefficient.

FIND:

Required gas side surface area.

SCHEMATIC:

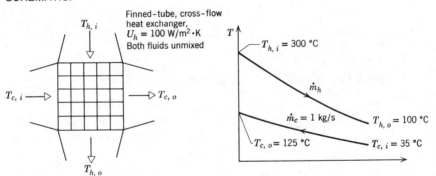

ASSUMPTIONS:

1. Negligible heat loss to the surroundings and kinetic and potential energy changes.
2. Constant properties.

PROPERTIES:

Table A.6, water ($\bar{T}_m = 80°C$): $c_p = 4197$ J/kg·K.
Exhaust gases: $c_p = 1000$ J/kg·K.

ANALYSIS:

The required surface area may be obtained from knowledge of the number of transfer units, which, in turn, may be obtained from knowledge of the ratio of heat capacity rates and the effectiveness. To determine the minimum heat capacity rate, we begin by computing

$$C_c = \dot{m}_c c_{p,c} = 1 \text{ kg/s} \times 4197 \text{ J/kg·K} = 4197 \text{ W/K}$$

Combining the overall energy balances, Equations 11.6 and 11.7, it follows that

$$C_h = \dot{m}_h c_{p,h} = C_c \frac{T_{c,o} - T_{c,i}}{T_{h,i} - T_{h,o}} = 4197 \frac{125 - 35}{300 - 100} = 1889 \text{ W/K}$$

Hence

$$C_{min} = C_h = 1889 \text{ W/K}$$

From Equation 11.19

$$q_{max} = C_{min}(T_{h,i} - T_{c,i}) = 1889 \text{ W/K} (300 - 35)^{\circ}\text{C}$$

or

$$q_{max} = 5.01 \times 10^5 \text{ W}$$

From Equation 11.7 the actual heat transfer rate is

$$q = \dot{m}_c c_{p,c}(T_{c,o} - T_{c,i}) = 1 \text{ kg/s} \times 4197 \text{ J/kg·K}(125 - 35)^{\circ}\text{C}$$

$$q = 3.77 \times 10^5 \text{ W}$$

Hence from Equation 11.20 the effectiveness is

$$\varepsilon = \frac{q}{q_{max}} = \frac{3.77 \times 10^5 \text{ W}}{5.01 \times 10^5 \text{ W}} = 0.75$$

With

$$\frac{C_{min}}{C_{max}} = \frac{1889}{4197} = 0.45$$

it follows from Figure 11.18 that

$$\text{NTU} = \frac{U_h A_h}{C_{min}} \approx 2.1$$

or

$$A_h = \frac{2.1(1889 \text{ W/K})}{100 \text{ W/m}^2 \text{·K}} = 39.7 \text{ m}^2$$

COMMENTS:

The desired heat transfer area may also be determined by using the LMTD method. From Equations 11.14 and 11.18,

$$A_h = \frac{q}{U_h F \Delta T_{\text{lm,CF}}}$$

With

$$P = \frac{t_o - t_i}{T_i - t_i} = \frac{125 - 35}{300 - 35} = 0.34$$

and

$$R = \frac{T_i - T_o}{t_o - t_i} = \frac{300 - 100}{125 - 35} = 2.22$$

it follows from Figure 11.12 that $F \approx 0.87$. From Equations 11.15 and 11.17,

$$\Delta T_{\text{lm,CF}} = \frac{(T_{h,i} - T_{c,o}) - (T_{h,o} - T_{c,i})}{\ln[(T_{h,i} - T_{c,o})/(T_{h,o} - T_{c,i})]} = 111°C$$

in which case

$$A_h = \frac{3.77 \times 10^5 \text{ W}}{100 \text{ W/m}^2 \cdot \text{K} \times 0.87 \times 111°C} = 39.1 \text{ m}^2 \qquad \lhd$$

Agreement of the two results is well within the accuracy associated with reading the F and ε, NTU, $C_{\text{min}}/C_{\text{max}}$ charts.

11.5 METHODOLOGY OF A HEAT EXCHANGER CALCULATION

We have developed two procedures for performing a heat exchanger analysis, the LMTD method and the NTU approach. For any problem both methods may be used to obtain equivalent results. However, depending on the nature of the problem, the NTU approach may be easier to implement than the LMTD approach.

Clearly, implementation of the LMTD method, Equations 11.14 and 11.15, is facilitated by knowledge of the hot and cold fluid inlet and outlet temperatures, since ΔT_{lm} may then be readily computed. Problems for which these temperatures are known may be classified as *heat exchanger design problems*. Typically, the fluid inlet temperatures and flowrates are specified, and a desired hot or cold fluid outlet temperature is prescribed. The design problem is then one of selecting an appropriate heat exchanger type and determining the size, that is, the heat transfer surface area A, required to achieve the desired outlet temperature. For example, consider an application for which $\dot{m}_c$, $\dot{m}_h$, $T_{c,i}$, and $T_{h,i}$ are known, and the objective is to specify a heat exchanger that will provide a desired value of $T_{c,o}$. The corresponding values of q and $T_{h,o}$ may be computed from the energy balances, Equations 11.7 and 11.6, respectively, and the value of ΔT_{lm} may be found from its definition, Equation 11.15. Using the rate equation, Equation 11.14, it is then a simple matter to determine the required value of A. Of course, the NTU method may also be used to obtain A by first calculating ε and (C_{min}/C_{max}). The appropriate chart (or equation) may then be used to obtain the NTU value, which in turn may be used to determine A.

Alternatively, problems arise for which the heat exchanger type and size are known and the objective is to determine the heat transfer rate and the fluid outlet temperatures for prescribed fluid flowrates and inlet temperatures. Although the LMTD method may be used for such a heat exchanger *performance calculation*, the computations would be tedious, requiring some kind of iteration. For example, a guess could be made for the value of $T_{c,o}$, and Equations 11.7 and 11.6 could be used to determine q and $T_{h,o}$, respectively. Knowing all fluid temperatures, ΔT_{lm} could be determined and Equation 11.14 could then be used to again compute the value of q. The original guess for $T_{c,o}$ would be correct, if the values of q obtained from Equations 11.7 and 11.14 were in agreement. Such agreement would be fortuitous, however, and it is likely that some iteration on the value of $T_{c,o}$ would be needed.

For Performance Calculations use NTU method

The iterative nature of the above solution could be eliminated by using the NTU method. From knowledge of the heat exchanger type and size and the fluid flowrates, the NTU and (C_{min}/C_{max}) values may be computed and ε may then be determined from the appropriate chart (or equation). Since q_{max} may also be computed from Equation 11.19, it is then a simple matter to determine the actual heat transfer rate from the requirement that $q = \varepsilon q_{max}$. Both fluid outlet temperatures may then be determined from Equations 11.6 and 11.7.

EXAMPLE 11.4

Consider the heat exchanger design of Example 11.3, that is, a finned-tube, cross-flow heat exchanger with a gas side heat transfer coefficient and area of $U_h = 100$ W/m²·K and $A_h = 40$ m², respectively. The water flowrate and inlet temperature remain at $\dot{m}_c = 1$ kg/s and $T_{c,i} = 35°$C. However, a change in operating conditions for the hot gas generator causes the gases to now enter the heat exchanger with a flowrate of $\dot{m}_h = 1.5$ kg/s and a temperature of $T_{h,i} = 250°$C. What is the rate of heat transfer by the exchanger, and what are the gas and water outlet temperatures?

SOLUTION

KNOWN:

Hot and cold fluid inlet conditions for a finned-tube, cross-flow heat exchanger of known surface area and overall heat transfer coefficient.

FIND:

Heat transfer rate and fluid outlet temperatures.

SCHEMATIC:

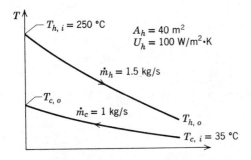

$T_{h,\,i} = 250\ °C$

$A_h = 40\ m^2$
$U_h = 100\ W/m^2 \cdot K$

$\dot{m}_h = 1.5\ kg/s$

$T_{c,\,o}$

$\dot{m}_c = 1\ kg/s$

$T_{h,\,o}$

$T_{c,\,i} = 35\ °C$

ASSUMPTIONS:

1. Negligible heat loss to surroundings and kinetic and potential energy changes.
2. Constant properties (unchanged from Example 11.3).

PROPERTIES:

Table A.6, water ($\bar{T}_m \approx 80°C$): $c_p = 4197\ J/kg \cdot K$.
Hot gases: $c_p = 1000\ J/kg \cdot K$.

ANALYSIS:

The problem may be classified as one requiring a heat exchanger *performance calculation*. Accordingly, it is expedient to base the calculations on the NTU method. The heat capacity rates are

$$C_c = \dot{m}_c c_{p,c} = 1\ kg/s \times 4197\ J/kg \cdot K = 4197\ W/K$$

$$C_h = \dot{m}_h c_{p,h} = 1.5\ kg/s \times 1000\ J/kg \cdot K = 1500\ W/K$$

in which case

$$C_{\min} = C_h = 1500 \text{ W/K}$$

and

$$\frac{C_{\min}}{C_{\max}} = \frac{1500}{4197} = 0.357$$

The number of transfer units is then

$$\text{NTU} = \frac{U_h A_h}{C_{\min}} = \frac{100 \text{ W/m}^2 \cdot \text{K} \times 40 \text{ m}^2}{1500 \text{ W/K}} = 2.67$$

From Figure 11.18 the heat exchanger effectiveness is then

$$\varepsilon \approx 0.82$$

and from Equation 11.19 the maximum possible heat transfer rate is

$$q_{\max} = C_{\min}(T_{h,i} - T_{c,i})$$

$$q_{\max} = 1500 \text{ W/K}(250 - 35)^\circ\text{C} = 3.23 \times 10^5 \text{ W}$$

Accordingly, from the definition of ε, Equation 11.20, the actual heat transfer rate is

$$q = \varepsilon \, q_{\max} = 0.82 \times 3.23 \times 10^5 \text{ W}$$

$$q = 2.65 \times 10^5 \text{ W} \qquad\qquad\qquad \triangleleft$$

It is now a simple matter to determine the outlet temperatures from the overall energy balances. From Equation 11.6

$$T_{h,o} = T_{h,i} - \frac{q}{\dot{m}_h c_{p,h}} = 250^\circ\text{C} - \frac{2.65 \times 10^5 \text{ W}}{1500 \text{ W/K}}$$

$$T_{h,o} = 73.3^\circ\text{C} \qquad\qquad\qquad\qquad \triangleleft$$

and from Equation 11.7

$$T_{c,o} = T_{c,i} + \frac{q}{\dot{m}_c c_{p,c}} = 35^\circ\text{C} + \frac{2.65 \times 10^5 \text{ W}}{4197 \text{ W/K}}$$

$$T_{c,o} = 98.1^\circ\text{C} \qquad\qquad\qquad\qquad \triangleleft$$

EXAMPLE 11.5

The condenser of a steam power plant is a heat exchanger in which steam is condensed to liquid water. Assume the condenser of a large power plant to be a *shell-and-tube* heat exchanger consisting of a single shell and 30,000 tubes, each executing two passes. The tubes are of thin wall construction with $D = 25$ mm, and steam condenses on their outer surface with an associated convection coefficient of $h_o = 11,000$ W/m²·K. The heat transfer rate that must be effected by the exchanger is $q = 2 \times 10^9$ W, and this is accomplished by passing cooling water through the tubes at a rate of 3×10^4 kg/s (the flowrate per tube is therefore 1 kg/s). The water enters at 20°C, while the steam condenses at 50°C.

1. What is the temperature of the cooling water emerging from the condenser?

2. What is the tube length L per pass required to achieve the above objectives?

SOLUTION

KNOWN:

Heat exchanger consisting of single shell and 30,000 tubes with two passes each.

FIND:

1. Outlet temperature of the cooling water.

2. Tube length per pass to achieve required heat transfer.

SCHEMATIC:

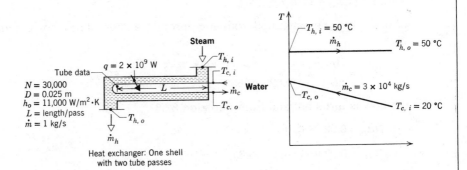

Tube data—
$N = 30,000$
$D = 0.025$ m
$h_o = 11,000$ W/m²·K
$L = $ length/pass
$\dot{m} = 1$ kg/s

Heat exchanger: One shell
with two tube passes

ASSUMPTIONS:

1. Negligible heat transfer between exchanger and surroundings and negligible kinetic and potential energy changes.

2. Tube internal flow and thermal conditions are fully developed.

3. Thermal resistance of tube material and fouling effects are negligible.

4. Constant properties.

PROPERTIES:

Table A.6, water (assume $\bar{T}_c \approx 27°C = 300$ K): $c_p = 4179$ J/kg·K, $\mu = 855 \times 10^{-6}$ N·s/m², $k = 0.613$ W/m·K, $Pr = 5.83$.

ANALYSIS:

1. The cooling water outlet temperature may be obtained from the overall energy balance, Equation 11.7. Accordingly,

$$T_{c,o} = T_{c,i} + \frac{q}{\dot{m}_c c_{p,c}} = 20°C + \frac{2 \times 10^9 \text{ W}}{3 \times 10^4 \text{ kg/s} \times 4179 \text{ J/kg·K}}$$

$$T_{c,o} = 36.0°C \qquad \lhd$$

2. The problem may be classified as one requiring a heat exchanger *design calculation*. Accordingly, either the LMTD or NTU method may be conveniently applied. Using the LMTD method, it follows from Equations 11.14 and 11.18 that

$$q = UAF\Delta T_{\text{lm,CF}}$$

where $A = N \times 2L \times \pi D$. From Equation 11.4 or 11.5

$$U = \frac{1}{(1/h_i) + (1/h_o)}$$

where h_i may be estimated from an appropriate internal flow correlation. With

$$Re_D = \frac{4\dot{m}}{\pi D \mu} = \frac{4 \times 1 \text{ kg/s}}{\pi(0.025 \text{ m})855 \times 10^{-6} \text{ N·s/m}^2} = 59,567$$

the flow is turbulent and from Equation 8.58

$$Nu_D = 0.023 Re_D^{4/5} Pr^{0.4}$$

$$Nu_D = 0.023(59,567)^{0.8}(5.83)^{0.4} = 308$$

Hence

$$h_i = Nu_D \frac{k}{D} = 308 \frac{0.613 \text{ W/m·K}}{0.025 \text{ m}} = 7552 \text{ W/m}^2 \text{·K}$$

and

$$U = \frac{1}{[(1/7552) + (1/11{,}000)]\ \text{m}^2 \cdot \text{K/W}} = 4478\ \text{W/m}^2 \cdot \text{K}$$

From Equations 11.15 and 11.17 the log mean temperature difference is

$$\Delta T_{\text{lm,CF}} = \frac{(T_{h,i} - T_{c,o}) - (T_{h,o} - T_{c,i})}{\ln[(T_{h,i} - T_{c,o})/(T_{h,o} - T_{c,i})]} = \frac{(50 - 36) - (50 - 20)}{\ln(14/30)}$$

$$\Delta T_{\text{lm,CF}} = 21^\circ \text{C}$$

The correction factor F may be obtained from Figure 11.10, where

$$P = \frac{t_o - t_i}{T_i - t_i} = \frac{(36 - 20)}{(50 - 20)} = 0.53$$

$$R = \frac{T_i - T_o}{t_o - t_i} = \frac{50 - 50}{36 - 20} = 0$$

Accordingly,

$$F = 1$$

It then follows that the tube length per pass is

$$L = \frac{q}{U(N2\pi D)F\Delta T_{\text{lm,CF}}}$$

$$L = \frac{2 \times 10^9\ \text{W}}{4478\ \text{W/m}^2 \cdot \text{K}\ (30{,}000 \times 2\pi \times 0.025\ \text{m}) \times 1 \times 21^\circ \text{C}}$$

$$L = 4.51\ \text{m} \qquad \qquad \triangleleft$$

COMMENTS:

1. Recognize that L is the tube length per pass, in which case the total tube length is 9.02 m.

2. Using the NTU method, $C_h = C_{\text{max}} = \infty$ and $C_{\text{min}} = \dot{m}_c c_{p,c} = 3 \times 10^4$ (kg/s) $\times$ 4179 (J/kg·K) = 1.25×10^8 W/K. It follows that $q_{\text{max}} = C_{\text{min}}(T_{h,i} - T_{c,i}) = 3.76 \times 10^9$ W and hence that $\varepsilon = 0.53$. From Figure 11.16 it also follows that NTU ≈ 0.75, from which it may then be shown that $L = 4.46$ m.

11.6 SUMMARY

The existence of many different industrial applications has stimulated a long history of research and development activities related to heat exchangers. Moreover, with increasing attention to energy conservation and alternative energy conversion schemes, sustained activity can be expected. In this chapter we have attempted to develop basic tools that allow you to effect approximate heat exchanger design and performance calculations. Results and methods of a more detailed nature, such as those that deal with variable fluid properties and convection coefficient effects, may be found in the literature.

REFERENCES

1. Fraas, A. P. and M. N. Ozisik, *Heat Exchanger Design*, Wiley, New York, 1965.
2. Kern, D. Q., *Process Heat Transfer*, McGraw-Hill, New York, 1950.
3. Kays, W. M. and A. L. London, *Compact Heat Exchangers*, 2nd Ed., McGraw-Hill, New York, 1964.
4. Coppage, J. E. and A. L. London, "The Periodic Flow Regenerator: A Summary of Design Theory," *Trans. ASME, 75*, 779, 1953.
5. Jakob, M., *Heat Transfer*, Vol. 2, Wiley, New York, 1957.
6. Sherwood, T. K., R. L. Pigford, and C. R. Wilkie, *Mass Transfer*, McGraw-Hill, New York, 1975.
7. Treybal, R. E., *Mass Transfer Operations*, 2nd Ed., McGraw-Hill, New York, 1968.
8. Geankoplis, C. J., *Mass Transport Phenomena*, Holt, Rinehart and Winston, New York, 1972.
9. *Standards of the Tubular Exchange Manufacturers Association*, 6th Ed., Tubular Exchanger Manufacturers Association, New York, 1978.
10. Bowman, R. A., A. C. Mueller, and W. M. Nagle, "Mean Temperature Difference in Design," *Trans. ASME, 62*, 283, 1940.

PROBLEMS

11.1 A shell and tube exchanger (two shells, four tube passes) is used to heat 10,000 kg/h of pressurized water from 35°C to 120°C with 5000 kg/h water entering the exchanger at 300°C. If the overall heat transfer coefficient is 1500 W/m²·K, determine the required heat exchanger area.

11.2 Consider the shell and tube heat exchanger of Problem 11.1. After several years of operation, it is observed that the outlet temperature of the cold water only reaches 95°C rather than the desired 120°C for the same flow rates and inlet temperatures of the fluids. Determine the fouling factor that is the cause of the poorer performance.

11.3 A finned-tube, cross-flow heat exchanger is to use the exhaust of a gas turbine to heat pressurized water. Laboratory measurements are performed on a prototype version of the exchanger, which has a surface area of 10 m², in order to determine the overall heat transfer coefficient as a function of operating conditions. Measurements made under particular conditions, for which $\dot{m}_h = 2$ kg/s, $T_{h,i} = 325°C$, $\dot{m}_c = 0.5$ kg/s, and $T_{c,i} = 25°C$, reveal a water outlet temperature of $T_{c,o}$

$= 150°C$. What is the overall heat transfer coefficient associated with the exchanger under these conditions?

11.4   Water at a rate of 45,500 kg/h is heated from 80°C to 150°C in a heat exchanger having two shell passes and eight tube passes with a total surface area of 925 m². Hot exhaust gases having approximately the same thermophysical properties as air enter at 350°C and exit at 175°C. Determine the overall heat transfer coefficient.

11.5   A thin-walled concentric tube heat exchanger is to be used to cool engine oil from an initial temperature of 160°C to 60°C, and water, which is available at 25°C, is to be used as the coolant. The oil and water flowrates are each 2 kg/s, and the diameter of the inner tube is 0.5 m. The corresponding value of the overall heat transfer coefficient is 250 W/m²·K. How long must the heat exchanger be to accomplish the desired cooling?

11.6   A single-pass, cross-flow heat exchanger uses hot exhaust gases (mixed) to heat water (unmixed) from 30°C to 80°C at a rate of 3 kg/s. The exhaust gases, having thermophysical properties similar to air, enter and exit the exchanger at 225°C and 100°C, respectively. If the overall heat transfer coefficient is 200 W/m²·K, estimate the required surface area.

11.7   Consider the fluid conditions of Problem 11.6 for a concentric tube heat exchanger operating in parallel flow where the thin-walled separator tube has a diameter of 100 mm.

   a) Determine the required length for the exchanger.

   b) Assuming the water flow in the inside of the separator tube is fully developed, estimate the convection heat transfer coefficient.

   c) If the length of the exchanger is twice the value found in part (a), determine the heat transfer rate and exit temperatures of the fluids assuming the overall heat transfer coefficient and the inlet temperatures of the fluids remain the same.

   d) If the exchanger is operated in counter flow, what would be the reduction in required length compared to that found in part (a)?

11.8   Steam at 0.14 bar is condensed in a shell and tube heat exchanger with one shell pass and two tube passes consisting of 130 brass tubes/pass of length 2 m. The tubes have inner and outer diameters of 13.4 mm and 15.9 mm, respectively. Cooling water enters the tubes at 20°C with a mean velocity of 1.25 m/s. The heat transfer convection coefficient for condensation on the outer surfaces of the tubes is 13,500 W/m²·K.

   a) Estimate the overall heat transfer coefficient for the heat exchanger.

   b) Determine the outlet temperature of the cooling water.

   c) Determine the condensation rate of the steam.

11.9   A two-fluid heat exchanger has inlet and outlet temperatures of 65°C and 40°C, respectively, for the hot fluid and 15°C and 30°C, respectively, for the cold fluid.

   a) Can you tell whether this exchanger is operating under counter flow or parallel flow conditions?

   b) If possible, determine the effectiveness of this exchanger when the cold fluid has the minimum capacity rate.

11.10 Water at a rate of 225 kg/h is to be heated from 35°C to 95°C by means of a concentric tube heat exchanger. Oil at a rate of 225 kg/h with a specific heat of 2095 J/kg·K and temperature of 210°C is to be used as the hot fluid. If the overall heat transfer coefficient based on the outer diameter of the inner tube is 550 W/m²·K, determine the length of the exchanger if the outer diameter of the tube is 100 mm.

11.11 A hot fluid enters a concentric pipe heat exchanger at a temperature of 150°C and is to be cooled to 100°C by a cold fluid entering at 35°C and heated to 65°C. Would you use parallel flow or counter flow for the most effective design?

11.12 A shell and tube heat exchanger is to be designed for heating 10,000 kg/h of water from 16°C to 84°C by hot engine oil flowing through the shell of the heat exchanger. The oil makes a single shell pass entering at 160°C and leaving at 94°C with an average heat transfer coefficient of 400 W/m²·K. The water flows through 11 brass tubes of 22.9 mm inside diameter and 25.4 mm outside diameter with each tube making four passes through the shell. Calculate the length of the tubes required for this heat exchanger assuming the water flow in the tubes is fully developed.

11.13 A shell and tube heat exchanger is comprised of 135 tubes in a double pass arrangement, each of 12.5 mm inner diameter with a total surface area 47.5 m². Hot exhaust gas at 200°C, having properties of air (1 atm) and flowing at a rate of 5 kg/s, is to be used for heating water entering at 15°C with a flow rate of 6.5 kg/s. The overall heat transfer coefficient for the exchanger has been estimated at 200 W/m²·K.

   a) Calculate the exchanger outlet temperatures of the gas and water.

   b) Assuming the flow conditions for the water to be fully developed, estimate the heat transfer coefficient for forced convection at the water-tube side of the exchanger.

11.14 An ocean thermal energy conversion system is being proposed for electrical power generation. Such a system is based on the standard power cycle for which the working fluid is evaporated, passed through a turbine, and subsequently condensed. The system is to be used in very special locations for which the temperature of the oceanic waters near the surface is approximately 300 K, while the temperature at reasonable depths beneath the surface is approximately 280 K. The higher temperature water is used as a heat source to evaporate the working fluid, while the lower temperature water is used as a heat sink for condensation of the fluid. Consider a power plant that is to generate 2 MW of electricity at an efficiency (electrical power output per heat input) of 3 percent. The evaporator is to be in the form of a heat exchanger consisting of a single shell with many tubes executing two passes. If the working fluid is to be evaporated at its phase change temperature of 290 K, with ocean water entering at 300 K and leaving at 292 K, what is the heat exchanger area required for the evaporator? What flowrate must be maintained for the water passing through the evaporator? The overall heat transfer coefficient may be approximated as 1200 W/m²·K.

11.15 The feedwater heater for a boiler supplies 10,000 kg/h of water at 65°C. The raw feedwater has an inlet temperature of 20°C and is to be heated in a single-shell, two-tube pass heat exchanger by condensing steam at 1.30 bars. The overall heat transfer coefficient is 2000 W/m$^2 \cdot$K.

a) Determine the required heat transfer area for the exchanger. Use both the LMTD method and the NTU approach to obtain your result.

b) Determine the condensation rate for the steam.

11.16 The oil in an engine is cooled by air in a cross-flow heat exchanger where both fluids are unmixed. Atmospheric air enters at 30°C and at a rate 0.53 kg/s. Oil at a rate of 0.026 kg/s enters the exchanger at 75°C and flows through a tube of 10 mm inner diameter.

a) Assuming fully developed conditions and constant wall flux heating, estimate the heat transfer convection coefficient on the oil side of the exchanger.

b) If the overall heat transfer convection coefficient is 53 W/m$^2 \cdot$K and the total heat transfer area is 1 m^2, determine the effectiveness of this exchanger.

c) What is the exit temperature of the oil?

11.17 A concentric tube heat exchanger uses water, which is available at 15°C, to cool ethylene glycol from 100°C to 60°C. The water and glycol flowrates are each 0.5 kg/s. What are the maximum possible heat transfer rate and effectiveness of the exchanger? Which is preferred, a parallel-flow or counterflow mode of operation?

11.18 Water is used for both fluids flowing through a single-pass, cross-flow heat exchanger with both fluids unmixed. The hot water enters at 90°C and at 10,000 kg/h, while the cold water enters at 10°C and at 20,000 kg/h. If the effectiveness of the exchanger is 60 percent, determine the cold water exit temperature.

11.19 A cross-flow heat exchanger arrangement is formed by a bundle of 32 tubes placed normal to the flow direction in a 0.6-m square duct. Hot water with a temperature of 150°C and mean velocity of 0.5 m/s enters the tubes having inner and outer diameters of 10.2 mm and 12.5 mm, respectively. Atmospheric air at 10°C enters the exchanger with a volumetric flow rate of 1.0 m^3/s. The convection heat transfer coefficient on the tube outer surfaces is 400 W/m$^2 \cdot$K. Estimate the fluid outlet temperatures.

11.20 Saturated steam at 100°C condenses in a shell-and-tube exchanger (single shell, two tube passes) with a surface area of 0.5 m^2 and overall heat transfer coefficient of 2000 W/m$^2 \cdot$K. Water at a flowrate of 0.5 kg/s enters at 15°C.

a) Determine the outlet temperature of the water.

b) Determine the rate of condensation of steam.

11.21 Consider a concentric tube heat exchanger that is characterized by a uniform overall heat transfer coefficient and operates under the following conditions:

	$\dot{m}$ (kg/s)	c_p (J/kg·K)	T_i (°C)	T_o (°C)
Cold fluid	0.125	4200	40	95
Hot fluid	0.125	2100	210	

a) What is the maximum possible heat transfer rate?

b) What is the heat exchanger effectiveness?

c) To minimize size and weight, should the heat exchanger be operated in parallel flow or in counterflow? What is the ratio of the required areas for these two flow conditions?

11.22 Consider a concentric tube heat exchanger with hot and cold water inlet temperatures of 200°C and 35°C, respectively. The flow rates of the hot and cold fluids are 42 kg/h and 84 kg/h, respectively. Assume the overall heat transfer coefficient is 180 W/m²·K.

a) What is the maximum heat transfer rate that could be achieved for the prescribed inlet conditions?

b) If the exchanger is operated in counterflow with a heat transfer area of 0.33 m², determine the outlet fluid temperatures.

c) What is the largest possible heat transfer rate that could be achieved for the prescribed inlet conditions if the exchanger is operated in parallel flow and its length is very long? What is the effectiveness of the exchanger in this configuration?

11.23 Hot water at 100°C enters a concentric tube counterflow heat exchanger having a total area of 23 m² at a rate of 2.5 kg/s. Cold water at 20°C enters at 5.0 kg/s, and the overall heat transfer coefficient is 1000 W/m²·K. Determine the total heat transfer rate, the outlet temperature of the hot fluid, and the outlet temperature of the cold fluid.

11.24 Hot exhaust gases are used in a shell-and-tube exchanger to heat 2.5 kg/s of water from 35 to 85°C. The gases, assumed to have the properties of air, enter at 200°C and leave at 93°C. The overall heat transfer coefficient is 180 W/m²·K. Using the NTU-effectiveness method, calculate the area of the heat exchanger.

11.25 A procedure for open heart surgery under hypothermic conditions involves cooling the patient's blood before the surgery and rewarming the blood following surgery. It is proposed that a concentric-tube, counterflow heat exchanger of length 0.5 m be used for this purpose with the thin-walled inner tube having a diameter of 55 mm. If water at a temperature of 60°C and a flow rate of 0.10 kg/s is used to heat blood entering the exchanger with a temperature of 18°C and a flow rate of 0.05 kg/s, what is the temperature of the blood leaving the exchanger? The overall heat transfer coefficient is 500 W/m²·K, and the specific heat of the blood is 3500 J/kg·K.

11.26 Ethylene glycol and water, which are at temperatures of 60°C and 10°C, respectively, enter a shell-and-tube heat exchanger for which the total heat transfer area is 15 m². With ethylene glycol and water flowrates of 2 kg/s and 5 kg/s, respectively, the overall heat transfer coefficient is known to be 800 W/m²·K. For these conditions, what is the rate of heat transfer for the exchanger and what are the fluid outlet temperatures?

12 Radiation: Processes and Properties

In our study of heat transfer by conduction and convection, we have come to recognize that such transfer requires the presence of a temperature gradient in some form of matter. In contrast, heat transfer by *thermal radiation* does not require the presence of any form of matter. It is an extremely important process, and in the physical sense it is the most interesting of the heat transfer modes. It is relevant to many industrial heating, cooling, and drying processes, as well as to energy conversion methods that involve fossil fuel combustion and solar radiation. Thermal radiation is also an inherent part of the natural environment.

In this chapter we consider the means by which thermal radiation is generated, the specific nature of the radiation, and the manner in which the radiation interacts with matter. We give particular attention to radiative interactions at a surface and to the properties which must be introduced to describe these interactions. In Chapter 13 we focus on the means for computing radiative exchange between two or more surfaces.

12.1 FUNDAMENTAL CONCEPTS

Consider a solid that is initially at a higher temperature, T_s, than that of its surroundings, T_{sur}, but around which there exists a vacuum (Figure 12.1). The presence of the vacuum precludes energy loss from the surface of the solid by conduction or convection. However, our intuition tells us that the solid will cool and eventually achieve thermal equilibrium with its surroundings. This cooling is associated with a reduction in the internal energy stored by the solid and is a direct consequence of the *emission* of thermal radiation from the surface. In turn, the surface will intercept and absorb radiation originating from the surroundings. However, if $T_s > T_{sur}$ the *net* heat transfer rate by radiation, $q_{rad,net}$, is *from* the surface, and the surface will cool until T_s reaches T_{sur}.

We associate thermal radiation with the rate at which energy is emitted by matter as a result of its finite temperature. At this moment thermal radiation is

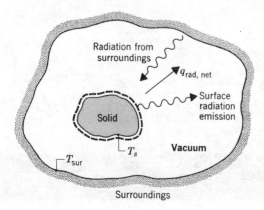

Figure 12.1 Radiation cooling of a heated solid.

being emitted by all of the matter that surrounds you: by the furniture and walls of the room, if you are indoors, or by the ground, buildings, and the atmosphere and sun above, if you are outdoors. The mechanism of emission is related to energy which is released due to oscillations or transitions of the many electrons which comprise matter. These oscillations are, in turn, sustained by the internal energy, and therefore the temperature, of the matter. Hence we associate the emission of thermal radiation with thermally excited conditions within the matter.

It is important to note that all forms of matter emit radiation. For gases and for semitransparent solids, such as glass and salt crystals, at elevated temperatures, emission is a *volumetric phenomenon*, as illustrated in Figure 12.2. That is, radiation emerging from a finite volume of matter is the integrated effect of local emission throughout the volume. However, in this text we concentrate on situations for which radiation is a *surface phenomenon*. In most solids and liquids, radiation emitted from interior molecules is strongly absorbed by adjoining molecules. Accordingly, radiation that is emitted from a solid or a liquid originates from molecules, that are within a distance of approximately 1 μm from the exposed surface. It is for this reason that emission from a solid or a liquid into an adjoining gas or a vacuum is viewed as a surface phenomenon.

We know that radiation originates due to emission by matter and that its subsequent transport does not require the presence of any matter. But what is the nature of this transport? One theory views radiation as the propagation of a collection of particles termed *photons* or *quanta*. Alternatively, radiation may be viewed as the propagation of *electromagnetic waves*. In any case we wish to attribute to radiation the standard wave properties of frequency, v, and wavelength, λ. For radiation propagating in a particular medium, the two properties are related by

$$\lambda = \frac{c}{v}$$

(12.1)

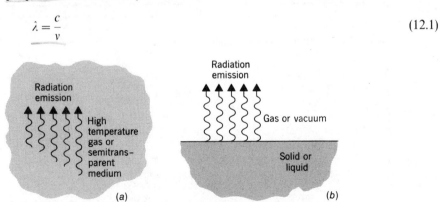

Figure 12.2 The emission process as a (a) volumetric phenomenon and (b) surface phenomenon.

where c is the speed of light in the medium. For propagation in a vacuum, $c_o = 2.998 \times 10^8$ m/s. The unit of wavelength is commonly the micrometer, μm, where 1 μm = 10^{-6} m = 10^4 Å (Angstroms).

The complete electromagnetic spectrum is delineated in Figure 12.3. The short wavelength gamma rays, X rays, and ultraviolet (UV) radiation are primarily of interest to the high-energy physicist and the nuclear engineer, while the long wavelength microwaves and radio waves are of concern to the electrical engineer. It is the intermediate portion of the spectrum, which extends from approximately 0.1 to 100 μm and includes a portion of the ultraviolet and all of the visible and infrared (IR), that is termed *thermal radiation* and is pertinent to heat transfer.

It is important to note that thermal radiation emitted by a surface encompasses a range of wavelengths. Moreover, as shown in Figure 12.4a, the magnitude of the radiation varies with wavelength, and the term *spectral* is used

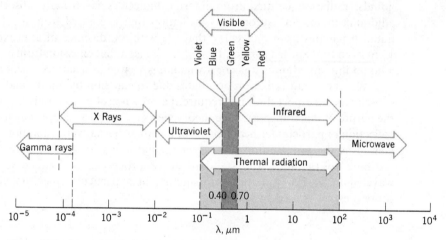

Figure 12.3 Spectrum of electromagnetic radiation.

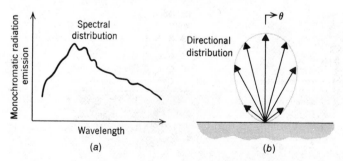

Figure 12.4 Radiation emitted by a surface. (a) Spectral distribution. (b) Directional distribution.

to refer to the nature of this dependence. In effect emitted radiation consists of a continuous, nonuniform distribution of *monochromatic* (single wavelength) components. As we will find, the magnitude of the radiation at any wavelength and the *spectral distribution* vary with the nature and temperature of the emitting surface.

The spectral nature of thermal radiation is one of two features that complicate its description. The second feature relates to its *directional* nature. That is, as shown in Figure 12.4*b*, a surface may emit preferentially in certain directions, creating a *directional distribution* of the emitted radiation. To properly quantify radiation heat transfer, we must therefore be able to treat both spectral and directional effects.

12.2 RADIATION INTENSITY

Radiation emitted by a surface propagates in all possible directions (Figure 12.4*b*), and we are often interested in knowing the directional distribution of this emission. Moreover, radiation incident on a surface may come from different directions, and the manner in which the surface responds to this radiation depends on the direction. Such directional effects may be treated by introducing the concept of *radiation intensity*.

12.2.1 Definitions

Consider emission in a particular direction from an element of area, dA_1, as shown in Figure 12.5*a*. This direction may be specified in terms of the zenith and

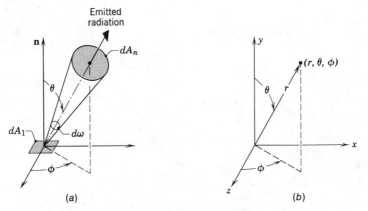

Figure 12.5 Directional nature of radiation. (*a*) Emission of radiation from a differential area dA_1 into a solid angle $d\omega$ subtended by dA_n at a point on dA_1. (*b*) The spherical coordinate system.

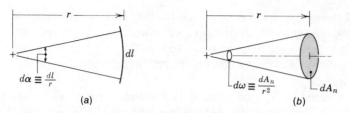

Figure 12.6   Definition of (a) plane and (b) solid angles.

azimuthal angles, θ and ϕ, respectively, of a spherical coordinate system, Figure 12.5b. Moreover, a differentially small surface in space, dA_n, through which this radiation passes subtends a solid angle, $d\omega$, when viewed from a point on dA_1. From Figure 12.6a we see that the differential plane angle $d\alpha$ is defined by a region between the rays of a circle and is measured as the ratio of the element of arc length dl on the circle to the radius r of the circle. Similarly, from Figure 12.6b the differential solid angle $d\omega$ is defined by a region between the rays of a sphere and is measured as the ratio of the element of area dA_n on the sphere to the square of the sphere's radius. Accordingly,

$$d\omega \equiv \frac{dA_n}{r^2} \tag{12.2}$$

Note that dA_n represents an area that is normal to the (θ, ϕ) direction. The differential solid angle may be related to the zenith and azimuthal angles by noting that, for a spherical surface, $dA_n = r^2 \sin\theta \, d\theta d\phi$, as shown in Figure 12.7.

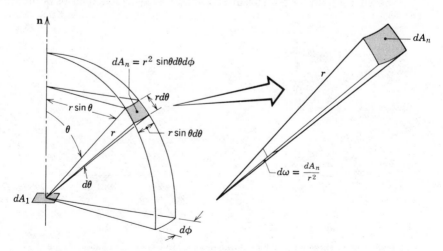

Figure 12.7   The solid angle subtended by dA_n at a point on dA_1 in the spherical coordinate system.

Accordingly,

$$dω = \sin θ \, dθdϕ$$
(12.3)

Note that, whereas the plane angle $dα$ has the unit of radians (rad), the unit of the solid angle is the steradian (sr).

Returning to Figure 12.5a, we may now consider the rate at which emission from dA_1 passes through dA_n. This quantity may be expressed in terms of the *spectral intensity*, $I_{λ,e}$, of the emitted radiation. We formally define $I_{λ,e}$ as the *rate at which radiant energy is emitted at the wavelength $λ$ in the $(θ,ϕ)$ direction, per unit area of the emitting surface normal to this direction, per unit solid angle about this direction, and per unit wavelength interval $dλ$ about $λ$.* Note that the area used to define the intensity is the component of dA_1 perpendicular to the direction of the radiation. From Figure 12.8 we see that this projected area is equal to $dA_1 \cos θ$. In effect it is how dA_1 would appear to an observer situated on dA_n. The spectral intensity, which has units of $\text{W/m}^2 \cdot \text{sr} \cdot μ\text{m}$, may then be expressed as

$$I_{λ,e}(λ,θ,ϕ) \equiv \frac{dq}{dA_1 \cos θ \cdot dω \cdot dλ}$$
(12.4)

where $(dq/dλ) \equiv dq_λ$ is the rate at which radiation of wavelength $λ$ leaves dA_1 and passes through dA_n. Rearranging Equation 12.4 it follows that

$$dq_λ = I_{λ,e}(λ,θ,ϕ)dA_1 \cos θ \, dω$$
(12.5)

where $dq_λ$ has the units of $\text{W}/μ\text{m}$. This important expression allows us to compute the rate at which radiation emitted by a surface propagates into the region of space defined by the solid angle $dω$ about the $(θ,ϕ)$ direction. However, to compute this rate, the spectral intensity, $I_{λ,e}$, of the emitted radiation must be known. The manner in which this quantity may be determined is discussed in subsequent sections. Expressing Equation 12.5 per unit area of the emitting surface and substituting from Equation 12.3, the spectral radiation *flux* as-

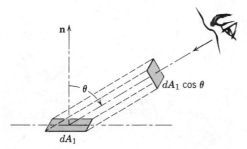

Figure 12.8 The projection of dA_1 normal to the direction of radiation.

sociated with dA_1 is

$$dq_\lambda'' = I_{\lambda,e}(\lambda,\theta,\phi)\cos\theta\sin\theta\,d\theta d\phi \tag{12.6}$$

If the spectral and directional distributions of $I_{\lambda,e}$ are known, that is $I_{\lambda,e}(\lambda,\theta,\phi)$, the heat flux associated with emission into any finite solid angle or over any finite wavelength interval may be determined by integrating Equation 12.6. For example, the spectral heat flux associated with emission into a hypothetical hemisphere above dA_1, as shown in Figure 12.9, may be expressed as

$$q_\lambda''(\lambda) = \int_0^{2\pi}\int_0^{\pi/2} I_{\lambda,e}(\lambda,\theta,\phi)\cos\theta\sin\theta\,d\theta d\phi \tag{12.7}$$

Note that the solid angle associated with the entire hemisphere may be obtained by integrating Equation 12.3 over the limits $\phi = 0$ to $\phi = 2\pi$ and $\theta = 0$ to $\theta = \pi/2$. It then follows that

$$\int_h d\omega = \int_0^{2\pi}\int_0^{\pi/2} \sin\theta\,d\theta d\phi = 2\pi\int_0^{\pi/2}\sin\theta\,d\theta = 2\pi \tag{12.8}$$

where the subscript h refers to integration over the entire hemisphere. The total heat flux associated with emission over all directions and at all wavelengths is then

$$q'' = \int_0^\infty q_\lambda''(\lambda)d\lambda \tag{12.9}$$

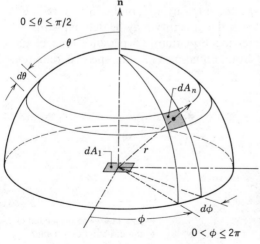

Figure 12.9 Emission from a differential element of area dA_1 into a hypothetical hemisphere centered at a point on dA_1.

12.2.2 Relation to Emission

The radiation intensity is related to several important radiation fluxes. Recall that radiation emission occurs from any surface that is at a finite temperature. The concept of *emissive power* is introduced to quantify the amount of radiation emitted per unit surface area. The *spectral, hemispherical emissive power*, E_λ ($W/m^2 \cdot \mu m$), is defined as the rate at which radiation of wavelength λ is emitted in *all directions* from a surface per unit wavelength $d\lambda$ about λ and per unit surface area. Accordingly, it is related to the spectral intensity of the emitted radiation by an expression of the form

$$E_\lambda(\lambda) = \int_0^{2\pi} \int_0^{\pi/2} I_{\lambda,e}(\lambda,\theta,\phi) \cos\theta \sin\theta \, d\theta d\phi \tag{12.10}$$

in which case it is equivalent to the heat flux given by Equation 12.7. Note that E_λ is a flux based on the *actual* surface area, whereas $I_{\lambda,e}$ is based on the *projected* area. The $\cos\theta$ term appearing in the above integrand is a consequence of this difference.

The *total, hemispherical emissive power*, E (W/m^2), is the rate at which radiation is emitted per unit area at all possible wavelengths and in all possible directions. Accordingly,

$$E = \int_0^\infty E_\lambda(\lambda) d\lambda \tag{12.11}$$

or from Equation 12.10

$$E = \int_0^\infty \int_0^{2\pi} \int_0^{\pi/2} I_{\lambda,e}(\lambda,\theta,\phi) \cos\theta \sin\theta \, d\theta d\phi d\lambda \tag{12.12}$$

Note that, since the term emissive power implies emission in all possible directions, the adjective hemispherical is redundant and is often dropped. One then speaks in terms of the *spectral emissive power*, E_λ, or the *total emissive power, E*.

Although the directional distribution of surface emission varies according to the nature of the surface, there is a special case that provides a reasonable approximation for many surfaces. We speak of a *diffuse emitter* as a surface for which the intensity of the emitted radiation is independent of direction, in which case $I_{\lambda,e}(\lambda,\theta,\phi) = I_{\lambda,e}(\lambda)$. Removing $I_{\lambda,e}$ from the integrand of Equation 12.10 and performing the integration, it follows that

$$E_\lambda(\lambda) = \pi I_{\lambda,e}(\lambda) \qquad \text{(Diffuse emitter)} \tag{12.13}$$

Similarly, from Equation 12.12

$$E = \pi I_e \qquad\qquad (12.14)$$

where I_e is the *total intensity* of the emitted radiation. Note that the constant appearing in the above expressions is π, not 2π, and has the unit sr.

EXAMPLE 12.1

A small surface of area $A_1 = 10^{-3}$ m^2 is known to emit diffusely, and from measurements the total intensity associated with emission in the normal direction is known to be $I_n = 7000$ W/m^2·sr.

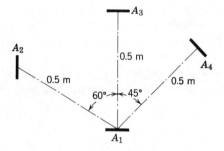

Radiation emitted from the surface is intercepted by three other surfaces of area $A_2 = A_3 = A_4 = 10^{-3}$ m^2, which are at a distance of 0.5 m from A_1 and are oriented as shown in the sketch.

1. What is the intensity associated with emission in each of the three directions?

2. What are the solid angles subtended by the three surfaces when viewed from A_1?

3. What is the rate at which radiation emitted by A_1 is intercepted by the three surfaces?

SOLUTION

KNOWN:

Normal intensity of diffuse emitter of area A_1 and orientation of three surfaces relative to A_1.

FIND:

1. Intensity of emission in each of the three directions.
2. Solid angles subtended by the three surfaces.
3. Rate at which radiation is intercepted by the three surfaces.

SCHEMATIC:

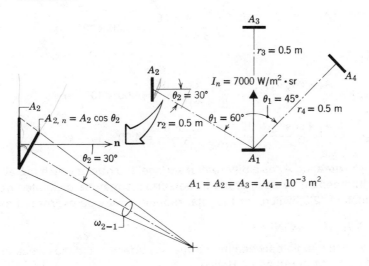

ASSUMPTIONS:

1. Surface A_1 emits diffusely.
2. A_1, A_2, A_3, and A_4 may be approximated as differential surfaces, $(A_j/r^2) \ll 1$.

ANALYSIS:

1. From the definition of a diffuse emitter, we know that the intensity of the emitted radiation is independent of direction. Hence

$$I = 7000 \text{ W/m}^2 \cdot \text{sr} \qquad \triangleleft$$

for each of the three directions.

2. Treating A_2, A_3, and A_4 as differential surface areas, the solid angles may be computed from Equation 12.2

$$d\omega \equiv \frac{dA_n}{r^2}$$

where dA_n is the projection of the surface normal to the direction of the radiation. Accordingly,

$$dA_{n,j} = dA_j \times \cos\theta_j$$

where θ_j is the angle between the surface normal and the direction of the radiation. The solid angle subtended by surface A_2 with respect to A_1 is then

$$\omega_{2-1} \approx \frac{A_2 \times \cos\theta_2}{r^2} = \frac{10^{-3}\ \mathrm{m}^2 \times \cos 30°}{(0.5\ \mathrm{m})^2} = 3.46 \times 10^{-3}\ \mathrm{sr} \qquad \triangleleft$$

Similarly,

$$\omega_{3-1} \approx \frac{A_3 \times \cos\theta_3}{r^2} = \frac{10^{-3}\ \mathrm{m}^2 \times \cos 0°}{(0.5\ \mathrm{m})^2} = 4.00 \times 10^{-3}\ \mathrm{sr} \qquad \triangleleft$$

$$\omega_{4-1} \approx \frac{A_4 \times \cos\theta_4}{r^2} = \frac{10^{-3}\ \mathrm{m}^2 \times \cos 0°}{(0.5\ \mathrm{m})^2} = 4.00 \times 10^{-3}\ \mathrm{sr} \qquad \triangleleft$$

3. Approximating A_1 as a differential surface, the rate at which radiation is intercepted by each of the three surfaces may be estimated from Equation 12.5, which, for the total radiation, may be expressed as

$$q_{1-j} \approx I \times A_1 \cos\theta_1 \times \omega_{j-1}$$

where θ_1 is the angle between the normal to surface 1 and the direction of propagation of the radiation. Hence

$$q_{1-2} \approx 7000\ \mathrm{W/m^2 \cdot sr}\ (10^{-3}\ \mathrm{m}^2 \times \cos 60°) 3.46 \times 10^{-3}\ \mathrm{sr}$$
$$= 12.1 \times 10^{-3}\ \mathrm{W} \qquad \triangleleft$$

$$q_{1-3} \approx 7000\ \mathrm{W/m^2 \cdot sr}\ (10^{-3}\ \mathrm{m}^2 \times \cos 0°)\ 4.00 \times 10^{-3}\ \mathrm{sr}$$
$$= 28.0 \times 10^{-3}\ \mathrm{W} \qquad \triangleleft$$

$$q_{1-4} \approx 7000\ \mathrm{W/m^2 \cdot sr}\ (10^{-3}\ \mathrm{m}^2 \times \cos 45°)\ 4.00 \times 10^{-3}\ \mathrm{sr}$$
$$= 19.8 \times 10^{-3}\ \mathrm{W} \qquad \triangleleft$$

COMMENTS:

1. Note the different values of θ_1 for the emitting surface and the values of θ_2, θ_3, and θ_4 for the receiving surfaces.

2. If the surfaces were not small relative to the square of the separation distance, the solid angles and radiation heat transfer rates would have to be obtained from complicated integrations involving Equations 12.3 and 12.5, respectively.

3. Any spectral component of the radiation rate could also be obtained

using the above procedures, so long as the spectral intensity, I_λ, is known.

4. Despite the fact that the intensity of the emitted radiation is independent of direction, the rate at which radiation is intercepted by the three surfaces differs significantly due to differences in the solid angles and projected areas.

12.2.3 Relation to Irradiation

Although we have focussed on radiation leaving a surface due to emission, the foregoing concepts may be adapted to radiation that is *incident* on a surface. This radiation may originate from emission and reflection occurring at other surfaces and will have specific directional and spectral distributions. As shown in Figure 12.10, the incident radiation can be characterized in terms of the spectral intensity, $I_{\lambda,i}$. It is defined as the rate at which radiant energy is incident at the wavelength λ from the (θ,ϕ) direction, per unit area of the *intercepting surface* normal to this direction, per unit solid angle about this direction, and per unit wavelength interval $d\lambda$ about λ.

The intensity of the incident radiation may be related to an important radiative flux, termed the *irradiation*, which encompasses radiation incident *from all directions*. The *spectral irradiation*, G_λ (W/m$^2 \cdot \mu$m), is defined as the rate at which radiation is incident on a surface per unit area of the surface at the wavelength λ, per unit wavelength interval $d\lambda$ about λ. Accordingly,

$$G_\lambda(\lambda) = \int_0^{2\pi}\int_0^{\pi/2} I_{\lambda,i}(\lambda,\theta,\phi)\cos\theta\sin\theta\,d\theta d\phi \qquad (12.15)$$

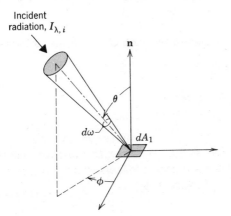

Incident radiation, $I_{\lambda,i}$

Figure 12.10 Directional nature of incident radiation.

where $\sin \theta \, d\theta d\phi$ is the unit solid angle. The $\cos \theta$ factor originates from the fact that G_λ is a flux based on the actual surface area, whereas $I_{\lambda,i}$ is defined in terms of the projected area. If the *total irradiation*, *G* (W/m²), represents the rate at which radiation is incident per unit area from all directions and at all wavelengths, it follows that

$$G = \int_0^\infty G_\lambda(\lambda) d\lambda \tag{12.16}$$

or from Equation 12.15

$$G = \int_0^\infty \int_0^{2\pi} \int_0^{\pi/2} I_{\lambda,i}(\lambda, \theta, \phi) \cos \theta \sin \theta \, d\theta d\phi d\lambda \tag{12.17}$$

If the incident radiation is diffuse, that is, the itercepting area is *irradiated diffusely*, $I_{\lambda,i}$ is independent of θ and ϕ and it follows that

$$G_\lambda(\lambda) = \pi I_{\lambda,i}(\lambda) \tag{12.18}$$

and

$$G = \pi I_i \tag{12.19}$$

EXAMPLE 12.2

What is the irradiation at surfaces A_2, A_3, and A_4 of Example 12.1 due to emission from A_1?

SOLUTION

KNOWN:

Rate at which radiation is intercepted by each of the three surfaces.

FIND:

Irradiation at each of the three surfaces.

ANALYSIS:

From the definition of irradiation as the rate at which radiation is incident on a surface per unit area of the surface, it follows that the irradiation of surface *j* due to emission from surface 1 is

$$G_j = \frac{q_{1-j}}{A_j}$$

With $A_2 = A_3 = A_4 = 10^{-3} \text{ m}^2$, it follows from the results of Example 12.1 that

$$G_2 = \frac{12.1 \times 10^{-3} \text{ W}}{10^{-3} \text{ m}^2} = 12.1 \text{ W/m}^2 \qquad \triangleleft$$

$$G_3 = \frac{28.0 \times 10^{-3} \text{ W}}{10^{-3} \text{ m}^2} = 28.0 \text{ W/m}^2 \qquad \triangleleft$$

$$G_4 = \frac{19.8 \times 10^{-3} \text{ W}}{10^{-3} \text{ m}^2} = 19.8 \text{ W/m}^2 \qquad \triangleleft$$

COMMENTS:

The irradiation could also be computed from Equation 12.15, which, for the present application, takes the form

$$G_j = I_i \cos \theta_j \omega_{1-j}$$

where $I_i = I = 7000 \text{ W/m}^2 \cdot \text{sr}$ and ω_{1-j} is the solid angle subtended by surface 1 with respect to j. For example, $G_2 = I \cos \theta_2 \omega_{1-2} = (7000 \text{ W/m}^2 \cdot \text{sr})(\cos 30°)(10^{-3} \text{ m}^2 \times \cos 60°/0.25 \text{ m}^2) \text{ sr} = 12.1 \text{ W/m}^2$.

EXAMPLE 12.3

The spectral distribution of radiation incident on a surface is of approximately the following form.

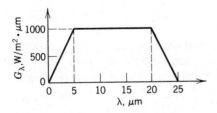

What is the total irradiation?

SOLUTION

KNOWN:

Spectral distribution of surface irradiation.

FIND:

Total irradiation.

ANALYSIS:

The total irradiation may be obtained from Equation 12.16

$$G = \int_0^\infty G_\lambda \, d\lambda$$

The integral is readily evaluated by breaking it into parts. That is,

$$G = \int_0^{5\ \mu m} G_\lambda d\lambda + \int_{5\ \mu m}^{20\ \mu m} G_\lambda d\lambda + \int_{20\ \mu m}^{25\ \mu m} G_\lambda d\lambda + \int_{25\ \mu m}^{\infty} G_\lambda d\lambda$$

or

$$G = 1/2 \ (1000 \ \text{W/m}^2 \cdot \mu m) \ (5 - 0)\mu m + (1000 \ \text{W/m}^2 \cdot \mu m) \ (20 - 5)\mu m$$
$$+ \ 1/2 \ (1000 \ \text{W/m}^2 \cdot \mu m)(25 - 20)\mu m + 0 = (2500 + 15000 + 2500) \ \text{W/m}^2$$

or

$$G = 20{,}000 \ \text{W/m}^2 \qquad\qquad\qquad \triangleleft$$

COMMENTS:

Generally, radiation sources would not provide such a regular spectral distribution for the irradiation. However, the procedure of computing the total irradiation from knowledge of the spectral distribution remains the same, although evaluation of the integral is likely to involve more detail.

12.2.4   Relation to Radiosity

The last radiative flux of interest, termed *radiosity*, accounts for *all* of the radiant energy leaving a surface. Since the radiation leaving a surface includes the *reflected* portion of the irradiation, as well as direct emission (Figure 12.11), the

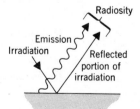

Figure 12.11   Surface radiosity.

radiosity is generally different from the emissive power. The *spectral radiosity, J_λ* ($W/m^2 \cdot \mu m$), represents the rate at which radiation leaves a unit area of the surface at the wavelength λ, per unit wavelength interval $d\lambda$ about λ. Since it accounts for radiation leaving in all directions, it is related to the intensity of the radiation associated with emission and reflection, $I_{\lambda,e+r}(\lambda,\theta,\phi)$, by an expression of the form

$$J_\lambda(\lambda) = \int_0^{2\pi} \int_0^{\pi/2} I_{\lambda,e+r}(\lambda,\theta,\phi)\cos\theta \sin\theta \, d\theta d\phi \qquad (12.20)$$

Hence the *total radiosity*, $J(W/m^2)$, associated with the entire spectrum is

$$J = \int_0^\infty J_\lambda(\lambda) d\lambda \qquad (12.21)$$

or

$$J = \int_0^\infty \int_0^{2\pi} \int_0^{\pi/2} I_{\lambda,e+r}(\lambda,\theta,\phi)\cos\theta \sin\theta \, d\theta d\phi d\lambda \qquad (12.22)$$

However, if the surface is both a *diffuse reflector* and a *diffuse emitter, $I_{\lambda,e+r}$* is independent of θ and ϕ, and it follows that

$$J_\lambda(\lambda) = \pi I_{\lambda,e+r} \qquad (12.23)$$

and

$$J = \pi I_{e+r} \qquad (12.24)$$

Again, it should be noted that the radiation flux, in this case the radiosity, is based on the actual surface area, while the intensity is based on the projected area.

12.3 BLACKBODY RADIATION

When describing the radiation characteristics of real surfaces, it is useful to introduce the concept of a *blackbody*. The blackbody is an *ideal* surface having the following properties.

1. *A blackbody absorbs all incident radiation, regardless of wavelength and direction.*

2. *For a prescribed temperature and wavelength, no surface can emit more energy than a blackbody.*

3. *Although the radiation emitted by a blackbody is a function of wavelength and temperature, it is independent of direction. That is, the blackbody is a diffuse emitter.*

As the perfect absorber and emitter, the blackbody therefore serves as a *standard* against which the radiative properties of actual surfaces may be compared.

Although closely approximated by some surfaces, it is important to note that there are no surfaces that have precisely the properties of a blackbody. The closest approximation to blackbody behavior is achieved by a *cavity* whose inner surface is at a uniform temperature. If radiation enters the cavity through a small aperture, as shown in Figure 12.12a, it is likely to experience many reflections prior to reemergence. Hence it is almost entirely absorbed by the cavity, and blackbody behavior is approximated. From thermodynamic principles it may then be argued that radiation leaving the aperture depends only on the surface temperature and corresponds to blackbody emission (Figure 12.12b). Since blackbody emission is diffuse, it follows that the spectral intensity, $I_{\lambda,b}$, of the radiation leaving the cavity is independent of direction. Moreover, since the radiation field in the cavity, which is the cumulative effect of emission and reflection from the cavity surface, must be of the same form as the radiation emerging from the aperture, it also follows that a blackbody radiation field exists within the cavity. Accordingly, any small surface in the cavity, Figure 12.12c, experiences irradiation for which $G_\lambda = E_{\lambda,b}(\lambda, T)$. This surface is diffusely irradiated, regardless of its orientation in the cavity. Note that *blackbody radiation exists within the cavity irrespective of the cavity surface condition.* That is, the surface may be highly reflecting or absorbing.

12.3.1 The Planck Distribution

The spectral distribution of the radiation intensity associated with blackbody emission is well known, having first been determined by Planck [1]. It is of the form

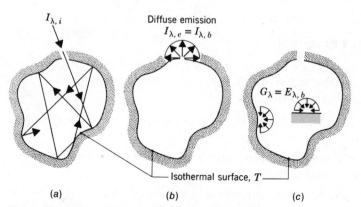

Figure 12.12 Characteristics of an isothermal blackbody cavity. (*a*) Complete absorption. (*b*) Diffuse emission from an aperture. (*c*) Diffuse irradiation of interior surfaces.

$$I_{\lambda,b}(\lambda,T) = \frac{2hc_o^2}{\lambda^5\left[\exp(hc_o/\lambda kT) - 1\right]} \qquad \text{BLACKBODY INTENSITY} \qquad (12.25)$$

where $h = 6.6256 \times 10^{-34}$ J·s and $k = 1.3805 \times 10^{-23}$ J/K are the universal Planck and Boltzmann constants, respectively, $c_o = 2.998 \times 10^8$ m/s is the speed of light in vacuum, and T is the *absolute* temperature of the blackbody (K). Since the blackbody is a diffuse emitter, it follows from Equation 12.13 that its spectral emissive power is of the form

$$E_{\lambda,b}(\lambda,T) = \pi I_{\lambda,b}(\lambda,T) = \frac{C_1}{\lambda^5\left[\exp(C_2/\lambda T) - 1\right]} \qquad \begin{array}{c}\text{BLACK BODY}\\ \text{EMISSIVE}\\ \text{POWER}\end{array} \qquad (12.26)$$

where the first and second radiation constants, C_1 and C_2, respectively, are $C_1 = 2\pi hc_o^2 = 3.742 \times 10^8$ W·μm⁴/m² and $C_2 = (hc_o/k) = 1.439 \times 10^4$ μm·K.

Equation 12.26, known as the *Planck distribution*, is plotted in Figure 12.13 for selected temperatures. Several important features should be noted.

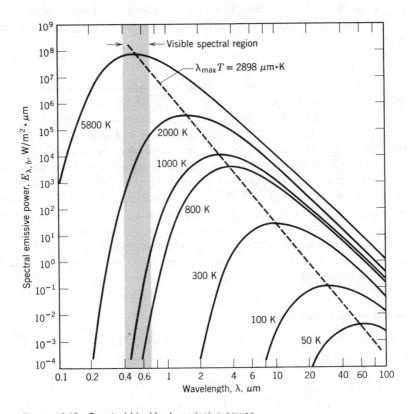

Figure 12.13 Spectral blackbody emissive power.

1. The emitted radiation varies *continuously* with wavelength.

2. At any wavelength the magnitude of the emitted radiation increases with increasing temperature.

3. The spectral region in which the radiation is concentrated depends on temperature, with *comparatively* more radiation appearing at shorter wavelengths as the temperature increases.

4. A significant fraction of the radiation emitted by the sun, which may be considered to be a blackbody at approximately 5800 K, is in the visible region of the spectrum. In contrast for $T \lesssim 800$ K, emission is predominantly in the infrared region of the spectrum and hence is not visible to the eye.

12.3.2 Wien's Displacement Law

From Figure 12.13 we see that the blackbody spectral distribution is characterized by a maximum and that the wavelength associated with this maximum, λ_{max}, depends on temperature. The nature of this dependence may be obtained by differentiating Equation 12.26 with respect to λ and setting the result equal to zero. In so doing, we obtain

$$\lambda_{max} T = C_3 \tag{12.27}$$

where the third radiation constant is $C_3 = 2897.6$ $\mu m \cdot K$.

Equation 12.27 is known as *Wien's displacement law*, and the locus of points described by the law is plotted as the dashed line of Figure 12.13. According to this result, the maximum spectral emissive power is displaced to shorter wavelengths with increasing temperature. This emission is in the middle of the visible spectrum ($\lambda \approx 0.50$ μm) for solar radiation, since the sun emits as a blackbody at approximately 5800 K. For a blackbody at 1000 K peak emission occurs at 2.90 μm, with some of the emitted radiation appearing visible to the eye as red light. With increasing temperature, shorter wavelengths become more prominent, until eventually significant emission occurs over the entire visible spectrum. For example, a tungsten filament lamp operating at 2900 K ($\lambda_{max} = 1$ μm) emits white light, although most of the emission remains in the infrared.

12.3.3 The Stefan-Boltzmann Law

Substituting the Planck distribution, Equation 12.26, into Equation 12.11, the total emissive power of a blackbody, E_b, may be expressed as

$$E_b = \int_0^\infty \frac{C_1}{\lambda^5 \left[\exp\left(C_2/\lambda T\right) - 1\right]} \, d\lambda$$

Performing the integration, it may be shown that

$$E_b = \sigma T^4 \qquad \text{TOTAL EMISSIVE POWER OF BLACKBODY} \tag{12.28}$$

where the *Stefan-Boltzmann* constant, which depends on C_1 and C_2, has the numerical value

$$\sigma = 5.670 \times 10^{-8} \ \text{W/m}^2 \cdot \text{K}^4$$

This simple, yet important, result is termed the *Stefan-Boltzmann law*. It enables calculation of the amount of radiation emitted in all directions and over all wavelengths simply from knowledge of the temperature of the blackbody. Because this emission is diffuse, it follows from Equation 12.14 that the total intensity associated with blackbody emission is

$$I_b = \frac{E_b}{\pi} \qquad \text{TOTAL INTENSITY OF BLACKBODY EMISSION} \tag{12.29}$$

12.3.4 Band Emission

It is often necessary to know the fraction of the total emission from a blackbody which is in a certain wavelength interval or *band*. For a prescribed temperature and the interval from 0 to λ, this fraction is determined by the ratio of the shaded section to the total area under the curve of Figure 12.14. Moreover, it may be expressed as

$$F_{(0 \to \lambda)} \equiv \frac{\int_0^\lambda E_{\lambda,b} \, d\lambda}{\int_0^\infty E_{\lambda,b} \, d\lambda} = \frac{\int_0^\lambda E_{\lambda,b} \, d\lambda}{\sigma T^4} = \int_0^{\lambda T} \frac{E_{\lambda,b}}{\sigma T^5} \, d(\lambda T) = f(\lambda T) \tag{12.30}$$

Since the integrand $(E_{\lambda,b}/\sigma T^5)$ is exclusively a function of the wavelength-temperature product, λT, the integral of Equation 12.30 may be evaluated to obtain $F_{(0 \to \lambda)}$ as a universal function of λT. The results are presented in Table 12.1 and Figure 12.15. They may also be used to obtain the fraction of the

FRACTION OF EMISSION FROM BLACKBODY at certain WAVELENGTH

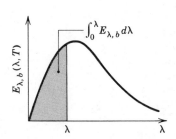

Figure 12.14 Radiation emission from a blackbody in the spectral band 0 to λ.

Table 12.1 Blackbody radiation functions

λT $(\mu\text{m}\cdot\text{K})$	$F_{(0\to\lambda)}$	$I_{\lambda,b}(\lambda,T)/\sigma T^5$ $(\mu\text{m}\cdot\text{K}\cdot\text{sr})^{-1}$	$\dfrac{I_{\lambda,b}(\lambda,T)}{I_{\lambda,b}(\lambda_{\max},T)}$
200	0.000000	0.375034×10^{-27}	0.000000
400	0.000000	0.490335×10^{-13}	0.000000
600	0.000000	0.104046×10^{-8}	0.000014
800	0.000016	0.991126×10^{-7}	0.001372
1,000	0.000321	0.118505×10^{-5}	0.016406
1,200	0.002134	0.523927×10^{-5}	0.072534
1,400	0.007790	0.134411×10^{-4}	0.186082
1,600	0.019718	0.249130	0.344904
1,800	0.039341	0.375568	0.519949
2,000	0.066728	0.493432	0.683123
2,200	0.100888	0.589649×10^{-4}	0.816329
2,400	0.140256	0.658866	0.912155
2,600	0.183120	0.701292	0.970891
2,800	0.227897	0.720239	0.997123
2,898	0.250108	0.722318×10^{-4}	1.000000
3,000	0.273232	0.720254×10^{-4}	0.997143
3,200	0.318102	0.705974	0.977373
3,400	0.361735	0.681544	0.943551
3,600	0.403607	0.650396	0.900429
3,800	0.443382	0.615225	0.851737
4,000	0.480877	0.578064	0.800291
4,200	0.516014	0.540394×10^{-4}	0.748139
4,400	0.548796	0.503253	0.696720
4,600	0.579280	0.467343	0.647004
4,800	0.607559	0.433109	0.599610
5,000	0.633747	0.400813	0.554898
5,200	0.658970	0.370580×10^{-4}	0.513043
5,400	0.680360	0.342445	0.474092
5,600	0.701046	0.316376	0.438002
5,800	0.720158	0.292301	0.404671
6,000	0.737818	0.270121	0.373965
6,200	0.754140	0.249723×10^{-4}	0.345724
6,400	0.769234	0.230985	0.319783
6,600	0.783199	0.213786	0.295973
6,800	0.796129	0.198008	0.274128
7,000	0.808109	0.183534	0.254090
7,200	0.819217	0.170256×10^{-4}	0.235708
7,400	0.829527	0.158073	0.218842
7,600	0.839102	0.146891	0.203360
7,800	0.848005	0.136621	0.189143
8,000	0.856288	0.127185	0.176079

Table 12.1 Continued

λT $(\mu m \cdot K)$	$F_{(0 \to \lambda)}$	$I_{\lambda, b}(\lambda, T)/\sigma T^5$ $(\mu m \cdot K \cdot sr)^{-1}$	$\dfrac{I_{\lambda, b}(\lambda, T)}{I_{\lambda, b}(\lambda_{max}, T)}$
8,500	0.874608	0.106772×10^{-4}	0.147819
9,000	0.890029	0.901463×10^{-5}	0.124801
9,500	0.903085	0.765338	0.105956
10,000	0.914199	0.653279	0.090442
10,500	0.923710	0.560522	0.077600
11,000	0.931890	0.483321×10^{-5}	0.066913
11,500	0.939959	0.418725	0.057970
12,000	0.945098	0.364394	0.050448
13,000	0.955139	0.279457	0.038689
14,000	0.962898	0.217641	0.030131
15,000	0.969981	0.171866×10^{-5}	0.023794
16,000	0.973814	0.137429	0.019026
18,000	0.980860	0.908240×10^{-6}	0.012574
20,000	0.985602	0.623310	0.008629
25,000	0.992215	0.276474	0.003828
30,000	0.995340	0.140469×10^{-6}	0.001945
40,000	0.997967	0.473891×10^{-7}	0.000656
50,000	0.998953	0.201605	0.000279
75,000	0.999713	0.418597×10^{-8}	0.000058
100,000	0.999905	0.135752	0.000019

Note: The radiation constants used to generate these blackbody functions are $C_1 = 3.7420 \times 10^8 \cdot \mu m^4/m^2$, $C_2 = 1.4388 \times 10^4 \, \mu m \cdot K$, and $\sigma = 5.670 \times 10^{-8} \, W/m^2 \cdot K$.

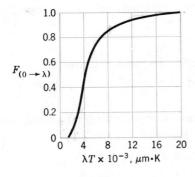

Figure 12.15 Fraction of the total blackbody emission in the spectral band from 0 to λ as a function of λT.

radiation between any two wavelengths λ_1 and λ_2, since

$$F_{(\lambda_1 \to \lambda_2)} = \frac{\int_0^{\lambda_2} E_{\lambda,b}\, d\lambda - \int_0^{\lambda_1} E_{\lambda,b}\, d\lambda}{\sigma T^4} = F_{(0 \to \lambda_2)} - F_{(0 \to \lambda_1)} \qquad (12.31)$$

Additional blackbody functions have been listed in the third and fourth columns of Table 12.1. The third column facilitates calculation of the spectral intensity for a prescribed wavelength and temperature. In lieu of computing this quantity from Equation 12.25, it may be obtained by simply multiplying the tabulated value of $I_{\lambda,b}/\sigma T^5$ by σT^5. The fourth column may be used to obtain a quick comparison of the ratio of the spectral intensity at any wavelength to that at λ_{max}.

EXAMPLE 12.4

Consider a large isothermal enclosure that is maintained at a uniform temperature of 2000 K.

1. Calculate the emissive power of the radiation that emerges from a small aperture on the enclosure surface.
2. What is the wavelength λ_1 below which 10 percent of the emission is concentrated? What is the wavelength λ_2 above which 10 percent of the emission is concentrated?
3. Determine the maximum spectral emissive power and the wavelength at which this emission occurs.
4. If a small object is placed within the enclosure, what is the irradiation incident on the object?

SOLUTION

KNOWN:

Large isothermal enclosure at a uniform temperature, $T = 2000$ K.

FIND:

1. Emissive power of a small aperture on the enclosure.
2. Wavelengths below which and above which 10 percent of the radiation is concentrated.
3. Maximum spectral emissive power and wavelength associated with maximum emission.
4. Irradiation on a small object within the enclosure.

SCHEMATIC:

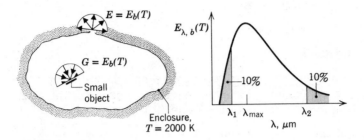

Enclosure,
$T = 2000$ K

ASSUMPTIONS:

Areas of aperture and object are very small relative to enclosure surface area.

ANALYSIS:

1. Emission from the aperture of any isothermal enclosure will have the characteristics of blackbody radiation. Hence, from Equation 12.28

$$E = E_b(T) = \sigma T^4 = 5.670 \times 10^{-8} \text{ W/m}^2 \cdot \text{K}^4 \text{ (2000 K)}^4$$

$$E = 9.07 \times 10^5 \text{ W/m}^2 \qquad \triangleleft$$

2. The wavelength λ_1 corresponds to the upper limit of the spectral band $(0 \rightarrow \lambda_1)$ containing 10 percent of the emitted radiation. With $F_{(0 \rightarrow \lambda_1)} = 0.10$ it follows from Table 12.1 that

$$\lambda_1 T \approx 2200 \ \mu\text{m} \cdot \text{K}$$

Hence

$$\lambda_1 = 1.1 \ \mu\text{m} \qquad \triangleleft$$

The wavelength λ_2 corresponds to the lower limit of the spectral band $(\lambda_2 \rightarrow \infty)$ containing 10 percent of the emitted radiation. With

$$F_{(\lambda_2 \rightarrow \infty)} = 1 - F_{(0 \rightarrow \lambda_2)} = 0.1$$

$$F_{(0 \rightarrow \lambda_2)} = 0.9$$

it follows from Table 12.1 that

$$\lambda_2 T = 9382 \ \mu\text{m} \cdot \text{K}$$

Hence

$$\lambda_2 = 4.69 \ \mu\text{m} \qquad \triangleleft$$

3. From Wien's law, Equation 12.27,

$$\lambda_{max} T = 2898 \ \mu m \cdot K$$

in which case, for $T = 2000$ K,

$$\lambda_{max} = 1.45 \ \mu m \qquad \lhd$$

The spectral emissive power associated with this wavelength may be computed from Equation 12.26 or from the third column of Table 12.1. For $\lambda_{max} T = 2898 \ \mu m \cdot K$ it follows from Table 12.1 that

$$I_{\lambda, b}(1.45 \ \mu m, \ T) = 0.722 \times 10^{-4} \sigma T^5$$

Hence

$$I_{\lambda, b}(1.45 \ \mu m, \ 2000 \ K) = 0.722 \times 10^{-4} \ (1/\mu m \cdot K \cdot sr) \ 5.67 \times 10^{-8} W/m^2 \cdot K^4 (2000 \ K)^5$$

$$I_{\lambda, b}(1.45 \ \mu m, \ 2000 \ K) = 1.31 \times 10^5 \ W/m^2 \cdot sr \cdot \mu m$$

Since the emission is diffuse it then follows from Equation 12.13 that

$$E_{\lambda, b} = \pi I_{\lambda, b} = 4.12 \times 10^5 \ W/m^2 \cdot \mu m \qquad \lhd$$

4. Irradiation of any small object within the enclosure may be approximated as being equal to emission from a blackbody at the enclosure surface temperature. Hence

$$G = E_b(T)$$

$$G = 9.07 \times 10^5 \ W/m^2 \qquad \lhd$$

EXAMPLE 12.5

A surface is known to emit as a blackbody at 1500 K. What is the rate per unit area (W/m^2) at which the surface emits radiation in directions corresponding to $0° \le \theta \le 60°$ and in the wavelength interval 2 $\mu m \le \lambda \le 4 \ \mu m$?

SOLUTION

KNOWN:

Temperature of a surface that emits as a blackbody.

FIND:

Rate of emission per unit area in directions between $\theta = 0°$ and $60°$ and at wavelengths between $\lambda = 2 \ \mu m$ and 4 μm.

SCHEMATIC:

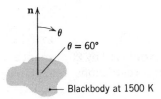

θ = 60°

—— Blackbody at 1500 K

ASSUMPTIONS:

Surface emits as a blackbody.

ANALYSIS:

The desired emission may be inferred from Equation 12.12, with the limits of integration restricted as follows

$$\Delta E = \int_2^4 \int_0^{2\pi} \int_0^{\pi/3} I_{\lambda,b} \cos\theta \sin\theta \, d\theta d\phi d\lambda$$

or, since a blackbody emits diffusely,

$$\Delta E = \int_2^4 I_{\lambda,b} \left(\int_0^{2\pi} \int_0^{\pi/3} \cos\theta \sin\theta \, d\theta d\phi \right) d\lambda$$

$$= \int_2^4 I_{\lambda,b} \left(2\pi \frac{\sin^2\theta}{2} \bigg|_0^{\pi/3} \right) d\lambda = 0.75 \int_2^4 \pi I_{\lambda,b} d\lambda$$

Substituting from Equation 12.13 and multiplying and dividing by E_b, this result may be put in a form which allows for use of Table 12.1 in evaluating the spectral integration. In particular

$$\Delta E = 0.75 E_b \int_2^4 \frac{E_{\lambda,b}}{E_b} d\lambda = 0.75 E_b [F_{(0\to4)} - F_{(0\to2)}]$$

where from Table 12.1

$$\lambda_1 T = 2\,\mu\text{m} \times 1500\,\text{K} = 3000\,\mu\text{m}\cdot\text{K}, \; F_{(0\to2)} = 0.273$$

$$\lambda_2 T = 4\mu\text{m} \times 1500\,\text{K} = 6000\,\mu\text{m}\cdot\text{K}, \; F_{(0\to4)} = 0.738$$

Hence

$$\Delta E = 0.75(0.738 - 0.273)E_b = 0.75(0.465)E_b$$

From Equation 12.28, it then follows that

$$\Delta E = 0.75(0.465)5.67 \times 10^{-8} \; \text{W/m}^2\cdot\text{K}^4 \; (1500\,\text{K})^4$$

or

$$\Delta E = 10^5 \ W/m^2 \qquad\qquad\qquad \lhd$$

COMMENTS:

The total hemispherical emissive power is reduced by 25 percent and 53.5 percent due to the directional and spectral restrictions, respectively.

12.4 SURFACE EMISSION

Having developed the notion of a blackbody to describe *ideal* surface behavior, we may now consider the behavior of *real* surfaces. Recall that the blackbody is an ideal emitter in the sense that no surface can emit more radiation than a blackbody at the same temperature. It is therefore convenient to choose the blackbody as a reference in describing emission from a real surface. A surface radiative property known as the *emissivity*[1] may then be defined as the *ratio* of the radiation emitted by the surface to the radiation emitted by a blackbody at the same temperature.

It is important to acknowledge that, in general, the spectral radiation emitted by a real surface differs from the Planck distribution, as shown in Figure 12.16a. Moreover, as shown in Figure 12.16b, the directional distribution may be other than diffuse. Hence the emissivity may assume different values according to whether one is interested in emission at a given wavelength, in a

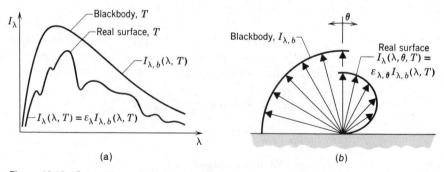

Figure 12.16 Comparison of blackbody and real surface emission. (a) Spectral distribution. (b) Directional distribution.

---

[1] In this text we use the *-ivity*, rather than the *-ance*, ending for material radiative properties (e.g., emissivity rather than emittance). Although efforts are being made to reserve the -ivity ending for optically smooth, uncontaminated surfaces, no such distinction is made in much of the literature, and it will not be made in this text.

given direction or in integrated averages over all possible wavelengths and directions.

We define the *spectral, directional emissivity*, $\varepsilon_{\lambda,\theta}$ (λ,θ,ϕ,T), of a surface at the temperature T as the ratio of the intensity of the radiation emitted at the wavelength λ and in the direction of θ and ϕ to the intensity of the radiation emitted by a blackbody at the same values of T and λ. Hence

$$\varepsilon_{\lambda,\theta}(\lambda,\theta,\phi,T) \equiv \frac{I_{\lambda,e}(\lambda,\theta,\phi,T)}{I_{\lambda,b}(\lambda,T)} \tag{12.32}$$

Note how the subscripts λ and θ are used to designate interest in a specific wavelength and direction for the emissivity. In contrast, terms appearing within a parenthesis designate functional dependence on wavelength, direction, and/or temperature. The absence of directional variables in the parenthesis of the denominator in the foregoing expression implies that the intensity is independent of direction, which is, of course, a characteristic of blackbody emission. In a like manner a *total, directional emissivity*, ε_θ, which represents a spectral average of $\varepsilon_{\lambda,\theta}$, may be defined as

$$\varepsilon_\theta(\theta,\phi,T) \equiv \frac{I_e(\theta,\phi,T)}{I_b(T)} \tag{12.33}$$

For most engineering calculations, it is desirable to work with surface properties which represent directional averages. A *spectral, hemispherical emissivity* is therefore defined as

$$\varepsilon_\lambda(\lambda,T) \equiv \frac{E_\lambda(\lambda,T)}{E_{\lambda,b}(\lambda,T)} \qquad \text{Directional Average emissivity} \tag{12.34}$$

It may be related to the directional emissivity, $\varepsilon_{\lambda,\theta}$, by substituting the expression for the spectral emissive power, Equation 12.10, to obtain

$$\varepsilon_\lambda(\lambda,T) = \frac{\int_0^{2\pi}\int_0^{\pi/2} I_{\lambda,e}(\lambda,\theta,\phi,T)\cos\theta\sin\theta\,d\theta d\phi}{\int_0^{2\pi}\int_0^{\pi/2} I_{\lambda,b}(\lambda,T)\cos\theta\sin\theta\,d\theta d\phi}$$

Note that, in contrast to Equation 12.10, the temperature dependence of emission is now acknowledged. From Equation 12.32 and the fact that $I_{\lambda,b}$ is independent of θ and ϕ, it then follows that

$$\varepsilon_\lambda(\lambda,T) = \frac{\int_0^{2\pi}\int_0^{\pi/2} \varepsilon_{\lambda,\theta}(\lambda,\theta,\phi,T)\cos\theta\sin\theta\,d\theta d\phi}{\int_0^{2\pi}\int_0^{\pi/2} \cos\theta\sin\theta\,d\theta d\phi} \tag{12.35}$$

Assuming $\varepsilon_{\lambda,\theta}$ to be independent of ϕ, which is a reasonable assumption for most surfaces, and evaluating the denominator, we then obtain

$$\varepsilon_\lambda(\lambda, T) = 2\pi \frac{\int_0^{\pi/2} \varepsilon_{\lambda,\theta}(\lambda,\theta,T)\cos\theta\sin\theta\, d\theta}{\pi}$$

$$\varepsilon_\lambda(\lambda, T) = 2 \int_0^{\pi/2} \varepsilon_{\lambda,\theta}(\lambda,\theta,T)\cos\theta\sin\theta\, d\theta \tag{12.36}$$

The *total, hemispherical emissivity*, which represents an average over all possible directions and wavelengths, is defined as

$$\varepsilon(T) \equiv \frac{E(T)}{E_b(T)} \tag{12.37}$$

Substituting from Equations 12.11 and 12.34, it then follows that

$$\varepsilon(T) = \frac{\int_0^\infty \varepsilon_\lambda(\lambda, T)\, E_{\lambda,b}(\lambda, T) d\lambda}{E_b(T)} \tag{12.38}$$

If the emissivities of a surface are known, it is a simple matter to compute its emission characteristics. For example, if $\varepsilon_\lambda(\lambda, T)$ is known, it may be used with Equations 12.26 and 12.34 to compute the spectral emissive power of the surface at any wavelength and temperature. Similarly, if $\varepsilon(T)$ is known, it may be used with Equations 12.28 and 12.37 to compute the total emissive power of the surface at any temperature. Measurements have been performed to determine these properties for many different materials and surface coatings.

The directional emissivity of a diffuse emitter is, of course, a constant, independent of direction. However, although this condition is often a reasonable *approximation*, all surfaces exhibit some departure from diffuse behavior. Representative variations of ε_θ with θ are shown schematically in Figure 12.17 for conducting and nonconducting materials. For conductors ε_θ is approximately constant over the range $\theta \lesssim 40°$, after which it increases with increasing θ but ultimately decays to zero. In contrast for nonconductors ε_θ is approximately constant for $\theta \lesssim 70°$, beyond which it decreases sharply with increasing θ. One implication of these variations is that, although there are preferential

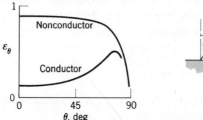

Figure 12.17 Representative directional distributions of the total, directional emissivity.

directions for emission, the hemispherical emissivity, ε, will not differ markedly from the normal emissivity, ε_n, corresponding to $\theta = 0$. In fact the ratio rarely falls outside the range $1.0 \leq (\varepsilon/\varepsilon_n) \leq 1.3$ for conductors and the range of $0.95 \leq (\varepsilon/\varepsilon_n) \leq 1.0$ for nonconductors. Hence to a reasonable approximation,

$$\varepsilon \approx \varepsilon_n \qquad\qquad (12.39)$$

Note that, although the foregoing statements were made for the total emissivity, they also apply to spectral components.

Since the spectral distribution of emission from real surfaces exhibits departures from the Planck distribution, Figure 12.16a, we do not expect the value of the spectral emissivity, ε_λ, to be independent of wavelength. Representative spectral distributions of ε_λ are shown in Figure 12.18. The manner in which ε_λ varies with λ depends on whether the solid is a conductor or nonconductor, as well as the nature of the surface coating.

Representative values of the total, normal emissivity, ε_n, are plotted in Figures 12.19 and 12.20 and listed in Table A.11. Several generalizations may be made.

1. The emissivity of metallic surfaces is generally small, achieving values as low as 0.02 for highly polished gold and silver.

2. The presence of oxide layers may significantly increase the emissivity of metallic surfaces. Contrast the value of 0.10 for lightly oxidized stainless steel with the value of approximately 0.50 for the heavily oxidized form.

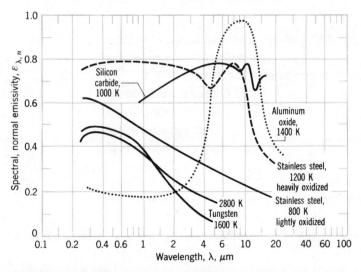

Figure 12.18 Spectral dependence of the spectral, normal emissivity, $\varepsilon_{\lambda,n}$, of selected materials.

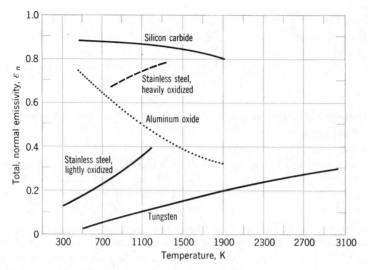

Figure 12.19 Temperature dependence of the total, normal emissivity, ε_n, of selected materials.

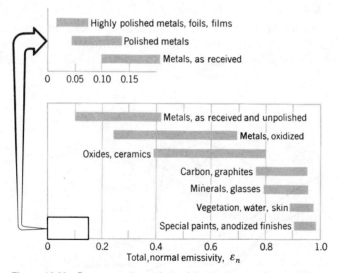

Figure 12.20 Representative values of the total, normal emissivity, ε_n.

3. The emissivity of nonconductors is comparatively large, generally exceeding 0.6.

4. The emissivity of conductors increases with increasing temperature; however, depending on the specific material, the emissivity of nonconductors

may either increase or decrease with increasing temperature. Note that the variations of ε_n with T shown in Figure 12.19 are consistent with the spectral distributions of $\varepsilon_{\lambda,n}$ shown in Figure 12.18. These trends follow from Equation 12.38. Although the spectral distribution of $\varepsilon_{\lambda,n}$ is approximately independent of temperature, there is proportionately more emission from a blackbody at lower wavelengths with increasing temperature. Hence, if $\varepsilon_{\lambda,n}$ increases with decreasing wavelength for a particular material, ε_n will increase with increasing temperature for that material.

It should be recognized that the emissivity depends strongly on the nature of the surface, which can be influenced by the method of fabrication, thermal cycling, and chemical reaction with its environment. More comprehensive surface emissivity compilations are provided by Gubareff et al. [2], Wood et al. [3], Touloukian [4], and Touloukian and DeWitt [5].

EXAMPLE 12.6

A diffuse surface at a temperature of 1600 K has the spectral, hemispherical emissivity shown below.

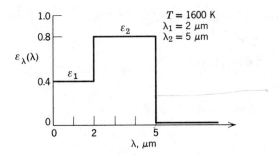

1. Determine the total, hemispherical emissivity.
2. Calculate the total emissive power.
3. At what wavelength will the spectral emissive power be a maximum?

SOLUTION

KNOWN:

The spectral, hemispherical emissivity, $\varepsilon_\lambda (\lambda)$, of a diffuse surface at $T = 1600$ K.

FIND:

1. Total, hemispherical emissivity.
2. Total emissive power.
3. Wavelength at which spectral emissive power will be a maximum.

ASSUMPTIONS:

Surface is a diffuse emitter.

ANALYSIS:

1. The total, hemispherical emissivity is given by Equation 12.38, where the integration may be performed in parts as follows.

$$\varepsilon = \frac{\int_0^\infty \varepsilon_\lambda E_{\lambda,b}\, d\lambda}{E_b} = \frac{\varepsilon_1 \int_0^2 E_{\lambda,b}\, d\lambda}{E_b} + \frac{\varepsilon_2 \int_2^5 E_{\lambda,b}\, d\lambda}{E_b}$$

or

$$\varepsilon = \varepsilon_1 F_{(0\to2\ \mu m)} + \varepsilon_2 [F_{(0\to5\ \mu m)} - F_{(0\to2\ \mu m)}]$$

From Table 12.1 we obtain

$$\lambda_1 T = 2\ \mu m \times 1600\ K = 3200\ \mu m\cdot K: F_{(0\to2\ \mu m)} = 0.318$$

$$\lambda_2 T = 5\ \mu m \times 1600\ K = 8000\ \mu m\cdot K: F_{(0\to5\ \mu m)} = 0.856$$

Hence

$$\varepsilon = 0.4 \times 0.318 + 0.8[0.856 - 0.318]$$

$$\varepsilon = 0.558 \qquad \triangleleft$$

2. From Equation 12.37 the total emissive power is

$$E = \varepsilon E_b = \varepsilon \sigma T^4$$

$$E = 0.558(5.67 \times 10^{-8}\ W/m^2\cdot K^4)(1600\ K)^4$$

$$E = 207\ kW/m^2 \qquad \triangleleft$$

3. If the surface emitted as a blackbody or if its emissivity were a constant, independent of λ, the wavelength corresponding to maximum spectral emission could be obtained from Wien's law. However, because ε_λ varies with λ, it is not immediately obvious where peak emission occurs. From Equation 12.27 we know that

$$\lambda_{max} = \frac{2898\ \mu m\cdot K}{1600\ K} = 1.81\ \mu m$$

The spectral emissive power at this wavelength may be obtained by using Equation 12.34 with Table 12.1. That is,

$$E_\lambda(\lambda_{max}, T) = \varepsilon_\lambda(\lambda_{max})E_{\lambda,b}(\lambda_{max}, T)$$

or, since the surface is a diffuse emitter,

$$E_\lambda(\lambda_{max}, T) = \pi\varepsilon_\lambda(\lambda_{max})I_{\lambda,b}(\lambda_{max}, T)$$

$$E_\lambda(\lambda_{max}, T) = \pi\varepsilon_\lambda(\lambda_{max}) \frac{I_{\lambda,b}(\lambda_{max}, T)}{\sigma T^5} \times \sigma T^5$$

$E_\lambda(1.81~\mu m,~1600~K) = \pi \times 0.4 \times 0.722 \times 10^{-4}(1/\mu m \cdot K \cdot sr)\,5.67 \times 10^{-8}~W/m^2 \cdot K^4 \times (1600~K)^5$

$E_\lambda(1.81~\mu m,~1600~K) = 54~kW/m^2 \cdot \mu m$

Since $\varepsilon_\lambda = 0.4$ from $\lambda = 0$ to $\lambda = 2~\mu m$, the foregoing result provides the maximum spectral emissive power for the region $\lambda < 2~\mu m$. However, with the change in ε_λ that occurs at $\lambda = 2~\mu m$, the value of E_λ at $\lambda = 2~\mu m$ may be larger than that for $\lambda = 1.81~\mu m$. To determine whether this is, in fact, the case, we compute

$$E_\lambda(\lambda_1, T) = \pi\varepsilon_\lambda(\lambda_1) \frac{I_{\lambda,b}(\lambda_1, T)}{\sigma T^5} \times \sigma T^5$$

where, for $\lambda_1 T = 3200~\mu m \cdot K$, $[I_{\lambda,b}(\lambda_1, T)/\sigma T^5] = 0.706 \times 10^{-4}~(\mu m \cdot K \cdot sr)^{-1}$. Hence

$E_\lambda(2\mu m,~1600~K) = \pi \times 0.80 \times 0.706 \times 10^{-4}~(1/\mu m \cdot K \cdot sr)\,5.67 \times 10^{-8}~W/m^2 \cdot K^4~(1600~K)^5$

$E_\lambda(2~\mu m,~1600~K) = 105.5~kW/m^2 \cdot \mu m > E_\lambda(1.81~\mu m,~1600~K)$

and peak emission occurs at

$$\lambda = \lambda_1 = 2~\mu m \qquad\qquad \triangleleft$$

COMMENTS:

For the prescribed spectral distribution of ε_λ, the spectral emissive power will vary with wavelength as shown in the following diagram.

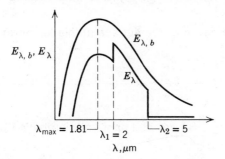

EXAMPLE 12.7

Measurements of the spectral, directional emissivity of a metallic surface at elevated temperatures and for $\lambda = 1.0\ \mu m$ reveal a directional distribution which may be approximated by the results shown below.

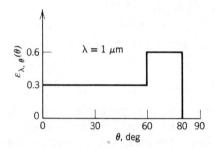

1. What are the spectral, normal emissivity and the spectral, hemispherical emissivity for $\lambda = 1\ \mu m$?

2. Consider the surface at 2000 K. What is the spectral intensity of radiation emitted at 1 μm in the normal direction? What is the spectral emissive power at 1 μm?

Additional measurements reveal that the spectral, hemispherical emissivity of the surface varies with wavelength in approximately the following manner.

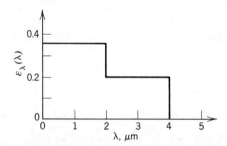

3. What is the total, hemispherical emissivity when the surface is at a temperature of 2000 K?

4. What is the total emissive power of the surface at this temperature?

SOLUTION

KNOWN:

Directional distribution of $\varepsilon_{\lambda,\theta}$ at $\lambda = 1\ \mu m$ and spectral distribution of ε_λ for a metallic surface.

FIND:

1. Spectral, normal emissivity, $\varepsilon_{\lambda,n}$, and spectral, hemispherical emissivity, ε_λ, at 1 μm.
2. Spectral, normal intensity, $I_{\lambda,n}$, and spectral emissive power, E_λ, at 1 μm for a surface temperature of 2000 K.
3. Total, hemispherical emissivity, ε, at 2000 K.
4. Total emissive power at 2000 K.

ANALYSIS:

1. From the measurement of $\varepsilon_{\lambda,\theta}$ at $\lambda = 1$ μm, we see that

$$\varepsilon_{\lambda,n} = \varepsilon_{\lambda,\theta}(1\ \mu m, 0°) = 0.3$$ ◁

From Equation 12.36, the spectral, hemispherical emissivity is

$$\varepsilon_\lambda(1\ \mu m) = 2\int_0^{\pi/2} \varepsilon_{\lambda,\theta}\cos\theta\sin\theta\,d\theta$$

or

$$\varepsilon_\lambda(1\ \mu m) = 2\left[0.3\int_0^{\pi/3}\cos\theta\sin\theta\,d\theta + 0.6\int_{\pi/3}^{4\pi/9}\cos\theta\sin\theta\,d\theta\right]$$

$$\varepsilon_\lambda(1\ \mu m) = 2\left[0.3\frac{\sin^2\theta}{2}\Big|_0^{\pi/3} + 0.6\frac{\sin^2\theta}{2}\Big|_{\pi/3}^{4\pi/9}\right] = 2\left[\frac{0.3}{2}(0.75)\right.$$

$$\left. + \frac{0.6}{2}(0.97 - 0.75)\right]$$

$$\varepsilon_\lambda(1\ \mu m) = 0.36$$ ◁

2. From Equation 12.32 the spectral intensity of radiation emitted at $\lambda = 1$ μm in the normal direction is

$$I_{\lambda,n}(1\ \mu m,\ 0°,\ 2000\ K) = \varepsilon_{\lambda,\theta}(1\ \mu m, 0°)I_{\lambda,b}(1\ \mu m,\ 2000\ K)$$

where

$$\varepsilon_{\lambda,\theta}(1\ \mu m,\ 0°) = 0.3$$

and $I_{\lambda,b}(1\ \mu m,\ 2000\ K)$ may be obtained from Table 12.1. For $\lambda T = 2000$ μm·K, $(I_{\lambda,b}/\sigma T^5) = 0.493 \times 10^{-4}(\mu m\cdot K\cdot sr)^{-1}$ and

$$I_{\lambda,b} = 0.493 \times 10^{-4}(\mu m\cdot K\cdot sr)^{-1} \times 5.67 \times 10^{-8}\ W/m^2\cdot K^4(2000\ K)^5$$

$$I_{\lambda,b} = 8.95 \times 10^4\ W/m^2\cdot\mu m\cdot sr$$

Hence

$$I_{\lambda,n}(1 \ \mu m, 0°, 2000 \ K) = 0.3 \times 8.95 \times 10^4 \ W/m^2 \cdot \mu m \cdot sr$$

$$I_{\lambda,n}(1 \ \mu m, 0°, 2000 \ K) = 2.69 \times 10^4 \ W/m^2 \cdot \mu m \cdot sr \qquad \triangleleft$$

From Equation 12.34 the spectral emissive power for $\lambda = 1 \ \mu m$ and $T = 2000 \ K$ is

$$E_\lambda(1 \ \mu m, 2000 \ K) = \varepsilon_\lambda(1 \ \mu m)E_{\lambda,b}(1 \ \mu m, 2000 \ K)$$

where

$$E_{\lambda,b}(1 \ \mu m, 2000 \ K) = \pi I_{\lambda,b}(1 \ \mu m, 2000 \ K)$$

$$E_{\lambda,b}(1 \ \mu m, 2000 \ K) = \pi sr \times 8.95 \times 10^4 \ W/m^2 \cdot \mu m \cdot sr$$
$$= 2.81 \times 10^5 \ W/m^2 \cdot \mu m$$

Hence

$$E_\lambda(1 \ \mu m, 2000 \ K) = 0.36 \times 2.81 \times 10^5 \ W/m^2 \cdot \mu m$$

or

$$E_\lambda(1 \ \mu m, 2000 \ K) = 1.01 \times 10^5 \ W/m^2 \cdot \mu m \qquad \triangleleft$$

3. From knowledge of $\varepsilon_\lambda(\lambda)$ and Equation 12.38,

$$\varepsilon(T) = \frac{\int_0^\infty \varepsilon_\lambda(\lambda)E_{\lambda,b}(\lambda,T)d\lambda}{E_b(T)}$$

$$\varepsilon(T) = 0.36 \int_0^{2 \ \mu m} \frac{E_{\lambda,b}(\lambda,T)}{E_b(T)} \, d\lambda + 0.2 \int_{2 \ \mu m}^{4 \ \mu m} \frac{E_{\lambda,b}(\lambda,T)}{E_b(T)} \, d\lambda$$

From Table 12.1,

$$\lambda_1 = 2 \ \mu m, \ T = 2000 \ K: \ \lambda_1 T = 4000 \ \mu m \cdot K, \ F_{(0 \rightarrow \lambda_1)} = 0.481$$

$$\lambda_2 = 4 \ \mu m, \ T = 2000 \ K: \ \lambda_2 T = 8000 \ \mu m \cdot K, \ F_{(0 \rightarrow \lambda_2)} = 0.856$$

Hence from Equations 12.30 and 12.31

$$\varepsilon(2000 \ K) = 0.36(0.481) + 0.2(0.856 - 0.481)$$

or

$$\varepsilon(2000 \ K) = 0.25 \qquad \triangleleft$$

4. From Equations 12.28 and 12.37, the total emissive power at 2000 K is

$$E(2000 \ K) = \varepsilon(2000 \ K)E_b(2000 \ K)$$

$$E(2000 \ K) = 0.25 \times 5.67 \times 10^{-8} \ W/m^2 \cdot K^4 \times (2000 \ K)^4$$

or

$$E(2000 \ K) = 2.27 \times 10^5 \ W/m^2 \qquad \triangleleft$$

12.5 SURFACE ABSORPTION, REFLECTION, AND TRANSMISSION

In Section 12.2.2 we defined the *spectral irradiation*, G_λ (W/m$^2 \cdot \mu$m), as the rate at which radiation of wavelength λ is incident on a surface per unit area of the surface and per unit wavelength interval $d\lambda$ about λ. It may be incident from all possible directions and originate from several different sources. The *total irradiation*, G (W/m^2), encompasses all spectral contributions and may be evaluated from Equation 12.16. In this section we consider the *processes* resulting from interception of this radiation by a solid (or liquid) medium.

The most general situation is one in which the irradiation interacts with a *semitransparent medium,* such as a layer of water or a glass plate. As shown in Figure 12.21 for a spectral component of the irradiation, portions of this radiation may be *reflected, absorbed* and *transmitted*. From a radiation balance on the medium, it follows that

$$G_\lambda = G_{\lambda,\text{ref}} + G_{\lambda,\text{abs}} + G_{\lambda,\text{tr}} \tag{12.40}$$

However, determination of the above components is a complex problem, depending on the upper and lower surface conditions, the wavelength of the radiation and the composition and thickness of the medium. In this case conditions are strongly influenced by *volumetric* effects occurring within the medium.

A simpler situation, which pertains to most engineering applications, is one for which the medium is said to be *opaque* to the incident radiation. In this case, $G_{\lambda,\text{tr}} = 0$ and the remaining absorption and reflection processes may be treated as *surface phenomena*. That is, reflection and absorption by the medium are controlled by processes occurring within a fraction of a micrometer from the surface being irradiated. It is therefore appropriate to speak of the irradiation being absorbed or reflected *by the surface*, with the relative magnitudes of $G_{\lambda,\text{abs}}$ and $G_{\lambda,\text{ref}}$ depending on λ and the nature of the surface material. Note that there

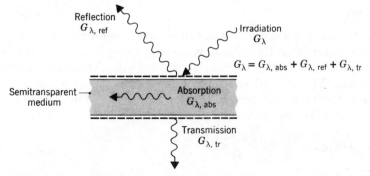

Figure 12.21 Absorption, reflection, and transmission processes associated with a semitransparent medium.

is no net effect of the reflection process on the medium, while absorption has the effect of increasing the internal thermal energy of the medium.

It is interesting to note that surface absorption and reflection characteristics are responsible for the *colors* that we see in the objects around us. Unless the surface is at a high temperature ($T_s \gtrsim 1000$ K), such that it is incandescent, the colors are in no way due to emission, which is concentrated in the infrared and hence imperceptible to the eye. The colors are instead due to the selective reflection and absorption of the visible portion of the irradiation which is incident from the sun or an artificial source of light. A shirt is "red" because it contains a pigment that preferentially absorbs the blue, green, and yellow components of the incident light. Hence the relative contributions of these components to the reflected light, which is seen, is diminished, and the red component is dominant. Similarly, a leaf is "green" because its cells contain chlorophyll, a pigment that shows strong absorption in the blue and the red and preferential reflection in the green. Similarly, a surface appears "black" if it absorbs all incident visible radiation, and it is "white" if it reflects this radiation. However, we must be careful in how we interpret such *visual* effects. For a prescribed irradiation, the "color" of a surface may not indicate its overall capacity as an absorber or reflector, since much of the irradiation may be in the infrared. A "white" surface such as snow, for example, is high reflective to visible radiation but strongly absorbs infrared radiation, thereby approximating blackbody behavior at long wavelengths.

In the foregoing section we introduced a property, which we called the emissivity, to characterize the process of surface emission. In the following subsections we shall introduce properties to characterize the absorption, reflection and transmission processes. In general these properties are a function of surface material and finish, surface temperature, and the wavelength and direction of the incident radiation.

12.5.1 Absorptivity

The absorptivity is a property which determines the fraction of the irradiation absorbed by a surface. Determination of the property is complicated by the fact that, like emission, it may be characterized by both a directional and spectral dependence. The *spectral, directional absorptivity*, $\alpha_{\lambda,\theta}(\lambda,\theta,\phi)$, of a surface is defined as the fraction of the spectral intensity incident in the direction of θ and ϕ that is absorbed by the surface. Hence,

$$\alpha_{\lambda,\theta}(\lambda,\theta,\phi) \equiv \frac{I_{\lambda,i,\text{abs}}(\lambda,\theta,\phi)}{I_{\lambda,i}(\lambda,\theta,\phi)} \tag{12.41}$$

Note that, in the foregoing expression, we have neglected any dependence of the absorptivity on the surface temperature. Such a dependence is generally small for all spectral radiative properties.

It is implicit in the foregoing result that surfaces may exhibit selective absorption with respect to the wavelength and direction of the incident radiation. For most engineering calculations, however, it is desirable to work with surface properties that represent directional averages. We therefore define a *spectral, hemispherical absorptivity*, $\alpha_\lambda(\lambda)$, as

$$\alpha_\lambda(\lambda) \equiv \frac{G_{\lambda,\text{abs}}(\lambda)}{G_\lambda(\lambda)} \tag{12.42}$$

which, from Equation 12.15 and 12.41, may be expressed as

$$\alpha_\lambda(\lambda) = \frac{\int_0^{2\pi} \int_0^{\pi/2} \alpha_{\lambda,\theta}(\lambda,\theta,\phi)\, I_{\lambda,i}(\lambda,\theta,\phi) \cos\theta \sin\theta\, d\theta d\phi}{\int_0^{2\pi} \int_0^{\pi/2} I_{\lambda,i}(\lambda,\theta,\phi) \cos\theta \sin\theta\, d\theta d\phi} \tag{12.43}$$

Hence α_λ depends on the directional distribution of the incident radiation, as well as on the wavelength of the radiation and the nature of the absorbing surface. Note that, if the incident radiation is diffusely distributed and $\alpha_{\lambda,\theta}$ is independent of ϕ, Equation 12.43 reduces to

$$\alpha_\lambda(\lambda) = 2 \int_0^{\pi/2} \alpha_{\lambda,\theta}(\lambda,\theta) \cos\theta \sin\theta\, d\theta \tag{12.44}$$

The *total, hemispherical absorptivity*, α, represents an integrated average over both direction and wavelength. It is defined as the fraction of the total irradiation absorbed by a surface

$$\alpha \equiv \frac{G_{\text{abs}}}{G} \tag{12.45}$$

and, from Equations 12.16 and 12.42, may be expressed as

$$\alpha = \frac{\int_0^\infty \alpha_\lambda(\lambda)\, G_\lambda(\lambda) d\lambda}{\int_0^\infty G_\lambda(\lambda) d\lambda} \tag{12.46}$$

Accordingly, α depends on the spectral distribution of the incident radiation, as well as on its directional distribution and the nature of the absorbing surface. Note that, although α is approximately independent of temperature, the same may not be said for the total, hemispherical emissivity, ε. From Equation 12.38 it is evident that this property is strongly temperature dependent. Note also that, because α depends on the spectral distribution of the irradiation, its value for a particular surface exposed to solar irradiation may differ appreciably from its value for the same surface exposed to longer wavelength irradiation originating from a source of lower temperature. Since the spectral distribution of solar irradiation is nearly proportional to that of emission from a blackbody at 5800 K, it follows from Equation 12.46 that the total absorptivity to solar irradiation, α_S, may be approximated as

$$\alpha_S \approx \frac{\int_0^\infty \alpha_\lambda(\lambda) E_{\lambda,b}(\lambda, 5800 \text{ K})d\lambda}{\int_0^\infty E_{\lambda,b}(\lambda, 5800 \text{ K})d\lambda} \tag{12.47}$$

Evaluation of the ratio of integrals appearing in Equation 12.47 may often be facilitated by using the blackbody radiation function, $F_{(0 \to \lambda)}$, of Table 12.1.

12.5.2 Reflectivity

The reflectivity is a property that determines the fraction of the incident radiation reflected by a surface. However, its specific definition may take several different forms, because the property is inherently *bidirectional* [6]. That is, in addition to depending on the direction of the incident radiation, it also depends on the direction of the reflected radiation. We shall avoid this complication by working exclusively with a reflectivity that represents an integrated average over the hemisphere associated with the reflected radiation and therefore provides no information concerning the directional distribution of this radiation. Accordingly, the *spectral, directional reflectivity*, $\rho_{\lambda,\theta}(\lambda,\theta,\phi)$, of a surface is defined as the fraction of the spectral intensity incident in the direction of θ and ϕ, which is reflected by the surface. Hence,

$$\rho_{\lambda,\theta}(\lambda,\theta,\phi) \equiv \frac{I_{\lambda,i,\text{ref}}(\lambda,\theta,\phi)}{I_{\lambda,i}(\lambda,\theta,\phi)} \tag{12.48}$$

The *spectral, hemispherical reflectivity*, $\rho_\lambda(\lambda)$, is then defined as the fraction of the spectral irradiation which is reflected by the surface. Accordingly,

$$\rho_\lambda(\lambda) \equiv \frac{G_{\lambda,\text{ref}}(\lambda)}{G_\lambda(\lambda)} \tag{12.49}$$

which is equivalent to

$$\rho_\lambda(\lambda) = \frac{\int_0^{2\pi} \int_0^{\pi/2} \rho_{\lambda,\theta}(\lambda,\theta,\phi) I_{\lambda,i}(\lambda,\theta,\phi) \cos\theta \sin\theta \, d\theta d\phi}{\int_0^{2\pi} \int_0^{\pi/2} I_{\lambda,i}(\lambda,\theta,\phi) \cos\theta \sin\theta \, d\theta d\phi} \tag{12.50}$$

The *total, hemispherical reflectivity*, ρ, is then defined as

$$\rho \equiv \frac{G_{\text{ref}}}{G} \tag{12.51}$$

in which case

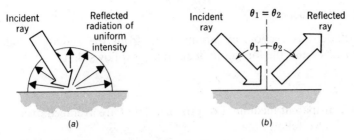

Figure 12.22 Diffuse and specular reflection.

$$\rho = \frac{\int_0^\infty \rho_\lambda(\lambda)G_\lambda(\lambda)d\lambda}{\int_0^\infty G_\lambda(\lambda)d\lambda} \tag{12.52}$$

We conclude this discussion by noting that surfaces may be idealized as *diffuse* or *specular*, according to the manner in which they reflect radiation (Figure 12.22). Diffuse reflection occurs if, regardless of the direction of the incident radiation, the intensity of the reflected radiation is independent of the reflection angle. In contrast, if all of the reflection is in the direction of θ_2, which equals the incident angle θ_1, specular reflection is said to occur. Although no surface is perfectly diffuse or specular, the latter condition is more closely approximated by polished, mirrorlike surfaces and the former condition by rough surfaces. The assumption of diffuse reflection is reasonable for most engineering applications.

12.5.3 Transmissivity

Although treatment of the response of a semitransparent material to incident radiation is a complicated problem [6], reasonable results may often be obtained through the use of hemispherical transmissivities defined as

$$\tau_\lambda = \frac{G_{\lambda,\text{tr}}}{G_\lambda} \tag{12.53}$$

and

$$\tau = \frac{G_{\text{tr}}}{G} \tag{12.54}$$

The total transmissivity, τ, is related to the spectral component, τ_λ, by an expression of the form

$$\tau = \frac{\int_0^\infty G_{\lambda,\text{tr}}(\lambda)d\lambda}{\int_0^\infty G_\lambda(\lambda)d\lambda} = \frac{\int_0^\infty \tau_\lambda(\lambda)G_\lambda(\lambda)d\lambda}{\int_0^\infty G_\lambda(\lambda)d\lambda} \tag{12.55}$$

12.5.4 Special Considerations

We conclude this section by noting that, from the radiation balance of Equation 12.40 and the foregoing definitions, it follows that

$$\rho_\lambda + \alpha_\lambda + \tau_\lambda = 1 \tag{12.56}$$

for a semitransparent medium. For properties that are averaged over the entire spectrum, it also follows that

$$\rho + \alpha + \tau = 1 \tag{12.57}$$

Of course, if the medium is opaque, there is no transmission, and absorption and reflection are surface processes for which

$$\alpha_\lambda + \rho_\lambda = 1 \tag{12.58}$$

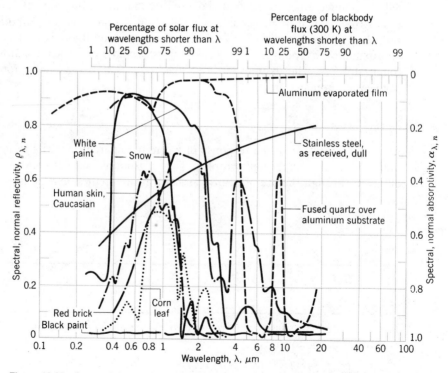

Figure 12.23 Spectral dependence of the spectral, normal absorptivity, $\alpha_{\lambda,n}$, and reflectivity, $\rho_{\lambda,n}$, of selected materials.

and

$$\alpha + \rho = 1 \tag{12.59}$$

Hence knowledge of one property implies knowledge of the other.

Spectral distributions of the normal reflectivity and absorptivity are plotted in Figure 12.23 for selected *opaque* surfaces. Note that a material such as glass or water, which is semitransparent at short wavelengths, becomes opaque at longer wavelengths. This behavior is shown in Figure 12.24, which presents the spectral transmissivity of several common semitransparent materials. Note that the transmissivity of glass is affected by its iron content and that the transmissivity of plastics, such as Tedlar, is greater than that of glass in the infrared region. These factors have an important bearing on the selection of cover plate materials for solar collector applications. Values for the total transmissivity to solar radiation of common collector cover plate materials are presented in Table A.12, along with surface solar absorptivities and low temperature emissivities.

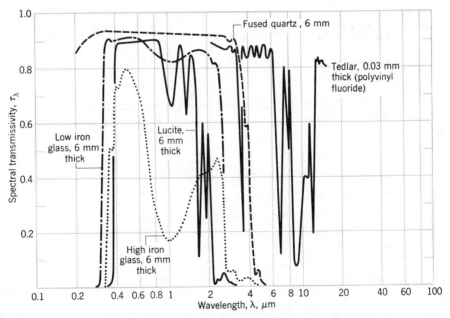

Figure 12.24 Spectral dependence of the spectral transmissivities, τ_λ, of selected materials.

EXAMPLE 12.8

The variation with wavelength of the spectral, hemispherical absorptivity of an opaque surface and the spectral irradiation at the surface are shown below.

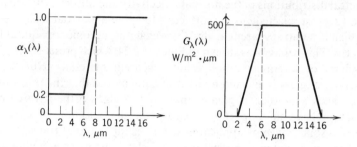

1. What is the spectral distribution of the spectral, hemispherical reflectivity?

2. What is the total, hemispherical absorptivity of the surface?

3. If the surface is in equilibrium at an initial temperature of 500 K and has a total, hemispherical emissivity of 0.8, will this temperature increase with time, decrease with time, or remain the same upon exposure to the irradiation?

SOLUTION

KNOWN:

Spectral, hemispherical absorptivity and irradiation of a surface. Surface temperature (500 K) and total, hemispherical emissivity (0.8).

FIND:

1. Spectral distribution of reflectivity.

2. Total, hemispherical absorptivity.

3. Time rate of surface temperature change.

SCHEMATIC:

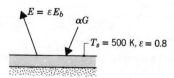

ASSUMPTIONS:

1. Surface is opaque.
2. Surface convection effects are negligible.

ANALYSIS:

1. From Equation 12.58, $\rho_\lambda = 1 - \alpha_\lambda$. Hence from knowledge of $\alpha_\lambda(\lambda)$, the spectral distribution of ρ_λ is as shown below.

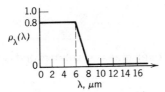

2. From Equations 12.45 and 12.46

$$\alpha = \frac{G_{abs}}{G} = \frac{\int_0^\infty \alpha_\lambda G_\lambda d\lambda}{\int_0^\infty G_\lambda d\lambda}$$

or, subdividing the integral into parts,

$$\alpha = \frac{0.2 \int_2^6 G_\lambda d\lambda + 500 \int_6^8 \alpha_\lambda d\lambda + 1.0 \int_8^{16} G_\lambda d\lambda}{\int_2^6 G_\lambda d\lambda + \int_6^{12} G_\lambda d\lambda + \int_{12}^{16} G_\lambda d\lambda}$$

$\alpha = \{0.2(1/2)500 \text{ W/m}^2 \cdot \mu\text{m } (6-2)\mu\text{m} + 500 \text{ W/m}^2 \cdot \mu\text{m } [0.2(8-6)\mu\text{m}$
$+ 1 - 0.2/2 \, (8-6)\mu\text{m}] + [1 \times 500 \text{ W/m}^2 \cdot \mu\text{m } (12-8)\mu\text{m}$
$+ 1(1/2)500 \text{ W/m}^2 \cdot \mu\text{m } (16-12)\mu\text{m}]\} - [(1/2)500 \text{ W/m}^2 \cdot \mu\text{m } (6-2)\mu\text{m}$
$+ 500 \text{ W/m}^2 \cdot \mu\text{m} (12-6)\mu\text{m} + (1/2)500 \text{ W/m}^2 \cdot \mu\text{m } (16-12) \, \mu\text{m}]$

Hence

$$\alpha = \frac{G_{abs}}{G} = \frac{(200 + 600 + 3000)\text{W/m}^2}{(1000 + 3000 + 1000)\text{W/m}^2} = \frac{3800 \text{ W/m}^2}{5000 \text{ W/m}^2}$$

$\alpha = 0.76$ ◁

3. Neglecting convection effects, the net heat flux *to* the surface is

$$q''_{net} = \alpha G - E = \alpha G - \varepsilon\sigma T^4$$

Hence

$$q''_{net} = 0.76(5000 \text{ W/m}^2) - 0.8 \times 5.67 \times 10^{-8} \text{ W/m}^2 \cdot \text{K}^4 (500 \text{ K})^4$$

$q''_{net} = 3800 - 2835 = 965 \text{ W/m}^2$

Since $q''_{net} > 0$, the surface temperature will *increase* with time.   ◁

EXAMPLE 12.9

The cover glass on a flat plate solar collector has a low iron content, and its spectral transmissivity may be approximated by the following distribution.

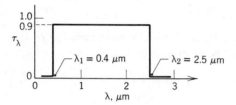

What is the total transmissivity of the cover glass to solar radiation?

SOLUTION

KNOWN:

Spectral transmissivity of solar collector cover glass.

FIND:

Total transmissivity of cover glass to solar radiation.

ASSUMPTIONS:

1. The prescribed spectral distribution is the appropriate directional average for the solar irradiation.
2. Solar irradiation of the cover glass has a spectral distribution which is proportional to that associated with emission from a blackbody at 5800 K.

ANALYSIS:

From Equation 12.55 the total transmissivity of the cover is

$$\tau = \frac{\int_0^\infty \tau_\lambda G_\lambda d\lambda}{\int_0^\infty G_\lambda d\lambda}$$

where the irradiation G_λ is due to solar emission. Having assumed that the sun emits as a blackbody at 5800 K, it follows that

$$G_\lambda(\lambda) \propto E_{\lambda,b}(5800 \text{ K})$$

With the proportionality constant canceling from the numerator and denominator of the expression for τ, we then obtain

$$\tau = \frac{\int_0^\infty \tau_\lambda E_{\lambda,b}(5800 \text{ K})d\lambda}{\int_0^\infty E_{\lambda,b}(5800 \text{ K})d\lambda}$$

or, for the prescribed distribution, $\tau_\lambda(\lambda)$,

$$\tau = 0.90 \frac{\int_{0.4}^{2.5} E_{\lambda,b}(5800 \text{ K})d\lambda}{E_b(5800 \text{ K})}$$

From Table 12.1

$\lambda_1 = 0.4 \ \mu m, \ T = 5800 \text{ K}: \ \lambda_1 T = 2320 \ \mu m \cdot K, \ F_{(0 \to \lambda_1)} = 0.1245$

$\lambda_2 = 2.5 \ \mu m, \ T = 5800 \text{ K}: \ \lambda_2 T = 14,500 \ \mu m \cdot K, \ F_{(0 \to \lambda_2)} = 0.9660$

Hence from Equation 12.31

$$\tau = 0.90 [F_{(0 \to \lambda_2)} - F_{(0 \to \lambda_1)}] = 0.90(0.9660 - 0.1245)$$

or

$$\tau = 0.76 \qquad \qquad \triangleleft$$

COMMENTS:

It is important to recognize that the irradiation at the cover plate is not equal to the emissive power of a blackbody at 5800 K, $G_\lambda \neq E_{\lambda,b}(5800 \text{ K})$. It is simply assumed to be proportional to this emissive power, in which case it is assumed to have a spectral distribution of the same form. With G_λ appearing in both the numerator and denominator of the expression for τ, it is then possible to replace G_λ by $E_{\lambda,b}$.

12.6 KIRCHHOFF'S LAW

In the foregoing sections we separately considered surface properties associated with emission and absorption. In this and the following section we consider conditions for which these properties are equal.

Consider a large, isothermal enclosure of surface temperature T_s, within which several small bodies are confined (Figure 12.25). Since these bodies are small relative to the enclosure, their presence has a negligible effect on the radiation field within the enclosure. Accordingly, this field is due to the cumulative effect of radiation that is emitted and subsequently reflected by the

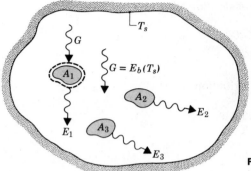

Figure 12.25 Radiative exchange in an isothermal enclosure.

enclosure surface. Recall that, regardless of the nature of its radiative properties, such a surface forms a *blackbody cavity*. Accordingly, regardless of its orientation, the irradiation experienced by any body in the cavity must be diffuse and equal in magnitude to emission from a blackbody at T_s. Hence

$$G = E_b(T_s) \tag{12.60}$$

Under steady-state conditions, *thermal equilibrium* must exist between the bodies and the enclosure. Hence $T_1 = T_2 = ... = T_s$, and the net rate of energy transfer to each surface must be zero. Applying an energy balance to a control surface about body 1, it follows that

$$\alpha_1 G A_1 - E_1(T_s) A_1 = 0$$

or, from Equation 12.60

$$\frac{E_1(T_s)}{\alpha_1} = E_b(T_s)$$

Since this result must apply to each of the confined bodies, we then obtain

$$\frac{E_1(T_s)}{\alpha_1} = \frac{E_2(T_s)}{\alpha_2} = ... = E_b(T_s) \tag{12.61}$$

This relation is known as *Kirchhoff's law*. A major implication is that, since $\alpha \leq 1$, $E(T_s) \leq E_b(T_s)$. Hence *no real surface can have an emissive power exceeding that of a black surface at the same temperature*, and the notion of the blackbody as an ideal emitter is confirmed.

From the definition of the total, hemispherical emissivity, Equation 12.37, an alternative form of Kirchhoff's law is given by

$$\frac{\varepsilon_1}{\alpha_1} = \frac{\varepsilon_2}{\alpha_2} = ... = 1$$

Hence, for any surface in the enclosure

$$\varepsilon = \alpha \qquad (12.62)$$

That is, the total, hemispherical emissivity of the surface is equal to its total, hemispherical absorptivity.

We will later find that calculations of radiative exchange between surfaces are greatly simplified if Equation 12.62 may be applied to each of the surfaces. However, the restrictive conditions inherent in its derivation should be remembered. In particular the surface irradiation has been assumed to correspond to emission from a blackbody at the same temperature as the surface. In the following section we consider other, less restrictive, conditions for which Equation 12.62 is applicable.

The preceding derivation may be repeated for spectral conditions, and for any surface in the enclosure it follows that

$$\varepsilon_\lambda = \alpha_\lambda \qquad (12.63)$$

Conditions associated with the use of Equation 12.63 are less restrictive than those associated with Equation 12.62. In particular it will be shown that Equation 12.63 is applicable if the irradiation is diffuse *or* if the surface is diffuse. The most general form of Kirchhoff's law, for which there are no restrictions, is expressed in terms of the spectral, directional properties and is of the form

$$\varepsilon_{\lambda,\theta} = \alpha_{\lambda,\theta} \qquad (12.64)$$

This equality is generally applicable because $\varepsilon_{\lambda,\theta}$ and $\alpha_{\lambda,\theta}$ are *inherent* surface properties. That is, respectively, they are independent of the spectral and directional distributions of the emitted and incident radiation.

More detailed developments of Kirchhoff's law are provided by Planck [1] and Siegel and Howell [6].

12.7   THE GRAY SURFACE

In Chapter 13 we will find that the problem of predicting radiant energy exchange between surfaces is greatly simplified if Equation 12.62 may be assumed to apply for each surface. It is therefore important to examine if this equality may be applied to conditions other than those for which it was derived, namely irradiation due to emission from a blackbody at the same temperature as the surface.

Accepting the fact that the spectral, directional emissivity and absorptivity are equal under any conditions, Equation 12.64, we begin by considering the conditions associated with using Equation 12.63. According to the definitions of the spectral, hemispherical properties, Equations 12.35 and 12.43, we are really asking under what conditions, if any, the following equality will hold.

$$\varepsilon_\lambda = \frac{\int_0^{2\pi}\int_0^{\pi/2} \varepsilon_{\lambda,\theta} \cos\theta \sin\theta \, d\theta d\phi}{\int_0^{2\pi}\int_0^{\pi/2} \cos\theta \sin\theta \, d\theta d\phi} \stackrel{?}{=} \frac{\int_0^{2\pi}\int_0^{\pi/2} \alpha_{\lambda,\theta} I_{\lambda,i} \cos\theta \sin\theta \, d\theta d\phi}{\int_0^{2\pi}\int_0^{\pi/2} I_{\lambda,i} \cos\theta \sin\theta \, d\theta d\phi} = \alpha_\lambda$$

(12.65)

Since $\varepsilon_{\lambda,\theta} = \alpha_{\lambda,\theta}$, it follows by inspection that Equation 12.63 is applicable if *either* one of the following conditions is satisfied:

1. The *irradiation is diffuse* ($I_{\lambda,i}$ is independent of θ and ϕ).
2. The *surface is diffuse* ($\varepsilon_{\lambda,\theta}$ and $\alpha_{\lambda,\theta}$ are independent of θ and ϕ).

The first of these conditions is a reasonable approximation for many engineering calculations; the second condition is reasonable for many surfaces, particularly for electrically nonconducting materials (Figure 12.17).

Assuming the existence of either diffuse irradiation or a diffuse surface, we now consider what *additional* conditions must be satisfied in order for Equation 12.62 to be valid. From Equations 12.38 and 12.46 it follows that the equality applies if

$$\varepsilon = \frac{\int_0^\infty \varepsilon_\lambda E_{\lambda,b}(\lambda,T)d\lambda}{E_b(T)} \stackrel{?}{=} \frac{\int_0^\infty \alpha_\lambda G_\lambda(\lambda)d\lambda}{G} = \alpha$$

(12.66)

Since $\varepsilon_\lambda = \alpha_\lambda$, it follows by inspection that Equation 12.62 applies if *either* one of the following conditions is satisfied.

1. The irradiation corresponds to emission from a blackbody at the surface temperature, T, in which case $G_\lambda(\lambda) = E_{\lambda,b}(\lambda,T)$ and $G = E_b(T)$.
2. The *surface is gray* (α_λ and ε_λ are independent of λ).

Note that the first condition corresponds to the major assumption required for the derivation of Kirchhoff's law (Section 12.6).

Because the total absorptivity of a surface depends on the spectral distribution of the irradiation, it cannot be stated unequivocally that $\alpha = \varepsilon$. For example, a particular surface may be highly absorbing to radiation in one spectral region and virtually nonabsorbing in another region, as shown in Figure 12.26a. Accordingly, for the two possible irradiation fields, $G_{\lambda,1}(\lambda)$ and $G_{\lambda,2}(\lambda)$, of Figure 12.26b, the values of α will differ drastically. In contrast the value of ε is independent of the irradiation. Hence, there is *no* basis for stating that α is, in general, equal to ε.

To assume gray surface behavior, and hence the validity of Equation 12.62, it is not necessary for α_λ and ε_λ to be independent of λ over the entire spectrum. Practically speaking, a *gray surface* may be defined as *one for which α_λ and ε_λ are independent of λ over the spectral regions of the irradiation and the surface*

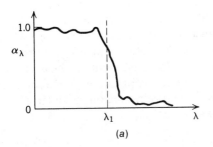
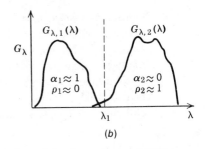

Figure 12.26 Spectral distribution of (a) the spectral absorptivity of a surface and (b) the spectral irradiation at the surface.

emission. From Equation 12.66 it is readily shown that gray surface behavior may be assumed for the conditions of Figure 12.27. That is, the irradiation and surface emission are concentrated in a region for which the spectral properties of the surface are approximately constant. Accordingly,

$$\varepsilon = \frac{\varepsilon_{\lambda,o} \int_{\lambda_1}^{\lambda_2} E_{\lambda,b}(\lambda, T)d\lambda}{E_b(T)} = \varepsilon_{\lambda,o}$$

and

$$\alpha = \frac{\alpha_{\lambda,o} \int_{\lambda_3}^{\lambda_4} G_\lambda(\lambda)d\lambda}{G} = \alpha_{\lambda,o}$$

in which case $\alpha = \varepsilon = \varepsilon_{\lambda,o}$. However, if the irradiation were in a spectral region corresponding to $\lambda < \lambda_1$ or $\lambda > \lambda_4$, gray surface behavior could not be assumed.

A surface for which $\alpha_{\lambda,\theta}$ and $\varepsilon_{\lambda,\theta}$ are independent of θ and λ is termed a

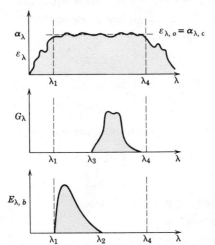

Figure 12.27 A set of conditions for which gray surface behavior may be assumed.

diffuse, gray surface (diffuse because of the directional independence and gray because of the wavelength independence). It is a surface for which both Equations 12.62 and 12.63 are satisfied. We assume such surface conditions in many of our subsequent considerations. However, although the assumption of a gray surface is reasonable for many practical applications, some caution should be exercised in its use, particularly if the spectral regions of the irradiation and emission are widely separated.

EXAMPLE 12.10

A diffuse, fire brick wall of temperature $T_s = 500$ K has a spectral emissivity as shown in the plot below and is exposed to a bed of coals at 2000 K.

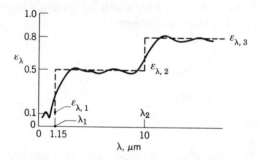

1. Determine the total, hemispherical emissivity, ε, of the fire brick wall.
2. Calculate the total emissive power of the fire brick wall.
3. Determine the total absorptivity, α, of the fire brick wall for irradiation due to emission from the bed of coals.

SOLUTION

KNOWN:

Brick wall of surface temperature $T_s = 500$ K and prescribed $\varepsilon_\lambda(\lambda)$ is exposed to coals at $T_c = 2000$ K.

FIND:

1. Total, hemispherical emissivity of the fire brick wall.
2. Total emissive power of the brick wall.
3. Absorptivity of the wall to irradiation from the coals.

SCHEMATIC:

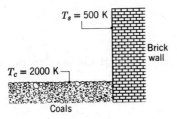

$T_s = 500$ K

Brick wall

$T_c = 2000$ K

Coals

ASSUMPTIONS:

1. Brick wall is opaque and diffuse.
2. Spectral distribution of irradiation at the brick wall approximates that due to emission from a blackbody at 2000 K.

ANALYSIS:

1. The total, hemispherical emissivity follows from Equation 12.38

$$\varepsilon(T_s) = \frac{\int_0^\infty \varepsilon_\lambda(\lambda) E_{\lambda,b}(\lambda, T_s)\, d\lambda}{E_b(T_s)}$$

Breaking the integral into parts,

$$\varepsilon(T_s) = \varepsilon_{\lambda,1} \frac{\int_0^{\lambda_1} E_{\lambda,b}\, d\lambda}{E_b} + \varepsilon_{\lambda,2} \frac{\int_{\lambda_1}^{\lambda_2} E_{\lambda,b}\, d\lambda}{E_b} + \varepsilon_{\lambda,3} \frac{\int_{\lambda_2}^\infty E_{\lambda,b}\, d\lambda}{E_b}$$

and introducing the blackbody functions, it follows that

$$\varepsilon(T_s) = \varepsilon_{\lambda,1} F_{(0\to\lambda_1)} + \varepsilon_{\lambda,2}[F_{(0\to\lambda_2)} - F_{(0\to\lambda_1)}] + \varepsilon_{\lambda,3}[1 - F_{(0\to\lambda_2)}]$$

From Table 12.1

$$\lambda_1 T_s = 1.5 \ \mu m \times 500 \ K = 750 \ \mu m \cdot K: F_{(0\to\lambda_1)} = 0.000$$
$$\lambda_2 T_s = 10 \ \mu m \times 500 \ K = 5000 \ \mu m \cdot K: F_{(0\to\lambda_2)} = 0.634$$

Hence

$$\varepsilon(T_s) = 0.1 \times 0 + 0.5 \times 0.634 + 0.8(1 - 0.634)$$

$$\varepsilon(T_s) = 0.610 \qquad \triangleleft$$

2. From Equations 12.28 and 12.37, the total emissive power of the brick is

$$E(T_s) = \varepsilon(T_s)E_b(T_s) = \varepsilon(T_s)\sigma T_s^4$$

$$E(T_s) = 0.61 \times 5.67 \times 10^{-8} \text{ W/m}^2 \cdot \text{K}^4 (500 \text{ K})^4$$

$$E(T_s) = 2161 \text{ W/m}^2 \qquad \qquad \triangleleft$$

3. From Equation 12.46, the total absorptivity of the wall to radiation from the coals is

$$\alpha = \frac{\int_0^\infty \alpha_\lambda(\lambda)G_\lambda(\lambda)d\lambda}{\int_0^\infty G_\lambda(\lambda)d\lambda}$$

Since the surface is diffuse, $\alpha_\lambda(\lambda) = \varepsilon_\lambda(\lambda)$. Moreover, since the spectral distribution of the irradiation approximates that due to emission from a blackbody at 2000 K, $G_\lambda(\lambda) \propto E_{\lambda,b}(\lambda,T_c)$. It follows that

$$\alpha = \frac{\int_0^\infty \varepsilon_\lambda(\lambda)E_{\lambda,b}(\lambda,T_c)d\lambda}{\int_0^\infty E_{\lambda,b}(\lambda,T_c)d\lambda}$$

Breaking the integral into parts and introducing the blackbody functions, we then obtain

$$\alpha = \varepsilon_{\lambda,1}F_{(0 \to \lambda_1)} + \varepsilon_{\lambda,2}[F_{(0 \to \lambda_2)} - F_{(0 \to \lambda_1)}] + \varepsilon_{\lambda,3}[1 - F_{(0 \to \lambda_3)}]$$

From Table 12.1

$$\lambda_1 T_c = 1.5 \text{ } \mu\text{m} \times 2000 \text{ K} = 3000 \text{ } \mu\text{m} \cdot \text{K}: F_{(0 \to \lambda_1)} = 0.273$$

$$\lambda_2 T_c = 10 \text{ } \mu\text{m} \times 2000 \text{ K} = 20{,}000 \text{ } \mu\text{m} \cdot \text{K}: F_{(0 \to \lambda_2)} = 0.986$$

Hence

$$\alpha = 0.1 \times 0.273 + 0.5(0.986 - 0.273) + 0.8(1 - 0.986)$$

$$\alpha = 0.395 \qquad \qquad \triangleleft$$

COMMENTS:

1. Recognize that the emissivity depends on the surface temperature, T_s, while the absorptivity depends on the spectral distribution of the irradiation, which in this case depends on the temperature of the source, T_c.

2. Note that the surface is not gray, $\alpha \neq \varepsilon$. This result is to be expected. Since emission is associated with $T_s = 500$ K, its spectral maximum occurs at $\lambda_{\text{max}} \approx 6$ μm. In contrast, since irradiation is associated with emission from a source at $T_c = 2000$ K, its spectral maximum occurs at $\lambda_{\text{max}} \approx 1.5$ μm. Because $\varepsilon_\lambda = \alpha_\lambda$ is not constant over the spectral ranges of both emission and irradiation, $\alpha \neq \varepsilon$.

EXAMPLE 12.11

A small, highly conducting, solid, metallic sphere has an opaque, diffuse coating with a spectral absorptivity as shown. This sphere, initially at a uniform temperature of $T_s = 300$ K, is inserted into a large furnace whose walls are at a temperature of $T_f = 1200$ K.

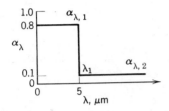

1. Calculate the total, hemispherical absorptivity of the coating for the initial condition described above.

2. Calculate the total, hemispherical emissivity of the coating for the initial condition described above.

3. Derive an expression for predicting the change in temperature with time of the sphere assuming that there are no gradients within the sphere and that there is no convection heat transfer at the surface.

4. What are the values of the absorptivity and emissivity of the coating after the sphere has been in the furnace for a very long period of time?

SOLUTION

KNOWN:

Small highly conductive, metallic sphere with spectrally selective absorptivity, initially at $T_s = 300$ K, is inserted into a large furnace at $T_f = 1200$ K.

FIND:

1. Total, hemispherical absorptivity of sphere coating for the initial condition.

2. Total, hemispherical emissivity of sphere coating for the initial condition.

3. Expression for predicting T_s as a function of time.

4. Values of α and ε after sphere has been in furnace a long time.

SCHEMATIC:

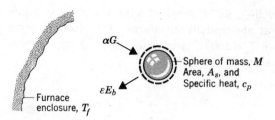

ASSUMPTIONS:

1. Opaque, diffuse coating.
2. Interior furnace surface is very much larger than surface of sphere.
3. Sphere is isothermal.

ANALYSIS:

1. From Equation 12.46 the total, hemispherical absorptivity is

$$\alpha = \frac{\int_0^\infty \alpha_\lambda(\lambda) G_\lambda(\lambda) d\lambda}{\int_0^\infty G_\lambda(\lambda) d\lambda}$$

or, with $G_\lambda = E_{\lambda,b}(T_f) = E_{\lambda,b}(\lambda, 1200\ \text{K})$

$$\alpha = \frac{\int_0^\infty \alpha_\lambda(\lambda) E_{\lambda,b}(\lambda, 1200\ \text{K}) d\lambda}{E_b(1200\ \text{K})}$$

Hence

$$\alpha = \alpha_{\lambda,1} \frac{\int_0^{\lambda_1} E_{\lambda,b}(\lambda, 1200\ \text{K}) d\lambda}{E_b(1200\ \text{K})} + \alpha_{\lambda,2} \frac{\int_{\lambda_1}^\infty E_{\lambda,b}(\lambda, 1200\ \text{K}) d\lambda}{E_b(1200\ \text{K})}$$

or,

$$\alpha = \alpha_{\lambda,1} F_{(0 \to \lambda_1)} + \alpha_{\lambda,2}[1 - F_{(0 \to \lambda_1)}]$$

From Table 12.1,

$$\lambda_1 T_f = 5\ \mu\text{m} \times 1200\ \text{K} = 6000\ \mu\text{m} \cdot \text{K}: F_{(0 \to \lambda_1)} = 0.738$$

Hence

$$\alpha = 0.8 \times 0.738 + 0.1(1 - 0.738)$$

$$\alpha = 0.62 \qquad \triangleleft$$

2. The total, hemispherical emissivity follows from Equation 12.38.

$$\varepsilon(T_s) = \frac{\int_0^\infty \varepsilon_\lambda E_{\lambda,b}(\lambda, T_s)d\lambda}{E_b(T_s)}$$

Since the surface is diffuse, $\varepsilon_\lambda = \alpha_\lambda$ and it follows that

$$\varepsilon = \alpha_{\lambda,1} \frac{\int_0^{\lambda_1} E_{\lambda,b}(\lambda, 300\ K)d\lambda}{E_b(300\ K)} + \alpha_{\lambda,2} \frac{\int_{\lambda_1}^\infty E_{\lambda,b}(\lambda, 300\ K)d\lambda}{E_b(300\ K)}$$

or,

$$\varepsilon = \alpha_{\lambda,1}F_{(0\to\lambda_1)} + \alpha_{\lambda,2}[1 - F_{(0\to\lambda_1)}]$$

From Table 12.1,

$$\lambda_1 T_s = 5\ \mu m \times 300\ K = 1500\ \mu m \cdot K: F_{(0\to\lambda_1)} = 0.014$$

Hence

$$\varepsilon = 0.8 \times 0.014 + 0.1(1 - 0.014)$$
$$\varepsilon = 0.11 \qquad \triangleleft$$

3. Performing an energy balance for a control volume about the sphere,

$$\dot{E}_{in} - \dot{E}_{out} = \dot{E}_{st}$$
$$(\alpha G)A_s - (\varepsilon \sigma T_s^4)A_s = Mc_p \frac{dT_s}{dt} \qquad \triangleleft$$

This differential equation could be solved to determine $T(t)$ for $t > 0$.

4. Because the spectral characteristics of the coating and the furnace temperature remain fixed, there is no change in the value of α with increasing time. However, as T_s increases with increasing time, the value of ε will change. After a sufficiently long period of time, $T_s = T_f$ and hence $\varepsilon = \alpha$ ($\varepsilon = 0.62$).

COMMENTS:

1. Note that the equilibrium condition that eventually exists ($T_s = T_f$) corresponds precisely to the condition for which Kirchhoff's law was derived. Hence α must equal ε.

2. The variation in ε that occurs with increasing time would have to be included in the solution to the energy balance equation of part 3.

12.8 ENVIRONMENTAL RADIATION

It would be inappropriate to conclude this chapter without commenting on the nature of the radiation that comprises our natural environment. Solar radiation is, of course, essential to all life on earth. Through the process of photosynthesis,

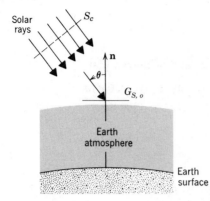

Figure 12.28 Directional nature of solar radiation outside the earth's atmosphere.

it satisfies our need for food, fiber, and fuel. Moreover, through thermal and photovoltaic processes, it has the potential to satisfy much of our demand for space heat, process heat, and electricity.

The sun is a nearly spherical radiation source that is 1.39×10^9 m in diameter and located 1.50×10^{11} m from the earth. With respect to the magnitude and the spectral and directional dependence of the incident solar radiation, it is necessary to distinguish between conditions at the earth's surface and at the outer edge of the earth's atmosphere. For a horizontal surface outside the earth's atmosphere, solar radiation appears as a beam of nearly *parallel rays* that form an angle θ, the *zenith angle*, relative to the surface normal (Figure 12.28). The *extraterrestrial* solar irradiation, $G_{S,o}$, depends on the geographic latitude, as well as the time of day and year. It may be determined from an expression of the form

$$G_{S,o} = S_c \cdot f \cdot \cos \theta \tag{12.67}$$

where S_c, the *solar constant*, is the flux of solar energy incident on a surface oriented *normal* to the sun's rays, when the earth is at its mean distance from the sun. It is known to have a value of $S_c = 1353$ W/m^2. The quantity f is a small correction factor to account for the eccentricity of the earth's orbit about the sun ($0.97 \lesssim f \lesssim 1.03$).

The spectral distribution of solar radiation is significantly different from that associated with emission by engineering surfaces. As shown in Figure 12.29, this distribution approximates that of a blackbody at 5800 K. The radiation is concentrated in the low wavelength region ($0.2 \lesssim \lambda \lesssim 3 \mu$m) of the thermal spectrum with the peak occurring at approximately 0.50 μm. It is this short wavelength concentration that frequently precludes the assumption of graybody behavior for solar irradiated surfaces, since emission is generally in the spectral region beyond 4 μm and it is unlikely that surface spectral properties would be constant over such a wide spectral range.

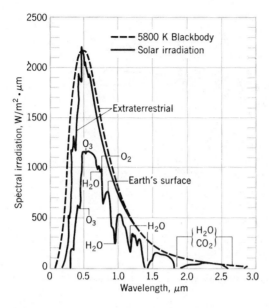

Figure 12.29 Spectral distribution of solar radiation.

As solar radiation passes through the earth's atmosphere, its magnitude and its spectral and directional distributions experience significant change. The change is due to *absorption* and *scattering* of the radiation by the atmospheric constituents. The effect of absorption by the atmospheric gases O_3 (ozone), H_2O, O_2, and CO_2 is shown by the lower curve of Figure 12.29. Absorption by ozone is strong in the UV, providing considerable attenuation below 0.4 μm and complete attenuation below 0.3 μm. In the visible there is some absorption by O_3 and O_2; and in the near and far infrared, absorption is dominated by water vapor. Throughout the solar spectrum, there is also continuous absorption of radiation by the dust and aerosol content of the atmosphere.

Atmospheric scattering provides for *redirection* of the sun's rays and consists of two kinds (Figure 12.30). *Rayleigh* (or *molecular*) *scattering* by the gas molecules provides for nearly uniform scattering of radiation in all directions. Hence approximately half of the scattered radiation is redirected back to space, while the remaining portion strikes the earth's surface. At any point on this surface, the scattered radiation is incident from all directions. In contrast *Mie scattering* by the dust and aerosol particles of the atmosphere is concentrated in directions that are close to that of the incident rays. Hence, virtually all of this radiation strikes the earth's surface in directions close to that of the sun's rays.

The cumulative effect of the foregoing scattering processes on the directional distribution of solar radiation striking the earth's surface is shown in Figure 12.31a. That portion of the radiation that has penetrated the atmosphere

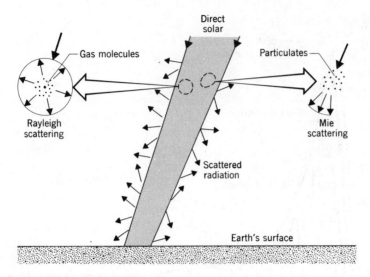

Figure 12.30 Scattering of solar radiation in the earth's atmosphere.

without having been scattered (or absorbed) is in the direction of the zenith angle and is termed the *direct radiation*. The scattered radiation is incident from all directions, although its intensity is largest for directions close to that of the direct radiation. However, because the radiation intensity is often assumed to be independent of direction, Figure 12.31*b*, the radiation is termed *diffuse*. The total solar radiation reaching the earth's surface may therefore be viewed as the sum of direct and diffuse contributions. The diffuse contribution may vary from

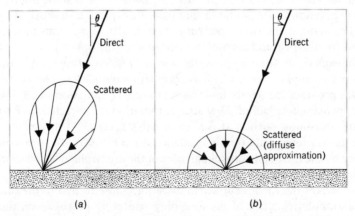

Figure 12.31 Directional distribution of solar radiation at the earth's surface. (*a*) Actual distribution. (*b*) Diffuse approximation.

approximately 10 percent of the total solar radiation on a clear day to nearly 100 percent on a totally overcast day.

The foregoing discussion has focused on the *nature* of solar radiation, an understanding of which is essential to the development of schemes for using this radiation. Moreover, heat transfer analyses related to solar energy utilization have been developed in many of the examples and problems of this text. Detailed consideration of solar energy technologies, however, is left to the literature [7–9].

Long-wave forms of environmental radiation include emission from the earth's surface, as well as emission from certain atmospheric constituents. The emissive power associated with the earth's surface may be computed in the conventional manner. That is,

$$E = \varepsilon \sigma T^4 \tag{12.68}$$

where ε and T are the surface emissivity and temperature, respectively. Emissivities are generally close to unity, with the emissivity of water, for example, approximately equal to 0.97, and temperatures are typically in the range from 250 to 320 K. Accordingly, emission is concentrated in the spectral region from approximately 4 to 40 μm, with the peak occurring at approximately 10 μm.

Atmospheric emission is largely due to emission from the CO_2 and H_2O molecules and is concentrated in the spectral regions from 5 to 8 μm and above 13 μm. Although the spectral distribution of atmospheric emission does not correspond to that of a blackbody, the contribution which it makes to irradiation of the earth's surface can be estimated using Equation 12.28. In particular earth irradiation due to atmospheric emission may be expressed in the form

$$G_{\text{atm}} = \sigma T_{\text{sky}}^4 \tag{12.69}$$

where T_{sky} is termed the effective sky temperature. Its value depends on atmospheric conditions, ranging from a low of 230 K under cold, clear sky conditions to a high of approximately 285 K under warm, cloudy conditions. At night atmospheric emission is the only source of earth irradiation, and when its value is small, as on a cool, clear night, water may freeze in spite of the air temperature exceeding 273 K.

We close by recalling that values of the spectral properties of a surface at short wavelengths may be appreciably different from values at long wavelengths (Figures 12.18 and 12.23). Since solar radiation is concentrated in the short wavelength region of the spectrum and surface emission is at much longer wavelengths, it follows that many surfaces may not be approximated as gray in their response to solar irradiation. In other words the solar absorptivity of a surface, α_S, may differ from its emissivity, ε. Values of α_S and the emissivity at moderate temperatures are presented in Table 12.2 for representative surfaces. Note that the ratio α_S/ε is an important engineering parameter. Small values are desired if the surface is intended to reject heat; large values are required if the surface is intended to collect solar energy.

Table 12.2 Solar absorptivity, α_S, and emissivity, ε, of surfaces whose spectral absorptivity is given in Figure 12.23

SURFACE	α_S	ε(300 K)	α_S/ε
Evaporated aluminium film	0.09	0.03	3.0
Fused quartz on aluminium film	0.19	0.81	0.24
White paint on metallic substrate	0.21	0.96	0.22
Black paint on metallic substrate	0.97	0.97	1.0
Stainless steel, as received, dull	0.50	0.21	2.4
Red brick	0.63	0.93	0.68
Human skin, Caucasian	0.62	0.97	0.64
Snow	0.28	0.97	0.29
Corn leaf	0.76	0.97	0.78

EXAMPLE 12.12

A flat plate solar collector with no cover plate has a selective absorber surface of emissivity 0.1 and solar absorptivity 0.95. At a given time of day the absorber surface temperature is $T_s = 120°C$ when the solar irradiation is 750 W/m², the effective sky temperature is $-10°C$, and the ambient air temperature is 30°C. Assume that the heat transfer convection coefficient for the calm day conditions can be estimated as

$$\bar{h} = 0.22(T_s - T_\infty)^{1/3} \text{ W/m}^2 \cdot \text{K}$$

Calculate the useful heat removal rate (W/m²) from the collector for these conditions. What is the efficiency of the collector for these conditions?

SOLUTION

KNOWN:

Operating conditions for a flat plate solar collector.

FIND:

1. Useful heat removal rate, q_u'' (W/m²).
2. Efficiency, η, of the collector.

SCHEMATIC:

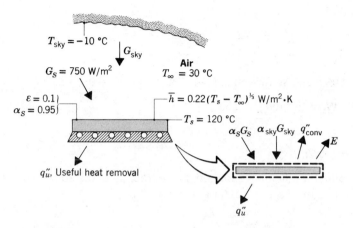

ASSUMPTIONS:

1. Steady-state conditions.
2. Bottom of collector is well insulated.
3. Absorber surface is diffuse.

ANALYSIS:

1. Performing an energy balance on the absorber,

$$\dot{E}_{in} - \dot{E}_{out} = 0$$

or

$$\alpha_S G_S + \alpha_{sky} G_{sky} - q''_{conv} - E - q''_u = 0$$

From Equation 12.69

$$G_{sky} = \sigma T^4_{sky}$$

Moreover, since the sky radiation is concentrated in approximately the same spectral region as that of the surface emission, it is reasonable to assume that

$$\alpha_{sky} \approx \varepsilon = 0.1$$

With

$$q''_{conv} = \bar{h}(T_s - T_\infty) = 0.22(T_s - T_\infty)^{4/3}$$

and

$$E = \varepsilon \sigma T_s^4$$

it follows that

$$q_u'' = \alpha_S G_S + \varepsilon \sigma T_{sky}^4 - 0.22(T_s - T_\infty)^{4/3} - \varepsilon \sigma T_s^4$$

$$q_u'' = \alpha_S G_S - 0.22(T_s - T_\infty)^{4/3} - \varepsilon \sigma (T_s^4 - T_{sky}^4)$$

$$q_u'' = 0.95 \times 750 \text{ W/m}^2 - 0.22(120 - 30)^{4/3} \text{ W/m}^2 - 0.1 \times 5.67$$

$$\times 10^{-8} \text{ W/m}^2 \text{ K}^4 (393^4 - 263^4) \text{K}^4$$

$$q_u'' = (712.5 - 88.7 - 108.1) \text{ W/m}^2$$

$$q_u'' = 516 \text{ W/m}^2 \qquad\qquad \triangleleft$$

2. The collector efficiency, defined as the fraction of the solar irradiation which is extracted as useful energy, is then

$$\eta = \frac{q_u''}{G_S} = \frac{516 \text{ W/m}^2}{750 \text{ W/m}^2} = 0.69 \qquad\qquad \triangleleft$$

COMMENTS:

1. Since the spectral range of G_{sky} is entirely different from that of G_S, it would be *incorrect* to assume that $\alpha_{sky} = \alpha_S$.

2. Recognize that the convection heat transfer coefficient is extremely low ($\bar{h} \approx 1$ W/m²·K). With a modest increase to $\bar{h} = 5$ W/m²·K, the useful heat flux and the efficiency are reduced to $q_u'' = 161$ W/m² and $\eta = 0.21$. Use of a cover plate can contribute significantly to the reduction of convection (and radiation) heat losses from the absorber plate.

EXAMPLE 12.13

Consider an opaque, gray surface whose directional absorptivity is 0.8 for $0° \le \theta \le 60°$ and 0.1 for $\theta > 60°$. The surface is horizontal and exposed to solar irradiation which is comprised of direct and diffuse components.

1. What is the surface absorptivity to direct solar radiation which is incident at an angle of 45° from the normal? What is the absorptivity to diffuse irradiation?

2. Neglecting convection heat transfer between the surface and the surrounding air, what would be the equilibrium temperature of the

surface if the direct and diffuse components of the irradiation are 600 W/m² and 100 W/m², respectively? The back side of the surface is insulated.

SOLUTION

KNOWN:

Directional distribution of α_θ for a horizontal, opaque, gray surface exposed to direct and diffuse irradiation.

FIND:

1. Absorptivity to direct radiation at $\theta = 45°$ and to diffuse radiation.
2. Equilibrium temperature for specified direct and diffuse irradiation components.

SCHEMATIC:

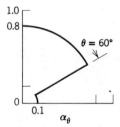
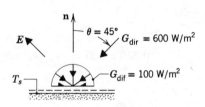

ASSUMPTIONS:

1. Steady-state conditions.
2. Opaque, gray surface behavior.
3. Negligible convection at top surface and perfectly insulated back surface.

ANALYSIS:

1. From knowledge of $\alpha_\theta(\theta)$, it is evident that the absorptivity of the surface to the direct radiation is

$$\alpha_{\text{dir}} = \alpha_{\theta = 45°} = 0.8 \qquad \triangleleft$$

The absorptivity to the diffuse radiation is the hemispherical absorptivity given by Equation 12.44. Hence dropping the subscript λ

$$\alpha_{\text{dif}} = 2 \int_0^{\pi/2} \alpha_\theta(\theta) \cos \theta \sin \theta \, d\theta$$

or

$$\alpha_{\text{dif}} = 2 \left(0.8 \left. \frac{\sin^2 \theta}{2} \right|_0^{\pi/3} + 0.1 \left. \frac{\sin^2 \theta}{2} \right|_{\pi/3}^{\pi/2} \right)$$

$$\alpha_{\text{dif}} = 0.625 \qquad \qquad \triangleleft$$

2. Performing a surface energy balance,

$$\dot{E}_{\text{in}} - \dot{E}_{\text{out}} = 0$$

or

$$\alpha_{\text{dir}} G_{\text{dir}} + \alpha_{\text{dif}} G_{\text{dif}} - \varepsilon \sigma T_s^4 = 0$$

The total, hemispherical emissivity may be obtained from Equation 12.36, where the subscript λ may again be deleted. Since this equation is of precisely the same form as Equation 12.44 and since $\alpha_\theta = \varepsilon_\theta$, it follows that

$$\varepsilon = \alpha_{\text{dif}} = 0.625$$

Hence

$$T_s^4 = \frac{(0.8 \times 600 + 0.625 \times 100) \text{W/m}^2}{0.625 \times 5.67 \times 10^{-8} \text{ W/m}^2 \cdot \text{K}^4} = 1.53 \times 10^{10} \text{ K}^4$$

or

$$T_s = 351.8 \text{ K} \qquad \qquad \triangleleft$$

COMMENTS:

In assuming gray surface behavior, all spectral effects may be neglected. It is implied that total properties and spectral properties are identical. In contrast the surface is not diffuse and the hemispherical and directional properties differ.

12.9 SUMMARY

Many new and important ideas have been introduced in this chapter, and at this stage you may well be confused, particularly by the terminology. However, the subject matter has been developed in a systematic fashion, and a careful rereading of the material should make you more comfortable with its application. A glossary of terms has been provided in Table 12.3 to assist you in assimilating the terminology.

Table 12.3 Glossary of radiative terms

TERM	DEFINITION
Absorption	The process of converting radiation intercepted by matter to internal thermal energy.
Absorptivity	Fraction of the incident radiation absorbed by matter. Equations 12.41, 12.42, and 12.45. Modifiers: *directional, hemispherical, spectral, total.*
Blackbody	The ideal emitter and absorber. Modifier referring to ideal behavior. Denoted by the subscript *b.*
Diffuse	Modifier referring to the directional independence of the intensity associated with emitted, reflected, or incident radiation.
Directional	Modifier referring to a particular direction. Denoted by the subscript θ.
Directional distribution	Refers to variation with direction.
Emission	The process of radiation production by matter at a finite temperature. Modifiers: *diffuse, blackbody, spectral.*
Emissivity	Ratio of the radiation emitted by a surface to the radiation emitted by a blackbody at the same temperature. Equations 12.32, 12.33, 12.34, and 12.37. Modifiers: *directional, hemispherical, spectral, total.*
Emissive power	Rate of radiant energy emitted by a surface in all directions per unit area of the surface. $E(W/m^2)$. Modifiers: *spectral, total, blackbody.*
Gray surface	A surface for which the spectral absorptivity and emissivity are independent of wavelength over the spectral regions of surface irradiation and emission.
Hemispherical	Modifier referring to all directions in the space above a surface.
Intensity	Rate of radiant energy propagation in a particular direction, per unit area normal to the direction, per unit solid angle about the direction. $I(W/m^2 \cdot sr)$. Modifier: *spectral.*
Irradiation	Rate at which radiation is incident on a surface from all directions per unit area of the surface. $G(W/m^2)$. Modifiers: *spectral, total, diffuse.*
Kirchhoff's law	Relation between emission and absorption properties for surfaces irradiated by a blackbody at the same temperature. Equations 12.61, 12.62, 12.63, and 12.64.
Planck's law	Spectral distribution of emission from a blackbody. Equations 12.25 and 12.26.
Radiosity	Rate at which radiation leaves a surface due to emission and reflection in all directions per unit area of the surface. $J(W/m^2)$. Modifiers: *spectral, total.*
Reflection	The process of redirection of radiation incident on a surface. Modifiers: *diffuse, specular.*

Table 12.3 Continued

TERM	DEFINITION
Reflectivity	Fraction of the incident radiation reflected by matter. Equations 12.48, 12.49, and 12.51. Modifiers: *directional, hemispherical, spectral, total.*
Semitransparent	Refers to a medium in which radiation absorption is a volumetric process.
Solid angle	Region subtended by an element of area on the surface of a sphere with respect to the origin of the sphere. $\omega(sr)$. Equations 12.2 and 12.3.
Spectral	Modifier referring to a single wavelength (monochromatic) component. Denoted by the subscript λ.
Spectral distribution	Refers to variation with wavelength.
Specular	Refers to a surface for which the angle of reflected radiation is equal to the angle of incident radiation.
Stefan-Boltzmann law	Emissive power of a blackbody. Equation 12.28.
Thermal radiation	Electromagnetic energy emitted by matter at a finite temperature and concentrated in the spectral region from approximately 0.1 to 100 μm.
Total	Modifier referring to all wavelengths.
Transmission	The process of thermal radiation passing through matter.
Transmissivity	Fraction of the incident radiation transmitted by matter. Equations 12.53 and 12.54. Modifiers: *hemispherical, spectral, total.*
Wien's law	Locus of the wavelength corresponding to peak emission by a blackbody. Equation 12.27.

You should be able to answer the following questions. What is the nature of radiation, and what position does *thermal radiation* occupy in *electromagnetic spectrum*? What are the physical origins of thermal radiation? What are meant by the terms *irradiation, emissive power,* and *radiosity*? What material property characterizes the ability of a surface to emit thermal radiation? What processes and associated material properties characterize the manner in which a surface responds to irradiation? What is an *opaque* surface? A *semitransparent* surface?

It is essential that you recognize the difference between *directional* and *spectral* radiation properties on the one hand and *hemispherical* and *total* properties on the other. Moreover, you must be able to proceed from knowledge of the former to determination of the latter. You must also be cognizant of what independent variables determine the values of the directional, spectral, hemispherical, and total properties.

You should appreciate the unique role of the *blackbody* in the description of thermal radiation. In what sense is the blackbody *perfect*? Why is it an idealization, and how may it be approximated in practice? What is the *Planck distribution*? How is it altered with increasing surface temperature? What are the *Wien* and *Stefan-Boltzmann* laws? For what purposes may the blackbody emission functions of Table 12.1 be used?

Relations between emissivity and absorptivity are often extremely important in radiative exchange calculations. What is *Kirchhoff's law* and what restrictive conditions are inherent in its derivation? What is a *gray surface*, and under what conditions might the assumption of a gray surface be particularly suspect? Finally, because of its considerable importance to life on earth, you should be familiar with the characteristics of *solar radiation*. In what region of the spectrum is this radiation concentrated, and how is it altered due to passage through the earth's atmosphere? What is the nature of its directional distribution at the earth's surface?

REFERENCES

1. Planck, M., *The Theory of Heat Radiation*, Dover Publications, New York, 1959.

2. Gubareff, G. G., J. E. Janssen, and R. H. Torberg, *Thermal Radiation Properties Survey*, 2nd Ed., Honeywell Research Center, Minneapolis, 1960.

3. Wood, W. D., H. W. Deem, and C. F. Lucks, *Thermal Radiative Properties*, Plenum Press, New York, 1964.

4. Touloukian, Y. S., *Thermophysical Properties of High Temperature Solid Materials*, Macmillan, New York, 1967.

5. Touloukian, Y. S. and D. P. DeWitt, *Thermal Radiative Properties*, Volumes 7, 8 and 9, from *Thermophysical Properties of Matter*, TPRC Data Series, Y. S. Touloukian and C. Y. Ho, Ed., IFI Plenum, New York, 1970–1972.

6. Siegel, R. and J. R. Howell, *Thermal Radiation Transfer*, McGraw-Hill, New York, 1972.

7. Duffie, J. A. and W. A. Beckman, *Solar Energy Thermal Processes*, Wiley, New York, 1974.

8. Meinel, A. B. and M. P. Meinel, *Applied Solar Energy: An Introduction*, Addison-Wesley, Reading, Mass., 1976.

9. Sayigh, A. A. M., Ed., *Solar Energy Engineering*, Academic Press, New York, 1977.

PROBLEMS

12.1   Consider a small surface of area $A_1 = 10^{-4}$ m^2, which emits diffusely with a total, hemispherical emissive power of $E_1 = 5 \times 10^4$ W/m^2. At what rate is this emission intercepted by the circular disc of area $A_2 = 5 \times 10^{-4}$ m^2, which is oriented as shown on the next page.

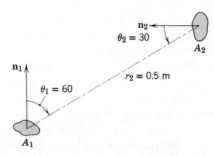

12.2   According to its directional distribution, solar radiation incident on the earth's surface may be divided into two components. The *direct* component consists of parallel rays incident at a fixed zenith angle θ, while the *diffuse* component consists of radiation that may be approximated as being diffusely distributed with θ.

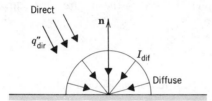

Consider clear sky conditions for which the direct radiation is incident at $\theta = 30°$, with a total flux (based on an area which is normal to the rays) of $q''_{dir} = 1000$ W/m^2, and the total intensity of the diffuse radiation is $I_{dif} = 70$ W/m$^2 \cdot$sr. What is the total solar irradiation at the earth's surface?

12.3   On an overcast day the directional distribution of the solar radiation incident on the earth's surface may be approximated by an expression of the form

$$I_i = I_n \cos \theta$$

where $I_n = 80$ W/m$^2 \cdot$sr is the total intensity of radiation directed normal to the surface and θ is the zenith angle. What is the solar irradiation at the earth's surface?

12.4   The spectral distribution of the radiation emitted by a diffuse surface may be approximated as follows.

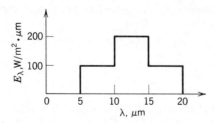

What is the total emissive power? What is the total intensity of the radiation emitted in the normal direction and at an angle of 30° from the normal?

12.5 A furnace with an aperture of 20 mm diameter and emissive power of 3.72×10^5 W/m² is used to calibrate a heat flux gage having a sensitive area of 1.6×10^{-5} m².

a) At what distance, measured along a normal from the aperture, should the gage be positioned in order to receive an irradiation of 1000 W/m²?

b) If the gage is tilted off normal by 20°, what will be the irradiation on the gage?

12.6 Determine the fraction of the total, hemispherical emissive power that leaves a diffuse surface in the directions $\pi/4 \leq \theta \leq \pi/2$ and $0 \leq \phi \leq \pi$.

12.7 What fraction of the total, hemispherical emissive power, E (W/m²), leaving a diffuse radiator will be in the directions $0 \leq \theta \leq 30°$?

12.8 Approximations to Planck's law for the spectral emissive power are the Wien and Rayleigh-Jeans spectral distributions, which are useful for the extreme low and high limits of the product λT, respectively.

a) Show that Planck's spectral distribution will have the form

$$E_{\lambda,b}(\lambda, T) \approx \frac{C_1}{\lambda^5} \exp(-C_2/\lambda T)$$

when $C_2/\lambda T \gg 1$ and determine the error, when compared to Planck's law for the condition $\lambda T = 2898 \ \mu m \cdot K$. This form is known as Wien's law.

b) Show that the Planck distribution will have the form

$$E_{\lambda,b}(\lambda, T) \approx \frac{C_1}{C_2} \frac{T}{\lambda^4}$$

when $C_2/\lambda T \ll 1$ and determine the error, when compared to Planck's law, for the condition $\lambda T = 100,000 \ \mu m \cdot K$. This form is known as the Rayleigh-Jeans law.

12.9 The fourth radiation constant, C_4, is defined from the relationship

$$I_{\lambda,b}(\lambda_{max}, T) = C_4 T^5$$

Using Planck's distribution for the spectral intensity of a blackbody, determine the value of C_4 in units $W/m^2 \cdot \mu m \cdot sr \cdot K^5$. Verify your result by using the appropriate values from Table 12.2.

12.10 Isothermal furnaces with small apertures approximating a blackbody are frequently used to calibrate heat flux gages, radiation thermometers and other radiometric devices. In such applications, it is necessary to control power to the furnace such that the variation of temperature and the spectral intensity of the aperture are within desired limits.

a) By considering the Planck spectral distribution, Equation 12.26, show that the ratio of the fractional change in the spectral intensity to the fractional change in the temperature of the furnace has the form

$$\frac{dI_\lambda/I_\lambda}{dT/T} = \frac{C_2}{\lambda T} \frac{1}{1 - \exp(-C_2/\lambda T)}$$

b) Using this relation, determine the allowable variation in temperature of the furnace operating at 2000 K in order that the spectral intensity at 0.65 μm will not vary by more than $\frac{1}{2}$ percent. What is the allowable variation for the spectral condition 10 μm?

12.11 Substitute the Planck distribution, Equation 12.26, into Equation 12.11 and perform the spectral integration to obtain Equation 12.28 for the total emissive power of a blackbody. Show that

$$\sigma = \left(\frac{\pi}{C_2}\right)^4 \frac{C_1}{15}$$

and calculate the value of the Stefan-Boltzmann constant using values of the radiation constants C_1 and C_2.

12.12 A spherical aluminum shell of inside diameter $D = 2$ m is evacuated and is used as a radiation test chamber. If the inner surface is coated with carbon black and maintained at 600 K, what is the irradiation on a small test surface placed in the chamber? If the inner surface is not coated and maintained at 600 K, what would the irradiation be?

12.13 An enclosure has an inside area of 100 m^2, and its inside surface is black and is maintained at a constant temperature. A small opening in the enclosure has an area of 0.02 m^2. The radiant power emitted from this opening is 70 W.

a) What is the temperature of the interior enclosure wall?

b) If the interior surface is maintained at this temperature, but is now polished such that its emissivity is 0.15, what will be the value of the radiant power emitted from the opening?

12.14 The energy flux associated with solar radiation incident on the outer surface of the earth's atmosphere has been accurately measured and is known to be 1353 W/m^2. The diameters of the sun and earth are 1.39×10^9 m and 1.29×10^7 m, respectively, and the distance between the sun and the earth is 1.5×10^{11} m.

a) What is the emissive power of the sun?

b) Approximating the sun's surface as black, what is the temperature of this surface?

c) At what wavelength is the spectral emissive power of the sun a maximum?

d) Assuming the earth's surface to be black and the sun to be the only source of energy for the earth, estimate the earth surface temperature.

12.15 Estimate the wavelength corresponding to maximum emission from each of the following surfaces: the sun, a tungsten filament at 2500 K, a heated metal at 1500 K, human skin at 305 K, and a cryogenically cooled metal surface at 60 K. Estimate the fraction of the solar emission which is in the following spectral regions: the ultraviolet, the visible, and the infrared.

12.16 What is the efficiency of a 100-W light source consisting of a filament that radiates as a blackbody at 2900 K? You may assume that the glass bulb transmits all of the incident visible radiation and that the filament is a thin rectangular strip 5 mm long by 2 mm wide.

12.17 Consider the radiation emerging from a small aperture of a furnace operating at 1000 K.

a) Calculate the emissive power of the aperture.

b) Determine the spectral intensity at 2 μm.

c) What is the ratio of the spectral intensity at 2 μm to the spectral intensity at 6 μm?

d) What fraction of the emissive power is in the spectral range 2 to 6 μm?

12.18 Determine the fraction of the radiation emitted by the sun, considered to be a blackbody at approximately 5800 K, which is in the visible region of the spectrum.

12.19 The spectral, hemispherical emissivity of tungsten may be approximated by the following distribution.

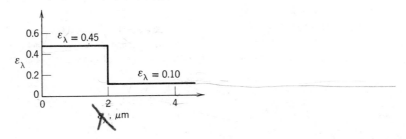

Consider a cylindrical tungsten filament that is of diameter $D = 0.8$ mm and length $L = 20$ mm. The filament is enclosed in an evacuated bulb and is heated by an electrical current to a steady-state temperature of 2900 K.

a) What is the initial rate of cooling (K/s) of the filament following discontinuation of the current?

b) Estimate the time required for the filament to cool to a temperature of 1300 K, below which the light is barely visible.

12.20 For materials A and B, whose spectral, hemispherical emissivities vary with wavelength as shown, how does the total, hemispherical emissivity vary with temperature? Explain briefly.

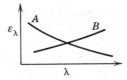

12.21 Various materials are to be evaluated for design of more energy efficient household cooking pans. Compare the emissive powers of the following materials at 200°C and explain what influence this quantity would have on energy consumption: (a) polished copper, (b) teflon-coated copper, (c) cleaned stainless steel, (d) pyrex, and (e) pyroceram.

12.22 The spectral, directional emissivity of a diffuse material at 2000 K has the following distribution.

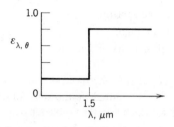

a) Determine the total, hemispherical emissivity of this surface at 2000 K.

b) Determine the emissive power over the spectral range 0.8 to 2.5 μm *and* for the directions $0 \leq \theta \leq 30°$.

12.23 Consider the directionally selective surface having a directional emissivity, ε_θ, as shown in the following sketch. Assuming the surface is isotropic in the ϕ direction, calculate the ratio of the normal emissivity, ε_n, to the hemispherical emissivity, ε_h.

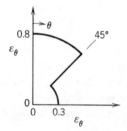

12.24 A sphere is suspended in air in a dark room and maintained at a uniform incandescent temperature. When first viewed with the naked eye, the sphere appears to be brighter around the rim. After several hours, however, it appears to be brighter in the center. Of what type material would you reason the sphere is made? Give plausible reasons for the nonuniformity of brightness of the sphere and for the changing appearance with time.

12.25 A radiation thermometer is a device that responds to a radiant flux within a prescribed spectral interval and is calibrated to indicate the temperature of a blackbody that produces the same flux.

a) When viewing a surface at some elevated temperature, T_s, and emissivity less than unity, the thermometer will indicate an apparent temperature referred to as the brightness or spectral radiance temperature, T_λ. Will this spectral radiance temperature, T_λ, be greater than, less than, or equal to the surface temperature, T_s?

b) Write an expression for the spectral emissive power of the surface in terms of Wien's spectral distribution (see Problem 12.8) and the spectral emissivity of the surface. Write the equivalent expression using the spectral radiance temperature of the surface and show that

$$\frac{1}{T_s} = \frac{1}{T_\lambda} + \frac{\lambda}{C_2} \ln \varepsilon_\lambda$$

where λ represents the wavelength at which the thermometer operates.

c) Consider a radiation thermometer that responds to a spectral flux centered about the wavelength 0.65 μm. What temperature will the thermometer indicate when viewing a surface with $\varepsilon_\lambda(0.65 \, \mu m) = 0.9$ and $T_s = 1000$ K? Verify that Wien's spectral distribution is a reasonable approximation to Planck's law for this situation.

12.26 A diffusely emitting surface is exposed to a radiant source causing the irradiation on the surface to be 1000 W/m². The intensity for emission is 143 W/m²·sr and the reflectivity of the surface is 0.8.

a) Determine the emissive power, $E(W/m^2)$, for the surface.

b) Calculate the radiosity, $J(W/m^2)$, for the surface.

c) What is the net heat flux to the surface by the radiation mode?

12.27 An opaque surface with the prescribed spectral, hemispherical reflectivity distribution is subjected to the spectral irradiation shown in the following sketch.

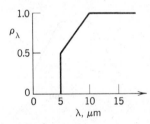
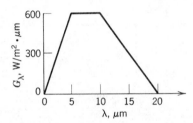

a) Sketch the spectral, hemispherical absorptivity distribution.

b) Determine the total irradiation on the surface.

c) Determine the radiant flux that is absorbed by the surface.

d) What is the total, hemispherical absorptivity of this surface?

12.28 The spectral transmissivity of a plain and tinted glass can be approximated as follows:

Plain glass: $\tau_\lambda = 0.9$ $0.3 \leq \lambda \leq 2.5 \, \mu m$
Tinted glass: $\tau_\lambda = 0.9$ $0.5 \leq \lambda \leq 1.5 \, \mu m$

Outside the specified wavelength ranges, the spectral transmissivity is zero for both glasses.

a) Compare the solar energy that could be transmitted through the glasses.

b) With solar irradiation on the glasses, compare the visible radiant energy that could be transmitted.

12.29 Referring to the distribution of the spectral transmissivity of low iron glass shown in Figure 12.24, describe briefly what is meant by the "greenhouse effect." That is, how does the glass influence energy transfer to and from the contents of the greenhouse?

12.30 The spectral, hemispherical absorptivity of an opaque surface is shown below.

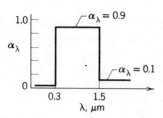

a) What is the solar absorptivity, α_S?

b) If it is assumed that $\varepsilon_\lambda = \alpha_\lambda$ and that the surface is at a temperature of 340 K, what is its total, hemispherical emissivity?

12.31 The spectral, hemispherical absorptivity of an opaque surface and the spectral distribution of radiation incident on the surface are as shown in the following diagram.

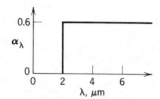
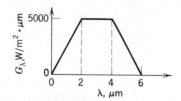

a) What is the total, hemispherical absorptivity of the surface?

b) If it is assumed that $\varepsilon_\lambda = \alpha_\lambda$ and that the surface is at a temperature of 1000 K, what is its total, hemispherical emissivity?

c) What is the net radiant heat flux to the surface?

12.32 Two small surfaces, designated as A and B, are placed inside an isothermal enclosure at a uniform temperature. The enclosure is known to provide an irradiation of 6300 W/m² to each of the surfaces, and from measurements, surfaces A and B are known to absorb incident radiation at rates of 5600 W/m² and 630 W/m², respectively. Consider conditions after a long period of time has expired.

a) What are the net heat transfer rates for each surface? What are their temperatures?

b) Determine the absorptivity of each surface.

c) What are the emissive powers of each surface?

d) Determine the emissivity of each surface.

12.33 Solar irradiation of 1100 W/m² is incident on a large, flat, horizontal metal roof on a day when the wind blowing over the roof causes a convection heat transfer coefficient of 25 W/m². The outside air temperature is 27°C, the metal surface absorptivity for incident solar radiation is 0.60, the metal surface emissivity is 0.20, and the roof is well insulated from below. Estimate the roof temperature under steady-state conditions.

12.34 The 50-mm peephole of a large furnace operating at 450°C is covered with a material having $\tau = 0.8$ and $\rho = 0$ for irradiation originating from the furnace. The material has an emissivity of 0.8 and is opaque to irradiation from a source at room temperature. The outer surface of the cover is exposed to surroundings and ambient air at 27°C with a convection heat transfer coefficient of 50 W/m²·K. Assuming that convection effects on the inner surface of the cover are negligible, calculate the heat loss by the furnace and the temperature of the cover.

12.35 Consider the evacuated tube solar collector described in Problem 1.19g of Chapter 1. In the interest of maximizing collector efficiency, what spectral radiative characteristics are desired for the outer tube and for the inner tube?

12.36 Solar flux of 900 W/m² is incident on the top side of a plate whose surface has a solar absorptivity of 0.9 and an emissivity of 0.1. The air and surroundings are at 17°C and the convection heat transfer coefficient between the plate and air is 20 W/m²·K. Assuming the bottom side of the plate is insulated, determine the steady-state temperature of the plate.

12.37 Consider an opaque horizontal plate that is well insulated on its backside. The irradiation on the plate is 2500 W/m², of which 500 W/m² is reflected. The plate is at a temperature of 227°C and has an emissive power of 1200 W/m². Air at a temperature of 127°C flows over the plate with a heat transfer convection coefficient of 15 W/m²·K.

a) Determine the emissivity, absorptivity, and radiosity of the plate.

b) What is the net heat transfer rate per unit area?

12.38 A horizontal, opaque surface at a steady-state temperature of 77°C is exposed to an air flow having a freestream temperature of 27°C with a convection heat transfer coefficient of 28 W/m²·K. The emissive power of the surface is 628 W/m², the irradiation incident on the surface is 1380 W/m², and the reflectivity of the surface is 0.40.

a) Determine the absorptivity of the surface.

b) Determine the net radiation heat transfer rate for this surface. Is this heat transfer to the surface or from the surface?

c) Determine the total heat transfer rate for the surface. Is this heat transfer to the surface or from the surface?

12.39 Consider small cylindrical shapes fabricated from materials of different composition and surface finishes as described below. These cylinders initially at 300 K are inserted into a large furnace at 600 K. Using the tabulated values for the emissivity of the materials, indicate which cylinder you expect to heat up faster. Briefly explain your reasoning.

a) Aluminum: (A) highly polished, film or (B) anodized.

b) Stainless steel: (A) typical, polished or (B) stably oxidized.

12.40 An apparatus commonly used for measuring the reflectivity of materials is shown in the sketch. A water-cooled sample, of 30 mm diameter and temperature T_s

$= 300$ K, is mounted flush with the inner surface of a large enclosure. The walls of the enclosure are gray/diffuse with an emissivity of 0.8 and having a uniform temperature $T_f = 1000$ K. A small aperture is located at the bottom of the enclosure to permit sighting of the sample or the enclosure wall.

The spectral reflectivity, ρ_λ, of an opaque, diffuse sample material is shown in the following diagram. The heat transfer coefficient for convection between the sample and the air within the cavity, which is also at 1000 K, has been estimated as $h = 10$ W/m$^2 \cdot$K.

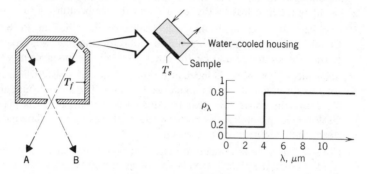

a) Calculate the absorptivity of the sample.
b) Calculate the emissivity of the sample.
c) Determine the heat removal rate (W) by the coolant.
d) The ratio of the radiation in the A direction to that in the B direction will give the reflectivity of the sample. Briefly explain why this is so.

12.41 The spectral absorptivity, α_λ, and spectral reflectivity, ρ_λ, for a spectrally selective, diffuse material are shown in these plots.

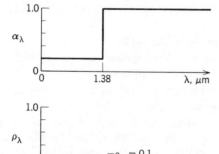

a) Sketch the spectral transmissivity, τ_λ.
b) If solar irradiation with $G_S = 750$ W/m^2, having a spectral distribution the same as a blackbody at 5800 K, is incident on this material, determine the fractions of the irradiation that are transmitted through the material, reflected by the material, and absorbed by the material.

c) If the temperature of this material is 350 K, determine the emissivity, ε.

d) Determine the net heat flux by radiation to the material.

12.42 The window of a large vacuum chamber is fabricated from a material having the spectral characteristics shown below.

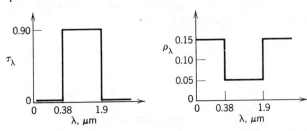

A collimated beam of radiant energy from a solar simulator is incident on the window and has a flux of 3000 W/m². The inside walls of the chamber, which are large compared to the window area, are maintained at 77 K. The outer surface of the window is subjected to surroundings and room air at 25°C, with a convection heat transfer coefficient of 15 W/m²·K.

a) Determine the transmissivity of the window material to the radiation from the solar simulator, which approximates the solar spectral distribution.

b) What is the steady-state temperature reached by the window, assuming it is insulated from its chamber mounting arrangement?

c) Calculate the net radiation transfer per unit area of the window to the vacuum chamber wall, excluding the transmitted simulated solar flux.

12.43 Two plates, one with a black paint surface and the other with a special coating (chemically oxidized copper) are in an earth orbit and are exposed to solar radiation. The solar rays make an angle of 30° with the normal to the plate. Estimate the equilibrium temperature of each plate assuming they are diffuse and that the solar flux is 1353 W/m². The spectral absorptivity of the black painted surface can be approximated by

$$\alpha_\lambda = 0.95, \, 0 \le \lambda \le \infty$$

and that of the special coating by

$$\alpha_\lambda = 0.95, \, 0 \le \lambda < 3 \, \mu m$$

$$\alpha_\lambda = 0.05, \, \lambda \ge 3 \, \mu m$$

12.44 A diffuse surface having the following spectral characteristics is maintained at 500 K when situated in a large furnace enclosure whose walls are maintained at 1500 K.

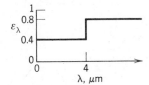

a) Sketch the spectral distribution of the surface emissive power, E_λ, and the emissive power, $E_{\lambda,b}$, which the surface would have if it were a blackbody.

b) Neglecting convection effects, what is the net heat flux to the surface for the prescribed conditions?

12.45 The spectral, hemispherical emissivity distributions for two diffuse panels to be used in a spacecraft are shown below.

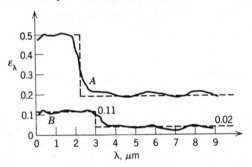

Assuming the backsides of the panels are insulated and that the panels are oriented normal to the solar flux at 1353 W/m², determine which of the two panels has the higher steady-state temperature.

12.46 Consider an opaque, diffuse surface whose spectral reflectivity varies with wavelength as shown. The surface is at 750 K, and irradiation on one side varies with wavelength as shown. The other side of the surface is insulated.

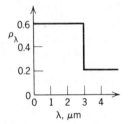 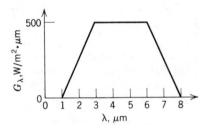

a) What is the total absorptivity of the surface?

b) What is the total emissivity of the surface?

c) What is the net radiative heat flux to the surface?

12.47 A very small sample of an opaque surface is initially at 1200 K and has the spectral, hemispherical absorptivity shown as follows.

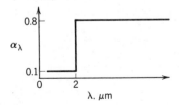

The sample is placed inside a very large enclosure whose walls have an emissivity of 0.2 and are maintained at 2400 K.

a) Determine the total, hemispherical absorptivity of the sample surface.

b) Determine the total, hemispherical emissivity of the sample surface.

c) What are the values of the absorptivity and emissivity after the sample has been in the enclosure a long time?

12.48 An opaque, gray surface at 27°C is exposed to irradiation of 1000 W/m², and 800 W/m² are reflected. Air at 17°C flows over the surface and the heat transfer convection coefficient is 15 W/m²·K. Determine the net heat flux from the surface.

12.49 A thermocouple whose surface is diffuse/gray with an emissivity of 0.6 indicates a temperature of 180°C when used to measure the temperature of a gas flowing through a large duct whose walls have an emissivity of 0.85 and a uniform temperature of 450°C. If the convection heat transfer coefficient between the thermocouple and the gas stream is 125 W/m²·K and there are negligible conduction losses from the thermocouple, determine the temperature of the gas.

12.50 A sphere of 30 mm diameter whose surface is diffuse/gray with emissivity of 0.8 is placed in a large oven whose walls are of uniform temperature at 600 K. The temperature of the air in the oven is 400 K and the convection heat transfer coefficient between the sphere and the oven air is 15 W/m²·K. Determine the net heat transfer rate to the sphere when its temperature is 300 K.

12.51 A thermograph is a device responding to the radiant power from the scene that reaches its radiation detector within the spectral region 9 to 12 μm. The thermograph provides an image of the scene, such as the side of a furnace as shown below, from which the surface temperature can be determined.

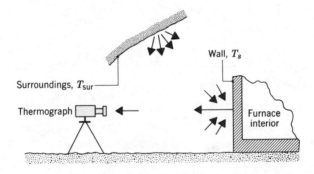

a) For a black surface at 60°C, determine the emissive power for the spectral region 9 to 12 μm.

b) Calculate the radiant power (W) received by the thermograph in the 9- to 12-μm range when viewing, in a normal direction, a small black wall area, 200 mm², at $T_s = 60°C$. The solid angle, ω, subtended by the aperture of the thermograph when viewed from the target is 0.001 sr.

c) Determine the radiant power (W) received by the thermograph for the same wall area (200 mm^2) and solid angle (0.001 sr) when the wall is a gray, opaque, diffuse material at $T_s = 60°C$ with emissivity 0.7 and the surroundings are black at $T_{sur} = 23°C$.

12.52 Consider a gray surface whose directional absorptivity varies with θ shown as follows.

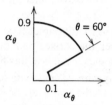

a) What is the ratio of the normal absorptivity, α_n, to the hemispherical emissivity of the surface?

b) Consider a plate with the above surface characteristics on both sides in orbit around the earth. If the solar flux incident on one side of the plate is $q''_S = 1353$ W/m^2, what equilibrium temperature will the plate assume if it is oriented normal to the sun's rays? What temperature will it assume if it is oriented at 75° to the sun's rays?

12.53 Four diffuse surfaces having the spectral characteristics shown below are at 300 K and are exposed to solar radiation.

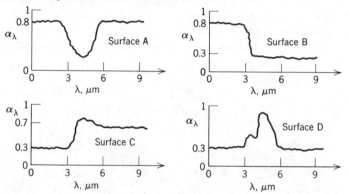

Which of the surfaces may be approximated as being gray?

12.54 A radiation detector has an aperture of area $A_d = 10^{-6}$ m^2 and is positioned at a distance of $r = 1$ m from a surface of area $A_s = 10^{-4}$ m^2. The angle formed by the normal to the detector and the surface normal is $\theta = 30°$.

The surface is at a temperature of 500 K, is opaque, and is diffuse/gray with an emissivity of $\varepsilon = 0.7$. If the surface irradiation is $G = 1500 \ \text{W/m}^2$, what is the rate at which the detector intercepts radiation from the surface?

12.55 Two special coatings are available for application to an absorber plate installed below the cover glass described in Example 12.9. Each coating is diffuse and is characterized by the spectral distributions shown below.

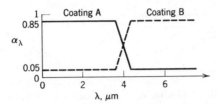

Which coating would you select for the absorber plate? Explain briefly. For the selected coating, what is the rate at which radiation is absorbed per unit area of the absorber plate if the total solar irradiation at the cover glass is $G_S = 1000 \ \text{W/m}^2$?

12.56 A small anodized aluminum block at 35°C is being heated in a large oven whose walls are diffuse and gray with $\varepsilon = 0.85$ and maintained at a uniform temperature 175°C. The anodized coating is also diffuse and gray with $\varepsilon = 0.92$. A radiation detector views the block through a small opening in the oven and receives the radiant energy from a small area, referred to as the target, A_t, on the block. The target has a diameter of 3 mm, and the detector receives radiation within a solid angle 0.001 sr centered about the normal from the block.

a) If the radiation detector views a small, but deep, hole drilled into the block, what is the total power (W) received by the detector?

b) If the radiation detector now views an area on the block surface, what is the total power (W) received by the detector?

12.57 A flat plate whose surface is gray, but has the total, directional emissivity, ε_θ, shown below, is in earth orbit where the solar flux is 1353 W/m².

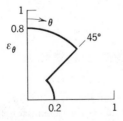

Determine the equilibrium temperatures for the plate when it is oriented normal to the solar flux and at 60° to the solar flux.

12.58 A contractor must select a roof covering material from the two diffuse, opaque coatings for which $\alpha_\lambda(\lambda)$ is shown in the following diagram. Which of the two coatings would result in a lower roof temperature? Which is preferred for summer use? For winter use? Sketch the spectral distribution of α_λ which would be ideal for summer use. For winter use.

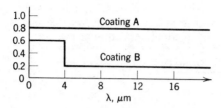

12.59 A radiator on a proposed satellite solar power station must dissipate heat being generated within the satellite by radiating it into space. The radiator surface has a solar absorptivity of 0.5 and an emissivity of 0.95. What is the equilibrium surface temperature when the solar irradiation is 1000 W/m² and the required heat dissipation is 1500 W/m²?

12.60 A gray plate of emissivity 0.6 insulated from the ground is exposed to solar irradiation of 1000 W/m². The effective sky temperature is −40°C and the heat transfer coefficient between the plate and the ambient air at 25°C is 6 W/m²·K. The temperature of the plate is 65°C at this particular time.

a) What is the radiosity of the plate?

b) What is the net heat flux between the plate and the environment?

c) In what manner will the temperature of the plate change with respect to time?

12.61 A thin sheet of glass is used on the roof of a greenhouse and is irradiated as shown.

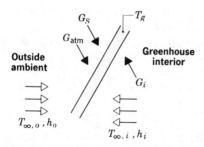

The irradiation is comprised of the total solar flux G_S, the flux G_{atm} due to atmospheric emission (sky radiation), and the flux G_i due to emission from interior surfaces. The solar flux may be assumed to have the spectral distribution of radiation emitted from a blackbody at 5800 K, and the fluxes G_{atm} and G_i are concentrated in the far infrared ($\lambda \gtrsim 8\ \mu$m). The glass may also exchange energy by convection with the outside and inside atmospheres. The glass may be assumed to be totally transparent for $\lambda < 1\ \mu$m ($\tau_\lambda = 1.0$ for $\lambda < 1\ \mu$m) and opaque, with $\alpha_\lambda = 1.0$, for $\lambda \geq 1\ \mu$m.

a) Assuming steady-state conditions, with all radiative fluxes uniformly distributed over the surfaces and the glass characterized by a uniform temperature T_g, write an appropriate energy balance for a unit area of the glass.

b) For the following conditions, what is the temperature of the greenhouse ambient air, $T_{\infty,i}$?

$T_g = 27°C$ $h_i = 10 \text{ W/m}^2 \cdot \text{K}$ $G_S = 1100 \text{ W/m}^2$

$T_{\infty,o} = 24°C$ $h_o = 55 \text{ W/m}^2 \cdot \text{K}$ $G_{atm} = 250 \text{ W/m}^2$

$G_i = 440 \text{ W/m}^2$

13 Radiation Exchange Between Surfaces

Having thus far restricted our attention to the radiative processes which occur at a *single surface*, we now consider the problem of radiative exchange between two or more surfaces. This exchange depends strongly on the surface geometries and orientations, in addition to their radiative properties and temperatures. We assume that the surfaces are separated by a *nonparticipating medium*. Since such a medium neither emits, absorbs nor scatters radiation, it has no effect on the transfer of radiation between surfaces. A vacuum meets these requirements exactly, and most gases meet them to an excellent approximation.

We initially focus on the geometrical features of the radiation exchange problem by developing the notion of a *view factor*. We then consider radiation exchange between *black surfaces* and follow with a treatment of exchange between *diffuse-gray surfaces*. We also consider the problem of radiation transfer in an *enclosure*, a term that describes the region surrounded by a collection of surfaces.

13.1 THE VIEW FACTOR

To compute the magnitude of radiation exchange between any two surfaces, we must first introduce the concept of a *view factor* (also called a *configuration* or *shape factor*). It is a geometrical quantity which is associated with the amount of radiation that leaves one surface and reaches a second surface.

13.1.1 The View Factor Integral

The view factor F_{ij} is defined as the *fraction of the radiation leaving surface i, which is intercepted by surface j*. To develop a general expression for this quantity, we consider the arbitrarily oriented surfaces, A_i and A_j, of Figure 13.1.

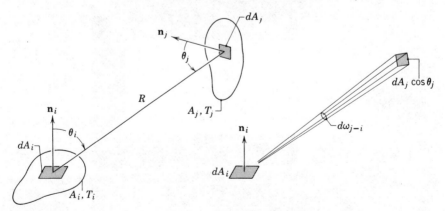

Figure 13.1 View factor associated with radiation exchange between elemental surfaces of area dA_i and dA_j.

Elemental areas on each surface, designated as dA_i and dA_j, are connected by a line of length R, which forms the polar angles θ_i and θ_j, respectively, with the surface normals $\mathbf{n}_i$ and $\mathbf{n}_j$. The values of R, θ_i, and θ_j vary with the position of the elemental areas on A_i and A_j.

From the definition of the radiation intensity, Section 12.2.1, and Equation 12.5 it follows that $dq_{i \to j}$, the rate at which radiation *leaves* dA_i and is *intercepted* by dA_j, may be expressed as

$$dq_{i \to j} = I_i \cos \theta_i \, dA_i \, d\omega_{j-i}$$

where I_i is the intensity of the radiation leaving surface i and $d\omega_{j-i}$ is the solid angle subtended by dA_j when viewed from dA_i. With $d\omega_{j-i} = (\cos \theta_j \, dA_j)/R^2$ from Equation 12.2, it follows that

$$dq_{i \to j} = I_i \frac{\cos \theta_i \cos \theta_j}{R^2} dA_i dA_j$$

Assuming that surface i emits and reflects *diffusely* and substituting from Equation 12.24, we then obtain

$$dq_{i \to j} = J_i \frac{\cos \theta_i \cos \theta_j}{\pi R^2} dA_i dA_j \qquad J_i = \pi I_i$$

The total rate at which radiation leaves surface i and is intercepted by j may then be obtained by integrating over the two surfaces. That is,

$$q_{i \to j} = J_i \int_{A_i} \int_{A_j} \frac{\cos \theta_i \cos \theta_j}{\pi R^2} dA_i dA_j$$

where it is assumed that the radiosity, J_i, is uniform over the surface A_i. From the definition of the view factor as the fraction of the radiation that leaves A_i and is intercepted by A_j,

$$F_{ij} = \frac{q_{i \to j}}{A_i J_i}$$

it follows that

$$F_{ij} = \frac{1}{A_i} \int_{A_i} \int_{A_j} \frac{\cos \theta_i \cos \theta_j}{\pi R^2} dA_i dA_j \qquad (13.1)$$

Similarly, the view factor F_{ji} is defined as the fraction of the radiation that leaves A_j and is intercepted by A_i. Following the same development it may then be shown that

$$F_{ji} = \frac{1}{A_j} \int_{A_i} \int_{A_j} \frac{\cos \theta_i \cos \theta_j}{\pi R^2} dA_i dA_j \qquad (13.2)$$

Either Equation 13.1 or 13.2 may be used to determine the view factor associated with any two surfaces that are *diffuse emitters* and *reflectors* and have *uniform radiosity*.

13.1.2 View Factor Relations

An important view factor relation is suggested by Equations 13.1 and 13.2. In particular, equating the integrals appearing in these equations, it follows that

$$A_i F_{ij} = A_j F_{ji} \tag{13.3}$$

This expression, termed the *reciprocity relation*, is useful in determining one view factor from knowledge of the other.

Another important view factor relation pertains to the surfaces of an *enclosure* (Figure 13.2). From the definition of the view factor, it follows that the *summation rule*

$$\sum_{j=1}^{N} F_{ij} = 1 \tag{13.4}$$

may be applied to each of the N surfaces in the enclosure. This rule follows from the conservation requirement that all of the radiation leaving surface i must be intercepted by the enclosure surfaces. The term F_{ii} appearing in this summation represents the fraction of the radiation that leaves surface i and is directly intercepted by i. That is, if the surface is concave, it *sees itself* and F_{ii} is nonzero. However, for a plane or convex surface, $F_{ii} = 0$.

To calculate radiation exchange in an enclosure of N surfaces, a total of N^2 view factors are needed. This requirement becomes evident when the view factors are arranged in the matrix form

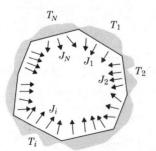

Figure 13.2 Radiation exchange in an enclosure.

$$
\begin{bmatrix}
F_{11} & F_{12}\ldots F_{1N} \\
F_{21} & F_{22}\ldots F_{2N} \\
\vdots & \vdots & \vdots \\
F_{N1} & F_{N2} & F_{NN}
\end{bmatrix}
$$

However, all of the view factors need not be calculated *directly*. A total of N view factors may be obtained by using the N equations resulting from application of the summation rule, Equation 13.4, to each of the surfaces in the enclosure. In addition $N(N-1)/2$ view factors may be obtained from the $N(N-1)/2$ applications of the reciprocity relation, Equation 13.3, which are possible for the enclosure. Accordingly, only $[N^2 - N - N(N-1)/2] = N(N-1)/2$ view factors need be determined directly. For example, in a three-surface enclosure this requirement corresponds to only $3(3-1)/2 = 3$ view factors. The remaining six view factors may be obtained by solving the six equations that result from use of Equations 13.3 and 13.4.

To illustrate the foregoing procedure, consider a simple, two-surface enclosure involving the spherical surfaces of Figure 13.3. Although the enclosure is characterized by $N^2 = 4$ view factors ($F_{11}, F_{12}, F_{21}, F_{22}$), only $N(N-1)/2 = 1$ view factor need be determined directly. In this case such a determination may be made by *inspection*. In particular, since all radiation leaving the inner surface must reach the outer surface, it follows that $F_{12} = 1$. The same may not be said of radiation leaving the outer surface, since this surface sees itself. However, from the reciprocity relation, Equation 13.3, which may only be applied once, we obtain

$$
F_{21} = (A_1/A_2)F_{12} = (A_1/A_2)
$$

Moreover, from the summation rule, which may be applied to each of the two surfaces, it follows that

$$
F_{11} + F_{12} = 1
$$

in which case $F_{11} = 0$, and

$$
F_{21} + F_{22} = 1
$$

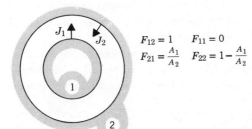

$F_{12} = 1 \quad F_{11} = 0$

$F_{21} = \dfrac{A_1}{A_2} \quad F_{22} = 1 - \dfrac{A_1}{A_2}$

Figure 13.3 View factors for the enclosure formed by two spheres.

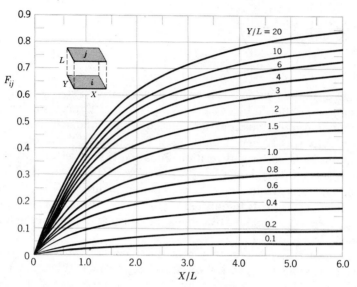

Figure 13.4   View factor for aligned parallel rectangles.

in which case

$$F_{22} = 1 - (A_1/A_2)$$

For more complicated geometries, direct determination of the view factor must involve use of the double integral of Equation 13.1. Because the integrations are generally difficult to effect, however, the specific details are left to the literature [1–5]. For particularly comprehensive compilations of view factor results, Hamilton and Morgan [2] and Siegel and Howell [5] should be consulted. The results are often presented in graphical form, and for three common configurations they are shown in Figures 13.4 to 13.6.

It is useful to note that the results of Figures 13.4 to 13.6 may be used to determine other view factors. For example, the view factor for an end surface of a cylinder (or a truncated cone) relative to the lateral surface may be obtained by using the results of Figure 13.5 with the summation rule, Equation 13.4. Moreover, Figures 13.4 and 13.6 may be used to obtain other useful results if two additional view factor relations are developed.

The first relation concerns the additive nature of the view factor for a subdivided surface and may be inferred from Figure 13.7. Considering radiation from surface i to surface j, which is divided into n components, it is evident that

$$F_{i(j)} = \sum_{k=1}^{n} F_{ik} \tag{13.5}$$

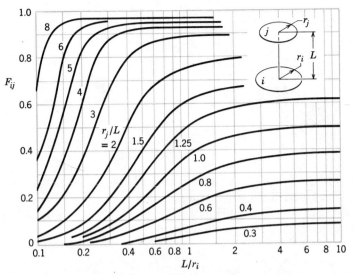

Figure 13.5 View factor for coaxial parallel disks.

where the parentheses around a subscript are used to indicate that it is a composite surface, in which case (j) is equivalent to $(1,2,...,k,...,n)$. This expression simply states that radiation reaching a composite surface is the sum of the radiation reaching its parts. Although it pertains to subdivision of the

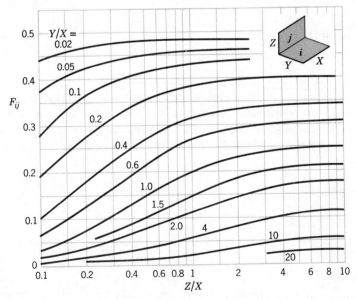

Figure 13.6 View factor for perpendicular rectangles with a common edge.

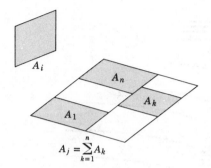

$$A_j = \sum_{k=1}^{n} A_k$$

Figure 13.7 Areas used to illustrate view factor relations.

receiving surface, it may also be used to obtain the second view factor relation of interest, which pertains to subdivision of the originating surface. Multiplying Equation 13.5 by A_i and applying the reciprocity relation, Equation 13.3, to each of the resulting terms, it follows that

$$A_j F_{(j)i} = \sum_{k=1}^{n} A_k F_{ki} \tag{13.6}$$

or

$$F_{(j)i} = \frac{\sum_{k=1}^{n} A_k F_{ki}}{\sum_{k=1}^{n} A_k} \tag{13.7}$$

Equations 13.6 and 13.7 may be applied when the originating surface is comprised of several parts.

EXAMPLE 13.1

Consider the two diffuse surfaces shown in the following sketch. One surface is a circular disk of diameter D and area A_j, while the other is a plane surface of area $A_i \ll A_j$. The surfaces are parallel to each other, and A_i is located at a distance L from the center of A_j. Obtain an expression for the view factor F_{ij}.

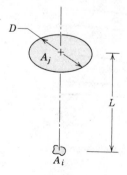

SOLUTION

KNOWN:

Orientation of a small surface relative to a large circular disk.

FIND:

View factor of small surface with respect to disk, F_{ij}.

SCHEMATIC:

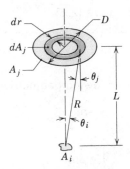

ASSUMPTIONS:

1. Diffuse surfaces.
2. $A_i \ll A_j$.

ANALYSIS:

The desired view factor may be obtained from Equation 13.1.

$$F_{ij} = \frac{1}{A_i} \int_{A_i} \int_{A_j} \frac{\cos \theta_i \cos \theta_j}{\pi R^2} \, dA_i \, dA_j$$

Recognizing that θ_i, θ_j, and R are approximately independent of position on A_i, this expression reduces to

$$F_{ij} = \int_{A_j} \frac{\cos \theta_i \cos \theta_j}{\pi R^2} \, dA_j$$

or, with $\theta_i = \theta_j \equiv \theta$,

$$F_{ij} = \int_{A_j} \frac{\cos^2 \theta}{\pi R^2} \, dA_j$$

With $R^2 = r^2 + L^2$, $\cos \theta = (L/R)$ and $dA_j = 2\pi r dr$, it follows that

$$F_{ij} = 2L^2 \int_0^{D/2} \frac{r\,dr}{(r^2 + L^2)^2}$$

Integrating

$$F_{ij} = \frac{D^2}{D^2 + 4L^2} \qquad \triangleleft \qquad (13.8)$$

COMMENTS:

The preceding geometry is one of the simplest cases for which the view factor may be obtained from Equation 13.1. Geometries involving more detailed integrations are considered in the literature [2, 5].

EXAMPLE 13.2

Determine the view factors F_{12} and F_{21} for the following geometries.

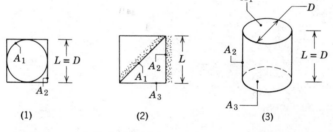

(1)   (2)   (3)

1. Sphere of diameter D inside a cubical box of length $L = D$.
2. Diagonal partition within a long square duct.
3. End and side of a circular tube of equal length and diameter.

SOLUTION

KNOWN:

Surface geometries.

FIND:

View factors.

ASSUMPTIONS:

Diffuse surfaces.

ANALYSIS:

The desired view factors may be obtained from inspection, the reciprocity rule, the summation rule and/or use of the charts.

1. Sphere within a cube:

    By inspection, $F_{12} = 1$ ◁

    By reciprocity, $F_{21} = \dfrac{A_1}{A_2} F_{12} = \dfrac{\pi D^2}{6L^2} \times 1 = \dfrac{\pi}{6}$ ◁

2. Partition within a square duct:

    From summation rule, $F_{11} + F_{12} + F_{13} = 1$

    where $\qquad\qquad F_{11} = 0$

    By symmetry, $\qquad F_{12} = F_{13}$

    Hence $\qquad\qquad F_{12} = 0.50$ ◁

    By reciprocity, $\qquad F_{21} = \dfrac{A_1}{A_2} F_{12} = \dfrac{\sqrt{2}\,L}{L} \times 0.5 = 0.71$ ◁

3. Circular tube:

    From Figure 13.5, with $(r_3/L) = 0.5$ and $(L/r_1) = 2$, $F_{13} \approx 0.17$

    From summation rule, $F_{11} + F_{12} + F_{13} = 1$

    or, with $F_{11} = 0$, $F_{12} = 1 - F_{13} = 0.83$ ◁

    From reciprocity, $F_{21} = \dfrac{A_1}{A_2} F_{12} = \dfrac{\pi D^2/4}{\pi D L} \times 0.83 = 0.21$ ◁

EXAMPLE 13.3

The reciprocity relation, the summation rule, and Equations 13.5 to 13.7 can be used to develop view factor relations that allow for applications of Figure 13.4 and/or 13.6 to more complex configurations. Consider the view factor F_{14} for surfaces 1 and 4 of the following geometry. These surfaces are perpendicular but do not share a common edge.

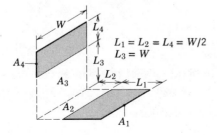

$L_1 = L_2 = L_4 = W/2$
$L_3 = W$

1. Obtain an expression for the view factor F_{14}.
2. If $L_1 = L_2 = L_4 = (W/2)$ and $L_3 = W$, what is the value of F_{14}?

SOLUTION

KNOWN:

Arrangement of perpendicular surfaces without a common edge.

FIND:

1. A relation for the view factor F_{14}.
2. The value of F_{14} for prescribed dimensions.

ASSUMPTIONS:

Diffuse surface behavior.

ANALYSIS:

1. To determine F_{14} it is convenient to work with the hypothetical surfaces, A_2 and A_3, shown in the sketch. From Equation 13.6 we may then write

$$(A_1 + A_2) \, F_{(1,2)(3,4)} = A_1 \, F_{1(3,4)} + A_2 \, F_{2(3,4)}$$

where $F_{(1,2)(3,4)}$ and $F_{2(3,4)}$ may be obtained from Figure 13.6. Substituting for $A_1 \, F_{1(3,4)}$ from Equation 13.5, which may be expressed as

$$A_1 \, F_{1(3,4)} = A_1 \, F_{13} + A_1 \, F_{14}$$

and solving for F_{14}, we then obtain

$$F_{14} = \frac{1}{A_1}[(A_1 + A_2) \, F_{(1,2)(3,4)} - A_1 \, F_{13} - A_{2(3,4)}]$$

Substituting for $A_1 \, F_{13}$ from Equation 13.6, which may be expressed as

$$(A_1 + A_2) \, F_{(1,2)3} = A_1 \, F_{13} + A_2 \, F_{23}$$

$$F_{14} = \frac{1}{A_1}[(A_1 + A_2) \, F_{(1,2)(3,4)} + A_2 \, F_{23} - (A_1 + A_2) \, F_{(1,2)3} - A_2 \, F_{2(3,4)}]$$

Since all view factors on the right-hand side of this expression may be obtained from Figure 13.6, F_{14} may be evaluated.

2. For the prescribed dimensions and Figure 13.6.

Surfaces $(1,2)(3,4)$:

$$(Y/X) = \frac{L_1 + L_2}{W} = 1 \qquad (Z/X) = \frac{L_3 + L_4}{W} = 1.45$$

$$F_{(1,2)(3,4)} = 0.22$$

Surfaces 23:

$$(Y/X) = \frac{L_2}{W} = 0.5 \qquad (Z/X) = \frac{L_3}{W} = 1 \qquad F_{23} = 0.35$$

Surfaces $(1,2)3$

$$(Y/X) = \frac{L_1 + L_2}{W} = 1 \qquad (Z/X) = \frac{L_3}{W} = 1 \qquad F_{(1,2)3} = 0.20$$

Surfaces $2(3,4)$

$$(Y/X) = \frac{L_2}{W} = 0.5 \qquad (Z/X) = \frac{L_3 + L_4}{W} = 1.5 \qquad F_{2(3,4)} = 0.37$$

Hence

$$F_{14} = \frac{1}{(WL_1)}[(WL_1 + WL_2)0.22 + (WL_2)0.35$$

$$- (WL_1 + WL_2)0.20 - (WL_2)0.37]$$

$$F_{14} = [2(0.22) + 1(0.35) - 2(0.20) - 1(0.37)]$$

$$F_{14} = 0.02$$

◁

13.2   BLACKBODY RADIATION EXCHANGE

In general, radiation may leave a surface due to both reflection and emission, and on reaching a second surface, experience reflection as well as absorption. However, matters are simplified for surfaces that may be approximated as blackbodies, since there is no reflection. Hence energy only leaves as a result of emission, and all incident radiation is absorbed.

Consider radiation exchange between two black surfaces of arbitrary shape (Figure 13.8). Defining $q_{i \to j}$ as the rate at which radiation *leaves* surface i and is *intercepted* by surface j, it follows that

$$q_{i \to j} = (A_i J_i)F_{ij} \tag{13.9}$$

or, since radiosity equals emissive power for a black surface ($J_i = E_{bi}$),

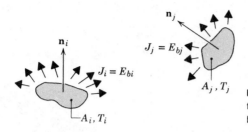

Figure 13.8 Radiation transfer between two surfaces that may be approximated as blackbodies.

$$q_{i \to j} = A_i F_{ij} E_{bi} \tag{13.10}$$

Similarly,

$$q_{j \to i} = A_j F_{ji} E_{bj} \tag{13.11}$$

We define the *net radiative exchange* between any two surfaces as

$$\boxed{q_{ij} = q_{i-j} - q_{j-i}} \tag{13.12}$$

where q_{i-j} is the rate at which radiation is *emitted* by surface i and *absorbed* by surface j and q_{j-i} is the rate at which radiation is emitted by j and absorbed by i. Equation 13.12 provides the *net* rate at which radiation leaves surface i due to its interaction with j, which is equal to the *net* rate at which j gains radiation due to its interaction with i. However, since $q_{i-j} = q_{i \to j}$ for black surfaces, Equations 13.10 and 13.11 may be substituted into Equation 13.12. Hence

$$\boxed{q_{ij} = A_i F_{ij} E_{bi} - A_j F_{ji} E_{bj}} \quad \text{BLACKBODY ONLY !!}$$

or from Equations 12.28 and 13.3, the net radiation exchange between any two black surfaces may then be expressed as

$$q_{ij} = A_i F_{ij} \sigma (T_i^4 - T_j^4) \tag{13.13}$$

The foregoing result may also be used to evaluate the net radiation transfer from any surface in an enclosure of black surfaces. With N surfaces maintained at different temperatures, the net transfer of radiation from surface i is due to exchange with the remaining surfaces and may be expressed as

$$q_i = \sum_{j=1}^{N} A_i F_{ij} \sigma (T_i^4 - T_j^4) \tag{13.14}$$

It is important to note the subtle distinctions between the various heat transfer rates that have been introduced in this section. For your convenience their definitions are summarized in Table 13.1.

Table 13.1 Surface radiative heat transfer rates

SYMBOL	DEFINITION
$q_{i \to j}$	Rate at which radiation *leaves* surface i and is *intercepted* by surface j, Equation 13.9 or Equation 13.10 for black surfaces
q_{i-j}	Rate at which radiation is *emitted* by surface i and is *absorbed* by surface j, Equation 13.12 for a black surface
q_{ij}	Net radiative exchange between surfaces i and j, Equation 13.20 or Equation 13.13 for black surfaces
q_i	Net radiative heat transfer *from* surface i, Equations 13.15, 13.17, 13.19, 13.20, 13.22, or Equation 13.14 for black surfaces

EXAMPLE 13.4

A furnace cavity, which is in the form of a cylinder of diameter $D = 75$ mm and length $L = 150$ mm, is open at one end to surroundings that are at a temperature of 27°C. The sides and bottom may be approximated as blackbodies, are heated electrically, are well insulated, and are maintained at temperatures of $T_1 = 1350$°C and $T_2 = 1650$°C, respectively.

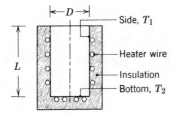

How much power is required to maintain the furnace at the prescribed conditions?

SOLUTION

KNOWN:

Cylindrical furnace with bottom and sides at different temperatures.

FIND:

Power required to maintain furnace at the prescribed temperatures.

SCHEMATIC:

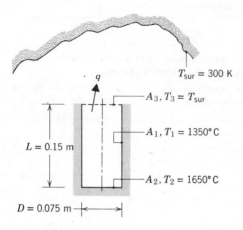

$T_{sur} = 300$ K

$A_3, T_3 = T_{sur}$

$A_1, T_1 = 1350°$ C

$L = 0.15$ m

$A_2, T_2 = 1650°$ C

$D = 0.075$ m

ASSUMPTIONS:

1. Interior surfaces behave as blackbodies.
2. Negligible heat transfer by convection.
3. Outer surface of furnace is adiabatic.

ANALYSIS:

The power needed to operate the furnace at the prescribed conditions must balance the heat losses from the furnace. Subject to the foregoing assumptions, the only heat loss is by radiation through the opening, which may be treated as a hypothetical surface of area A_3. Moreover, because the surroundings are large, the effect that this surface has on radiation exchange between the furnace and the surroundings may be treated by approximating the surface as a blackbody at $T_3 = T_{sur}$. The heat loss may then be expressed as

$$q = q_{13} + q_{23}$$

or, from Equation 13.13,

$$q = A_1 F_{13} \sigma (T_1^4 - T_3^4) + A_2 F_{23} \sigma (T_2^4 - T_3^4)$$

From Figure 13.5, it follows that, with $(r_j/L) = (0.0375 \text{ m}/0.15 \text{ m}) = 0.25$ and $(L/r_i) = (0.15 \text{ m}/0.0375 \text{ m}) = 4$,

$$F_{23} = 0.06$$

From the summation rule

$$F_{21} = 1 - F_{23} = 1 - 0.06 = 0.94$$

and from reciprocity

$$F_{12} = \frac{A_2}{A_1} F_{21} = \frac{\pi(0.075 \text{ m})^2/4}{\pi(0.075 \text{ m})(0.15 \text{ m})} \times 0.94 = 0.118$$

Hence, since $F_{13} = F_{12}$ from symmetry,

$$q = (\pi \times 0.075 \text{ m} \times 0.15 \text{ m})0.118 \times 5.67 \times 10^{-8} \text{ W/m}^2 \cdot \text{K}^4 \, [(1623 \text{ K})^4 - (300 \text{ K})^4]$$

$$+ \frac{\pi(0.075 \text{ m})^2}{4} \times 0.06 \times 5.67 \times 10^{-8} \text{ W/m}^2 \cdot \text{K}^4 \, [(1923 \text{ K})^4 - (300 \text{ K})^4]$$

$$q = 1639 \text{ W} + 205 \text{ W}$$

$$q = 1844 \text{ W} \qquad\qquad \triangleleft$$

13.3 RADIATION EXCHANGE BETWEEN DIFFUSE-GRAY SURFACES IN AN ENCLOSURE

Although useful to a point, the foregoing results are limited by the assumption of blackbody behavior. The blackbody is, of course, an idealization, which, although closely approximated by some surfaces, is never precisely achieved. A major complication associated with radiation exchange between nonblack surfaces is due to surface reflection. In an enclosure, such as that of Figure 13.9a, radiation may experience multiple reflections off all surfaces, with partial absorption occurring at each.

Analyzing radiation exchange in an enclosure may be simplified by making certain assumptions. Each surface of the enclosure is assumed to be *isothermal* and to be characterized by a *uniform radiosity* and *irradiation*. *Opaque, diffuse-gray* surface behavior is also assumed, and the medium within the enclosure is taken to be *nonparticipating*. The problem is generally one in which the temperature, T_i, associated with each of the surfaces is known, and the objective is to determine the *net radiative heat flux, q_i''*, from each surface.

13.3.1 Net Radiation Exchange at a Surface

The term q_i, which is the *net* rate at which radiation *leaves* surface i, represents the net effect of radiative interactions occurring at the surface, Figure 13.9b. It is the rate at which energy would have to be transferred to the surface by other means in order to maintain it at a constant temperature. It is equal to the difference between the surface radiosity and irradiation and may be expressed as

$$q_i = A_i(J_i - G_i) \qquad\qquad (13.15)$$

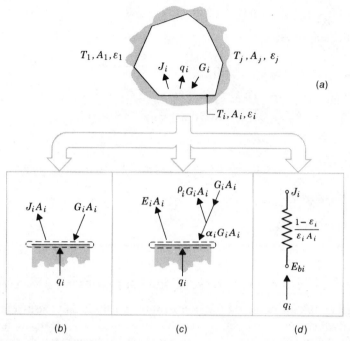

Figure 13.9 Radiation exchange in an enclosure of diffuse-gray surfaces with a nonparticipating medium. (a) Schematic of the enclosure. (b) Radiative balance according to Equation 13.15. (c) Radiative balance according to Equation 13.17. (d) Network element representing the net radiation transfer from a surface.

From Figure 13.9c and the definition of the radiosity, J_i,

$$J_i \equiv E_i + \rho_i G_i \qquad (13.16)$$

it is evident that the net radiative transfer from the surface may also be expressed in terms of the surface emissive power and the absorbed irradiation.

$$q_i = A_i(E_i - \alpha_i G_i) \qquad (13.17)$$

Substituting from Equation 12.37 and recognizing that $\rho_i = 1 - \alpha_i = 1 - \varepsilon_i$ for an opaque, diffuse-gray surface, the radiosity may also be expressed as

$$J_i = \varepsilon_i E_{bi} + (1 - \varepsilon_i)G_i \qquad (13.18)$$

Solving for G_i and substituting into Equation 13.15, it follows that

$$q_i = A_i\left(J_i - \frac{J_i - \varepsilon_i E_{bi}}{1 - \varepsilon_i}\right)$$

or

✳ $$q_i = \frac{E_{bi} - J_i}{(1 - \varepsilon_i)/\varepsilon_i A_i} \qquad (13.19)$$

Equation 13.19 provides a convenient representation for the net radiative heat transfer rate from a surface. This transfer, which may be represented by the network element of Figure 13.9d, is associated with the driving potential, $(E_{bi} - J_i)$, and a *surface radiative resistance* of the form $(1 - \varepsilon_i)/\varepsilon_i A_i$. Hence if the emissive power that the surface would have if it were black exceeds its radiosity, there is net radiation heat transfer from the surface; if the inverse is true, the net transfer is to the surface.

If $E_{bi} - J_i$ is neg, q_i is neg

13.3.2 Radiation Exchange Between Surfaces

To use Equation 13.19 the surface radiosity, J_i, must be known. To determine this quantity, it is necessary to consider radiation exchange between the surfaces of the enclosure.

The irradiation of surface i can be evaluated from the radiosities of all of the surfaces in the enclosure. In particular from the definition of the view factor, it follows that the total rate at which radiation reaches surface i from all surfaces, including i, is

$$A_i G_i = \sum_{j=1}^{N} F_{ji} A_j J_j$$

or from the reciprocity relation, Equation 13.3,

$$A_i G_i = \sum_{j=1}^{N} A_i F_{ij} J_j$$

Canceling the area, A_i, and substituting into Equation 13.15 for G_i, it follows that

$$q_i = A_i \left(J_i - \sum_{j=1}^{N} F_{ij} J_j \right)$$

or, from the summation rule, Equation 13.4,

$$q_i = A_i \left(\sum_{j=1}^{N} F_{ij} J_i - \sum_{j=1}^{N} F_{ij} J_j \right)$$

Hence

✳ $$q_i = \sum_{j=1}^{N} A_i F_{ij} (J_i - J_j) = \sum_{j=1}^{N} q_{ij} \qquad (13.20)$$

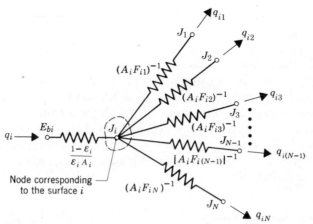

Figure 13.10 Network representation of radiative exchange between surface i and the remaining surfaces of an enclosure.

This result suggests that the net rate of radiation transfer from surface i, q_i, may be represented as a sum of components, q_{ij}, related to radiative exchange with the other surfaces. Each component may be represented by a network element for which $(J_i - J_j)$ is the driving potential and $(A_i F_{ij})^{-1}$ is a *space* or *geometrical resistance* (Figure 13.10).

Combining Equations 13.19 and 13.20, we then obtain

$$\frac{E_{bi} - J_i}{(1 - \varepsilon_i)/\varepsilon_i A_i} = \sum_{j=1}^{N} \frac{J_i - J_j}{(A_i F_{ij})^{-1}} \tag{13.21}$$

As shown in Figure 13.10 this expression represents a radiation balance for the i node associated with the i surface of radiosity J_i. The rate of radiation transfer (current flow) to i through its surface resistance must equal the rate of radiation transfer (current flows) from i to all other surfaces through the corresponding geometrical resistances.

Note that Equation 13.21 is especially useful when the surface temperature T_i (and hence E_{bi}) is known. Although this situation is typical, it is not always the case. In particular, situations may arise for which the net radiation transfer rate at the surface, q_i, rather than the temperature, T_i, is known. In such cases the preferred form of the radiation balance is Equation 13.20, rearranged as

$$q_i = \sum_{j=1}^{N} \frac{J_i - J_j}{(A_i F_{ij})^{-1}} \tag{13.22}$$

Use of network representations to solve enclosure radiation problems was first suggested by Oppenheim [6]. The method provides a useful tool for visualizing radiation exchange in the enclosure and, at least for simple enclosures, may be used as the basis for predicting this exchange. However, a more direct approach, which lends itself to use with a computer, simply involves working with Equations 13.21 and 13.22. Equation 13.21 is written for each surface at which T_i is known, and Equation 13.22 is written for each surface at which q_i is known. The resulting set of N linear, algebraic equations are then solved for the N unknowns, $J_1, J_2, ..., J_N$. With knowledge of the J_i, Equation 13.19 may then be used to determine the net radiation heat transfer rate, q_i, at each surface of known T_i or the value of T_i at each surface of known q_i.

For any number, N, of surfaces in the enclosure, the foregoing problem may be readily solved by using matrix inversion. For each of the N surfaces Equation 13.21 or 13.22 may be rearranged to obtain the following system of N equations

$$\begin{bmatrix} a_{11}J_1 + ... + a_{1i}J_i + ... + a_{1N}J_N = C_1 \\ \vdots \\ a_{i1}J_1 + ... + a_{ii}J_i + ... + a_{iN}J_N = C_i \\ \vdots \\ a_{N1}J_1 + ... + a_{Ni}J_i + ... + a_{NN}J_N = C_N \end{bmatrix}$$

where the coefficients a_{ij} and C_i are known quantities. In matrix form these equations may be expressed as

$$[A][J] = [C]$$

where

$$[A] = \begin{bmatrix} a_{11}...a_{1i}...a_{1N} \\ \cdot \quad \cdot \quad \cdot \\ a_{i1} ...a_{ii} ...a_{iN} \\ \cdot \quad \cdot \quad \cdot \\ a_{N1}...a_{Ni}...a_{NN} \end{bmatrix} \quad [J] = \begin{bmatrix} J_1 \\ \vdots \\ J_i \\ \vdots \\ J_N \end{bmatrix} \quad [C] = \begin{bmatrix} C_1 \\ \vdots \\ C_i \\ \vdots \\ C_N \end{bmatrix}$$

Expressing the unknown radiosities as

$$\begin{bmatrix} J_1 = b_{11}C_1 + ... + b_{1i}C_i + ... + b_{1N}C_N \\ \vdots \\ J_i = b_{i1}C_1 + ... + b_{ii}C_i + ... + b_{iN}C_N \\ \vdots \\ J_N = b_{N1}C_1 + ... + b_{Ni}C_i + ... + b_{NN}C_N \end{bmatrix}$$

it follows that they may be found by obtaining the inverse of $[A]$, $[A]^{-1}$, such that

$$[J] = [A]^{-1}[C]$$

where

$$[A]^{-1} = \begin{bmatrix} b_{11} \ldots b_{1i} \ldots b_{1N} \\ \cdot \quad \cdot \quad \cdot \quad \cdot \\ b_{i1} \ldots b_{ii} \ldots b_{iN} \\ \cdot \quad \cdot \quad \cdot \quad \cdot \\ b_{N1} \ldots b_{Ni} \ldots b_{NN} \end{bmatrix}$$

The foregoing matrix inversion may be readily obtained by using any of numerous computer routines available for this purpose or, for simpler problems, programmable hand calculators.

EXAMPLE 13.5

In manufacturing, the special coating on a curved solar absorber surface of area $A_2 = 15 \text{ m}^2$ is cured by exposing it to an infrared heater of width $W = 1 \text{ m}$. The absorber and heater are each of length $L = 10 \text{ m}$ and are separated by a distance of $H = 1 \text{ m}$.

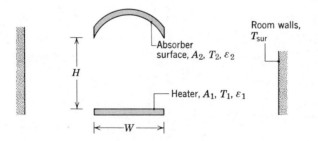

The heater is at a temperature of $T_1 = 1000 \text{ K}$ and has an emissivity of $\varepsilon_1 = 0.9$, while the absorber is at a temperature of $T_2 = 600 \text{ K}$ and has an emissivity of $\varepsilon_2 = 0.5$. The system is located in a large room whose walls are at a temperature of 300 K. What is the net rate of heat transfer to the absorber surface?

SOLUTION

KNOWN:

A curved, solar absorber surface with a special coating is being cured by use of an infrared heater in a large room.

FIND:

Net rate of heat transfer to the absorber surface.

SCHEMATIC:

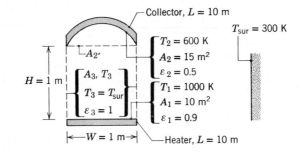

Collector, $L = 10$ m

$T_{sur} = 300$ K

$H = 1$ m

$A_{2'}$

$\begin{cases} T_2 = 600 \text{ K} \\ A_2 = 15 \text{ m}^2 \\ \varepsilon_2 = 0.5 \end{cases}$

$\begin{cases} A_3, T_3 \\ T_3 = T_{sur} \\ \varepsilon_3 = 1 \end{cases}$

$\begin{cases} T_1 = 1000 \text{ K} \\ A_1 = 10 \text{ m}^2 \\ \varepsilon_1 = 0.9 \end{cases}$

$\leftarrow W = 1 \text{ m} \rightarrow$

Heater, $L = 10$ m

ASSUMPTIONS:

1. Steady-state conditions.
2. Convection effects are negligible.
3. Absorber and heater surfaces are diffuse-gray.
4. The surroundings may be represented by a hypothetical surface A_3, which completes the enclosure and which is approximated as a blackbody of temperature $T_3 = 300$ K.

ANALYSIS:

The problem may be viewed as a three-surface enclosure problem for which we are specifically interested in obtaining the net rate of radiation transfer to surface 2. From Equation 13.19 this quantity is given by

$$q_2 = \frac{E_{b2} - J_2}{(1 - \varepsilon_2)/\varepsilon_2 A_2} \tag{1}$$

where all quantities are known except J_2. In any enclosure problem all unknown radiosities must be determined simultaneously, and the matrix inversion procedure described in Section 13.3.2 is well suited for this purpose. In the present problem, however, matters are simplified by the fact that the hypothetical surface is being approximated as a blackbody of known temperature. Hence $J_3 = E_{b3}$ is known, and the only unknowns are J_1 and J_2. Since T_1 and T_2 are known, J_1 and J_2 may be obtained by expressing Equation 13.21 for the heater and absorber surfaces. The network representation of the system is shown below, and the analysis proceeds as follows.

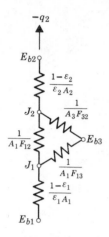

Expressing Equation 13.21 for the absorber surface, it follows that

$$\frac{E_{b2} - J_2}{(1 - \varepsilon_2)/\varepsilon_2 A_2} = \frac{J_2 - J_1}{(1/A_2 F_{21})} + \frac{J_2 - J_3}{(1/A_2 F_{23})}$$

where $J_3 = E_{b3} = \sigma T_3^4 = 459$ W/m^2 and $E_{b2} = \sigma T_2^4 = 7348$ W/m^2. The quantity $A_2 F_{21}$ may be obtained by recognizing that, from reciprocity,

$$A_2 F_{21} = A_1 F_{12}$$

where F_{12} may be obtained, from Figure 13.4. That is, $F_{12} = F_{12'}$, where $A_{2'}$ is simply the rectangular base of the absorber surface. Hence with $Y/L = 10/1 = 10$ and $X/L = 1/1 = 1$,

$$F_{12} = 0.39$$

$$F_{21} = \frac{A_1}{A_2} F_{12} = \frac{1 \text{ m} \times 10 \text{ m}}{15 \text{ m}^2} \times 0.39 = 0.26$$

From the summation rule, it also follows that

$$F_{13} = 1 - F_{12} = 1 - 0.39 = 0.61$$

and from reciprocity

$$F_{31} = \frac{A_1}{A_3} F_{13} = \frac{1 \text{ m} \times 10 \text{ m}}{(10 + 10 + 1 + 1) \text{ m}^2} \times 0.61 = 0.28$$

But from symmetry, $F_{31} = F_{32'} = F_{32}$. Hence

$$F_{23} = \frac{A_3}{A_2} F_{32} = \frac{22 \text{ m}^2}{15 \text{ m}^2} \times 0.28 = 0.41$$

Canceling the area A_2, the radiation balance for surface 2 is then

$$\frac{7348 - J_2}{(1 - 0.5)/0.5} = \frac{J_2 - J_1}{(1/0.26)} + \frac{J_2 - 459}{(1/0.41)}$$

or

$$7348 - J_2 = 0.26J_2 - 0.26J_1 + 0.41J_2 - 188$$

$$0.26J_1 - 1.67J_2 = -7536 \qquad\qquad\qquad (2)$$

Expressing Equation 13.21 for the heater surface, it follows that

$$\frac{E_{b1} - J_1}{(1 - \varepsilon_1)/\varepsilon_1 A_1} = \frac{J_1 - J_2}{(1/A_1 F_{12})} + \frac{J_1 - J_3}{(1/A_1 F_{13})}$$

where $E_{b1} = \sigma T_1^4 = 56{,}700 \text{ W/m}^2$. Hence canceling the area A_1,

$$\frac{56{,}700 - J_1}{(1 - 0.9)/0.9} = \frac{J_1 - J_2}{(1/0.39)} + \frac{J_1 - 459}{(1/0.61)}$$

or

$$510{,}300 - 9J_1 = 0.39J_1 - 0.39J_2 + 0.61J_1 - 280$$

$$- 10J_1 + 0.39J_2 = -510{,}002 \qquad\qquad\qquad (3)$$

From Equation 3, we then obtain

$$J_1 = 51{,}000 + 0.039J_2$$

and substituting into Equation 2

$$0.26(51{,}000 + 0.039J_2) - 1.67J_2 = -7536$$

$$13{,}260 + 0.010J_2 - 1.67J_2 = -7536$$

$$- 1.66J_2 = -20{,}796$$

$$J_2 = 12{,}528 \text{ W/m}^2$$

Substituting into Equation 1 we then obtain

$$q_2 = \frac{(7348 - 12{,}528) \text{ W/m}^2}{(1 - 0.5)/0.5 \times 15 \text{ m}^2}$$

$$q_2 = -77.7 \text{ kW} \qquad \qquad \lhd$$

COMMENTS:

Recognize the utility associated with the use of hypothetical surfaces, in one case to complete the enclosure with A_3 and in the other case to simplify the evaluation of the shape factor with $A_{2'}$.

13.3.3 The Two-Surface Enclosure

The simplest example of an enclosure is one involving two surfaces which exchange radiation only with each other. Such a two-surface enclosure is shown schematically in Figure 13.11*a*. Since there are only two surfaces, the net rate of radiation transfer *from* surface 1, q_1, must equal the net rate of radiation transfer *to* surface 2, $-q_2$, and both quantities must equal the net rate at which radiation is exchanged between 1 and 2. Accordingly,

$$q_1 = -q_2 = q_{12}$$

The radiation transfer rate may be determined by applying Equation 13.21 to surfaces 1 and 2 and solving the resulting two equations for J_1 and J_2. The results could then be used with Equation 13.19 to determine q_1 (or q_2). However, in this case the desired result is more readily obtained by working with the network representation of the enclosure shown in Figure 13.11*b*.

From Figure 13.11*b* we see that the total resistance to radiation exchange between surfaces 1 and 2 is comprised of the two surface resistances and the geometrical resistance. Hence, substituting from Equation 12.28, the net radiation exchange between surfaces may be expressed as

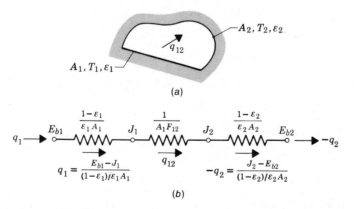

(a)

(b)

Figure 13.11 The two-surface enclosure. (a) Schematic. (b) Network representation.

$$q_{12} = q_1 = -q_2 = \frac{\sigma(T_1^4 - T_2^4)}{\dfrac{1-\varepsilon_1}{\varepsilon_1 A_1} + \dfrac{1}{A_1 F_{12}} + \dfrac{1-\varepsilon_2}{\varepsilon_2 A_2}} \tag{13.23}$$

The foregoing result may be used for any two diffuse-gray surfaces *that form an enclosure*. Important special cases are summarized in Figure 13.12.

13.3.4 Radiation Shields

Radiation shields constructed from low emissivity (high reflectivity) materials can be used to reduce the net radiation transfer between two surfaces. Consider placing a radiation shield, surface 3, between the two large, parallel planes of Figure 13.13a. Without the radiation shield, the net rate of radiation transfer between surfaces 1 and 2 is given by Equation 13.24. However, with the

Large (Infinite) Parallel Planes

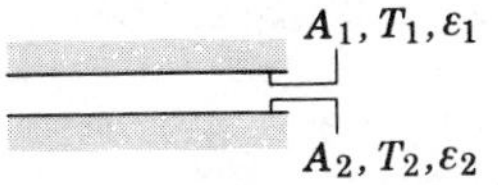

A_1, T_1, ε_1 $A_1 = A_2 = A$

A_2, T_2, ε_2 $F_{12} = 1$

$$q_{12} = \frac{A\sigma(T_1^4 - T_2^4)}{\dfrac{1}{\varepsilon_1} + \dfrac{1}{\varepsilon_2} - 1} \tag{13.24}$$

Long (Infinite) Concentric Cylinders

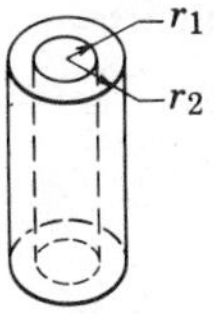

$\dfrac{A_1}{A_2} = \dfrac{r_1}{r_2}$

$F_{12} = 1$

$$q_{12} = \frac{\sigma A_1(T_1^4 - T_2^4)}{\dfrac{1}{\varepsilon_1} + \dfrac{1-\varepsilon_2}{\varepsilon_2}\left(\dfrac{r_1}{r_2}\right)} \tag{13.25}$$

Concentric Spheres

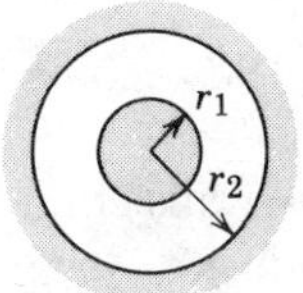

$\dfrac{A_1}{A_2} = \dfrac{r_1^2}{r_2^2}$

$F_{12} = 1$

$$q_{12} = \frac{\sigma A_1(T_1^4 - T_2^4)}{\dfrac{1}{\varepsilon_1} + \dfrac{1-\varepsilon_2}{\varepsilon_2}\left(\dfrac{r_1}{r_2}\right)^2} \tag{13.26}$$

Small Convex Object in a Large Cavity

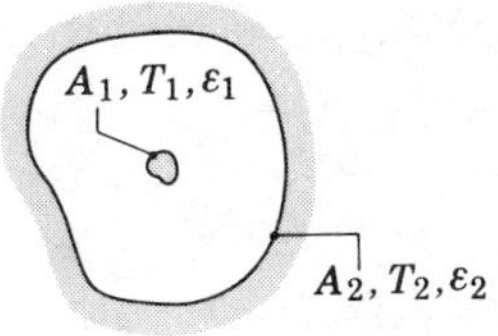

A_1, T_1, ε_1

$\dfrac{A_1}{A_2} \approx 0$

A_2, T_2, ε_2 $F_{12} = 1$

$$q_{12} = \sigma A_1 \varepsilon_1 (T_1^4 - T_2^4) \tag{13.27}$$

Figure 13.12 Special diffuse-gray, two-surface enclosures.

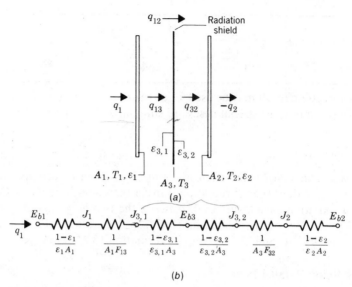

Figure 13.13   Radiation exchange between large parallel planes with a radiation shield. (a) Schematic. (b) Network representation.

radiation shield, additional resistances are present, as shown in Figure 13.13b, and the heat transfer rate is reduced. Note that the emissivity associated with one side of the shield ($\varepsilon_{3,1}$) may differ from that associated with the opposite side ($\varepsilon_{3,2}$) and the radiosities will always differ. Summing the resistances and recognizing that $F_{13} = F_{32} = 1$, it follows that

$$q_{12} = \frac{A_1 \sigma (T_1^4 - T_2^4)}{\dfrac{1}{\varepsilon_1} + \dfrac{1}{\varepsilon_2} + \dfrac{1 - \varepsilon_{3,1}}{\varepsilon_{3,1}} + \dfrac{1 - \varepsilon_{3,2}}{\varepsilon_{3,2}}} \tag{13.28}$$

Note that the resistances associated with the radiation shield become very large when the emissivities, $\varepsilon_{3,1}$ and $\varepsilon_{3,2}$, are very small.

Equation 13.28 may be used to determine the net heat transfer rate if T_1 and T_2 are known. From knowledge of q_{12} and the fact that $q_{12} = q_{13} = q_{32}$, the value of T_3 may then be determined by expressing Equation 13.24 for q_{13} or q_{32}.

The foregoing procedure may readily be extended to problems involving multiple radiation shields. In the special case for which all of the emissivities are equal, it may be shown that, with N shields,

$$(q_{12})_N = \frac{1}{N + 1} (q_{12})_0 \tag{13.29}$$

where $(q_{12})_0$ is the radiation transfer rate with no shields ($N = 0$).

EXAMPLE 13.6

A cryogenic fluid flows through a long tube of diameter $D_1 = 20$ mm, the outer surface of which is diffuse-gray with $\varepsilon_1 = 0.02$ and $T_1 = 77$ K. This tube is concentric with a larger tube of diameter $D_2 = 50$ mm, the inner surface of which is diffuse-gray with $\varepsilon_2 = 0.05$ and $T_2 = 300$ K. The space between the surfaces is evacuated.

1. Calculate the heat gain by the cryogenic fluid per unit length of tubes.
2. If a thin radiation shield of diameter $D_3 = 35$ mm and emissivity $\varepsilon_3 = 0.02$ (both sides) is inserted midway between the inner and outer surfaces, calculate the change (percent) in heat gain per unit length of the tubes.

SOLUTION

KNOWN:

Concentric tube arrangement with diffuse-gray surfaces of different emissivities and temperatures.

FIND:

1. Heat gain by the cryogenic fluid passing through the inner tube.
2. Percentage change in heat gain with radiation shield inserted midway between inner and outer tubes.

SCHEMATIC:

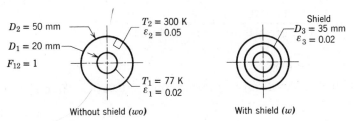

Without shield (*wo*)   With shield (*w*)

ASSUMPTIONS:

1. Surfaces are diffuse gray.
2. Space between tubes is evacuated.
3. Negligible conduction resistance for radiation shield.
4. Concentric tubes form a two-surface enclosure (end effects are negligible).

ANALYSIS:

1. The network representation of the system without the shield is shown in Figure 13.11, and the desired heat rate may be obtained from Equation 13.25, where

$$q = \frac{\sigma(\pi D_1 L)(T_1^4 - T_2^4)}{\dfrac{1}{\varepsilon_1} + \dfrac{1 - \varepsilon_2}{\varepsilon_2}\left(\dfrac{D_1}{D_2}\right)}$$

Hence

$$q' = \frac{q}{L} = \frac{5.67 \times 10^{-8} \text{ W/m}^2 \cdot \text{K}^4 \ (\pi \times 0.02 \text{ m})[(77 \text{ K})^4 - (300 \text{ K})^4]}{\dfrac{1}{0.02} + \dfrac{1 - 0.05}{0.05}\left(\dfrac{0.02 \text{ m}}{0.05 \text{ m}}\right)}$$

$$q' = -0.50 \text{ W/m} \qquad \triangleleft$$

2. The network representation of the system with the shield is shown in Figure 13.13, and the desired heat rate is now

$$q = \frac{E_{b1} - E_{b2}}{R_{\text{tot}}} = \frac{\sigma(T_1^4 - T_2^4)}{R_{\text{tot}}}$$

where

$$R_{\text{tot}} = \frac{1 - \varepsilon_1}{\varepsilon_1(\pi D_1 L)} + \frac{1}{(\pi D_1 L)F_{13}} + 2\left[\frac{1 - \varepsilon_3}{\varepsilon_3(\pi D_3 L)}\right] + \frac{1}{(\pi D_3 L)F_{32}} + \frac{1 - \varepsilon_2}{\varepsilon_2(\pi D_2 L)}$$

or

$$R_{\text{tot}} = \frac{1}{L}\left\{\frac{1 - 0.02}{0.02\,(\pi \times 0.02 \text{ m})} + \frac{1}{(\pi \times 0.02 \text{ m})1}\right.$$

$$\left. + 2\left[\frac{1 - 0.02}{0.02(\pi \times 0.035 \text{ m})}\right] + \frac{1}{(\pi \times 0.035 \text{ m})1} + \frac{1 - 0.05}{0.05(\pi \times 0.05 \text{ m})}\right\}$$

$$R_{\text{tot}} = \frac{1}{L}(779.9 + 15.9 + 891.3 + 9.1 + 121.0) = \frac{1817}{L}\left(\frac{1}{\text{m}^2}\right)$$

Hence

$$q' = \frac{q}{L} = \frac{5.67 \times 10^{-8} \text{ W/m}^2 \cdot \text{K}^4 [(77 \text{ K})^4 - (300 \text{ K})^4]}{1817 \ (1/\text{m})}$$

$$q' = -0.25 \text{ W/m} \qquad \triangleleft$$

The percentage change in the heat gain is then

$$\frac{q'_w - q'_{wo}}{q'_{wo}} \times 100 = \frac{(-0.25 \text{ W/m}) - (-0.50 \text{ W/m})}{-0.50 \text{ W/m}} \times 100 = -50 \text{ percent}$$

That is, there is a 50 percent reduction in the heat rate.

13.3.5 The Reradiating Surface

Reradiating surfaces are common to many industrial applications. They are simply surfaces that are well insulated and hence may be approximated as adiabatic $(q_i = 0)$. Accordingly, from Equation 13.19, it follows that $E_{bi} = J_i$. Hence, if the radiosity of a reradiating surface is known, its temperature is readily determined. Moreover, from Equation 13.15 it also follows that $J_i = G_i$. In an enclosure the equilibrium temperature of a reradiating surface is determined by its interaction with the other surfaces, and it is *independent of the value of emissivity*. The presence of such a surface strongly influences the radiation environment associated with the enclosure.

A three-surface enclosure, for which the third surface, surface R, is reradiating, is shown in Figure 13.14a, and the corresponding network is shown in Figure 13.14b. Since surface R is adiabatic, the net radiation *transfer* from

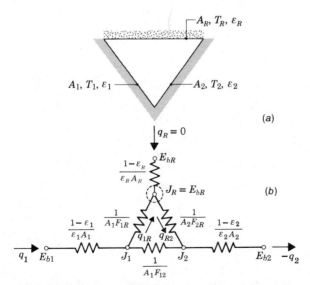

Figure 13.14 A three-surface enclosure with one surface reradiating. (*a*) Schematic. (*b*) Network representation.

surface 1 must equal the net radiation transfer to surface 2. The network is a simple series-parallel arrangement, and from its analysis it is readily shown that

$$q_1 = -q_2$$

$$= \frac{E_{b1} - E_{b2}}{\dfrac{1-\varepsilon_1}{\varepsilon_1 A_1} + \dfrac{1}{A_1 F_{12} + [(1/A_1 F_{1R}) + (1/A_2 F_{2R})]^{-1}} + \dfrac{1-\varepsilon_2}{\varepsilon_2 A_2}} \qquad (13.30)$$

With knowledge of $q_1 = -q_2$, Equation 13.19 may then be applied to surfaces 1 and 2 to determine their radiosities, J_1 and J_2, respectively. From knowledge of J_1, J_2, and the geometrical resistances, the radiosity of the reradiating surface, J_R, may then be determined from application of a radiation balance to this surface. Accordingly,

$$\frac{J_1 - J_R}{(1/A_1 F_{1R})} - \frac{J_R - J_2}{(1/A_2 F_{2R})} = 0 \qquad (13.31)$$

The temperature of the reradiating surface may then be determined from the requirement that $\sigma T_R^4 = J_R$.

Note that the general procedure described in Section 13.3.2 may be applied to enclosures with reradiating surfaces. For each such surface, it is appropriate to use Equation 13.22 with $q_i = 0$.

EXAMPLE 13.7

A paint baking oven is in the form of a long, triangular duct in which one surface is maintained at an elevated temperature of 1200 K, and another surface is insulated. Painted panels, which are maintained at 500 K, occupy the third surface. The triangle is of width $W = 1$ m on a side, and the heated and insulated surfaces have an emissivity of 0.8. The emissivity of the panels is $\varepsilon = 0.4$. During steady-state operation, at what rate must energy be supplied to the heated side per unit length of the duct to maintain its temperature at 1200 K? What is the temperature of the insulated surface?

SOLUTION

KNOWN:

Surface properties of a long triangular duct that is insulated on one side and heated and cooled on the other sides.

FIND:

1. Rate at which heat must be supplied per unit length of duct.

2. Temperature of the insulated surface.

SCHEMATIC:

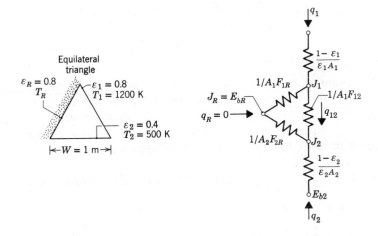

$\varepsilon_R = 0.8$
T_R

Equilateral triangle

$\varepsilon_1 = 0.8$
$T_1 = 1200$ K

$\varepsilon_2 = 0.4$
$T_2 = 500$ K

$\leftarrow W = 1 \text{ m} \rightarrow$

ASSUMPTIONS:

1. Steady-state conditions.

2. All surfaces are opaque, diffuse gray, and of uniform radiosity.

3. Negligible convection effects.

4. Surface R is reradiating (adiabatic).

5. End effects are negligible.

ANALYSIS:

1. The system may be modeled as a three-surface enclosure with one surface reradiating. The rate at which energy must be supplied to the heated surface may then be obtained from Equation 13.30

$$q_1 = \frac{E_{b1} - E_{b2}}{\dfrac{1 - \varepsilon_1}{\varepsilon_1 A_1} + \dfrac{1}{A_1 F_{12} + [(1/A_1 F_{1R}) + (1/A_2 F_{2R})]^{-1}} + \dfrac{1 - \varepsilon_2}{\varepsilon_2 A_2}}$$

From symmetry, $F_{12} = F_{1R} = F_{2R} = 0.5$.

Also

$$A_1 = A_2 = W \cdot L$$

where L is the duct length. Hence

$$q_1' = \frac{q_1}{L} = \frac{5.67 \times 10^{-8} \text{ W/m}^2 \cdot \text{K}^4 \, (1200^4 - 500^4) \text{K}^4}{\dfrac{1 - 0.8}{0.8 \times 1 \text{ m}} + \dfrac{1}{1 \text{ m} \times 0.5 + (2 + 2)^{-1} \text{ m}} + \dfrac{1 - 0.4}{0.4 \times 1 \text{ m}}}$$

or

$$q_1' = 37 \text{ kW/m} = -q_2' \qquad \lhd$$

2. The temperature of the insulated surface may be obtained from the requirement that $J_R = E_{bR}$, where J_R may be obtained from Equation 13.31. However, to use this expression J_1 and J_2 must be known. Applying the surface energy balance equation, Equation 13.19, to surfaces 1 and 2, it follows that

$$J_1 = E_{b1} - \frac{1 - \varepsilon_1}{\varepsilon_1 W} q_1' = 5.67 \times 10^{-8} \text{ W/m}^2 \cdot \text{K}^4 \, (1200 \text{ K})^4 - \frac{1 - 0.8}{0.8 \times 1 \text{ m}}$$

$$\times 37,000 \text{ W/m}$$

$$J_1 = 108,323 \text{ W/m}^2$$

$$J_2 = E_{b2} - \frac{1 - \varepsilon_2}{\varepsilon_2 W} q_2' = 5.67 \times 10^{-8} \text{ W/m}^2 \cdot \text{K}^4 \, (500 \text{ K})^4$$

$$- \frac{1 - 0.4}{0.4 \times 1 \text{ m}} (-37,000 \text{ W/m})$$

$$J_2 = 59,043 \text{ W/m}^2$$

From the energy balance for the reradiating surface, Equation 13.31,

$$\frac{J_1 - J_R}{(1/A_1 F_{1R})} - \frac{J_R - J_2}{(1/A_2 F_{2R})} = 0$$

it follows that

$$\frac{108,323 - J_R}{\dfrac{1}{W \times L \times 0.5}} - \frac{J_R - 59,043}{\dfrac{1}{W \times L \times 0.5}} = 0$$

or

$$2J_R = 167,366 \text{ W/m}^2$$

Hence

$$J_R = E_{bR} = \sigma T_R^4 = 83{,}683 \text{ W/m}^2$$

$$T_R = \left(\frac{83{,}683 \text{ W/m}^2}{5.67 \times 10^{-8} \text{ W/m}^2 \cdot \text{K}^4} \right)^{1/4}$$

$$T_R = 1102 \text{ K} \qquad \triangleleft$$

COMMENTS:

1. Note that the temperature and radiosity discontinuities cannot exist at the corners, and the assumptions of uniform temperature and radiosity are weakest in these regions.

2. The above results are independent of the value of ε_R.

3. This problem may also be solved using the matrix inversion method. The solution would involve first determining the three unknown radiosities, J_1, J_2, and J_R. The governing equations would be obtained by writing Equation 13.21 for each of the two surfaces of known temperature, 1 and 2, and Equation 13.22 for surface R, whose heat rate is known ($q_R = 0$). The three equations are of the form

$$\frac{E_{b1} - J_1}{(1 - \varepsilon_1)/\varepsilon_1 A_1} = \frac{J_1 - J_2}{(A_1 F_{12})^{-1}} + \frac{J_1 - J_R}{(A_1 F_{1R})^{-1}}$$

$$\frac{E_{b2} - J_2}{(1 - \varepsilon_2)/\varepsilon_2 A_2} = \frac{J_2 - J_1}{(A_2 F_{21})^{-1}} + \frac{J_2 - J_R}{(A_2 F_{2R})^{-1}}$$

$$0 = \frac{J_R - J_1}{(A_R F_{R1})^{-1}} + \frac{J_R - J_2}{(A_R F_{R2})^{-1}}$$

Canceling the area A_1, the first of these equations reduces to

$$\frac{117{,}573 - J_1}{0.25} = \frac{J_1 - J_2}{2} + \frac{J_1 - J_R}{2}$$

or

$$940{,}584 - 8J_1 = J_1 - J_2 + J_1 - J_R$$

Hence

$$10J_1 - J_2 - J_R = 940{,}584 \qquad (1)$$

Similarly, for surface 2,

$$\frac{3544 - J_2}{1.50} = \frac{J_2 - J_1}{2} + \frac{J_2 - J_R}{2}$$

or

$$-J_1 + 3.33J_2 - J_R = 4725 \qquad (2)$$

and for the reradiating surface

$$0 = \frac{J_R - J_1}{2} + \frac{J_R - J_2}{2}$$

or

$$-J_1 - J_2 + 2J_R = 0 \qquad (3)$$

From Equations 1, 2, and 3 the coefficient matrices are the form

$$\begin{vmatrix} a_{11} & a_{12} & a_{1R} \\ a_{21} & a_{22} & a_{2R} \\ a_{R1} & a_{R2} & a_{RR} \end{vmatrix} = \begin{vmatrix} 10 & -1 & -1 \\ -1 & 3.33 & -1 \\ -1 & -1 & +2 \end{vmatrix}$$

$$\begin{vmatrix} C_1 \\ C_2 \\ C_R \end{vmatrix} = \begin{vmatrix} 940{,}584 \\ 4725 \\ 0 \end{vmatrix}$$

Using a calculator with the capability to solve a system of simultaneous equations by matrix inversion, the results are

$$J_1 = 108{,}328 \ \text{W/m}^2 \qquad J_2 = 59{,}018 \ \text{W/m}^2 \qquad \text{and} \qquad J_R = 83{,}673 \ \text{W/m}^2$$

Recognizing that

$$J_R = \sigma T_R^4$$

it follows that

$$T_R = \left(\frac{J_R}{\sigma}\right)^{1/4} = \left(\frac{83{,}673 \ \text{W/m}^2}{5.67 \times 10^{-8} \ \text{W/m}^2 \cdot \text{K}^4}\right)^{1/4} = 1102 \ \text{K}$$

As expected, these results agree with those obtained from Equation 13.30 and appropriate radiation balances.

13.4   ADDITIONAL EFFECTS

Although we have developed means for predicting radiation exchange between surfaces, it is important to be cognizant of the inherent limitations. Recall that we have considered *isothermal, opaque, gray* surfaces that *emit* and *reflect diffusely* and that are characterized by *uniform* surface *radiosity* and *irradiation*. For enclosures we have also considered the medium that separates the surfaces to be *nonparticipating*; that is, it neither absorbs nor scatters the surface radiation, and it emits no radiation.

The foregoing conditions and the related equations may often be used to obtain reliable first estimates and, in some cases, highly accurate results for radiation transfer in an enclosure. However, situations do arise for which the assumptions are grossly inappropriate and more refined prediction methods are needed to obtain reliable results. Although beyond the scope of this text, the methods are discussed in more advanced treatments of radiation transfer. Radiation between nongray, nondiffuse surfaces with a nonparticipating medium is discussed by Siegel and Howell [5] and Sparrow [7]. The effect of a participating gas (one which emits and absorbs) is treated by Dunkle [8] for diffuse surfaces and again by Dunkle [9] for specular surfaces.

In this text we have said little about gaseous radiation, having confined our attention principally to radiation exchange at the surface of an opaque solid or liquid. For many of the commonly occurring *nonpolar* gases, such as O_2 or N_2, such neglect is justified, since these gases do not emit radiation and are essentially transparent to incident thermal radiation. However, the same may not be said for polar molecules, such as CO_2, H_2O (vapor), NH_3, and hydrocarbon gases, which emit and absorb over a wide temperature range. For such gases matters are complicated by the fact that unlike radiation from a solid or a liquid, which is distributed continuously with wavelength, gaseous radiation is concentrated in specific *wavelength intervals* (called bands). Moreover, gaseous radiation is not a surface phenomenon, but is instead a *volumetric* phenomenon.

13.4.1 Volumetric Absorption

Spectral radiation absorption in a gas (or in a semitransparent liquid or solid) is a function of the absorption coefficient, κ_λ (1/m), and the thickness, L, of the medium (Figure 13.15). If a monochromatic beam of intensity $I_{\lambda,0}$ is incident on

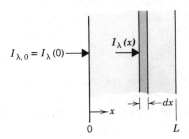

$I_{\lambda,0} = I_\lambda(0)$

$I_\lambda(x)$

dx

x

0 L **Figure 13.15** Absorption in a gas or liquid layer.

the medium, the intensity is reduced due to absorption, and the reduction occurring in an infinitesimal layer of thickness dx may be expressed as

$$dI_\lambda(x) = -\kappa_\lambda I_\lambda(x)dx \qquad (13.32)$$

Separating variables and integrating over the entire layer, we obtain

$$\int_{I_{\lambda,0}}^{I_{\lambda,L}} \frac{dI_\lambda(x)}{I_\lambda(x)} = -\kappa_\lambda \int_0^L dx$$

where κ_λ is assumed to be independent of x. It follows that

$$\boxed{\frac{I_{\lambda,L}}{I_{\lambda,0}} = e^{-\kappa_\lambda L}} \qquad (13.33)$$

This exponential decay is termed *Beer's law*, and it is a useful tool in radiation analysis.

The foregoing result may be used to infer the spectral absorptivity of the medium. In particular with its transmissivity defined as

$$\tau_\lambda = \frac{I_{\lambda,L}}{I_{\lambda,0}} = e^{-\kappa_\lambda L} \qquad (13.34)$$

it follows that the absorptivity is

$$\alpha_\lambda = 1 - \tau_\lambda = 1 - e^{-\kappa_\lambda L} \qquad (13.35)$$

If Kirchhoff's law is assumed to be valid, $\alpha_\lambda = \varepsilon_\lambda$, and Equation 13.35 also provides the spectral emissivity of the medium.

13.4.2 Gaseous Emission and Absorption

A common engineering calculation is one that requires determination of the radiant heat flux from a gas to an adjoining surface. Despite the complicated spectral and directional effects inherent in such calculations, a simplified procedure may be used. The method was developed by Hottel [1] and involves determining radiation emission from a hemispherical gas mass of temperature T_g to a surface element, dA_1, which is located at the center of the hemisphere's base. Emission from the gas per unit area of the surface is expressed as

$$E_g = \varepsilon_g \sigma T_g^4 \qquad (13.36)$$

where the gas emissivity, ε_g, was determined by correlating available data. In particular ε_g was correlated in terms of the temperature T_g and total pressure p of the gas, the partial pressure p_g of the radiating species, and the radius L of the hemisphere.

Results for the emissivity of water vapor are plotted in Figure 13.16 as a function of the gas temperature, for a total pressure of one atmosphere and for different values of the product of the vapor partial pressure and the hemisphere radius. To evaluate the emissivity for total pressures other than one atmosphere, the emissivity from Figure 13.16 must be multiplied by the correction factor C_w from Figure 13.17. Similar results were obtained for carbon dioxide and are presented in Figures 13.18 and 13.19.

The foregoing results apply when water vapor or carbon dioxide appear separately in a mixture with other species that are nonradiating. However, the results may readily be extended to situations in which water vapor and carbon dioxide appear together in a mixture with other nonradiating gases. In particular the total gas emissivity may be expressed as

$$\varepsilon_g = \varepsilon_w + \varepsilon_c - \Delta\varepsilon \tag{13.37}$$

where the correction factor $\Delta\varepsilon$ is presented in Figure 13.20 for different values of the gas temperature. This factor accounts for the reduction in emission associated with the mutual absorption of radiation between the two species.

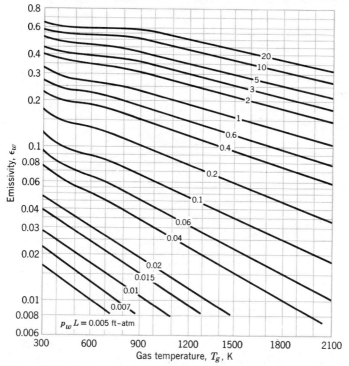

Figure 13.16 Emissivity of water vapor in a mixture with nonradiating gases at 1 atm total pressure and of hemispherical shape [1]. Used with permission.

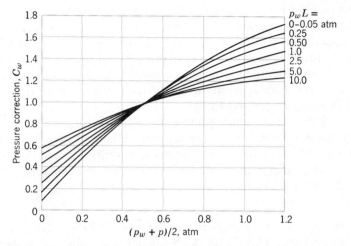

Figure 13.17 Correction factor for obtaining water vapor emissivities at pressures other than 1 atm ($\varepsilon_{w,\,p\neq1\text{ atm}} = C_w\varepsilon_{w,\,p=1\text{ atm}}$)[1]. Used with permission.

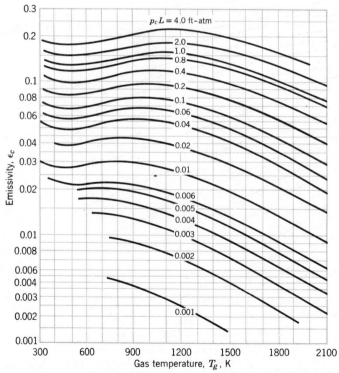

Figure 13.18 Emissivity of carbon dioxide in a mixture with non-radiating gases at 1 atm total pressure and of hemispherical shape [1]. Used with permission.

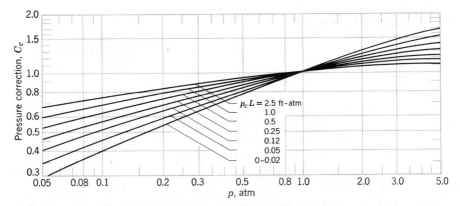

Figure 13.19 Correction factor for obtaining carbon dioxide emissivities at pressures other than 1 atm ($\varepsilon_{c,\,p \neq 1 \text{ atm}} = C_c\varepsilon_{c,\,p=1 \text{ atm}}$)[1]. Used with permission.

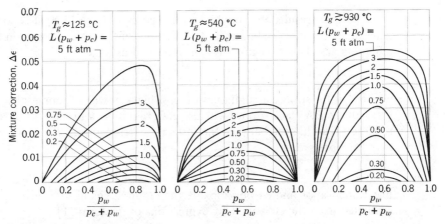

Figure 13.20 Correction factor associated with mixtures of water vapor and carbon dioxide [1]. Used with permission.

Recall that the foregoing results provide the emissivity of a hemispherical gas mass of radius L radiating to an element of area at the center of its base. However, the results may be extended to other gas geometries by introducing the concept of a *mean beam length*, L_e. This quantity may be interpreted as the radius of a hemispherical gas mass whose emissivity is equivalent to that for the geometry of interest. Its value has been determined for numerous gas shapes [1], and representative results are listed in Table 13.2. Replacing L by L_e in Figures 13.16 to 13.20, the emissivity associated with the geometry of interest may then be determined.

Table 13.2 Mean beam lengths, L_e, for various gas geometries

GEOMETRY	CHARACTERISTIC LENGTH	L_e
Sphere (radiation to surface)	Diameter (D)	$0.65D$
Infinite circular cylinder (radiation to curved surface)	Diameter (D)	$0.95D$
Semiinfinite circular cylinder (radiation to base)	Diameter (D)	$0.65D$
Circular cylinder of equal weight and diameter (radiation to entire surface)	Diameter (D)	$0.60D$
Infinite parallel planes (radiation to planes)	Spacing between planes (L)	$1.80L$
Cube (radiation to any surface)	Side (L)	$0.66L$
Arbitrary shape of volume V, (radiation to surface of Area A)	Volume to area ratio (V/A)	$3.6V/A$

Using the results of Table 13.2 with Figures 13.16 to 13.20, it is possible to determine the rate of radiant heat transfer to a surface due to emission from an adjoining gas. This heat rate may be expressed as

$$q = \varepsilon_g A_s \sigma T_g^4 \tag{13.38}$$

where A_s is the surface area. If the surface is black, it will, of course, absorb all of this radiation. A black surface will also emit radiation, and the net rate at which radiation is exchanged between the surface at a temperature T_s and the gas at T_g may be expressed as

$$q_{\text{net}} = A_s \sigma (\varepsilon_g T_g^4 - \alpha_g T_s^4) \tag{13.39}$$

For water vapor and carbon dioxide the required gas absorptivity α_g may be evaluated from the emissivity by expressions of the form [1]

$$\text{Water:} \quad \alpha_w = C_w \left(\frac{T_g}{T_s}\right)^{0.45} \times \varepsilon_w \left(T_s, p_w L_e \frac{T_s}{T_g}\right) \tag{13.40}$$

$$\text{Carbon Dioxide:} \quad \alpha_c = C_c \left(\frac{T_g}{T_s}\right)^{0.45} \times \varepsilon_c \left(T_s, p_c L_e \frac{T_s}{T_g}\right) \tag{13.41}$$

where ε_w and ε_c are evaluated from Figures 13.16 and 13.18, respectively, and C_w and C_c are evaluated from Figures 13.17 and 13.19, respectively. Note, however, that in using Figures 13.16 and 13.18, T_g is replaced by T_s and $p_w L_e$ or $p_c L_e$ is replaced by $p_w L_e (T_s/T_g)$ or $p_c L_e (T_s/T_g)$, respectively. Note also that, in the presence of both water vapor and carbon dioxide, the total gas absorptivity may be expressed as

$$\alpha_g = \alpha_w + \alpha_c - \Delta\alpha \qquad\qquad (13.42)$$

where $\Delta\alpha = \Delta\varepsilon$ is obtained from Figure 13.20.

13.5 SUMMARY

In this chapter we focused on the analysis of radiation exchange between the surfaces of an enclosure, and in treating this exchange we introduced the notion of a *view factor*. Since knowledge of this geometrical quantity is essential to determining radiation exchange between any two diffuse surfaces, you should be familiar with the means by which it may be determined. You should also be adept at performing radiation calculations for an enclosure of *isothermal*, *opaque*, *diffuse*, and *gray* surfaces of *uniform radiosity* and *irradiation*. Moreover, you should be familiar with the results that apply to simple cases such as the two-surface enclosure or the three-surface enclosure with a reradiating surface.

REFERENCES

1. Hottel, H. C., "Radiant-Heat Transmission" in W. H. McAdams, *Heat Transmission*, 3rd Ed., McGraw-Hill, New York, 1954.

2. Hamilton, D. C. and W. R. Morgan, "Radiant Interchange Configuration Factors," NACA TN 2836, 1952.

3. Leuenberger, H. and R. A. Pearson, "Compilation of Radiant Shape Factors for Cylindrical Assemblies," ASME Paper 56-A-144, ASME Annual Meeting, New York, 1956.

4. Eckert, E. R. G., "Radiation: Relations and Properties" in W. M. Rohsenow and J. P. Hartnett, Eds., *Handbook of Heat Transfer*, McGraw-Hill, New York, 1973.

5. Siegel, R. and J. R. Howell, *Thermal Radiation Heat Transfer*, McGraw-Hill, New York, 1972.

6. Oppenheim, A. K., *Trans. ASME*, 65, 725, 1956.

7. Sparrow, E. M., "Radiant Interchange between Surfaces Separated by Nonabsorbing and Nonemitting Media" in W. M. Rohsenow and J. P. Hartnett, Eds., *Handbook of Heat Transfer*, McGraw-Hill, New York, 1973.

8. Dunkle, R. V., "Radiation Exchange in an Enclosure with a Participating Gas" in W. M. Rohsenow and J. P. Hartnett, Eds., *Handbook of Heat Transfer*, McGraw-Hill, New York, 1973.

9. Dunkle, R. V., "Radiant Interchange in an Enclosure with Specular Surfaces and Enclosures with Windows or Diathermous Walls" in C. H. A. Johnson, Ed., *Heat Transfer, Thermodynamics, and Education*, Boelter Anniversary Volume, McGraw-Hill, New York, 1964, p. 133.

PROBLEMS

13.1   Determine F_{12} and F_{21} for the following configurations using the reciprocity theorem and other basic shape factor relations. Do not use tables or charts.

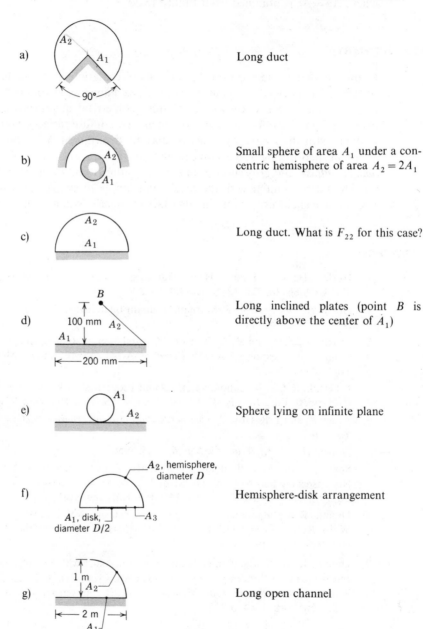

a)   Long duct

b)   Small sphere of area A_1 under a concentric hemisphere of area $A_2 = 2A_1$

c)   Long duct. What is F_{22} for this case?

d)   Long inclined plates (point B is directly above the center of A_1)

e)   Sphere lying on infinite plane

f)   Hemisphere-disk arrangement

g)   Long open channel

13.2 Consider the following grooves, each of width W, that have been machined from a solid block of material.

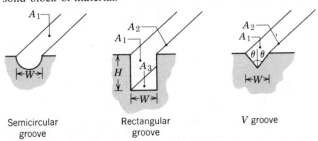

| Semicircular groove | Rectangular groove | V groove |

a) For each case obtain an expression for the view factor of the groove with respect to the surroundings outside the groove.

b) For the V groove, obtain an expression for the view factor F_{12}, where A_1 and A_2 are opposite surfaces.

c) If $H = 2W$ in the rectangular groove, what is the view factor F_{12}?

13.3 Consider an enclosure that is in the shape of a rectangular parallelepiped.

a) How many view factors are involved and how many of these view factors need be determined directly?

b) Acknowledging symmetry and the fact that all surfaces are planar, how many of the view factors need be determined directly?

c) If the enclosure is cubical, how many view factors need be determined directly?

13.4 Consider the two diffuse surfaces shown in Example 13.1. Beginning with the general definition of $F_{ij} = q''_{i \to j}/J_i$, show that when $L \gg D$, the shape factor for a small area element (i) to a disk (j) is $F_{ij} = D^2/4L^2$. How does this compare with the result of Example 13.1?

13.5 Determine the shape factor F_{12} for the perpendicular rectangles shown.

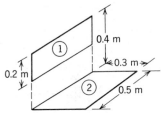

13.6 Determine the shape factor F_{12} for the arrangement of perpendicular rectangles shown below.

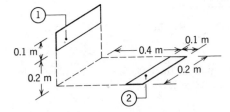

13.7   Determine the shape factor, F_{12}, for the rectangles shown in the figures below.
a) Perpendicular rectangles without a common edge.

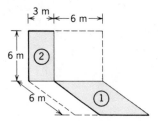

b) Parallel rectangles of unequal areas.

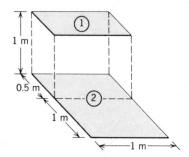

13.8   Determine the radiant power from the tubular heater that is incident on the disk. The inner surface of the heater is black and at a uniform temperature of 1000 K.

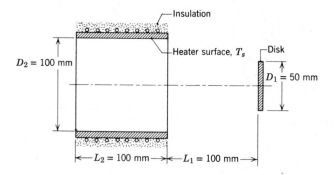

13.9   Consider the arrangement of the three black surfaces as shown in the sketch, where A_1 is small compared to A_2 or A_3.

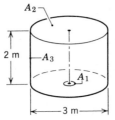

a) Determine the value of F_{13}.

b) Calculate the net radiation heat transfer from A_1 to A_3 if $A_1 = 0.05$ m^2, $T_1 = 1000$ K and $T_2 = 500$ K.

13.10 Two perfectly black, parallel disks 1 m in diameter are separated a distance of 0.25 m. One disk is maintained at 60°C and the other at 20°C. The disks are located in a large room whose walls are 40°C. Assume that the exterior surfaces of the disks (the surfaces that do not face each other) are very well insulated.

a) Determine the net radiation exchange between the disks.

b) Determine the net radiation exchange between the disks and the room.

13.11 A drying oven consists of a long duct of semicircular cross-section and diameter $D = 1$ m as shown below.

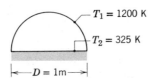

Materials to be dried cover the base of the oven, while the walls of the oven are maintained at an elevated temperature of $T_1 = 1200$ K. What is the drying rate per unit length of the oven (kg/s·m), if a water coated layer of material maintains a temperature of 325 K during the drying process? Blackbody behavior may be assumed for the water surface and for the oven wall.

13.12 A circular disk of diameter $D_1 = 20$ mm is located at the base of an enclosure that has a cylindrical sidewall and a hemispherical dome. The enclosure is of diameter $D = 0.5$ m, and the height of the cylindrical section is $L = 0.3$ m. The disk and the enclosure surface are black and at temperatures of $T_1 = 1000$ K and $T = 300$ K, respectively. (Figure appears on next page.)

a) What is the net rate of radiation exchange between the disk and the hemispherical dome?

b) What is the net rate of radiation exchange between the disk and the top one-third portion of the cylindrical section?

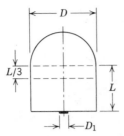

13.13 Consider the cylindrical cavity of diameter D and length L having the lateral (A_1) and bottom (A_2) surfaces maintained at temperatures T_1 and T_2, respectively. Develop an expression for the emissive power of the cavity opening A_3 in terms of T_1, T_2, and the shape factor F_{13}.

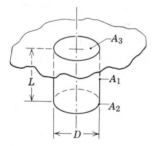

13.14 A cylindrical cavity of diameter D and depth L is machined in a metal block, and conditions are such that the base and side surfaces of the cavity are maintained at temperatures of $T_1 = 1000$ K and $T_2 = 700$ K, respectively. Approximating the surfaces as black, determine the emissive power of the cavity if $L = 20$ mm and $D = 10$ mm. The emissive power is defined as the rate at which radiation leaves the cavity opening per unit area of the opening.

13.15 Two parallel plates 1 m × 1 m in size, insulated on their backsides and separated by 1 m, may be approximated as blackbodies at temperatures of 500 K and 750 K. The plates are located in a large room whose walls are maintained at 300 K. Determine the net radiative heat transfer from each plate and the net radiative heat transfer to the room walls.

13.16 The arrangement shown below is to be used to calibrate a heat flux gage. The gage has a black surface that is 10 mm in diameter and maintained at 17°C by means of a water-cooled backing plate. The heater, 200 mm in diameter, has a black surface maintained at 800 K and is located 0.5 m from the gage. The surroundings and the air are at 27°C and a convection heat transfer coefficient between the gage and the air is 15 W/m²·K.

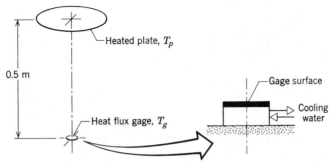

a) Determine the net radiation exchange between the heater and the gage.

b) Determine the net transfer of radiation to the gage per unit area of the gage.

c) What is the net heat transfer rate to the gage per unit area of the gage?

d) If the gage is constructed according to the description of Problem 3.28, what heat flux will it indicate?

13.17 A manufacturing process calls for heating long copper rods, which are coated with a thin film having $\varepsilon = 1$, by placing them in a large evacuated oven whose surface is maintained at a temperature of 1650 K. If the rods are of 10 mm diameter and are placed in the oven with an initial temperature of 300 K, what is the initial rate of change of the rod temperature?

13.18 Consider the very long, inclined black surfaces (A_1, A_2) oriented as shown in the following sketch and maintained at uniform temperatures of $T_1 = 1000$ K and $T_2 = 800$ K.

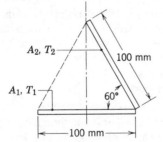

a) Determine the net radiation exchange between the surfaces per unit length of the surfaces.

b) Consider the configuration when a black surface (A_3), whose backside is insulated, is positioned along the dashed line shown in the figure. Calculate the net radiation transfer to surface A_2 per unit length of the surface and determine the temperature of the insulated surface A_3.

13.19 Two plane coaxial disks, as shown below, are separated by a distance $L = 0.20$ m. The lower disk (A_1) is solid with an outer diameter $D_o = 0.80$ m and a temperature $T_1 = 300$ K. The upper disk (A_2), at temperature $T_2 = 1000$ K, has the same outer diameter but is ring shaped with an inner diameter $D_i = 0.40$ m. Assuming the disks to be blackbodies, calculate the net radiative heat exchange between them.

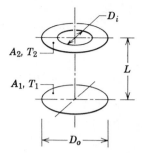

13.20 A circular ice rink 25 m in diameter is enclosed by a hemispherical dome 35 m in diameter. If the ice and dome surfaces may be approximated as blackbodies and are at temperatures of 0°C and 15°C, respectively, what is the net rate of radiative transfer from the dome to the rink?

13.21 A round tube with a diameter of 0.75 m and a length of 0.33 m has an electrical heater wrapped around the outside, and a heavy layer of insulation is wrapped over the tube-heater combination. The tube is open at both ends and is suspended in a very large vacuum chamber whose walls are at an ambient temperature 27°C. The inside surface of the tube is black and is maintained at a steady uniform temperature of 127°C. Determine the electrical power that is supplied to the heater.

13.22 A flat-bottomed hole 6 mm in diameter is drilled to a depth of 24 mm in a diffuse-gray material having an emissivity of 0.8 and a uniform temperature of 1000 K.

a) Determine the radiant power leaving the opening of the cavity.

b) The effective emissivity, ε_e, of a cavity is defined as the ratio of the radiant power leaving the cavity to that from a blackbody having the area of the cavity opening and a temperature of the inner surfaces of the cavity. Calculate the effective emissivity of the cavity described above.

c) If the depth of the hole were increased, would ε_e increase or decrease? What is the limit of ε_e as the depth increases?

13.23 Consider a cylindrical cavity 100 mm in diameter and 50 mm in depth whose walls are diffuse gray with an emissivity of 0.6 and are of uniform temperature at 1500 K. Assuming the surroundings of the cavity opening are very large and at 300 K, calculate the net radiation transfer from the cavity.

13.24 A long, thin-walled horizontal tube 100 mm in diameter is maintained at 120°C by the passage of steam through its interior. A radiation shield is installed around the tube, providing an air gap of 10 mm between the tube and the shield, and reaches a surface temperature of 35°C. The tube and shield are diffuse-gray surfaces with emissivities of 0.80 and 0.10, respectively. What is the radiant heat transfer from the tube per unit length?

13.25 A very long electrical conductor 10 mm in diameter is concentric with a cooled cylindrical tube 50 mm in diameter whose surface is diffuse gray with an emissivity of 0.9 and temperature of 27°C. The electrical conductor has a diffuse-gray surface with an emissivity of 0.6 and is dissipating 6.0 watts per meter of length. Assuming the space between the two surfaces is evacuated, calculate the surface temperature of the conductor.

13.26 Under steady-state operation a 50-W incandescent light bulb has a surface temperature of 135°C when the room air is at a temperature of 25°C. If the bulb may be approximated as a 60-mm diameter sphere with a diffuse-gray surface of emissivity 0.8, what is the radiant heat transfer from the bulb surface to its surroundings?

13.27 An arrangement for direct thermophotovoltaic conversion of thermal energy to electrical power is shown in the sketch. The inner cylinder of diameter $D_i = 25$ mm is heated internally by a combustion process that brings the ceramic cylinder ($\varepsilon_i = 0.9$) to a surface temperature of $T_i = 1675°C$. The outer cylinder of diameter $D_o = 0.38$ m consists of a semiconductor material ($\varepsilon_o = 0.5$) which converts absorbed incident irradiation to electrical current; the backing material for the semiconductor is a highly conducting metal that is water cooled to 20°C. The converter is assumed to be very long compared to the outer diameter. The space between the two concentric cylinders is evacuated so there is no convection.

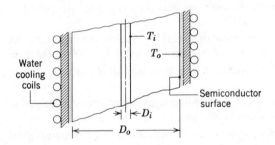

a) Assuming diffuse-gray surfaces, determine the heat transfer rate per unit area of the outer cylinder.

b) The electrical output of the semiconductor surface is 10 percent of the absorbed irradiation in the wavelength region 0.6 to 2.0 μm. For the above conditions, determine the power generation for this converter in watts per unit outer surface area.

13.28 Liquid oxygen is stored in a thin-walled, spherical container 0.8 m in diameter, which is enclosed within a second thin-walled, spherical container 1.2 m in diameter. The opaque, diffuse-gray container surfaces have an emissivity of 0.05 and are separated by an evacuated space. If the outer surface is at 280 K and the inner surface is at 95 K, what is the mass rate of oxygen lost due to evaporation? (The latent heat of vaporization of oxygen is 2.13×10^5 J/kg.)

13.29 The end of a cylindrical liquid cryogenic propellant tank in free space is to be protected from external (solar) radiation by placing a thin metallic shield in front of the tank as illustrated in the sketch below. Assume the view factor, F_{ts}, between the tank and the shield is unity; all surfaces are diffuse and gray with emissivities as indicated; and the space surroundings are black at 0 K.

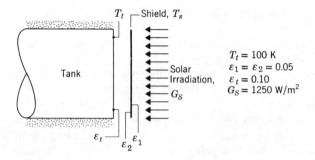

$T_t = 100$ K
$\varepsilon_1 = \varepsilon_2 = 0.05$
$\varepsilon_t = 0.10$
$G_S = 1250$ W/m^2

a) Find the temperature of the shield, T_s, for the conditions given.

b) Determine the heat flux (W/m^2) to the end of the tank.

13.30 Consider two large, diffuse-gray, parallel surfaces separated by a small distance. If the surface emissivities are 0.8, what emissivity should a thin radiation shield have in order to reduce the radiation heat transfer rate between the two surfaces by a factor of 10?

13.31 Two concentric spheres of diameters $D_1 = 0.8$ m and $D_2 = 1.2$ m are separated by an airspace and have surface temperatures of $T_1 = 400$ K and $T_2 = 300$ K.

a) If their surfaces are black, what is the net rate of radiation exchange between the spheres?

b) What is the net rate of radiation exchange between the surfaces, if they are diffuse gray with $\varepsilon_1 = 0.5$ and $\varepsilon_2 = 0.05$?

c) What is the net rate of radiation exchange if D_2 is increased to 20 m, with $\varepsilon_2 = 0.05$, $\varepsilon_1 = 0.5$ and $D_1 = 0.8$ m? What error would be introduced by assuming blackbody behavior for the outer surface ($\varepsilon_2 = 1$), with all other conditions remaining the same?

13.32 A cryogenic fluid flows through a tube 20 mm in diameter, the outer surface of which is diffuse gray with an emissivity of 0.02 and temperature of 77 K. This tube is concentric with a larger tube of 50 mm diameter, the inner surface of which is diffuse gray with an emissivity of 0.05 and temperature of 300 K. The space between the surface is evacuated.

a) Determine the heat gain by the cryogenic fluid per unit length of the inner tube (W/m).

b) If a thin-walled radiation shield that is diffuse gray with an emissivity of 0.02 (both sides) is inserted midway between the inner and outer surfaces, calculate the change (percent) in heat gain per unit length of the inner tube.

13.33 At the bottom of a very large vacuum chamber whose walls are at 300 K, a black panel 0.1 m in diameter is maintained at 77 K. To reduce the heat gain to this panel, a radiation shield, of the same diameter D, and with an emissivity of 0.05 is placed very close to the panel. Calculate the net heat gain to the panel (shown on the next page).

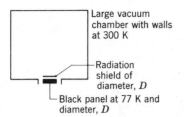

Large vacuum
chamber with walls
at 300 K

Radiation
shield of
diameter, D

Black panel at 77 K and
diameter, D

13.34 Heat transfer by radiation occurs between two large parallel plates, which are maintained at temperatures $T_1 > T_2$. To reduce the rate of heat transfer between the plates, it is proposed that they be separated by a thin shield which is known to have different emissivities on opposite surfaces. In particular one surface is known to have the emissivity $\varepsilon_s < 0.5$, while the opposite surface has an emissivity of $2\varepsilon_s$.

   a) How should the shield be oriented to provide the largest reduction in heat transfer between the plates? That is, should the surface of emissivity ε_s or that of emissivity $2\varepsilon_s$ be oriented towards the plate at T_1?

   b) What orientation will result in the larger value of the shield temperature, T_s?

13.35 A large slab of meat 0.15 m × 0.30 m in area is taken from a refrigerator at 4°C and is placed on a wire grill 0.15 m above and parallel to a bed of hot coals at 850°C with approximately the same area. Assume that the meat and the coals are essentially black and neglect convection.

   a) What is the initial heat transfer rate between the coals and the meat?

   b) By what percentage would this rate increase if well-insulated side walls were placed around the perimeter of the system?

   c) What average temperature would the insulated wall assume if steady-state conditions exist with the temperatures of the meat and coals at 4°C and 850°C, respectively.

13.36 A long cylindrical heater element of diameter $D = 10$ mm, temperature $T_1 = 1500$ K, and emissivity $\varepsilon_1 = 1$ is used in the furnace shown as follows. The bottom area, A_2, is a diffuse-gray surface with $\varepsilon_2 = 0.6$ and is maintained at $T_2 = 500$ K. The side and top walls are fabricated from an insulating, refractory brick that is diffuse gray with $\varepsilon = 0.9$. The length of the furnace normal to the page, l, is very large compared to the width, w, and height, h.

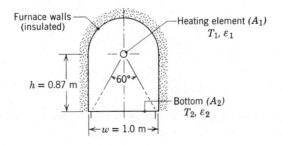

Furnace walls (insulated)

Heating element (A_1)
T_1, ε_1

$h = 0.87$ m

60°

Bottom (A_2)
T_2, ε_2

$w = 1.0$ m

a) Neglecting convection and treating the furnace walls as an isothermal area, determine the power per unit length (W/m) that must be provided to the heating element in order to maintain steady-state conditions.

b) Calculate the temperature of the furnace wall area.

13.37 Consider two aligned, parallel, square planes (0.4 m × 0.4 m) spaced 0.8 m apart and maintained at $T_1 = 500$ K and $T_2 = 800$ K. Calculate the net radiative heat transfer *from surface 1* for the following special conditions.

a) Both planes are black.

b) Both planes are black with connecting, reradiating walls.

c) Both planes are diffuse gray with $\varepsilon_1 = 0.6$, $\varepsilon_2 = 0.8$, and radiation-free surroundings at 0 K.

d) Both planes are diffuse gray ($\varepsilon_1 = 0.6$ and $\varepsilon_2 = 0.8$) with connecting, reradiating walls.

13.38 Two parallel, diffuse-gray surfaces, each 1 m × 2 m, face each other and are separated by a distance of 1 m. Each surface has an emissivity of 0.2. One surface has a temperature of 27°C and the other has a temperature of 277°C.

a) Calculate the net radiation transfer from the hotter surface with radiation-free surroundings at 0 K.

b) If well-insulated, diffuse-gray side walls are added to this configuration, what will be the net radiation heat transfer rate from the hotter surface?

c) Review the assumptions you made in performing the calculations and briefly explain what effect they could have on the accuracy of your results.

13.39 Consider a room that is 4 m long by 3 m wide with a floor-to-ceiling distance of 2.5 m. The four walls of the room are well insulated, while the surface of the floor is maintained at a uniform temperature of 30°C by means of electric resistance heaters. Heat loss occurs through the ceiling, which has a surface temperature of 12°C. If all surfaces have an emissivity of 0.9, what is the rate of heat loss by radiation from the room?

13.40 Two parallel, aligned disks, 0.4 m in diameter and separated by 0.1 m, are located in a large room whose walls are maintained at 300 K. One of the disks is maintained at a uniform temperature of 500 K with an emissivity 0.6 while the backside of the second disk is well insulated. If the disks are diffuse-gray surfaces, determine the temperature of the insulated disk.

13.41 Consider the very long duct constructed with walls that are diffuse gray and of 1 m width.

$A_1, T_1 = 1000\text{K}, \varepsilon_1 = 0.33$ $A_2, T_2 = 700$ K, $\varepsilon_2 = 0.5$

$A_3, T_3, \varepsilon_3 = 0.8$

a) Determine the net radiation transfer from surface A_1 per unit length of the duct.

b) Determine the temperature of the insulated surface A_3.

c) Comment on what effect changing the value of ε_3 would have on your result. Also, after considering your assumptions, comment on whether you expect your results to be exact.

13.42 A solar collector consists of a long duct through which air is blown and whose cross section forms an equilateral triangle 1 m in length on a side. One side of the collector consists of a glass cover of emissivity $\varepsilon_1 = 0.9$, while the other two sides are absorber plates of emissivity $\varepsilon_2 = \varepsilon_3 = 1.0$.

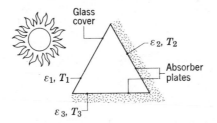

During operation the surface temperatures are known to be $T_1 = 25°C$, $T_2 = 60°C$, and $T_3 = 70°C$. What is the net rate at which radiation is transferred to the cover due to exchange with the absorber plates?

13.43 Consider a circular furnace that is 0.3 m in length and 0.3 m in diameter. The two ends have diffuse-gray surfaces that are maintained at 400 K and 500 K with emissivities of 0.4 and 0.5, respectively. The lateral surface is also diffuse gray with an emissivity of 0.8 and temperature of 800 K. Determine the net radiative heat transfer from each of the surfaces.

13.44 Consider the cylindrical cavity closed at the bottom with an opening in the top surface as shown in cross-section as follows.

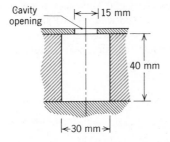

For the conditions identified below, calculate the rate at which radiation passes through the cavity opening. Also determine the effective emissivity of the cavity as defined in Problem 13.22.

a) All interior surfaces are black at 600 K.

b) The bottom surface of the cavity is diffuse gray with an emissivity of 0.6 while all other interior surfaces are reradiating. All surfaces are at 600 K.

c) All the interior surfaces are diffuse gray with an emissivity of 0.6 and uniform temperature of 600 K.

13.45 The enclosure analysis of diffuse-gray surfaces presented in Section 13.3 requires that the irradiation over any surfaces must be uniform. When the geometry of the enclosure precludes this requirement, it is necessary to divide the isothermal surface into zones over which the irradiation is approximately uniform. In the cylindrical cavity shown below, the irradiation along the lateral surface and on the base are likely to be different.

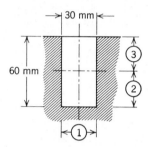

Dividing the cavity into three zones, each having the same surface temperature, perform an enclosure analysis to determine the radiative heat transfer leaving the cavity opening. The surfaces of the cavity are diffuse gray with an emissivity of 0.8 and with a uniform temperature of 1000 K. Compare your result to the calculation treating the cavity walls (base plus lateral side) as a single surface.

13.46 A room is represented by the enclosure shown as follows where the ceiling (1) has an emissivity of 0.8 and is maintained at 40°C by imbedded electrical heating elements. Heaters are also used to maintain the floor (2) with an emissivity of 0.9 at 50°C. The right wall (3) with an emissivity of 0.7 reaches a temperature of 15°C on a cold, winter day. The left wall (4) and end walls (5A, 5B) are very well insulated. To simplify the analysis, treat the two end walls as a single surface (5). Assuming all the surfaces are diffuse gray, find the net radiation heat transfer from each surface.

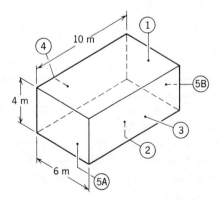

13.47 Consider the diffuse-gray, four-surface enclosure with all sides equal as shown in the following sketch. The temperatures of three surfaces are specified while the fourth surface is well insulated and can be treated as a reradiating surface.

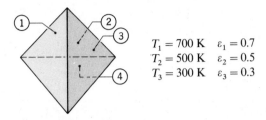

$$T_1 = 700 \text{ K} \quad \varepsilon_1 = 0.7$$
$$T_2 = 500 \text{ K} \quad \varepsilon_2 = 0.5$$
$$T_3 = 300 \text{ K} \quad \varepsilon_3 = 0.3$$

Determine the temperature of the reradiating surface (4).

13.48 A furnace having a spherical cavity of 0.5 m diameter contains a gas mixture at 1 atm and 1400 K. The mixture consists of CO_2 with a partial pressure of 0.25 atm and nitrogen with a partial pressure of 0.75 atm. If the cavity wall is black, what is the cooling rate needed to maintain its temperature at 500 K?

13.49 A gas turbine combustion chamber may be approximated as a long tube of 0.4 m diameter. The combustion gas is at a pressure and temperature of 1 atm and 1000°C, respectively, while the chamber surface temperature is 500°C. If the combustion gas contains carbon dioxide and water vapor, each with a mole fraction of 0.15, what is the net radiative heat flux between the gas and the chamber surface, which may be approximated as a blackbody?

13.50 A flue gas at 1 atm total pressure and a temperature of 1400 K contains CO_2 and water vapor at partial pressures of 0.05 atm and 0.10 atm, respectively. If the gas flows through a long flue of 1 m diameter and 400 K surface temperature, determine the net radiative heat flux from the gas to the surface. Blackbody behavior may be assumed for the surface.

13.51 A furnace consists of two large parallel plates separated by a distance of 0.75 m. A gas mixture comprised of O_2, N_2, CO_2, and H_2O (v), with mole fractions of 0.20, 0.50, 0.15 and 0.15, respectively, flows between the plates at a total pressure of 2 atm and a temperature 1300 K. If the plates may be approximated as blackbodies and are maintained at 500 K, what is the net radiative heat flux to the plates?

14 Multimode Heat and Mass Transfer

In Chapter 1 we introduced the different modes of heat and mass transfer and stressed the importance of being able to identify which modes are important in any given problem. In subsequent chapters, however, we isolated each mode in order to develop the computational tools needed for its treatment. In Chapters 2 through 5, for example, we focused on the means for determining temperatures and heat transfer rates in solids experiencing heat transfer by conduction. Although we recognized that convection heat transfer occurring at the surface of a solid strongly influences conditions within the solid, we were forced to presume that the convection coefficient was known. In Chapters 6 through 11 we considered means of determining the convection coefficient for a variety of flow conditions. However, in most cases we presumed the surface temperature of the adjoining solid to be known, when, in fact, this temperature could be strongly influenced by conduction within the solid. In Chapters 2 through 11 we also gave little attention to the important role that radiation exchange at a surface or between surfaces can have on thermal conditions. The situation was rectified in Chapters 12 and 13, but with little attention given to conduction and convection effects.

The time has come to recognize and to give serious consideration to the fact that most technically important problems involve *multimode heat transfer* effects. For example, conduction may occur within a solid, while convection and radiation occur at its boundaries. Also, radiation exchange may occur between two or more solid surfaces, while heat is transferred by convection at the surfaces and by conduction within the solids. To solve such problems, it may be necessary to simultaneously work with the heat equation, surface energy balances, overall energy balances and appropriate rate equations for conduction, convection and radiation heat transfer. Moreover, the problems may involve several unknowns and require the use of iterative solution methods.

In this chapter we consider several multimode problems that illustrate the manner in which previously developed tools may be integrated to obtain solutions. In the process we hope that you will begin to develop a state of mind that allows you to identify relevant transport processsesses and to implement appropriate rate equations and conservation principles.

EXAMPLE 14.1

A horizontal pipe of 0.17 m outside diameter is used to transport steam through a building in which the air and walls are at a temperature of 25°C. If the pipe is insulated with a layer of calcium silicate insulation 25 mm thick and the inner surface of the insulation is at 650 K, what is the heat loss per unit length of pipe?

SOLUTION

KNOWN:

Surface temperature and dimensions of an insulated steam pipe.

FIND:

Heat loss per unit length of pipe.

SCHEMATIC:

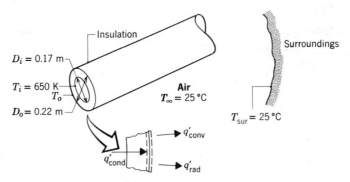

ASSUMPTIONS:

1. One-dimensional, steady-state conditions.
2. Constant properties.
3. Room air is quiescent.
4. Insulation outer surface is diffuse gray, and surroundings are large.
5. Negligible contact resistance between insulation and pipe wall.

PROPERTIES:

Assume $T_o \approx 350$ K.
Table A.3, calcium silicate $[(T_i + T_o)/2 \approx 500$ K]: $k_s = 0.072$ W/m·K.
Table A.4, air $[(T_o + T_\infty)/2 \approx 324$ K]: $k_f = 0.028$ W/m·K, $\alpha = 26.1 \times 10^{-6}$ m²/s, $\nu = 18.3 \times 10^{-6}$ m²/s, $\beta = 3.09 \times 10^{-3}$ K^{-1}, $Pr = 0.71$.
Table A.11, Assume the emissivity of calcium silicate to be that of standard building materials such as asbestos or gypsum: $\varepsilon \approx 0.9$.

ANALYSIS:

The heat loss may be obtained by first computing the outer surface temperature of the insulation. By applying an energy balance to this surface, it follows that

$$q'_{cond} = q'_{conv} + q'_{rad}$$

From Equations 3.26 and 13.27

$$q'_{cond} = \frac{2\pi k_s(T_i - T_o)}{\ln(D_o/D_i)}$$

$$q'_{rad} = \pi D_o \varepsilon \sigma (T_o^4 - T_{sur}^4)$$

The free convection heat transfer rate is

$$q'_{conv} = \bar{h}(\pi D_o)(T_o - T_\infty)$$

where $\bar{h}$ may be obtained from Equation 9.34.

$$\bar{h} = \frac{k_f}{D_o}\left\{ 0.60 + \frac{0.387 Ra_D^{1/6}}{[1 + (0.559/Pr)^{9/16}]^{8/27}} \right\}^2$$

where

$$Ra_D = \frac{g\beta(T_o - T_\infty)D_o^3}{\nu\alpha}$$

$$Ra_D = \frac{9.8 \text{ m/s}^2 \times 3.09 \times 10^{-3}\,\text{K}^{-1}\,(T_o - T_\infty)(0.22\text{ m})^3}{18.3 \times 10^{-6}\text{ m}^2/\text{s} \times 26.1 \times 10^{-6}\text{ m}^2/\text{s}}$$

$$= 6.75 \times 10^5\,(T_o - T_\infty)$$

Substituting into the energy balance, it follows that

$$\frac{2\pi k_s(T_i - T_o)}{\ln(D_o/D_i)} = \frac{k_f}{D_o}\left\{ 0.60 + \frac{0.387[6.75 \times 10^5(T_o - T_\infty)]^{1/6}}{[1 + (0.559/Pr)^{9/16}]^{8/27}} \right\}^2$$

$$\times \pi D_o(T_o - T_\infty) + \pi D_o \varepsilon \sigma (T_o^4 - T_{sur}^4)$$

Hence

$$\frac{2\pi(0.072 \text{ W/m·K})(650 \text{ K} - T_o)}{\ln(0.22 \text{ m}/0.17 \text{ m})}$$

$$= \frac{0.028 \text{ W/m·K}}{0.22 \text{ m}}\left\{ 0.60 + \frac{0.387[6.75 \times 10^5(T_o - 298 \text{ K})]^{1/6}}{[1 + (0.559/0.71)^{9/16}]^{8/27}} \right\}^2$$

$$\times (\pi \times 0.22 \text{ m})(T_o - 298 \text{ K}) + \pi(0.22 \text{ m})(0.9)5.67 \times 10^{-8} \text{ W/m}^2\text{·K}^4$$

$$\times [T_o^4 - (298 \text{ K})^4]$$

or

$$1.75(650 - T_o) = 0.127[0.60 + 3.01(T_o - 298)^{1/6}]^2(0.691)(T_o - 298)$$

$$+ 3.527 \times 10^{-8}(T_o^4 - 298^4)$$

The outer surface temperature of the insulation may be obtained from a trial-and-error solution, which gives

$T_o = 357$ K

Accordingly,

$$q' = q'_{cond} = \frac{2\pi k_s (T_i - T_o)}{\ln(D_o/D_i)} = \frac{2\pi(0.072 \text{ W/m·K})(650 - 357)\text{K}}{\ln(0.22 \text{ m}/0.17 \text{ m})}$$

or

$q' = 514$ W/m   ◁

COMMENTS:

1. An outside surface temperature of 357 K exceeds the maximum allowable for safety reasons (by approximately 20 K), and additional insulation should be used.

2. The assumption of $T_o = 350$ K is reasonable for evaluation of the properties.

3. The convective and radiative contributions to the heat loss are approximately 220 W/m and 294 W/m, respectively, for the above conditions. If a polished aluminum foil ($\varepsilon = 0.07$) is applied to the outer surface, the convective and radiative heat losses become 408 W/m and 45 W/m, respectively, with $q' = 453$ W/m and $T_o = 392$ K.

EXAMPLE 14.2

The rear window of an automobile is of thickness $L = 8$ mm and height $H = 0.5$ m and contains fine-meshed heating wires that can induce nearly volumetric heating, $\dot{q}$ (W/m³). Consider steady-state conditions for which the interior surface of the window is exposed to air at 10°C, while the exterior surface is exposed to air at -10°C and a very cold sky with an effective temperature of -33°C. The air moves in parallel flow over the surface with a velocity of 20 m/s. Determine the volumetric heating rate needed to maintain the interior window surface at $T_{s,i} = 15$°C.

SOLUTION

KNOWN:

Boundary conditions associated with a rear window experiencing uniform volumetric heating.

FIND:

Volumetric heating rate, $\dot{q}$, needed to maintain inner surface temperature at $T_{s,i} = 15°C$.

SCHEMATIC:

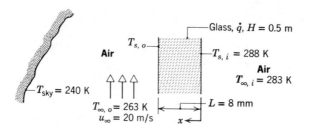

ASSUMPTIONS:

1. Steady-state conditions.
2. One-dimensional conduction in window.
3. Constant properties.
4. Uniform volumetric heating in window.
5. Convection heat transfer from interior surface of window to interior air may be approximated as free convection from a vertical plate. Radiation exchange with automobile interior is negligible.
6. Heat transfer from outer surface is due to forced convection over a flat plate and radiation exchange with the sky.
7. Negligible boundary layer separation on the outer surface.

PROPERTIES:

Table A.3, glass (300 K): $k = 1.4$ W/m·K.
Table A.11, glass (300 K): $\varepsilon \approx 0.90$.
Table A.4, air ($T_{f,i} = 12.5°C$): $\nu = 14.6 \times 10^{-6}$ m²/s, $k = 0.0251$ W/m·K, $\alpha = 20.59 \times 10^{-6}$ m²/s, $\beta = (1/285.5) = 3.503 \times 10^{-3}$ K⁻¹, $Pr = 0.711$. Air ($T_f \approx 0°C$): $\nu = 13.49 \times 10^{-6}$ m²/s, $k = 0.0241$ W/m·K, $Pr = 0.714$.

ANALYSIS:

1. The temperature distribution in the glass is governed by the appropriate form of the heat equation, Equation 3.38, whose general solution is given by Equation 3.39.

$$T(x) = -\frac{\dot{q}}{2k}x^2 + C_1 x + C_2$$

The constants of integration may be evaluated by applying appropriate boundary conditions at $x = 0$. In particular, with $T(0) = T_{s,i}$

$$C_2 = T_{s,i}$$

Applying an energy balance to the inner surface, it also follows that

$$q''_{cond} = q''_{conv,i}$$

or

$$-k\frac{dT}{dx}\Big|_{x=0} = \bar{h}_i(T_{\infty,i} - T_{s,i})$$

$$-k\left(-\frac{\dot{q}}{k}x + C_1\right)\Big|_{x=0} = \bar{h}_i(T_{\infty,i} - T_{s,i})$$

$$C_1 = -\frac{\bar{h}_i}{k}(T_{\infty,i} - T_{s,i})$$

Hence

$$T(x) = -\frac{\dot{q}}{2k}x^2 - \frac{\bar{h}_i(T_{\infty,i} - T_{s,i})}{k}x + T_{s,i} \tag{1}$$

The required generation may then be obtained by formulating an energy balance at the outer surface. That is, at $x = L$

$$q''_{cond} = q''_{conv,o} + q''_{rad}$$

or

$$-k\frac{dT}{dx}\Big|_{x=L} = \bar{h}_o(T_{s,o} - T_{\infty,o}) + \varepsilon\sigma(T_{s,o}^4 - T_{sky}^4) \tag{2}$$

From Equation 1 it follows that

$$-k\frac{dT}{dx}\Big|_{x=L} = -k\left(-\frac{\dot{q}L}{k}\right) + \bar{h}_i(T_{\infty,i} - T_{s,i})$$

$$-k\frac{dT}{dx}\Big|_{x=L} = \dot{q}L + \bar{h}_i(T_{\infty,i} - T_{s,i}) \tag{3}$$

Substituting Equation 3 into Equation 2, the energy balance becomes

$$\dot{q}L = \bar{h}_o(T_{s,o} - T_{\infty,o}) + \bar{h}_i(T_{s,i} - T_{\infty,i}) + \varepsilon\sigma(T_{s,o}^4 - T_{sky}^4) \tag{4}$$

where $T_{s,o}$ may be evaluated by applying Equation 1 at $x = L$.

$$T_{s,o} = -\frac{\dot{q}L^2}{2k} - \frac{\bar{h}_i(T_{\infty,i} - T_{s,i})}{k}L + T_{s,i} \tag{5}$$

The required generation rate may now be obtained by substituting appropriate numerical values into Equations 4 and 5 and solving for $\dot{q}$ by trial and error.

The inside convection coefficient may be obtained from Equation 9.26

$$\overline{Nu}_H = \left\{0.825 + \frac{0.387Ra_H^{1/6}}{[1 + (0.492/Pr)^{9/16}]^{8/27}}\right\}^2$$

where

$$Ra_H = \frac{g\beta(T_{s,i} - T_{\infty,i})H^3}{\nu\alpha}$$

$$Ra_H = \frac{9.8 \text{ m/s}^2 (3.503 \times 10^{-3} \text{ K}^{-1})(15 - 10)\text{K}(0.5 \text{ m})^3}{14.60 \times 10^{-6} \text{ m}^2/\text{s} \times 20.59 \times 10^{-6} \text{ m}^2/\text{s}}$$

$$Ra_H = 7.137 \times 10^7$$

Hence

$$\overline{Nu}_H = \left\{0.825 + \frac{0.387(7.137 \times 10^7)^{1/6}}{[1 + (0.492/0.711)^{9/16}]^{8/27}}\right\}^2$$

$$\overline{Nu}_H = 56$$

and

$$\bar{h}_i = \overline{Nu}_H\frac{k}{H} = \frac{56 \times 0\,0251 \text{ W/m·K}}{0.5 \text{ m}}$$

$$\bar{h}_i = 2.81 \text{ W/m}^2\text{·K}$$

The outside convection coefficient may be obtained by first evaluating the Reynolds number. With

$$Re_H = \frac{u_\infty H}{\nu} = \frac{20 \text{ m/s} \times 0.5 \text{ m}}{13.49 \times 10^{-6} \text{ m}^2/\text{s}} = 7.413 \times 10^5$$

and $Re_{x,c}$ assumed to equal 5×10^5, mixed boundary layer conditions exist on the outer surface. Hence, from Equation 7.21,

$$\overline{Nu}_H = (0.037Re_H^{4/5} - 871)Pr^{1/3}$$

$$\overline{Nu}_H = [0.037(7.413 \times 10^5)^{4/5} - 871](0.714)^{1/3}$$

$$\overline{Nu}_H = 864$$

and

$$\bar{h}_o = \overline{Nu}_H \frac{k}{H} = \frac{864 \times 0.0241 \text{ W/m} \cdot \text{K}}{0.5 \text{ m}}$$

$$\bar{h}_o = 41.6 \text{ W/m}^2 \cdot \text{K}$$

Equation 5 may now be expressed as

$$T_{s,o} = -\frac{\dot{q}(0.008 \text{ m})^2}{2(1.4 \text{ W/m} \cdot \text{K})} - \frac{2.81 \text{ W/m}^2 \cdot \text{K} (10 - 15)\text{K}}{1.4 \text{ W/m} \cdot \text{K}} \times 0.008 \text{ m} + 288 \text{ K}$$

$$T_{s,o} = -2.286 \times 10^{-5} \dot{q} + 288.1 \text{ K}$$

or, solving for $\dot{q}$,

$$\dot{q} = -43745(T_{s,o} - 288.1) \tag{6}$$

and substituting into Equation 4,

$$\begin{aligned} -43745 \ (T_{s,o} - 288.1) \ (0.008 \text{ m}) &= 41.6 \ \text{W/m}^2 \cdot \text{K} \ (T_{s,o} - 263 \text{ K}) \\ &+ 2.81 \ \text{W/m}^2 \cdot \text{K} \ (288 \text{ K} - 283 \text{ K}) + 0.90 \times 5.67 \\ &\times 10^{-8} \ \text{W/m}^2 \cdot \text{K}^4 \ (T_{s,o}^4 - 240^4)\text{K}^4 \end{aligned}$$

or

$$\begin{aligned} -350 \ (T_{s,o} - 288.1 \text{ K}) &= 41.6(T_{s,o} - 263) + 14.05 \\ &+ 5.10 \times 10^{-8} \ (T_{s,o}^4 - 240^4) \end{aligned}$$

From trial-and-error calculations it follows that

$$T_{s,o} \approx 285 \text{ K}$$

in which case, from Equation 6

$$\dot{q} \approx 136 \text{ kW/m}^3 \qquad \triangleleft$$

EXAMPLE 14.3

Cylindrical dry-bulb and wet-bulb thermometers are installed in a large diameter duct for the purpose of obtaining the temperature, T_∞, and the relative humidity, ϕ_∞, of moist air flowing through the duct at a velocity V. The dry-bulb thermometer has a bare glass surface of diameter, D_{db}, and emissivity, ε_g. The wet-bulb thermometer is covered with a thin wick that is saturated with water flowing continuously by capillary action from a bottom reservoir. Its diameter and emissivity are designated as D_{wb} and ε_w, respectively. The duct inside surface is at a known temperature, T_s, which is less than T_∞. Develop expressions that may be used to obtain T_∞ and ϕ_∞

from knowledge of the dry-bulb and wet-bulb temperatures, T_{db} and T_{wb}, and the foregoing parameters. Determine the values of T_∞ and ϕ_∞ under conditions for which $T_{db} = 45°C$, $T_{wb} = 25°C$, $T_s = 35°C$, $p = 1$ atm, $V = 5$ m/s, $D_{db} = 3$ mm, $D_{wb} = 4$ mm, and $\varepsilon_g = \varepsilon_w = 0.95$.

SOLUTION

KNOWN:

Dry- and wet-bulb temperatures associated with a moist air flow of prescribed velocity through a large diameter duct of prescribed surface temperature.

FIND:

Temperature and relative humidity of airflow.

SCHEMATIC:

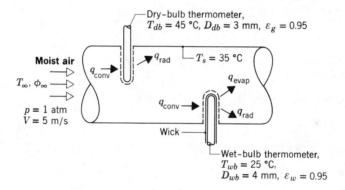

ASSUMPTIONS:

1. Steady-state conditions.
2. Negligible heat transfer by conduction along the thermometers.
3. Duct wall forms a large enclosure about the thermometers.
4. Thermometer surfaces are diffuse gray.

PROPERTIES:

As a first approximation, evaluate the dry-bulb air properties at 45°C and the wet-bulb properties at 25°C.
Table A.4, air (318 K): $v = 17.7 \times 10^{-6}$ m^2/s, $k = 0.0276$ W/m·K, $Pr = 0.70$.

Table A.4, air (298 K): $v = 15.7 \times 10^{-6}$ m²/s, $k = 0.0261$ W/m·K, $Pr = 0.71$.
Table A.6, saturated water vapor (298 K): $v_g = 44.3$ m³/kg, $h_{fg} = 2442$ kJ/kg. Saturated water vapor (318.5 K): $v_g = 15.5$ m³/kg.
Table A.8, water vapor–air (298 K): $D_{AB} = 0.26 \times 10^{-4}$ m²/s, $Sc = 0.60$.
Table A.11, glass and water: $\varepsilon_g \approx \varepsilon_w \approx 0.95$.

ANALYSIS:

Dry-Bulb Thermometer
Since $T_{db} > T_s$, there is net radiation transfer from the surface of the dry-bulb thermometer to the duct wall. Hence to maintain steady-state conditions, the thermometer temperature must be less than that of the air ($T_{db} < T_\infty$) to allow for convection heat transfer from the air. Hence, from application of a surface energy balance to the thermometer

$$q_{conv} = q_{rad}$$

or, from Equations 6.4 and 13.27,

$$\bar{h}A_{db}(T_\infty - T_{db}) = \varepsilon_g A_{db}\sigma(T_{db}^4 - T_s^4)$$

The air temperature is then

$$T_\infty = T_{db} + \frac{\varepsilon_g \sigma}{\bar{h}}(T_{db}^4 - T_s^4) \tag{1}$$

where $\bar{h}$ may be obtained from Equation 7.31.

Wet-Bulb Thermometer
The relative humidity may be obtained by performing an energy balance on the wet-bulb thermometer. In this case convection heat transfer to the wick is balanced by evaporative and radiative heat losses from the wick. Hence

$$q_{conv} = q_{evap} + q_{rad}$$

where

$$q_{evap} = n_A'' A_{wb}h_{fg} = \bar{h}_m[\rho_{A,sat}(T_{wb}) - \phi_\infty \rho_{A,sat}(T_\infty)]A_{wb}h_{fg}$$

Hence

$$\bar{h}A_{wb}(T_\infty - T_{wb}) = \bar{h}_m[\rho_{A,sat}(T_{wb}) - \phi_\infty \rho_{A,sat}(T_\infty)]A_{wb}h_{fg}$$
$$+ \varepsilon_w A_{wb}\sigma(T_{wb}^4 - T_s^4)$$

or

$$\phi_\infty = \frac{\rho_{A,sat}(T_{wb}) + [\varepsilon_w \sigma(T_{wb}^4 - T_s^4) - \bar{h}(T_\infty - T_{wb})]/h_{fg}\bar{h}_m}{\rho_{A,sat}(T_\infty)} \tag{2}$$

where $\bar{h}_m$ may be determined from the mass transfer analog of Equation 7.31.

Convection Calculations

For the prescribed conditions the Reynolds number associated with the dry-bulb thermometer is

$$Re_{D(db)} = \frac{VD_{db}}{v} = \frac{5 \text{ m/s} \times 0.003 \text{ m}}{17.7 \times 10^{-6} \text{ m}^2/\text{s}} = 847$$

Approximating the Prandtl number ratio as unity, it follows from Equation 7.31 and Table 7.3 that

$$\overline{Nu}_{D(db)} = CRe_{D(db)}^m Pr^n = 0.51(847)^{0.5}(0.70)^{0.37} = 13.01$$

Hence

$$\bar{h} = 13.01 \frac{k}{D_{db}} = 13.01 \frac{0.0276 \text{ W/m}\cdot\text{K}}{0.003 \text{ m}} = 120 \text{ W/m}^2\cdot\text{K}$$

From Equation 1 the air temperature is then

$$T_\infty = 45°\text{C} + \frac{0.95 \times 5.67 \times 10^{-8} \text{ W/m}^2\cdot\text{K}^4}{120 \text{ W/m}^2\cdot\text{K}}(318^4 - 308^4)\text{K}^4$$

or

$$T_\infty = 45°\text{C} + 0.55°\text{C} = 45.6°\text{C} \qquad \triangleleft$$

The realative humidity may now be obtained from Equation 2. The Reynolds number associated with the wet-bulb thermometer is

$$Re_{D(wb)} = \frac{VD_{wb}}{v} = \frac{5 \text{ m/s} \times 0.004 \text{ m}}{15.7 \times 10^{-6} \text{ m}^2/\text{s}} = 1274$$

From Equation 7.31 and Table 7.3 it follows that

$$\overline{Nu}_{D(wb)} = 0.26(1274)^{0.6}(0.71)^{0.37} = 16.71$$

Hence

$$\bar{h} = 16.71 \frac{k}{D_{wb}} = 16.71 \frac{0.0261 \text{ W/m}\cdot\text{K}}{0.004 \text{ m}} = 109 \text{ W/m}^2\cdot\text{K}$$

Using the mass transfer analog of Equation 7.31, it also follows that

$$\overline{Sh}_{D(wb)} = 0.26Re_{D(wb)}^{0.6}Sc^{0.37} = 0.26(1274)^{0.6}(0.6)^{0.37} = 15.7$$

Hence the convection mass transfer coefficient is

$$\bar{h}_m = 15.7 \frac{D_{AB}}{D_{wb}} = 15.7 \frac{0.26 \times 10^{-4} \text{ m}^2/\text{s}}{0.004 \text{ m}} = 0.102 \text{ m/s}$$

Also,

$$\rho_{A,\,sat}(T_{wb}) = v_g(298\text{ K})^{-1} = (44.3\text{ m}^3/\text{kg})^{-1} = 0.0226\text{ kg/m}^3$$

$$\rho_{A,\,sat}(T_\infty) = v_g(318.5\text{ K})^{-1} = (15.5\text{ m}^3/\text{kg})^{-1} = 0.0645\text{ kg/m}^3$$

Hence the relative humidity is, from Equation 2

$$\phi_\infty = \cfrac{0.0226\text{ kg/m}^3 + \cfrac{\{[0.95 \times 5.67 \times 10^{-8}\text{ W/m}^2\cdot\text{K}^4(298^4 - 308^4)\text{K}^4 \\ -109\text{ W/m}^2\cdot\text{K}(45.55 - 25)\text{K}]\}}{(2.442 \times 10^6\text{ J/kg})(0.102\text{ m/s})}}{0.0645\text{ kg/m}^3}$$

or

$$\phi_\infty = 0.21 \qquad \triangleleft$$

COMMENTS:

1. The effect of radiation exchange between the duct wall and the thermometers is small. For this reason $T_\infty \approx T_{db}$.

2. The evaporative heat loss is significant due to the small value of ϕ_∞, causing T_{wb} to be significantly less than T_∞.

EXAMPLE 14.4

The composite insulation described in Problem 1.19d of Chapter 1 and shown as follows is being considered as a roofing material.

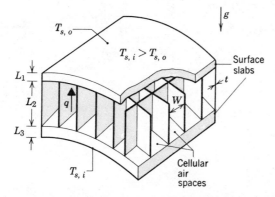

It is proposed that the outer and inner slabs be made from low-density particle board of thicknesses $L_1 = L_3 = 12.5$ mm and that the honeycomb core be constructed from a high-density particle board. The square cells of the core are to be of length $L_2 = 50$ mm, width $W = 10$ mm and of wall thickness $t = 2$ mm. The emissivity of both particle boards is known to be approximately 0.85, and the honeycomb cells are filled with air at 1 atm pressure. To assess the effectiveness of the insulation, its total thermal resistance must be evaluated under representative operating conditions for which the bottom (inner) surface temperature is $T_{s,i} = 25°C$ and the top (outer) surface temperature is $T_{s,o} = -10°C$.

SOLUTION

KNOWN:

Dimensions of a composite insulation consisting of a honeycomb core sandwiched between solid slabs.

FIND:

Total thermal resistance.

SCHEMATIC:

Because of the repetitive nature of the honeycomb core, the cell side walls will be adiabatic. That is, there is no lateral heat transfer from cell to cell, and it suffices to consider the heat transfer across a single cell.

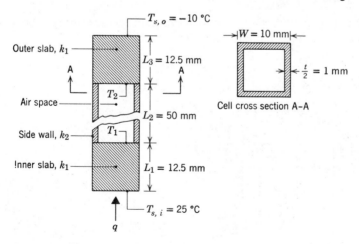

ASSUMPTIONS:

1. One-dimensional, steady-state conditions.
2. Equivalent conditions for each cell.

3. Constant properties.
4. Diffuse-gray surface behavior.

PROPERTIES:

Table A.3, particle board (low density): $k_1 = 0.078$ W/m·K. Particle board (high density): $k_2 = 0.170$ W/m·K.
For both board materials, $\varepsilon = 0.85$.
Table A.4, air ($\bar{T} \approx 7.5°C$): $\nu = 14.15 \times 10^{-6}$ m²/s, $k = 0.0247$ W/m·K, $\alpha = 19.9 \times 10^{-6}$ m²/s, $Pr = 0.71$, $\beta = 3.57 \times 10^{-3}$ K⁻¹.

ANALYSIS:

The total resistance of the composite is determined by conduction, convection and radiation processes occurring within the honeycomb and by conduction across the inner and outer slabs. The corresponding thermal circuit is shown as follows.

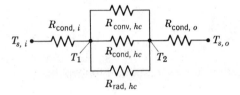

The total resistance of the composite is

$$R = R_{\text{cond},i} + R_{\text{eq}} + R_{\text{cond},o}$$

where the equivalent resistance for the honeycomb is

$$R_{\text{eq}}^{-1} = (R_{\text{cond}}^{-1} + R_{\text{conv}}^{-1} + R_{\text{rad}}^{-1})_{hc}$$

The component resistances may be evaluated as follows.

The inner and outer slabs are plane walls, for which the thermal resistance is given by Equation 3.6. Hence, since $L_1 = L_3$ and the slabs are constructed from low-density particle board,

$$R_{\text{cond},i} = R_{\text{cond},o} = \frac{L_1}{k_1 W^2} = \frac{0.0125 \text{ m}}{0.078 \text{ W/m·K } (0.01 \text{ m})^2} = 1603 \text{ K/W}$$

Similarly, applying Equation 3.6 to the sidewalls of the cell,

$$R_{\text{cond},hc} = \frac{L_2}{k_2[W^2 - (W-t)^2]} = \frac{L_2}{k_2(2Wt - t^2)}$$

$$R_{\text{cond}, hc} = \frac{0.050 \text{ m}}{0.170 \text{ W/m} \cdot \text{K} \, [2 \times 0.01 \text{ m} \times 0.002 \text{ m} - (0.002 \text{ m})^2]}$$
$$= 8170 \text{ K/W}$$

From Equation 3.9 the convection resistance associated with the cellular airspace may be expressed as

$$R_{\text{conv}, hc} = \frac{1}{h(W - t)^2}$$

The cell forms an enclosure that may be classified as a horizontal cavity heated from below, and the appropriate form of the Rayleigh number is

$$Ra_L = \frac{g\beta(T_1 - T_2)L_2^3}{\alpha \nu}$$

To evaluate this parameter, however, it is necessary to *assume* a value of the cell temperature difference. As a first approximation we choose

$$(T_1 - T_2) = [15^\circ\text{C} - (-5^\circ\text{C})] = 20^\circ\text{C}$$

in which case

$$Ra_L = \frac{9.8 \text{ m/s}^2 (3.57 \times 10^{-3} \text{ K}^{-1})(20 \text{ K})(0.05 \text{ m})^3}{19.9 \times 10^{-6} \text{ m}^2/\text{s} \times 14.15 \times 10^{-6} \text{ m}^2/\text{s}} = 3.11 \times 10^5$$

Applying Equation 9.37 as a first approximation, it follows that

$$h = \frac{k}{L_2} [0.069 \, Ra_L^{1/3} Pr^{0.074}]$$

$$h = \frac{0.0247 \text{ W/m} \cdot \text{K}}{0.05 \text{ m}} [0.069(3.11 \times 10^5)^{1/3}(0.71)^{0.074}] = 2.25 \text{ W/m}^2 \cdot \text{K}$$

The convection resistance is then

$$R_{\text{conv}, hc} = \frac{1}{2.25 \text{ W/m}^2 \cdot \text{K} \, (0.01 \text{ m} - 0.002 \text{ m})^2} = 6944 \text{ K/W}$$

The resistance to heat transfer by radiation may be obtained by first noting that the cell forms a three-surface enclosure for which the sidewalls are reradiating. The net radiation heat transfer between the end surfaces of the cell is then given by Equation 13.30. With $\varepsilon_1 = \varepsilon_2 = \varepsilon$ and $A_1 = A_2 = (W - t)^2$, the equation reduces to

$$q_{\text{rad}} = \frac{(W - t)^2 \sigma(T_1^4 - T_2^4)}{2\left(\dfrac{1}{\varepsilon} - 1\right) + \dfrac{1}{F_{12} + [(F_{1R} + F_{2R})/F_{1R}F_{2R}]^{-1}}}$$

However, with $F_{1R} = F_{2R} = (1 - F_{12})$ it follows that

$$q_{rad} = \frac{(W - t)^2 \sigma (T_1^4 - T_2^4)}{2\left(\dfrac{1}{\varepsilon} - 1\right) + \left[F_{12} + \dfrac{(1 - F_{12})^2}{2(1 - F_{12})}\right]^{-1}}$$

or

$$q_{rad} = \frac{(W - t)^2 \sigma (T_1^4 - T_2^4)}{2\left(\dfrac{1}{\varepsilon} - 1\right) + \dfrac{2}{1 + F_{12}}}$$

The view factor F_{12} may be obtained from Figure 13.4

where

$$\frac{X}{L} = \frac{Y}{L} = \frac{(W - t)}{L_2} = \frac{(10 \text{ mm} - 2 \text{ mm})}{50 \text{ mm}} = 0.16$$

Hence

$$F_{12} \approx 0.01$$

Defining the radiation resistance as

$$R_{rad,hc} = \frac{(T_1 - T_2)}{q_{rad}}$$

it follows that

$$R_{rad,hc} = \frac{2\left(\dfrac{1}{\varepsilon} - 1\right) + \dfrac{2}{1 + F_{12}}}{(W - t)^2 \sigma (T_1^2 + T_2^2)(T_1 + T_2)}$$

where

$$(T_1^4 - T_2^4) = (T_1^2 + T_2^2)(T_1 + T_2)(T_1 - T_2).$$

Accordingly,

$$R_{rad,hc} = \left[2\left(\frac{1}{0.85} - 1\right) + \frac{2}{1 + 0.01}\right]\Bigg/$$

$$\{(0.01 \text{ m} - 0.002 \text{ m})^2 \times 5.67 \times 10^{-8} \text{ W/m}^2 \cdot \text{K}^4 \, [(288 \text{ K})^2$$
$$+ (268 \text{ K})^2](288 + 268)\text{K}\}$$

where, again, it is *assumed* that $T_1 = 15°\text{C}$ and $T_2 = -5°\text{C}$. From the above expression it follows that

$$R_{rad,hc} = \frac{0.353 + 1.980}{3.123 \times 10^{-4}} = 7471 \ \text{K/W}$$

In summary the component resistances are then

$$R_{cond,i} = R_{cond,o} = 1603 \ \text{K/W}$$

$$R_{cond,hc} = 8170 \ \text{K/W} \qquad R_{conv,hc} = 6944 \ \text{K/W} \qquad R_{rad,hc} = 7471 \ \text{K/W}$$

The equivalent resistance is then

$$R_{eq} = \left(\frac{1}{8170} + \frac{1}{6944} + \frac{1}{7471} \right)^{-1} = 2498 \ \text{K/W}$$

and the total resistance is

$$R = 1603 + 2498 + 1603 = 5704 \ \text{K/W} \qquad\qquad \triangleleft$$

COMMENTS:

1. Note that the problem is iterative in nature, since values of T_1 and T_2 were assumed in order to calculate $R_{conv,hc}$ and $R_{rad,hc}$. To check the validity of the assumed values, we first obtain the heat transfer rate q from the expression

$$q = \frac{T_{s,1} - T_{s,2}}{R} = \frac{25°C - (-10°C)}{5704 \ \text{K/W}} = 6.14 \times 10^{-3} \ \text{W}$$

Hence

$$T_1 = T_{s,i} - qR_{cond,i} = 25°C - 6.14 \times 10^{-3} \ \text{W} \times 1603 \ \text{K/W} = 15.2°C$$

$$T_2 = T_{s,o} + qR_{cond,o} = -10°C + 6.14 \times 10^{-3} \ \text{W} \times 1603 \ \text{K/W} = -0.2°C$$

Using these values of T_1 and T_2, $R_{conv,hc}$ and $R_{rad,hc}$ should be recomputed and the process repeated until satisfactory agreement is obtained between the initial and computed values of T_1 and T_2.

2. The resistance of a section of low-density particle board 75 mm thick $(L_1 + L_2 + L_3)$ of area W^2 is 9615 K/W, which exceeds the total resistance of the composite by approximately 70 percent. Accordingly, use of the honeycomb structure offers no advantages as an insulating material. Its effectiveness as an insulator could be improved (R_{eq} increased) by reducing the wall thickness t to increase R_{cond}, evacuating the cell to increase R_{conv}, and/or by decreasing ε to increase R_{rad}. A significant increase in $R_{rad,hc}$ could be achieved by aluminizing the top and bottom surfaces of the cell.

PROBLEMS

14.1 A square plate of pure aluminum, 0.5 m on a side and 16 mm thick, is initially at 300°C and is suspended to cool in a large chamber. The walls of the chamber and the enclosed air are each maintained at 27°C.

a) If the surface emissivity of the plate is 0.25, what is the initial cooling rate?

b) Is it reasonable to assume a uniform plate temperature during the cooling process?

14.2 A square (0.25 m × 0.25 m) channel is used to transport preheated atmospheric air for an industrial process application. The channel is constructed from steel, except for a special glass viewing panel, which is of width $w = 80$ mm and extends from $x = L_1 = 0.8$ m to $x = L_2 = 1$ m from the channel entrance. The glass is mounted flush with the steel on the inside surface of the channel, and its sides are separated from the steel by insulating strips.

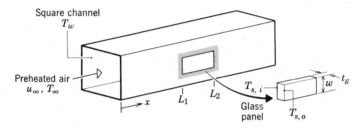

The glass has a thickness of $t_g = 50$ mm, an emissivity of $\varepsilon_g = 0.8$, and a thermal conductivity of $k_g = 2$ W/m·K. It may be assumed to be opaque to radiation of wavelengths greater than 1 μm. During operation air at a velocity of $u_\infty = 15$ m/s and a temperature of $T_\infty = 800$ K passes through the channel, while the outer surface of the glass is uniformly cooled to a temperature of $T_{s,o} = 300$ K. Steady-state conditions exist, and it may be assumed that the inner surface of the glass exchanges radiation only with the steel walls, which have an emissivity of $\varepsilon_w = 1.0$ and a temperature of $T_w = 800$ K. Estimate the heat loss through the glass panel and the glass inner surface temperature, $T_{s,i}$.

14.3 Hot air flows from a furnace through a 0.15 m diameter, thin-walled steel duct with a velocity of 3 m/s. The duct passes through the crawl space of a house, and its uninsulated exterior surface is exposed to quiescent air and surroundings at 0°C. At a location in the duct for which the mean air temperature is 70°C, determine the heat loss per unit duct length and the duct wall temperature.

14.4 A radiant heating oven is one whose walls are maintained at an elevated surface temperature, T_s, while a gas at a lower temperature, T_∞, flows through the oven. A product placed in the oven is then heated to an equilibrium temperature, T_p, which is less than T_s and larger than T_∞. (See figure on next page.)

Consider the product to be in the form of a solid cylinder whose length is much larger than its diameter, $D = 50$ mm. The surface emissivity of the solid is known to be $\varepsilon = 0.8$. If the solid is placed in an oven whose surfaces are maintained at T_s

$= 650$ K and atmospheric nitrogen is in cross flow over the cylinder with T_∞ $= 350$ K and $V = 3$ m/s, what is the equilibrium temperature, T_p, of the product?

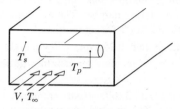

14.5 A thermocouple junction is inserted in a large duct to measure the temperature of hot gases flowing through the duct.

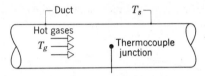

a) If the duct surface temperature T_s is less than the gas temperature T_g, will the thermocouple sense a temperature that is less than, equal to, or greater than T_g? Justify your answer on the basis of a simple analysis.

b) A thermocouple junction in the shape of a 2-mm-diameter sphere with a surface emissivity of 0.60 is placed in a gas stream moving at 3 m/s. If the thermocouple senses a temperature of 320°C when the duct surface temperature is 175°C, what is the actual gas temperature? The gas may be assumed to have the properties of air at atmospheric pressure.

14.6 Consider temperature measurement in a gas stream using the thermocouple junction described in the foregoing problem ($D = 2$ mm, $\varepsilon = 0.60$). If the gas velocity and temperature are 3 m/s and 500°C, respectively, what temperature will be indicated by the thermocouple if the duct surface temperature is 200°C? The gas may be assumed to have the properties of atmospheric air. What temperature will be indicated by the thermocouple if the gas pressure is doubled and all other conditions remain the same?

14.7 The temperature of flue gases flowing through the stack of a boiler is measured by means of a thermocouple enclosed within a cylindrical tube as shown.

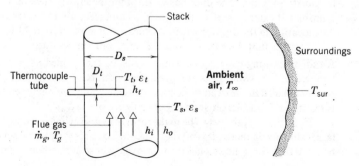

The tube axis is oriented normal to the gas flow, and the thermocouple senses a temperature T_t corresponding to that of the tube surface. The gas flow rate and temperature are designated as $\dot{m}_g$ and T_g, respectively, and the gas flow may be assumed to be fully developed. The stack is fabricated from sheet metal that is at a uniform temperature T_s and is exposed to ambient air at T_∞ and surroundings at T_{sur}. The convection coefficient associated with the outer surface of the duct is designated as h_o, while those associated with the inner surface of the duct and the tube surface are designated as h_i and h_t, respectively. The tube and duct surface emissivities are designated as ε_t and ε_s, respectively.

a) Neglecting conduction losses along the thermocouple tube, develop an analysis that could be used to predict the error $(T_g - T_t)$ in the temperature measurement.

b) Assuming the flue gas to have the properties of atmospheric air, evaluate the error for $T_t = 300°C$, $D_s = 0.6$ m, $D_t = 10$ mm, $\dot{m}_g = 1$ kg/s, $T_\infty = T_{\text{sur}} = 27°C$, $\varepsilon_t = \varepsilon_s = 0.8$ and $h_o = 25$ W/m$^2 \cdot$K.

14.8   Hot coffee is contained in a cylindrical thermos bottle which is of length $L = 0.3$ m and is laying on its side (horizontally).

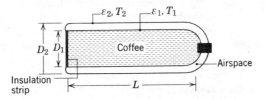

The coffee container consists of a glass flask of diameter $D_1 = 0.07$ m, which is separated from an aluminum housing of diameter $D_2 = 0.08$ m by air at atmospheric pressure. The outer surface of the flask and the inner surface of the housing are silver coated to provide emissivities of $\varepsilon_1 = \varepsilon_2 = 0.25$. If these surfaces are at temperatures of $T_1 = 75°C$ and $T_2 = 35°C$, respectively, what is the heat loss from the coffee?

14.9   A room heater panel consists of a vertical array of N copper pipes of diameter D

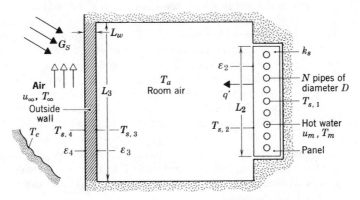

imbedded in a solid panel of thermal conductivity, k_s. Hot water flows through the pipes with a mean velocity and temperature of u_m and T_m, respectively, maintaining the pipe wall at a temperature $T_{s,1} < T_m$. The surface of the panel is at a uniform temperature $T_{s,2}$, which exceeds the mean temperature T_a of the air in the room. The outside wall of the room, which is of thermal conductivity, k_w, and without windows, has uniform inner and outer surface temperatures of $T_{s,3}$ and $T_{s,4}$, respectively. The outer surface is exposed to a cold ambient air flow of temperature T_∞ and velocity u_∞, to solar irradiation of magnitude, G_S, and to surroundings characterized by an effective temperature, T_e. The room is long in the direction normal to the sketch; and the floor, ceiling, and rear wall, which houses the panel, are well insulated.

a) Assuming steady-state conditions, identify all pertinent heat transfer processes.

b) For prescribed values of u_m, T_m, u_∞, T_∞, T_e and G_S, develop whatever equations are needed to determine the heat flow per unit length (normal to the sketch) through the panel, q', and the air temperature, T_a.

14.10 A liquid enters a long, thin-walled circular tube of diameter D_i at a temperature $T_{m,i}$ and a mass flowrate $\dot{m}$. Fully developed turbulent flow may be assumed for the liquid. A second tube, whose surface is maintained at a uniform, elevated temperature T_h, surrounds the first tube, and the space between the two tubes is evacuated. Both tube surfaces may be approximated as blackbodies. Obtain expressions that could be used to determine the variation of the liquid mean temperature T_m and the inner tube surface temperature T_s as a function of distance x along the tube. Express your results in terms of the foregoing variables and appropriate liquid thermophysical properties.

14.11 A vertical air space in the wall of a home has a thickness of 0.1 m and a height of 3 m. The air separates a brick exterior from a plaster board wall, with each surface having an emissivity of 0.9. Consider conditions for which the temperatures of the brick and plaster surfaces exposed to the air are $-10°C$ and $18°C$, respectively.

a) What is the rate of heat loss per unit surface area?

b) What would be the heat loss per unit area if the space were filled with urethane foam?

14.12 To estimate the energy savings associated with using double-pane (thermopane) versus single-pane window construction, consider a window that is oriented vertically and is 0.6 m × 0.6 m on a side. The inner surface of the window is exposed to the air and walls of a room that are maintained at a temperature of T_i, while the outer surface is exposed to ambient air at T_∞ and to surroundings (the sky, ground, etc.) characterized by an effective temperature T_{sur}. The convection coefficient associated with the outer surface and the ambient air is designated as h_o. The window is constructed of glass that is opaque to longwave radiation and of known emissivity ε_g. The single-pane window is of thickness t_1, while each pane of the thermopane window is of thickness t_2 with an intervening airspace of thickness t_a.

a) Assuming steady-state, nighttime conditions and negligible temperature gradients within the glass, develop a model that could be used to determine the rate of heat transfer through the window and the glass temperature(s) for both single pane and thermopane constructions.

b) Determine the heat transfer rate and the glass temperatures when $T_i = 18°C$, $T_\infty = -20°C$, $T_{sur} = -30°C$, $h_o = 25 \text{ W/m}^2 \cdot \text{K}$, $\varepsilon_g = 0.9$, $t_1 = 10$ mm, $t_2 = 5$ mm, and $t_a = 15$ mm.

14.13 In recent years there has been considerable interest in developing flat plate solar collectors for use in satisfying home heating needs. A typical collector configuration is shown below. It consists of an absorbing surface that is thermally insulated on the edges and on the back side. Solar radiation is transmitted through one or more glass *cover plates* before reaching the absorber. The collected energy is removed by circulating water through tubes that are in good thermal contact with the absorber.

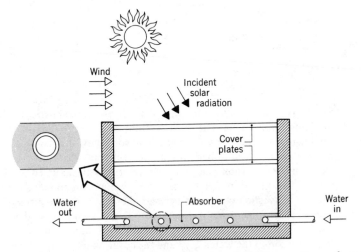

Identify all of the energy exchange processes that influence the performance of the collector. Suggest design features which would increase collection efficiency.

14.14 Consider a flat plate solar collector, such as that shown schematically in the foregoing problem, but with one, rather than two, cover plates. The following quantities are presumed to be known.

G_S The solar irradiation
$\alpha_{cp,S}$ The absorptivity of the cover plate to solar radiation
$\tau_{cp,S}$ The transmissivity of the cover plate to solar radiation
$\alpha_{ap,S}$ The absorptivity of the absorber plate to solar radiation
$\varepsilon_{cp}, \varepsilon_{ap}$ Emissivities of the cover and absorber plates, respectively
h_i Air space convection coefficient based on the temperature difference, $T_{ap} - T_{cp}$.
h_o Convection coefficient between cover plate and ambient air
T_{ap} Absorber plate temperature
T_∞ Ambient air temperature
T_{sky} Effective sky temperature

Assuming the absorber-plate tube assembly to be perfectly insulated from below, develop expressions that could be used to obtain the rate at which useful energy is collected per unit area of the absorber plate surface, q_u'', under steady-state conditions.

14.15 The performance of wastewater treatment systems is known to diminish during the winter, when water temperatures drop below the level required for maximum treatment rates. However, it has been suggested that it may be economically feasible to maintain the desired water temperature with water heaters, if a thin, semitransparent cover is installed over the water surface. Consider the system shown as follows.

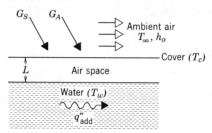

Shortwave solar irradiation, G_S, and longwave atmospheric irradiation, G_A, are incident on the system. The cover is semitransparent to the solar radiation, with a total hemispherical transmissivity and absorptivity of τ_S and α_S, respectively. It is opaque to the atmospheric radiation, with an absorptivity α_A ($\tau_A = 0$). The emissivity of the water is ε_w, and the water may be assumed to absorb all of the solar radiation transmitted by the cover. The ambient air is at a temperature of T_∞, and the convection coefficient between the cover and the air is designated as h_o. For a prescribed water temperature $T_w > T_\infty$ and prescribed environmental conditions (G_S, G_A, T_∞, and h_o), the unknowns of the system are the required rate of heat addition to the water, q_{add}'', and the cover temperature, T_c. Develop equations that could be used to solve for q_{add}'' and T_c, expressing your results exclusively in terms of the designated variables and appropriate thermophysical properties. Note that the water is well insulated from below and is separated from the cover by a distance L.

14.16 The following solar collector configuration involves passing a working fluid (water) at a flowrate of $\dot{m}_w$ through a rectangular channel bounded from above by the absorber plate and from below by a thick layer of insulation. Solar irradiation, G_S, and longwave atmospheric irradiation, G_A, are incident on the cover plate, which is exposed to ambient air at a temperature T_∞ with a convection coefficient of $h_{c\infty}$. The airspace between the cover and absorber plates is characterized by a convection coefficient, h_{ac}, which, when multiplied by the difference in temperatures between the two plates, $T_{ap} - T_{cp}$, provides the convection heat flux across the space. Heat transfer by convection occurs from the absorber plate to the water, which enters at a mean temperature $T_{w,i}$ and for which the convection coefficient is h_{aw}.

Expressing your results in terms of the foregoing parameters and the collector surface area, A_c, formulate a set of equations that could be used to solve for the water outlet temperature, $T_{w,o}$, and the absorber and cover plate temperatures, T_{ap}

and T_{cp}. You may assume the following: the solar irradiation is completely transmitted by the cover plate and completely absorbed by the absorber plate; the atmospheric irradiation is completely absorbed by the cover plate; the cover and absorber plate temperatures are uniform.

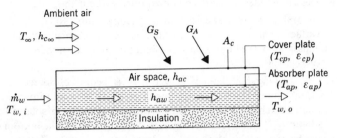

14.17 It is known that on clear nights the air temperature need not drop below 0°C before a thin layer of water on the ground will freeze. Consider such a layer of water on a clear night for which the effective sky temperature is −30°C and the convection heat transfer coefficient due to wind motion is $h = 25$ W/m²·K. The water may be assumed to have an emissivity of 1.0 and to be insulated from the ground as far as conduction is concerned.

a) Neglecting evaporation, determine the lowest temperature that the air can have without the water freezing.

b) Accounting now for the effect of evaporation, what is the lowest temperature that the air can have without the water freezing? Assume the air to be dry.

14.18 A shallow layer of water is exposed to the natural environment shown as follows:

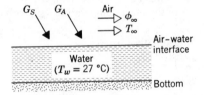

Consider conditions for which the solar and atmospheric irradiations are $G_S = 600$ W/m² and $G_A = 300$ W/m², respectively, and the air temperature and relative humidity are $T_\infty = 27$°C and $\phi_\infty = 0.50$, respectively. The reflectivities of the water surface to the solar and atmospheric irradiation are known to be $\rho_S = 0.3$ and $\rho_A = 0$, respectively, while the surface emissivity is $\varepsilon = 0.97$. The convection heat transfer coefficient at the air–water interface is $h = 25$ W/m²·K. If the water is at a temperature of 27°C, will this temperature increase or decrease with time?

14.19 A roof cooling system, which operates by maintaining a thin film of water on the roof surface, may be used to reduce air conditioning costs or to maintain a cooler environment in nonconditioned buildings. To determine the effectiveness of such a system, consider a sheet metal roof for which the solar absorptivity is $\alpha_S = 0.50$ and the hemispherical emissivity is $\varepsilon = 0.3$. Representative conditions correspond to a surface convection coefficient of $h = 20$ W/m²·K, a solar irradiation of $G_S = 700$ W/m², a sky temperature of $-10°C$, and an atmospheric temperature and relative humidity of 30°C and 65 percent respectively. The roof may be assumed to be well insulated from below.

a) Determine the roof surface temperature without the water film.

b) Assuming the film and roof surface temperatures to be equal, determine the surface temperature with the film. The solar absorptivity and hemispherical emissivity of the film–surface combination are known to be $\alpha_S = 0.8$ and $\varepsilon = 0.9$, respectively.

14.20 A long rectangular channel of width w is being considered for solar energy collection. As envisioned, a shallow layer of water would flow at a mass rate $\dot{m}$ through the channel and would be exposed to uniform solar and atmospheric irradiation, G_S and G_A, respectively, at its free interface. The bottom of the channel would be painted flat black and insulated from below. The free interface would be exposed to ambient air at a temperature T_∞ and relative humidity ϕ_∞. Convection heat transfer across the air–water interface would be characterized by a convection coefficient h presumed to be known.

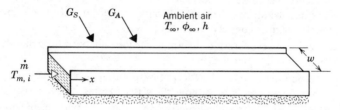

a) Assuming water to enter the channel with a mean temperature of $T_{m,i}$, derive an expression for the variation of the mean temperature $T_{m,x}$ with distance along the channel.

b) Defining an accumulated, local collector efficiency as

$$\eta(x) = \frac{\dot{m}c_p[T_{m(x)} - T_{m,i}]}{(G_S + G_A)wx}$$

sketch the anticipated variation of η with x.

14.21 A climate control system consists of a long rectangular channel of width w and length L. The top surface is heated to a uniform temperature, T_s, while the bottom surface holds a shallow, stagnant layer of water whose temperature, T_w, varies with x. The entire system is well insulated from its surroundings.

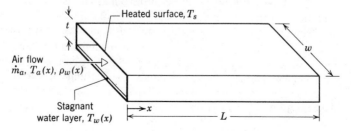

Air enters the system with a mass flowrate $\dot{m}_a$ and moves in laminar flow between the heated surface and the water layer. The separation t between the heated surface and the water may be assumed to be small compared to w and L but large relative to any inside boundary layer thickness. Blackbody behavior may be assumed for the heated surface and the water.

a) Derive differential equations that determine the variation of the air temperature, T_a, and water vapor density in the air, ρ_w, with x along the channel.

b) Derive an algebraic equation that could be used to determine the temperature of the water as a function of x.

14.22 The thermal pollution problem is associated with discharging warm water from an electrical power plant or from an industrial source to a natural body of water. Methods for alleviating this problem involve cooling the warm water before allowing the discharge to occur. Two such methods, involving wet cooling towers or spray ponds, rely on heat transfer from the warm water in droplet form to the surrounding atmosphere. To develop an understanding of the mechanisms that contribute to this cooling, consider a spherical droplet of diameter D and temperature T, which is moving at a velocity V relative to air at a temperature T_∞ and relative humidity ϕ_∞. The surroundings are characterized by a temperature, T_{sur}.

a) Develop expressions for the droplet evaporation and cooling rates.

b) Calculate the evaporation rate (kg/s) and cooling rate (K/s) when $D = 3$ mm, $V = 7$ m/s, $T = 40°C$, $T_\infty = 25°C$, $T_{sur} = 15°C$, and $\phi_\infty = 0.60$.

14.23 The radiation generated in an evacuated 60-W light bulb may be approximated as having the spectral distribution of radiation emitted by a blackbody at 2900 K.

a) If the radiative properties of the glass envelope are as follows,

$$0 < \lambda < 1 \ \mu m \qquad \tau_\lambda = 1, \ \alpha_\lambda = 0$$

$$1 \ \mu m < \lambda \qquad \tau_\lambda = 0, \ \alpha_\lambda = 1$$

what fraction of the radiation emitted by the filament is transmitted to the surroundings of the bulb?

b) Neglecting conduction losses through the socket, what will be the temperature of the glass bulb, which may be approximated as a sphere of diameter $D = 50$ mm, when it is situated in a room whose walls and air are at $20°C$?

14.24 Four high-powered silicon-controlled rectifiers (SCR), each dissipating 150 W, are mounted on a water-cooled heat sink as shown in the following sketch.

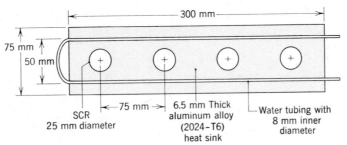

Water at 15°C is supplied at a rate of 4 l/min. The thermal contact resistance between an SCR and the heat sink is estimated to be 0.1°C/W. Estimate the operating temperature of the SCRs by performing the following series of calculations.

a) Determine the temperature rise of the water coolant. What is the average mean temperature of the coolant?

b) Using the proper correlation for forced convection inside a tube, estimate the heat transfer coefficient between the water and the tube wall. What is the approximate temperature rise between the coolant and the tube wall?

c) Determine the shape factor for the heat sink material by using the two-dimensional graphical flux plotting method. Be sure to recognize symmetry of the system. What is the temperature rise between the tube wall and the location where an SCR contacts the heat sink?

d) Draw the thermal resistance circuit and identify each of the elements present. Label the temperatures at each node and indicate the operating temperature of the SCRs.

e) Summarize briefly the assumptions that were made in arriving at the SCR operating temperatures.

14.25 It is desired to estimate the effectiveness of a horizontal, straight fin of rectangular cross section when applied to a surface operating at 45°C in an environment when the surroundings and ambient air are at 25°C. The fin is to be fabricated from aluminum alloy (2024-T6) with an anodized finish and is 2 mm thick and 10 mm long.

a) Considering only free convection from the fin surface and estimating an average heat transfer coefficient, determine the effectiveness of the fin.

b) Estimate the effectiveness of the fin including the influence of radiation exchange with the surroundings.

c) Using a numerical method, develop the finite-difference equations and detail the procedure for obtaining the fin effectiveness. The free convection and radiation exchange modes should be based on local, rather than average, values for the entire fin.

15 Diffusion Mass Transfer

We have learned that heat transfer will occur whenever there exists a temperature difference in a medium. Similarly, whenever there exists a difference in the concentration or density of some chemical species in a mixture, mass transfer *must* occur.

> *Mass transfer is mass in transit as the result of a species concentration difference in a mixture.*

Hence, just as a *temperature gradient* constitutes the *driving potential* for heat transfer, the existence of a *concentration gradient* for some species in a mixture provides the *driving potential* for transport of that species.

It is important to clearly understand the context in which the term *mass transfer* is used. Although mass is certainly transferred whenever there is bulk fluid motion, this is not what we have in mind. For example, we would *not* use the term *mass transfer* to describe the motion of air that is induced by a fan or the motion of water being forced through a pipe. In both cases, there is gross or bulk fluid motion due to mechanical work. We would, however, use the term *mass transfer* to describe the relative motion of species in a mixture due to the presence of concentration gradients. One example would be the dispersion of oxides of sulphur released from a powerplant smokestack into the environment. Another example would be the transfer of water vapor into dry air, as in a home humidifier.

There are *modes* of *mass transfer* that are similar to the conduction and convection modes of heat transfer. In Chapters 6 through 8 we considered mass transfer by convection, which is similar to convection heat transfer and will occur whenever the concentration of some species at a surface differs from its concentration in a gas moving over the surface. In this chapter we consider mass transfer by diffusion, which is similar to conduction heat transfer and occurs whenever there is a species concentration gradient in a stationary medium.

15.1   PHYSICAL ORIGINS AND RATE EQUATIONS

From the standpoint of physical origins and the governing rate equations, strong analogies exist between heat and mass transfer by diffusion.

15.1.1   Physical Origins

Both conduction heat transfer and mass diffusion are transport processes which originate from atomic and molecular activity. Consider a chamber in which two different gas species at the same temperature and pressure are initially separated by a partition. If the partition is removed, both species will be transported by diffusion. Figure 15.1 shows the situation as it might exist shortly after removal of the partition. Since a higher concentration means more molecules per unit

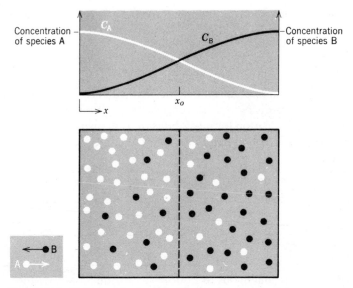

Figure 15.1   Mass transfer by diffusion in a binary gas mixture.

volume, it is evident that at the instant shown the concentration of species A (open circles) decreases with increasing x while the concentration of B increases with x. Mass diffusion will be in the direction of decreasing concentration, and there will be a net transport of species A to the right and of species B to the left. The physical mechanism becomes more evident if we look at the imaginary plane shown as a dashed line at x_o. Since the molecular motion is random, there is equal probability for motion of any molecule to the left or to the right. Accordingly, more molecules of species A will cross the plane from the left (since this is the side of higher A concentration) than from the right. Similarly, the concentration of B molecules is higher to the right of the plane than to the left, in which case random motion will provide for a *net* transfer of species B to the left. Of course, after a sufficient period of time, uniform concentrations of A and B would be achieved, and mass diffusion would cease. Under this condition there is no *net* transport of species A or B across the imaginary plane.

Mass diffusion may occur in liquids and solids, as well as in gases. However, since mass transfer is strongly influenced by molecular spacing, diffusion occurs more readily in gases than in liquids and more readily in liquids than in solids. Mass diffusion occurs in many practical situations. An example of diffusion in a gas mixture is the transport of nitrous oxide from an automobile exhaust through a stagnant atmosphere. There will be transport of the nitrous oxide in the direction away from the source, where its concentration is highest. Also very common and technically important is the diffusion of a single species through a stationary solid. For example, helium can diffuse through pyrex or carbon dioxide through rubber.

15.1.2 Fick's Law of Diffusion

Since the same physical mechanism is associated with heat and mass transfer by diffusion, it is not surprising that the corresponding rate equations are of the same form. The rate equation for mass diffusion is known as *Fick's law*, and for the transfer of species A in a binary mixture of *A* and *B*, it may be expressed in vector form as

$$\mathbf{j}_A = -\rho D_{AB} \nabla m_A \tag{15.1}$$

or

$$\mathbf{J}_A^* = -C D_{AB} \nabla x_A \tag{15.2}$$

These expressions are analogous to Fourier's law, Equation 2.3. Moreover, just as Fourier's law serves to define one important transport property, the thermal conductivity, Fick's law defines a second important transport property, namely the *binary diffusion coefficient* or *mass diffusivity*, D_{AB}.

The quantity $\mathbf{j}_A$ (kg/s $\cdot$m^2) is defined as the mass flux of species A. It is the amount of species A that is transferred per unit time and per unit area perpendicular to the direction of transfer, and it is proportional to the mixture mass density, $\rho = \rho_A + \rho_B$ (kg/m^3), and to the gradient in the mass fraction of species A, $m_A = \rho_A/\rho$. The species flux may also be evaluated on a molar basis, where $\mathbf{J}_A^*$ (kmol/s$\cdot$m^2) is the molar flux of species A. It is proportional to the total molar concentration of the mixture, $C = C_A + C_B$ (kmol/m^3), and to the gradient in the mole fraction of species A, $x_A = C_A/C$.[1] Note that the foregoing forms of Fick's law may be simplified in situations for which the total mass density, ρ, or the total molar concentration, C, are independent of x.

15.1.3 Restrictive Conditions

Despite the existence of several similarities, it is important to note that mass diffusion is a good deal more complicated than conduction. There are two restrictive conditions inherent in Equations 15.1 and 15.2. First, these equations are valid only when mass diffusion occurs as the result of a concentration gradient. It should be noted that mass diffusion may occur as the result of a

---

[1] Do not confuse x_i, the mole fraction of species i, with the spatial coordinate x. The former variable will always be subscripted with the species designation.

temperature gradient, a pressure gradient, or an external force, *as well* as originating due to a concentration gradient. In using Equations 15.1 and 15.2 to specify the transfer rates, we are assuming that these additional effects are not present or are negligible. In most practical problems of interest, this is in fact the case, and the dominant driving potential is the species concentration gradient. This condition is referred to as *ordinary diffusion*. Treatment of the other (higher order) effects is presented by Bird et al. [1–3].

The second restrictive condition on Equations 15.1 and 15.2 is that the fluxes (mass or molar) are measured relative to *coordinates which move with some average velocity for the mixture*. If the mass or molar flux of a species is expressed relative to a *fixed set of coordinates*, Equations 15.1 and 15.2 are not generally valid.

To obtain an expression for the mass flux relative to a fixed coordinate system, consider species A in a binary mixture of A and B. For species A, the mass flux, n_A'', relative to a fixed coordinate system is related to the species absolute velocity, v_A, by the expression

$$\mathbf{n}_A'' \equiv \rho_A \mathbf{v}_A \tag{15.3}$$

A value of v_A may be associated with any point in the mixture, and it is interpreted as the average velocity of all of the A particles in a small volume element about the point. An average, or aggregate, velocity may also be associated with the particles of species B, in which case

$$\mathbf{n}_B'' \equiv \rho_B \mathbf{v}_B \tag{15.4}$$

A *mass-average velocity for the mixture* may then be obtained from the requirement that

$$\rho \mathbf{v} = \mathbf{n}'' = \mathbf{n}_A'' + \mathbf{n}_B'' = \rho_A \mathbf{v}_A + \rho_B \mathbf{v}_B \tag{15.5}$$

giving

$$\mathbf{v} = m_A \mathbf{v}_A + m_B \mathbf{v}_B \tag{15.6}$$

It is important to note that we have defined the velocities $(\mathbf{v}_A, \mathbf{v}_B, \mathbf{v})$ and the fluxes $(\mathbf{n}_A'', \mathbf{n}_B'', \mathbf{n}'')$ as *absolute* quantities. That is, they are referred to axes that are fixed in space. The mass-average velocity $\mathbf{v}$ is a useful parameter of the binary mixture, since it need only be multiplied by the total mass density to obtain the total mass flux with respect to fixed axes. Moreover, since the velocities $\mathbf{v}_A$, $\mathbf{v}_B$, and $\mathbf{v}$ are averages associated with *aggregates* of particles, the fluxes $\mathbf{n}_A'', \mathbf{n}_B''$, and $\mathbf{n}''$ may be associated with transport due to *bulk* or *gross* motion.

We may now define the *mass flux of species A relative to the mixture mass-average velocity* as

$$\mathbf{j}_A \equiv \rho_A (\mathbf{v}_A - \mathbf{v}) \tag{15.7}$$

Whereas $\mathbf{n}_A''$ is the *absolute flux* of species A, $\mathbf{j}_A$ is the *relative* or *diffusive flux* of the species. It represents the motion of the species relative to the average motion of the mixture. It then follows from Equation 15.3 that

$$\mathbf{n}_A'' = \mathbf{j}_A + \rho_A \mathbf{v} \tag{15.8}$$

This expression indicates that there are two distinct contributions to the absolute flux of species A: a contribution due to *diffusion* (i.e., due to the motion of A *relative to the mass-average motion* of the mixture) and a contribution due to motion of A *with the mass-average motion* of the mixture. Substituting from Equations 15.1 and 15.5, we then obtain

$$\mathbf{n}_A'' = -\rho D_{AB} \nabla m_A + m_A (\mathbf{n}_A'' + \mathbf{n}_B'') \tag{15.9}$$

At this point it is useful to recap what we have done by noting the alternative formulations for the mass flux of species A. The form given by Equation 15.1 determines the transport of A relative to the mixture mass-average velocity, whereas the form given by Equation 15.9 determines the absolute transport of A. Note that, if the second term on the right-hand side of Equation 15.9 is finite in a mass transfer situation, the expression for the *absolute* flux of species A, Equation 15.9, is not analogous to that for the heat flux, Equation 2.3. We will later identify special situations for which the equations become analogous.

The foregoing considerations may now be extended to species B. The *mass flux of* B *relative to the mixture mass-average velocity* (the *diffusive flux*) is

$$\mathbf{j}_B \equiv \rho_B (\mathbf{v}_B - \mathbf{v}) \tag{15.10}$$

where

$$\mathbf{j}_B = -\rho D_{BA} \nabla m_B \tag{15.11}$$

It then follows from Equations 15.5, 15.7, and 15.10 that the diffusive fluxes in a binary mixture are related by

$$\mathbf{j}_A + \mathbf{j}_B = 0 \tag{15.12}$$

Moreover, if Equations 15.1 and 15.11 are substituted into Equation 15.12 and it is recognized that $\nabla m_A = -\nabla m_B$, since $m_A + m_B = 1$ for a binary mixture, it follows that

$$D_{BA} = D_{AB} \tag{15.13}$$

Hence, as in Equation 15.9, the *absolute* flux of species B may be expressed as

$$\mathbf{n}_B'' = -\rho D_{AB} \nabla m_B + m_B (\mathbf{n}_A'' + \mathbf{n}_B'') \tag{15.14}$$

Although the foregoing expressions are for binary diffusion on a *mass* basis, the same procedures can be used to obtain results on a *molar* basis. The absolute molar fluxes for species A and B may be expressed as

$$N''_A \equiv C_A v_A \qquad N''_B \equiv C_B v_B \tag{15.15}$$

and a *molar-average velocity for the mixture*, v^*, is obtained from the requirement that

$$N'' = N''_A + N''_B = C v^* = C_A v_A + C_B v_B \tag{15.16}$$

giving

$$v^* = x_A v_A + x_B v_B \tag{15.17}$$

The significance of the molar-average velocity is that, when multiplied by the total molar concentration, C, it provides the total molar flux, N'', with respect to fixed axes. Equation 15.15 provides the *absolute molar flux* of species A and B. In contrast, the molar flux of A relative to the mixture molar average velocity, J^*_A, termed the *diffusive flux*, may be obtained from Equation 15.2 or from the expression

$$J^*_A \equiv C_A(v_A - v^*) \tag{15.18}$$

To determine an expression similar in form to Equation 15.9, we combine Equations 15.15 and 15.18 to obtain

$$N''_A = J^*_A + C_A v^* \tag{15.19}$$

or, from Equations 15.2 and 15.16

$$N''_A = -C D_{AB} \nabla x_A + x_A(N''_A + N''_B) \tag{15.20}$$

Compare the molar fluxes given by Equations 15.2 and 15.20. In the first case the molar flux of A is designated relative to the mixture molar-average velocity, and in the second case it is the absolute molar flux that is specified. Note also that Equation 15.20 represents the absolute molar flux as the sum of a diffusive flux and a flux due to the bulk motion of the mixture. For the binary mixture, it also follows that

$$J^*_A + J^*_B = 0 \tag{15.21}$$

A *special case* for which the *absolute flux* of a species is *equal* to the *diffusive flux* pertains to what is termed a *stationary* medium. In terms of mass units, it is a medium for which $v = 0$, in which case $j_A = n''_A$. In terms of molar units, it is a medium for which $v^* = 0$ and hence $J^*_A = N''_A$. For this special case, the analogy between heat and mass transfer is complete since the rate equations then have the same physical form regardless of the reference frame.

EXAMPLE 15.1

Gaseous hydrogen is stored at elevated pressure in a rectangular container having steel walls of 10 mm thickness. The molar concentration of hydrogen in the steel at the inner surface is known to be $C_{A,1} = 1$ kmol/m^3, while the concentration of hydrogen in the steel at the outer surface is negligible, $C_{A,2} \approx 0$ kmol/m^3. The binary diffusion coefficient for hydrogen in steel is $D_{AB} = 0.26 \times 10^{-12}$ m^2/s. What is the molar diffusion flux for hydrogen through the steel?

SOLUTION

KNOWN:

Molar concentrations of hydrogen at inner and outer surfaces of a steel wall.

FIND:

Hydrogen molar diffusion flux.

SCHEMATIC:

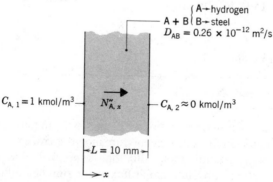

ASSUMPTIONS:

1. Steady-state conditions.
2. One-dimensional species diffusion through a plane wall.
3. Plane wall may be approximated as a stationary medium.
4. The molar concentration of hydrogen is much less than that of the steel $(C_A \ll C_B)$ and the total molar concentration, $C = C_A + C_B$, is uniform.
5. No chemical reactions of hydrogen in the steel.

ANALYSIS:

The molar diffusion flux of hydrogen with respect to a fixed reference frame is given by Equation 15.20. If the plane wall were, in fact, a stationary medium, $N_A'' + N_B''$, and hence the second term on the right-hand side of Equation 15.20, would be identically equal to zero. Although this condition does not apply exactly, the second term on the right-hand side of Equation 15.20 may still be neglected to an excellent approximation. In particular $N_B'' = 0$, and since $x_A \ll 1$, $(x_A N_A'') \ll N_A''$. Hence Equation 15.20 reduces to

$$N_{A,x}'' = -CD_{AB}\frac{dx_A}{dx}$$

In addition, since the total molar concentration is approximately constant,

$$C\frac{dx_A}{dx} = \frac{dC_A}{dx}$$

and

$$N_{A,x}'' = -D_{AB}\frac{dC_A}{dx}$$

With steady-state, one-dimensional conditions and no chemical reactions, it also follows from the species conservation requirement that $N_{A,x}''$ is independent of x. Hence dC_A/dx is constant and the species flux may be expressed as

$$N_{A,x}'' = -D_{AB}\frac{C_{A,2} - C_{A,1}}{L} = D_{AB}\frac{C_{A,1}}{L}$$

Hence,

$$N_{A,x}'' = 0.26 \times 10^{-12}\text{ m}^2/\text{s}\,\frac{1\text{ kmol/m}^3}{0.01\text{ m}}$$

$$N_{A,x}'' = 2.6 \times 10^{-11}\text{ kmol/s}\cdot\text{m}^2 \qquad \triangleleft$$

COMMENTS:

1. It is the existence of the conditions, $N_B'' = 0$ and $x_A \ll 1$, that will be implied in future use of the *stationary medium approximation*.

2. The mass flux of hydrogen is $n_{A,x}'' = \mathcal{M}_A N_{A,x}'' = 2$ kg/kmol $\times 2.6 \times 10^{-11}$ kmol/s·m² $= 5.2 \times 10^{-11}$ kg/s·m².

15.1.4   Mass Diffusion Coefficient

Considerable attention has been given to predicting the mass diffusion coefficient, D_{AB}, for the binary mixture of two gases, A and B. Assuming ideal gas behavior, kinetic theory may be used to show that

$$D_{AB} \sim p^{-1} T^{3/2}$$

This relation applies for restricted pressure and temperature ranges and is useful for estimating values of the diffusion coefficient at conditions other than those for which data is available. Bird et al. [1–3] provide detailed discussions of available theoretical treatments and comparisons with experiment.

For binary liquid mixtures, it is necessary to rely exclusively on experimental measurements. For small concentrations of liquid A (the solute) in liquid B (the solvent), D_{AB} is known to increase with increasing temperature. The mechanism of diffusion of gases, liquids and solids in solid media is extremely complicated and generalized theories are not available. Furthermore, only limited experimental results are available in the literature.

Data for binary diffusion in selected mixtures are presented in Table A.8. Skelland [4] and Ried and Sherwood [5] provide more recent treatments of this subject.

15.2   CONSERVATION OF SPECIES

Just as the first law of thermodynamics (the law of *conservation of energy*) plays an important role in heat transfer analyses, the law of *conservation of atomic species* plays an important role in the analysis of mass transfer problems. In this section we consider a general statement of this law, as well as application of the law to species diffusion in a stationary medium.

15.2.1   Conservation of Species for a Control Volume

Just as a general formulation of the energy conservation requirement, Equation 1.10, was expressed for the control volume of Figure 1.6, we may express an analogous species mass conservation requirement for the control volume of Figure 15.2.

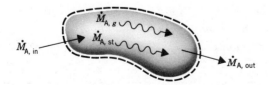

Figure 15.2   Conservation of mass for a control volume.

The rate at which the mass of some species enters a control volume minus the rate at which this species mass leaves the control volume must equal the rate at which the species mass is stored in the control volume.

For example, any species A may enter and leave the control volume due to both fluid motion and diffusion across the control surface; these processes are *surface* phenomena represented by $\dot{M}_{A,in}$ and $\dot{M}_{A,out}$. The same species A may also be generated, $\dot{M}_{A,g}$, and accumulated or stored, $\dot{M}_{A,st}$, *within* the control volume. The conservation equation may then be expressed on a rate basis as

$$\dot{M}_{A,in} + \dot{M}_{A,g} - \dot{M}_{A,out} = \dot{M}_{A,st} \tag{15.22}$$

Species generation will exist when there are chemical reactions occurring within the system. There would, for example, be a net production of species A if a dissociation reaction of the form $AB \rightarrow A + B$ were occurring.

15.2.2 The Mass Diffusion Equation

The foregoing result may be used to obtain a mass, or species, diffusion equation which is analogous to the heat equation of Chapter 2. We will consider a homogeneous medium that is a binary mixture of species A and B and that is *stationary*. That is, the mass average or molar average velocity of the mixture is everywhere zero and mass transfer may occur only by diffusion. The resulting equation could be solved for the species concentration distribution, which could, in turn, be used with Fick's law to determine the species diffusion rate at any point in the medium.

Allowing for concentration gradients in each of the x, y, and z coordinate directions, we first define a differential control volume, $dx\,dy\,dz$, within the medium (Figure 15.3) and then consider the processes that influence the distribution of species A. With the concentration gradients, diffusion must result in the transport of species A through the control surfaces. Moreover, relative to stationary coordinates, the species transport rates at opposite surfaces must be related by

$$n''_{A,x+dx}\,dy\,dz = n''_{A,x}\,dy\,dz + \frac{\partial[n''_{A,x}\,dy\,dz]}{\partial x}\,dx \tag{15.23a}$$

$$n''_{A,y+dy}\,dx\,dz = n''_{A,y}\,dx\,dz + \frac{\partial[n''_{A,y}\,dx\,dz]}{\partial y}\,dy \tag{15.23b}$$

$$n''_{A,z+dz}\,dx\,dy = n''_{A,z}\,dx\,dy + \frac{\partial[n''_{A,z}\,dx\,dy]}{\partial z}\,dz \tag{15.23c}$$

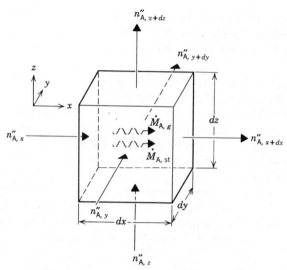

Figure 15.3 Differential control volume, *dxdydz*, for species diffusion analysis in cartesian coordinates.

In addition there may also be volumetric (homogeneous) chemical reactions occurring throughout the medium. The rate at which species A is generated within the control volume due to such reactions may be expressed as

$$\dot{M}_{A,g} = \dot{n}_A \, dxdydz \tag{15.24}$$

where $\dot{n}_A$ is the rate of increase of the mass of species A *due to chemical reactions* per unit volume of the mixture (kg/s·m³). Finally, the above processes may cause a change in the mass of species A stored within the control volume, and the rate at which this occurs may be expressed as

$$\dot{M}_{A,st} = \frac{\partial \rho_A}{\partial t} \, dxdydz \tag{15.25}$$

Recall that the general form of the conservation of species requirement on a rate basis is

$$\dot{M}_{A,in} + \dot{M}_{A,g} - \dot{M}_{A,out} = \dot{M}_{A,st} \tag{15.22}$$

Note also that the transfer rates, $n''_{A,x}$, $n''_{A,y}$ and $n''_{A,z}$, determine the mass inflow rate, $\dot{M}_{A,in}$, and Equations 15.23 determine the outflow rate, $\dot{M}_{A,out}$. Hence substituting from Equations 15.23, 15.24, and 15.25, canceling terms, and dividing by *dxdydz*, we obtain

$$-\frac{\partial n''_A}{\partial x} - \frac{\partial n''_A}{\partial y} - \frac{\partial n''_A}{\partial z} + \dot{n}_A = \frac{\partial \rho_A}{\partial t}$$

For a *stationary* medium, the mass average velocity, **v**, is zero, and from Equation 15.8 it follows that $\mathbf{n}_A'' = \mathbf{j}_A$. Hence, substituting the x, y, and z components of Equation 15.1, we obtain

$$\frac{\partial}{\partial x}\left(\rho D_{AB}\frac{\partial m_A}{\partial x}\right) + \frac{\partial}{\partial y}\left(\rho D_{AB}\frac{\partial m_A}{\partial y}\right) + \frac{\partial}{\partial z}\left(\rho D_{AB}\frac{\partial m_A}{\partial z}\right)$$

$$+ \dot{n}_A = \frac{\partial \rho_A}{\partial t} \tag{15.26a}$$

In terms of the molar concentration, the expression becomes

$$\frac{\partial}{\partial x}\left(C D_{AB}\frac{\partial x_A}{\partial x}\right) + \frac{\partial}{\partial y}\left(C D_{AB}\frac{\partial x_A}{\partial y}\right) + \frac{\partial}{\partial z}\left(C D_{AB}\frac{\partial x_A}{\partial z}\right)$$

$$+ \dot{N}_A = \frac{\partial C_A}{\partial t} \tag{15.27a}$$

In subsequent treatments of species diffusion phenomena, we shall work with simplified versions of the foregoing equations. In particular, if D_{AB} and ρ are constant, Equation 15.26a may be expressed as

$$\frac{\partial^2 \rho_A}{\partial x^2} + \frac{\partial^2 \rho_A}{\partial y^2} + \frac{\partial^2 \rho_A}{\partial z^2} + \frac{\dot{n}_A}{D_{AB}} = \frac{1}{D_{AB}}\frac{\partial \rho_A}{\partial t} \tag{15.26b}$$

Similarly, if D_{AB} and C are constant, Equation 15.27a may be expressed as

$$\frac{\partial^2 C_A}{\partial x^2} + \frac{\partial^2 C_A}{\partial y^2} + \frac{\partial^2 C_A}{\partial z^2} + \frac{\dot{N}_A}{D_{AB}} = \frac{1}{D_{AB}}\frac{\partial C_A}{\partial t} \tag{15.27b}$$

The foregoing equations are analogous to the heat equation, Equation 2.15. Hence it follows that for like boundary conditions, the solution to Equation 15.26b for $\rho_A(x,y,z,t)$ or to Equation 15.27b for $C_A(x,y,z,t)$ is of the same form as the solution to Equation 2.15 for $T(x,y,z,t)$.

The foregoing species diffusion equations may also be expressed in cylindrical and spherical coordinates. These alternative forms can be inferred from the analogous expressions for heat transfer, Equations 2.18 and 2.19, and in terms of the molar concentration have the form

Cylindrical Coordinates

$$\frac{1}{r}\frac{\partial}{\partial r}\left(C D_{AB} r \frac{\partial x_A}{\partial r}\right) + \frac{1}{r^2}\frac{\partial}{\partial \phi}\left(C D_{AB}\frac{\partial x_A}{\partial \phi}\right)$$

$$+ \frac{\partial}{\partial z}\left(C D_{AB}\frac{\partial x_A}{\partial z}\right) + \dot{N}_A = \frac{\partial C_A}{\partial t} \tag{15.28}$$

Spherical Coordinates

$$\frac{1}{r^2}\frac{\partial}{\partial r}\left(CD_{AB}r^2\frac{\partial x_A}{\partial r}\right) + \frac{1}{r^2\sin^2\theta}\frac{\partial}{\partial\phi}\left(CD_{AB}\frac{\partial x_A}{\partial\phi}\right)$$

$$+ \frac{1}{r^2\sin\theta}\frac{\partial}{\partial\theta}\left(CD_{AB}\sin\theta\frac{\partial x_A}{\partial\theta}\right) + \dot{N}_A = \frac{\partial C_A}{\partial t} \tag{15.29}$$

Simpler forms are, of course, associated with the absence of chemical reactions ($\dot{n}_A = \dot{N}_A = 0$) and with one-dimensional, steady-state conditions.

15.3 BOUNDARY AND INITIAL CONDITIONS

To determine the species concentration distribution in a stationary medium, it is necessary to solve the appropriate diffusion equation. However, such a solution depends on the physical conditions existing at the *boundaries* of the medium and, if the situation is time dependent, on conditions existing in the medium at some *initial time*. Two kinds of boundary conditions are commonly encountered in mass, or species, diffusion problems and are *analogous* to the first two heat transfer conditions of Table 2.1. The first condition corresponds to a situation for which the species concentration at the surface is maintained at a fixed or constant value, $x_{A,s}$. Expressing this condition on a molar basis for the surface at $x = 0$, we have

$$x_A(0,t) = x_{A,s} \tag{15.30}$$

The second condition corresponds to the existence of a fixed or constant species flux, $J_{A,s}^*$, at the surface and, using Fick's law, Equation 15.2, may be expressed as

$$-CD_{AB}\frac{\partial x_A}{\partial x}\bigg|_{x=0} = J_{A,s}^* \tag{15.31}$$

A special case of this condition corresponds to the *impermeable surface* for which $\partial x_A/\partial x|_{x=0} = 0$. At this point it is left to the student to envision physical situations that correspond to the above conditions.

With regard to the application of Equation 15.30, it should be noted that we are often interested in knowing the concentration of a species A in a liquid or a solid (species B) at the interface with a gas containing species A (Figure 15.4). For situations in which species A is only weakly soluble in a *liquid*, species B, the following form of Henry's law may be used to relate the mole fraction of A *in the liquid* B to the partial pressure of A in the gas phase outside the liquid.

$$x_A(0) = p_A/H \tag{15.32}$$

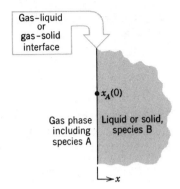

Figure 15.4 Species concentration at the gas–liquid or gas–solid interface.

The coefficient H is known as *Henry's constant*, and values for selected aqueous solutions are listed in Table A.9. Although H depends strongly on temperature, its dependence on pressure may generally be neglected for values of p up to 5 bars.

A similar equation may also be used to determine conditions at a *gas–solid* interface. We will confine our attention to problems in which, to a good approximation, the gas, species A, dissolves in the solid, species B, and a homogeneous solution is formed. In such cases gas transport in the solid is independent of the structure of the solid and may be treated as a diffusion process. In contrast there are many situations for which the porosity of a solid strongly influences gas transport through the solid. Treatment of such cases is left to more advanced texts on the subject [2, 4].

Treating the solid as a uniform substance, the concentration of the gas in the solid at the interface may be obtained through use of a property known as the *solubility*, S. It is defined by the expression

$$C_A(0) = S\,p_A \tag{15.33}$$

where p_A is the partial pressure (bar) of the gas adjoining the interface. The molar concentration of A in the solid at the interface, $C_A(0)$, is in units of kmol of A per cubic meter of solid, in which case the units of S must be *kmol of* A per cubic meter *of solid per* bar *partial pressure of* A. Values of S for several gas–solid combinations are given in Table A.10.

Note that the foregoing considerations relate to transfer across an interface for which a gas solute differs from the liquid or solid solvent. Another interfacial condition that has numerous applications, but is fundamentally different from the above, involves transfer of a species into a gas stream due to evaporation or sublimation from a liquid or solid surface, respectively. In this case, although there are two phases, they both involve the same species. The concentration (or partial pressure) of the vapor at the interface may readily be determined by assuming thermodynamic equilibrium. The partial pressure of the vapor will then correspond to saturation at the temperature of the interface and may be determined from standard thermodynamic tables.

EXAMPLE 15.2

Helium gas is stored at 20°C in a spherical container of fused silica (SiO$_2$), which has a diameter of 0.20 m and a wall thickness of 2 mm. If the container is charged to an initial pressure of 4 bar, what is the rate at which this pressure decreases with time?

SOLUTION

KNOWN:

Initial pressure of helium in a spherical, fused silica container of prescribed diameter, D, and wall thickness, L.

FIND:

The rate of change of the helium pressure, dp_A/dt.

SCHEMATIC:

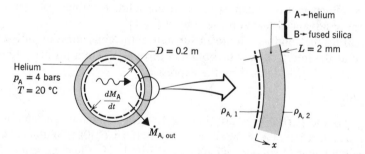

ASSUMPTIONS:

1. Since $D \gg L$, diffusion may be approximated as being one-dimensional through a plane wall.
2. Quasi-steady diffusion (pressure variation is sufficiently slow to permit assuming steady-state conditions for diffusion through the fused silica at any instant).
3. Fused silica is a stationary medium of uniform density ρ.
4. Pressure of helium in outside air is negligible.

PROPERTIES:

Table A.8, helium-fused silica mixture (293 K): $D_{AB} = 0.4 \times 10^{-13}$ m²/s.
Table A.10, helium-fused silica mixture (293 K): $S = 0.45 \times 10^{-3}$ kmol/m³·bar.

ANALYSIS:

The rate of change of the helium pressure may be obtained by applying the species conservation requirement, Equation 15.22, to a control volume about the helium. It follows that

$$-\dot{M}_{A,\text{out}} = \dot{M}_{A,\text{st}}$$

or, since the helium outflow is due to diffusion through the fused silica,

$$\dot{M}_{A,\text{out}} = n''_{A,x}A$$

and the change in mass storage is

$$\dot{M}_{A,\text{st}} = \frac{dM_A}{dt} = \frac{d(\rho_A V)}{dt}$$

the species balance reduces to

$$-n''_{A,x}A = \frac{d(\rho_A V)}{dt}$$

Recognizing that $\rho_A = \mathscr{M}_A C_A$ and applying the perfect gas law

$$C_A = \frac{p_A}{\mathscr{R}T}$$

the species balance becomes

$$\frac{dp_A}{dt} = -\frac{\mathscr{R}T}{\mathscr{M}_A V} A \, n''_{A,x}$$

For a stationary medium the absolute flux of species A through the fused silica is equal to the diffusion flux, $n''_{A,x} = j_{A,x}$, in which case, from Fick's law, Equation 15.1

$$n''_{A,x} = -\rho D_{AB} \frac{dm_A}{dx} = -D_{AB} \frac{d\rho_A}{dx}$$

Since diffusion through the fused silica is assumed to be characterized by steady-state conditions, assumption 2, it follows that

$$n''_{A,x} = D_{AB} \frac{\rho_{A,1} - \rho_{A,2}}{L}$$

The species densities, $\rho_{A,1}$ and $\rho_{A,2}$, pertain to conditions *within* the fused silica at its inner and outer surfaces, respectively, and may be evaluated from knowledge of the solubility through Equation 15.33. Hence with $\rho_A = \mathcal{M}_A C_A$,

$$\rho_{A,1} = \mathcal{M}_A S p_{A,i} = \mathcal{M}_A S p_A$$

$$\rho_{A,2} = \mathcal{M}_A S p_{A,o} = 0$$

where $p_{A,i}$ and $p_{A,o}$ represent the helium pressures at the inner and outer surfaces, respectively. Hence

$$n''_{A,x} = \frac{D_{AB}\mathcal{M}_A S p_A}{L}$$

and substituting into the species balance it follows that

$$\frac{dp_A}{dt} = -\frac{\mathcal{R}TAD_{AB}S}{LV}p_A$$

or with $A = \pi D^2$ and $V = \pi D^3/6$

$$\frac{dp_A}{dt} = -\frac{6\mathcal{R}TD_{AB}S}{LD}p_A$$

Substituting numerical values, the rate of change of the pressure is

$$\frac{dp_A}{dt} = \left[-6\left(0.08314 \ \frac{m^3 \cdot bar}{kmol \cdot K}\right) 293 \ K \left(0.4 \times 10^{-13} \ \frac{m^2}{s}\right) 0.45 \times 10^{-3} \right.$$
$$\left. \frac{kmol}{m^3 \cdot bar} \times 4 \ bar \right] \Big/ [0.002 \ m \ (0.2 \ m)]$$

$$\frac{dp_A}{dt} = -2.63 \times 10^{-8} \ bars/s \qquad \triangleleft$$

COMMENTS:

The foregoing result provides the initial (maximum) leakage rate for the system. The leakage rate decreases as the inside pressure decreases.

15.4   MASS DIFFUSION WITHOUT HOMOGENEOUS CHEMICAL REACTIONS

There are numerous situations for which species transfer occurs under one-dimensional, steady-state conditions. Moreover, there are many special cases for which results are entirely analogous to those obtained in Chapter 3 for heat transfer. In this section we will consider some of these special cases, restricting ourselves to situations that do not involve homogeneous chemical reactions. A *homogeneous* reaction is one that occurs *within* the medium. It is a *volumetric phenomenon* whose magnitude may vary from point to point in the medium. Homogeneous chemical reactions involve *species generation* and are therefore analogous to internal sources of heat generation. In contrast a *heterogeneous* chemical reaction is one which results from contact between the reactants and a surface. It is therefore a *surface phenomenon* that can be treated as a *boundary condition*. Heterogeneous chemical reactions are therefore analogous to the constant surface heat flux condition commonly encountered in heat diffusion problems.

In Section 15.4.1 we will consider species transfer in stationary media for which the species concentrations may be specified at the boundaries. In Section 15.4.2 this treatment will be extended to situations characterized by heterogeneous reactions at a surface boundary. Two special cases, both of which involve the gas phase and are without chemical reactions, are considered in Sections 15.4.3 and 15.4.4. One situation involves equimolar counterdiffusion in a gas mixture, and the other involves evaporation from a liquid surface in a column. Homogeneous chemical reactions are treated in Section 15.5. For each of the above cases, we are interested in determining the species concentration distribution and the corresponding species flux.

15.4.1 Stationary Media with Specified Surface Concentrations

In Section 15.1.3 we defined a *stationary medium* as one for which the molar (or mass) average velocity of the mixture is zero, in which case $N_A'' = J_A^*$ (or $n_A'' = j_A$). That is, the absolute flux of species A is equivalent to the diffusive flux, and from Equations 15.1 and 15.2 we have

$$n_A'' = -\rho D_{AB} \nabla m_A \tag{15.34}$$

$$N_A'' = -C D_{AB} \nabla x_A \tag{15.35}$$

From Equation 15.9 it is evident that Equation 15.34 will apply to a good approximation in nonstationary media for which $m_A \ll 1$ and $n_B'' \approx 0$. (Recall Example 15.1.) Similarly, from Equation 15.20 we see that Equation 15.35 serves as a good approximation in media for which $x_A \ll 1$ and $N_B'' \approx 0$. Such conditions often characterize dilute gas mixtures and liquid solutions (m_A or x_A is small) in which the motion of the solvent is negligible (n_B'' or N_B'' is small). Both of these conditions must be satisfied in order that the species flux due to bulk motion of the mixture be negligible compared to the diffusive flux. This situation can also occur for species diffusion in a solid, and satisfactory results may be obtained from the use of Equation 15.34 or 15.35. Examples include the diffusion of oxygen through rubber or helium through pyrex.

Let us begin by considering the one-dimensional diffusion of some species A through a planar medium of A and B, as shown in Figure 15.5. For steady-state conditions with no homogeneous chemical reactions, the molar form of the species diffusion equation, Equation 15.27a reduces to

$$\frac{d}{dx}\left(CD_{AB}\frac{dx_A}{dx}\right) = 0 \tag{15.36}$$

The standard solution method involves first obtaining the general solution for the species concentration distribution, $x_A(x)$, in the medium. The boundary conditions are then invoked to determine the arbitrary constants, and Equation 15.35 is used to obtain the species diffusion flux. Assuming the total molar concentration and the binary diffusion coefficient to be constant, it is easily verified that the distribution and the flux are of the form

$$x_A(x) = (x_{A,s2} - x_{A,s1})\frac{x}{L} + x_{A,s1} \tag{15.37}$$

and

$$N''_{A,x} = -CD_{AB}\frac{(x_{A,s2} - x_{A,s1})}{L} \tag{15.38}$$

Multiplying by the surface area A and substituting for $x_A \equiv C_A/C$, the molar rate is then

$$N_{A,x} = \frac{D_{AB}A}{L}(C_{A,s1} - C_{A,s2}) \tag{15.39}$$

From this expression we can define a resistance to species transfer by diffusion in a planar medium as

$$R_{m,\text{dif}} = \frac{(C_{A,s1} - C_{A,s2})}{N_{A,x}} = \frac{L}{D_{AB}A} \tag{15.40}$$

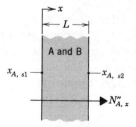

Figure 15.5   Mass transfer in a stationary planar medium.

Comparing the foregoing results with those obtained for one-dimensional, steady-state heat diffusion in a plane wall with no generation (Section 3.1), it is evident that, for the prescribed conditions, a direct analogy exists between heat and mass transfer by diffusion.

The analogy also applies to cylindrical and spherical systems. For steady-state conditions with no homogeneous reactions and one-dimensional diffusion in the radial direction of a cylindrical medium, Equation 15.28 reduces to

$$\frac{d}{dr}\left(r\, C D_{AB} \frac{dx_A}{dr}\right) = 0 \tag{15.41}$$

Similarly, from Equation 15.29 the appropriate form of the mass diffusion

Table 15.1 Summary of species diffusion solutions for stationary media with specified surface concentrations[a]

GEOMETRY	SPECIES CONCENTRATION DISTRIBUTION, $x_A(x)$ or $x_A(r)$	SPECIES DIFFUSION RESISTANCE, $R_{m,\,\mathrm{dif}}$
	$x_A(x) = (x_{A,s2} - x_{A,s1})\dfrac{x}{L} + x_{A,s1}$	$R_{m,\,\mathrm{dif}} = \dfrac{L}{D_{AB}A}$ [b]
	$x_A(r) = \dfrac{x_{A,s1} - x_{A,s2}}{\ln(r_1/r_2)}\ln\left(\dfrac{r}{r_2}\right) + x_{A,s2}$	$R_{m,\,\mathrm{dif}} = \dfrac{\ln(r_2/r_1)}{2\pi L\, D_{AB}}$ [c]
	$x_A(r) = \dfrac{x_{A,s1} - x_{A,s2}}{\left(\dfrac{1}{r_1} - \dfrac{1}{r_2}\right)}\left(\dfrac{1}{r} - \dfrac{1}{r_2}\right) + x_{A,s2}$	$R_{m,\,\mathrm{dif}} = \dfrac{1}{4\pi D_{AB}}\left(\dfrac{1}{r_1} - \dfrac{1}{r_2}\right)$ [c]

[a] Assuming C and D_{AB} are constant.
[b] $N_{A,x} = (C_{A,s1} - C_{A,s2})/R_{m,\,\mathrm{dif}}$.
[c] $N_{A,r} = (C_{A,s1} - C_{A,s2})/R_{m,\,\mathrm{dif}}$.

equation for the spherical system is

$$\frac{1}{r^2}\frac{d}{dr}\left(r^2 C D_{AB}\frac{dx_A}{dr}\right) = 0 \tag{15.42}$$

Note that Equations 15.41 and 15.42, as well as Equation 15.36, dictate that the molar transfer rate, $N_{A,r}$ or $N_{A,x}$, is constant in the direction of transfer (r or x). Assuming C and D_{AB} to be constant, it is a simple matter to obtain the general solution to Equations 15.41 and 15.42. For prescribed surface species concentrations, the corresponding solution and the diffusion resistance are summarized in Table 15.1.

EXAMPLE 15.3

Hydrogen gas is maintained at pressures of 3 bar and 1 bar on opposite sides of a plastic membrane which is 0.3 mm thick. The temperature is 25°C throughout, and the binary diffusion coefficient of hydrogen in the plastic is 8.7×10^{-8} m²/s. The solubility of hydrogen in the membrane is 1.5×10^{-3} kmol/m³·bar. What is the mass diffusion flux of hydrogen through the membrane?

SOLUTION

KNOWN:

Pressure of hydrogen gas on opposite sides of a plastic membrane.

FIND:

The hydrogen mass diffusion flux, $n_{A,x}''$ (kg/s·m²).

SCHEMATIC:

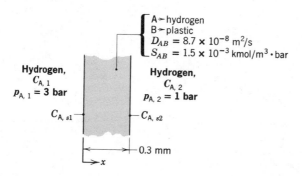

ASSUMPTIONS:

1. Steady-state conditions.
2. One-dimensional diffusion.
3. Membrane is a stationary medium.
4. $x_A \ll 1$.
5. Uniform total molar concentration.
6. No homogeneous chemical reactions.

ANALYSIS:

The general rate expression, Equation 15.20, reduces to Equation 15.35 because of assumptions 3 and 4. For the prescribed conditions, Equation 15.35 reduces to Equation 15.38, which may be expressed as

$$N''_{A,x} = C D_{AB} \frac{x_{A,s1} - x_{A,s2}}{L} = \frac{D_{AB}}{L}(C_{A,s1} - C_{A,s2})$$

The molar concentrations of hydrogen at the surfaces of the membrane may be obtained from Equation 15.33, which is of the form

$$C_{A,s} = S_{AB} p_A$$

Hence

$$C_{A,s1} = 1.5 \times 10^{-3} \, \text{kmol/m}^3 \cdot \text{bar} \times 3 \, \text{bar} = 4.5 \times 10^{-3} \, \text{kmol/m}^3$$

$$C_{A,s2} = 1.5 \times 10^{-3} \, \text{kmol/m}^3 \cdot \text{bar} \times 1 \, \text{bar} = 1.5 \times 10^{-3} \, \text{kmol/m}^3$$

$$N''_{A,x} = \frac{8.7 \times 10^{-8} \, \text{m}^2/\text{s}}{0.3 \times 10^{-3} \, \text{m}} (4.5 \times 10^{-3} - 1.5 \times 10^{-3}) \, \text{kmol/m}^3$$

$$N''_{A,x} = 8.7 \times 10^{-7} \, \text{kmol/s} \cdot \text{m}^2$$

On a mass basis,

$$n''_{A,x} = N''_{A,x} \mathcal{M}_A$$

where the molecular weight of hydrogen is 2 kg/kmol. Hence

$$n''_{A,x} = 8.7 \times 10^{-7} \, \text{kmol/s} \cdot \text{m}^2 \times 2 \, \text{kg/kmol}$$

$$n''_{A,x} = 1.74 \times 10^{-6} \, \text{kg/s} \cdot \text{m}^2 \qquad \triangleleft$$

COMMENTS:

Note that the molar concentrations of hydrogen in the gas phase, $C_{A,1}$ and $C_{A,2}$, differ from the surface concentrations in the membrane and may be calculated from the perfect gas equation of state

$$c_A = \frac{p_A}{\mathcal{R} T}$$

where $\mathcal{R} = 8.314 \times 10^{-2}$ m^3·bar/kmol·K. It follows that $c_{A,1} = 0.121$ kmol/m^3 and $c_{A,2} = 0.040$ kmol/m^3. Even though $C_{A,s2} < C_{A,2}$, hydrogen transport will occur from the membrane to the gas at $p_{A,2} = 1$ bar. This seemingly anomalous result may be explained by recognizing that the two concentrations are based on *different* volumes; in one case the concentration is per unit volume of the membrane and in the other case it is per unit volume of the adjoining gas phase. For this reason it is not possible to infer the direction of hydrogen transport from a simple comparison of the numerical values of $C_{A,s2}$ and $C_{A,2}$.

15.4.2 Stationary Media with Catalytic Surface Reactions

Many mass transfer problems involve specification of the species flux, rather than the species concentration, at a surface. One such problem relates to the process of catalysis, which involves the use of special surfaces to promote chemical reactions. Often a one-dimensional diffusion analysis may be used to *approximate* the performance of a catalytic reactor.

Consider the system of Figure 15.6. A catalytic surface is placed in a gas stream for the purpose of promoting a heterogeneous chemical reaction involving species A. Assume that the reaction results in the production of species A at a rate given by $\dot{N}_A''$, which is defined as the molar rate of production of species A per unit surface area of the catalyst. Such a reaction enhances the

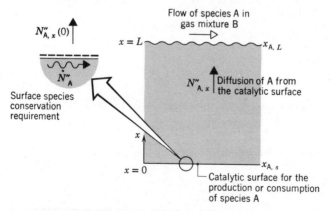

Figure 15.6 One-dimensional diffusion with heterogeneous catalysis.

surface concentration of A, and in order to maintain steady-state conditions it is necessary that the rate of species transfer from the surface, $N''_{A,x}$, be equal to the surface reaction rate. That is,

$$N''_{A,x}(0) = \dot{N}''_A \qquad (15.43)$$

It may be assumed that species A leaves the surface as a result of one-dimensional transfer through a thin film of thickness L and that no reactions occur within the film itself. The mole fraction of A at $x = L$, $x_{A,L}$, corresponds to conditions in the mainstream of the mixture and is presumed to be known. Although in practice bulk motion will contribute to the transfer of A through the film, it is reasonable to assume, as a first estimate, that such effects are negligible and that transfer occurs exclusively by diffusion. Assuming that the remaining species of the mixture are represented as a single species B and that the medium is stationary, Equation 15.27a reduces to

$$\frac{d}{dx}\left(C D_{AB} \frac{dx_A}{dx}\right) = 0 \qquad (15.44)$$

where D_{AB} is the binary diffusion coefficient for A in B and B may be a multicomponent mixture. Assuming C and D_{AB} to be constant, Equation 15.44 may be solved subject to the conditions that

$$x_A(L) = x_{A,L}$$

and

$$N''_{A,x}(0) = -C D_{AB} \frac{dx_A}{dx}\bigg|_{x=0} = \dot{N}''_A \qquad (15.45)$$

This expression follows from Equation 15.43 and the substitution of Fick's law, Equation 15.35.

For a catalytic surface, the surface reaction rate, $\dot{N}''_A$, generally depends upon the surface concentration, $C_A(0)$. For a *first-order reaction* that results in species consumption at the surface, the reaction rate is of the form

$$\dot{N}''_A = -k''_1 C_A(0) \qquad (15.46)$$

where k''_1 (m/s) is the reaction rate constant. Accordingly, the surface boundary condition, Equation 15.45 reduces to

$$-D_{AB} \frac{dx_A}{dx}\bigg|_{x=0} = -k''_1 x_A(0) \qquad (15.47)$$

Solving Equation 15.44 subject to the above conditions, it is readily verified that the concentration distribution is linear and of the form

$$\frac{x_A(x)}{x_{A,L}} = \frac{1 + (xk''_1/D_{AB})}{1 + (Lk''_1/D_{AB})} \qquad (15.48)$$

At the catalytic surface this result reduces to

$$\frac{x_A(0)}{x_{A,L}} = \frac{1}{1 + (Lk_1''/D_{AB})}$$

(15.49)

and the molar flux is

$$N_A''(0) = -CD_{AB}\frac{dx_A}{dx}\bigg|_{x=0} = -k_1'' C x_A(0)$$

or

$$N_A''(0) = -\frac{k_1'' C x_{A,L}}{1 + (Lk_1''/D_{AB})}$$

(15.50)

The negative sign implies mass transfer *to* the surface (in the negative x direction).

Two limiting cases of the foregoing results are of special interest. For the limit $k_1'' \to 0$, $(Lk_1''/D_{AB}) \ll 1$ and Equations 15.49 and 15.50 reduce to

$$\frac{x_{A,s}}{x_{A,L}} \approx 1$$

and

$$N_A''(0) \approx -k_1'' C x_{A,L}$$

In such cases the rate of reaction is controlled by the finite surface reaction rate, and the limitation due to diffusion is negligible. The process is said to be *reaction limited*. Conversely, for the limit $k_1'' \to \infty$, $(Lk_1''/D_{AB}) \gg 1$ and Equations 15.49 and 15.50 reduce to

$$x_{A,s} \approx 0$$

and

$$N_{A,s}'' \approx -\frac{CD_{AB} x_{A,L}}{L}$$

In this case the reaction is controlled by the rate of diffusion to the surface, and the process is said to be *diffusion limited*.

15.4.3 Equimolar Counterdiffusion

A special case of the conditions considered in Section 15.4.1 is shown in Figure 15.7. A channel connecting two large reservoirs contains an isothermal, perfect gas mixture of species A and B. The species concentrations are maintained constant in each of the reservoirs, such that $x_{A,0} > x_{A,L}$ and $x_{B,0}$

$< x_{B,L}$, while the total pressure $p = p_A + p_B$ is uniform throughout. The species concentration gradients cause the diffusion of A molecules in the direction of increasing x and the diffusion of B molecules in the opposite direction. However, under steady-state conditions, it is also necessary for the diffusion of the species to occur at equal and opposite rates, in which case the total molar flux must be zero relative to stationary coordinates. This requirement is expressed as

$$N''_{A,x} + N''_{B,x} = 0 \tag{15.51}$$

and is necessitated by the fact that, with p and T constant, the total molar concentration C must also be constant throughout the system. This condition may only be maintained if the molar flux of A to the right is balanced by the molar flux of B to the left, in which case the process is referred to as *equimolar counterdiffusion*. From Equations 15.16 and 15.51, it is evident that the mixture molar average velocity, v_x^*, must be zero. The mixture is therefore *stationary* and is, in fact, a special case of the situation considered in Section 15.4.1. From Equation 15.20 it therefore follows that

$$N''_{A,x} = -CD_{AB}\frac{dx_A}{dx} \tag{15.52}$$

Similarly

$$N''_{B,x} = -CD_{BA}\frac{dx_B}{dx} \tag{15.53}$$

From Equations 15.13 and 15.51 it is also evident that

$$\frac{dx_A}{dx} = -\frac{dx_B}{dx} \tag{15.54}$$

Since the mole fraction of either species may be expressed as a ratio of the species partial pressure to the total pressure, that is, $x_i = (p_i/p)$, it follows that

$$\frac{dp_A}{dx} = -\frac{dp_B}{dx} \tag{15.55}$$

Since we are dealing with one-dimensional, steady-state conditions involving no homogeneous chemical reactions, the species diffusion rates are independent of x. Assuming D_{AB} to be constant and recalling that C is independent of x, Equation 15.52 may be expressed as

$$N_{A,x}\int_0^L \frac{dx}{A} = -CD_{AB}\int_{x_{A,0}}^{x_{A,L}} dx_A \tag{15.56}$$

Assuming the channel cross-sectional area, A, to be constant, we then obtain for the species A molar diffusion rate

$$N_{A,x} = C D_{AB} A \frac{x_{A,0} - x_{A,L}}{L} = D_{AB} A \frac{C_{A,0} - C_{A,L}}{L} \tag{15.57}$$

Alternatively, substituting from the equation of state,

$$p_A = C_A \mathscr{R} T \tag{15.58}$$

we obtain

$$N_{A,x} = \frac{D_{AB} A}{\mathscr{R} T} \frac{p_{A,0} - p_{A,L}}{L} \tag{15.59}$$

Similar results may be obtained for species B. Note that the foregoing results imply that the species mole fraction gradients, and therefore the species pressure gradients, are linear and of opposite sign (Figure 15.7).

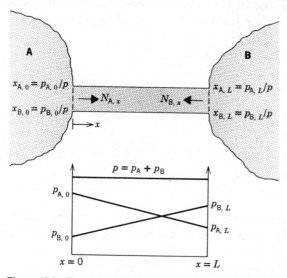

Figure 15.7 Equimolar counterdiffusion in a binary, isothermal, perfect gas mixture.

EXAMPLE 15.4

To avoid overpressurization as well as to maintain a pressure close to one atmosphere, an industrial pipeline containing ammonia gas is vented to ambient air. Venting is achieved by tapping the pipe and inserting a 3-mm-diameter tube, which extends for 20 m into the atmosphere. With the entire system operating at 25°C, determine:

1. The mass rate of ammonia lost to the atmosphere in kg/h.
2. The mass rate of contamination of the pipe with air in kg/h.
3. The mole and mass fractions of air in the pipe, downstream of the tube, when the ammonia flowrate is 5 kg/h.

SOLUTION

KNOWN:

Conditions in an ammonia pipeline vented to atmosphere.

FIND:

1. Mass rate of ammonia lost through the vent, n_A.
2. Mass rate of air contamination in the pipe, n_B.
3. Mole and mass fractions of air in the pipe.

SCHEMATIC:

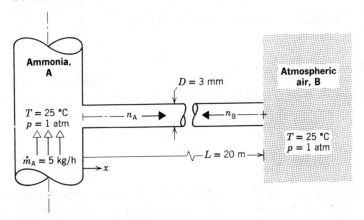

ASSUMPTIONS:

1. Steady-state conditions.
2. One-dimensional axial diffusion in the tube.
3. Constant properties.
4. Uniform temperature and total pressure, $p = p_A + p_B$, in the tube.
5. Negligible mole fraction of air in the pipe, $x_{B,0} \ll 1$, and negligible mole fraction of ammonia in the atmosphere, $x_{A,L} \ll 1$.
6. No chemical reactions in the tube.

PROPERTIES:

Table A.8, ammonia-air mixture (298 K): $D_{AB} = 0.28 \times 10^{-4}$ m^2/s.

ANALYSIS:

1. The prescribed conditions are precisely those that provide for equimolar counterdiffusion. Hence the mass rate of ammonia lost to the atmosphere may be obtained from Equation 15.59 where

$$n_A = \mathscr{M}_A N_A$$

with

$$N_A = \frac{D_{AB} A}{\mathscr{R} T} \frac{p_{A,0} - p_{A,L}}{L}$$

From assumption 5, $p_{A,0} = p$ and $p_{A,L} = 0$. Hence

$$N_A = \frac{0.28 \times 10^{-4}\ \text{m}^2/\text{s} \times \dfrac{\pi}{4}(0.003\ \text{m})^2}{8.205 \times 10^{-2}\ \text{m}^3 \cdot \text{atm/kmol} \cdot \text{K}\ (298\ \text{K})} \times \frac{(1 - 0)\ \text{atm}}{20\ \text{m}} \times 3600\ \text{s/h}$$

or

$$N_A = 1.46 \times 10^{-9}\ \text{kmol/h}$$

Hence, with a molecular weight for the ammonia of $\mathscr{M}_A = 17$ kg/kmol,

$$n_A = 17\ \text{kg/kmol} \times 1.46 \times 10^{-9}\ \text{kmol/h}$$

or

$$n_A = 2.48 \times 10^{-8}\ \text{kg/h} \qquad \triangleleft$$

2. The mass rate of contamination of the pipe by air is determined from the equimolar diffusion requirement, Equation 15.51 in which case

$$N_B = -N_A = -1.46 \times 10^{-9}\ \text{kmol/h}$$

Hence, with the molecular weight of air given by $\mathscr{M}_B = 28.97$ kg/kmol

$$n_B = \mathscr{M}_B N_B = -28.97\ \text{kg/kmol} \times 1.46 \times 10^{-9}\ \text{kmol/h}$$

or

$$n_B = -4.23 \times 10^{-8}\ \text{kg/h} \qquad \triangleleft$$

3. For an ammonia flowrate of $\dot{m}_A = 5$ kg/h, the mass fraction of air in the pipe is

$$m_{B,0} = \frac{-n_B}{\dot{m}_A} = \frac{4.23 \times 10^{-8} \text{ kg/h}}{5 \text{ kg/h}}$$

or

$$m_{B,0} = 0.85 \times 10^{-8} \qquad \lhd$$

With the molar flowrate of ammonia in the pipe being $\dot{m}_A / \mathcal{M}_A$ = 5 kg/h/17 kg/kmol, it follows that the mole fraction of air in the pipe is

$$x_{B,0} = \frac{-N_B}{\dot{m}_A / \mathcal{M}_A} = \frac{1.46 \times 10^{-9} \text{ kmol/h}}{5 \text{ kg/h/17 kg/kmol}}$$

or

$$x_{B,0} = 4.96 \times 10^{-9} \qquad \lhd$$

COMMENTS:

The foregoing results for $m_{B,0}$ and $x_{B,0}$ are based on assumption 5, which presumes that $C \approx C_A$ (or $\rho \approx \rho_A$) in the pipe. The validity of this assumption is confirmed by the small value of $x_{B,0}$ (or $m_{B,0}$).

15.4.4   Evaporation in a Column

Let us now consider diffusion in the binary gas mixture of Figure 15.8. Fixed species concentrations, $x_{A,L}$ and $x_{B,L}$, are maintained at the top of a beaker containing a liquid layer of species A, and the system is at constant pressure and temperature. Since equilibrium exists between the vapor and liquid phases at the

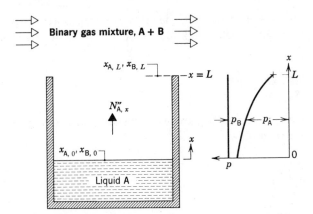

Figure 15.8   Evaporation of liquid A into a binary gas mixture, $A + B$.

liquid interface, the vapor concentration is determined from the saturated vapor properties available in standard thermodynamic tables. Assuming that $x_{A,0} > x_{A,L}$, species A *evaporates* from the liquid interface and is transferred upward by diffusion. Moreover for steady-state, one-dimensional conditions with no chemical reactions, the absolute molar flux of A must be a constant throughout the column. Hence

$$\frac{dN''_{A,x}}{dx} = 0 \tag{15.60}$$

Since p and T are constant, it follows that the total molar concentration, $C = C_A + C_B$, is also constant, in which case $x_A + x_B = 1$ throughout the column. Having assumed that $x_{A,0} > x_{A,L}$, we conclude that $x_{B,L} > x_{B,0}$ and therefore that species B must diffuse from the top of the column toward the liquid interface. However, if species B is insoluble in liquid A, steady-state conditions can only be maintained if the downward diffusion of B is balanced by an upward bulk motion. That is, the absolute flux of species B must be zero ($N''_{B,x} = 0$). A major implication of this condition is that we are no longer dealing with a stationary medium, and hence Equation 15.35 does not apply. However, an appropriate expression for $N''_{A,x}$ may be obtained by substituting the requirement that $N''_{B,x} = 0$ into Equation 15.20, giving

$$N''_{A,x} = -C D_{AB} \frac{dx_A}{dx} + x_A N''_{A,x} \tag{15.61}$$

or, from Equation 15.16,

$$N''_{A,x} = -C D_{AB} \frac{dx_A}{dx} + C_A v_x^* \tag{15.62}$$

From this equation it is evident that the diffusive transport of species A $[-C D_{AB}(dx_A/dx)]$ is augmented by bulk motion ($C_A v_x^*$). Rearranging Equation 15.61, we obtain

$$N''_{A,x} = -\frac{C D_{AB}}{1 - x_A} \frac{dx_A}{dx} \tag{15.63}$$

For constant p and T, C and D_{AB} are also constant. Substituting Equation 15.63 into (15.60), we then obtain

$$\frac{d}{dx}\left(\frac{1}{1 - x_A} \frac{dx_A}{dx}\right) = 0$$

Integrating twice, we have

$$-\ln(1 - x_A) = C_1 x + C_2$$

and applying the conditions $x_A(0) = x_{A,0}$ and $x_A(L) = x_{A,L}$, the constants of integration may be evaluated and the concentration distribution becomes

$$\frac{1 - x_A}{1 - x_{A,0}} = \left(\frac{1 - x_{A,L}}{1 - x_{A,0}}\right)^{x/L} \tag{15.64}$$

Since $1 - x_A = x_B$, we also obtain

$$\frac{x_B}{x_{B,0}} = \left(\frac{x_{B,L}}{x_{B,0}}\right)^{x/L} \tag{15.65}$$

To determine the evaporation rate of species A, Equation 15.64 is first used to evaluate the concentration gradient (dx_A/dx). Substituting the result into Equation 15.63, it follows that

$$N''_{A,x} = \frac{C D_{AB}}{L} \ln\left(\frac{1 - x_{A,L}}{1 - x_{A,0}}\right) \tag{15.66}$$

15.5 MASS DIFFUSION WITH HOMOGENEOUS CHEMICAL REACTIONS

Just as heat diffusion may be influenced by internal sources of heat, species transfer by diffusion may be influenced by the existence of homogeneous chemical reactions. We restrict our attention to stationary media, in which case Equation 15.34 or 15.35 provides a good approximation to the absolute species flux. Accordingly, if we also assume steady, one-dimensional transfer in the x direction and that D_{AB} and C are constant, Equation 15.27b reduces to

$$D_{AB}\frac{d^2 C_A}{dx^2} + \dot{N}_A = 0 \tag{15.67}$$

If there are no homogeneous chemical reactions involving species A, the volumetric species *production* rate, $\dot{N}_A$, is zero. When reactions do occur, they are often of the form

Zero-order reaction: $\dot{N}_A = k_0$

First-order reaction: $\dot{N}_A = k_1 C_A$

That is, the reaction may occur at a constant rate (zero order) or at a rate which is proportional to the local concentration (first order). Note that the units of k_0 and k_1 are kmol/s·m^3 and 1/s, respectively. Remember that if $\dot{N}_A$ is positive, the reaction results in the generation of species A; if it is negative, it results in the consumption of A.

In many applications the species of interest is being converted to another form through a first-order chemical reaction, and Equation 15.67 becomes

$$D_{AB}\frac{d^2C_A}{dx^2} - k_1 C_A = 0 \tag{15.68}$$

This linear, homogeneous differential equation has a general solution of the form, case 7 of Appendix B.1,

$$C_A(x) = C_1 e^{mx} + C_2 e^{-mx} \tag{15.69}$$

where $m = (k_1/D_{AB})^{1/2}$ and the constants C_1 and C_2 depend on the prescribed boundary conditions. The form of this equation is identical to that which characterizes heat conduction in an extended surface.

Consider the situation illustrated in Figure 15.9. Gas A is soluble in liquid B, where it is transferred by diffusion and experiences a first-order chemical reaction. The solution is dilute, and the concentration of A in the liquid at the interface is a known constant $C_{A,0}$. Moreover, the bottom of the liquid container is impermeable to A. The appropriate boundary conditions are then

$$C_A(0) = C_{A,0}$$

$$dC_A/dx|_{x=L} = 0$$

Using these boundary conditions with Equation 15.69, it may be shown, after some manipulation, that

$$C_A(x) = C_{A,0}(\cosh mx - \tanh mL \sinh mx) \tag{15.70}$$

In such a problem, quantities of special interest would likely be the concentration of A at the bottom and the flux of A across the gas–liquid interface. Applying Equation 15.70 at $x = L$, we obtain

$$C_A(L) = C_{A,0}\frac{(\cosh^2 mL - \sinh^2 mL)}{\cosh mL} = \frac{C_{A,0}}{\cosh mL} \tag{15.71}$$

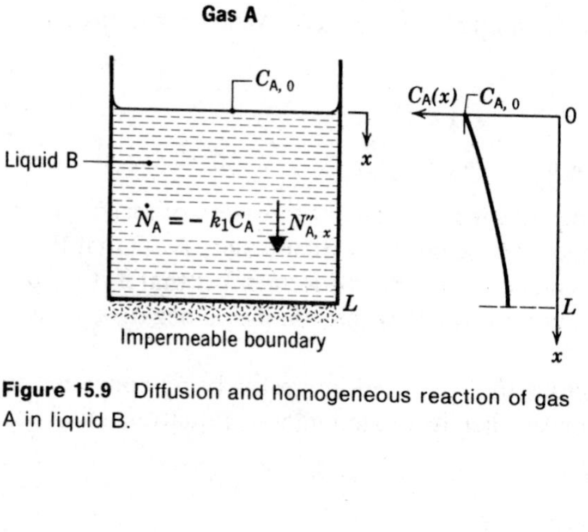

Figure 15.9 Diffusion and homogeneous reaction of gas A in liquid B.

Moreover,

$$N''_{A,x}(0) = -D_{AB}\, dC_A/dx\,|_{x=0}$$

$$= -D_{AB}C_{A,0}\, m(\sinh mx - \tanh mL \cosh mx)|_{x=0}$$

or,

$$N''_{A,x}(0) = D_{AB}C_{A,0}\, m \tanh mL$$

EXAMPLE 15.5

A solid waste treatment system operates on the principle of aerobic fermentation to decompose organic matter into its basic chemical constituents. That is, bacteria distributed within the organic matter will utilize gaseous oxygen originating from the atmosphere in the decomposition process. Consider a plane layer of organic matter that has thickness L and is supported by a concrete slab. The top of the layer is exposed to atmospheric air that maintains a fixed molar concentration of oxygen, $C_{A,0}$, in the layer (at the exposed surface). The diffusion coefficient, D_{AB}, of oxygen in the organic matter is known, as is the volumetric rate at which the oxygen is consumed by biochemical reactions. This consumption depends on the local concentration of oxygen and may be expressed as

$$\dot{N}_A = -k_1 C_A \,(\text{kmol/s} \cdot \text{m}^3)$$

1. From consideration of a differential control volume within the organic matter, derive a differential equation which could be solved for the local concentration of oxygen, $C_A(x)$. Assume one-dimensional, steady-state conditions. Write the general solution to this equation.

2. Write appropriate boundary conditions that could be used to obtain the constants in the general solution, and determine the specific form of $C_A(x)$.

SOLUTION

KNOWN:

Oxygen transport by diffusion with a homogeneous, first-order chemical reaction.

FIND:

1. Differential equation for the concentration distribution, $C_A(x)$, and general solution.

2. Appropriate boundary conditions and final solution form.

SCHEMATIC:

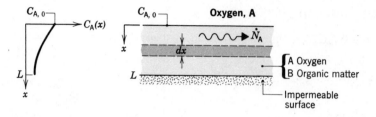

ASSUMPTIONS:

1. Steady-state conditions.
2. One-dimensional diffusion.
3. Medium B is stationary.
4. $x_A \ll 1$.
5. Homogeneous chemical reaction is present.
6. Constant properties and uniform total molar concentration.
7. Impermeable bottom.

ANALYSIS:

1. Applying a species mass balance to the differential control volume of the schematic, it follows from Equation 15.22 that, on a molar basis,

$$\dot{N}_{A, \text{in}} + \dot{N}_{A, g} - \dot{N}_{A, \text{out}} = \dot{N}_{A, \text{st}}$$

where

$$\dot{N}_{A, \text{in}} = N_{A, x} = -D_{AB} A \frac{dC_A}{dx}$$

$$\dot{N}_{A, \text{out}} = N_{A, x} + \frac{dN_{A, x}}{dx} \, dx = -D_{AB} A \frac{dC_A}{dx} - D_{AB} A \frac{d^2 C_A}{dx^2} \, dx$$

$$\dot{N}_{A, g} = \dot{N}_A A \, dx$$

$$\dot{N}_{A, \text{st}} = 0$$

Substituting the rate equations into the species balance and dividing by the area A, it follows that

$$D_{AB} \frac{d^2 C_A}{dx^2} - k_1 C_A = 0 \qquad \qquad \triangleleft$$

The foregoing differential equation is of second order with constant coefficients, and its general solution, case 7 of Appendix B.1, is

$$C_A(x) = C_1 e^{-mx} + C_2 e^{mx} \qquad \lhd$$

where

$$m \equiv (k_1/D_{AB})^{1/2}$$

2. Appropriate boundary conditions are of the form

$$C_A(0) = C_{A,0} \qquad \lhd$$

$$\left. \frac{dC_A}{dx} \right|_{x=L} = 0 \qquad \lhd$$

Applying these conditions to the general solution, we obtain

$$C_{A,0} = C_1 + C_2$$

$$\left. \frac{dC_A}{dx} \right|_{x=L} = -mC_1 e^{-mL} + mC_2 e^{mL} = 0$$

Hence, from the second condition

$$C_1 = C_2 e^{2mL}$$

and from the first condition

$$C_{A,0} = C_2(e^{2mL} + 1)$$

Hence

$$C_2 = \frac{C_{A,0}}{e^{2mL} + 1}$$

$$C_1 = \frac{C_{A,0} e^{2mL}}{e^{2mL} + 1}$$

$$C_A(x) = \frac{C_{A,0}}{e^{2mL} + 1} [e^{m(2L-x)} + e^{mx}] \qquad \lhd$$

15.6 TRANSIENT DIFFUSION

Results analogous to those of Chapter 5 may be obtained for the transient diffusion of a dilute species A in a stationary medium. Assuming no homogeneous reactions, constant D_{AB} and C, and one-dimensional transfer in the x direction, Equation 15.27b reduces to

$$D_{AB} \frac{\partial^2 C_A}{\partial x^2} = \frac{\partial C_A}{\partial t} \tag{15.73}$$

Assuming an initial uniform concentration,

$$C_A(x,0) = C_{A,i} \tag{15.74}$$

Equation 15.73 may be solved for boundary conditions that depend on the particular geometry and surface conditions. If, for example, the geometry is that of a plane wall of thickness $2L$ with surface convection, the boundary conditions may be expressed as

$$\left. \frac{\partial C_A}{\partial x} \right|_{x=0} = 0 \tag{15.75}$$

$$C_A(L,t) = C_{A,s} \tag{15.76}$$

Equation 15.75 describes the symmetry requirement at the midplane. Equation 15.76 represents the surface convection condition corresponding to the assumption that the *mass transfer Biot number*, $Bi_m = h_m L/D_{AB}$, is much larger than unity. This situation corresponds to the requirement that the resistance to species transfer by diffusion in the medium be much larger than the resistance to species transfer by convection at the surface of the medium. If this situation is taken to the limit of $Bi_m \to \infty$, or $Bi_m^{-1} \to 0$, it follows that the freestream species concentration, $C_{A,\infty}$, may be replaced by the surface concentration, $C_{A,s}$, in a transient diffusion analysis. Note, however, that $C_{A,s}$ represents the species concentration in the medium and hence must be determined by using Equation 15.32 or 15.33.

The analogy between heat and mass transfer may be more conveniently applied if we nondimensionalize the above equations. Introducing a dimensionless concentration and time, as follows,

$$\gamma^* \equiv \frac{\gamma}{\gamma_i} = \frac{C_A - C_{A,s}}{C_{A,i} - C_{A,s}} \tag{15.77}$$

$$t_m^* \equiv \frac{D_{AB}t}{L^2} \equiv Fo_m \tag{15.78}$$

and substituting into Equation 15.73, we obtain

$$\frac{\partial^2 \gamma^*}{\partial x^{*2}} = \frac{\partial \gamma^*}{\partial Fo_m} \tag{15.79}$$

where $x^* = x/L$. Similarly, the initial and boundary conditions may be expressed as

$$\gamma^*(x^*,0) = 1 \tag{15.80}$$

$$\left.\frac{\partial \gamma^*}{\partial x^*}\right|_{x^*=0} = 0 \tag{15.81}$$

and

$$\gamma^*(1, t_m^*) = 0 \tag{15.82}$$

One need only compare Equations 15.79 through 15.82 with Equations 5.22 through 5.24 and Equation 5.25 for the case of $Bi \to \infty$ to confirm the existence of the analogy. Note that for $Bi \to \infty$, Equation 5.25 reduces to $\theta^*(1, t^*) = 0$, which is analogous to Equation 15.82. Hence the two systems of equations must have equivalent solutions.

The correspondence between variables for transient heat and mass diffusion problems is summarized in Table 15.2. From this correspondence it is possible to use many of the previous heat transfer results to solve transient mass diffusion problems. For example, replacing θ_o^* and Fo by γ_o^* and Fo_m, Figure 5.6 could be used, for $Bi^{-1} = 0$, to determine the midplane concentration, $C_{A,o}$. The remaining Heisler charts may be applied in a similar fashion, along with results obtained for the semiinfinite solid.

Table 15.2 Correspondence between heat and mass transfer variables for transient diffusion

HEAT TRANSFER	MASS TRANSFER
$\theta^* = \dfrac{T - T_\infty}{T_i - T_\infty}$	$\gamma^* = \dfrac{C_A - C_{A,s}}{C_{A,i} - C_{A,s}}$
$1 - \theta^* = \dfrac{T - T_i}{T_\infty - T_i}$	$1 - \gamma^* = \dfrac{C_A - C_{A,i}}{C_{A,s} - C_{A,i}}$
$Fo = \dfrac{\alpha t}{L^2}$	$Fo_m = \dfrac{D_{AB} t}{L^2}$
$Bi = \dfrac{hL}{k}$	$Bi_m = \dfrac{h_m L}{D_{AB}}$
$\dfrac{x}{2\sqrt{\alpha t}}$	$\dfrac{x}{2\sqrt{D_{AB} t}}$

EXAMPLE 15.6

A slab of salt (NaCl) of thickness L is used to support a deep layer of water. The salt dissolves in the water, maintaining a fixed mass density, $\rho_{A,s}$ (kg/m^3), at the water-salt interface.

1. If the salt density in the water is initially everywhere zero, how does this density vary with position and time after contact between the solid salt and the water is made?

2. What is the surface recession rate, dL/dt, and how does the magnitude of the surface recession vary with time? If the mass density of solid salt is $\rho_A(S) = 2165$ kg/m^3 and its density in solution at the surface is $\rho_{A,s} = 380$ kg/m^3, by how much will the surface recede after 24 hours? The saltwater diffusion coefficient is $D_{AB} = 1.2 \times 10^{-9}$ m^2/s.

SOLUTION

KNOWN:

Salt from a slab of thickness L is diffusing into a deep layer of water under transient conditions.

FIND:

1. Density distribution of salt in the water, $\rho_A(x,t)$.

2. Surface recession rate, dL/dt, and magnitude of the surface recession as a function of time.

SCHEMATIC:

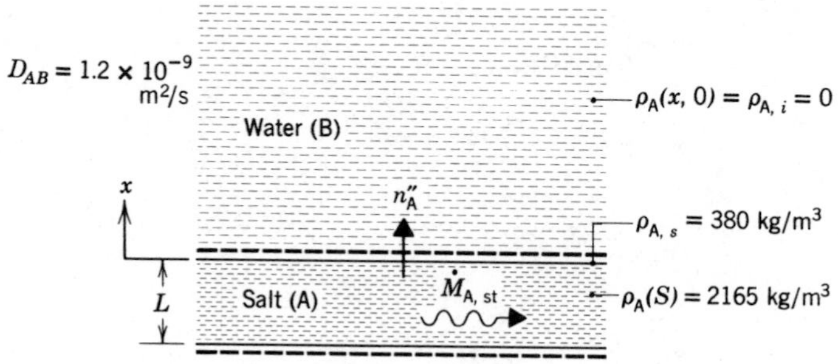

ASSUMPTIONS:

1. One-dimensional species diffusion in x.

2. No chemical reactions.

3. The water is stagnant ($n_B'' = 0$) and the solution is dilute ($m_A \ll 1$).

4. The water is a semiinfinite medium.

5. Constant properties, including the total density ρ of the saltwater solution.

ANALYSIS:

1. With the foregoing assumptions, $n''_A = j_A$ and Equation 15.26b reduces to

$$\frac{\partial^2 \rho_A}{\partial x^2} = \frac{1}{D_{AB}}\frac{\partial \rho_A}{\partial t}$$

which is analogous to Equation 5.14. Moreover, the initial condition

$$\rho_A(x,0) = \rho_{A,i} = 0$$

and the boundary conditions

$$\rho_A(\infty,t) = \rho_{A,i} = 0 \qquad \rho_A(0,t) = \rho_{A,s}$$

are analogous to Equations 5.15, 5.28, and 5.29, respectively, which correspond to case 1 of Figure 5.15. By analogy to Equation 5.30, the density distribution of salt in the water is then

$$\frac{\rho_A(x,t) - \rho_{A,s}}{\rho_{A,i} - \rho_{A,s}} = \mathrm{erf}\left[\frac{x}{2(D_{AB}t)^{1/2}}\right]$$

or

$$\rho_A(x,t) = \rho_{A,s}\left\{1 - \mathrm{erf}\left[\frac{x}{2(D_{AB}t)^{1/2}}\right]\right\} \qquad \triangleleft$$

2. Performing a mass balance on a control volume about the slab, it follows from Equation 15.22 that

$$-\dot{M}_{A,\mathrm{out}} = \dot{M}_{A,\mathrm{st}}$$

or, for a unit surface area,

$$-n''_{A,s} = \frac{d[\rho_A(S)L]}{dt}$$

where, by analogy to Equation 5.31, the species mass flux at the surface is

$$n''_{A,s} = \frac{D_{AB}\rho_{A,s}}{(\pi D_{AB}t)^{1/2}} = \left(\frac{D_{AB}}{\pi t}\right)^{1/2}\rho_{A,s}$$

Hence

$$\frac{dL}{dt} = -\left(\frac{D_{AB}}{\pi t}\right)^{1/2}\frac{\rho_{A,s}}{\rho_A(S)}$$

Integrating

$$\int_0^{\Delta L} dL = -\left(\frac{D_{AB}}{\pi}\right)^{1/2} \frac{\rho_{A,s}}{\rho_A(S)} \int_0^t \frac{dt}{t^{1/2}}$$

and the magnitude of the surface recession that has occurred after the time t is

$$\Delta L = -2\frac{\rho_{A,s}}{\rho_A(S)}\left(\frac{D_{AB}t}{\pi}\right)^{1/2} \qquad \triangleleft$$

For the prescribed conditions, it follows that

$$\Delta L = -2\frac{380 \text{ kg/m}^3}{2165 \text{ kg/m}^3}\left(\frac{1.2 \times 10^{-9} \text{ m}^2/\text{s} \times 24 \text{ h} \times 3600 \text{ s/h}}{\pi}\right)^{1/2}$$

or

$$\Delta L = 2.02 \times 10^{-3} \text{ m} = 2.02 \text{ mm} \qquad \triangleleft$$

COMMENTS:

1. Recognize that D_{AB} is analogous to α through their appearance in the diffusion equations. In contrast D_{AB} is analogous to k through their appearance in the rate equations (Fick's and Fourier's laws). Note how D_{AB} has been substituted for both α and k in the foregoing implementation of the analogy.

2. The above results will apply to only a first approximation, since the large salt concentration near the surface precludes the existence of a dilute solution in this region.

REFERENCES

1. Bird, R. B., "Theory of Diffusion" in *Advances in Chemical Engineering*, *1*, 170, 1956.
2. Bird, R. B., W. E. Stewart, and E. N. Lightfoot, *Transport Phenomena*, Wiley, New York, 1960.
3. Hirschfelder, J. O., C. F. Curtiss, and R. B. Bird, *Molecular Theory of Gases and Liquids*, Wiley, New York, 1954.
4. Skelland, A. H. P., *Diffusional Mass Transfer*, Wiley, New York, 1974.
5. Ried, R. C. and T. K. Sherwood, *The Properties of Gases and Liquids*, McGraw-Hill, New York, 1966.

PROBLEMS

15.1 Assuming air to be composed exclusively of O_2 and N_2, with their partial pressures in the ratio 0.21 to 0.79, what are their mass fractions?

15.2 A mixture of CO_2 and N_2 is in a container at 25°C, with each species having a partial pressure of 1 bar. Calculate the molar concentration, the mass density, the mole fraction, and the mass fraction of each species.

15.3 Consider a perfect gas mixture of n species.

a) Derive a general equation for determining the mass fraction of species i from knowledge of the mole fraction and the molecular weight of each of the n species. Derive a general equation for determining the mole fraction of species i from knowledge of the mass fraction and the molecular weight of each of the n species.

b) In a mixture containing equal mole fractions of O_2, N_2, and CO_2, what is the mass fraction of each species? In a mixture containing equal mass fractions of O_2, N_2, and CO_2, what is the mole fraction of each species?

15.4 Consider air in a closed, cylindrical container with its axis vertical and with opposite ends maintained at different temperatures. Assume that the total pressure of the air is uniform throughout the container.

a) If the bottom surface is colder than the top surface, what is the nature of conditions within the container? For example, will there be vertical gradients of the species (O_2 and N_2) concentrations? Is there any motion of the air? Does mass transfer occur?

b) What is the nature of conditions within the container if it is inverted (i.e., the warm surface is now at the bottom)?

15.5 Gaseous hydrogen at 10 bars and 27°C is stored in a 100-mm-diameter spherical tank having a steel wall 2 mm thick. The molar concentration of hydrogen in the steel is known to be 1.50 $kmol/m^3$ at the inner surface and negligible at the outer surface, while the diffusion coefficient of hydrogen in steel is approximately 0.3 $\times 10^{-12}$ m^2/s.

a) What is the initial rate of mass loss of hydrogen by diffusion through the tank wall?

b) What is the initial rate of pressure drop within the tank?

15.6 A thin plastic membrane is used in a process that involves the separation of helium from a gas stream. Under steady-state conditions the concentration of helium in the membrane is known to be 0.02 $kmol/m^3$ and 0.005 $kmol/m^3$ at the inner and outer surfaces, respectively. If the membrane is 1 mm thick and the binary diffusion coefficient of helium with respect to the plastic is 10^{-9} m^2/s, what is the diffusion flux?

15.7 Estimate the values of the mass diffusion coefficients, D_{AB}, for the binary mixtures of these gases at 350 K and 1 atm: ammonia–air and hydrogen–air.

15.8 Beginning with a properly defined differential control volume element, derive the diffusion equation, on a molar basis, for species A in a three-dimensional (cartesian coordinates), stationary medium (B) considering species generation with constant properties. Compare your result with Equation 15.27b.

15.9 Consider the radial diffusion of a gaseous species (A) through the wall of a plastic tube (B) and allow for the occurrence of chemical reactions that provide for the depletion of A at a rate $\dot{N}_A$ (kmol/s·m³). Derive a differential equation that governs the molar concentration of species A in the plastic.

15.10 Beginning with a properly defined differential control volume element, derive the diffusion equation, on a molar basis, for species A in a one-dimensional, spherical, stationary medium (B) considering species generation. Compare your result with Equation 15.29.

15.11 Oxygen gas is maintained at pressures of 2 bars and 1 bar on opposite sides of a rubber membrane that is 0.5 mm thick, and the entire system is at 25°C.

a) What is the molar diffusion flux of O_2 through the membrane?

b) What are the molar concentrations of O_2 on both sides of the membrane (outside the rubber)?

15.12 Insulation is known to degrade (experience an increase in thermal conductivity) if it is subjected to water vapor condensation, and the problem may occur in home insulation during cold periods of the year. It occurs when vapor in a humidified room diffuses through the dry wall (plaster board) and condenses in the adjoining insulation. Estimate the mass diffusion rate for a 3 m × 5 m wall, under conditions for which the vapor pressure is 0.03 bar in the room air and 0.0 bar in the insulation. The dry wall is 10 mm thick, and the solubility of water vapor in the wall material is approximately 5×10^{-3} kmol/m³·bar. The binary diffusion coefficient for water vapor in the dry wall is approximately 10^{-9} m²/s.

15.13 A rubber plug that is 20 mm thick and has a surface area of 300 mm² is used to contain CO_2 at 25°C and a pressure of 5 bars in a 10-liter vessel. What is the rate of mass loss of CO_2 from the vessel? What is the reduction in pressure that would be experienced over a 24-hour period?

15.14 Helium gas at 25°C and a pressure of 4 bars is contained in a glass cylinder of 100 mm inside diameter and 5 mm thickness. What is the rate of mass loss per unit length of the cylinder?

15.15 Helium gas at 25°C and a pressure of 4 bars is stored in a spherical pyrex container of 200 mm inside diameter and 10 mm thickness. What is the rate of mass loss from the container?

15.16 The reduction of nitric oxide (NO) emissions from an automobile exhaust can be promoted by using a catalytic converter, and the following reaction is known to occur at the catalytic surface

$$NO + CO \rightarrow \tfrac{1}{2}N_2 + CO_2$$

Hence the concentration of NO may be reduced by passing the exhaust gases over the surface. The rate at which the reduction of NO occurs at the catalyst is known to be governed by a first-order reaction of the form given by Equation 15.46. Moreover, as a first approximation it may be assumed that NO reaches the surface as a result of one-dimensional diffusion through a thin gaseous film of thickness L that adjoins the surface.

Referring to Figure 15.6 consider a situation for which the exhaust gas is at 500°C and 1.2 bars and the mole fraction of NO is $x_{A,L} = 0.15$. If $D_{AB} = 10^{-4}$ m²/s, $k_1'' = 0.05$ m/s, and the film thickness may be approximated as $L = 1$ mm, what is the mole fraction of NO at the catalytic surface and what is the NO removal rate for a surface of area $A = 200$ cm^2?

15.17 Carbon dioxide and nitrogen experience equimolar counterdiffusion in a circular tube whose length and diameter are 1 m and 50 mm, respectively. The system is at a total pressure of 1 atm and a temperature of 25°C. The ends of the tubes are connected to large chambers in which the species concentrations are maintained at fixed values. The partial pressure of CO_2 at one end is known to be 100 mm Hg, while at the other end it is 50 mm Hg. What is the mass transfer rate of CO_2 through the tube?

15.18 Consider evaporation in a column, where the vapor A is transferred through a gas B. Which of the following two limiting cases is characterized by the largest evaporation rate? (a) Gas B has unlimited solubility in liquid A. (b) Gas B is completely insoluble in liquid A. What is the ratio of the evaporation rate for (a) to that of (b) if the vapor pressure is zero at the top of the column and if the saturated vapor pressure is one tenth of the total pressure?

15.19 An open pan of diameter 0.2 m and height 80 mm contains water at 27°C and is exposed to ambient air at 27°C and 25 percent relative humidity.

a) Determine the evaporation rate assuming only mass diffusion occurs.

b) Determine the evaporation rate considering convective or bulk motion effects.

15.20 As an employee of the Los Angeles Air Quality Commission you have been asked to develop a model for computing the distribution of NO_2 in the atmosphere. The situation you are to consider is one in which the molar flux of NO_2 at ground level, $N''_{A,o}$, is presumed known. This flux is attributed to automobile and smokestack emissions. It is also known that the concentration of NO_2 at a distance well above ground level is zero and that NO_2 reacts chemically in the atmosphere. In particular, NO_2 reacts with unburned hydrocarbons (in a process which is activated by sunlight) to produce PAN (peroxyacetylnitrate), the final product of photochemical smog. The reaction is first order, and the local rate at which it occurs may be expressed as $\dot{N}_A = -k_1 C_A$.

a) Assuming steady-state conditions and a stagnant atmosphere, obtain an expression for the vertical distribution, $C_A(x)$, of the molar concentration of NO_2 in the atmosphere.

b) If an NO_2 partial pressure of $p_A = 2 \times 10^{-6}$ bars is sufficient to cause pulmonary damage, what is the value of the ground level molar flux for which you would issue a smog alert? You may assume an isothermal atmosphere at $T = 300$ K, a reaction coefficient of $k_1 = 0.03$ (1/s), and an NO_2–air diffusion coefficient of $D_{AB} = 0.15 \times 10^{-4}$ m^2/s.

15.21 Consider the problem of oxygen transfer from the interior lung cavity, across the lung tissue, to the network of blood vessels on the opposite side. The lung tissue (species B) may be approximated as a plane wall of thickness L. The inhalation process may be assumed to maintain a constant molar concentration, $C_A(0)$, of oxygen (species A) in the tissue at its inner surface ($x = 0$), and assimilation of oxygen by the blood may be assumed to maintain a constant molar concentration, $C_A(L)$, of oxygen in the tissue at its outer surface ($x = L$). There is oxygen consumption in the tissue due to metabolic processes, and the reaction is zero order, with $\dot{N}_A = -k_0$. Obtain expressions for the distribution of the oxygen concentration in the tissue and for the rate of assimilation of oxygen by the blood per unit tissue surface area.

15.22 In Problem 15.20 NO_2 transport by diffusion in a stagnant atmosphere was considered under the assumption of steady-state conditions. However, the problem is actually time dependent, and a more realistic approach would be one that attempts to account for transient effects. Consider the ground level emission of NO_2 to begin in the early morning (at $t = 0$), when the NO_2 concentration in the atmosphere is everywhere zero. Emission occurs throughout the day at a constant flux $N''_{A,0}$, and the NO_2 again experiences a first-order photochemical reaction in the atmosphere whose rate may be expressed as $\dot{N}_A = -k_1 C_A$.

a) From consideration of a differential element in the atmosphere, derive a differential equation that could be used to determine the molar concentration $C_A(x,t)$. State appropriate initial and boundary conditions.

b) Obtain an expression for $C_A(x,t)$ under the special condition for which photochemical reactions may be neglected. For this condition what is the molar concentration of NO_2 at ground level and at an elevation of 100 m three hours after the start of the emissions, if $N''_{A,0} = 3 \times 10^{-11}$ kmol/s·m² and $D_{AB} = 0.15 \times 10^{-4}$ m²/s?

15.23 The presence of CO_2 in solution is essential to the growth of aquatic plant life in any body of water. The CO_2 is used as a reactant in the photosynthesis process.

Consider a stagnant body of water in which the concentration of CO_2 (ρ_A) is everywhere zero. At time $t = 0$, the water is exposed to a source of CO_2, which maintains the surface ($x = 0$) concentration at a fixed value $\rho_{A,0}$. For time $t > 0$, CO_2 will begin to accumulate in the water, but the accumulation is inhibited by CO_2 consumption due to photosynthesis. The time rate at which this consumption occurs per unit volume is equal to the product of a reaction rate constant k_1 and the local CO_2 concentration $\rho_A(x,t)$.

a) Write (do not derive) a differential equation that could be used to determine $\rho_A(x,t)$ in the water. What does each term in the equation represent physically? Assume one-dimensional conditions.

b) Write appropriate boundary conditions that could be used to obtain a particular solution, assuming a "deep" body of water. What would be the form of this solution for the special case of negligible CO_2 consumption ($k_1 \approx 0$)?

15.24 Steel is carburized in a high-temperature process that depends on the transfer of carbon by diffusion. The value of the diffusion coefficient is strongly temperature dependent and may be approximated as

$$D_{c-s}(m^2/s) \approx 2 \times 10^{-5} \exp\left[-17{,}000/T(K)\right]$$

If the process is effected at $1000°C$ and a carbon mole fraction of 0.02 is maintained at the surface of the steel, how much time is required to elevate the carbon content of the steel from an initial value of 0.1 percent to a value of 1.0 percent at a depth of 1 mm?

Appendix A
Thermophysical Properties
of Matter[1]

[1] The convention used to present numerical values of the properties is illustrated by this example

T	$v \cdot 10^7$	$k \cdot 10^3$
(K)	(m^2/s)	(W/m·K)
300	0.349	521

where $v = 0.349 \times 10^{-7}$ m^2/s and $k = 521 \times 10^{-3} = 0.521$ W/m·K at 300 K.

Table A.1 Thermophysical Properties of Selected Metallic Solids[a]

| | MELTING POINT K | PROPERTIES AT 300 K | | | | PROPERTIES AT VARIOUS TEMPERATURES (K) k, W/m·K / c_p, J/kg·K | | | | | | | | | |
		ρ kg/m³	c_p J/kg·K	k W/m·K	$\alpha \cdot 10^6$ m²/s	100	200	400	600	800	1000	1200	1500	2000	2500
Aluminum Pure	933	2702	903	237	97.1	302	237	240	231	218					
						482	798	949	1033	1146					
Alloy 2024-T6 (4.5% Cu, 1.5% Mg, 0.6% Mn)	775	2770	875	177	73.0	65	163	186	186						
						473	787	925	1042						
Alloy 195, Cast (4.5% Cu)		2790	883	168	68.2			174	185						
								—	—						
Beryllium	1550	1850	1825	200	59.2	990	301	161	126	106	90.8	78.7			
						203	1114	2191	2604	2823	3018	3227	3519		
Bismuth	545	9780	122	7.86	6.59	16.5	9.69	7.04							
						112	120	127							
Boron	2573	2500	1107	27.0	9.76	190	55.5	16.8	10.6	9.60	9.85				
						128	600	1463	1892	2160	2338				
Cadmium	594	8650	231	96.8	48.4	203	99.3	94.7							
						198	222	242							
Chromium	2118	7160	449	93.7	29.1	159	111	90.9	80.7	71.3	65.4	61.9	57.2	49.4	
						192	384	484	542	581	616	682	779	937	
Cobalt	1769	8862	421	99.2	26.6	167	122	85.4	67.4	58.2	52.1	49.3	42.5		
						236	379	450	503	550	628	733	674		
Copper Pure	1358	8933	385	401	117	482	413	393	379	366	352	339			
						252	356	397	417	433	451	480			
Commercial bronze (90% Cu, 10% Al)	1293	8800	420	52	14		42	52	59						
							785	460	545						
Phosphor gear bronze (89% Cu, 11% Sn)	1104	8780	355	54	17		41	65	74						
							—	—	—						
Cartridge brass (70% Cu, 30% Zn)	1188	8530	380	110	33.9	75	95	137	149						
						—	360	395	425						
Constantan (55% Cu, 45% Ni)	1493	8920	384	23	6.71	17	19								
						237	362								

[a]Adapted from References 1 7.

Table A.1 Continued

COMPOSITION	MELTING POINT K	PROPERTIES AT 300 K				PROPERTIES AT VARIOUS TEMPERATURES (K) k, W/m·K / c_p, J/kg·K									
		ρ kg/m³	c_p J/kg·K	k W/m·K	$\alpha \cdot 10^6$ m²/s	100	200	400	600	800	1000	1200	1500	2000	2500
Germanium	1211	5360	322	59.9	34.7	232 / 190	96.8 / 290	43.2 / 337	27.3 / 348	19.8 / 357	17.4 / 375	17.4 / 395			
Gold	1336	19300	129	317	127	327 / 109	323 / 124	311 / 131	298 / 135	284 / 140	270 / 145	255 / 155			
Iridium	2720	22500	130	147	50.3	172 / 90	153 / 122	144 / 133	138 / 138	132 / 144	126 / 153	120 / 161	111 / 172		
Iron															
Pure	1810	7870	447	80.2	23.1	134 / 216	94.0 / 384	69.5 / 490	54.7 / 574	43.3 / 680	32.8 / 975	28.3 / 609	32.1 / 654		
Armco (99.75% pure)		7870	447	72.7	20.7	95.6 / 215	80.6 / 384	65.7 / 490	53.1 / 574	42.2 / 680	32.3 / 975	28.7 / 609	31.4 / 654		
Carbon steels															
Plain carbon (Mn ≤ 1%, Si ≤ 0.1%)		7854	434	60.5	17.7			56.7 / 487	48.0 / 559	39.2 / 685	30.0 / 1169				
AISI 1010		7832	434	63.9	18.8			58.7 / 487	48.8 / 559	39.2 / 685	31.3 / 1168				
Carbon-silicon (Mn ≤ 1%, 0.1% < Si ≤ 0.6%)		7817	446	51.9	14.9			49.8 / 501	44.0 / 582	37.4 / 699	29.3 / 971				
Carbon-manganese-silicon (1% < Mn ≤ 1.65%, 0.1% < Si ≤ 0.6%)		8131	434	41.0	11.6			42.2 / 487	39.7 / 559	35.0 / 685	27.6 / 1090				
Chromium (low) steels															
½Cr-¼Mo-Si (0.18% C, 0.65% Cr, 0.23% Mo, 0.6% Si)		7822	444	37.7	10.9			38.2 / 492	36.7 / 575	33.3 / 688	26.9 / 969				
1Cr-½Mo (0.16% C, 1% Cr, 0.54% Mo, 0.39% Si)		7858	442	42.3	12.2			42.0 / 492	39.1 / 575	34.5 / 688	27.4 / 969				
1Cr-V (0.2% C, 1.02% Cr, 0.15% V)		7836	443	48.9	14.1			46.8 / 492	42.1 / 575	36.3 / 688	28.2 / 969				

Composition	Melting Point (K)	ρ (kg/m³)	cₚ (J/kg·K)	k (W/m·K)	α·10⁶ (m²/s)	100	200	400	600	800	1000	1200	1500	2000	2500
Stainless steels AISI 302	1670	8055	480	15.1	3.91			17.3 / 512	20.0 / 559	22.8 / 585	25.4 / 606				
AISI 304		7900	477	14.9	3.95	9.2 / 272	12.6 / 402	16.6 / 515	19.8 / 557	22.6 / 582	25.4 / 611	28.0 / 640	31.7 / 682		
AISI 316		8238	468	13.4	3.48			15.2 / 504	18.3 / 550	21.3 / 576	24.2 / 602				
AISI 347		7978	480	14.2	3.71			15.8 / 513	18.9 / 559	21.9 / 585	24.7 / 606				
Lead	601	11340	129	35.3	24.1	39.7 / 118	36.7 / 125	34.0 / 132	31.4 / 142						
Magnesium	923	1740	1024	156	87.6	169 / 649	159 / 934	153 / 1074	149 / 1170	146 / 1267					
Molybdenum	2894	10240	251	138	53.7	179 / 141	143 / 224	134 / 261	126 / 275	118 / 285	112 / 295	105 / 308	98 / 330	90 / 380	86 / 459
Nickel Pure	1728	8900	444	90.7	23.0	164 / 232	107 / 383	80.2 / 485	65.6 / 592	67.6 / 530	71.8 / 562	76.2 / 594	82.6 / 616		
Nichrome (80% Ni, 20% Cr)	1672	8400	420	12	3.4			14 / 480	16 / 525	21 / 545					
Inconel X-750 (73% Ni, 15% Cr, 6.7% Fe)	1665	8510	439	11.7	3.1	8.7 / –	10.3 / 372	13.5 / 473	17.0 / 510	20.5 / 546	24.0 / 626	27.6 / –	33.0 / –		
Niobium	2741	8570	265	53.7	23.6	55.2 / 188	52.6 / 249	55.2 / 274	58.2 / 283	61.3 / 292	64.4 / 301	67.5 / 310	72.1 / 324	79.1 / 347	
Palladium	1827	12020	244	71.8	24.5	76.5 / 168	71.6 / 227	73.6 / 251	79.7 / 261	86.9 / 271	94.2 / 281	102 / 291	110 / 307		
Platinum Pure	2045	21450	133	71.6	25.1	77.5 / 100	72.6 / 125	71.8 / 136	73.2 / 141	75.6 / 146	78.7 / 152	82.6 / 157	89.5 / 165	99.4 / 179	
Alloy 60Pt-40Rh (60% Pt, 40% Rh)	1800	16630	162	47	17.4			52 / –	59 / –	65 / –	69 / –	73 / –	76 / –		
Rhenium	3453	21100	136	47.9	16.7	58.9 / 97	51.0 / 127	46.1 / 139	44.2 / 145	44.1 / 151	44.6 / 156	45.7 / 162	47.8 / 171	51.9 / 186	
Rhodium	2236	12450	243	150	49.6	186 / 147	154 / 220	146 / 253	136 / 274	127 / 293	121 / 311	116 / 327	110 / 349	112 / 376	
Silicon	1685	2330	712	148	89.2	884 / 259	264 / 556	98.9 / 790	61.9 / 867	42.2 / 913	31.2 / 946	25.7 / 967	22.7 / 992		

Table A.1 Continued

COMPOSITION	MELTING POINT K	PROPERTIES AT 300 K				PROPERTIES AT VARIOUS TEMPERATURES (K) k, W/m·K / c_p, J/kg·K									
		ρ kg/m³	c_p J/kg·K	k W/m·K	$\alpha \cdot 10^6$ m²/s	100	200	400	600	800	1000	1200	1500	2000	2500
Silver	1235	10500	235	429	174	444 / 187	430 / 225	425 / 239	412 / 250	396 / 262	379 / 277	361 / 292			
Tantalum	3269	16600	140	57.5	24.7	59.2 / 110	57.5 / 133	57.8 / 144	58.6 / 146	59.4 / 149	60.2 / 152	61.0 / 155	62.2 / 160	64.1 / 172	65.6 / 189
Thorium	2023	11700	118	54.0	39.1	59.8 / 99	54.6 / 112	54.5 / 124	55.8 / 134	56.9 / 145	56.9 / 156	58.7 / 167			
Tin	505	7310	227	66.6	40.1	85.2 / 188	73.3 / 215	62.2 / 243							
Titanium	1953	4500	522	21.9	9.32	30.5 / 300	24.5 / 465	20.4 / 551	19.4 / 591	19.7 / 633	20.7 / 675	22.0 / 620	24.5 / 686		
Tungsten	3660	19300	132	174	68.3	208 / 87	186 / 122	159 / 137	137 / 142	125 / 145	118 / 148	113 / 152	107 / 157	100 / 167	95 / 176
Uranium	1406	19070	116	27.6	12.5	21.7 / 94	25.1 / 108	29.6 / 125	34.0 / 146	38.8 / 176	43.9 / 180	49.0 / 161			
Vanadium	2192	6100	489	30.7	10.3	35.8 / 258	31.3 / 430	31.3 / 515	33.3 / 540	35.7 / 563	38.2 / 597	40.8 / 645	44.6 / 714	50.9 / 867	
Zinc	693	7140	389	116	41.8	117 / 297	118 / 367	111 / 402	103 / 436						
Zirconium	2125	6570	278	22.7	12.4	33.2 / 205	25.2 / 264	21.6 / 300	20.7 / 322	21.6 / 342	23.7 / 362	26.0 / 344	28.8 / 344	33.0 / 344	

Table A.2 Thermophysical Properties of Selected Nonmetallic Solids[a]

COMPOSITION	MELTING POINT K	PROPERTIES AT 300 K ρ (kg/m³)	c_p (J/kg·K)	k (W/m·K)	α·10⁶ (m²/s)	100	200	400	600	800	1000	1200	1500	2000	2500
Aluminum oxide, sapphire	2323	3970	765	46	15.1	450	82	32.4	18.9	13.0	10.5	6.55	5.66	6.00	
(c_p)								940	1110	1180	1225				
Aluminum oxide, polycrystalline	2323	3970	765	36.0	11.9	133	55	26.4	15.8	10.4	7.85				
(c_p)								940	1110	1180	1225				
Beryllium oxide	2725	3000	1030	272	88.0			196	111	70	47	33	21.5	15	
(c_p)								1350	1690	1865	1975	2055	2145	2750	
Boron	2573	2500	1105	27.6	9.99	190	52.5	18.7	11.3	8.1	6.3	5.2			
(c_p)								1490	1880	2135	2350	2555			
Boron fiber epoxy (30% vol) composite	590	2080													
k, ‖ to fibers			2.29			2.10	2.23	2.28							
k, ⊥ to fibers			0.59			0.37	0.49	0.60							
c_p		1122				364	757	1431							
Carbon Amorphous	1500	1950	—	1.60	—	0.67	1.18	1.89	2.19	2.37	2.53	2.84	3.48		
Diamond, type IIa insulator	—	3500	509	2300		10000	4000	1540							
(c_p)						21	194	853							
Graphite, pyrolytic	2273	2210													
k, ‖ to layers			1950			4970	3230	1390	892	667	534	448	357	262	
k, ⊥ to layers			5.70			16.8	9.23	4.09	2.68	2.01	1.60	1.34	1.08	0.81	
c_p		709				136	411	992	1406	1650	1793	1890	1974	2043	
Graphite fiber epoxy (25% vol) composite	450	1400													
k, heat flow ‖ to fibers			11.1			5.7	8.7	13.0							
k, heat flow ⊥ to fibers			0.87			0.46	0.68	1.1							
c_p		935				337	642	1216							
Pyroceram, Corning 9606	1623	2600	808	3.98	1.89	5.25	4.78	3.64	3.28	3.08	2.96	2.87	2.79		
(c_p)								908	1038	1122	1197	1264	1498		
Silicon carbide	3100	3160	675	490	230		—	—	—	—	87	58	30		
(c_p)								880	1050	1135	1195	1243	1310		

[a] Adapted from References 1, 2, 3, and 6.

Table A.2 Continued

COMPOSITION	MELTING POINT K	PROPERTIES AT 300 K				PROPERTIES AT VARIOUS TEMPERATURES (K)									
		ρ kg/m³	c_p J/kg·K	k W/m·K	$\alpha \cdot 10^6$ m²/s	k, W/m·K / c_p, J/kg·K									
						100	200	400	600	800	1000	1200	1500	2000	2500
Silicon dioxide, crystalline (quartz)	1883	2650													
k, ∥ to c axis				10.4		39	16.4	7.6	5.0	4.2					
k, ⊥ to c axis				6.21		20.8	9.5	4.70	3.4	3.1					
c_p			745					885	1075	1250					
Silicon dioxide, polycrystalline (fused silica)	1883	2220	745	1.38	0.834	0.69	1.14	1.51	1.75	2.17	2.87	4.00			
								905	1040	1105	1155	1195			
Silicon nitride	2173	2400	691	16.0	9.65			13.9	11.3	9.88	8.76	8.00	7.16	6.20	
							578	778	937	1063	1155	1226	1306	1377	
Sulphur	392	2070	708	0.206	0.141	0.165	0.185								
						403	606								
Thorium dioxide	3573	9110	235	13	6.1			10.2	6.6	4.7	3.68	3.12	2.73	2.5	
								255	274	285	295	303	315	330	
Titanium dioxide, polycrystalline	2133	4157	710	8.4	2.8			7.01	5.02	3.94	3.46	3.28			
								805	880	910	930	945			

Table A.3 Thermophysical Properties of Common Materials[a]

Structural Building Materials

DESCRIPTION/COMPOSITION	TYPICAL PROPERTIES AT 300 K		
	DENSITY, ρ kg/m^3	THERMAL CONDUCTIVITY, k W/m·K	SPECIFIC HEAT, c_p J/kg·K
Building Boards			
Asbestos-cement board	1,920	0.58	—
Gypsum or plaster board	800	0.17	—
Plywood	545	0.12	1,215
Sheathing, regular density	290	0.055	1,300
Acoustic tile	290	0.058	1,340
Hardboard, siding	640	0.094	1,170
Hardboard, high density	1,010	0.15	1,380
Particle board, low density	590	0.078	1,300
Particle board, high density	1,000	0.170	1,300
Woods			
Hardwoods (oak, maple)	720	0.16	1,255
Softwoods (fir, pine)	510	0.12	1,380
Masonry Materials			
Cement mortar	1,860	0.72	780
Brick, common	1,920	0.72	835
Brick, face	2,083	1.3	—
Clay tile, hollow			
1 cell deep, 10 cm thick	—	0.52	—
3 cells deep, 30 cm thick	—	0.69	—
Concrete block, 3 oval cores			
sand/gravel, 20 cm thick	—	1.0	—
cinder aggregate, 20 cm thick	—	0.67	—
Concrete block, rectangular core			—
2 core, 20 cm thick, 16 kg	—	1.1	—
same with filled cores	—	0.60	—
Plastering Materials			
Cement plaster, sand aggregate	1,860	0.72	—
Gypsum plaster, sand aggregate	1,680	0.22	1,085
Gypsum plaster, vermiculite aggregate	720	0.25	—

[a]Adapted from References 1, 8–13.

Table A.3 Continued

Insulating Materials and Systems

DESCRIPTION/COMPOSITION	TYPICAL PROPERTIES AT 300 K		
	DENSITY, ρ kg/m³	THERMAL CONDUCTIVITY, k W/m·K	SPECIFIC HEAT, c_p J/kg·K
Blanket and Batt			
Glass fiber, paper faced	16	0.046	—
	28	0.038	—
	40	0.035	—
Glass fiber, coated; duct liner	32	0.038	835
Board and Slab			
Cellular glass	145	0.058	1,000
Glass fiber, organic bonded	105	0.036	795
Polystyrene, expanded			
extruded (R-12)	55	0.027	1,210
molded beads	16	0.040	1,210
Mineral fiberboard; roofing material	265	0.049	—
Wood, shredded/cemented	350	0.087	1,590
Cork	120	0.039	1,800
Loose Fill			
Cork, granulated	160	0.045	—
Diatomaceous silica, coarse powder	350	0.069	—
	400	0.091	—
Diatomaceous silica, fine powder	200	0.052	—
	275	0.061	—
Glass fiber, poured or blown	16	0.043	835
Vermiculite, flakes	80	0.068	835
	160	0.063	1,000
Formed/Foamed-in-Place			
Mineral wool granules with asbestos/inorganic binders, sprayed	190	0.046	—
Polyvinyl acetate cork mastic; sprayed or troweled	—	0.100	—
Urethane, two-part mixture; rigid foam	70	0.026	1,045
Reflective			
Aluminum foil separating fluffy glass mats; 10-12 layers; evacuated; for cryogenic application (150 K)	40	0.00016	—
Aluminum foil and glass paper laminate; 75-150 layers; evacuated; for cryogenic application (150 K)	120	0.000017	—
Typical silica powder, evacuated	160	0.0017	—

Table A.3 Thermophysical Properties of Common Materials Continued

Industrial Insulation

DESCRIPTION/COMPOSITION	MAX SERVICE TEMP, K	TYPICAL DENSITY kg/m³	TYPICAL THERMAL CONDUCTIVITY, k(W/m·K), AT VARIOUS TEMPERATURES (K)													
			200	215	230	240	255	270	285	300	310	365	420	530	645	750
Blankets																
Blanket, mineral fiber, metal reinforced	920	96-192									0.038	0.046	0.056	0.078		
	815	40-96									0.035	0.045	0.058	0.088		
Blanket, mineral fiber, glass; fine fiber, organic bonded	450	10				0.036	0.038	0.040	0.043	0.048	0.052	0.076				
		12				0.035	0.036	0.039	0.042	0.046	0.049	0.069				
		16				0.033	0.035	0.036	0.039	0.042	0.046	0.062				
		24				0.030	0.032	0.033	0.036	0.039	0.040	0.053				
		32				0.029	0.030	0.032	0.033	0.036	0.038	0.048				
		48				0.027	0.029	0.030	0.032	0.033	0.035	0.045				
Blanket, alumina-silica fiber	1,530	48												0.071	0.105	0.150
		64												0.059	0.087	0.125
		96												0.052	0.076	0.100
		128												0.049	0.068	0.091
Felt, semi-rigid; organic bonded	480	50-125						0.035	0.036	0.038	0.039	0.051	0.063			
	730	50						0.030	0.032	0.033	0.035	0.051	0.079			
Felt, laminated; no binder	920	120	0.023	0.025	0.026	0.027	0.029						0.051	0.065	0.087	
Blocks, Boards, and Pipe Insulations																
Asbestos paper, laminated and corrugated																
4-ply	420	190								0.078	0.082	0.098				
6-ply	420	255								0.071	0.074	0.085				
8-ply	420	300								0.068	0.071	0.082				
Magnesia, 85%	590	185									0.051	0.055	0.061			
Calcium silicate	920	190									0.055	0.059	0.063	0.075	0.089	0.104
Cellular glass	700	145			0.046	0.048	0.051	0.052	0.055	0.058	0.062	0.069	0.079			
Diatomaceous silica	1,145	345												0.092	0.098	0.104
	1,310	385												0.101	0.100	0.115
Polystyrene, rigid																
Extruded (R-12)	350	56	0.023	0.023	0.022	0.023	0.023	0.025	0.026	0.027	0.029					
Extruded (R-12)	350	35	0.023	0.023	0.025	0.023	0.025	0.026	0.027	0.029						
Molded beads	350	16	0.026	0.029	0.030	0.033	0.035	0.036	0.038	0.040						
Rubber, rigid foamed	340	70						0.029	0.030	0.032	0.033					

Table A.3 Continued

Industrial Insulation Continued

DESCRIPTION/COMPOSITION	MAX SERVICE TEMP, K	TYPICAL DENSITY kg/m³	TYPICAL THERMAL CONDUCTIVITY, k(W/m·K), AT VARIOUS TEMPERATURES (K)													
			200	215	230	240	255	270	285	300	310	365	420	530	645	750
Insulating Cement																
Mineral fiber (rock, slag or glass)																
with clay binder	1,255	430									0.071	0.079	0.088	0.105	0.123	
with hydraulic setting binder	922	560									0.108	0.115	0.123	0.137		
Loose Fill																
Cellulose, wood or paper pulp	—	45							0.038	0.039	0.042					
Perlite, expanded	—	105	0.036	0.039	0.042	0.043	0.046	0.049	0.051	0.053	0.056					
Vermiculite, expanded	—	122			0.056	0.058	0.061	0.063	0.065	0.068	0.071					
		80			0.049	0.051	0.055	0.058	0.061	0.063	0.066					

Table A.3   Thermophysical Properties of Common Materials   Continued

Other Materials

DESCRIPTION/COMPOSITION	TEMPERATURE K	DENSITY, ρ kg/m^3	THERMAL CONDUCTIVITY, k W/m·K	SPECIFIC HEAT, c_p J/kg·K
Asphalt	300	2,115	0.062	920
Bakelite	300	1,300	1.4	1,465
Brick, refractory				
Carborundum	872	—	18.5	—
	1,672	—	11.0	—
Chrome brick	473	3,010	2.3	835
	823		2.5	
	1,173		2.0	
Diatomaceous silica, fired	478	—	0.25	—
	1,145	—	0.30	
Fire clay, burnt 1600 K	773	2,050	1.0	960
	1,073	—	1.1	
	1,373	—	1.1	
Fire clay, burnt 1725 K	773	2,325	1.3	960
	1,073		1.4	
	1,373		1.4	
Fire clay brick	478	2,645	1.0	960
	922		1.5	
	1,478		1.8	
Magnesite	478	—	3.8	1,130
	922	—	2.8	
	1,478		1.9	
Clay	300	1,460	1.3	880
Coal, Anthracite	300	1,350	0.26	1,260
Concrete (stone mix)	300	2,300	1.4	880
Cotton	300	80	0.06	1,300
Foodstuffs				
Banana (75.7% water content)	300	980	0.481	3,350
Apple, red (75% water content)	300	840	0.513	3,600
Cake, batter	300	720	0.223	—
Cake, fully done	300	280	0.121	—
Chicken meat, white	198	—	1.60	—
(74.4% water content)	233	—	1.49	
	253		1.35	
	263		1.20	
	273		0.476	
	283		0.480	
	293		0.489	

Table A.3 Continued

Other Materials Continued

DESCRIPTION/COMPOSITION	TEMPERATURE K	DENSITY, ρ kg/m^3	THERMAL CONDUCTIVITY, k W/m·K	SPECIFIC HEAT, c_p J/kg·K
Glass				
Plate (soda lime)	300	2,500	1.4	750
Pyrex	300	2,225	1.4	835
Ice	273	920	0.188	2,040
	253	—	0.203	1,945
Leather (sole)	300	998	0.013	—
Paper	300	930	0.011	1,340
Paraffin	300	900	0.020	2,890
Rock				
Granite, Barre	300	2,630	2.79	775
Limestone, Salem	300	2,320	2.15	810
Marble, Halston	300	2,680	2.80	830
Quartzite, Sioux	300	2,640	5.38	1,105
Sandstone, Berea	300	2,150	2.90	745
Rubber, vulcanized				
Soft	300	1,100	0.012	2,010
Hard	300	1,190	0.013	—
Sand	300	1,515	0.027	800
Soil	300	2,050	0.52	1,840
Snow	273	110	0.049	—
		500	0.190	—
Teflon	300	2,200	0.35	—
	400		0.45	—
Tissue, human				
Skin	300	—	0.37	—
Fat layer (adipose)	300	—	0.2	—
Muscle	300	—	0.41	—
Wood, cross grain				
Balsa	300	140	0.055	—
Cypress	300	465	0.097	—
Fir	300	415	0.11	2,720
Oak	300	545	0.17	2,385
Yellow pine	300	640	0.15	2,805
White pine	300	435	0.11	—
Wood, radial				
Oak	300	545	0.19	2,385
Fir	300	420	0.14	2,720

Table A.4 Thermophysical Properties of Gases at Atmospheric Pressure[a]

T K	ρ kg/m^3	c_p kJ/kg·K	$\mu \cdot 10^7$ N·s/m^2	$v \cdot 10^6$ m^2/s	$k \cdot 10^3$ W/m·K	$\alpha \cdot 10^6$ m^2/s	Pr
Air							
100	3.5562	1.032	71.1	2.00	9.34	2.54	0.786
150	2.3364	1.012	103.4	4.426	13.8	5.84	0.758
200	1.7458	1.007	132.5	7.590	18.1	10.3	0.737
250	1.3947	1.006	159.6	11.44	22.3	15.9	0.720
300	1.1614	1.007	184.6	15.89	26.3	22.5	0.707
350	0.9950	1.009	208.2	20.92	30.0	29.9	0.700
400	0.8711	1.014	230.1	26.41	33.8	38.3	0.690
450	0.7740	1.021	250.7	32.39	37.3	47.2	0.686
500	0.6964	1.030	270.1	38.79	40.7	56.7	0.684
550	0.6329	1.040	288.4	45.57	43.9	66.7	0.683
600	0.5804	1.051	305.8	52.69	46.9	76.9	0.685
650	0.5356	1.063	322.5	60.21	49.7	87.3	0.690
700	0.4975	1.075	338.8	68.10	52.4	98.0	0.695
750	0.4643	1.087	354.6	76.37	54.9	109	0.702
800	0.4354	1.099	369.8	84.93	57.3	120	0.709
850	0.4097	1.110	384.3	93.80	59.6	131	0.716
900	0.3868	1.121	398.1	102.9	62.0	143	0.720
950	0.3666	1.131	411.3	112.2	64.3	155	0.723
1000	0.3482	1.141	424.4	121.9	66.7	168	0.726
1100	0.3166	1.159	449.0	141.8	71.5	195	0.728
1200	0.2902	1.175	473.0	162.9	76.3	224	0.728
1300	0.2679	1.189	496.0	185.1	82	238	0.719
1400	0.2488	1.207	530	213	91	303	0.703
1500	0.2322	1.230	557	240	100	350	0.685
1600	0.2177	1.248	584	268	106	390	0.688
1700	0.2049	1.267	611	298	113	435	0.685
1800	0.1935	1.286	637	329	120	482	0.683
1900	0.1833	1.307	663	362	128	534	0.677
2000	0.1741	1.337	689	396	137	589	0.672
2100	0.1658	1.372	715	431	147	646	0.667
2200	0.1582	1.417	740	468	160	714	0.655
2300	0.1513	1.478	766	506	175	783	0.647
2400	0.1448	1.558	792	547	196	869	0.630
2500	0.1389	1.665	818	589	222	960	0.613
3000	0.1135	2.726	955	841	486	1570	0.536
Ammonia, NH$_3$							
300	0.6894	2.158	101.5	14.7	24.7	16.6	0.887
320	0.6448	2.170	109	16.9	27.2	19.4	0.870
340	0.6059	2.192	116.5	19.2	29.3	22.1	0.872
360	0.5716	2.221	124	21.7	31.6	24.9	0.872
380	0.5410	2.254	131	24.2	34.0	27.9	0.869

[a]Adapted from References 8, 14, and 15.

Table A.4 Continued

T K	ρ kg/m^3	c_p kJ/kg·K	$\mu \cdot 10^7$ N·s/m^2	$\nu \cdot 10^6$ m^2/s	$k \cdot 10^3$ W/m·K	$\alpha \cdot 10^6$ m^2/s	Pr
Ammonia, NH$_3$	Continued						
400	0.5136	2.287	138	26.9	37.0	31.5	0.853
420	0.4888	2.322	145	29.7	40.4	35.6	0.833
440	0.4664	2.357	152.5	32.7	43.5	39.6	0.826
460	0.4460	2.393	159	35.7	46.3	43.4	0.822
480	0.4273	2.430	166.5	39.0	49.2	47.4	0.822
500	0.4101	2.467	173	42.2	52.5	51.9	0.813
520	0.3942	2.504	180	45.7	54.5	55.2	0.827
540	0.3795	2.540	186.5	49.1	57.5	59.7	0.824
560	0.3708	2.577	193	52.0	60.6	63.4	0.827
580	0.3533	2.613	199.5	56.5	63.8	69.1	0.817
Carbon Dioxide, CO$_2$							
280	1.9022	0.830	140	7.36	15.20	9.63	0.765
300	1.7730	0.851	149	8.40	16.55	11.0	0.766
320	1.6609	0.872	156	9.39	18.05	12.5	0.754
340	1.5618	0.891	165	10.6	19.70	14.2	0.746
360	1.4743	0.908	173	11.7	21.2	15.8	0.741
380	1.3961	0.926	181	13.0	22.75	17.6	0.737
400	1.3257	0.942	190	14.3	24.3	19.5	0.737
450	1.1782	0.981	210	17.8	28.3	24.5	0.728
500	1.0594	1.02	231	21.8	32.5	30.1	0.725
550	0.9625	1.05	251	26.1	36.6	36.2	0.721
600	0.8826	1.08	270	30.6	40.7	42.7	0.717
650	0.8143	1.10	288	35.4	44.5	49.7	0.712
700	0.7564	1.13	305	40.3	48.1	56.3	0.717
750	0.7057	1.15	321	45.5	51.7	63.7	0.714
800	0.6614	1.17	337	51.0	55.1	71.2	0.716
Carbon Monoxide, CO							
200	1.6888	1.045	127	7.52	17.0	9.63	0.781
220	1.5341	1.044	137	8.93	19.0	11.9	0.753
240	1.4055	1.043	147	10.5	20.6	14.1	0.744
260	1.2967	1.043	157	12.1	22.1	16.3	0.741
280	1.2038	1.042	166	13.8	23.6	18.8	0.733
300	1.1233	1.043	175	15.6	25.0	21.3	0.730
320	1.0529	1.043	184	17.5	26.3	23.9	0.730
340	0.9909	1.044	193	19.5	27.8	26.9	0.725
360	0.9357	1.045	202	21.6	29.1	29.8	0.725
380	0.8864	1.047	210	23.7	30.5	32.9	0.729
400	0.8421	1.049	218	25.9	31.8	36.0	0.719
450	0.7483	1.055	237	31.7	35.0	44.3	0.714
500	0.67352	1.065	254	37.7	38.1	53.1	0.710

Table A.4 Continued

T K	ρ kg/m^3	c_p kJ/kg·K	$\mu \cdot 10^7$ N·s/m^2	$\nu \cdot 10^6$ m^2/s	$k \cdot 10^3$ W/m·K	$\alpha \cdot 10^6$ m^2/s	Pr
Carbon Monoxide, CO Continued							
550	0.61226	1.076	271	44.3	41.1	62.4	0.710
600	0.56126	1.088	286	51.0	44.0	72.1	0.707
650	0.51806	1.101	301	58.1	47.0	82.4	0.705
700	0.48102	1.114	315	65.5	50.0	93.3	0.702
750	0.44899	1.127	329	73.3	52.8	104	0.702
800	0.42095	1.140	343	81.5	55.5	116	0.705
Helium, He							
100	0.4871	5.193	96.3	19.8	73.0	28.9	0.686
120	0.4060	5.193	107	26.4	81.9	38.8	0.679
140	0.3481	5.193	118	33.9	90.7	50.2	0.676
160	—	5.193	129	—	99.2	—	—
180	0.2708	5.193	139	51.3	107.2	76.2	0.673
200	—	5.193	150	—	115.1	—	--
220	0.2216	5.193	160	72.2	123.1	107	0.675
240	—	5.193	170	—	130	—	—
260	0.1875	5.193	180	96.0	137	141	0.682
280	—	5.193	190	—	145	—	—
300	0.1625	5.193	199	122	152	180	0.680
350	—	5.193	221	—	170	—	—
400	0.1219	5.193	243	199	187	295	0.675
450		5.193	263	—	204	—	—
500	0.09754	5.193	283	290	220	434	0.668
550	—	5.193	—	—	—	—	—
600	—	5.193	320	—	252	—	—
650	—	5.193	332	—	264	—	—
700	0.06969	5.193	350	502	278	768	0.654
750	—	5.193	364	—	291	—	—
800	—	5.193	382	—	304	—	—
900	—	5.193	414	—	330	—	—
1000	0.04879	5.193	·446	914	354	1400	0.654
Hydrogen, H$_2$							
100	0.24255	11.23	42.1	17.4	67.0	24.6	0.707
150	0.16156	12.60	56.0	34.7	101	49.6	0.699
200	0.12115	13.54	68.1	56.2	131	79.9	0.704
250	0.09693	14.06	78.9	81.4	157	115	0.707
300	0.08078	14.31	89.6	111	183	158	0.701
350	0.06924	14.43	98.8	143	204	204	0.700
400	0.06059	14.48	108.2	179	226	258	0.695
450	0.05386	14.50	117.2	218	247	316	0.689
500	0.04848	14.52	126.4	261	266	378	0.691
550	0.04407	14.53	134.3	305	285	445	0.685

Table A.4 Continued

T K	ρ kg/m^3	c_p kJ/kg·K	$\mu \cdot 10^7$ N·s/m^2	$\nu \cdot 10^6$ m^2/s	$k \cdot 10^3$ W/m·K	$\alpha \cdot 10^6$ m^2/s	Pr
Hydrogen, H$_2$	**Continued**						
600	0.04040	14.55	142.4	352	305	519	0.678
700	0.03463	14.61	157.8	456	342	676	0.675
800	0.03030	14.70	172.4	569	378	849	0.670
900	0.02694	14.83	186.5	692	412	1030	0.671
1000	0.02424	14.99	201.3	830	448	1230	0.673
1100	0.02204	15.17	213.0	966	488	1460	0.662
1200	0.02020	15.37	226.2	1120	528	1700	0.659
1300	0.01865	15.59	238.5	1279	568	1955	0.655
1400	0.01732	15.81	250.7	1447	610	2230	0.650
1500	0.01616	16.02	262.7	1626	655	2530	0.643
1600	0.0152	16.28	273.7	1801	697	2815	0.639
1700	0.0143	16.58	284.9	1992	742	3130	0.637
1800	0.0135	16.96	296.1	2193	786	3435	0.639
1900	0.0128	17.49	307.2	2400	835	3730	0.643
2000	0.0121	18.25	318.2	2630	878	3975	0.661
Nitrogen, N$_2$							
100	3.4388	1.070	68.8	2.00	9.58	2.60	0.768
150	2.2594	1.050	100.6	4.45	13.9	5.86	0.759
200	1.6883	1.043	129.2	7.65	18.3	10.4	0.736
250	1.3488	1.042	154.9	11.48	22.2	15.8	0.727
300	1.1233	1.041	178.2	15.86	25.9	22.1	0.716
350	0.9625	1.042	200.0	20.78	29.3	29.2	0.711
400	0.8425	1.045	220.4	26.16	32.7	37.1	0.704
450	0.7485	1.050	239.6	32.01	35.8	45.6	0.703
500	0.6739	1.056	257.7	38.24	38.9	54.7	0.700
550	0.6124	1.065	274.7	44.86	41.7	63.9	0.702
600	0.5615	1.075	290.8	51.79	44.6	73.9	0.701
700	0.4812	1.098	321.0	66.71	49.9	94.4	0.706
800	0.4211	1.122	349.1	82.90	54.8	116	0.715
900	0.3743	1.146	375.3	100.3	59.7	139	0.721
1000	0.3368	1.167	399.9	118.7	64.7	165	0.721
1100	0.3062	1.187	423.2	138.2	70.0	193	0.718
1200	0.2807	1.204	445.3	158.6	75.8	224	0.707
1300	0.2591	1.219	466.2	179.9	81.0	256	0.701
Oxygen, O$_2$							
100	3.945	0.962	76.4	1.94	9.25	2.44	0.796
150	2.585	0.921	114.8	4.44	13.8	5.80	0.766
200	1.930	0.915	147.5	7.64	18.3	10.4	0.737
250	1.542	0.915	178.6	11.58	22.6	16.0	0.723
300	1.284	0.920	207.2	16.14	26.8	22.7	0.711

Table A.4   Continued

T K	ρ kg/m^3	c_p kJ/kg·K	$\mu \cdot 10^7$ N·s/m^2	$\nu \cdot 10^6$ m^2/s	$k \cdot 10^3$ W/m·K	$\alpha \cdot 10^6$ m^2/s	Pr
Oxygen, O$_2$ Continued							
350	1.100	0.929	233.5	21.23	29.6	29.0	0.733
400	0.9620	0.942	258.2	26.84	33.0	36.4	0.737
450	0.8554	0.956	281.4	32.90	36.3	44.4	0.741
500	0.7698	0.972	303.3	39.40	41.2	55.1	0.716
550	0.6998	0.988	324.0	46.30	44.1	63.8	0.726
600	0.6414	1.003	343.7	53.59	47.3	73.5	0.729
700	0.5498	1.031	380.8	69.26	52.8	93.1	0.744
800	0.4810	1.054	415.2	86.32	58.9	116	0.743
900	0.4275	1.074	447.2	104.6	64.9	141	0.740
1000	0.3848	1.090	477.0	124.0	71.0	169	0.733
1100	0.3498	1.103	505.5	144.5	75.8	196	0.736
1200	0.3206	1.115	532.5	166.1	81.9	229	0.725
1300	0.2960	1.125	588.4	188.6	87.1	262	0.721
Water Vapor (steam)							
380	0.5863	2.060	127.1	21.68	24.6	20.4	1.06
400	0.5542	2.014	134.4	24.25	26.1	23.4	1.04
450	0.4902	1.980	152.5	31.11	29.9	30.8	1.01
500	0.4405	1.985	170.4	38.68	33.9	38.8	0.998
550	0.4005	1.997	188.4	47.04	37.9	47.4	0.993
600	0.3652	2.026	206.7	56.60	42.2	57.0	0.993
650	0.3380	2.056	224.7	66.48	46.4	66.8	0.996
700	0.3140	2.085	242.6	77.26	50.5	77.1	1.00
750	0.2931	2.119	260.4	88.84	54.9	88.4	1.00
800	0.2739	2.152	278.6	101.7	59.2	100	1.01
850	0.2579	2.186	296.9	115.1	63.7	113	1.02

Table A.5   Thermophysical Properties of Saturated Liquids[a]

T K	ρ kg/m^3	c_p kJ/kg·K	$\mu \cdot 10^2$ N·s/m^2	$\nu \cdot 10^6$ m^2/s	$k \cdot 10^3$ W/m·K	$\alpha \cdot 10^7$ m^2/s	Pr	$\beta \cdot 10^3$ K^{-1}
Engine Oil (unused)								
273	899.1	1.796	385	4,280	147	0.910	47,000	0.70
280	895.3	1.827	217	2,430	144	0.880	27,500	0.70
290	890.0	1.868	99.9	1,120	145	0.872	12,900	0.70
300	884.1	1.909	48.6	550	145	0.859	6,400	0.70
310	877.9	1.951	25.3	288	145	0.847	3,400	0.70
320	871.8	1.993	14.1	161	143	0.823	1,965	0.70
330	865.8	2.035	8.36	96.6	141	0.800	1,205	0.70
340	859.9	2.076	5.31	61.7	139	0.779	793	0.70
350	853.9	2.118	3.56	41.7	138	0.763	546	0.70
360	847.8	2.161	2.52	29.7	138	0.753	395	0.70
370	841.8	2.206	1.86	22.0	137	0.738	300	0.70
380	836.0	2.250	1.41	16.9	136	0.723	233	0.70
390	830.6	2.294	1.10	13.3	135	0.709	187	0.70
400	825.1	2.337	0.874	10.6	134	0.695	152	0.70
410	818.9	2.381	0.698	8.52	133	0.682	125	0.70
420	812.1	2.427	0.564	6.94	133	0.675	103	0.70
430	806.5	2.471	0.470	5.83	132	0.662	88	0.70
Ethylene Glycol, $C_2H_4(OH)_2$								
273	1,130.8	2.294	6.51	57.6	242	0.933	617	0.65
280	1,125.8	2.323	4.20	37.3	244	0.933	400	0.65
290	1,118.8	2.368	2.47	22.1	248	0.936	236	0.65
300	1,111.4	2.415	1.57	14.1	252	0.939	151	0.65
310	1,103.7	2.460	1.07	9.65	255	0.939	103	0.65
320	1,096.2	2.505	0.757	6.91	258	0.940	73.5	0.65
330	1,089.5	2.549	0.561	5.15	260	0.936	55.0	0.65
340	1,083.8	2.592	0.431	3.98	261	0.929	42.8	0.65
350	1,079.0	2.637	0.342	3.17	261	0.917	34.6	0.65
360	1,074.0	2.682	0.278	2.59	261	0.906	28.6	0.65
370	1,066.7	2.728	0.228	2.14	262	0.900	23.7	0.65
373	1,058.5	2.742	0.215	2.03	263	0.906	22.4	0.65
Glycerin, $C_3H_5(OH)_3$								
273	1,276.0	2.261	1,060	8,310	282	0.977	85,000	0.47
280	1,271.9	2.298	534	4,200	284	0.972	43,200	0.47
290	1,265.8	2.367	185	1,460	286	0.955	15,300	0.48
300	1,259.9	2.427	79.9	634	286	0.935	6,780	0.48
310	1,253.9	2.490	35.2	281	286	0.916	3,060	0.49
320	1,247.2	2.564	21.0	168	287	0.897	1,870	0.50

[a] Adapted from References 15 and 16.

Table A.5 Continued

T K	ρ kg/m^3	c_p kJ/kg·K	$\mu \cdot 10^2$ N·s/m^2	$v \cdot 10^6$ m^2/s	$k \cdot 10^3$ W/m·K	$\alpha \cdot 10^7$ m^2/s	Pr	$\beta \cdot 10^3$ K^{-1}
Freon (refrigerant-12), CCl_2F_2								
230	1,528.4	0.8816	0.0457	0.299	68	0.505	5.9	1.85
240	1,498.0	0.8923	0.0385	0.257	69	0.516	5.0	1.90
250	1,469.5	0.9037	0.0354	0.241	70	0.527	4.6	2.00
260	1,439.0	0.9163	0.0322	0.224	73	0.554	4.0	2.10
270	1,407.2	0.9301	0.0304	0.216	73	0.558	3.9	2.25
280	1,374.4	0.9450	0.0283	0.206	73	0.562	3.7	2.35
290	1,340.5	0.9609	0.0265	0.198	73	0.567	3.5	2.55
300	1,305.8	0.9781	0.0254	0.195	72	0.564	3.5	2.75
310	1,268.9	0.9963	0.0244	0.192	69	0.546	3.4	3.05
320	1,228.6	1.0155	0.0233	0.190	68	0.545	3.5	3.5
Mercury, Hg								
273	13,595	0.1404	0.1688	0.1240	8,180	42.85	0.0290	0.181
300	13,529	0.1393	0.1523	0.1125	8,540	45.30	0.0248	0.181
350	13,407	0.1377	0.1309	0.0976	9,180	49.75	0.0196	0.181
400	13,287	0.1365	0.1171	0.0882	9,800	54.05	0.0163	0.181
450	13,167	0.1357	0.1075	0.0816	10,400	58.10	0.0140	0.181
500	13,048	0.1353	0.1007	0.0771	10,950	61.90	0.0125	0.182
550	12,929	0.1352	0.0953	0.0737	11,450	65.55	0.0112	0.184
600	12,809	0.1355	0.0911	0.0711	11,950	68.80	0.0103	0.187

Table A.6   Thermophysical Properties of Saturated Water[a]

TEMP. (K) T	PRESSURE (BAR)[c] p	SPECIFIC VOLUME (m³/kg) $v_f \cdot 10^3$	v_g	HEAT OF VAPORIZATION (kJ/kg) h_{fg}	SPECIFIC HEAT (kJ/kg·K) $c_{p,f}$	$c_{p,g}$	VISCOSITY (N·s/m²) $\mu_f \cdot 10^6$	$\mu_g \cdot 10^6$	THERMAL CONDUCTIVITY (W/m·K) $k_f \cdot 10^3$	$k_g \cdot 10^3$	PRANDTL NO. Pr_f	Pr_g	SURFACE TENSION (N/m) $\sigma_f \cdot 10^3$	EXPANSION COEFFICIENT (K⁻¹) $\beta_f \cdot 10^6$	TEMP. (K) T
273.15	0.00611	1.000	206.3	2502	4.217	1.854	1750	8.02	659	18.2	12.99	0.815	75.5	−68.05	273.15
275	0.00697	1.000	181.7	2497	4.211	1.855	1652	8.09	574	18.3	12.22	0.817	75.3	−32.74	275
280	0.00990	1.000	130.4	2485	4.198	1.858	1422	8.29	582	18.6	10.26	0.825	74.8	46.04	280
285	0.01387	1.000	99.4	2473	4.189	1.861	1225	8.49	590	18.9	8.81	0.833	74.3	114.1	285
290	0.01917	1.001	69.7	2461	4.184	1.864	1080	8.69	598	19.3	7.56	0.841	73.7	174.0	290
295	0.02617	1.002	51.94	2449	4.181	1.868	959	8.89	606	19.5	6.62	0.849	72.7	227.5	295
300	0.03531	1.003	39.13	2438	4.179	1.872	855	9.09	613	19.6	5.83	0.857	71.7	276.1	300
305	0.04712	1.005	27.90	2426	4.178	1.877	769	9.29	620	20.1	5.20	0.865	70.9	320.6	305
310	0.06221	1.007	22.93	2414	4.178	1.882	695	9.49	628	20.4	4.62	0.873	70.0	361.9	310
315	0.08132	1.009	17.82	2402	4.179	1.888	631	9.69	634	20.7	4.16	0.883	69.2	400.4	315
320	0.1053	1.011	13.98	2390	4.180	1.895	577	9.89	640	21.0	3.77	0.894	68.3	436.7	320
325	0.1351	1.013	11.06	2378	4.182	1.903	528	10.09	645	21.3	3.42	0.901	67.5	471.2	325
330	0.1719	1.016	8.82	2366	4.184	1.911	489	10.29	650	21.7	3.15	0.908	66.6	504.0	330
335	0.2167	1.018	7.09	2354	4.186	1.920	453	10.49	656	22.0	2.88	0.916	65.8	535.5	335
340	0.2713	1.021	5.74	2342	4.188	1.930	420	10.69	660	22.3	2.66	0.925	64.9	566.0	340
345	0.3372	1.024	4.683	2329	4.191	1.941	389	10.89	668	22.6	2.45	0.933	64.1	595.4	345
350	0.4163	1.027	3.846	2317	4.195	1.954	365	11.09	668	23.0	2.29	0.942	63.2	624.2	350
355	0.5100	1.030	3.180	2304	4.199	1.968	343	11.29	671	23.3	2.14	0.951	62.3	652.3	355
360	0.6209	1.034	2.645	2291	4.203	1.983	324	11.49	674	23.7	2.02	0.960	61.4	697.9	360
365	0.7514	1.038	2.212	2278	4.209	1.999	306	11.69	677	24.1	1.91	0.969	60.5	707.1	365
370	0.9040	1.041	1.861	2265	4.214	2.017	289	11.89	679	24.5	1.80	0.978	59.5	728.7	370
373.15	1.0133	1.044	1.679	2257	4.217	2.029	279	12.02	680	24.8	1.76	0.984	58.9	750.1	373.15
375	1.0815	1.045	1.574	2252	4.220	2.036	274	12.09	681	24.9	1.70	0.987	58.6	761	375
380	1.2869	1.049	1.337	2239	4.226	2.057	260	12.29	683	25.4	1.61	0.999	57.6	788	380
385	1.5233	1.053	1.142	2225	4.232	2.080	248	12.49	685	25.8	1.53	1.004	56.6	814	385
390	1.794	1.058	0.980	2212	4.239	2.104	237	12.69	686	26.3	1.47	1.013	55.6	841	390
400	2.455	1.067	0.731	2183	4.256	2.158	217	13.05	688	27.2	1.34	1.033	53.6	896	400
410	3.302	1.077	0.553	2153	4.278	2.221	200	13.42	688	28.2	1.24	1.054	51.5	952	410
420	4.370	1.088	0.425	2123	4.302	2.291	185	13.79	688	29.8	1.16	1.075	49.4	1010	420
430	5.699	1.099	0.331	2091	4.331	2.369	173	14.14	685	30.4	1.09	1.10	47.2		430
440	7.333	1.110	0.261	2059	4.36	2.46	162	14.50	682	31.7	1.04	1.12	45.1		440
450	9.319	1.123	0.208	2024	4.40	2.56	152	14.85	678	33.1	0.99	1.14	42.9		450
460	11.71	1.137	0.167	1989	4.44	2.68	143	15.19	673	34.6	0.95	1.17	40.7		460
470	14.55	1.152	0.136	1951	4.48	2.79	136	15.54	667	36.3	0.92	1.20	38.5		470
480	17.90	1.167	0.111	1912	4.53	2.94	129	15.88	660	38.1	0.89	1.23	36.2		480

T	p														T
490	21.83	1.184	0.0922	1870	4.59	3.10	124	16.23	651	40.1	0.87	1.25	33.9	—	490
500	26.40	1.203	0.0766	1825	4.66	3.27	118	16.59	642	42.3	0.86	1.28	31.6	—	500
510	31.66	1.222	0.0631	1779	4.74	3.47	113	16.95	631	44.7	0.85	1.31	29.3	—	510
520	37.70	1.244	0.0525	1730	4.84	3.70	108	17.33	621	47.5	0.84	1.35	26.9	—	520
530	44.58	1.268	0.0445	1679	4.95	3.96	104	17.72	608	50.6	0.85	1.39	24.5	—	530
540	52.38	1.294	0.0375	1622	5.08	4.27	101	18.1	594	54.0	0.86	1.43	22.1	—	540
550	61.19	1.323	0.0317	1564	5.24	4.64	97	18.6	580	58.3	0.87	1.47	19.7	—	550
560	71.08	1.355	0.0269	1499	5.43	5.09	94	19.1	563	63.7	0.90	1.52	17.3	—	560
570	82.16	1.392	0.0228	1429	5.68	5.67	91	19.7	548	76.7	0.94	1.59	15.0	—	570
580	94.51	1.433	0.0193	1353	6.00	6.40	88	20.4	528	76.7	0.99	1.68	12.8	—	580
590	108.3	1.482	0.0163	1274	6.41	7.35	84	21.5	513	84.1	1.05	1.84	10.5	—	590
600	123.5	1.541	0.0137	1176	7.00	8.75	81	22.7	497	92.9	1.14	2.15	8.4	—	600
610	137.3	1.612	0.0115	1068	7.85	11.1	77	24.1	467	103	1.30	2.60	6.3	—	610
620	159.1	1.705	0.0094	941	9.35	15.4	72	25.9	444	114	1.52	3.46	4.5	—	620
625	169.1	1.778	0.0085	858	10.6	18.3	70	27.0	430	121	1.65	4.20	3.5	—	625
630	179.7	1.856	0.0075	781	12.6	22.1	67	28.0	412	130	2.0	4.8	2.6	—	630
635	190.9	1.935	0.0066	683	16.4	27.6	64	30.0	392	141	2.7	6.0	1.5	—	635
640	202.7	2.075	0.0057	560	26	42	59	32.0	367	155	4.2	9.6	0.8	—	640
645	215.2	2.351	0.0045	361	90	—	54	37.0	331	178	12	26	0.1	—	645
647.3[b]	221.2	3.170	0.0032	0	∞	∞	45	45.0	238	238	∞	∞	0.0	—	647.3[b]

[a] Adapted from References 16 and 17.
[b] Critical temperature.
[c] 1 bar = 10^5 N/m².

Table A.7   Thermophysical Properties of Liquid Metals[a]

COMPOSITION	MELTING POINT K	T K	ρ kg/m^3	c_p kJ/kg·K	$\nu \cdot 10^7$ m^2/s	k W/m·K	$\alpha \cdot 10^5$ m^2/s	Pr
Bismuth	544	589	10,011	0.1444	1.617	16.4	0.138	0.0142
		811	9,739	0.1545	1.133	15.6	1.035	0.0110
		1033	9,467	0.1645	0.8343	15.6	1.001	0.0083
Lead	600	644	10,540	0.159	2.276	16.1	1.084	0.024
		755	10,412	0.155	1.849	15.6	1.223	0.017
		977	10,140	—	1.347	14.9	—	—
Potassium	337	422	807.3	0.80	4.608	45.0	6.99	0.0066
		700	741.7	0.75	2.397	39.5	7.07	0.0034
		977	674.4	0.75	1.905	33.1	6.55	0.0029
Sodium	371	366	929.1	1.38	7.516	86.2	6.71	0.011
		644	860.2	1.30	3.270	72.3	6.48	0.0051
		977	778.5	1.26	2.285	59.7	6.12	0.0037
NaK, (45%/55%)	292	366	887.4	1.130	6.522	25.6	2.552	0.026
		644	821.7	1.055	2.871	27.5	3.17	0.0091
		977	740.1	1.043	2.174	28.9	3.74	0.0058
NaK, (22%/78%)	262	366	849.0	0.946	5.797	24.4	3.05	0.019
		672	775.3	0.879	2.666	26.7	3.92	0.0068
		1033	690.4	0.883	2.118	—	—	—
PbBi (44.5%/55.5%)	398	422	10,524	0.147	—	9.05	0.586	—
		644	10,236	0.147	1.496	11.86	0.790	0.189
		922	9,835	—	1.171	—	—	—
Mercury	234	See table A.5						

[a]Adapted from *Liquid Materials Handbook*, 23rd Ed., the Atomic Energy Commission, Department of the Navy, Washington, D.C., 1952.

Table A.8 Binary Diffusion Coefficients at One Atmosphere[a]

SUBSTANCE A	SUBSTANCE B	T (K)	D_{AB} (m^2/s)
Gases			
NH_3	Air	298	0.28×10^{-4}
H_2O	Air	298	0.26×10^{-4}
CO_2	Air	298	0.16×10^{-4}
H_2	Air	298	0.41×10^{-4}
O_2	Air	298	0.21×10^{-4}
Acetone	Air	273	0.11×10^{-4}
Benzene	Air	298	0.88×10^{-5}
Napthalene	Air	300	0.62×10^{-5}
Ar	N_2	293	0.19×10^{-4}
H_2	O_2	273	0.70×10^{-4}
H_2	N_2	273	0.68×10^{-4}
H_2	CO_2	273	0.55×10^{-4}
CO_2	N_2	293	0.16×10^{-4}
CO_2	O_2	273	0.14×10^{-4}
O_2	N_2	273	0.18×10^{-4}
Dilute Solutions			
Caffeine	H_2O	298	0.63×10^{-9}
Ethanol	H_2O	298	0.12×10^{-8}
Glucose	H_2O	298	0.69×10^{-9}
Glycerol	H_2O	298	0.94×10^{-9}
Acetone	H_2O	298	0.13×10^{-8}
CO_2	H_2O	298	0.20×10^{-8}
O_2	H_2O	298	0.24×10^{-8}
H_2	H_2O	298	0.63×10^{-8}
N_2	H_2O	298	0.26×10^{-8}
Solids			
O_2	Rubber	298	0.21×10^{-9}
N_2	Rubber	298	0.15×10^{-9}
CO_2	Rubber	298	0.11×10^{-9}
He	SiO_2	293	$0.4 \ \times 10^{-13}$
H_2	Fe	293	0.26×10^{-12}
Cd	Cu	293	0.27×10^{-18}
Al	Cu	293	0.13×10^{-33}

[a]Adapted with permission from References 18, 19, and 20.

Table A.9 Henry's Constant for Selected Gases in Water at Moderate Pressure[a]

$H = p_{A,i}/x_{A,i}$ (bar)

T (K)	NH_3	Cl_2	H_2S	SO_2	CO_2	CH_4	O_2	H_2
273	21	265	260	165	710	22,880	25,500	58,000
280	23	365	335	210	960	27,800	30,500	61,500
290	26	480	450	315	1300	35,200	37,600	66,500
300	30	615	570	440	1730	42,800	45,700	71,600
310	—	755	700	600	2175	50,000	52,500	76,000
320	—	860	835	800	2650	56,300	56,800	78,600
323	—	890	870	850	2870	58,000	58,000	79,000

[a]Adapted with permission from Reference 21.

Table A.10 The Solubility of Selected Gases and Solids[a]

GAS	SOLID	T (K)	$S = C_{A,i}/p_{A,i}$ (kmol/m^3·bar)
O_2	Rubber	298	3.12×10^{-3}
N_2	Rubber	298	1.56×10^{-3}
CO_2	Rubber	298	40.15×10^{-3}
He	SiO_2	293	0.45×10^{-3}
H_2	Ni	358	9.01×10^{-3}

[a]Adapted with permission from Reference 20.

Table A.11 Total, Normal (n) or Hemispherical (h) Emissivity of Selected Surfaces
Metallic Solids and Their Oxides[a]

DESCRIPTION/COMPOSITION		ε_n or ε_h at various temperatures (K)										
		100	200	300	400	600	800	1000	1200	1500	2000	2500
Aluminum												
Highly polished, film	(h)	0.02	0.03	0.04	0.05	0.06						
Foil, bright	(h)	0.06	0.06	0.07								
Anodized	(h)			0.82	0.76							
Chromium												
Polished or plated	(n)	0.05	0.07	0.10	0.12	0.14						
Copper												
Highly polished	(h)			0.03	0.03	0.04	0.04	0.04				
Stably oxidized	(h)					0.50	0.58	0.80				
Gold												
Highly polished or film	(h)	0.01	0.02	0.03	0.03	0.04	0.05	0.06				
Foil, bright	(h)	0.06	0.07	0.07								
Molybdenum												
Polished	(h)						0.08	0.10	0.12	0.15	0.21	0.26
Shot-blasted, rough	(h)					0.25	0.28	0.31	0.35	0.42		
Stably oxidized	(h)					0.80	0.82					
Nickel												
Polished	(h)					0.09	0.11	0.14	0.17			
Stably oxidized	(h)					0.40	0.49	0.57				
Platinum												
Polished	(h)						0.10	0.13	0.15	0.18		
Silver												
Polished	(h)			0.02	0.02	0.03	0.05	0.08				
Stainless steels												
Typical, polished	(n)			0.17	0.17	0.19	0.23	0.30				
Typical, cleaned	(n)			0.22	0.22	0.24	0.28	0.35				
Typical, lightly oxidized	(n)						0.33	0.40				
Typical, highly oxidized	(n)						0.67	0.70	0.76			
AISI 347, stably oxidized	(n)					0.87	0.88	0.89	0.90			
Tantalum												
Polished	(h)							0.11		0.17	0.23	0.28
Tungsten												
Polished	(h)							0.10	0.13	0.18	0.25	0.29

[a]Adapted from Reference 1.

Table A.11   Total, Normal (*n*) or Hemispherical (*h*) Emissivity of Selected Materials
Nonmetallic Solids[b]

DESCRIPTION/COMPOSITION		TEMPERATURE (K)	EMISSIVITY ε
Aluminum oxide	(*n*)	600	0.69
		1000	0.55
		1500	0.41
Asphalt pavement	(*h*)	300	0.85–0.93
Building materials			
Asbestos sheet	(*h*)	300	0.93–0.96
Brick, red	(*h*)	300	0.93–0.96
Gypsum or plaster board	(*h*)	300	0.90–0.92
Wood	(*h*)	300	0.82–0.92
Cloth	(*h*)	300	0.75–0.90
Concrete	(*h*)	300	0.88–0.93
Glass, window	(*h*)	300	0.90–0.95
Ice	(*h*)	273	0.95–0.98
Paints			
Black (Parsons)	(*h*)	300	0.98
White, acrylic	(*h*)	300	0.90
White, zinc oxide	(*h*)	300	0.92
Paper, white	(*h*)	300	0.92–0.97
Pyrex	(*n*)	300	0.82
		600	0.80
		1000	0.71
		1200	0.62
Pyroceram	(*n*)	300	0.85
		600	0.78
		1000	0.69
		1500	0.57
Refractories (furnace liners)			
Alumina brick	(*n*)	800	0.40
		1000	0.33
		1400	0.28
		1600	0.33
Magnesia brick	(*n*)	800	0.45
		1000	0.36
		1400	0.31
		1600	0.40
Kaolin insulating brick	(*n*)	800	0.70
		1200	0.57
		1400	0.47
		1600	0.53

[a]Adapted from References 1, 9, 22, and 23.

Table A.11   Continued

Nonmetallic Solids   Continued

DESCRIPTION/COMPOSITION		TEMPERATURE (K)	EMISSIVITY ε
Sand	(h)	300	0.90
Silicon carbide	(n)	600	0.87
		1000	0.87
		1500	0.85
Skin	(h)	300	0.95
Snow	(h)	273	0.82–0.90
Soil	(h)	300	0.93–0.96
Rocks	(h)	300	0.88–0.95
Teflon	(h)	300	0.85
		400	0.87
		500	0.92
Vegetation	(h)	300	0.92–0.96
Water	(h)	300	0.96

Table A.12 Solar Radiative Properties for Selected Materials[a]

DESCRIPTION/COMPOSITION	α_S	ε[b]	α_S/ε	τ_S
Aluminum				
Polished	0.09	0.03	3.0	
Anodized	0.14	0.84	0.17	
Quartz overcoated	0.11	0.37	0.30	
Foil	0.15	0.05	3.0	
Brick, red (Purdue)	0.63	0.93	0.68	
Concrete	0.60	0.88	0.68	
Galvanized sheet metal				
Clean, new	0.65	0.13	5.0	
Oxidized, weathered	0.80	0.28	2.9	
Glass, 3.2 mm thickness				
Float or tempered				0.79
Low iron oxide type				0.88
Metals, plated				
Black sulphide	0.92	0.10	9.2	
Black cobalt oxide	0.93	0.30	3.1	
Black nickel oxide	0.92	0.08	11	
Black chrome	0.87	0.09	9.7	
Mylar, 0.13 mm thickness				0.87
Paints				
Black (Parsons)	0.98	0.98	1.0	
White, acrylic	0.26	0.90	0.29	
White, zinc oxide	0.16	0.93	0.17	
Plexiglas, 3.2 mm thickness				0.90
Snow				
Fine particles, fresh	0.13	0.82	0.16	
Ice granules	0.33	0.89	0.37	
Tedlar, 0.10 mm thickness				0.92
Teflon, 0.13 mm thickness				0.92

[a]Adapted with permission from Reference 23.
[b]The emissivity values in this table correspond to a surface temperature of approximately 300 K.

REFERENCES

1. Touloukian, Y. S. and C. Y. Ho, Eds., *Thermophysical Properties of Matter*, Plenum Press, New York, Vol. 1, *Thermal Conductivity of Metallic Solids*; Vol. 2, *Thermal Conductivity of Nonmetallic Solids*; Vol. 4, *Specific Heat of Metallic Solids*; Vol. 5, *Specific Heat of Nonmetallic Solids*; Vol. 7, *Thermal Radiative Properties of Metallic Solids*; Vol. 8, *Thermal Radiative Properties of Nonmetallic Solids*; Vol. 9, *Thermal Radiative Properties of Coatings*; 1972.

2. Touloukian, Y. S. and C. Y. Ho, Eds., *Thermophysical Properties of Selected Aerospace Materials*, Part I: Thermal Radiative Properties; Part II: Thermophysical Properties of Seven Materials; Thermophysical and Electronic Properties Information Analysis Center, CINDAS, Purdue University, West Lafayette, Ind., 1976.

3. Ho, C. Y., R. W. Powell, and P. E. Liley, "Thermal Conductivity of the Elements: A Comprehensive Review," *J. Phys. and Chem. Reference Data, 3*, Supplement 1, 1974.

4. Desai, P. D., T. K. Chu, R. H. Bogaard, M. W. Ackermann, and C. Y. Ho, Part I: Thermophysical Properties of Carbon Steels, Part II: Thermophysical Properties of Low Chromium Steels, Part III: Thermophysical Properties of Nickel Steels, Part IV: Thermophysical Properties of Stainless Steels, CINDAS Special Report, September 1976.

5. American Society for Metals, *Metals Handbook*, Vol. 1, "Properties and Selection of Metals," 8th Ed., 1961.

6. Hultgren, R., P. D. Desai, D. T. Hawkins, M. Gleiser, K. K. Kelley, and D. D. Wagman, *Selected Values of the Thermodynamic Properties of the Elements*, American Society of Metals, 1973.

7. Hultgren, R., P. D. Desai, D. T. Hawkins, M. Gleiser, and K. K. Kelley, *Selected Values of the Thermodynamic Properties of Binary Alloys*, American Society of Metals, 1973.

8. American Society of Heating, Refrigerating and Air Conditioning Engineers, *ASHRAE Handbook of Fundamentals*, 1972.

9. Mallory, J. F., *Thermal Insulation*, Van Nostrand Rheinhold Co., New York, 1969.

10. Hanley, E. J., D. P. DeWitt, and R. E. Taylor, "The Thermal Transport Properties at Normal and Elevated Temperature of Eight Representative Rocks," *Proceedings of the Seventh Symposium on Thermophysical Properties*, American Society of Mechanical Engineers, 1977.

11. Sweat, V. E., "A Miniature Thermal Conductivity Probe for Foods," American Society of Mechanical Engineers, Paper 76-HT-60, August 1976.

12. Kothandaraman, C. P. and S. Subramanyan, *Heat and Mass Transfer Data Book*, Halsted Press, Wiley, New York, 1975.

13. Chapman, A. J., *Heat Transfer*, 3rd Ed., Macmillan, New York, 1974.

14. Vargaftik, N. B., *Tables of Thermophysical Properties of Liquids and Gases*, 2nd Ed., Hemisphere Publishing Corp., Washington, D.C., 1975.

15. Eckert, E. R. G. and R. M. Drake, *Analysis of Heat and Mass Transfer*, McGraw Hill, New York, 1972.

16. Vukalovich, M. P., A. I. Ivanov, L. R. Fokin, and A. T. Yakovelev, *Thermophysical Properties of Mercury*, State Committee on Standards, State Service for Standards and Handbook Data, Monograph Series No. 9, Izd. Standartov, Moscow, 1971.

17. Liley, P. E., Steam Tables in SI Units, Private communication, School of Mechanical Engineering, Purdue University, December 1978.

18. Perry, J. H., Ed., *Chemical Engineer's Handbook*, 4th Ed., McGraw-Hill, New York, 1963.

19. Geankoplis, C. J., *Mass Transport Phenomena*, Holt, Rinehart and Winston, New York, 1972.

20. Barrer, R. M., *Diffusion In and Through Solids*, Macmillan, New York, 1941.

21. Spalding, D. B., *Convective Mass Transfer*, McGraw-Hill, New York, 1963.

22. Gubareff, G. G., J. E. Janssen, and R. H. Torborg, *Thermal Radiation Properties Survey*, Minneapolis-Honeywell Regulator Company, Minneapolis, Minn., 1960.

23. Kreith, F. and J. F. Kreider, *Principles of Solar Energy*, Hemisphere Publishing Corp., Washington, D.C., 1978.

Appendix B
Mathematical Relations
and Functions

B.1 GENERAL SOLUTIONS TO SELECTED DIFFERENTIAL EQUATIONS

Case 1 Second Order — Rectangular Coordinates

$$\frac{d^2y}{dx^2} = 0$$

$$y = C_1 x + C_2$$

Case 2 Second Order with Source Term, S — Rectangular Coordinates

$$\frac{d^2y}{dx^2} + S = 0$$

$$y = C_1 x + C_2 - \tfrac{1}{2} S x^2$$

Case 3 Second Order — Cylindrical Coordinate

$$\frac{1}{r}\frac{d}{dr}\left(r\frac{dy}{dr}\right) = 0$$

$$y = C_1 \ln r + C_2$$

Case 4   Second Order With Source Term, S — Cylindrical Coordinate

$$\frac{1}{r}\frac{d}{dr}\left(r\frac{dy}{dr}\right)+S=0$$

$$y=C_1\ln r+C_2-\tfrac{1}{4}Sr^2$$

Case 5   Second Order — Spherical Coordinate

$$\frac{d}{dr}\left(r^2\frac{dy}{dr}\right)=0$$

$$y=\frac{C_1}{r}+C_2$$

Case 6   Second Order with Source Term, S — Spherical Coordinate

$$\frac{1}{r^2}\frac{d}{dr}\left(r^2\frac{dy}{dr}\right)+S=0$$

$$y=\frac{C_1}{r}+C_2-\tfrac{1}{6}Sr^2$$

Case 7   Second Order with Constant Coefficients

$$\frac{d^2y}{dx^2}-\lambda^2y=0$$

$$y=C_1e^{+\lambda x}+C_2e^{-\lambda x}$$

Case 8   Second Order with Constant Coefficients and Source Term, S

$$\frac{d^2y}{dx^2}-\lambda^2y+S=0$$

$$y=C_1e^{+\lambda x}+C_2e^{-\lambda x}+\frac{S}{\lambda^2}$$

Case 9   Second Order with Constant Coefficients

$$\frac{d^2y}{dx^2}+\lambda^2y=0$$

$$y=C_1\cos\lambda x+C_2\sin\lambda x$$

B.2 TABLES OF SELECTED MATHEMATICAL FUNCTIONS

Hyperbolic Functions [1]

x	sinh x	cosh x	tanh x	x	sinh x	cosh x	tanh x
0.00	0.0000	1.0000	0.00000	2.00	3.6269	3.7622	0.96403
0.10	0.1002	1.0050	0.09967	2.10	4.0219	4.1443	0.97045
0.20	0.2013	1.0201	0.19738	2.20	4.4571	4.5679	0.97574
0.30	0.3045	1.0453	0.29131	2.30	4.9370	5.0372	0.98010
0.40	0.4108	1.0811	0.37995	2.40	5.4662	5.5569	0.98367
0.50	0.5211	1.1276	0.46212	2.50	6.0502	6.1323	0.98661
0.60	0.6367	1.1855	0.53705	2.60	6.6947	6.7690	0.98903
0.70	0.7586	1.2552	0.60437	2.70	7.4063	7.4735	0.99101
0.80	0.8881	1.3374	0.66404	2.80	8.1919	8.2527	0.99263
0.90	1.0265	1.4331	0.71630	2.90	9.0596	9.1146	0.99396
1.00	1.1752	1.5431	0.76159	3.00	10.018	10.068	0.99505
1.10	1.3356	1.6685	0.80050	3.50	16.543	16.573	0.99818
1.20	1.5095	1.8107	0.83365	4.00	27.290	27.308	0.99933
1.30	1.6984	1.9709	0.86172	4.50	45.003	45.014	0.99975
1.40	1.9043	2.1509	0.88535	5.00	74.203	74.210	0.99991
1.50	2.1293	2.3524	0.90515	6.00	201.71	201.72	0.99999
1.60	2.3756	2.5775	0.92167	7.00	548.32	548.32	1.0000
1.70	2.6456	2.8283	0.93541	8.00	1490.5	1490.5	1.0000
1.80	2.9422	3.1075	0.94681	9.00	4051.5	4051.5	1.0000
1.90	3.2682	3.4177	0.95624	10.00	11013	11013	1.0000

[1] The hyperbolic functions are defined as

$$\sinh x = \tfrac{1}{2}(e^x - e^{-x}) \qquad \cosh x = \tfrac{1}{2}(e^x + e^{-x})$$

$$\tanh x = \frac{e^x - e^{-x}}{e^x + e^{-x}} = \frac{\sinh x}{\cosh x}$$

The derivatives of the hyperbolic functions of the variable u are given as

$$\frac{d}{dx}(\sinh u) = (\cosh u)\frac{du}{dx} \qquad \frac{d}{dx}(\cosh u) = (\sinh u)\frac{du}{dx}$$

$$\frac{d}{dx}(\tanh u) = \left(\frac{1}{\cosh^2 u}\right)\frac{du}{dx}$$

Gaussian Error Function[1]

w	erf w	w	erf w	w	erf w
0.00	0.00000	0.36	0.38933	1.04	0.85865
0.02	0.02256	0.38	0.40901	1.08	0.87333
0.04	0.04511	0.40	0.42839	1.12	0.88679
0.06	0.06762	0.44	0.46622	1.16	0.89910
0.08	0.09008	0.48	0.50275	1.20	0.91031
0.10	0.11246	0.52	0.53790	1.30	0.93401
0.12	0.13476	0.56	0.57162	1.40	0.95228
0.14	0.15695	0.60	0.60386	1.50	0.96611
0.16	0.17901	0.64	0.63459	1.60	0.97635
0.18	0.20094	0.68	0.66378	1.70	0.98379
0.20	0.22270	0.72	0.69143	1.80	0.98909
0.22	0.24430	0.76	0.71754	1.90	0.99279
0.24	0.26570	0.80	0.74210	2.00	0.99532
0.26	0.28690	0.84	0.76514	2.20	0.99814
0.28	0.30788	0.88	0.78669	2.40	0.99931
0.30	0.32863	0.92	0.80677	2.60	0.99976
0.32	0.34913	0.96	0.82542	2.80	0.99992
0.34	0.36936	1.00	0.84270	3.00	0.99998

[1] The Gaussian error function is defined as

$$\text{erf } w \equiv \frac{2}{\sqrt{\pi}} \int_0^w e^{-v^2}\, dv$$

The complementary error function is defined as

$$\text{erfc } w \equiv 1 - \text{erf } w$$

B.3 INDEFINITE INTEGRALS

A constant of integration should be added to the result of each integration. The symbols u and v represent functions of x while the symbol a represents an arbitrary constant.

$$\int u\,dv = uv - \int v\,du$$

$$\int e^u\,du = e^u$$

$$\int \sin u\,du = -\cos u$$

$$\int \cos u\,du = \sin u$$

$$\int \cos u \sin u\,du = \tfrac{1}{2}\sin^2 u$$

$$\int \cos^2 u \sin u\,du = -\tfrac{1}{3}\cos^3 u$$

$$\int \frac{dx}{(a^4 - x^4)} = \frac{1}{4a^3}\log\left|\frac{a+x}{a-x}\right| + \frac{1}{2a^3}\tan^{-1}\frac{x}{a}$$

Appendix C
Two-Dimensional, Steady-State
Conduction: An Exact Solution

To obtain an appreciation for how the method of separation of variables may be used to solve two-dimensional conduction problems, we consider the system of Figure C.1. Three sides of the rectangular plate are maintained at a constant temperature T_1, while the fourth side is maintained at a constant temperature T_2 $\neq T_1$. We are interested in the temperature distribution $T(x, y)$, but to simplify the solution we introduce the transformation

$$\theta \equiv \frac{T - T_1}{T_2 - T_1} \tag{C.1}$$

For steady-state conditions with no generation and constant thermal conductivity, the appropriate form of the heat equation is given by Equation 4.1. Substituting from Equation C.1 it may be expressed as

$$\frac{\partial^2 \theta}{\partial x^2} + \frac{\partial^2 \theta}{\partial y^2} = 0 \tag{C.2}$$

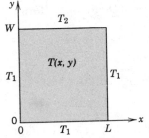

Figure C.1 Two-dimensional conduction in a rectangular plate.

Since the differential equation is second order in both x and y, two boundary conditions are needed for each of the coordinates. They are

$$\theta(0, y) = 0 \qquad \theta(x,0) = 0$$

$$\theta(L, y) = 0 \qquad \theta(x, W) = 1$$

Note that through the transformation of Equation C.1, we have been able to make three of the four boundary conditions homogeneous.

We now apply the separation of variables technique by assuming that the desired solution can be expressed as the product of two functions, one of which depends only on x and the other depends only on y. That is, we assume the existence of a solution of the form

$$\theta(x, y) = X(x) \cdot Y(y) \tag{C.3}$$

Substituting into Equation C.2 and dividing by XY, we then obtain

$$-\frac{1}{X}\frac{d^2 X}{dx^2} = \frac{1}{Y}\frac{d^2 Y}{dy^2} \tag{C.4}$$

and it is evident that the differential equation is, in fact, separable. Since the left-hand side of the equation depends only on x and the right-hand side depends only on y, the equality can only apply in general if both sides are equal to the same constant. Identifying this, as yet unknown, *separation constant* as λ^2, we then have

$$\frac{d^2 X}{dx^2} + \lambda^2 X = 0 \tag{C.5}$$

$$\frac{d^2 Y}{dy^2} - \lambda^2 Y = 0 \tag{C.6}$$

and the partial differential equation has been reduced to two ordinary differential equations. Note that the designation of λ^2 as a positive constant was not arbitrary. Were a negative value selected or were a value of $\lambda^2 = 0$ chosen, it can be readily shown that it would be impossible to obtain a solution which satisfies the prescribed boundary conditions.

The general solutions to Equations C.5 and C.6 are of the form

$$X = C_1 \cos \lambda x + C_2 \sin \lambda x$$

$$Y = C_3 e^{-\lambda y} + C_4 e^{+\lambda y}$$

in which case the general form of the two-dimensional solution is

$$\theta = (C_1 \cos \lambda x + C_2 \sin \lambda x)(C_3 e^{-\lambda y} + C_4 e^{\lambda y}) \tag{C.7}$$

Applying the condition that $\theta(0, y) = 0$, it is evident that $C_1 = 0$. In addition from the requirement that $\theta(x,0) = 0$, we obtain

$$C_2 \sin \lambda x (C_3 + C_4) = 0$$

which may only be satisfied if $C_3 = -C_4$.[1] If we now invoke the requirement that $\theta(L, y) = 0$, we obtain

$$C_2 C_4 \sin \lambda L (e^{\lambda y} - e^{-\lambda y}) = 0$$

The only way in which this condition may be satisfied (and still have an acceptable solution) is by requiring that λ assume discrete values for which $\sin \lambda L = 0$. These values must then be of the form

$$\lambda = \frac{n\pi}{L}, \qquad n = 1,2,3,\ldots \tag{C.8}$$

where the integer $n = 0$ is precluded since it provides an unacceptable solution. The desired solution may now be expressed as

$$\theta = C_2 C_4 \sin \frac{n\pi x}{L} (e^{n\pi y/L} - e^{-n\pi y/L}) \tag{C.9}$$

Combining constants and acknowledging that the new constant may depend on n, we obtain

$$\theta(x, y) = C_n \sin \frac{n\pi x}{L} \sinh \frac{n\pi y}{L}$$

where we have also used the fact that $(e^{n\pi y/L} - e^{-n\pi y/L}) = 2 \sinh(n\pi y/L)$. In the above form we have really obtained an infinite number of solutions that satisfy the original differential equation and the boundary conditions. However, since the problem is linear, a more general solution may be obtained from a superposition of the form

$$\theta(x, y) = \sum_{n=1}^{\infty} C_n \sin \frac{n\pi x}{L} \sinh \frac{n\pi y}{L} \tag{C.10}$$

To determine C_n we now apply the remaining boundary condition, which is of the form

$$\theta(x, W) = 1 = \sum_{n=1}^{\infty} C_n \sin \frac{n\pi x}{L} \sinh \frac{n\pi W}{L} \tag{C.11}$$

Although Equation C.11 would seem to be an extremely complicated relation for evaluating C_n, a standard method is available. It involves writing an

---

[1] The requirement could also be satisfied by having $C_2 = 0$, but this would totally eliminate the x dependence and hence provide an unacceptable solution.

analogous infinite series expansion in terms of *orthogonal functions*. An infinite set of functions $g_1(x), g_2(x), ..., g_n(x), ...$ is said to be orthogonal in the domain $a \leq x \leq b$ if

$$\int_a^b g_m(x)\, g_n(x)\, dx = 0 \qquad (m \neq n) \tag{C.12}$$

Many functions exhibit orthogonality, including the trigonometric functions $\sin n\pi x/L$ and $\cos n\pi x/L$ for $0 \leq x \leq L$. Their utility in the present problem rests with the fact that any function $f(x)$ may be expressed in terms of an infinite series of orthogonal functions

$$f(x) = \sum_{n=1}^{\infty} A_n g_n(x) \tag{C.13}$$

The form of the coefficients A_n in this series may be determined by multiplying each side of the equation by $g_n(x)$ and integrating between the limits a and b.

$$\int_a^b f(x)\, g_n(x) dx = \int_a^b g_n(x) \sum_{n=1}^{\infty} A_n\, g_n(x) dx \tag{C.14}$$

However, from Equation C.12 it is evident that all but one of the terms on the right-hand side of (C.14) must be zero, leaving us with

$$\int_a^b f(x) g_n(x) dx = A_n \int_a^b g_n^2(x) dx$$

Hence

$$A_n = \frac{\int_a^b f(x) g_n(x) dx}{\int_a^b g_n^2(x) dx} \tag{C.15}$$

The properties of orthogonal functions may be used to solve Equation C.11 for C_n by formulating an *analogous* infinite series for the appropriate form of $f(x)$. From Equation C.11 it is evident that we should choose $f(x) = 1$ and the orthogonal function $g_n(x) = \sin n\pi x/L$. Substituting into Equation C.15 we obtain

$$A_n = \frac{\displaystyle\int_0^L \sin \frac{n\pi x}{L}\, dx}{\displaystyle\int_0^L \sin^2 \frac{n\pi x}{L}\, dx}$$

$$= \frac{2}{\pi} \frac{(-1)^{n+1} + 1}{n}$$

Hence from Equation C.13, we have

$$1 = \sum_{n=1}^{\infty} \frac{2}{\pi} \frac{(-1)^{n+1} + 1}{n} \sin \frac{n\pi x}{L} \tag{C.16}$$

which is simply the expansion of unity in a Fourier series. Comparing Equations C.11 and C.16 we obtain

$$C_n = \frac{2[(-1)^{n+1} + 1]}{n\pi \sinh (n\pi\, W/L)}, \qquad n = 1, 2, 3, \ldots \tag{C.17}$$

Substituting Equation C.17 into C.10 we then obtain for the final solution

$$\theta(x, y) = \frac{2}{\pi} \sum_{n=1}^{\infty} \frac{(-1)^{n+1} + 1}{n} \sin \frac{n\pi x}{L} \frac{\sinh (n\pi y/L)}{\sinh (n\pi W/L)} \tag{C.18}$$

The foregoing infinite series converges, and the results are of the form shown in Figure C.2.

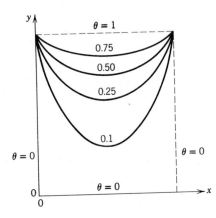

Figure C.2   Isotherms for two-dimensional conduction in a rectangular plate.

Appendix D
An Exact Laminar Boundary
Layer Solution for Parallel Flow
Over a Flat Plate

To obtain an appreciation for the nature of a *boundary layer solution*, we consider the simple case of parallel flow over a flat surface (Figures 6.3 to 6.5). We assume *steady, incompressible laminar flow* with *constant fluid properties* and *negligible viscous dissipation*. We also assume that, although species transfer may occur, it has a negligible effect on conditions in the velocity and thermal boundary layers. The boundary layer equations, Equations 6.51, 6.52, 6.54, and 6.55, are then

Continuity:
$$\frac{\partial u}{\partial x} + \frac{\partial v}{\partial y} = 0 \qquad (D.1)$$

Momentum:
$$u\frac{\partial u}{\partial x} + v\frac{\partial u}{\partial y} = \nu\frac{\partial^2 u}{\partial y^2} \qquad (D.2)$$

Energy:
$$u\frac{\partial T}{\partial x} + v\frac{\partial T}{\partial y} = \alpha\frac{\partial^2 T}{\partial y^2} \qquad (D.3)$$

Species:
$$u\frac{\partial \rho_A}{\partial x} + v\frac{\partial \rho_A}{\partial y} = D_{AB}\frac{\partial^2 \rho_A}{\partial y^2} \qquad (D.4)$$

We first note that for constant properties the velocity (hydrodynamic) boundary layer development is independent of thermal effects. Hence we may begin by solving the hydrodynamic problem, Equations D.1 and D.2, to the exclusion of Equations D.3 and D.4. Once the hydrodynamic problem is solved, the solutions to Equations D.3 and D.4, which depend on the variables u and v, may be obtained.

The classical hydrodynamic solution follows the method of Blasius [1, 2]. The velocity components are defined in terms of a stream function, $\psi(x, y)$,

$$u \equiv \frac{\partial \psi}{\partial y} \qquad v \equiv -\frac{\partial \psi}{\partial x} \tag{D.5}$$

such that Equation D.1 is automatically satisfied and hence is no longer needed. New dependent and independent variables, f and η, respectively, are then defined such that

$$f(\eta) \equiv \frac{\psi}{u_\infty \sqrt{vx/u_\infty}} \tag{D.6}$$

$$\eta \equiv y \sqrt{u_\infty/vx} \tag{D.7}$$

As we will find, use of these variables simplifies matters by reducing the partial differential equation, Equation D.2, to an ordinary differential equation.

The Blasius solution is termed a *similarity solution*, and η is a *similarity variable*. This terminology is used because, despite the growth of the boundary layer with distance x from the leading edge, the velocity profile, u/u_∞, remains *geometrically similar*. This similarity is of the form

$$\frac{u}{u_\infty} = \phi\left(\frac{y}{\delta}\right)$$

where δ is the boundary layer thickness. Assuming this thickness to vary as $(vx/u_\infty)^{1/2}$, it follows that

$$\frac{u}{u_\infty} = \phi(\eta) \tag{D.8}$$

Hence, the velocity profile is assumed to be uniquely determined by the similarity variable η, which depends on both x and y.

From Equations D.5 through D.7 we obtain

$$u = \frac{\partial \psi}{\partial y} = \frac{\partial \psi}{\partial \eta}\frac{\partial \eta}{\partial y} = u_\infty \sqrt{vx/u_\infty}\,\frac{df}{d\eta}\sqrt{u_\infty/vx} = u_\infty \frac{df}{d\eta} \tag{D.9}$$

and

$$v = -\frac{\partial \psi}{\partial x} = -\left(u_\infty \sqrt{vx/u_\infty}\,\frac{\partial f}{\partial x} + \frac{u_\infty}{2}\sqrt{v/u_\infty x}\,f\right)$$

$$= \tfrac{1}{2}\sqrt{vu_\infty/x}\left(\eta\frac{df}{d\eta} - f\right) \tag{D.10}$$

By differentiating the velocity components, it may also be shown that

$$\frac{\partial u}{\partial x} = -\frac{u_\infty}{2x}\eta\frac{d^2f}{d\eta^2} \tag{D.11}$$

$$\frac{\partial u}{\partial y} = u_\infty \sqrt{u_\infty/vx}\, \frac{d^2 f}{d\eta^2} \tag{D.12}$$

$$\frac{\partial^2 u}{\partial y^2} = \frac{u_\infty^2}{vx}\, \frac{d^3 f}{d\eta^3} \tag{D.13}$$

Substituting these expressions into Equation D.2, we then obtain

$$2\frac{d^3 f}{d\eta^3} + f\frac{d^2 f}{d\eta^2} = 0 \tag{D.14}$$

Hence the hydrodynamic boundary layer problem is reduced to one of solving a nonlinear, third-order ordinary differential equation. The appropriate boundary conditions for the flat plate are

$$u(x,0) = v(x,0) = 0$$

$$u(x,\infty) = u_\infty$$

in which case

$$\left.\frac{df}{d\eta}\right|_{\eta=0} = f(0) = 0$$

$$\left.\frac{df}{d\eta}\right|_{\eta=\infty} = 1 \tag{D.15}$$

The solution to Equation D.14, subject to the conditions of Equation D.15, is not easily obtained. In fact, a closed form solution does not exist. However, an accurate solution may still be obtained by a series expansion [2] or by numerical methods [3], and selected results are presented in Table D.1. Several important results may be extracted from this table. We first note that, to a good approximation, $(u/u_\infty) = 0.99$ for $\eta = 5.0$. Defining the boundary layer thickness, δ, as that value of y for which $(u/u_\infty) = 0.99$, it follows from Equation D.7 that

$$\delta = \frac{5.0}{\sqrt{u_\infty/vx}} = \frac{5x}{\sqrt{Re_x}} \tag{D.16}$$

From Equation D.16 it is clear that δ increases with x but decreases with increasing u_∞ (the larger the freestream velocity, the *thinner* the boundary layer). In addition, from Equation D.12 the wall shear stress may be expressed as

$$\tau_s = \mu\left.\frac{\partial u}{\partial y}\right|_{y=0} = \mu u_\infty \sqrt{u_\infty/vx}\left.\frac{d^2 f}{d\eta^2}\right|_{\eta=0}$$

Hence, from Table D.1

$$\tau_s = 0.332 u_\infty \sqrt{\rho\mu u_\infty/x}$$

Table D.1   Flat plate laminar boundary layer functions [3]

$\eta = y \sqrt{\dfrac{u_\infty}{vx}}$	f	$\dfrac{df}{d\eta} = \dfrac{u}{u_\infty}$	$\dfrac{d^2 f}{d\eta^2}$
0	0	0	0.332
0.4	0.027	0.133	0.331
0.8	0.106	0.265	0.327
1.2	0.238	0.394	0.317
1.6	0.420	0.517	0.297
2.0	0.650	0.630	0.267
2.4	0.922	0.729	0.228
2.8	1.231	0.812	0.184
3.2	1.569	0.876	0.139
3.6	1.930	0.923	0.098
4.0	2.306	0.956	0.064
4.4	2.692	0.976	0.039
4.8	3.085	0.988	0.022
5.2	3.482	0.994	0.011
5.6	3.880	0.997	0.005
6.0	4.280	0.999	0.002
6.4	4.679	1.000	0.001
6.8	5.079	1.000	0.000

The *local* friction coefficient is then

$$C_{f,x} = \frac{\tau_s}{\rho u_\infty^2 / 2} = \frac{0.664}{\sqrt{Re_x}} \tag{D.17}$$

From knowledge of conditions in the velocity boundary layer, the energy and species continuity equations may now be solved. To solve Equation D.3 we nondimensionalize the temperature according to Equation 6.58, and we assume a similarity solution of the form $T^* = T^*(\eta)$. Making the necessary substitutions, Equation D.3 reduces to

$$\frac{d^2 T^*}{d\eta^2} + \frac{Pr}{2} f \frac{dT^*}{d\eta} = 0 \tag{D.18}$$

Note the dependence of the thermal solution on hydrodynamic conditions through appearance of the variable f in Equation D.18. The appropriate boundary conditions are

$$T^*(0) = 0 \qquad T^*(\infty) = 1 \tag{D.19}$$

Equation D.18 has been solved subject to the conditions of Equation D.19 for variable Prandtl number [2, 4]. Since the details are somewhat involved, however, only the key results will be discussed. One important result is the expression obtained for $dT^*/d\eta|_{\eta=0}$. In particular, since

$$dT^*/d\eta|_{\eta=0} = 0.332 Pr^{1/3}$$

and the local convection coefficient is

$$h = \frac{q_s''}{T_s - T_\infty} = \frac{-k(T_\infty - T_s)}{(T_s - T_\infty)} \partial T^*/\partial y|_{y=0}$$

$$= k\sqrt{u_\infty/vx}\; dT^*/d\eta|_{\eta=0}$$

it follows that

$$Nu_x \equiv \frac{hx}{k} = 0.332 Re_x^{1/2}\, Pr^{1/3} \tag{D.20}$$

This result applies to a good approximation for any $Pr \gtrsim 0.6$. Another important result obtained from the solution to Equation D.18 concerns the ratio of the velocity and thermal boundary layer thicknesses. In particular

$$\frac{\delta}{\delta_t} \approx Pr^{1/3} \tag{D.21}$$

The concentration boundary layer solution is obtained in a similar manner. Introducing a normalized species density $\rho_A^* \equiv [(\rho_A - \rho_{A,s})/(\rho_{A,\infty} - \rho_{A,s})]$ and assuming a similarity solution, Equation D.4 becomes

$$\frac{d^2\rho_A^*}{d\eta^2} + \frac{Sc}{2} f \frac{d\rho_A^*}{d\eta} = 0 \tag{D.22}$$

with

$$\rho_A^*(0) = 0 \qquad \rho_A^*(\infty) = 1 \tag{D.23}$$

From the corresponding solution, it is known that

$$\frac{d\rho_A^*}{d\eta}\bigg|_{\eta=0} = 0.332 Sc^{1/3}$$

in which case

$$h_m \equiv \frac{n_A''}{\rho_{A,s} - \rho_{A,\infty}} = D_{AB}\frac{\partial\rho_A^*}{\partial y}\bigg|_{y=0} = D_{AB}\sqrt{u_\infty/vx}\,\frac{d\rho_A^*}{d\eta}\bigg|_{\eta=0}$$

and

$$Sh_x \equiv \frac{h_m x}{D_{AB}} = 0.332 Re_x^{1/2} Sc^{1/3} \tag{D.24}$$

Equation D.24 applies to a good approximation for $Sc \gtrsim 0.6$. It may also be shown that

$$\frac{\delta}{\delta_c} \approx Sc^{1/3} \tag{D.25}$$

Note that Equations D.20 and D.24, as well as Equations D.21 and D.25, are of precisely the same form, and hence confirm the heat and mass transfer analogy (Section 6.8.1). Note also that the boundary layer thickness ratios are proportional to $Pr^{1/3}$ and $Sc^{1/3}$.

REFERENCES

1. Blasius, H., *Z. Math. u. Phys.*, *56*, 1, 1908. English translation in NACA Tech. Memo. No. 1256.
2. Schlichting, H., *Boundary Layer Theory*, 4th Ed., 1960.
3. Howarth, L., *Proc. Roy. Soc. Lond.*, Ser. A, *164*, 1938, p. 547.
4. Pohlhausen, E., *Z. Angew. Math. Mech.*, *1*, 1921, p. 115.

Appendix E
An Integral Laminar Boundary Layer Solution for Parallel Flow Over a Flat Plate

An alternative approach to solving the boundary layer equations involves the use of an approximate *integral* method. The approach was originally proposed by von Karman [1] in 1921 and first applied by Pohlhausen [2]. It is without the mathematical complications inherent in the *exact* method of Appendix D; yet it can be used to obtain reasonably accurate results for the key boundary layer parameters (δ, δ_t, δ_c, C_f, h, and h_m). Although the method has been used with some success for a variety of flow conditions, we restrict our attention to parallel flow over a flat plate, subject to the same restrictions enumerated in Appendix D, that is, *incompressible laminar flow* with *constant fluid properties* and *negligible viscous dissipation*.

To use the method, the boundary layer equations, Equations D.1 to D.4, must be cast in integral form. These forms are obtained by integrating the equations in the y direction across the boundary layer. For example, integrating Equation D.1, we obtain

$$\int_0^\delta \frac{\partial u}{\partial x}\,dy + \int_0^\delta \frac{\partial v}{\partial y}\,dy = 0 \tag{E.1}$$

or, since $v = 0$ at $y = 0$,

$$v(y = \delta) = -\int_0^\delta \frac{\partial u}{\partial x}\,dy \tag{E.2}$$

Integrating Equation D.2 we obtain

$$\int_0^\delta u\frac{\partial u}{\partial x}\,dy + \int_0^\delta v\frac{\partial u}{\partial y}\,dy = v\int_0^\delta \frac{\partial}{\partial y}\left(\frac{\partial u}{\partial y}\right)dy$$

or, integrating the second term on the left-hand side by parts,

$$\int_0^\delta u \frac{\partial u}{\partial x} dy + uv \Big|_0^\delta - \int_0^\delta u \frac{\partial v}{\partial y} dy = v \frac{\partial u}{\partial y}\Big|_0^\delta$$

Substituting from Equations D.1 and E.2 we obtain

$$\int_0^\delta u \frac{\partial u}{\partial x} dy - u_\infty \int_0^\delta \frac{\partial u}{\partial x} dy + \int_0^\delta u \frac{\partial u}{\partial x} dy = -v \frac{\partial u}{\partial y}\Big|_{y=0}$$

or

$$u_\infty \int_0^\delta \frac{\partial u}{\partial x} dy - \int_0^\delta 2u \frac{\partial u}{\partial x} dy = v \frac{\partial u}{\partial y}\Big|_{y=0}$$

Therefore

$$\int_0^\delta \frac{\partial}{\partial x} (u_\infty \cdot u - u \cdot u) dy = v \frac{\partial u}{\partial y}\Big|_{y=0}$$

Rearranging, we then obtain

$$\frac{d}{dx}\left[\int_0^\delta (u_\infty - u)u\, dy \right] = v \frac{\partial u}{\partial y}\Big|_{y=0} \tag{E.3}$$

Equation E.3 is the integral form of the boundary layer momentum equation.

In a similar fashion the following integral forms of the boundary layer energy and species equations may be obtained.

$$\frac{d}{dx}\left[\int_0^{\delta_t} (T_\infty - T)u\, dy \right] = \alpha \frac{\partial T}{\partial y}\Big|_{y=0} \tag{E.4}$$

$$\frac{d}{dx}\left[\int_0^{\delta_c} (\rho_{A,\infty} - \rho_A)u\, dy \right] = D_{AB} \frac{\partial \rho_A}{\partial y}\Big|_{y=0} \tag{E.5}$$

Equations E.3 to E.5 satisfy the x momentum, the energy, and the species conservation requirements in an *integral* (or *average*) fashion over the entire boundary layer. In contrast the original conservation equations, Equations D.2 to D.4, satisfy the conservation requirements *locally*, that is, at each point in the boundary layer.

The integral equations can be used to obtain *approximate* boundary layer solutions. The procedure involves first *assuming* reasonable functional forms for the unknown variables u, T, and ρ_A in terms of the corresponding (*unknown*) boundary layer thicknesses. The assumed forms must satisfy the appropriate boundary conditions. Substituting these forms into the integral equations, expressions for the boundary layer thicknesses may be determined and the assumed functional forms may then be completely specified. Although this method is approximate, it frequently leads to accurate results for the surface parameters.

Consider the hydrodynamic boundary layer, for which the relevant boundary conditions are

$$u(y=0) = \frac{\partial u}{\partial y}\bigg|_{y=\delta} = 0$$

$$u(y=\delta) = u_\infty$$

Moreover, from Equation D.2 it is evident that, since $u = v = 0$ at $y = 0$,

$$\frac{\partial^2 u}{\partial y^2}\bigg|_{y=0} = 0$$

With four such conditions, which must be satisfied, we could then approximate the velocity profile as a third-degree polynomial of the form

$$\frac{u}{u_\infty} = a_1 + a_2\left(\frac{y}{\delta}\right) + a_3\left(\frac{y}{\delta}\right)^2 + a_4\left(\frac{y}{\delta}\right)^3$$

and apply the conditions to determine the coefficients a_1 to a_4. It is easily verified that $a_1 = a_3 = 0$, $a_2 = \frac{3}{2}$ and $a_4 = -\frac{1}{2}$, in which case

$$\frac{u}{u_\infty} = \frac{3}{2}\frac{y}{\delta} - \frac{1}{2}\left(\frac{y}{\delta}\right)^3 \tag{E.6}$$

The velocity profile is then specified in terms of the unknown boundary layer thickness δ. This unknown may be determined by substituting Equation E.6 into E.3 and integrating over y to obtain

$$\frac{d}{dx}\left(\frac{39}{280}u_\infty^2 \delta\right) = \frac{3}{2}\frac{vu_\infty}{\delta}$$

Separating variables and integrating over x, we then obtain

$$\frac{\delta^2}{2} = \frac{140}{13}\frac{vx}{u_\infty} + \text{const.}$$

However, since $\delta = 0$ at the leading edge of the plate ($x = 0$), the integration constant must be zero and

$$\delta = 4.64(vx/u_\infty)^{1/2} = 4.64x/Re_x^{1/2} \tag{E.7}$$

Substituting Equation E.7 into Equation E.6 and evaluating $\tau_s = \mu(\partial u/\partial y)_s$, we also obtain

$$C_{f,x} = \frac{\tau_s}{\rho u_\infty^2/2} = \frac{0.646}{Re_x^{1/2}} \tag{E.8}$$

Despite the approximate nature of the foregoing procedure, Equations E.7 and

E.8 compare quite well with results obtained from the exact solution, Equations D.16 and D.17.

In a similar fashion one could assume a temperature profile of the form

$$T^* = \frac{T - T_s}{T_\infty - T_s} = b_1 + b_2\left(\frac{y}{\delta_t}\right) + b_3\left(\frac{y}{\delta_t}\right)^2 + b_4\left(\frac{y}{\delta_t}\right)^3$$

and determine the coefficients from the conditions

$$T^*(y=0) = \left.\frac{\partial T^*}{\partial y}\right|_{y=\delta_t} = 0$$

$$T^*(y=\delta_t) = 1$$

as well as

$$\left.\frac{\partial^2 T^*}{\partial y^2}\right|_{y=0} = 0$$

which is inferred from the energy equation, Equation D.3. We then obtain

$$T^* = \frac{3}{2}\frac{y}{\delta_t} - \frac{1}{2}\left(\frac{y}{\delta_t}\right)^3 \tag{E.9}$$

Substituting Equations E.6 and E.9 into Equation E.4, we obtain, after some manipulation and assuming $Pr \gtrsim 1$,

$$\frac{\delta_t}{\delta} = \frac{Pr^{-1/3}}{1.026} \tag{E.10}$$

This result is in good agreement with that obtained from the exact solution, Equation D.21. Moreover, the heat transfer coefficient may be then computed from

$$h = \frac{-k\,\partial T/\partial y|_{y=0}}{T_s - T_\infty} = \frac{3}{2}\frac{k}{\delta_t}$$

and substituting from Equations E.7 and E.10 we obtain

$$Nu_x = \frac{hx}{k} = 0.332 Re_x^{1/2}\, Pr^{1/3} \tag{E.11}$$

This result agrees precisely with that obtained from the exact solution, Equation D.20. Using the same procedures, analogous results may also be obtained for the concentration boundary layer.

REFERENCES

1. von Kárman, T., Z. Angew. Math. Mech., 1, 232, 1921.
2. Pohlhausen, K., Z. Angew. Math. Mech., 1, 252, 1921.

Index